Harold L Wade

Distillation

McGraw-Hill Chemical Engineering Series

Editorial Advisory Board

James J. Carberry, *Professor of Chemical Engineering, University of Notre Dame*

James R. Fair, *Director, Engineering, Technology, Monsanto Company, Missouri*

Max S. Peters, *Dean of Engineering, University of Colorado*

William R. Schowalter, *Professor of Chemical Engineering, Princeton University*

James Wei, *Professor of Chemical Engineering, Massachusetts Institute of Technology*

Building the Literature of a Profession

Fifteen prominent chemical engineers first met in New York more than fifty years ago to plan a continuing literature for their rapidly growing profession. From industry came such pioneer practitioners as Leo H. Baekeland, Arthur D. Little, Charles L. Reese, John V. N. Dorr, M. C. Whitaker, and R. S. McBride. From the universities came such eminent educators as William H. Walker, Alfred H. White, D. D. Jackson, J. H. James, Warren K. Lewis, and Harry A. Curtis. H. C. Parmelee, then editor of *Chemical and Metallurgical Engineering*, served as chairman and was joined subsequently by S. D. Kirkpatrick as consulting editor.

After several meetings, this committee submitted its report to the McGraw-Hill Book Company in September, 1925. In the report were detailed specifications for a correlated series of more than a dozen texts and reference books which have since become the McGraw-Hill Chemical Engineering Series and which became the cornerstone of the chemical engineering curriculum.

From this beginning there has evolved a series of texts surpassing by far the scope and longevity envisioned by the founding Editorial Board. The McGraw-Hill Chemical Engineering Series stands as a unique historical record of the development of chemical engineering education and practice. In the series one finds the milestones of the subject's evolution: industrial chemistry, stoichiometry, unit operations and processes, thermodynamics, kinetics, and transfer operations.

Chemical engineering is a dynamic profession, and its literature continues to evolve. McGraw-Hill and its consulting editors remain committed to a publishing policy that will serve, and indeed lead, the needs of the chemical engineering profession during the years to come.

The Series

Bailey and Ollis—*Biochemical Engineering Fundamentals*
Bennett and Myers—*Momentum, Heat, and Mass Transfer*
Beveridge and Schechter—*Optimization: Theory and Practice*
Carberry—*Chemical and Catalytic Reaction Engineering*
Churchill—*The Interpretation and Use of Rate Data—The Rate Concept*
Clarke and Davidson—*Manual for Process Engineering Calculations*
Coughanowr and Koppel—*Process Systems Analysis and Control*
Danckwerts—*Gas Liquid Reactions*
Harriott—*Process Control*
Johnson—*Automatic Process Control*
Johnstone and Thring—*Pilot Plants, Models, and Scale-up Methods in Chemical Engineering*
Katz, Cornell, Kobayashi, Poettmann, Vary, Elenbaas and Weinaug—*Handbook of Natural Gas Engineering*
King—*Separation Processes*
Knudsen and Katz—*Fluid Dynamics and Heat Transfer*
Lapidus—*Digital Computation for Chemical Engineers*
Luyben—*Process Modeling, Simulation, and Control for Chemical Engineers*
McCabe and Smith, J.C.—*Unit Operations of Chemical Engineering*
Mickley, Sherwood and Reed—*Applied Mathematics in Chemical Engineering*
Nelson—*Petroleum Refinery Engineering*
Perry and Chilton (Editors)—*Chemical Engineers' Handbook*
Peters—*Elementary Chemical Engineering*
Peters and Timmerhaus—*Plant Design and Economics for Chemical Engineers*
Reed and Gubbins—*Applied Statistical Mechanics*
Reid, Prausnitz, and Sherwood—*The Properties of Gases and Liquids*
Sherwood, Pigford, and Wilke—*Mass Transfer*
Slattery—*Momentum, Energy, and Mass Transfer in Continua*
Smith, B. D.—*Design of Equilibrium Stage Processes*
Smith, J. M.—*Chemical Engineering Kinetics*
Smith, J. M., and Van Ness—*Introduction to Chemical Engineering Thermodynamics*
Thompson and Ceckler—*Introduction to Chemical Engineering*
Treybal—*Liquid Extraction*
Treybal—*Mass Transfer Operations*
Van Winkel—*Distillation*
Volk—*Applied Statistics for Engineers*
Walas—*Reaction Kinetics for Chemical Engineers*
Whitwell and Toner—*Conservation of Mass and Energy*

DISTILLATION

Matthew Van Winkle

Professor of Chemical Engineering
The University of Texas

McGraw-Hill Book Company
New York St. Louis San Francisco
Toronto London Sydney

DISTILLATION

Copyright © 1967 by McGraw-Hill, Inc. All Rights Reserved. Printed in the United States of America. No part of this publication may be reproduced, stored in a retrieval system, or transmitted, in any form or by any means, electronic, mechanical, photocopying, recording, or otherwise, without the prior written permission of the publisher.
Library of Congress Catalog Card Number 67-13203
07-067195-8

11 12 13 14 15 16 KPKP 8321

Dedicated to my wife Louise

Preface

This book was written to serve as a guide to fractional distillation design. The attempt has been made to treat the subject matter in such a manner that the "why" as well as the "how" is clearly delineated. In addition, design methods in common use are given along with working-size charts so that a process fractionater design can be made with a minimum of cross-reference consultation with other books and source material. Recommended design methods for each of the principal types of plate design are given, accompanied by the recommended limits of the design variables.

Distillation in general has been the subject of much research and this has resulted in a large accumulation of literature on the subject. Obviously, in one book, it is impossible to incorporate even a minimum of discussion of all of the information available. Therefore, the selection of material covered must lie with the author, and the selection has been made on the basis of the author's industrial, consulting, and academic experience, with regard to both the understanding of the subject and the production of workable and economical designs.

It is hoped that the book will serve as a text where the unit operation of distillation is to be taught and, certainly, as a reference for practicing engineers who are concerned with designing distillation equipment.

The many design equations given here can be programmed and used with various machine-computation methods for fractional distillation design. Also, many ready-made programs for such calculations and some very good books on the subject are available elsewhere. As an aid to those who wish to utilize machine computation, available literature on the subject has been referenced and included in the text.

The author wishes to express appreciation to former students who are in the field of industrial distillation design and engineering for their suggestions, to present graduate students who have given constructive suggestions regarding the presentation of the subject matter, and to friends in industry and in the academic field who have given help and encouragement in this work.

Matthew Van Winkle

Contents

Preface ix
Introduction 1

PART 1 VAPOR-LIQUID EQUILIBRIA 6

CHAPTER 1 PHASE EQUILIBRIA—BINARY SYSTEMS 7

1.1. Rules and Laws Describing Equilibrium 7
1.2. Vapor-Liquid Equilibria—Binary Systems 16
1.3. Prediction of Binary Vapor-Liquid Equilibrium Data from Pure-component Properties 30
1.4. Consistency Tests for Binary Vapor-Liquid Equilibrium Data 31
Problems 39
Nomenclature 42
References 43

CHAPTER 2 MULTICOMPONENT VAPOR-LIQUID EQUILIBRIA 46

2.1. Ideal Multicomponent Systems 46
2.2. Ideal Ternary Systems 49
2.3. Nonideal Vapor-Liquid Equilibria—Multicomponent Systems 55
Appendix: Relationship between Li and Coull and Bonham Vapor-Liquid Equilibrium Relations 117
Problems 121
Nomenclature 123
References 124

CHAPTER 3 COMPLEX SYSTEM VAPOR-LIQUID EQUILIBRIA 127

3.1. True-boiling-point Curve 127
3.2. ASTM or Engler-type Distillation 129
3.3. Equilibrium Flash Vaporization (EFV) Curves 130
3.4. Interrelation between EFV, TBP, and Engler Distillation Curves 131
3.5. Determination of EFV Curves—Superatmospheric Pressure 132

3.6. Atmospheric and Subatmospheric EFV Relationships from ASTM and TBP Data 136
3.7. Characteristics of Equilibrium Vapor and Liquid 153
Nomenclature 157
References 158

CHAPTER 4 EQUILIBRIUM AND SIMPLE DISTILLATION 160

4.1. Equilibrium Distillation 160
4.2. Differential Vaporization—Condensation 170
Problems 187
Nomenclature 188
References 189

PART 2 FRACTIONAL DISTILLATION 192

CHAPTER 5 GENERAL CONSIDERATIONS 193

5.1. Reflux 194
5.2. The Fractionating Column 195
5.3. The Contact Stage—Plate or Tray 196
5.4. Design Considerations in Fractionating Processes 197
5.5. Equilibrium Stages—General Considerations 202
Problems 243
Nomenclature 245
References 247

CHAPTER 6 BINARY FRACTIONATION—NUMBER OF EQUILIBRIUM STAGES 248

6.1. Analytical Determination of Number of Equilibrium Stages 249
6.2. Graphical Methods 256
Problems 280
Nomenclature 284
References 285

CHAPTER 7 TERNARY AND MULTICOMPONENT SYSTEM FRACTIONATION—NUMBER OF STAGES 286

7.1. Preliminary Calculations 287
7.2. Design Procedure 290
7.3. Short-cut Methods 303

7.4. Computer Rigorous Calculation of Number of Theoretical Plates 304
7.5. Graphical Evaluation of the Number of Theoretical Stages for Ternary Systems 321
Problems 329
Nomenclature 330
References 332

CHAPTER 8 COMPLEX SYSTEM FRACTIONATION—ESTIMATION OF NUMBER OF EQUILIBRIUM STAGES 334

8.1. Pseudocomponent Design Method 336
Problems 376
Nomenclature 377
References 378

PART 3 AZEOTROPIC AND EXTRACTIVE FRACTIONAL DISTILLATION 380

CHAPTER 9 TECHNIQUES OF SEPARATION OF AZEOTROPES AND OTHER DIFFICULT TO SEPARATE MIXTURES BY FRACTIONAL DISTILLATION 381

9.1. Techniques for Separation 382
Nomenclature 389

CHAPTER 10 SELECTION OF ADDITION AGENTS FOR AZEOTROPIC AND EXTRACTIVE DISTILLATION 390

10.1. Relative Volatility as an Economic Factor 390
10.2. Prediction of Vapor-Liquid Equilibrium Data 392
Nomenclature 431
References 432

CHAPTER 11 AZEOTROPIC AND EXTRACTIVE FRACTIONAL DISTILLATION 434

11.1. Mechanism of Relative Volatility Change 434
11.2. Choice of Entrainers or Solvents 436
11.3. Selection of an Azeotropic or Extractive Process 437

11.4. Design of an Azeotropic Distillation Process 437
11.5. Design of an Extractive Distillation Process 446
11.6. Solvent Recovery 469
Problems 473
Nomenclature 474
References 475

PART 4 FRACTIONATING PLATE AND COLUMN DESIGN 478

CHAPTER 12 FRACTIONATION DEVICES—EQUILIBRIUM STAGES 479

12.1. The Bubble-cap Plate 480
12.2. Perforated Plate 488
12.3. Variable-orifice Perforated Plates 492
12.4. Other Equilibrium-stage Trays 492
12.5. Plate Components 494
References 501

CHAPTER 13 PLATE AND COLUMN HYDRAULICS AND EFFICIENCY 502

13.1. Flow-energy Losses in Plate and Column 502
13.2. Evaluation of Pressure Drop and Liquid Backup in Downcomers 506
13.3. Flow Conditions Causing Inoperability 521
13.4. Efficiency Evaluation and Prediction 533
13.5. Prediction of Efficiency 550
Problems 561
Nomenclature 562
References 566

CHAPTER 14 PLATE-FRACTIONATING-COLUMN DESIGN METHODS 570

14.1. General Column Design 571
14.2. Bubble-cap Tray Design 575
14.3. Perforated-tray and Column Design 584
14.4. Value-tray Design Methods 598
Problems 600
Nomenclature 600
References 602

CHAPTER 15 PACKED COLUMN DESIGN 604

15.1. Packings 604
15.2. "Pseudoequilibrium Stage" Devices 613
15.3. Uses of Packed Columns 615
15.4. Differential Mass Transfer 616
15.5. Empirical Design Relationships 620
15.6. Capacity Design of Packed Towers 630
15.7. Design of Packed Towers for Distillation 638
Problems 643
Nomenclature 644
References 646
Appendix A: Tables 649
Appendix B: Vapor Pressures at Various Temperatures 653

Name Index 669

Subject Index 675

Introduction

The use of distillation as a separation and purification process is very old. Egloff and Lowry [8]† and Underwood [17] delved into the early historical writings pertaining to alchemy and science of the past to find where the process of distillation originated and when, and how it developed to its present state of use. Through the writings of early historians (Hoefer [10], Kopp [11], Bertholet [2]) it was indicated that the first recorded description of a batch distillation occurred in Cleopatra's time in Egypt around 50 B.C. However, because of other, older historical descriptions of products, essential oils, perfumes, medicines, beverages, etc., it can be deduced that distillation in some form was known probably as much as 1000 to 2000 years before that time. Fresh water was produced by distillation of sea water by using a sponge as a condenser around 300 A.D., and about the same time turpentine was distilled from rosin oil and condensed in a wool mat.

A simple condenser consisting of a long tube exposed to the air and leading to a receiver was described in the fourth century, and in the eleventh century the first record of alcohol by distillation was made. The beverage alcohol process became the first industrial distillation process during the period between the eleventh and fourteenth century, and it only became significant when the problems of sealing the joints in the still and devising a water-cooled condenser were solved. This was necessary to obtain reasonable yields of products containing high alcohol contents.

In the early sixteenth century distillation was being applied to separation and recovery of alcohol, water, vinegar, essences, oils, and numerous other products. Books appeared on the subject of distillation in the early part of the sixteenth century. Brunschwig [4] and Andrew [1] included information on distillation processes developed up to that time, which essentially covered small-scale batch distillations having practically no reflux but using straight or, in some cases, coiled tubes cooled with water for a condenser. Lonicer [13] described an air-cooled still head with water-cooled condenser.

It is interesting to note that Libavius [12] described a still provided with an air-cooled head which enabled simultaneous removal of five side streams.

† Numbers in brackets pertain to references given at end of each chapter.

Many refining engineers today consider the side-stream topping column to be a modern invention. In 1624 Donato d'Eremita [7] described a batch still with a fractionating still head, and Libavius [12] described batch stills with reboilers and fractionating heads. Boyle [3] distilled wood alcohol and vinegar and recovered various fractions according to their boiling points. This was probably the first analytical distillation.

Up until the nineteenth century the stills were of the differential batch type with little reflux. They were small, 18 to 30 in. in diameter, and 3 to 4 ft high including accessories. The yields were apparently reasonably high and the products usable. Cooper [6] described the "modern" distilling devices used in his time.

Near the end of the eighteenth century innovations were introduced which advanced the state of the distillation art measurably. In 1800 Rumford [15] used open steam as a heating agent. The forerunner of the present-day bubble cap was invented in 1822 by Perrier [14]. These caps were used as contact devices for steam introduced below the plate on which the caps were placed, wine was circulated around the caps on the plate, and the steam stripped the alcohol from the wine. In 1830 Coffee [5] developed a continuous still which used perforated plates, feed preheat, and internal reflux.

As an interesting sidelight, in 1822 it was estimated that approximately 10,000,000 gal of illicit liquor were made in the British Isles while only 3,000,000 gal were being taxed.

Distillation obviously was an art over the period of its early development, which encompassed about 3500 to 4000 years. The invention of bubble caps, perforated plates, water-cooled condensers, reflux, feed preheat, and the adaptation of the process to continuous operation took place over the relatively short period of 300 years. During the entire period, however, there was apparently no attempt to systematize or apply quantitative principles to distillation processes.

Late in the nineteenth century Hausbrand [9] and Sorel [16] introduced the first recorded quantitative mathematical discussions applied to fractionating still design. Sorel developed and applied mathematical relations to the fractional separation of binary mixtures—primarily those comprised of alcohol and water, and incorporated such considerations as variable boilup and overflow, variable molal enthalpy, heat losses from the column in his heat balances, compositions, rates, reflux, and pressure effects.

Thus, at the beginning of the present century, Sorel and Hausbrand evolved the basic relations upon which all modern distillation design calculations are based, and all methods used today are expansions and modifications of those original methods.

The rationalization and organization of mathematical applications to process fractionation design pointed up the necessity for study in several related areas. These were the thermodynamics of vapor-liquid equilibria and phase behavior in binary and multicomponent nonideal systems; contact stage improvement in design to obtain the best possible approach to an equilibrium stage; plate and column hydraulics study and the study of the variables which affect efficiency and operability of the fractionating column; and the rationalization of design methods to enable optimization of fractionating process design from the standpoint of economics.

These subjects form the basis for the discussion in the following chapters.

REFERENCES

1. Andrew, L. (translator): "The Vertuose Boke of Distyllacyon of the Waters of all Manner of Horbes," Brunshuig, 1527.
2. Bertholet, D.: "Histoire des sciences—la chémie au moyen age," Paris, 1893; "Les Origines de l'alchemie," Paris, 1885.
3. Boyle, R.: "Philosophical Works," vol. III, London, 1738.
4. Brunschwig, H.: "Liber de arte destillandi de simplicibus," Strasburg, 1500; "Liber de arte destillandi de composition," Strasburg, 1507.
5. Coffee, A.: British Patent No. 5974, 1830.
6. Cooper, A.: "The Complete Distiller," 2d ed., London, 1761.
7. Donato d'Eremita: "Dell elixir vitae," Naples, 1624.
8. Egloff, G., and C. D. Lowry: *Ind. Eng. Chem.*, **21**:920 (1929).
9. Hausbrand, E.: "Die Wirkungsweise der Rectifier und Distillin Apparate," Berlin, 1893.
10. Hoefer, F.: "Histoire de la chémie," Paris, 1866.
11. Kopp, H.: "Geschichte der Chemie," Braunschweig, 1847.
12. Libavius, A.: "Alchymia," Frankfurt, 1606.
13. Lonicer, A.: "Künstliche Conterfeytunge," Frankfurt, 1573.
14. Perrier, Sir A.: British Patent No. 4694, 1822.
15. Rumford, B. T.: "Essays, Political, Economical, and Philosophical," vol. III, London, 1802.
16. Sorel, E.: "La Rectification de l'alcool," Paris, 1894; "Distillation et rectification industrielles," Paris, 1899.
17. Underwood, A. J. V.: *Trans. Inst. Chem. Engrs. (London)*, **13**:34 1935.

part 1

Vapor-liquid equilibria

chapter 1

Phase equilibria—binary systems

In the distillation process the separation of a mixture of materials to obtain one or more desired products is achieved by selection of conditions of temperature and pressure so that at least a vapor and a liquid phase coexist and a difference in relative concentration of the materials to be separated in the two phases is attained. When the two (or more) phases are in a state of physical equilibrium, the maximum relative difference in concentration of the materials in the phases occurs. Therefore, attainment of equilibrium conditions is desirable in the distillation process, and most design methods use equilibrium as one of the boundary conditions for quantitative design calculations.

Because of this, equilibrium data must be available for such calculations to be made. Such data may be obtained from technical literature, they may be determined experimentally, or they may be obtained through the use of correlations, theoretical or empirical, relating pure-component and mixture properties.

1.1. RULES AND LAWS DESCRIBING EQUILIBRIUM

A number of quantitative rules and laws have been devised to describe and define systems in a state of physical equilibrium. Most thermodynamics and physical chemistry books give detailed development of the rules cited here and, therefore, complete discussions will not be included.

Gibbs Phase Rule

J. Willard Gibbs [16] described equilibrium mathematically in terms of the number of components, the number of phases coexisting, and the conditions necessary to establish equilibrium. In its simplest form, representing a system at a definite pressure and temperature, the phase rule is expressed as follows:

$$V = C + 2 - P \tag{1.1}$$

where V = the degrees of freedom or degrees of variance of the system in terms of the independent variables, concentration, temperature, and pressure, which must be fixed in order to define the system at equilibrium

C = the number of independent components appearing in all phases at equilibrium where the number of components represent the fewest variable constituents from which the composition of each phase can be expressed directly or in terms of an equation

P = the number of phases wherein a phase is a physically distinct part of the system having the same composition throughout and which is separated from the other parts of the system by means of an interface

Thus for systems at equilibrium, composed of a vapor phase and of a liquid phase, fixing two variables defines a binary system equilibrium, three variables a ternary system equilibrium, and n variables an n-component system equilibrium. For a system composed of three phases, fixing $n - 1$ variables establishes equilibrium for an n-component system.

Ideal Mixtures

Ideal gases are those whose behavior is described by the ideal-gas law, which can be stated mathematically as

$$V = \frac{RT}{P} \tag{1.2}$$

wherein the volume of one mole of gas is related to the conditions of temperature and pressure through the ideal-gas-law proportionality factor R. Ideal gases do not interact chemically by forming new molecular species or associate physically because of differences in molecular volumes and nonspecific molecular attractive forces. The same criteria may be said to apply to ideal liquids.

The concepts of "ideal" gas, "ideal" liquid, "ideal" gas mixtures, and "ideal" liquid mixtures have formed the basis for many quantitative relations describing equilibrium. Of principal interest in the field of distillation are Dalton's law of partial pressures and Raoult's law relating the pressure exerted by a component in the vapor phase of a gaseous mixture to its concentration in the liquid phase and its vapor pressure.

Dalton's [11] law states that the total pressure of a mixture of gases is equal to the sum of the partial pressures of the mixed gases. Thus,

$$P_t = \sum_1^n p_i = p_1 + p_2 + p_3 + \cdots + p_n \tag{1.3}$$

Also Dalton postulated that the partial pressure of an ideal gas in a gaseous mixture is proportional to the relative number of molecules of that gas in the mixture (or to its mole fraction). Thus

$$p_i = y_i P_t \tag{1.4}$$

Raoult's [36] law, relating the partial pressure in the vapor phase to the liquid phase composition, is expressed as

$$p_i = x_i P_i \tag{1.5}$$

Combining Dalton's and Raoult's laws results in an expression describing mixtures of ideal vapors and liquids in equilibrium.

$$P_t = \sum_1^n p_i = \sum_1^n y_i P_t = \sum_1^n x_i P_i \tag{1.6}$$

and for a single component,

$$y_i P_t = x_i P_i \tag{1.7}$$

The *vapor pressure* P_i of a component is a unique property of the component and is a direct function of temperature. Thus, it increases with an increase in temperature, and a material having a higher vapor pressure at a given temperature than another component is said to be *more volatile.*

Vapor pressure and temperature are commonly related by means of the Antoine [2] equation

$$\log P = A - \frac{B}{C + t} \tag{1.8}$$

where A, B, and C are constants for a particular compound over a relatively narrow temperature range (usually not over 100°C). Values of these constants for various compounds and families of compounds and the temperature ranges for which the constants apply appear in a number of references. Dreisbach [12], API Project Report No. 44 [3], Perry [34], and others present either the Antoine constants, tabular vapor pressure data, or both.

Example 1.1

The following vapor pressure-temperature data are available.

Pressure, mm Hg	Temperature, °C	
	Ethylbenzene	Ethyl cyclohexane
760	136.19	131.78
100	74.1	69.04
30	46.7	41.50

Determine the boiling temperatures for both compounds at 400 mm Hg pressure by the Antoine equation.

Solution

The three points of data permit the direct solution for a set of Antoine constants by algebraic means. An alternative method described in Dreisbach [12] can be used. The constant C is solved for first by the empirical formula:

$$C = 239 - 0.19t_B$$

where t_B is the normal boiling point. The three linear equations can then be solved for the best values of A and B.

For ethylbenzene,

$$\log P = A - \frac{B}{t + C}$$

$$C = 239 - 0.19(136.19) = 213.12$$

$$B_{12} = \frac{\log (P_2/P_1)}{1/(t_1 + C) - 1/(t_2 + C)} = \frac{0.52288}{1/259.82 - 1/287.22} = 1424.1$$

$$B_{13} = 1423.7 \qquad B_{23} = 1423.3 \qquad B_{\text{av}} = 1423.7$$

Substituting directly for A,

$$A_1 = 6.95668 \qquad A_2 = 6.95683 \qquad A_3 = 6.95656 \qquad A_{\text{av}} = 6.9567$$

For ethyl cyclohexane the solution is identical.

Tabulated Results

Source or method	Ethylbenzene Antoine constants			BP, 400 mm Hg, °C
	A	*B*	*C*	
Direct algebraic solution	9.093	3007.6	384.722	114.7
Dreisbach's method	6.9567	1423.7	213.12	113.82
Published values	6.95719	1424.255	213.206	113.82

Source or method	Ethyl cyclohexane Antoine constants			BP, 400 mm Hg, °C
	A	*B*	*C*	
Direct algebraic solution	8.941	2900	347.6	109.9
Dreisbach's method	6.8525	1373.2	213.96	109.11
Published values	6.87041	1384.036	215.128	109.12

For convenience in interpolation, vapor pressure charts in the form of Cox [10] and Katz et al. [25], among many others, are commonly used.

Relative volatility is a widely used relation in distillation calculations since it is a measure of separability; the larger the value of α_{ij}, the easier the separation. It is defined by

$$\alpha_{ij} = \frac{y_i/x_i}{y_j/x_j} \tag{1.9}$$

For an ideal mixture it is equal to the ratio of the vapor pressures.

$$\alpha_{ij} = \frac{P_i}{P_j} \tag{1.10}$$

Nonideal-Gas Mixtures

Actual gases and their mixtures generally do not follow the ideal-gas law because of the formation of different molecular species and because of intermolecular attractive forces. Deviations from the ideal-gas law can be expressed in a number of ways, and probably the most useful from a distillation standpoint are those of *compressibility factor* and *fugacity coefficient.*

$$Z = \frac{PV}{nRT} \tag{1.11}$$

where Z is the compressibility factor representing the ratio of the actual gas volume to that of an ideal gas under the same conditions of temperature and pressure. It approaches a value of unity at zero pressure and is essentially unity for most gases at and around atmospheric pressure. However, extreme deviations can be encountered at higher pressures.

Example 1.2

Assume that the ethylbenzene–ethyl cyclohexane system behaves ideally in a liquid mixture and determine the relative volatility of ethyl cyclohexane referred to ethylbenzene at (*a*) 50°C, (*b*) 150°C.

Components	Antoine constants		
	A	B	C
Ethylbenzene	6.95719	1424.255	213.206
Ethyl cyclohexane	6.87041	1384.036	215.128

Solution

$$\log P = A - \frac{B}{t + C}$$

Components	t	$t + C$	$B/(t + C)$	$\log P$	P, mm Hg
(1) Ethylbenzene	50	263.206	5.41118	1.54601	35.16
	150	363.206	3.92134	3.03585	1086.00
(2) Ethyl cyclohexane	50	265.128	5.22026	1.65015	44.68
	150	365.128	3.79055	3.07986	1201.90

At 50°C, $\alpha_{21} = 1.271$; at 150°C, $\alpha_{21} = 1.107$.

Most thermodynamics books give methods of determining compressibility factors using critical constants and correlations based on the theorem of corresponding states. For example, Hougen et al. [22] present a compressibility chart relating compressibility factor, reduced temperature, and reduced pressure for materials having a critical compressibility $Z_c = 0.27$. By means of tabular data and a correlating equation, the compressibility for compounds having critical compressibility factors other than 0.27 can be determined. This method resulted from the work by Lyderson et al. [31].

Fugacity coefficient ν relates the partial pressure exerted by an actual gas to that which it would exert if it behaved ideally. It is defined as follows:

$$\nu = \frac{f}{p} \tag{1.12}$$

where f is the *escaping tendency* or fugacity defined by Lewis and Randall [26] by

$$\left(\frac{\partial G}{\partial p}\right)_T = RT\, d \ln f \tag{1.13}$$

At lower pressures, around atmospheric and below, the fugacity is essentially equal to the partial pressure of the gas and the fugacity coefficient is essentially unity, whereas at higher pressures it deviates widely from unity. Hougen et al. [22] presented a fugacity coefficient chart based on the work of Lyderson et al. [31] relating the coefficient with reduced temperature and pressure for compounds having a critical compressibility $Z_c = 0.27$. Tabular data and a correction equation enable evaluation of the fugacity coefficient for other materials having different critical compressibility factors.

There are many compressibility and fugacity coefficient charts in the literature. These range in type from generalized relationships to specific compound charts correlated on the basis of true reduced conditions, pseudo-reduced conditions, and in some cases on actual temperatures and pressures. For specific compounds the specific charts are more accurate; for general use the author prefers to use the Hougen et al. [22] generalized charts and tables.

For a nonideal-gas mixture the fugacity coefficient enters into Dalton's-law equation as

$$\nu_i = \frac{f_i}{p_i} = \frac{y_i P_t}{p_i} \tag{1.14}$$

Nonideal-Liquid Mixtures

In actual liquids and liquid mixtures the mean distances between molecules are much less than those in the gaseous state and the forces of attraction between them are much greater. Nonideal behavior is evidenced by non-additivity of volumes when compounds are mixed in solution, and by heats of mixing when the pure components are mixed at constant temperature and pressure. The extent of deviation from nonideality of components in liquid mixtures is measured by the *activity coefficient* γ. Again the reader is referred to standard thermodynamics texts for derivations and definitions of this factor in terms of activity and fugacity.

Applying this correction factor to Raoult's law results in

$$p_i = \gamma_i x_i P_i \tag{1.15}$$

Incorporating both the fugacity coefficient and activity coefficient into a corrected Raoult's law results in Eq. (1.16), which can be used as

$$\nu_i y_i P_t = \gamma_i x_i P_i \tag{1.16}$$

a basis for calculation of vapor-liquid equilibria if the interrelationships of the variables are known or can be predicted. (Some thermodynamicists prefer to define three or more factors to correct for nonideal behavior. Here the two described are considered sufficient.)

Relative volatility in nonideal systems is shown by

$$\alpha_{ij} = \frac{y_i/x_i}{y_j/x_j} = \frac{\gamma_i P_i \nu_j}{\gamma_j P_j \nu_i} \tag{1.17}$$

It may increase, decrease, or remain constant with increase in temperature, depending upon the nature of the system. Because relative volatility varies less with temperature than does vapor pressure, it is used quite extensively in distillation calculations.

Phase Diagrams

Phase diagrams are used to describe two-component systems by plotting two of the three independent variables, composition, temperature, and pressure at a constant value of the remaining one. Composition versus enthalpy (a function of temperature) at constant pressure, and vapor composition versus liquid composition plots at constant pressure are useful in describing binary vapor-liquid equilibria and in quantitative distillation calculations. Figures 1.1 and 1.2 illustrate common types of binary phase diagrams.

In Fig. 1.1, the *a*, *e*, *i* diagrams are typical of *regular* or *normal* systems. The *b*, *f*, *j* diagrams are typical of *minimum boiling* homogeneous azeotropes, the *c*, *g*, *k* diagrams of *maximum boiling* homogeneous azeotropes, and the *d*, *h*, *l* diagrams of *minimum boiling* heterogeneous azeotropes. In the first three systems only one liquid phase exists, whereas in the fourth, two liquid phases can exist at and below the azeotrope temperature.

Effect of Pressure on Phase Equilibria

As the total pressure is increased on a binary system, the boiling temperatures of the pure components increase and the boiling points of their mixtures also increase. In addition, the two-phase area on the *tx* plot decreases until

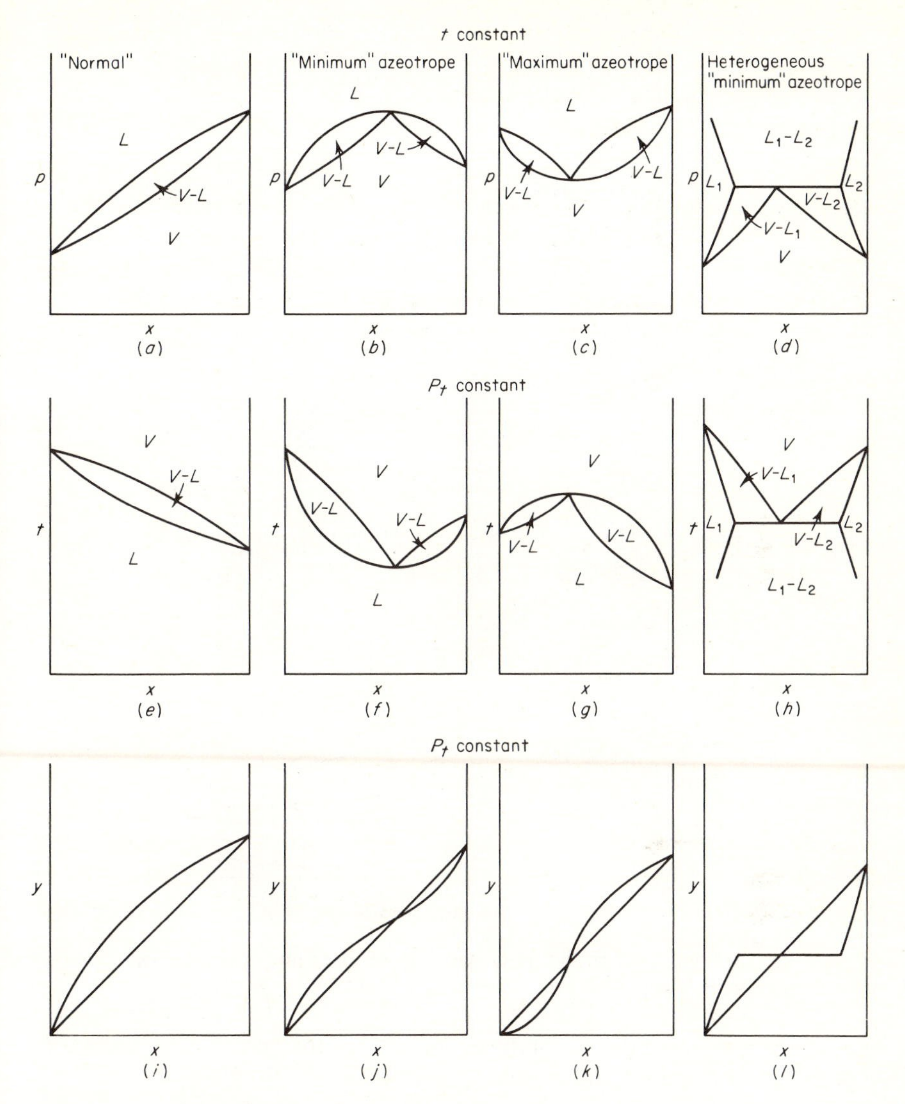

FIGURE 1.1 Phase diagrams for various types of binary systems.

it disappears at the critical temperature and pressure of the mixture. This is indicated in Fig. 1.3*a*.

A similar behavior is noted on the P_tx plot when the temperature is increased, as shown in Fig. 1.3*b*. It has been observed that, as the pressure increased, the difference in boiling points of the two components may decrease, increase, or remain essentially constant, depending on the character

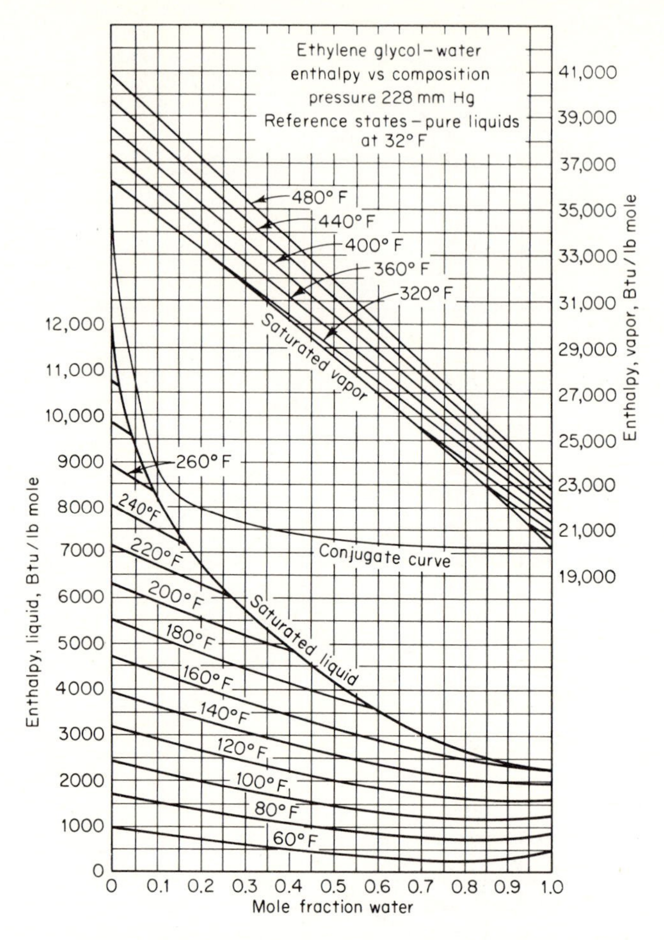

FIGURE 1.2 Enthalpy-concentration diagram. [*From Chu and Wang, Ind. Eng. Chem.*, ***42**:273* (*1950*).]

of the components in the mixtures, as long as the critical temperature (or pressure) of neither component is exceeded.

1.2. VAPOR-LIQUID EQUILIBRIA—BINARY SYSTEMS

Ideal Vapors and Liquids

The experimental criteria for ideal behavior of mixtures are:

1. There is no net change in volume when the components are mixed in the liquid or vapor state, i.e., the volumes are additive.

2. The heat of mixing is zero.

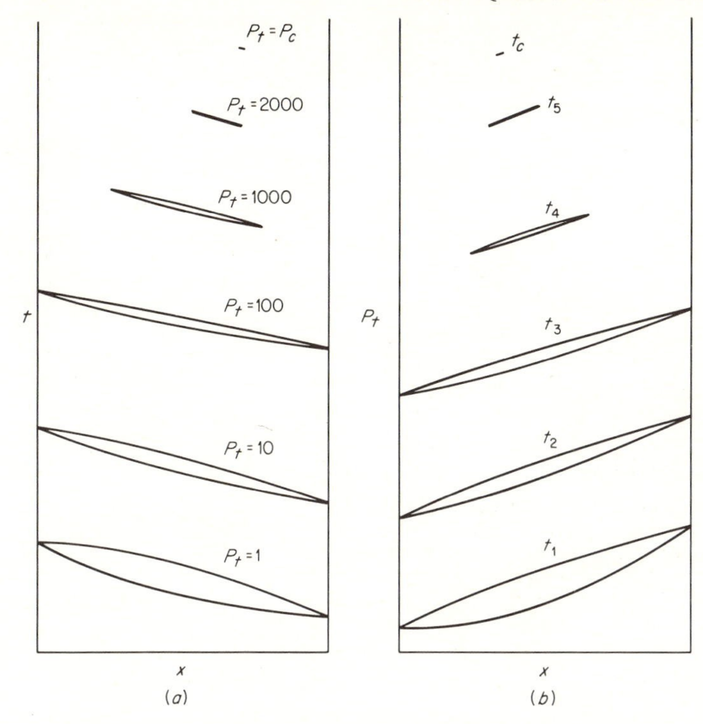

FIGURE 1.3 Binary phase diagrams at different pressures.

3. There is no change in molecular aggregation with respect to the components mixed and unmixed.

While in most cases nonideal mixtures can be detected on the basis of the above criteria, it is possible for molecular complexes between like molecules to be broken and others formed between unlike molecules with a zero net volume change. Also it is possible for physical interactions with endothermic heat effects and chemical interactions with exothermic heat effects to balance one another with no net effect. The chances of both occurring in a mixture are remote, but it is recommended that the ideality of mixtures be tested by more than one criterion.

The vapor-liquid equilibrium composition data may be calculated for a binary system assumed to be ideal from the vapor pressure-temperature data of the pure components at and around atmospheric pressure. A series of temperatures ranging from the boiling temperature of one component to that of the other are selected, and the vapor pressures of the components are evaluated at those temperatures from data, charts, or from Antoine constants. The vapor pressure data are then used in Eqs. (1.18) and (1.19) to calculate the vapor and liquid compositions in equilibrium at the temperature selected. From the combined Dalton's and Raoult's law Eq. (1.6), applied to a binary system, the following equations result:

$$x_1 = \frac{P_t - P_2}{P_1 - P_2} \tag{1.18}$$

$$y_1 = \frac{P_1 x_1}{P_t} \tag{1.19}$$

The liquid-phase boundary line on a Px diagram (at constant temperature) is a straight line for ideal systems described by

$$x_1(P_1 - P_2) + P_2 = P_t \tag{1.20}$$

In ideal binary systems and in actual systems whose behavior approximates ideality, the relative volatility is constant or essentially constant, and the equilibrium data may be calculated by

$$y_1 = \frac{\alpha_{12} x_1}{1 - (1 - \alpha_{12}) x_1} \tag{1.21}$$

and

$$x_1 = \frac{y_1}{\alpha_{12} + y_1(1 - \alpha_{12})} \tag{1.22}$$

Thus the entire range of equilibrium data can be derived from the constant relative volatility or from an average relative volatility, if the α values at the component boiling temperatures do not differ more than a few percent.

Nonideal Vapors and Liquids

Although many binary systems, both in the vapor and liquid states, approximate ideal behavior, the greater majority of systems encountered in practice are nonideal in either or both states. For distillation calculations it is necessary to have reasonably accurate vapor-liquid equilibrium composition data. These may be obtained by a variety of methods:

1. They may be determined experimentally.

2. The system may be assumed to be ideal in its behavior, and the data may be calculated from the ideal equations.

3. The data may be calculated from a few experimental points by utilizing empirical equations.

4. They may be roughly estimated from physical data on the pure components alone with the use of empirical relations.

Since no method of calculating vapor-liquid equilibrium data is completely accurate, actual experimental data should be used if at all possible. For systems known to be approximately ideal, the use of ideal equations may be satisfactory within the limits of engineering accuracy. Where the system is known to be nonideal, it is necessary to use method 3 if complete experimental data are not available. Method 4 is to be used only for rough approximations.

Because the fugacity coefficient is a quantitative correction factor showing the extent of the departure of actual gases from ideal behavior and because the activity coefficient is a similar correction factor relating the behavior of actual liquids to that of ideal liquids, evaluation of these coefficients as a function of temperature, pressure, and composition is essential to the determination of vapor-liquid equilibrium data. Also, because distillations are practically always conducted at "constant" pressure conditions, only the isobaric data are useful in direct application to design calculations. Thus, the relationship between the coefficients, temperature, and composition at constant pressure is of particular interest. Furthermore, most distillations are conducted at lower pressures where the behavior of the vapor phase is essentially ideal and the fugacity coefficients are approximately unity. Because of this, more effort has been extended to the study of solution nonideality through activity coefficient correlation with composition at constant pressure.

Variables Affecting Activity Coefficient

Composition The Gibbs-Duhem [15, 13] equation was derived from basic thermodynamic relations for constant temperature and constant pressure, and neglecting all internal energy contributions of surface, magnetic, electrical, etc., effects. (For derivation, see chemical engineering thermodynamics texts.) In terms of activity coefficient and composition, the general equation is

$$x_1 \left(\frac{\partial \ln \gamma_1}{\partial x_1}\right)_{T,P_t} + x_2 \left(\frac{\partial \ln \gamma_2}{\partial x_1}\right)_{T,P_t} + \cdots + x_n \left(\frac{\partial \ln \gamma_n}{\partial x_1}\right)_{T,P_t} = 0 \quad (1.23)$$

Because of the conditions assumed in the derivation, this equation is strictly applicable only to systems at constant temperature and pressure. It cannot be integrated unless some simplifying assumptions are introduced which are valid for the particular system under consideration. Otherwise the composition data calculated from the integrated equation can be seriously in error.

Ibl and Dodge [23] modified the Gibbs-Duhem equation for a binary system at constant pressure and obtained

$$x_1\, d \ln \gamma_1 + x_2\, d \ln \gamma_2 = Z\, dx_1 \quad (1\ 24)$$

where

$$Z = -\frac{\Delta H}{RT^2}\left(\frac{dT}{dx_1}\right)_{P_t} \tag{1.25}$$

and

$$\Delta H = H - x_1 H_1^\circ - x_2 H_2^\circ \tag{1.26}$$

where H, H_1°, H_2° are molal enthalpies of mixture and components, respectively, as liquids at the temperature and pressure of the equilibrium.

The effect of pressure on activity coefficient can be shown by

$$\ln \frac{\gamma_{i,P_{t1}}}{\gamma_{i,P_{t2}}} = -\frac{1}{RT}(\bar{V}_i - V_i^\circ)_{\text{av}}(P_{t1} - P_{t2}) \tag{1.27}$$

The variation of activity coefficient with temperature is described by

$$\ln \frac{\gamma_{i,T1}}{\gamma_{i,T2}} = -\left(\frac{\bar{H}_i - H_i^\circ}{R}\right)_{\text{av}} \int_{T_1}^{T_2} \frac{dT}{T^2} = \frac{k}{R}\,\frac{T_1 - T_2}{T_1 T_2} \tag{1.28}$$

Example 1.3

Assuming ideal behavior of both liquid and vapor for the binary system, *n*-octane–ethylbenzene at 200 mm Hg, compute the following data and plot *tx*, *ty*, and *xy*.

Components	Antoine constants		
	A	*B*	*C*
1,*n*-Octane	6.92377	1355.13	209.52
2-Ethylbenzene	6.95719	1424.255	213.206

Solution

(*a*) Vapor pressure data (from Antoine equation):

Temperature, °C	Pressure, mm Hg	
	n-Octane	Ethylbenzene
83.62	200.0	144.2
85	210.2	150.6
87	225.8	162.1
89	242.3	175.5
91	259.7	188.5
92.67	275.1	200.0

(*b*) Ideal compositions: $x_1 = (200 - P_2)/(P_1 - P_2)$, $y_1 = P_1x_1/200$.

Temperature, °C	$200 - P_2$	$P_1 - P_2$	x_1	x_1P_1	y_1
83.62	55.8	55.8	1.000	200	1.0
85	49.4	59.6	0.829	174.25	0.871
87	37.9	63.7	0.595	134.35	0.672
89	24.5	66.8	0.367	88.92	0.445
91	11.5	71.2	0.162	41.94	0.210
92.67	0.0	75.1	0.0	0.0	0.0

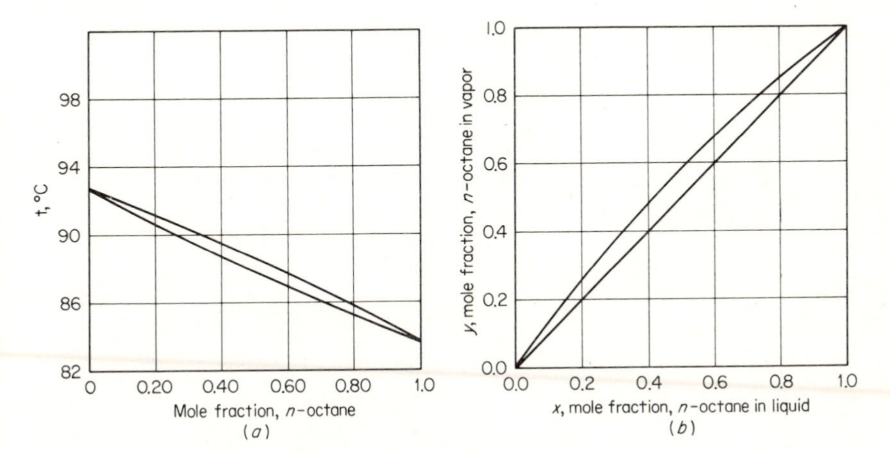

EXAMPLE 1.3 *n*-Octane–ethylbenzene system at 200 mm Hg. (*a*) Temperature-composition diagram. (*b*) Equilibrium diagram.

The van Laar Equations

Van Laar [44] attempted to relate activity coefficients and compositions by making the following assumptions regarding the thermodynamic changes occurring in mixing pure liquids. He assumed (*a*) that the change in the excess partial entropy of mixing of a component is zero or the actual change in partial entropy of the component upon mixing is equal to the change in partial entropy of the component upon mixing if it formed an ideal solution, i.e., $\Delta S_i^E = 0$; (*b*) that the partial molal volumes of the components remain constant upon mixing; (*c*) that the van der Waals [43] equation of state applies to both components and mixture as vapors or liquids.

Van Laar used the basic thermodynamic relation

$$\left(\frac{\partial U}{\partial V}\right)_T = T\left(\frac{\partial P}{\partial T}\right)_V - P$$

and the van der Waals equation of state.

$$P = \frac{RT}{V - b} - \frac{a}{V^2}$$

Differentiation yielded

$$T\left(\frac{\partial P}{\partial T}\right)_V = \frac{RT}{V - b} \tag{1.29}$$

and substitution gave

$$\left(\frac{\partial U}{\partial V}\right)_T = \frac{a}{V^2} \tag{1.30}$$

Integration from vapor at zero pressure to the liquid state yielded

$$U_L - U_V^\circ = -\frac{a}{V_L} \tag{1.31}$$

Using the assumption that $V = b$ and that the constants of the mixture could be obtained by

$$a_{\text{mix}} = (x_1 a_1^{0.5} + x_2 a_2^{0.5} + \cdots + x_n a_n^{0.5})^2 \tag{1.32}$$

and

$$b_{\text{mix}} = x_1 b_1 + x_2 b_2 + \cdots + x_n b_n \tag{1.33}$$

he derived

$$\ln \gamma_1 = \frac{\dfrac{b_1}{RT}\left(\dfrac{a_1^{0.5}}{b_1} - \dfrac{a_2^{0.5}}{b_2}\right)^2}{\left(1 + \dfrac{b_1}{b_2}\dfrac{x_1}{x_2}\right)^2} \tag{1.34}$$

for component 1 in a binary system. Defining

$$A = \frac{b_1}{b_2} \tag{1.35}$$

and

$$B = \frac{b_1}{R}\left(\frac{a_1^{0.5}}{b_1} - \frac{a_2^{0.5}}{b_2}\right)^2 \tag{1.36}$$

the classic van Laar equations resulted:

$$T \ln \gamma_1 = \frac{B}{[1 + A(x_1/x_2)]^2} \tag{1.37}$$

$$T \ln \gamma_2 = \frac{AB}{(A + x_2/x_1)^2} \tag{1.38}$$

Wohl Equations

From basic thermodynamic equations it can be shown that the activity coefficient of any component can be related to partial excess free energy and composition of a solution as follows:

$$\left(\frac{\partial \Delta G^E}{\partial n_i}\right)_{T,P_t,n \neq i} = RT \ln \gamma_i \tag{1.39}$$

Wohl [46, 47] related excess free energy and composition in the liquid phase by the expansion of the excess free-energy equation

$$\frac{\Delta \bar{G}^E}{2.303RT} \frac{1}{\sum_1^n q_i x_i} = \sum_{ij} z_i z_j a_{ij} + \sum_{ijk} z_i z_j z_k a_{ijk} + \cdots \tag{1.40}$$

where

$$\Delta \bar{G}^E = \frac{\Delta G^E}{n_i} \qquad \text{(partial molar excess free energy)} \tag{1.41}$$

and q_i, q_j, etc., are effective molal volumes of components i, j, etc.

For example,

$$z_i = \frac{q_i x_i}{\sum_j q_j x_j} \qquad \text{(effective volume fraction of } i) \tag{1.42}$$

and a_{ij}, a_{ijk}, etc., are constants relating force interaction between combinations of unlike molecules.

For a binary,

$$\sum_{ij} z_i z_j a_{ij} = z_1 z_2 a_{12} + z_2 z_1 a_{21} = 2 z_1 z_2 a_{12} \tag{1.43}$$

Since

$$a_{12} = a_{21} \tag{1.44}$$

the greater the number of terms in Eq. (1.40) the more accurate the fit to actual data. Equations involving, for example, a_{121} are third-order equations and those involving, for example, a_{1211} are fourth-order equations.

Third-order equations for a binary system were developed by Wohl as follows:

$$\frac{\Delta \bar{G}^E}{2.303RT} \frac{1}{q_1 x_1 + q_2 x_2} = 2 z_1 z_2 a_{12} + 3 z_1^2 z_2 a_{112} + 3 z_1 z_2^2 a_{122} \tag{1.45}$$

Since the sum of the effective volume fractions is equal to 1.0, any or all terms can be multiplied by $z_1 + z_2$:

$$\frac{\Delta \bar{G}^E}{2.303RT} = \left(x_1 + \frac{q_2}{q_1} x_2\right) z_1 z_2 [z_1 q_1 (2a_{12} + 3a_{112}) + z_2 q_1 (2a_{12} + 3a_{122})] \tag{1.46}$$

Letting

$$A = q_1(2a_{12} + 3a_{122}) \tag{1.47}$$

and

$$B = q_2(2a_{12} + 3a_{112}) \tag{1.48}$$

$$\frac{\Delta \bar{G}^E}{2.303RT} = \left(x_1 + \frac{q_2}{q_1} x_2\right) z_1 z_2 \left(z_1 \frac{Bq_1}{q_2} + z_2 A\right) \tag{1.49}$$

Differentiating Eq. (1.39),

$$\log \gamma_1 = \frac{\partial(\Delta \bar{G}^E/2.303RT)}{\partial n_1} = \frac{\partial[(n_1 + n_2)\Delta \bar{G}^E/2.303RT]}{\partial n_1} \tag{1.50}$$

$$x_1 = \frac{n_1}{n_1 + n_2} \tag{1.51}$$

$$x_2 = \frac{n_2}{n_1 + n_2} \tag{1.52}$$

$$z_1 = \frac{n_1}{n_1 + n_2(q_2/q_1)} \qquad z_2 = \frac{n_2(q_2/q_1)}{n_1 + n_2(q_2/q_1)} \tag{1.53}$$

$$\log \gamma_1 = \frac{n_2^2(q_2/q_1)^2[-n_1 A + 2n_1 B(q_1/q_2) + n_2 A(q_2/q_1)]}{[n_1 + n_2(q_1/q_2)]^3} \tag{1.54}$$

In terms of effective volume fraction z, the Wohl equations become

$$\log \gamma_1 = z_2^2 \left[A + 2z_1\left(B\frac{q_1}{q_2} - A\right)\right] \tag{1.55}$$

$$\log \gamma_2 = z_1^2 \left[B + 2z_2\left(A\frac{q_2}{q_1} - B\right)\right] \tag{1.56}$$

Other Equations Correlating Binary Vapor-Liquid Equilibria

Carlson and Colburn [9] modified the van Laar equations to eliminate the temperature variable. Since van Laar assumed q_1 and q_2 were the van der Waals volume constants b_1 and b_2, the ratio of the effective molal volumes becomes equal to the ratio of the modified constants. Thus, $q_2/q_1 = B'/A'$ and

$$\log \gamma_1 = \frac{A' x_2^2}{[(A'/B')x_1 + x_2]^2} \tag{1.57}$$

$$\log \gamma_2 = \frac{B' x_1^2}{[x_1 + (B'/A')x_2]^2} \tag{1.58}$$

Generally the "normal" binary systems show positive values of $\log \gamma$ over the whole composition range. The minimum boiling azeotropic systems have positive values throughout the composition range, and the maximum boiling azeotropic systems exhibit negative values over the entire range of composition. The constants A' and B' in the Carlson-Colburn modified van Laar equation are related to the corresponding constants in the original van Laar equations as follows:

$$A' = \frac{B}{2.303T} \tag{1.59}$$

$$B' = \frac{B}{2.303AT} \tag{1.60}$$

where A', B' are the Carlson-Colburn constants and A, B are the original van Laar constants.

The constants A' and B' in the foregoing equations can be evaluated from experimental data. For a binary system only one point of data would be required, theoretically, for simultaneous solution of pairs of equations. This requires an accurately determined point such as that represented by an azeotrope composition and temperature. Other points may be used if they are accurately evaluated and those determined in the middle range of concentration, i.e., $x = 0.4$ to 0.6, are usually more reliable.

A common method [41] utilizing the Carlson-Colburn modified van Laar equations for evaluation of isobaric binary vapor-liquid equilibrium data from one experimental point without the use of trial-and-error calculations is outlined as follows. Solving for the constants in terms of composition and activity coefficients and utilizing

$$A' = \log \gamma_1 \left(1 + \frac{x_2 \log \gamma_2}{x_1 \log \gamma_1}\right)^2 \tag{1.61}$$

$$B' = \log \gamma_2 \left(1 + \frac{x_1 \log \gamma_1}{x_2 \log \gamma_2}\right)^2 \tag{1.62}$$

and using

$$\gamma_i = \frac{y_i P_t}{x_i P_i}$$

the method is as follows:

1. From the experimental point x, y, t, P_t determine γ_1, γ_2, P_1, and P_2.
2. Calculate the values of A' and B' from Eqs. (1.61) and (1.62).
3. Compute $\gamma_1 x_1$ and $\gamma_2 x_2$ for the whole concentration range and plot.
4. Calculate tx and ty data by the following steps:

(*a*) Assume that Eq. (1.63) will express the vapor pressure relation of both components with temperature.

$$P_1 = a' + b' P_2 \tag{1.63}$$

(This is valid for most binary systems over their range of boiling temperature.)

$$P_1 = \frac{P_t}{\gamma_1 x_1} - \frac{\gamma_2 x_2}{\gamma_1 x_1} P_2 \tag{1.64}$$

Let

$$a = \frac{P_t}{\gamma_1 x_1} \tag{1.65}$$

and

$$b = -\frac{\gamma_2 x_2}{\gamma_1 x_1} \tag{1.66}$$

$$P_1 = a + bP_2 \tag{1.67}$$

(*b*) Solving Eq. (1.63) for P_2 and substituting in Eq. (1.67),

$$P_1 = \frac{ab' - a'b}{b' - b} \tag{1.68}$$

b' is obtained from the vapor pressures of the two components at two different temperatures (within the boiling range between that of the two components):

$$b' = \frac{P_{1,t1} - P_{1,t2}}{P_{2,t1} - P_{2,t2}} \tag{1.69}$$

$$a' = P_{1,t1} - b'P_{2,t1} \tag{1.70}$$

a' and b' can be considered constant for a given system.

(*c*) Calculate a and b for assumed values of x and calculate a' and b'. From Eq. (1.68) evaluate P_1 corresponding to the different values of x. From the vapor pressure data determine the temperatures corresponding to the different values of P_1.

From the limit equations

$$A' = \lim_{x_1 \to 0} \log \gamma_1 \tag{1.71}$$

$$B' = \lim_{x_2 \to 0} \log \gamma_2 \tag{1.72}$$

A' and B' can be determined also by extrapolation of the log γ versus curves to $x_1 = 0$ and $x_2 = 0$. This method is subject to possible error because the experimental points at low concentration are difficult to evaluate accurately.

Example 1.4

For a total pressure of 100 mm Hg calculate tx, ty, yx, γx curves for the system *n*-dodecane–butyl carbitol using the Carlson-Colburn modified van Laar equations. The azeotrope temperature was found to be 142.6°C and the composition is $x = 0.65$ mole fraction dodecane.

Solution

Vapor pressures are conveniently shown as data calculated by the Antoine equation:

Boiling Points at Several Pressures, °C

Pressure, mm Hg	Component 1, *n*-dodecane	Component 2, butyl carbitol
50	127.54	141.4
80	139.90	153.3
100	146.14	159.8
150	158.10	172.3
200	167.14	181.2

At the azeotrope:

$$\gamma_1 = \frac{P_t}{P_1} = \frac{100}{88.2} = 1.134 \qquad \gamma_2 = \frac{100}{53} = 1.887$$

$$\log \gamma_1 = 0.05461 \qquad \log \gamma_2 = 0.27577$$

The constants A' and B' of the Carlson-Colburn modification of the van Laar equations are solved for by Eqs. (1.61) and (1.62)

$$A' = \log \gamma_1 \left(1 + \frac{x_2 \log \gamma_2}{x_1 \log \gamma_1}\right)^2 = 0.05461 \left[1 + \frac{(0.35)(0.27577)}{(0.65)(0.05461)}\right]^2$$

$$= 0.75538 \qquad \text{or} \qquad 0.755$$

$$B' = \log \gamma_2 \left(1 + \frac{x_1 \log \gamma_1}{x_2 \log \gamma_2}\right)^2 = 0.27577 \left[1 + \frac{(0.65)(0.05461)}{(0.35)(0.27577)}\right]^2$$

$$= 0.51590 \qquad \text{or} \qquad 0.516$$

A comparison of experimental and calculated points is as follows:

Temperature, °C	*n*-Dodecane, mole %		
	Experimental		Calculated
	x	*y*	*y*
155.5	4.6	31.0	30.55
150.0	12.5	44.5	44.0
145.2	38.0	54.0	54.2
143.2	46.0	59.0	59.0
142.7	58.5	62.5	62.5
142.5	74.3	71.0	70.6
143.6	89.5	83.2	83.5

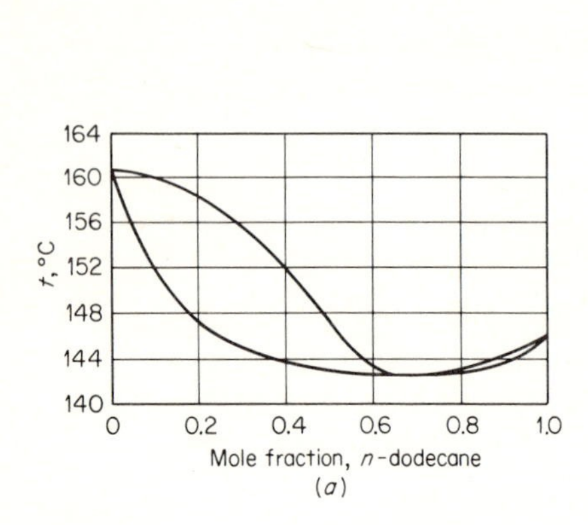

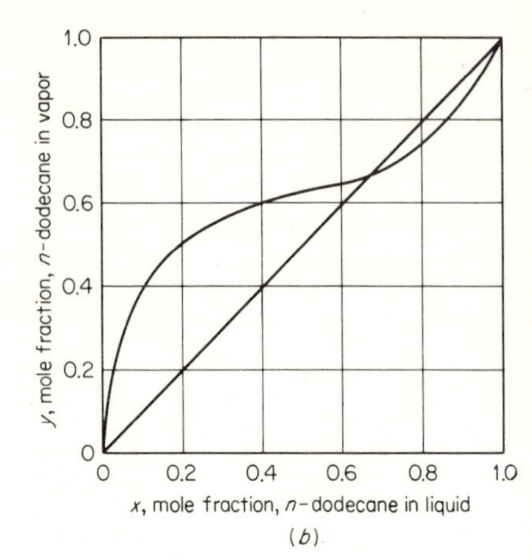

EXAMPLE 1.4 *n*-Dodecane–butyl carbitol system at 100 mm Hg. Equilibrium boiling point diagram. [*From Houser and Van Winkle, J. Chem. Eng. Data*, **2:**12 (*1957*).]

Many equations utilizing basic thermodynamic relations and based upon various simplifying assumptions have been developed to describe the interrelationship between activity coefficient and composition of nonideal binary systems for the purpose of extending partial data or correlating data.

Scatchard and Hamer [39] assumed the effective molal volume of the component was equal to its molal volume, i.e., $q_1 = V_1$ and $q_2 = V_2$. Margules [32] assumed the effective molal volumes of the components to be equal, i.e., $q_2/q_1 = 1.0$, resulting in the following frequently used equations:

$$\log \gamma_1 = x_2^2[A + 2x_1(B - A)] \tag{1.73}$$

$$\log \gamma_2 = x_1^2[B + 2x_2(A - B)] \tag{1.74}$$

Hala et al. [19] assumed ideal-vapor-phase behavior and derived from the Redlich and Kister [37] equations a set of equations involving vapor pressures of the pure components. White [45], Yu and Coull [49], Li and Coull [27], Bonham [7], Black [4, 5], Black et al. [6], and others have derived equations of the van Laar type. In addition there have been many other methods proposed for prediction of vapor-liquid equilibrium data—from total pressure measurements [35], from data on other but similar systems [48], from relative volatilities [17, 18], and from pure-component properties for certain classes of systems [29]. Binary azeotrope temperature and composition can be predicted at different pressures from data at one pressure [24].

Example 1.5

With the use of the Margules equations and the following experimental point, determine the activity coefficient composition curves for the ethyl alcohol (component 1)–methyl ethyl ketone (component 2) binary system.

Azeotrope: 50.0 mole % ethanol at 165.2°F (760 mm Hg total pressure).

Components	Antoine constants		
	A	B	C
(1) Ethanol	8.04494	1554.3	222.65
(2) MEK	6.97421	1209.6	216.00

Solution

At the azeotrope, $x_i = y_i$ and, consequently, $\gamma_1 = P_t/P_1$,

$$\gamma_1 = 760/643.3 = 1.181 \qquad \log \gamma_1 = 0.07225$$

$$\gamma_2 = 760/639.2 = 1.189 \qquad \log \gamma_2 = 0.07518$$

The Margules equations are:

$$\log \gamma_1 = x_2^2[A + 2x_1(B - A)]$$

$$\log \gamma_2 = x_1^2[B + 2x_2(A - B)]$$

Substituting the values of composition and activity coefficient at the azeotrope results in

$$0.07225 = 0.25[A + (B - A)]$$

$$0.07518 = 0.25[B + (A - B)]$$

Solution for A and B is direct at the midpoint:

$$A = 0.3007 \qquad B = 0.2890$$

Repeated solution of the Margules equations gives the following set of data.

x_1	$\log \gamma_1$	γ_1	$\log \gamma_2$	γ_2
0.0	0.30072	1.997	0.00	1.000
0.2	0.18946	1.547	0.01231	1.029
0.4	0.10488	1.273	0.04849	1.118
0.6	0.04587	1.111	0.10742	1.281
0.8	0.01128	1.026	0.18796	1.542
1.0	0.00	1.0	0.28900	1.945

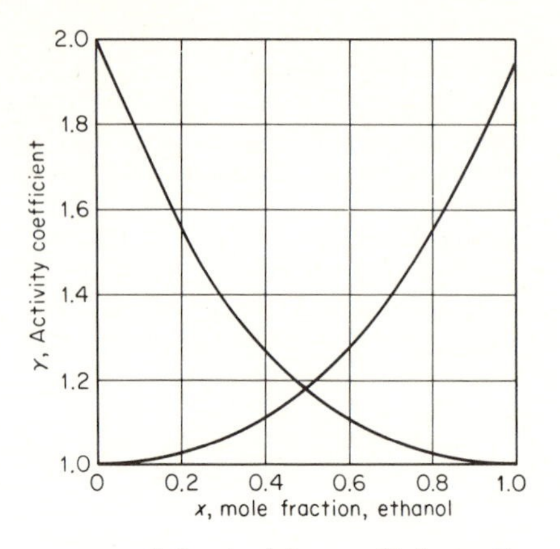

EXAMPLE 1.5 Activity coefficient—liquid composition relation for the methyl ethyl ketone—ethyl alcohol system at 760 mm Hg.

1.3. PREDICTION OF BINARY VAPOR-LIQUID EQUILIBRIUM DATA FROM PURE-COMPONENT PROPERTIES

The prediction of mixture properties from properties of the components composing the mixture is the most desirable method because extremely large quantities of pure-component data are available compared to mixture property data. For very gross approximations, the values of the a and b constants of the van Laar equation can be estimated from

$$a = \frac{27R^2T_c^2}{64P_c} \tag{1.75}$$

$$b = \frac{RT_c}{8P_c} \tag{1.76}$$

Critical data can be estimated by the method of Lyderson [30]. When the data are substituted into the van Laar equations for A and B, the following results are obtained:

$$A = \frac{b_1}{b_2} = \frac{T_{c_1}P_{c_2}}{T_{c_2}P_{c_1}} \tag{1.77}$$

$$B = 3.375T_{c_1}\left[1 - \left(\frac{P_{c_2}}{P_{c_1}}\right)^{0.5}\right]^2 \tag{1.78}$$

At best these approximations could be applicable only to binary mixtures showing no chemical interaction effects or hydrogen bonding.

Finch and Van Winkle [14], using Scatchard's and Hamer's [39, 40] modification of the van Laar equations, Hildebrand's [21] approximation for internal pressure and postulates by van Arkel [42] and London [28], developed a method for prediction of vapor-liquid equilibria for binary systems from pure-component properties. For nonpolar–nonpolar systems, satisfactory agreement with experimental data was obtained. For polar–nonpolar systems the agreement was less satisfactory, and for polar–polar systems the method failed to predict data within required engineering accuracy.

1.4. CONSISTENCY TESTS FOR BINARY VAPOR-LIQUID EQUILIBRIUM DATA

Distillation process design generally involves the use of some experimental vapor-liquid equilibrium data. These data may be reported in the literature as binary, ternary, or multicomponent data and may be complete, partially complete, or consist of only a few points. Because the sources of such data vary widely with regard to methods used, techniques involved, and even in skill of the individuals determining them, it may be necessary for the engineer, who proposes to use the data, to apply some kind of test to determine, approximately at least, their quality. Although this testing procedure may indicate only whether or not the data are internally consistent, such an indication is of considerable value in establishing the probable accuracy of calculations in which the data are involved. In addition, there may be more than one set of data reported on one system which differ in numerical values of temperature or composition. Consistency tests will generally enable the selection of the data which are probably more correct.

Most of the consistency tests have been derived from thermodynamic relations for binary data, and most of the tests are based upon the Gibbs-Duhem [15, 13] relation.

Rearrangement of the Gibbs-Duhem expression written for a binary system results in

$$\left(\frac{\partial \ln \gamma_1/\partial x_1}{\partial \ln \gamma_2/\partial x_1}\right)_{T,P_t} = -\frac{x_2}{x_1} \tag{1.79}$$

This states that the ratio of the slopes of the activity coefficient curves versus liquid composition is equal to the negative ratios of the mole fractions at any point. This is called a *thermodynamic consistency test* since the Gibbs-Duhem equation was developed from basic thermodynamic definitions. To check the consistency of any set of data, the curves are plotted as indicated and the slopes at selected compositions are evaluated and compared with the ratios of the liquid compositions.

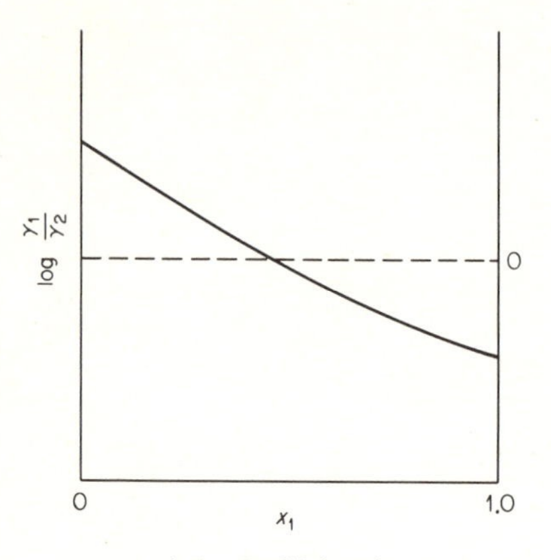

FIGURE 1.4 Redlich-Kister test.

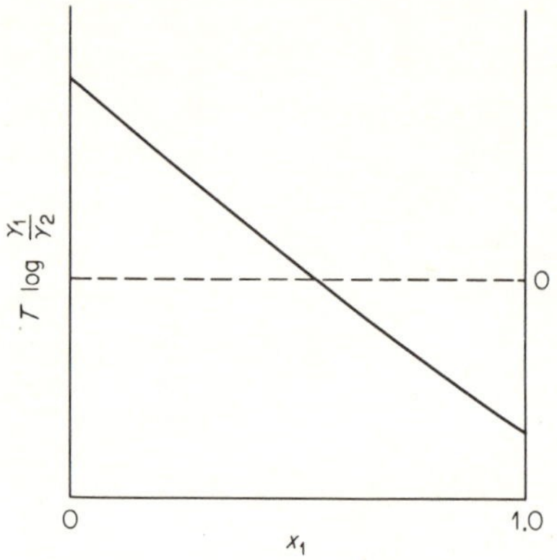

FIGURE 1.5 Broughton-Brearley test.

Redlich and Kister [38] recommended plotting the ratio of $\log(\gamma_1/\gamma_2)$ versus the liquid composition x_1, and determining the areas above and below the line where $\log(\gamma_1/\gamma_2) = 0$, and extending to the curve. If the areas are equal, the data are considered to be consistent. The extent of deviation from equality is a measure of the inconsistency. Figure 1.4 illustrates the method. Broughton and Brearley [8] developed a consistency test based upon the assumption that the term $T \log \gamma$ is independent of temperature. They recommended plotting $T \log(\gamma_1/\gamma_2)$ versus x_1 and measuring the areas above and below the line $T \log(\gamma_1/\gamma_2) = 0$ and the plotted line. This test is shown in Fig. 1.5.

A different type of consistency test based on the van Laar equations was proposed by Black [4] wherein the values of $(\log \gamma_1)^{0.5}$ are plotted against the values of $(\log \gamma_2)^{0.5}$, as shown in Fig. 1.6. If a straight line results, the data are said to be consistent with the van Laar relations. The Norrish and Twigg [33] consistency test used the following relations:

$$z = \ln\left(\frac{y_1}{y_2^k}\frac{x_2^k}{x_1}\right) \tag{1.80}$$

where k is the ratio of the molal latent heat of vaporization of the lower boiling component to that of the higher boiling component. They claim that the data are consistent when a straight line results from plotting z against the liquid composition x_1, as indicated in Fig. 1.7. Thus:

$$z = mx + C \tag{1.81}$$

where m and C are constants for a particular binary system.

Adler et al. [1] proposed a thermodynamic consistency test for binary systems when the more volatile component is above its critical temperature.

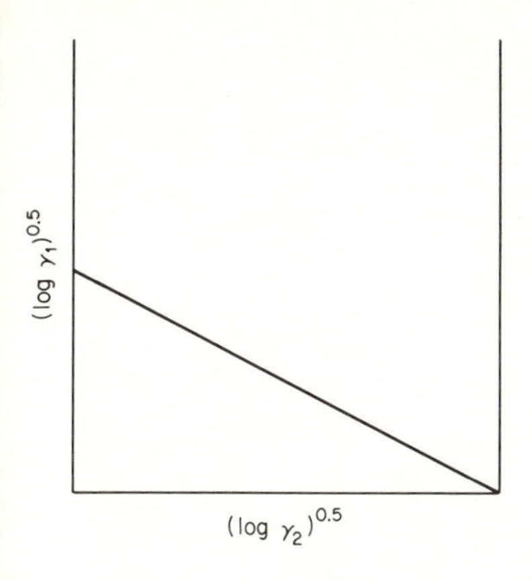

FIGURE 1.6 Black test.

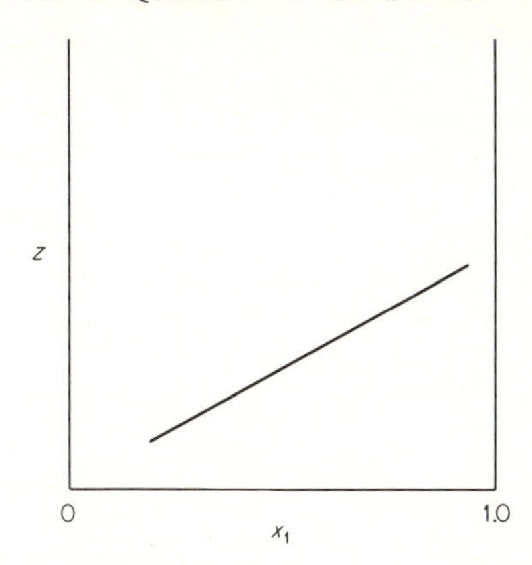

FIGURE 1.7 Norrish and Twigg test.

The following equations are presented:

$$x_1 d\ln(y_2 P_t \nu_{1,P_t}) + x_2\, d\ln(y_2 P_t \nu_{2,P_t}) = Z_L\, d\ln P_t \tag{1.82}$$

where

$$Z_L = \frac{V_L P_t}{RT} \tag{1.83}$$

where V_L = volume of liquid, cu ft/mole.

$$x_1\, d\ln(K_1 P_T \nu_{1,P_t}) + x_2\, d\ln(K_2 P_T \nu_{2,P_t}) = Z_L\, d\ln P_t \tag{1.84}$$

where

$$K = \frac{y}{x} \tag{1.85}$$

$$x_1\, d\ln K_1 + x_2\, d\ln K_2 = Z_L + Z_{1_V} y_1 \left(\frac{1}{K_2} - \frac{1}{K_1}\right) - \frac{Z_{2_V}}{K_2}\, d\ln P_T \tag{1.86}$$

Example 1.6

Two sets of experimental equilibrium data, *A* and *B*, are available on the binary system methyl propyl ketone–ethyl alcohol. P_t = 760 mm Hg.

A			*B*		
t, °F	x_{EtOH}	y_{EtOH}	*t*, °F	x_{EtOH}	y_{EtOH}
173	0.985	0.983	173	0.99	0.995
172.7	0.969	0.967	175	0.61	0.765
175.5	0.624	0.738	178	0.55	0.755
181.4	0.301	0.518	180	0.33	0.56
186.8	0.13	0.347	185	0.15	0.35

EtOH		MPK	
Temperature, °C	Vapor pressure, mm Hg	Temperature, °C	Vapor pressure, mm Hg
8.0	20	28.5	20
26.0	60	47.3	60
48.4	200	71.0	200
63.5	400	86.8	400
78.4	760	103.3	760

Determine which set of data is the more internally consistent by (*a*) the Redlich and Kister test and (*b*) the Broughton and Brearley test.

Solution

Activity coefficients are calculated by

$$\gamma_i = \frac{y_i P_t}{x_i P_i}$$

(ethanol is component 1).

System A

x_1	t, °C	P_1	P_2	γ_1	γ_2	$\log(\gamma_1/\gamma_2)$	$T\log(\gamma_1/\gamma_2)$
0.985	78.51	765.4	280.8	0.991	3.067	−0.4907	−172.5
0.969	78.34	760.3	278.7	0.998	2.903	−0.4639	−163.1
0.624	79.89	808.0	298.2	1.112	1.776	−0.2033	−71.76
0.301	83.17	917.4	343.4	1.426	1.526	−0.0294	−10.48
0.130	86.17	1028.0	389.7	1.973	1.464	0.1297	46.60

System B

x_1	t, °C	P_1	P_2	γ_1	γ_2	$\log(\gamma_1/\gamma_2)$	$T\log(\gamma_1/\gamma_2)$
0.99	78.51	765.4	280.8	0.998	1.352	−0.1319	−46.39
0.61	79.62	799.6	292.3	1.192	1.567	−0.1193	−42.09
0.55	81.28	869.0	316.7	1.200	1.307	−0.0370	−13.11
0.33	82.40	890.7	332.4	1.448	1.502	−0.0158	−5.62
0.15	85.17	989.8	373.7	1.792	1.555	0.0614	22.00

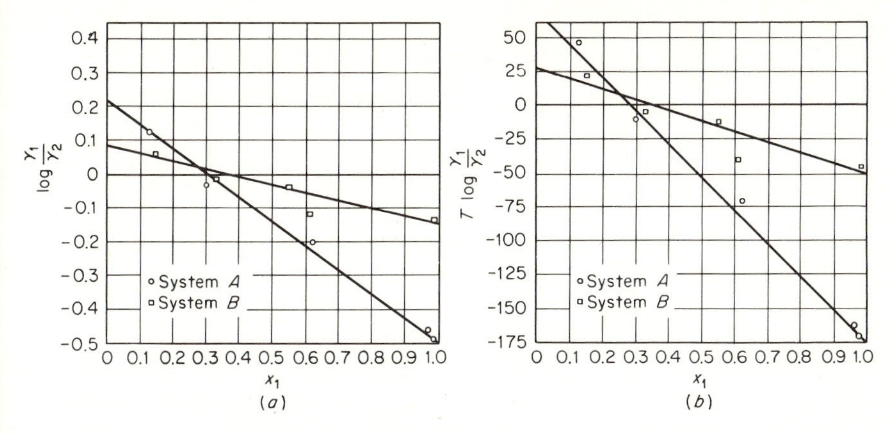

EXAMPLE 1.6 (*a*) Redlich and Kister thermodynamic consistency test. (*b*) Broughton and Brearley thermodynamic consistency test.

The results are plotted on the accompanying figures. The ratios of area above the abscissa to area (absolute value) below are:

System	Redlich and Kister	Broughton and Brearley
A	0.20	0.21
B	0.40	0.30

Neither set of data can be regarded as consistent by these two tests, although system *B* is the better of the two. The data of system *A* are noted to be the smoother of the two sets.

Example 1.7

The following data on the binary system *n*-octane–Cellosolve are to be tested for thermodynamic consistency by (*a*) the Black test; (*b*) the Norrish and Twigg test.

The vapor-liquid equilibrium data for the *n*-octane–Cellosolve system at 760 mm Hg are:

t, °C	$x_{n\text{-}0}$	$y_{n\text{-}0}$
133.6	0.010	0.064
127.75	0.054	0.261
123.45	0.108	0.386
118.4	0.302	0.540
117.1	0.651	0.630
119.0	0.880	0.746
122.6	0.948	0.869

The Antoine constants for *n*-octane are:

A	B	C
6.9237	1335.13	209.52

The temperature–vapor-pressure data for Cellosolve are:

T, °C	P, mm Hg
116.9	412.5
122.7	505
128	602
133	707

Solution

(*a*) The activity coefficients are solved for as in the previous example. Component 1 is *n*-octane.

t, °C	x_1	γ_1	γ_2	$\log \gamma_1$	$(\log \gamma_1)^{0.5}$	$\log \gamma_2$	$(\log \gamma_2)^{0.5}$
133.6	0.010	5.161	0.997	0.71273	0.8442	−0.00130	
127.75	0.054	4.598	0.993	0.66257	0.8140	−0.00303	
123.45	0.108	3.815	1.011	0.58149	0.7626	0.00475	0.0689
118.4	0.302	2.196	1.150	0.34163	0.5845	0.06070	0.2464
117.1	0.651	1.237	1.942	0.09237	0.3039	0.28825	0.5369
119.0	0.880	1.023	3.622	0.00988	0.0994	0.55895	0.7476
122.6	0.948	1.000	3.763	0.00000	0.0000	0.57553	0.7586

The data are not isothermal and, consequently, a straight-line curve would not be anticipated. The smoothness of the curve joining the points indicates, however, that the boiling points were precisely measured.

(*b*) The Norrish-Twigg method requires the evaluation of

$$k = \frac{(\Delta H_{V_1})_L}{(\Delta H_{V_2})_H}$$

The Clausius-Clapeyron equation is used for the evaluation of latent beat of vaporization.

$$\Delta H_V = T(V_g - V_l)\frac{dP}{dt}$$

It will be sufficiently accurate to assume that the ratio $(V_g - V_l)_1/(V_g - V_l)_2 = 1$ for these materials. Hence,

$$k = \frac{\Delta H_{V_1}}{\Delta H_{V_2}} = \frac{(dP/dt)_1}{(dP/dt)_2}$$

From the Antoine equation,

$$\frac{dP}{dt} = 2.303B\frac{P}{(t + C)^2}$$

and

$$k = \frac{B_1P_1}{B_2P_2}\left(\frac{t + C_2}{t + C_1}\right)^2$$

where the comparison is made at a fixed temperature.

A temperature in the middle of the range, 128°C, is chosen. A set of Antoine constants for Cellosolve is found for a narrow region around 128°C; namely, $\log P = 7.6412 - 1658/(t + 213)$. Then

$$k = \frac{(1355.1)(810.3)}{(1658)(602)}\left(\frac{341}{(337.5)}\right)^2 = 1.123$$

$$z = \ln\left(\frac{y_1}{y_2{}^k}\frac{y_2{}^k}{x_1}\right) = \ln\frac{y_1}{x_1} - k\ln\frac{y_2}{x_2}$$

x_1	z
0.010	0.8337
0.054	0.8046
0.108	0.7355
0.302	0.4557
0.651	−0.0425
0.880	−0.4377
0.948	−0.4883

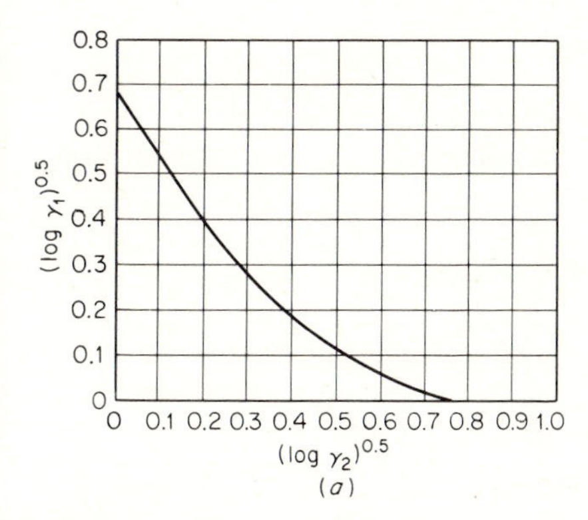

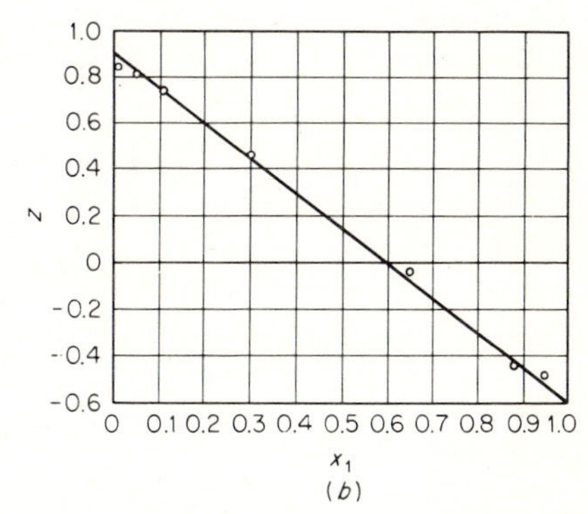

EXAMPLE 1.7 (*a*) Black thermodynamic consistency test. (*b*) Norrish-Twigg thermodynamic consistency test.

A very good straight line can be drawn through the plot of z versus x indicating thermodynamic consistency.

Equations (1.82) to (1.86) are graphically integrated with the use of available data or generalized correlations if the data are not at hand. Figures 1.8 and 1.9 schematically illustrate the integration.

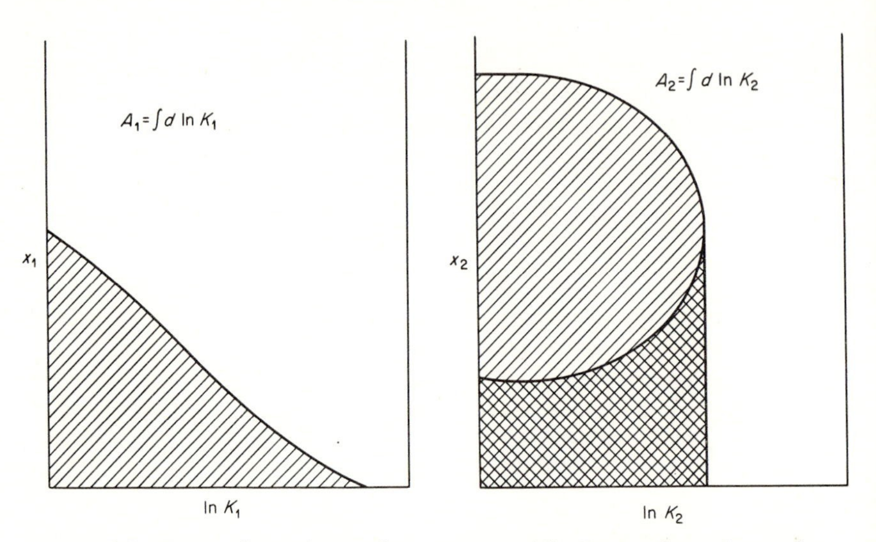

FIGURE 1.8 Integration of equation. FIGURE 1.9 Integration of equation.

Herington [20] devised an empirical test which claims to be a consistency test for binary data. He recommends the following procedure.

1. Calculate the value of $\log (\gamma_1/\gamma_2)$ for a series of compositions and plot these values against the values of x_1.

2. Evaluate the integral I as follows:

$$I = \int_0^1 \left(\log \alpha_{12} - \log \frac{P_1}{P_2}\right) dx \tag{1.87}$$

From Fig. 1.10,

$$I = A - B \tag{1.88}$$

Also,

$$\Sigma = A + B \tag{1.89}$$

3. Calculate the percentage deviation D by

$$D = \frac{100\,|I|}{\Sigma} \tag{1.90}$$

The value of I is taken with a positive sign.

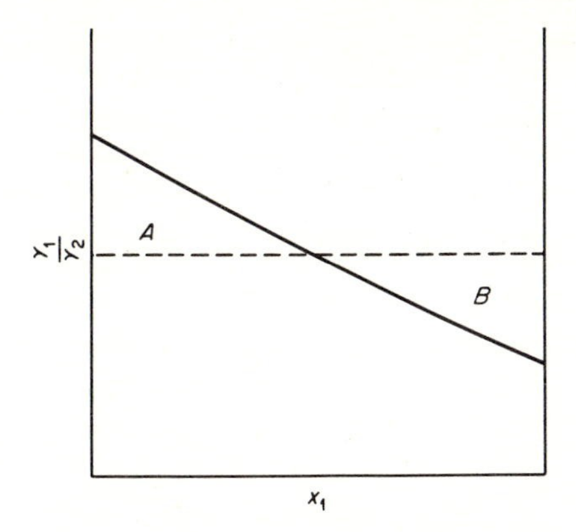

FIGURE 1.10 Areas used in Herington test.

4. Determine the lowest boiling point in the binary system t_i, which is either the boiling point of the lowest boiling component or that of the minimum boiling azeotrope and the maximum boiling range θ.

5. From Eq. (1.91) determine J,

$$J = \frac{150\,|\theta|}{T_i} \tag{1.91}$$

The data are considered consistent when

$$D < J \tag{1.92}$$

PROBLEMS

1. Experimental vapor pressure data for four alcohols are as follows:

P, mm Hg	Methanol	Ethanol	1-Propanol	2-Propanol
	t, °C at P, mm Hg			
40	5.0	19.0	36.4	23.8
100	21.2	34.9	52.8	39.5
400	49.9	63.5	82.0	67.8
2 atm	84.0	97.5	117.0	101.3
5 atm	112.5	126.0	149.0	130.2

(*a*) Determine the Antoine constants for the alcohols covering the range of pressures from 40 to 400 mm Hg.

(*b*) Determine the Antoine constants for the alcohols for the pressure range of 400 mm to 3800 mm Hg.

(*c*) Draw a Cox chart including on the chart all of the alcohols listed.

(*d*) Predict the vapor pressure temperature line for 1-butanol and draw it on the chart.

2. Determine the relative volatility of methanol relative to each of the other three alcohols listed at 45°C and at 120°C.

3. Determine the vapor-liquid equilibrium data for $x = 20, 40, 60, 80$ mole % for the binary system ethyl alcohol-benzene at a total pressure of 760 mm Hg from fugacity and activity coefficient calculations. $A'_{EB} = 0.845$, $B'_{BE} = 0.699$.

4. Calculate the (*a*) van Laar constants; (*b*) Margules constants; (*c*) Carlson and Colburn modified van Laar constants for the binary system toluene–2-methyl heptane which shows an azeotrope at $x_{Tol} = 0.82$ and 110.3°C.

5. Determine the tx, ty, γx data for the system methanol–ethyl acetate at 760 mm pressure.

Components	Antoine constants		
	A	*B*	*C*
MeOH	7.8786	1473.11	230.00
EtOAc	7.0981	1238.71	217.00

One experimental data point is available $x_{MeOH} = 0.4975$, $y_{MeOH} = 0.5970$, $t = 63.25$°C.

6. At 400 mm Hg pressure calculate the tx, ty, xy, and γx curves for the binary system ethyl cyclohexane–ethylbenzene from the following data

$x_{ECH} = 0.50 \qquad t = 110.1°C$

$\gamma_{ECH} = 1.05 \qquad \gamma_{EB} = 1.03$

Components	Antoine constants		
	A	*B*	*C*
EB	6.95722	1424.255	213.206
ECH	6.87041	1384.036	215.128

7. (*a*) Assuming the chloroform-acetone system to be an "ideal" system (no azeotrope), calculate the data for the tx, ty diagram. Apply the Redlich-Kister test to your calculated data.

(*b*) Using the following experimental data apply the same test.

Acetone–Chloroform System at 760 mm Hg

t, °C	x_A	y_A
62.1	0.056	0.039
63.2	0.149	0.113
64.4	0.275	0.250
63.65	0.521	0.579
62.0	0.645	0.731
58.75	0.839	0.905

The vapor pressure data are:

Components	t, °C at P, mm Hg			
	200	400	760	1520
Acetone	22.7	39.5	56.5	78.6
Chloroform	25.9	42.7	61.3	83.9

8. Apply the Black test to the following data on the 1,3-butadiene–chloroprene system.

t, °C	x_B	y_B
41.0	0.10	0.52
20.5	0.30	0.833
9.2	0.50	0.922
2.4	0.70	0.963
−2.8	0.90	0.987

The vapor pressure data are:

1,3-Butadiene		Chloroprene	
P, mm Hg	t, °C	P, mm Hg	t, °C
30	−65.7	185	20.5
100	−47.0	405	41.0
760	−4.4	760	59.4

9. Apply the Norrish and Twigg consistency test to the data in Prob. 8.

10. Apply the tests indicated to the following data on the cyclohexane–toluene system.

t, °C	x_{CH}	y_{CH}
108.25	0.041	0.102
99.5	0.217	0.422
94.2	0.368	0.596
90.6	0.504	0.702
84.8	0.763	0.864
82.7	0.874	0.926

Components	Antoine constants		
	A	B	C
Cyclohexane	6.845	1203.53	222.86
Toluene	6.9533	1343.94	219.38

(*a*) Apply the Redlich and Kister test.
(*b*) Apply the Herington test.

Nomenclature

a = constant, van der Waals equation of state, or molecular force constant
A = Antoine constant, or constant defined in activity coefficient equations
A = van Laar constant
A' = Carlson and Colburn modified van Laar constant
b = constant, van der Waals equation
B = Antoine constant or constant defined in activity coefficient equations
B = van Laar constant
B' = Carlson and Colburn modified van Laar constant
C = Antoine constant or number of phase-rule components
d = differential
D = defined quantity, Eq. (1.90)
E = (superscript) denotes excess property
f = fugacity
G = Gibbs molal free energy
$\bar{G}$ = Gibbs partial molal free energy
G^E = excess Gibbs free energy
H = enthalpy of mixture
H° = molar enthalpy of component
I = defined quantity, Eqs. (1.87) and (1.88)
i = subscript, any component
J = defined quantity, Eq. (1.91)
j = subscript, any component
k = ratio of molal heats of vaporization of lower to that of higher boiling component
K = constant
m = constant
n = subscript, nth component
p = absolute pressure
P = number of phases

p_c = critical pressure
P_i = vapor pressure of component i
P_t = total pressure
p_i = partial pressure of component i
q = effective molal volume of component
R = ideal-gas-law proportionality factor
S = molal entropy
$\bar{S}^E$ = excess molal entropy, $S_{\text{ideal}} - S_{\text{actual}}$
T_c = critical temperature
T = absolute temperature
t = temperature
t_B = normal boiling point
U = molal internal energy
U° = internal energy at zero pressure
$\bar{V}$ = partial molal volume
V = molal volume
V = degrees of freedom
V° = molal volume of pure component
x = mole fraction of component in liquid
y = mole fraction of component in vapor
Z = compressibility factor
z = effective volume fraction; defined quantity, Eq. (1.80)
Z_c = critical compressibility factor, $Z_c = \dfrac{P_c V_c}{RT_c}$

Greek letters

α = relative volatility
γ = activity coefficient
Δ = difference
∂ = differential operator
ν = fugacity coefficient
θ = constant
Σ = summation or defined by Eq. (1.89)

Subscripts

g = gas
H = higher boiling component
L = lower boiling component
l = liquid

REFERENCES

1. Adler, S. B., Leo Friend, R. L. Pigford, and C. M. Rosseli: *A.I.Ch.E. J.*, **6**:104 (1960).
2. Antoine, C.: *Compt. Rend. Acad. Sci., Paris*, **107**:681, 836, 1143 (1888).

3. API Project Report No. 44: "Selected Values of Physical and Thermodynamic Properties of Hydrocarbons and Related Compounds," Carnegie Press, Pittsburgh, 1953.
4. Black, Cline: *Ind. Eng. Chem.*, **50**:403 (1958).
5. Black, Cline: *A.I.Ch.E. J.*, **5**:249 (1959).
6. Black, Cline, E. L. Derr, and M. N. Papadopoulos: *Ind. Eng. Chem.*, **55**(8):40–49, 1963; (9):38–47, 1963.
7. Bonham, M. S.: M.S. Thesis in Chemical Engineering, MIT, 1941.
8. Broughton, D. B., and C. S. Brearley: *Ind. Eng. Chem.*, **47**:838 (1955).
9. Carlson, H. D., and A. P. Colburn: *Ind. Eng. Chem.*, **34**:581 (1942).
10. Cox, E. R.: *Ind. Eng. Chem.*, **15**:592 (1923).
11. Dalton, J.: "New System of Chemical Philosophy," vol. 1, i, p. 191, Manchester, England, 1808.
12. Dreisbach, R. R.: Physical Properties of Chemical Compounds I, II and III, *Am. Chem. Soc. Advan. Chem. Ser.*, nos. 15 (1955); 22 (1959); 29 (1961).
13. Duhem, P.: *Compt. Rend. Acad. Sci., Paris*, **102**:1449 (1886).
14. Finch, R. N., and M. Van Winkle: *A.I.Ch.E. J.* **8**:455–60 (1962).
15. Gibbs, J. W.: *Trans. Conn. Acad. Sci.*, **3**:152 (1876).
16. Gibbs, J. W.: "Collected Works," vol. 1, New York, Longmans 1928.
17. Gilmont, R., D. Zudkevitch, and D. F. Othmer: *Ind. Eng. Chem.*, **53**:223 (1961).
18. Gilmont, R., E. A. Weinman, F. Kramer, E. Miller, F. Hashmall, and D. F. Othmer: *Ind. Eng. Chem.*, **42**:120 (1950).
19. Hala, Edward, J. Pick, V. Fried, and O. Vilim: "Vapor Liquid Equilibrium" (Translated by G. Standart), Pergamon Press, New York, 1958.
20. Herington, E. F. G.: *J. Inst. Petrol.*, **37**:457 (1951).
21. Hildebrand, J. H., and R. L. Scott: "Solubility of Non-Electrolytes," 3d ed., Am. Chem. Soc. Monograph Series, Reinhold, New York, 1955; Dover, 1964.
22. Hougen, O. A., K. M. Watson, and R. A. Ragatz: "Chemical Process Principles—Part II: Thermodynamics," 2d ed., Wiley, New York, 1959.
23. Ibl, N. V., and B. F. Dodge: *Chem. Eng. Sci.*, **2**:120 (1953).
24. Joffe, J.: *Ind. Eng. Chem.*, **47**:2553 (1955).
25. Katz, D. L., D. Cornell, R. Kobayashi, F. H. Poettmann, C. F. Weinaug, J. A. Vary, and J. R. Elenbaas: "Handbook of Natural Gas Engineering," McGraw-Hill, New York, 1959.
26. Lewis, G. N., and M. Randall: *J. Am. Chem. Soc.*, **43**:233 (1921).
27. Li, Y. M., and J. Coull: *J. Inst. Petrol.*, **34**:692 (1948).
28. London, F.: *Trans. Faraday Soc.*, **33**:8 (1937).
29. Lu, B. C-Y, and W. F. Graydon: *Ind. Eng. Chem.*, **49**:1058 (1958).
30. Lyderson, A. L.: Critical Constants of Pure Compounds, *Univ. Wisconsin Eng. Exp. Sta. Rept.* no. 3 (April, 1955).
31. Lyderson, A. L., R. A. Greenkorn, and O. A. Hougen: Generalized Thermodynamic Properties of Pure Fluids, *Univ. Wisconsin Eng. Expt. Sta. Rept.* no. 4 (October, 1955).
32. Margules, M: *Sitzber. Akad. Wiss. Wien. Math. Naturw. Kl. II*, **104**:1243 (1895).
33. Norrish, R. S., and G. H. Twigg: *Ind. Eng. Chem.*, **46**:201 (1954).
34. Perry, J. H.: "Chemical Engineers' Handbook," 3d ed., McGraw-Hill, 1950.

35. Prengle, H. W., and M. A. Pike, Jr.: *J. Chem. Eng. Data*, **6**:400 (1961).
36. Raoult, F. M.: *Z. Physik. Chemie*, **2**:353 (1888).
37. Redlich, O., and A. T. Kister: *Ind. Eng. Chem.*, **40**:341 (1948).
38. Redlich, O., and A. T. Kister: *Chem. Rev.*, **10**:149 (1952).
39. Scatchard, G., and W. J. Hamer: *J. Am. Chem. Soc.*, **57**:1805 (1935).
40. Scatchard, G., *Chem. Rev.*, **8**:321 (1931).
41. Schechter, R. S., and M. Van Winkle: *Petrol. Refiner*, p. 301 (September, 1957).
42. Van Arkel, A. E.: *Trans. Faraday Soc.*, **42B**:81 (1946).
43. van der Waals, J. D.: "Over de continuet van den gassen vloeistoftoestand," Leiden, 1873.
44. Van Laar, J. J.: *Z. Phys. Chem.*, **185**:35 (1929).
45. White, R. R.: *Trans. A.I.Ch.E.*, **41**:546 (1945).
46. Wohl, Kurt: *Trans. A.I.Ch.E.*, **42**:215 (1946).
47. Wohl, Kurt: *Chem. Eng. Progr.*, **49**:218 (1953).
48. Wood, S. E.: *Ind. Eng. Chem.*, **42**:660 (1958).
49. Yu, K. T., and J. Coull: *Chem. Eng. Progr. Symp. Ser.*, **48**(2):38 (1952).

chapter 2

Multicomponent vapor-liquid equilibria

A multicomponent system is defined here as a system composed of more than two identifiable actual compounds or pseudocompounds or materials to which physical properties can be reasonably assigned. As in the case of binary systems, multicomponent systems are encountered which essentially behave ideally under conditions of low pressure and normal distilling temperatures. Those composed of mixtures of hydrocarbons, mixtures of isomeric compounds, or mixtures of homologous compounds may approximate ideal behavior in both vapor and liquid states. Where mixtures of materials of highly dissimilar nature are encountered or the conditions of pressure and temperature are severe, extreme nonideal behavior is evidenced.

2.1. IDEAL MULTICOMPONENT SYSTEMS

Multicomponent vapor-liquid equilibrium data for those systems which approximate ideal behavior may be calculated by Raoult's and Dalton's laws in the same manner as those for binary systems.

$$y_i P_t = p_i = x_i P_i \tag{2.1}$$

$$P_t = \sum_1^n P_i x_i = \sum_i^n p_i \tag{2.2}$$

$$y_i = \frac{x_i P_i}{\sum_1^n x_i P_i} \tag{2.3}$$

For a five-component system with two phases in equilibrium according to Gibbs phase rule, there are five degrees of freedom, four independent composition variables, and temperature or pressure. If the liquid composition is known and the temperature selected, the vapor composition can be calculated from Eq. (2.4). (If four liquid compositions are known, the fifth is also known because $\sum_1^n x_i = 1.0$.)

$$y_i = \frac{x_i P_i}{x_1 P_1 + x_2 P_2 + x_3 P_3 + x_4 P_4 + x_5 P_5} \tag{2.4}$$

and the total pressure is evaluated by means of

$$P_t = x_1 P_1 + x_2 P_2 + x_3 P_3 + x_4 P_4 + x_5 P_5 \tag{2.5}$$

If the temperature and vapor composition is known, the equilibrium composition of the liquid is also fixed but must be determined by trial-and-error calculation. This may be accomplished by assuming a total pressure and calculating a total pressure by

$$\sum_1^n y_i P_t = P_t \tag{2.6}$$

This may also be done by a *dew-point* pressure calculation. By definition the dew-point pressure is that pressure at which the first drop of liquid is formed as the pressure on the vapor mixture is increased at constant temperature. Mathematically,

$$\sum_1^n x_i = 1.0 = \sum_1^n \frac{y_i P_t}{P_i} \tag{2.7}$$

for an ideal mixture. In either calculation, plotting the calculated total pressures versus the assumed total pressures for three or four trials usually will enable the correct pressure to be evaluated from the plot with sufficient accuracy. Then the liquid compositions can be calculated by using the correctly determined pressure in

$$x_i = \frac{y_i P_t}{P_i} \tag{2.8}$$

Considering a five-component system of a fixed constant pressure with coexisting vapor and liquid phases, there are four remaining degrees of freedom, four independent compositions, and the temperature is fixed. If the liquid compositions are known, a *bubble-point* temperature calculation usually is made to determine the temperature. By definition, the bubble-point temperature is that temperature at which the first bubble of vapor is formed on heating a liquid at constant pressure. Thus, quantitatively,

$$\sum_1^n y_i = \sum_1^n \frac{x_i P_i}{P_t} = 1.0 \tag{2.9}$$

A minimum of three temperatures is assumed, the corresponding vapor pressures are obtained, and Eq. (2.9) is used to calculate the summation of y_i. The sum is plotted versus temperature and, at the point where the sum equals unity, the correct bubble-point temperature is established. By means of this temperature the individual values of y are calculated, using the correspondingly correct vapor pressures.

When the vapor compositions and total pressure are known, the temperature and the composition of the liquid in equilibrium may be calculated by a *dew-point temperature* calculation. The dew-point temperature is defined as the temperature at which the first drop of liquid is formed upon cooling a vapor mixture at constant pressure. Mathematically it is expressed by Eq. (2.7). At least three values of temperature are selected, the correct vapor pressures are obtained, and the summation is calculated by Eq. (2.7). The calculated sums are plotted versus the assumed temperatures, and where $\Sigma x_i = 1.0$ occurs the correct equilibrium (dew-point) temperature is established. The liquid compositions may then be calculated by means of Eq. (2.8).

Relative volatilities may be used where the temperature range (the difference in boiling points of the lightest and heaviest components) of the multicomponent equilibrium data is not great. Usually the relative volatilities are based on the heaviest component in the mixture. Arithmetic average relative volatilities (at the highest and lowest temperatures) may be used or, if the variation in relative volatility with temperature is slight, they may be evaluated at one temperature and assumed to be constant over the temperature range.

Thus,

$$\alpha_{15} = \frac{P_1}{P_5} \qquad \alpha_{25} = \frac{P_2}{P_5} \qquad \text{etc.}$$

for the ideal five-component mixture. Divide numerator and denominator of Eq. (2.4) by P_5. The composition of component 1 in the vapor is

$$y_1 = \frac{x_1(P_1/P_5)}{x_1 \frac{P_1}{P_5} + x_2 \frac{P_2}{P_5} + x_3 \frac{P_3}{P_5} + x_4 \frac{P_4}{P_5} + x_5 \frac{P_5}{P_5}} \tag{2.10}$$

and this in terms of relative volatility is

$$y_1 = \frac{x_1 \alpha_{15}}{x_1 \alpha_{15} + x_2 \alpha_{25} + x_3 \alpha_{35} + x_4 \alpha_{45} + x_5} \tag{2.11}$$

Similarly,

$$x_1 = \frac{y_1/\alpha_{15}}{y_1/\alpha_{15} + y_2/\alpha_{25} + y_3/\alpha_{35} + y_4/\alpha_{45} + y_5} \tag{2.12}$$

In terms of any number of components n,

$$y_i = \frac{x_i\alpha_{in}}{\sum_1^n x_i\alpha_{in}} \tag{2.13}$$

$$x_i = \frac{y_i/\alpha_{in}}{\sum_1^n y_i/\alpha_{in}} \tag{2.14}$$

2.2. IDEAL TERNARY SYSTEMS

Ternary systems represent a somewhat special case since many distillation problems can be reduced to consideration of essentially a ternary system, and the calculations are somewhat simplified as the result. From the phase rule there are three degrees of freedom for a ternary system in two co-existing phases. Raoult's and Dalton's laws are followed, i.e., Eqs. (2.1) to (2.3) apply and bubble-point and dew-point calculations are made to determine the phase compositions, temperature, and pressure as needed. Because

$$x_1 + x_2 + x_3 = 1.0 \tag{2.15}$$

and

$$y_1 + y_2 + y_3 = 1.0 \tag{2.16}$$

$$\sum_1^3 y_i = \frac{x_1P_1}{P_t} + \frac{x_2P_2}{P_t} + \frac{(1 - x_1 - x_2)P_3}{P_t} = 1.0 \tag{2.17}$$

and

$$\sum_1^3 x_i = \frac{y_1P_t}{P_1} + \frac{y_2P_t}{P_2} + \frac{(1 - y_1 - y_2)P_t}{P_3} = 1.0 \tag{2.18}$$

In many cases, particularly in graphical studies, it is useful to determine isothermal vapor and liquid composition lines. This involves fixing t and P_t and enables direct calculation of the compositions (fixing one and calculating the others) from the following relations:

$$p_3 = P_3x_3 = P_3(1 - x_1 - x_2) \tag{2.19}$$

$$p_3 = P_t - P_1x_1 - P_2x_2 \tag{2.20}$$

$$P_t - P_3 = x_1(P_1 - P_3) + x_2(P_2 - P_3) \tag{2.21}$$

$$x_1 = \frac{P_t - P_3}{P_1 - P_3} - x_2\frac{P_2 - P_3}{P_1 - P_3} \tag{2.22}$$

Thus, with temperature and pressure fixed, the vapor pressures are fixed, and the variation of x_1 as a function of x_2 can be calculated. For ideal ternary systems this is a straight line on an x_1 versus x_2 plot.

Example 2.1

Calculate the isothermal vapor and liquid lines for the ternary system toluene–ethylbenzene–*p*-xylene, at a total pressure of 800 mm Hg and for a composition $x_{Tol} = 0.30$, $x_{EB} = 0.45$. The system may be assumed to behave ideally for this calculation.

Components	Antoine constants		
	A	B	C
(1) Toluene	6.95334	1343.943	219.38
(2) Ethylbenzene	6.95719	1424.255	213.21
(3) *p*-Xylene	6.99052	1453.43	215.31

Solution

The liquid composition line is found from

$$x_1 = \frac{P_t - P_3}{P_1 - P_3} - x_2 \frac{P_2 - P_3}{P_1 - P_3}$$

The temperature at which $x_1 = 0.3$, $x_2 = 0.45$ at $P_t = 800$ mm Hg is found by trial and error.

Vapor pressure, mm Hg			Boiling point t, °C
(3) *p*-Xylene	(2) Ethylbenzene	(1) Toluene	
500	531.5	1077	123.36
600	637	1270	129.73
700	792.2	1459	135.30

Values of x_1 are calculated at each temperature for $x_2 = 0.45$:

t, °C	$\frac{P_t - P_3}{P_1 - P_3}$	$\frac{P_2 - P_3}{P_1 - P_3}$	x_1, calculated
123.36	0.5199	0.0546	0.4953
129.73	0.2985	0.0552	0.2737
135.30	0.1318	0.0556	0.1068

With the use of a graph of x_1 (calculated) versus P_3, the final value of $P_3 = 587$ is located. At this pressure, $t = 129°C$, $P_1 = 1244$ mm Hg, $P_2 = 623$ mm Hg.

$$x_1 = 0.3242 - 0.0551x_2$$

From Eq. (2.23),

$$y_1 = 0.5041 - 0.110y_2$$

Similarly, the isothermal vapor composition line can be calculated by

$$y_1 = \frac{P_1}{P_3}\frac{P_t - P_3}{P_1 - P_3} - y_2\frac{P_1}{P_2}\frac{P_2 - P_3}{P_1 - P_3} \tag{2.23}$$

(This represents the equation of the solid line in Fig. 2.1.)

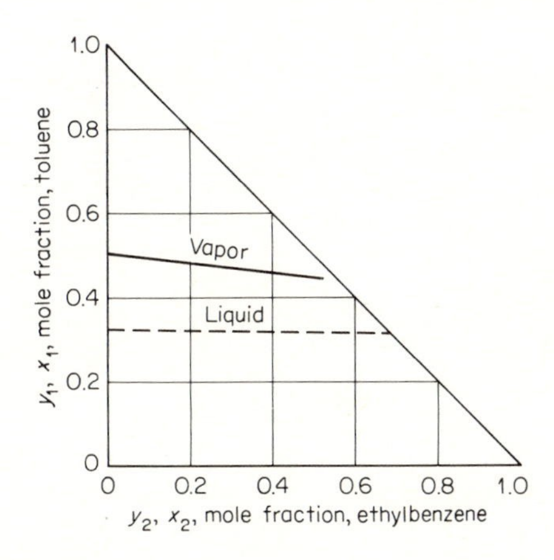

FIGURE 2.1 Isothermal vapor and liquid lines for Example 2.1.

When one composition is selected, the remaining vapor and liquid compositions can be calculated or obtained from the isothermal vapor or isothermal liquid composition plots at the fixed total pressure. In terms of relative volatilities, the following equations for ideal ternary mixtures can be used:

$$y_1 = \frac{\alpha_{13}x_1}{1 + (\alpha_{13} - 1)x_1 + (\alpha_{23} - 1)x_2} \tag{2.24}$$

$$y_2 = \frac{\alpha_{23}x_2}{1 + (\alpha_{13} - 1)x_1 + (\alpha_{23} - 1)x_2} \tag{2.25}$$

$$x_1 = \frac{y_1/\alpha_{13}}{y_2(1/\alpha_{23} - 1) + y_1(1/\alpha_{13} - 1) + 1} \tag{2.26}$$

$$x_2 = \frac{y_2/\alpha_{23}}{y_2(1/\alpha_{23} - 1) + y_1(1/\alpha_{13} - 1) + 1} \tag{2.27}$$

Qualitative Graphical Study

Visualizing the variance of composition, temperature, and pressure in a binary system is relatively easy since the number of variables is such that the variation can be shown on a two-dimensional plot. In the case of ternary systems there is difficulty in visualizing the variation of composition with temperature at constant pressure, unless solid figures of the type shown in Figs. 2.2, 2.3, and 2.4 are constructed or plastic sheet models such as shown in Figs. 2.5 to 2.7 are available. These are extremely useful and informative, although it may not be feasible to spend the necessary time and effort on this kind of three-dimensional plot.

On many occasions the engineer is interested in estimating qualitatively the temperature-composition relationship before the actual ternary data are determined or calculated. In such cases a fairly good representation of the tx and ty surfaces can be estimated from the characteristics of the binary systems alone. To do this, diagrams of the type shown in Figs. 2.8 to 2.11

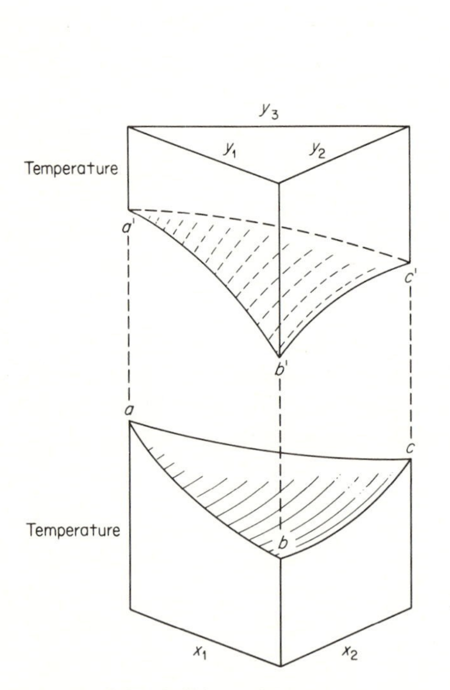

FIGURE 2.2 Solid model of a ternary temperature-composition phase diagram showing liquid and vapor surfaces—three "normal" binary systems.

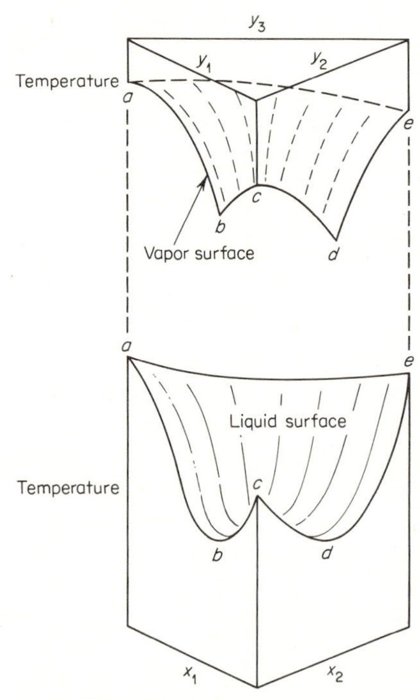

FIGURE 2.3 Solid model of a ternary temperature-composition phase diagram showing liquid and vapor surfaces—two binary azeotropes.

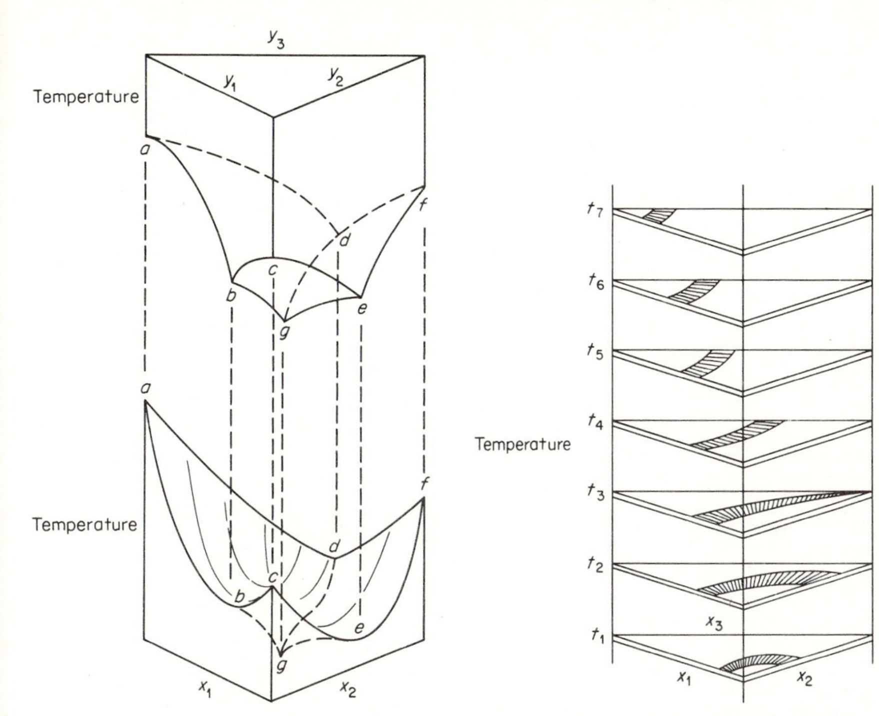

FIGURE 2.4 Solid model of a ternary temperature-composition phase diagram showing vapor and liquid surfaces—three binary azeotropes.

FIGURE 2.5 Plastic sheet model of ternary temperature-composition diagram—three "normal" binary systems. Each sheet shows the two-phase region at that temperature and representative tie lines.

are drawn up. These consist of conventional ternary plots using triangular coordinates or right angle, triangular plots using rectangular coordinates. Extensions are drawn from the corners of the triangle at right angles to the sides, giving three two-dimensional plots—one on each side of the triangle. The scale on the lines perpendicular to the sides of the triangle can be either temperature or pressure. Temperature is usually selected as the variable since practically all distillations are carried out at constant pressure.

On each of the binary plots the *tx* and *ty* curves are plotted for the binary system indicated. Then a series of lines at a number of selected temperatures are drawn on the binary diagrams. The temperatures are selected such that isotherms intersect the binary diagrams at intervals ranging from the lowest temperature at which the two-phase vapor-liquid region occurs to the highest boiling temperature on the diagram. The intersections of the vapor and liquid lines bound the isothermal two-phase region at that temperature. The points, vapor-vapor and liquid-liquid, are connected with lines, and the areas between the lines are the approximate isothermal, ternary, two-phase regions.

The tie lines (lines connecting vapor-liquid equilibrium compositions) for the binaries lie in the sides of the ternary plot, and in the ternary diagram all tie lines generally tend to point to a corner of the diagram. If binary azeotropes are present, the tie lines tend to converge at the binary azeotrope compositions.

Such diagrams are useful in characterizing the vapor and liquid surfaces of a ternary system. Since the tie lines sketched in for each isothermal two-phase region represent (at their intersections with the liquid-phase and vapor-phase boundaries) the liquid and vapor compositions in equilibrium at that temperature, continuous lines can be drawn from one composition in one part of the diagram to another composition in another part of the diagram. This line represents the locus of the tangency points of the curve with the equilibrium tie lines throughout the whole composition-temperature range. Such lines, as we shall see later, represent possible distillation paths.

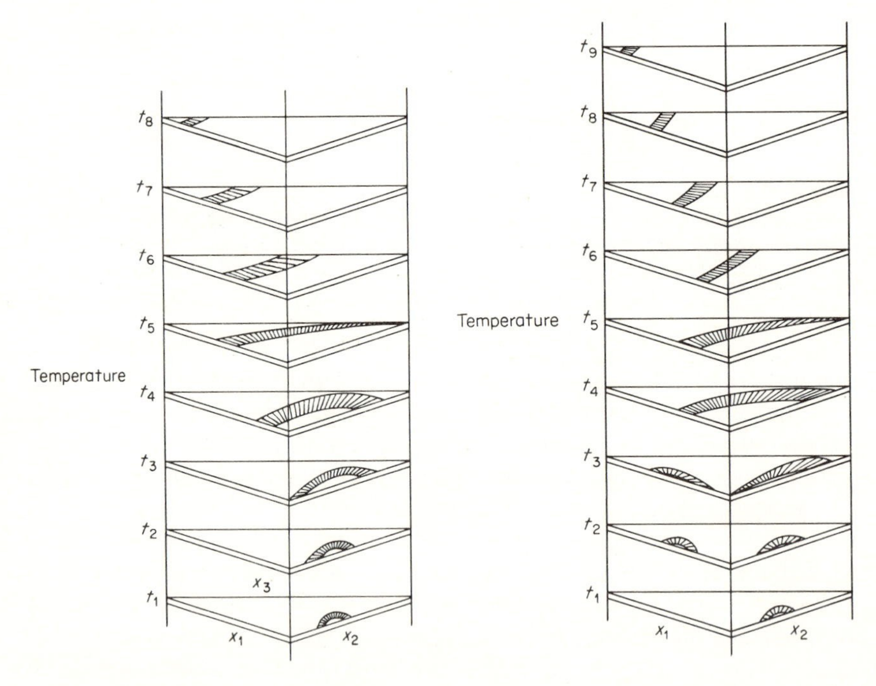

FIGURE 2.6 Plastic sheet model of ternary temperature-composition diagram—two "normal" binary systems and one binary azeotrope. Each sheet shows the two-phase region at that temperature and representative tie lines.

FIGURE 2.7 Plastic sheet model of a ternary temperature-composition diagram—two binary azeotropes and one "normal binary" system. Each sheet shows the two-phase region at that temperature and representative tie lines.

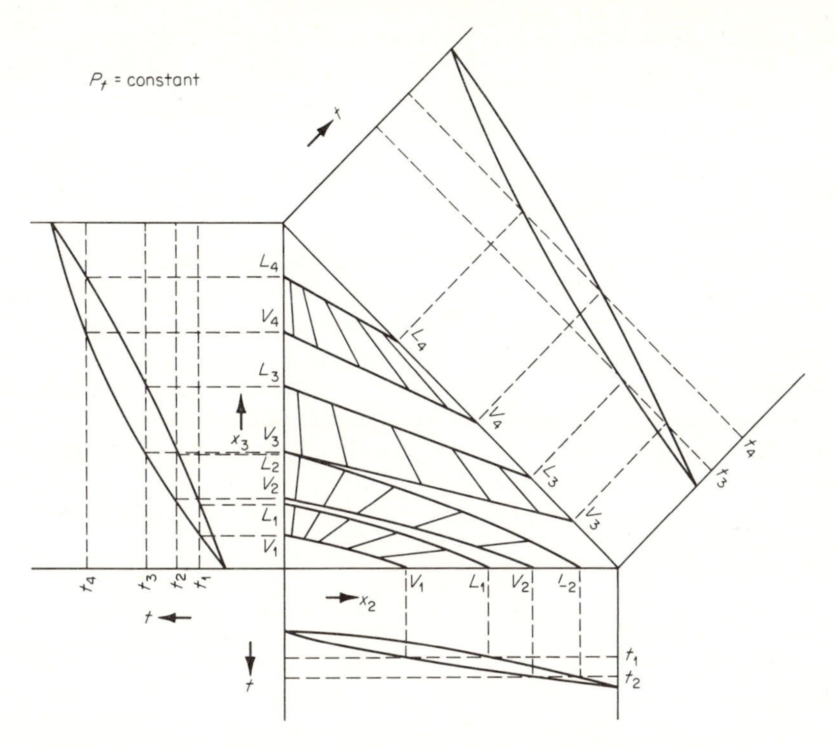

FIGURE 2.8 Isothermal two-phase ternary diagrams from binaries—normal ternary system.

Also diagrams of this sort enable reasonable prediction of the existence of valleys and hills or ridges on the surfaces which aid in selecting distillation conditions to obtain better advantages with regard to maximum changes in composition from plate to plate.

2.3. NONIDEAL VAPOR-LIQUID EQUILIBRIA—MULTICOMPONENT SYSTEMS

In the discussion of nonideal vapor-liquid equilibria for multicomponent systems, the subject naturally falls into two categories: nonideal ternary-system vapor-liquid equilibria and nonideal vapor-liquid equilibria of systems composed of more than three components. The reason for this is principally the degree of complexity of the equations. There have been many correlations proposed for the calculation of ternary vapor-liquid equilibria relating activity coefficients and composition, whereas relatively only a few have been proposed for systems having a greater number of components. Therefore, the less complex ternary material will be considered first.

No attempt is made to give a thorough discussion of correlations for multicomponent data because of space considerations. The reader is invited to refer to other books on this subject, e.g., Hala et al. [24].

Nonideal Ternary Systems

The nonideal ternary data needed for ternary distillation calculations may be experimentally determined or calculated from other information. Experimental evaluation of ternary vapor-liquid data is time consuming and expensive, because approximately a minimum of 100 points of experimental data are necessary to characterize completely an extremely nonideal ternary vapor-liquid system. If correlations are available by means of which the necessary data may be predicted, and sufficient information for use in the correlations is at hand or can be reasonably obtained, the ternary data are

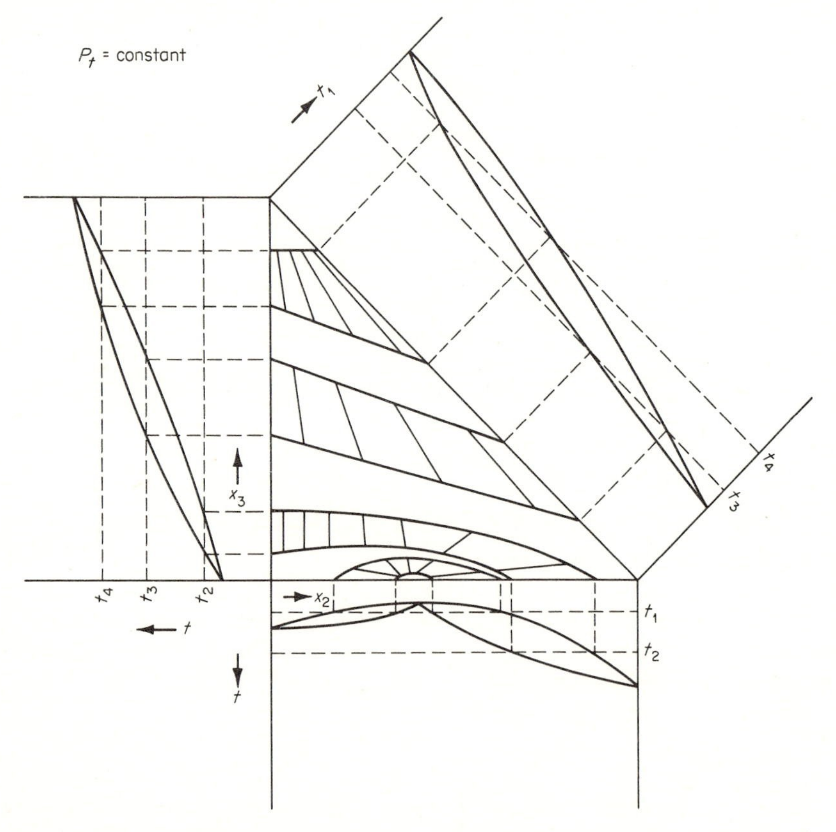

FIGURE 2.9 Isothermal two-phase ternary diagrams from binaries—two normal binaries, one binary azeotrope.

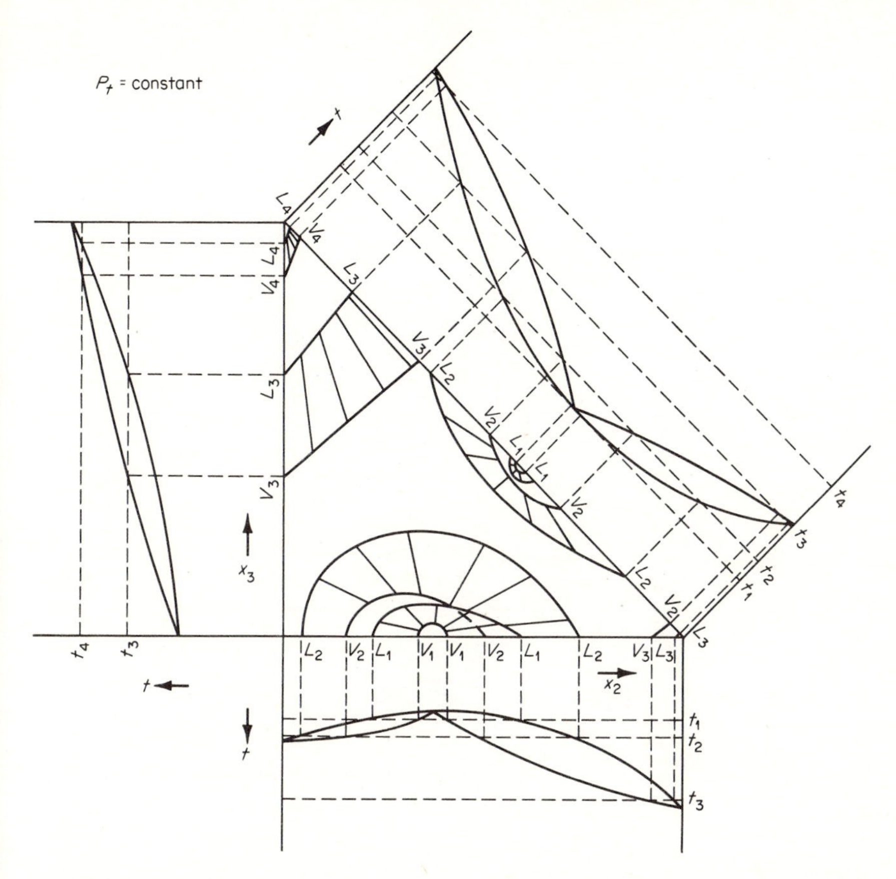

FIGURE 2.10 Isothermal two-phase ternary diagrams from binaries—one normal binary, two binary azeotropes.

usually predicted even at the loss of some accuracy. The ideal situation would be one in which ternary vapor-liquid equilibrium data could be predicted from some correlation or correlations from pure-component data, because most of the available information is pure-component data whereas only a comparatively small amount of data are available on mixture properties involving two or more components. This cannot be done at present with the necessary precision.

Calculation of Nonideal Ternary Vapor-Liquid Equilibrium Data

There have been a number of methods proposed for the calculation of ternary vapor-liquid equilibrium data with the use of equations derived from thermodynamic consideration of excess free energy of mixing. These

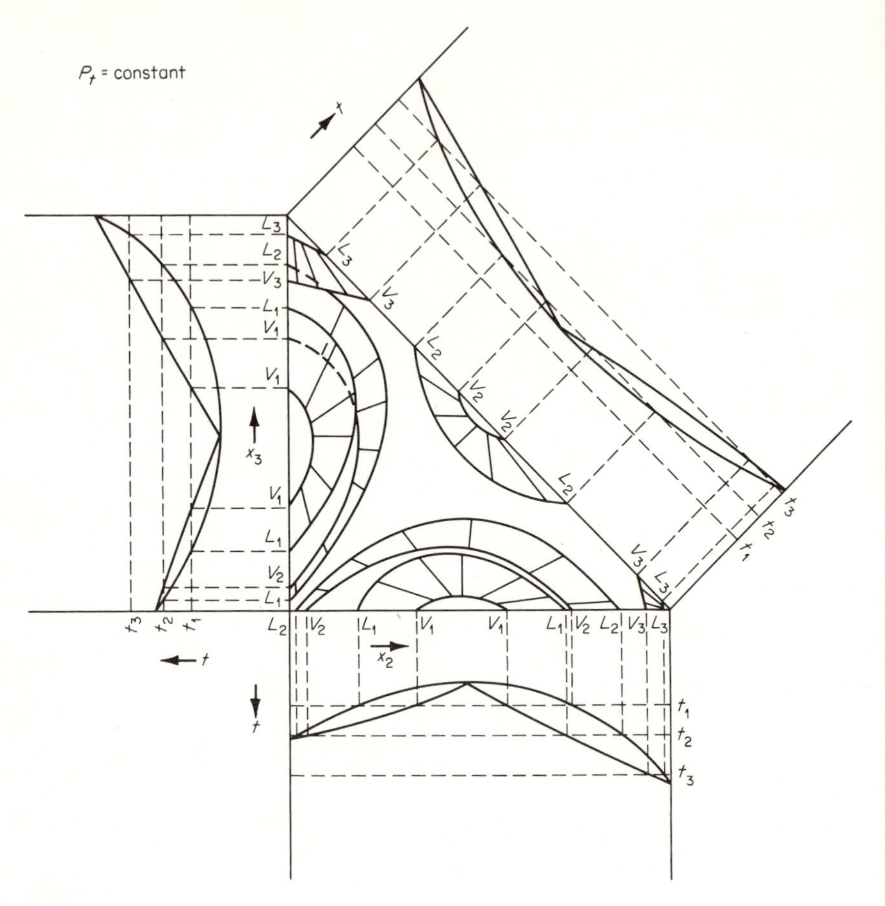

FIGURE 2.11 Isothermal ternary diagrams from binaries—three binary azeotropes.

equations range from purely empirical relationships to semitheoretical relations which differ only because of certain assumptions made in their derivation. The various methods may require the use of only binary data, on one hand, to relatively large amounts of ternary data on the other.

van Laar type of equations The original van Laar equations [50] were developed only for binary systems. However, several investigators have extended the van Laar type of equations to ternary systems, basing their equations on essentially the same assumptions as those of van Laar but using constants of different form.

Wohl [57, 58] through his expansion of the molal excess free-energy equations developed van Laar type of equations of the second, third, and fourth order for ternary systems.

His second-order or two-suffix equation for $\log \gamma_1$ in terms of binary van Laar constants and composition is

$$\log \gamma_1 = \frac{x_2^2 A_{12}\left(\frac{A_{21}}{A_{12}}\right)^2 + x_3^2 A_{13}\left(\frac{A_{31}}{A_{13}}\right)^2 + x_2 x_3 \frac{A_{21}A_{31}}{A_{12}A_{13}}\left(A_{12} + A_{13} - A_{32}\frac{A_{13}}{A_{31}}\right)}{[x_1 + x_2(A_{21}/A_{12}) + x_3(A_{31}/A_{13})]^2} \tag{2.28}$$

where A_{12} = Carlson-Colburn modification, van Laar A for system 1–2
A_{21} = Carlson-Colburn modification, van Laar B for system 1–2
A_{23} = Carlson-Colburn modification, van Laar A for system 2–3, etc.

(Equations for the other components, 2 and 3, can be written by rotating the subscripts as follows:

$$\begin{array}{ccc} & 1 & \\ \nearrow & & \searrow \\ 3 & \longleftarrow & 2 \end{array}$$

Thus, wherever 1 appears, 2 is substituted, 3 is substituted for 2, and 1 is substituted for 3.)

These binary constants are defined as follows:

$$A_{12} = A' = \log \gamma_1 \left(1 + \frac{x_2 \log \gamma_2}{x_1 \log \gamma_1}\right)^2 \tag{2.29}$$

$$A_{21} = B' = \log \gamma_2 \left(1 + \frac{x_1 \log \gamma_1}{x_2 \log \gamma_2}\right)^2 \tag{2.30}$$

The equations will describe many "normal" ternary systems using only the constants derived from component binary data. However, they do not adequately describe highly nonideal systems. For such systems the higher order, four- and five-suffix equations, should be used along with sufficient multicomponent data to enable evaluation of the interaction terms.

The Li and Coull [35] ternary equations of the van Laar type are:

$$T \log \gamma_1 = \frac{\frac{k_{12}}{b_1}\frac{b_2}{b_1}x_2^2 + \frac{k_{13}}{b_1}\frac{b_3}{b_1}x_3^2 + \left(\frac{k_{12}}{b_1}\frac{b_3}{b_1} + \frac{k_{13}}{b_1}\frac{b_2}{b_1} - \frac{k_{23}}{b_1}\right)x_2 x_3}{[x_1 + (b_2/b_1)x_2 + (b_3/b_1)x_3]^2} \tag{2.31}$$

$$T \log \gamma_2 = \frac{\frac{k_{12}}{b_2}\frac{b_1}{b_2}x_1^2 + \frac{k_{23}}{b_2}\frac{b_3}{b_2}x_3^2 + \left(\frac{k_{12}}{b_2}\frac{b_3}{b_2} + \frac{k_{23}}{b_2}\frac{b_1}{b_2} - \frac{k_{13}}{b_2}\right)x_1 x_3}{[x_2 + (b_1/b_2)x_1 + (b_3/b_2)x_3]^2} \tag{2.32}$$

$$T \log \gamma_3 = \frac{\frac{k_{13}}{b_3}\frac{b_1}{b_3}x_1^2 + \frac{k_{23}}{b_3}\frac{b_2}{b_3}x_2^2 + \left(\frac{k_{13}}{b_3}\frac{b_2}{b_3} + \frac{k_{23}}{b_3}\frac{b_1}{b_3} - \frac{k_{12}}{b_3}\right)x_1 x_2}{[x_3 + (b_1/b_3)x_1 + (b_2/b_3)x_2]^2} \tag{2.33}$$

Example 2.2

(*a*) Determine the binary van Laar constants from pure-component data for the system *n*-octane–ethylbenzene–Cellosolve at 760 mm Hg pressure.

(*b*) Determine the ternary activity coefficients using Eqs. (2.36), (2.37), and (2.38) and the calculated values from part (*a*) for $x_{n\text{-O}} = 0.25$, $x_{\text{EB}} = 0.52$, $T = 125°\text{C}$.

Solution

The van der Waals *a* and *b* are calculated from the following:

$$a_i = \frac{27R^2T_{c_i}^{\,2}}{64P_{c_i}} \qquad b_i = \frac{RT_{c_i}}{8P_{ci}}$$

$$A_{ij} = \frac{b_i}{b_j} = \frac{T_{c_i}P_{c_j}}{T_{c_j}P_{c_i}}$$

$$\sqrt{B_{ij}} = \sqrt{\frac{b_i}{R}}\left(\frac{\sqrt{a_i}}{b_i} - \frac{\sqrt{a_j}}{b_j}\right) = \sqrt{\frac{3.375T_{c_i}}{P_{c_i}}}\,(P_{c_i}^{\,1/2} - P_{c_j}^{\,1/2})$$

The critical properties of *n*-octane and ethylbenzene are reported in Dreisbach [15]. The critical properties of Cellosolve (ethylene glycol–monoethyl ether) were calculated from group contributions by the method of Lydersen.†

Components	Critical properties			
	T_c, °K	$(T_c)^{1/2}$, °K	P_c, mm Hg	$P_c^{1/2}$, mm Hg
(1) *n*-Octane	569.4	23.86	18,726	136.84
(2) Ethylbenzene	619.6	24.892	28,120	167.69
(3) Cellosolve	570.1	23.88	32,210	179.47

The A_{ij} and $\sqrt{B_{ij}}$ are readily calculated.

$$A_{12} = 1.3800 \qquad \sqrt{B_{12}} = -9.883$$

$$A_{13} = 1.7180 \qquad \sqrt{B_{13}} = -13.657$$

$$A_{23} = 1.2449 \qquad \sqrt{B_{23}} = -3.212$$

$$A_{21} = 0.7246 \qquad \sqrt{B_{21}} = 8.416$$

$$A_{31} = 0.5821 \qquad \sqrt{B_{31}} = 10.419$$

$$A_{32} = 0.8033 \qquad \sqrt{B_{32}} = 2.879$$

† As reported in Reid and Sherwood [44].

Calculations are to be carried out at $x_1 = 0.25$, $x_2 = 0.52$, $x_3 = 0.23$, $T = 398.2°K$.

$$T \ln \gamma_1 = \left[\frac{0.52(-9.883) + (0.23)(0.8033)(-13.657)}{0.25(1.3800) + 0.52 + 0.23(0.8033)}\right]^2$$

$$T \ln \gamma_1 = 53.275 \qquad \gamma_1 = 1.14$$

$$T \ln \gamma_2 = 4.842 \qquad \gamma_2 = 1.01$$

$$T \ln \gamma_3 = 23.526 \qquad \gamma_3 = 1.06$$

The constants b and k are evaluated from the binary data with the use of the binary equations:

$$(T \log \gamma_1)^{-0.5} = \frac{b_1}{b_2}\left(\frac{b_2}{k_{12}}\right)^{0.5} \frac{x_1}{x_2} + \left(\frac{b_2}{k_{12}}\right)^{0.5} \tag{2.34}$$

$$(T \log \gamma_2)^{-0.5} = \frac{b_2}{b_1}\left(\frac{b_1}{k_{12}}\right)^{0.5} \frac{x_2}{x_1} + \left(\frac{b_1}{k_{12}}\right)^{0.5} \tag{2.35}$$

The constants may be evaluated by plotting $(T \log \gamma_1)^{-0.5}$ versus x_1/x_2. The slope of the line is $(b_1/b_2)(b_2/k_{12})^{0.5}$ and the intercept is $(b_2/k_{12})^{0.5}$. The constants may be determined for the other binary systems in a similar manner by plotting the correct terms. These equations correlate the data on a wide range of systems including hydrocarbons, alcohols, ketones, and water. White [52] presented equations similar in nature to those of Li and Coull, utilizing the same type of binary system constants for evaluation of ternary data.

The Bonham [9, 45] ternary equations are of the van Laar type:

$$T \ln \gamma_1 = \frac{(x_2\sqrt{B_{12}} + x_3A_{32}\sqrt{B_{13}})^2}{(x_1A_{12} + x_2 + x_3A_{32})^2} \tag{2.36}$$

$$T \ln \gamma_2 = \frac{(x_1A_{12}\sqrt{B_{21}} + x_3A_{32}\sqrt{B_{23}})^2}{(x_1A_{12} + x_2 + x_3A_{32})^2} \tag{2.37}$$

$$T \ln \gamma_3 = \frac{(x_1A_{12}\sqrt{B_{31}} + x_2\sqrt{B_{32}})^2}{(x_1A_{12} + x_2 + x_3A_{32})^2} \tag{2.38}$$

The general form for these equations is as follows:

$$T \ln \gamma_k = \frac{[\Sigma x_i A_{ij}(B_{ki}^{0.5})]^2}{[\Sigma(x_i A_{ij})]^2} \tag{2.39}$$

where k = component whose activity coefficient is being evaluated
j = reference component, usually the heavy key
i = any component from 1 to n

$$A_{12} = \frac{b_1}{b_2} \qquad A_{21} = \frac{b_2}{b_1} \tag{2.40}$$

$$\sqrt{B_{21}} = \sqrt{\frac{b_2}{R}}\left(\frac{\sqrt{a_2}}{b_2} - \frac{\sqrt{a_1}}{b_1}\right) \tag{2.41a}$$

$$\sqrt{B_{12}} = \sqrt{\frac{b_1}{R}}\left(\frac{\sqrt{a_1}}{b_1} - \frac{\sqrt{a_2}}{b_2}\right) \tag{2.41b}$$

where a and b are the van der Waals constants (see Chap. 1). The Bonham and Li and Coull equations are identical when compared on the basis of the same assumptions (see appendix to Chap. 2). They require only binary data for solution, and describe many of the types of nonideal ternary systems.

Margules type of equations Wohl [57, 58] in his work on the thermodynamics of vapor-liquid equilibria also developed equations of the two-, three-, and four-suffix Margules type. The three-suffix equation relating the activity coefficient for component 1 with composition in ternary systems that are symmetrical and require binary and some ternary data for the evaluation of the constants is

$$\begin{aligned}\log \gamma_1 = {} & x_2^2[A_{12} + 2x_1(A_{21} - A_{12})] + x_3^2[A_{13} + 2x_1(A_{31} - A_{13})] \\ & + x_2x_3[\tfrac{1}{2}(A_{21} + A_{12} + A_{31} + A_{13} - A_{23} - A_{32}) \\ & + x_1(A_{21} - A_{12} + A_{31} - A_{13}) + (x_2 - x_3)(A_{23} - A_{32}) - (1 - 2x_1)C^*]\end{aligned} \tag{2.42}$$

Equations for the activity coefficients for the other two components in terms of composition and binary and ternary constants may be written by using the rotation principle for subscripts.

$$C^* = \tfrac{3}{2}(a_{112} + a_{122} + a_{113} + a_{133} + a_{223} + a_{233}) - 6a_{123} \tag{2.43}$$

In simple, normal systems the value of $C^* = 0$ and the ternary data can be found by using only the binary constants derived from the binary data. Normal systems are those in which the component characteristics are similar and there is no chemical interaction. When the system is extremely nonideal, ternary data are needed for evaluation of the ternary constant C^*. Severns et al. [47] proposed a method whereby a minimum of ternary data are required for evaluation of the ternary constant C^* which simplifies the complete evaluation of the ternary data.

The defining equations for the Margules binary constants are

$$A = \frac{x_2 - x_1}{x_2^2} \log \gamma_1 + \frac{2 \log \gamma_2}{x_1} \tag{2.44}$$

$$B = \frac{x_1 - x_2}{x_1^2} \log \gamma_2 + \frac{2 \log \gamma_1}{x_2} \tag{2.45}$$

Benedict et al. [6] used a somewhat different expansion of the excess free-energy equation which led to equations essentially of the Margules type. These equations which relate activity coefficient with liquid composition may be expanded to the desired order.

The two-suffix equations are as follows:

$$RT \ln \gamma_1 = 2(x_2 - x_1x_2)A'_{12} + 2(x_3 - x_1x_3)A'_{13} - 2x_2x_3A'_{23} \tag{2.46}$$

$$RT \ln \gamma_2 = 2(x_1 - x_1x_2)A'_{12} - 2x_1x_3A'_{13} + 2(x_3 - x_2x_3)A'_{23} \tag{2.47}$$

$$RT \ln \gamma_3 = -2(x_1x_2A'_{12}) + 2(x_1 - x_1x_3)A'_{13} + 2(x_2 - x_2x_3)A'_{23} \tag{2.48}$$

The authors indicate that only a few systems are represented by the two-suffix equations. Usually four-suffix equations are needed, and these require evaluation of a number of points of ternary data in order to establish the value of the constants.

Empirical Prediction of Ternary Vapor-Liquid Equilibria

Edwards et al. [17] used the correlating equations developed by Gilmont et al. [20] to predict ternary and multicomponent data from relative volatility data related to activity coefficient. He used an empirical form of equation to evaluate the integral form of equation developed.

Scheibel and Friedland [46] developed an empirical method for predicting ternary vapor-liquid equilibria which classed all ternary systems into three groups: (1) Ternary systems made up of three binary systems all having the same sign of deviation from Raoult's law; (2) ternary systems made up of one ideal and two nonideal binary systems all showing the same sign of deviation; and (3) ternary systems made up of one nonideal binary system having the opposite sign deviation from those of the other two binaries. A different graphical method was proposed for each type of system. For the systems tested, a high degree of accuracy was claimed.

TABLE 2.1 **Calculation of γ_i from Margules Three-suffix Equation for Example 2.3**

Col. 9 $= x_j^2[A_{ij} + 2x_i(A_{ji} - A_{ij})]$

$i\ j\ k$	(1) x_j	(2) x_j^2	(3) A_{ji}	(4) A_{ij}	(5) 3–4	(6) x_i	(7) $2.0 \times 5 \times 6$	(8) $4 + 7$	(9) 2×8
1 2 3	0.52	0.27040	0.085	0.085	0.0	0.25	0.0	0.085	0.02298
2 3 1	0.23	0.0529	0.455	0.385	0.070	0.52	0.0728	0.4578	0.02422
3 1 2	0.25	0.0625	0.700	0.715	−0.015	0.23	−0.0069	0.7081	0.04426

Col. 18 $= x_k^2[A_{ik} + 2x_i(A_{ki} - A_{ik})]$

$i\ j\ k$	(10) x_k	(11) x_k^2	(12) A_{ki}	(13) A_{ik}	(14) 12–13	(15) x_i	(16) $2.0 \times 14 \times 15$	(17) $13 + 16$	(18) 11×17
1 2 3	0.23	0.0529	0.715	0.700	0.015	0.25	0.00750	0.70750	0.03743
2 3 1	0.25	0.0625	0.085	0.085	0.0	0.52	0.0	0.08500	0.00531
3 1 2	0.52	0.2704	0.385	0.455	−0.070	0.23	−0.03220	0.42280	0.11433

Col. 26 $= \frac{1}{2}(A_{ji} + A_{ij} + A_{ki} + A_{ik} - A_{jk} - A_{kj})$

$i\ j\ k$	(19) A_{ji}	(20) A_{ij}	(21) A_{ki}	(22) A_{ik}	(23) $-A_{jk}$	(24) $-A_{kj}$	(25) $19 + \cdots + 24$	(26) $\frac{1}{2} \times 25$
1 2 3	0.085	0.085	0.715	0.700	−0.385	−0.455	0.745	0.3725
2 3 1	0.455	0.385	0.085	0.085	−0.715	−0.700	−0.405	−0.2025
3 1 2	0.700	0.715	0.385	0.455	−0.085	−0.085	2.085	1.0425

Col. 33 = $x_i(A_{ji} - A_{ij} + A_{ki} - A_{ik})$

$i\ j\ k$	(27) A_{ji}	(28) $-A_{ij}$	(29) A_{ki}	(30) $-A_{ik}$	(31) $27 + \cdots + 30$	(32) x_i	(33) 31×32	(34) x_j	(35) x_k	(36) $x_j x_k$
1 2 3	0.085	−0.085	0.715	−0.700	0.015	0.25	0.00375	0.52	0.23	0.11960
2 3 1	0.455	−0.385	0.085	−0.085	0.070	0.52	0.03640	0.23	0.25	0.05750
3 1 2	0.700	−0.715	0.385	−0.455	−0.085	−0.23	−0.01955	0.25	0.52	0.13000

Col. 43 = $(x_j - x_k)(A_{jk} - A_{kj})$ | Col. 46 = $-(1 - 2x_i)C^*$

$i\ j\ k$	(37) x_i	(38) x_k	(39) 37–38	(40) A_{jk}	(41) A_{kj}	(42) 40–41	(43) 39×42	(44) x_i	(45) $(1 - 2x_i)$	(46) $-45 \times C^*$
1 2 3	0.52	0.23	0.29	0.385	0.455	−0.070	−0.02030	0.25	0.5	0.015
2 3 1	0.23	0.25	−0.02	0.715	0.700	0.015	−0.00030	0.52	−0.04	−0.0012
3 1 2	0.25	0.52	−0.27	0.085	0.085	0.0	0.00	0.27	0.46	0.0138

Col. 50 = γ_i for $C^* = 0$ | Col. 54 = γ_i for $C^* = -0.03$

$i\ j\ k$	(47) $26 + 33 + 43$	(48) 36×47	(49) $9 + 18 + 48$	(50) γ_i	(51) $46 + 47$	(52) 36×51	(53) $9 + 18 + 52$	(54) γ_i
1 2 3	0.35595	0.04257	0.10298	1.268	0.37095	0.04437	0.10478	1.273
2 3 1	−0.16640	−0.00957	0.01996	1.047	−0.16760	−0.00964	0.01989	1.047
3 1 2	1.02295	0.13298	0.29157	1.957	1.03675	0.13478	0.29337	1.965

Example 2.3

Experimental Margules binary constants were determined at 760 mm Hg for the system *n*-octane–ethylbenzene–Cellosolve as follows:

System	Margules constants	
	A	B
n-Octane–ethylbenzene	0.085	0.085
n-Octane–Cellosolve	0.700	0.715
Ethylbenzene–Cellosolve	0.385	0.455

(*a*) Compute the ternary activity coefficients at $x_{n\text{-O}} = 0.25$, $x_{\text{EB}} = 0.52$, using the Wohl form of the Margules equation.

(*b*) Compute the ternary activity coefficients if the ternary constant $C^* = -0.03$.

Solution

(*a*) In order to apply Eq. (2.42) and the corresponding equations for activity coefficients for components 2 and 3, the components are designated 1, 2, 3 (*n*-octane, ethylbenzene, Cellosolve) and the Margules constants are suffixed accordingly.

$$A_{12} = 0.085 \qquad A_{13} = 0.700 \qquad A_{23} = 0.385$$
$$A_{21} = 0.085 \qquad A_{31} = 0.715 \qquad A_{32} = 0.455$$
$$x_1 = 0.25 \qquad x_2 = 0.52 \qquad x_3 = 0.23$$

In order to carry out the work in a systematic manner, a table is set up. The rotation of subscripts is utilized to set up column headings applicable to each component. Result ($C^* = 0$):

$$\gamma_1 = 1.268 \qquad \gamma_2 = 1.047 \qquad \gamma_3 = 1.957$$

(*b*) For $C^* = -0.03$,

$$\gamma_1 = 1.273 \qquad \gamma_2 = 1.047 \qquad \gamma_3 = 1.965$$

Example 2.4

Assuming the constants given in Example 2.3 were van Laar constants, determine the ternary activity coefficients for the composition $x_{n\text{-O}} = 0.25$, $x_{\text{EB}} = 0.52$ at $P_T = 760$ mm Hg, using the Wohl ternary two-suffix van Laar equations. Assume $t = 125°\text{C}$.

Solution

Equation (2.28) and corresponding equations can be expressed by means of the following general equation, with a specific form to be obtained by rotation of subscripts:

$$\log \gamma_i = \frac{x_j^2 A_{ij}\left(\dfrac{A_{ji}}{A_{ij}}\right)^2 + x_k^2 A_{ik}\left(\dfrac{A_{ki}}{A_{ik}}\right)^2 + x_j x_k \dfrac{A_{ji}A_{ki}}{A_{ij}A_{ik}}\left(A_{ij} + A_{ik} - A_{kj}\dfrac{A_{ik}}{A_{ki}}\right)}{[x_i + x_j(A_{ji}/A_{ij}) + x_k(A_{ki}/A_{ik})]^2}$$

The results of the tabulated calculations are

$$\gamma_1 = 1.265 \qquad \gamma_2 = 1.039 \qquad \gamma_3 = 1.968$$

TABLE 2.2 **Calculation of γ_i from van Laar Two-suffix Equation for Example 2.4**

$$\text{Col. } 7 = x_j^2 A_{ij} \frac{A_{ji}}{A_{ij}}$$

$i\ j\ k$	(1) x_j	(2) x_j^2	(3) A_{ji}	(4) A_{ij}	(5) $3 \div 4$	(6) 5×5	(7) $2 \times 4 \times 6$	(8) x_k	(9) x_k^2
1 2 3	0.52	0.27040	0.085	0.085	1.0	1.0	0.02298	0.23	0.0529
2 3 1	0.23	0.0529	0.455	0.385	1.18182	1.39670	0.02845	0.25	0.0625
3 1 2	0.25	0.0625	0.700	0.715	0.97902	0.95848	0.04283	0.52	0.2704

$$\text{Col. } 14 = x_k^2 A_{ik} \left(\frac{A_{ki}}{A_{ik}}\right)^2$$

$i\ j\ k$	(10) A_{ki}	(11) A_{ik}	(12) $10 \div 11$	(13) 12×12	(14) $9 \times 11 \times 13$	(15) $11 \div 10$	(16) $-15 \times A_{kj}$	(17) $4 + 11 + 16$	(18) 1×8
1 2 3	0.715	0.700	1.02143	1.04332	0.03863	0.97902	−0.44545	0.33955	0.11960
2 3 1	0.085	0.085	1.0	1.0	0.00531	1.0	−0.700	−0.230	0.05750
3 1 2	0.385	0.455	0.85615	0.71597	0.08809	1.18182	0.10045	1.06955	0.13000

$$\text{Col. } 20 = x_j x_k \frac{A_{ji}A_{ki}}{A_{ij}A_{ik}} \left(A_{ij} + A_{ik} - A_{kj} \frac{A_{ik}}{A_{ki}}\right)$$

$i\ j\ k$	(19) $5 \times 12 \times 18$	(20) 17×19	(21) $7 + 14 + 20$	(22) 1×5	(23) 8×12	(24) x_i	(25) $22 + 23 + 24$	(26) 25×25	(27) $21 \div 26$	(28) γ_i
1 2 3	0.12216	0.04148	0.10309	0.52	0.23493	0.25	1.00493	1.00988	0.10208	1.265
2 3 1	0.06795	−0.01563	0.01813	0.27182	0.25	0.52	1.04182	1.08539	0.01670	1.039
3 1 2	0.10769	0.11518	0.24610	0.24476	0.44000	0.23	0.91476	0.83679	0.29410	1.968

Example 2.5

Using the equations of Benedict et al., determine the ternary activity coefficients for the *n*-octane–ethylbenzene–Cellosolve system, $x_{n\text{-}O} = 0.25$, $x_{EB} = 0.52$.

Solution

Since Eqs. (2.46) through (2.48) should reduce to the symmetrical Margules equation for the binaries ($x_k = 0$),

$$\ln \gamma_i = \frac{2(x_j - x_i x_j)}{RT} A'_{ij} = \frac{2x_j^2}{RT} A'_{ij} = x_j^2 A_{ij}$$

where the extreme right-hand expression is the symmetrical Margules equation for $\ln \gamma_i$. Then

$$A_{ij} = \frac{2A'_{ij}}{RT}$$

The data given for the system are unsymmetrical in the sense that $A_{ij} \neq A_{ji}$. As an approximation it will be necessary to use an average, as

$$\bar{A}_{ij} = \frac{A_{ij} + A_{ji}}{2}$$

Equations (2.46) through (2.48) then reduce to

$$\ln \gamma_i = (x_j - x_i x_j)\bar{A}_{ij} + (x_k - x_i x_k)\bar{A}_{ik} - x_j x_k \bar{A}_{jk}$$

The data are:

$$\bar{A}_{12} = 0.085 \qquad \bar{A}_{13} = 0.7075 \qquad \bar{A}_{23} = 0.420$$

$$x_1 = 0.25 \qquad x_2 = 0.52 \qquad x_3 = 0.23$$

Solving,

$$\ln \gamma_1 = (0.52 - 0.13)(0.085) + (0.23 - 0.0575)(0.7075) - (0.1196)(0.420)$$

$$\ln \gamma_1 = 0.10496 \qquad \gamma_1 = 1.273$$

$$\ln \gamma_2 = 0.01589 \qquad \gamma_2 = 1.037$$

$$\ln \gamma_3 = 0.29331 \qquad \gamma_3 = 1.965$$

Chao and Hougen [12] applied the Ibl and Dodge [28] correction to the Gibbs-Duhem equation for isobaric conditions and combined this with the expression for excess free energy. This resulted in equations of the form

$$\begin{aligned} \ln \frac{\gamma_2}{\gamma_1} = &- b_{12}(x_1 - x_2) + c_{12}[2x_1x_2 - (x_1 - x_2)^2] \\ &+ x_3\{b_{31} - b_{23} - c_{23}(2x_2 - x_3) + c_{31}(x_3 - 2x_1) - b_{12}(x_1 - x_2) \\ &- c_1[x_1(2x_2 - x_3) + x_2(x_3 - x_2)] - c_2[x_1(2x_2 - x_1) + x_3(x_1 - x_2)] \\ &+ c_3[2x_1x_2 - (x_1 - x_2)^2] + \cdots\} \end{aligned} \tag{2.49}$$

The rotation principle can be used to get other ratios of $\ln(\gamma_i/\gamma_j)$. The Redlich and Kister [43] fourth-order equations for ternary systems are similar in nature:

$$\log \frac{\gamma_1}{\gamma_2} = b_{12}(x_2 - x_1) - c_{12}[(x_1 - x_2)^2 - 2x_1x_2] + d_{12}(x_2 - x_1)$$
$$\times [(x_1 - x_2)^2 - 4x_1x_2] + x_3[b_{13} + c_{13}(2x_1 - x_3) + d_{13}(x_1 - x_3)(3x_1 - x_3)$$
$$- b_{23} - c_{23}(2x_2 - x_3) - d_{23}(x_2 - x_3)(3x_2 - x_3) + c(x_2 - x_1)$$
$$+ D_1x_1(2x_2 - x_3) + D_2x_2(x_2 - 2x_1)] \quad (2.50)$$

For correlation of ternary data, the Wohl three- and four-suffix equations utilizing some ternary data to evaluate the ternary constants are reasonably accurate. Where no ternary data are available, the Li-Coull and Bonham equations have proved satisfactory and reliable for most ternary systems, using only binary data. The Redlich-Kister and Chao-Hougen equations require evaluation of some ternary constants to obtain accurate reproduction of experimental data. Finally, the Bonham type of equations are good for very rough approximation only, using pure-component data to evaluate the binary constants.

Presentation of Data

Ternary data may be presented in tabular form or in graphical form. The tabular data are exact, but this advantage is lost when interpolation is required, and therefore graphical presentation is generally more useful.

Such graphs may take the form of activity coefficient of a component in solution versus the ternary compositions. Three graphs are necessary such as shown in Fig. 2.12.

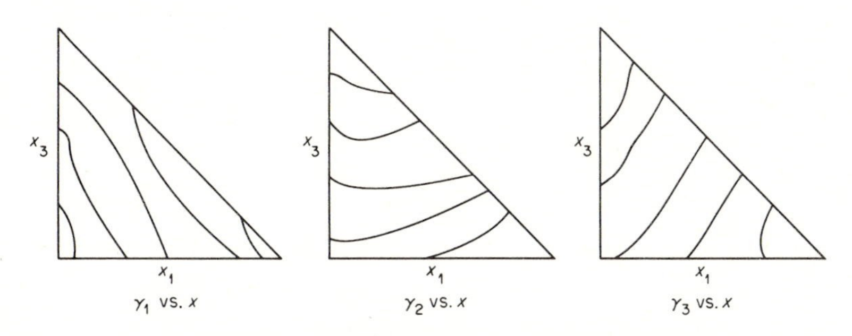

FIGURE 2.12 Activity-coefficient composition curves.

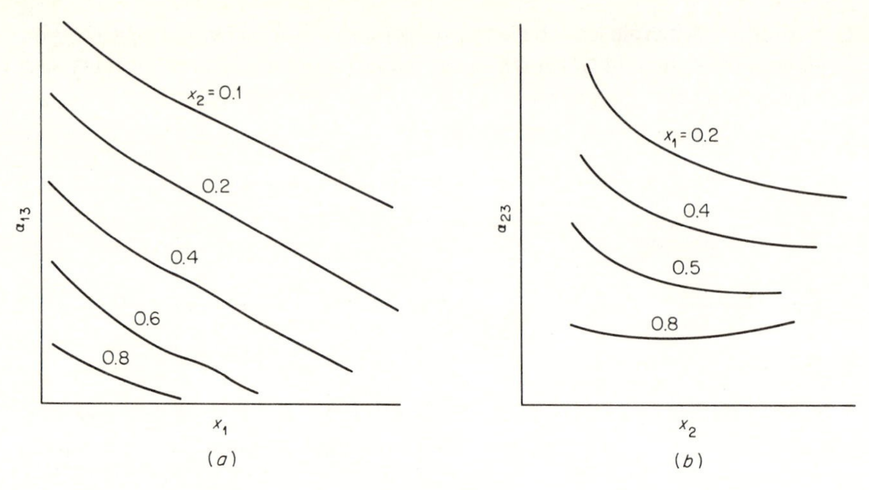

FIGURE 2.13 Relative volatility composition curves.

The relative volatility of two components based on the third may be plotted for constant values of composition of the second component (see Fig. 2.13). As a third alternative the actual y versus x diagram may be plotted on a ternary diagram with parameters of x_1, x_2 versus y_1 and y_2 (Fig. 2.14).

Nonideal Multicomponent Systems

Black Equations

Black [7, 8] utilized modifications of the van Laar equations to relate activity coefficients and composition for multicomponent systems. He

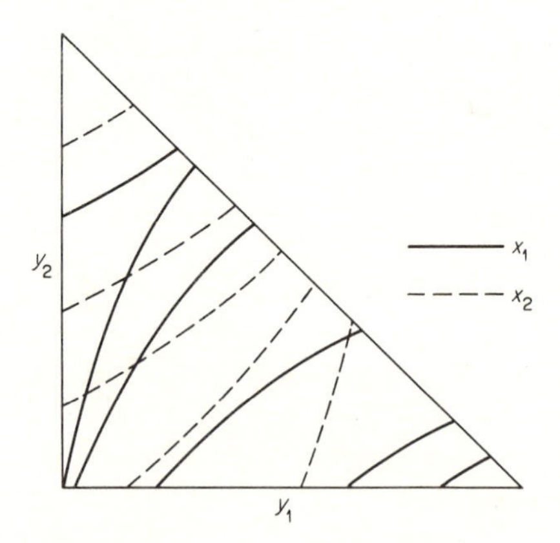

FIGURE 2.14 Ternary vapor-liquid compositions.

distinguished between classes of components ("class" constituting a homologous series made up of a number of components), class variables, and component variables. The general equation for activity coefficient in terms of composition and class designation is

$$\log \gamma_i = \frac{\sum_j (a_{ij}R_{j2}x_j)^2 + \sum_{jk} (B_{i,jk}x_jx_k)}{\left(\sum_k x_kR_{k2}\right)^2} + E_{si} \tag{2.51}$$

where

$$E_{si} = \sum_R [(X_S - X_R)^2(\sum c_{ir}x_r)] + 2\sum_R \left[(X_S - X_R)\left(\sum_{sr} c_{sr}x_sx_r\right)\right]$$
$$- \frac{3}{2}\sum_{RM}\left[(X_R - X_M)^2\left(\sum_{rm} c_{rm}x_rx_m\right)\right] \tag{2.52}$$

and where

$$B_{i,jk} = R_{j2}R_{k2}[(a_{ij} - a_{ik})^2 - a_{jk}{}^2R_{ij}] \tag{2.53}$$

The terms R and M refer to two classes of compounds, and r and m are variables denoting compounds within classes R and M. S refers to a particular class and i, j, and k refer to specific compounds within classes. The compositions x_r and x_m refer to mole fractions of the individual components in classes R and M, respectively. The sums of compositions within the classes are X_R and X_M, respectively.

Black established a number of rules for determining the values of the variables and coefficients in the equations for handling the different situations by use of his equations. For the various systems considered which included both ternary and multicomponent systems and hydrocarbons, alcohols, ketones and water, the method gave very good results when the calculated data are compared with the experimental data.

For complete details concerning the method, see Black [7, 8].

Wilson Equation

A method of calculating multicomponent vapor-liquid equilibria by means of the Wilson [56] equation, which extended the theoretical equation of Flory [18] and Huggins [27], has been proposed by Orye and Prausnitz [39]. They utilize the excess free-energy concept to express solution nonideality and present a generalized equation (2.54) for multicomponent systems:

$$\ln \gamma_k = -\ln\left(\sum_{j=1}^{n} x_j\Lambda_{kj}\right) + 1 - \sum_{i=1}^{n} \frac{x_i\Lambda_{ik}}{\sum_{j=1}^{n} x_j\Lambda_{ij}} \tag{2.54}$$

where

$$\Lambda_{ij} \equiv \frac{V_{jL}}{V_{iL}} e^{-[(\lambda_{ij}-\lambda_{ii})/RT]} \tag{2.55}$$

$$\Lambda_{ji} = \frac{V_{iL}}{V_{jL}} e^{-[(\lambda_{ji}-\lambda_{jj}/RT]} \tag{2.56}$$

Only binary interaction terms are required in the generalized equation. Therefore, activity coefficients for the compounds in multicomponent systems can be calculated from the binary activity coefficients determined experimentally.

TABLE 2.3 **Summary of Activity-Coefficient Equations for Multicomponent System**

Reference	Type and variables	Application	Data needed
Wohl [57, 58]	van Laar γ, x, T	Ternary	Binary, some ternary
Bonham [9, 45]	van Laar γ, x, T	Ternary, multicomponent	Binary
Li-Coull [35]	van Laar γ, x, T	Ternary	Binary
Wohl [57, 58]	Margules γ, x, T	Ternary, multicomponent	Binary, some ternary, multicomponent
Redlich-Kister [43]	Margules γ, x	Ternary	Binary, some ternary
Benedict et al. [6]	Margules γ, x, T	Ternary	Binary, some ternary
Black [7, 8]	van Laar γ, x	Ternary, multicomponent	Binary, some multicomponent
Wilson [56, 39]	Flory-Huggins γ, x	Ternary, multicomponent	Binary
White [52]	van Laar γ, x, T	Ternary	Binary
Weimer-Prausnitz [51]	Hildebrand $\gamma^{\infty}, V_L, x, \phi, \lambda, \gamma, T$	Binary, ternary, multicomponent	Pure component, binary
Edwards et al. [17]	Empirical γ, α, P, x	Ternary, multicomponent	Binary, some multicomponent
Scheibel-Friedland [46]	Empirical γ, x	Ternary	Binary
Chao-Hougen [12]	Empirical γ, x	Ternary	Binary, some ternary

Although theoretically only one point of x, y, T, and P data is required to evaluate Λ_{ij} and Λ_{ji}, the proponents of the method state that a series of data points gives more reliable results.

For a binary solution the activity-coefficient equations are

$$\ln \gamma_1 = -\ln (x_1 + \Lambda_{12}x_2) + x_2\left(\frac{\Lambda_{12}}{x_1 + \Lambda_{12}x_2} - \frac{\Lambda_{21}}{\Lambda_{21}x_1 + x_2}\right) \tag{2.57}$$

$$\ln \gamma_2 = -\ln (x_2 + \Lambda_{21}x_1) - x_1\left(\frac{\Lambda_{12}}{x_1 + \Lambda_{12}x_2} - \frac{\Lambda_{21}}{\Lambda_{21}x_1 + x_2}\right) \tag{2.58}$$

Using the values of Λ_{ij} and Λ_{ji} determined from binary data, the activity coefficients for the components in a multicomponent mixture can be calculated from the component binaries.

The authors predicted the vapor-liquid data on some 65 systems using the Wilson parameters and found that the predicted and experimental data agreed very well. Comparison with the predicted data from the van Laar and Margules equations showed that those derived from the Wilson equation were as good in most cases and better in many others.

The disadvantage of this type of correlation is that it does not apply to partially miscible systems, whereas the Black method is applicable. However, the number of constants needed for this correlation is less than required for the Black correlation, and thus fewer data are needed.

Equilibrium Vaporization Ratios

Equilibrium vaporization ratios or K values have proved to be extremely useful in hydrocarbon distillation, absorption, and stripping calculations. K is defined as

$$K_i = \frac{y_i}{x_i} \tag{2.59}$$

In terms of vapor pressure, activity coefficient, and fugacity coefficient,

$$K_i = \frac{\gamma_i P_i}{\nu_i P_t} \tag{2.60}$$

When the system can be considered ideal, $\gamma_i = 1.0$ and $\nu_i = 1.0$, and the equilibrium vaporization ratio reduces to

$$K_i = \frac{P_i}{P_t} \tag{2.61}$$

K values for a particular compound are a function of temperature, pressure, composition of the mixture, and the amount of the compound in the mixture.

There are in general two principal methods for obtaining K values—the thermodynamic method and the empirical method. The thermodynamic method involves the use of equations which describe all of the thermodynamic properties of the phases including the fugacity, while the empirical approach involves the prediction of K values without particular regard to thermodynamic properties which represent the best fit to the experimental data available for derivation of the correlation.

Thermodynamic Methods for K

The thermodynamic method normally requires an equation of state to describe phase behavior of mixtures. Benedict et al. [2–5] developed a practical equation describing the behavior of light hydrocarbons and their mixtures. This equation is similar in form to the Beattie-Bridgeman [1] equation of state. The equation includes eight constants evaluated from vapor pressure, critical constants, and pressure, volume, and temperature properties of the pure component. The coefficients describing the mixture behavior were calculated from the individual pure-component constants. The equation, while accurate to a high degree, must be solved by a complicated trial-and-error solution for each point and requires the use of a high-speed digital computer.

A practical correlation for primarily paraffinic hydrocarbon K values was developed by Benedict et al. by using molal average boiling point as a bulk phase parameter. Fugacity coefficients and activity coefficients were determined over wide ranges for the different variables, and were plotted on usable charts as functions of temperature, pressure, and molal average boiling point of the two phases. These comprise the Kellogg [30] charts.

De Priester [14] presented the same fugacity functions on pressure, temperature, and composition charts for a pressure range of 0 to 1000 psia. He also constructed nomographs to show average K's for first trial calculations. The De Priester charts were an improvement because only two charts were needed for each hydrocarbon and the entire pressure range was included on each chart. However, use of the charts for exact determination of K values requires trial-and-error calculations.

Edmister and Ruby [16] developed generalized correlations of activity coefficients based on the fugacity functions developed by Benedict et al. The activity coefficients for both liquid and vapor phases were correlated with reduced temperature, reduced pressure, and ratio of component boiling point to the molal average boiling point of the phases. This is a further improvement in determining K values but still trial-and-error calculations are required. The Edmister-Ruby correlations cover a pressure range to 3600 psia.

Chao and Seader [13] developed a generalized correlation for K values in hydrocarbon mixtures using three factors: ϕ_i, the fugacity coefficient of component i in the vapor mixture; ν_1°, the fugacity coefficient of pure liquid i at system conditions; and γ_i, the activity coefficient of component i in liquid solution.

$$K_i = \frac{\nu_i \gamma_i}{\phi_i} \tag{2.62}$$

Methods and equations for evaluation of the three factors in Eq. (2.62) are given in the original reference and are suited to solution using a high-speed digital computer. The authors claim an over-all average deviation of calculated from experimental values of 8.7%. This method is claimed to apply to mixtures of paraffin, aromatic, naphthenic, and olefinic hydrocarbons and inert gases such as hydrogen.

Hoffman et al. [25] presented a method for evaluation of fugacities of the components in a multicomponent liquid mixture to aid in predicting vaporization equilibrium ratios for multicomponent mixtures.

Prausnitz et al. [42] used the Hildebrand solubility parameter to calculate K values for light hydrocarbons in paraffinic, naphthenic, and aromatic absorption oils with reasonably good results.

The foregoing methods represent only a few of the many which utilize some of the various equations of state to evaluate individual and mixture properties from which K values for various materials can be calculated. Readers interested in this aspect of K-value determination should refer to modern thermodynamics books and the copious literature on the subject. Extended discussion of this phase of the subject is considered beyond the scope of this book.

Empirical Methods for K

The empirical methods vary greatly in detail but generally are based upon fitting curves to experimental data points to get the best agreement and to devise suitable means of extrapolating and interpolating for other conditions while maintaining the same relative accuracy. Essentially these correlations involve modifying the pure-component fugacity relations by means of a "convergence pressure" parameter which corrects for the behavior of the components in a mixture and thus includes composition effects.

A number of such correlations have been devised [10, 11, 21–23, 29, 36–38, 40, 41, 48, 49, 54, 55].

Table 2.4 lists some of the general equilibrium vaporization ratio, or K, correlations. In most cases these are for paraffinic and olefinic hydrocarbons, but may present K values for hydrogen, hydrogen sulfide, sulfur dioxide, carbon dioxide, nitrogen, water, as well as aromatic and saturated

TABLE 2.4 **Equilibrium Vaporization Ratios**

Reference	Presented form	System	T, °F	Pressure range
Winn [54]	Nomographs	HC	40–800	Atm.–10,000 psia
Winn [55]	Nomographs	HC	40–800	Atm.–10,000 psia
Hadden-Grayson [23]	Nomographs	HC and Pet. fr.	−260–800	Atm.–10,000 psia
Smith-Smith [48]	Graphs	HC and Pet. fr.	Almost to T_c	Atm.–p_c
Smith-Watson [49]	Graphs	HC	0–1000	To 10,000 psia
		HC	−200–500	To 10,000 psia
N.G.P.A. [37]	Graphs	Pet. fr.	100–1000	1.0–1000 psia
Myers-Lenoir [36]	Graphs	HC	−120–350	1.0–10,000 psia
Kellogg [30]	Graphs	HC	−100–400	To 3500 psia
De Priester [14]	Graphs		−100–400	To 1000 psia
	Nomographs	HC	−100–400	To 1000 psia
Organick-Brown [38]	Graphs	HC	80–500	Up to 10,000 psia
Poettman-Mayland [40]	Graphs	Pet. fr	100–1000	100–1000 psia
Prausnitz-Duffin [41]	Graphs	C_4—C_{10} HC in H	50–400	To 2500 psia
Brown [10]	Graphs	HC	40–500	To 1000 psia
Hadden [22]	Nomographs	HC	−30–600	3–500 psia
Katz-Hackmuth [29]	Graphs	HC and crude oil	−30–270	To 3000 psia
Cajander et al. [11]	Graphs	HC	−200–900 (max)	10–10,000 psia
Lenoir-Hipkin [32]	Graphs	H_2 in HC	−200–303	To 10,000 psia

cyclic hydrocarbons. Only two of the correlations in the table are used specifically for heavy oil fractions [40, 49]. In addition to those listed in the table, there are numerous K charts reported in the literature which are applicable to specific compounds in specific mixtures. These, in general, are very accurate for the particular compound and conditions for which they were devised because they do not attempt to generalize for all compounds.

Convergence Pressure

As pointed out in the foregoing, the correlating parameter most commonly used which introduces the mixture effect is convergence pressure. If the K values for two hydrocarbons are plotted on a log K versus log P plot at constant temperature, as shown in Fig. 2.15, the point of convergence with respect to pressure of the two lines at $K = 1.0$ is the *convergence* pressure. If the temperature of the plot was the critical temperature, then and only then does the convergence pressure correspond to the critical pressure.

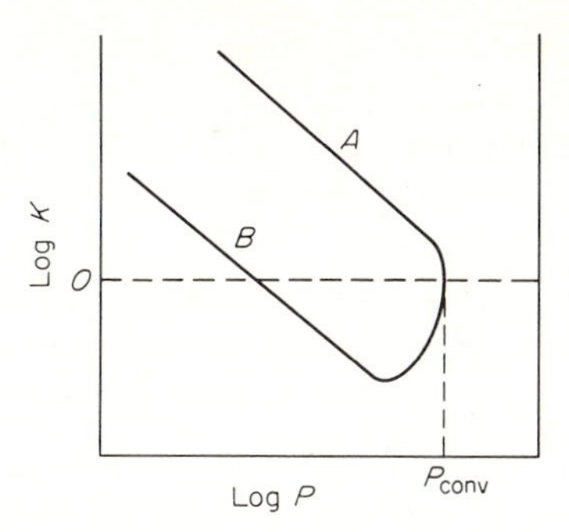

FIGURE 2.15 Convergence pressure.

Example 2.6

Determine K values for the following compounds at 350°F and 100 psia by assuming ideality in the vapor and liquid.

Solution

$P_t = 5171$ mm Hg, $t = 176.8$°C.

Component	Antoine constants			Vapor pressure, mm Hg	$K = P/P_t$
	A	B	C		
Hexane	7.31938	1483.1	265.9	9317	1.802
Heptane	7.3270	1581.7	257.6	4852	0.938
Benzene	7.42912	1628.32	279.56	7262	1.404
Toluene	6.95334	1343.943	219.38	3640	0.704
Styrene	6.92409	1420.0	206.0	1639	0.317
Ethylbenzene	6.95719	1424.26	213.21	2020	0.391

The convergence pressure will vary both with composition of the mixture and the temperature. Therefore, it is considered a mixture parameter and capable of characterizing the mixture. Unfortunately, convergence pressure is difficult to determine accurately. Hadden [21], Lenoir and White [33, 34], Myers and Lenoir [36], and White and Brown [53] all used the convergence-pressure parameter to account for mixture properties, but the methods used in determining it varied considerably as did the calculated results for the same mixtures.

Lenoir and White [34] derived a relatively simple but reliable method for determination of convergence pressures of hydrocarbon systems and hydrocarbon systems containing other components by considering the system to be composed of pseudobinary mixtures. From the liquid-phase composition, effective boiling points (EBP's) for the pseudolight component and for the pseudoheavy component are determined. Correlating factors are determined from tables, and the convergence pressure is read from one of a number of charts—each for a different value of constant EBP of the light component—on which system temperature and convergence pressure are related with parameters of constant EBP of the heavy pseudocomponent.

Figures 2.16 and 2.17 give the necessary information for determining the EBP for the light and heavy components.

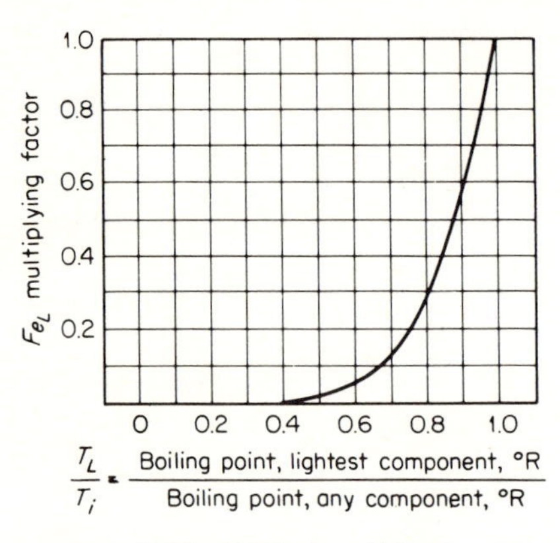

FIGURE 2.16 EBP for lightest component. [*From Lenoir and White, Petrol. Refiner* (*March, 1958*), *p. 173. Copyright Gulf Publishing Company, Houston, Texas* (*1958*).]

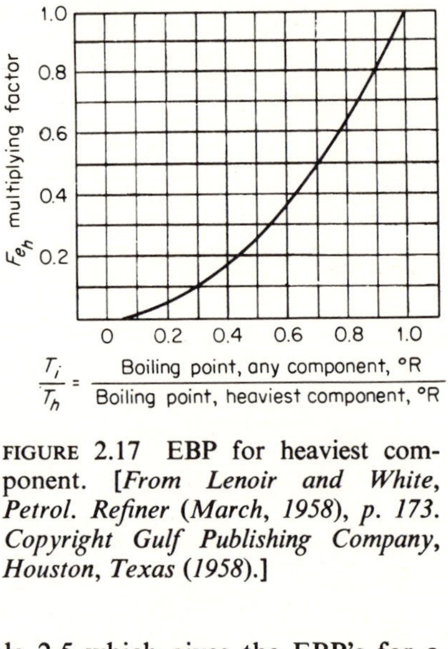

FIGURE 2.17 EBP for heaviest component. [*From Lenoir and White, Petrol. Refiner* (*March, 1958*), *p. 173. Copyright Gulf Publishing Company, Houston, Texas* (*1958*).]

T_L/T_i and T_i/T_H are read from Table 2.5 which gives the EBP's for a number of compounds. The EBP for a paraffin or olefin compound is the same as its normal boiling point. Aromatics, acetylenes, naphthenes, and nonhydrocarbons have EBP's, the same as the normal boiling point of the hypothetical paraffin compound having the same critical temperature as the nonparaffin under consideration.

Once the multiplying factors F_e are determined for each component, the mole fraction of each component in the mixture is multiplied with the corresponding F_e and the component EBP from Table 2.5, giving values of $F_e xT$. This is done for both the light and heavy components. Then the EBP of the pseudolight and pseudoheavy components are determined by

$$\text{EBP}_L = \frac{\Sigma F_{e_L} x_i T_i}{\Sigma F_{e_L} x_i} \tag{2.63}^*$$

* Sum for all components lighter than EBP_H.

TABLE 2.5 **Effective Properties for Convergence Pressures [34]**

Component	EBP		Pressure function P_F, psia
	t_i, °F	T_i, °R	
n-Paraffins			
Methane	−259	201	—
Ethane	−128	332	—
Propane	−44	416	—
Butane	31	491	—
Pentane	97	557	—
Hexane	156	616	—
Heptane	209	669	—
Octane	258	718	—
Nonane	303	763	—
Decane	345	805	—
Undecane	384	844	—
Dodecane	421	881	—
i-Paraffins			
i-Butane	11	471	—
i-Pentane	82	542	—
i-Hexane	141	601	—
Olefins			
Ethylene	−155	305	—
Propylene	−54	406	—
Isobutylene	20	480	—
Butylene-1	21	481	—
cis-2-Butylene	39	499	—
trans-2-Butylene	34	494	—
1-Pentylene	86	546	—
1,3-Butadiene	24	484	—
Naphthenes			
Cyclopentane	176	636	225
Methylcyclopentane	216	676	142
Cyclohexane	232	692	205
Methylcyclohexane	269	729	140
Dimethylcyclohexane	306	766	140
Ethylcyclohexane	324	784	140
Propylcyclohexane	365	825	140
Aromatics			
Benzene	246	706	335
Toluene	301	761	272
o-Xylene	362	822	231
m,p-Xylene	352	812	199
Styrene	364	824	200
Ethylbenzene	347	807	238
Propylbenzene	389	849	164
Isopropylbenzene	376	836	164
Butyl benzene	432	892	158
Acetylenes			
Acetylene	−120	340	205

TABLE 2.5 **Contd.**

Component	EBP		Pressure function P_F, psia
	t_i, °F	T_i, °R	
Nonhydrocarbons			
Air	−300	160	—
Carbon dioxide	−130	330	363
Carbon monoxide	−288	172	—
Hydrogen	−393	67	—
Hydrogen sulfide	−40	420	686
Nitrogen	−330	130	—
Water	397	857	2726

and

$$\text{EBP}_H = \frac{\Sigma F_{e_H} x_i T_i}{\Sigma F_{e_H} x_i} \qquad (2.64)^*$$

where

$$F_{e_L} = \left(\frac{T_L}{T_i}\right)^m \quad \text{and} \quad F_{e_H} = \left(\frac{T_i}{T_H}\right)^n \qquad (2.65)$$

Figures 2.18 through 2.29 are used to determine the convergence pressure. Cross plotting to construct figures of intermediate values of $(\text{EBP})_L$ can be resorted to or logarithmic interpolation may be used.

The authors recommend accurate calculation of the $(\text{EBP})_L$ (rather than estimation) when the system pressure is equal to or above 0.6 of the convergence pressure. They suggest for estimating the $(\text{EBP})_L$ the use of the molal average of the EBP values of the material composing the lightest 7% of the mixture, and for the $(\text{EBP})_H$ the molal average of the EBP values of the materials comprising the heaviest 40% of the mixture. When the system temperature is more than 40°F below the critical temperature of the light or pseudolight component, Lenoir and White recommend the quasi-convergence pressure be estimated to be 5000 psia.

Where nonaliphatic hydrocarbons and nonhydrocarbon materials exist in the mixture, it is necessary to use another parameter, namely, the pressure function P_F.

$$P_F = \frac{\Sigma F_{e_i} x_i P_{F_i}}{\Sigma F_{e_i} x_i}. \qquad (2.66)$$

The values of P_{F_i} are read from Table 2.5. Where the light component is aliphatic and the heavy component nonaliphatic, the value of P_F is added to the base-line pressure isothermally, as shown in Fig. 2.30, and the critical locus of the mixture is constructed as shown by the heavy line. This line AD is used to determine the convergence pressure.

* Sum for all components heavier than EBP_L.

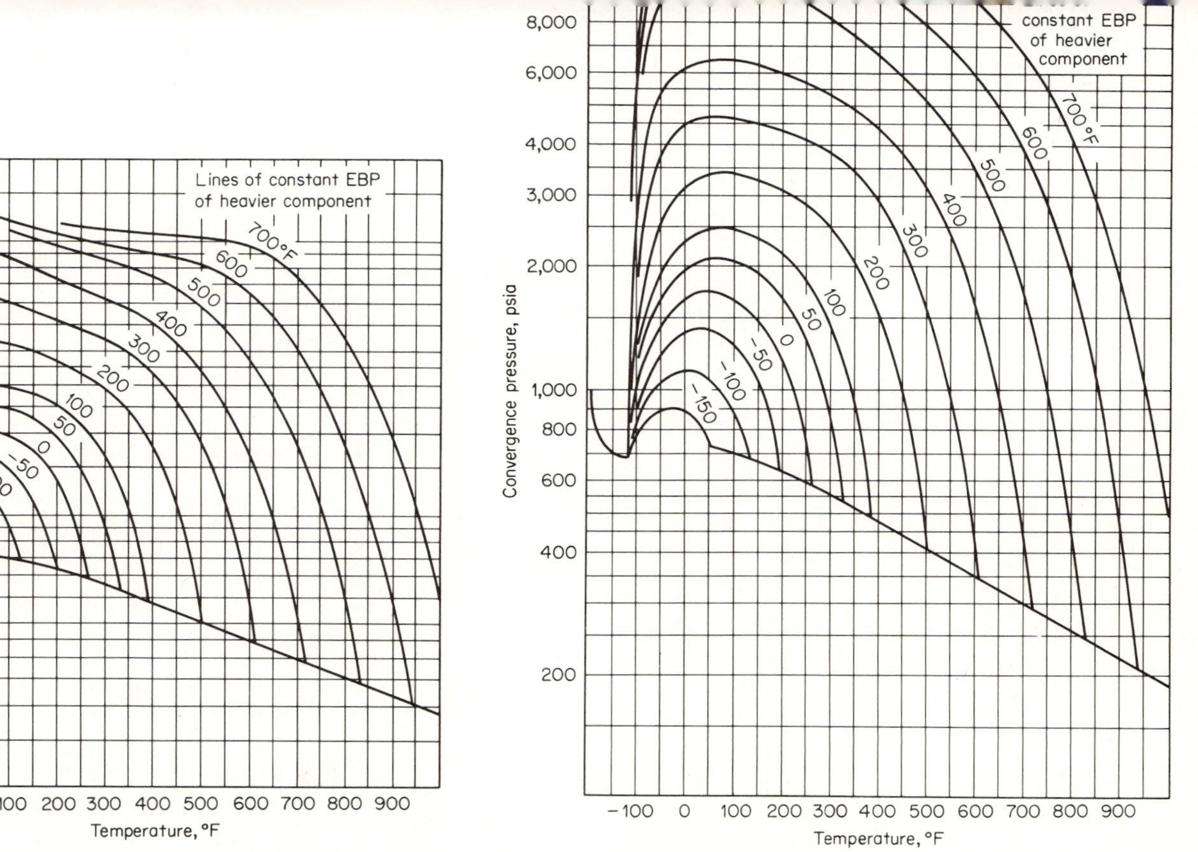

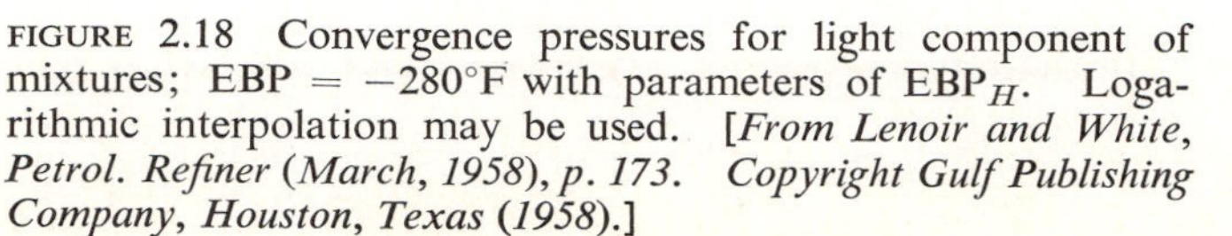

FIGURE 2.18 Convergence pressures for light component of mixtures; EBP = −280°F with parameters of EBP_H. Logarithmic interpolation may be used. [*From Lenoir and White, Petrol. Refiner* (*March, 1958*), *p. 173. Copyright Gulf Publishing Company, Houston, Texas* (*1958*).]

FIGURE 2.19 Convergence pressures for light component of mixtures; EBP = −259°F with parameters of EBP_H. [*From Lenoir and White, Petrol. Refiner* (*March, 1958*), *p. 173. Copyright Gulf Publishing Company, Houston, Texas* (*1958*).]

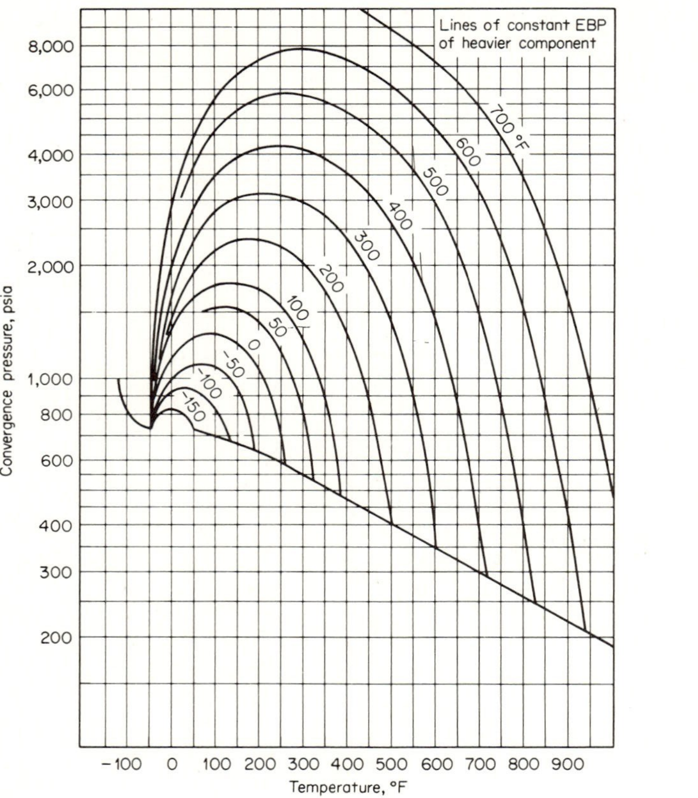

FIGURE 2.20 Convergence pressures for light component of mixtures; EBP = −220°F with parameters of EBP_H. [*From Lenoir and White, Petrol. Refiner (March, 1958), p. 173. Copyright Gulf Publishing Company, Houston, Texas (1958).*]

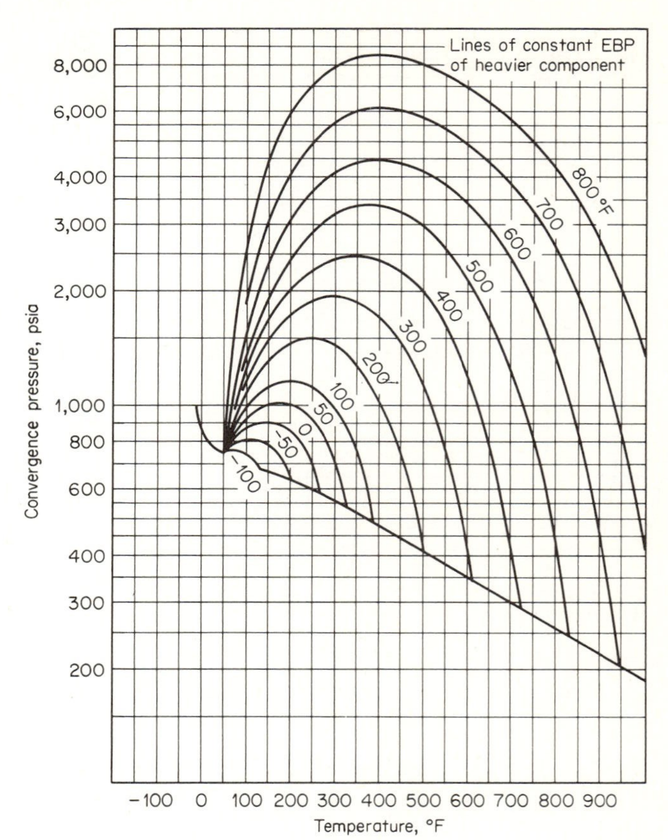

FIGURE 2.21 Convergence pressures for light component of mixtures; EBP = −155°F with parameters of EBP_H. [*From Lenoir and White, Petrol. Refiner (March, 1958), p. 173. Copyright Gulf Publishing Company, Houston, Texas (1958).*]

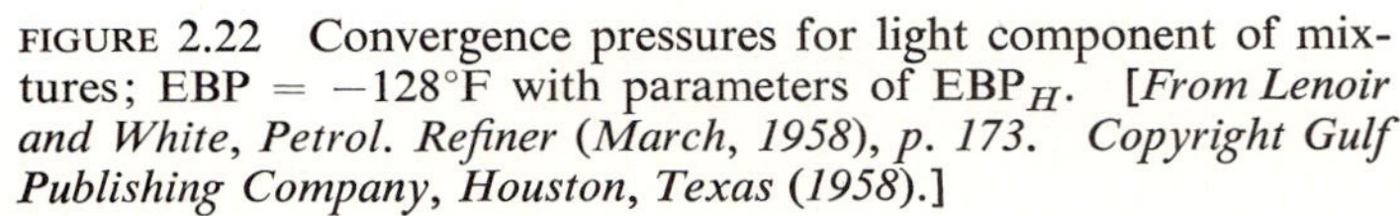

FIGURE 2.22 Convergence pressures for light component of mixtures; EBP = −128°F with parameters of EBP_H. [*From Lenoir and White, Petrol. Refiner (March, 1958), p. 173. Copyright Gulf Publishing Company, Houston, Texas (1958).*]

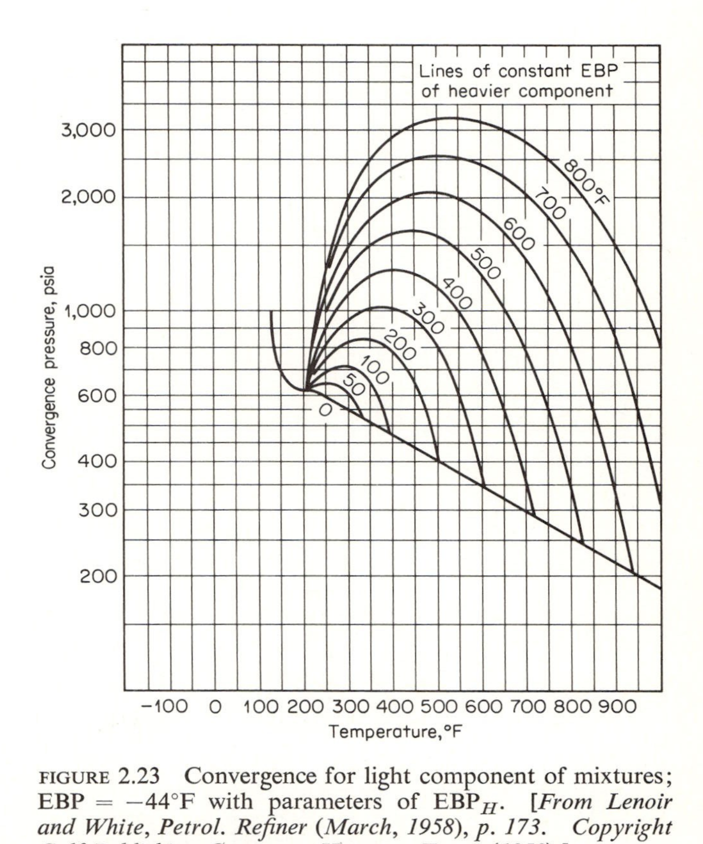

FIGURE 2.23 Convergence for light component of mixtures; EBP = −44°F with parameters of EBP_H. [*From Lenoir and White, Petrol. Refiner (March, 1958), p. 173. Copyright Gulf Publishing Company, Houston, Texas (1958).*]

FIGURE 2.24 Convergence pressures for light component of mixtures; EBP = 31°F with parameters of EBP_H. [*From Lenoir and White, Petrol. Refiner (March, 1958), p. 173. Copyright Gulf Publishing Company, Houston, Texas (1958).*]

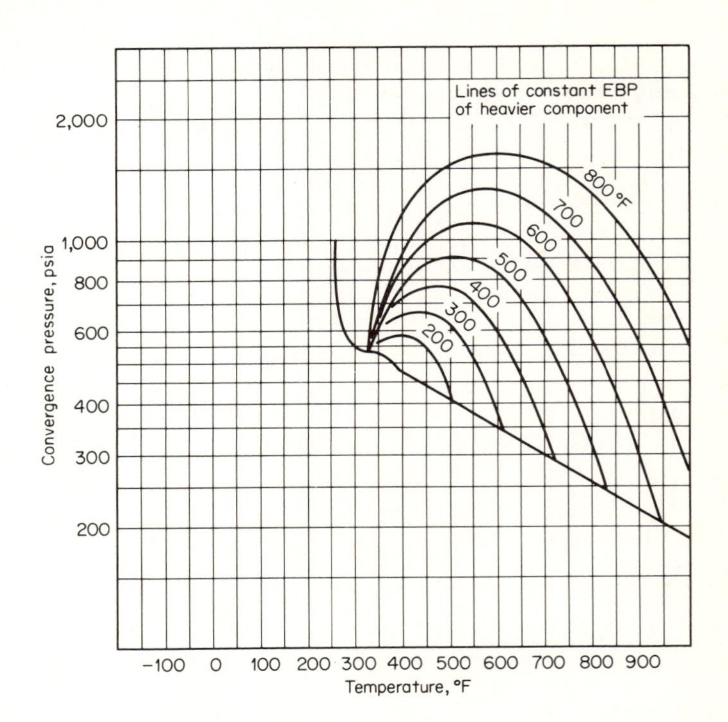

FIGURE 2.25 Convergence pressures for light component of mixtures; EBP = 50°F with parameters of EBP_H. [*From Lenoir and White, Petrol. Refiner (March, 1958), p. 173. Copyright Gulf Publishing Company, Houston, Texas (1958).*]

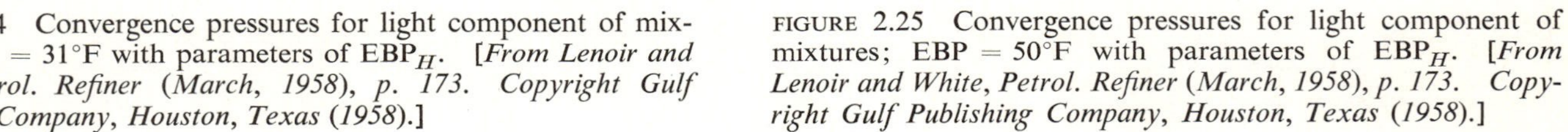

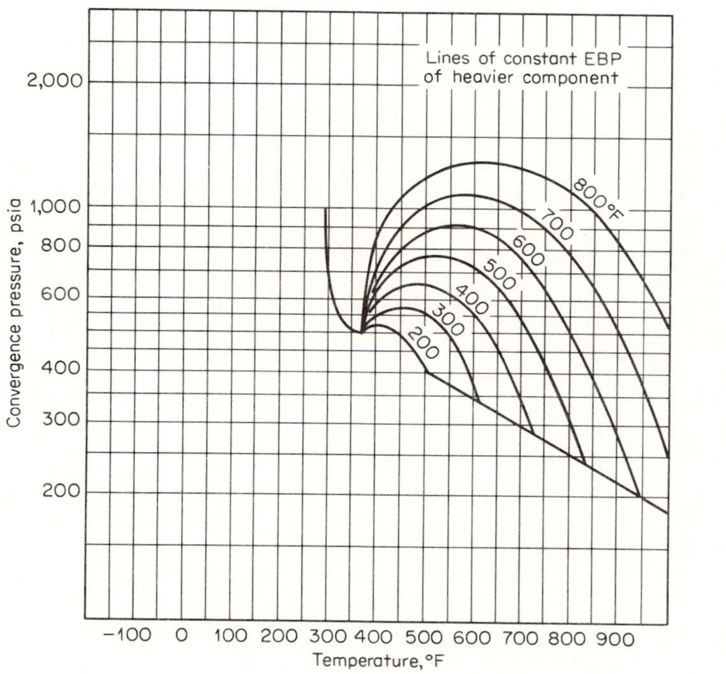

FIGURE 2.26 Convergence pressures for light component of mixtures; EBP = 80°F with parameters of EBP_H. [*From Lenoir and White, Petrol. Refiner (March, 1958), p. 173. Copyright Gulf Publishing Company, Houston, Texas (1958).*]

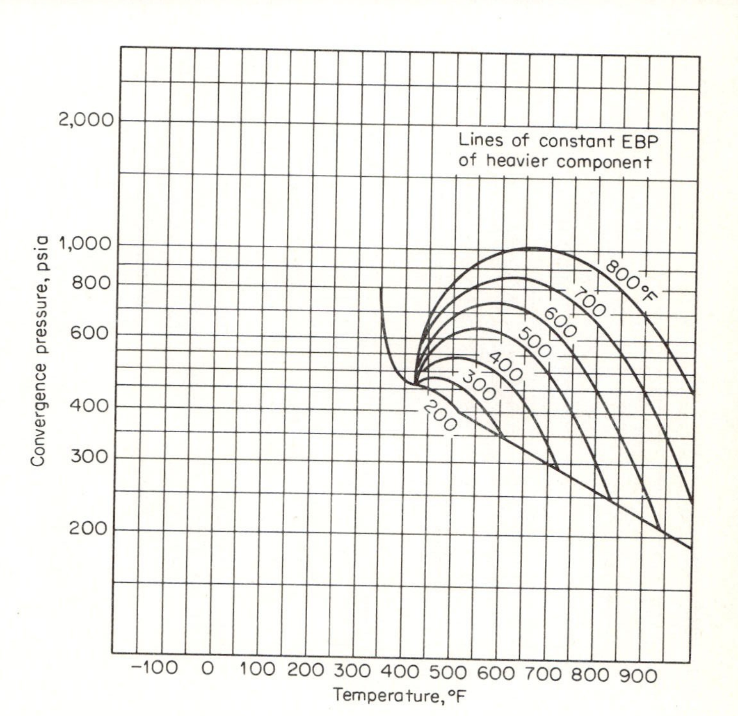

FIGURE 2.27 Convergence pressures for light component of mixtures; EBP = 120°F with parameters of EBP_H. [*From Lenoir and White, Petrol. Refiner (March, 1958), p. 173. Copyright Gulf Publishing Company, Houston, Texas (1958).*]

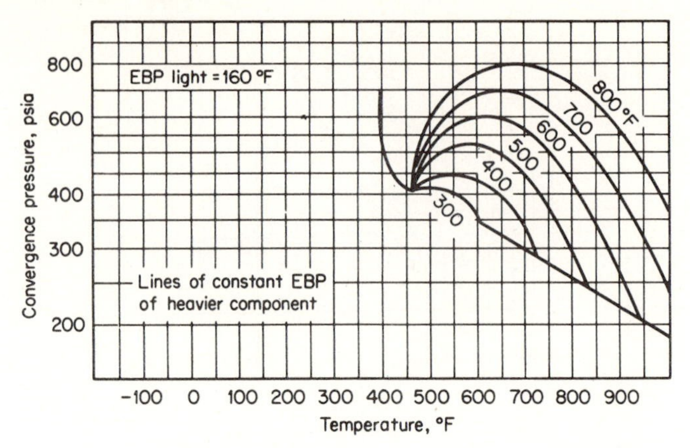

FIGURE 2.28 Convergence pressures for light component of mixtures; EBP = 160°F with parameters of EBP_H. (*From Lenoir and White, Petrol. Refiner* (*March, 1958*), *p. 173. Copyright Gulf Publishing Company, Houston, Texas* (*1958*).]

Where the light component is nonaliphatic and the heavy one is aliphatic, the construction shown in Fig. 2.31 is used. The value of P_F is added to point A. The aliphatic curve passes through A and C. The partially aliphatic curve passes through D and C. A curve is found on Figs. 2.18 through 2.29 which does this, and is plotted as shown on the heavy line.

Where both components are nonaliphatic, the construction shown on Fig. 2.32 is used. The nonaliphatic curve passes through DE.

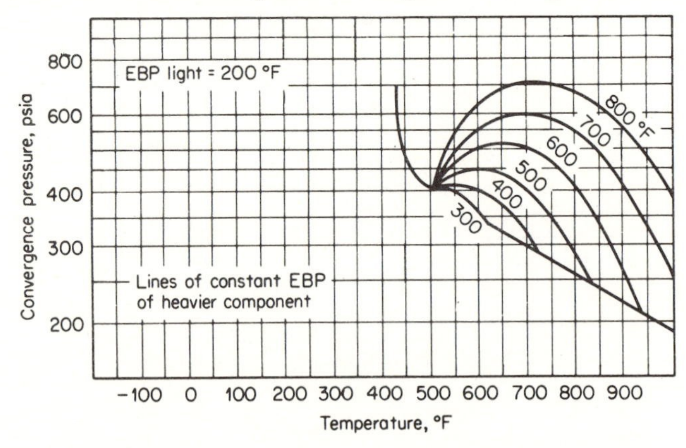

FIGURE 2.29 Convergence pressures for light component of mixtures; EBP = 200°F with parameters of EBP_H. [*From Lenoir and White, Petrol. Refiner* (*March, 1958*), *p. 173. Copyright Gulf Publishing Company, Houston, Texas* (*1958*).]

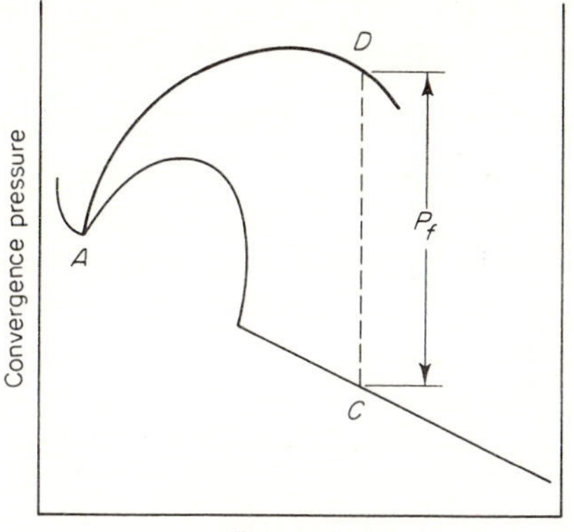

FIGURE 2.30 Construction for aliphatic light component, nonaliphatic heavy component. [*From Lenoir and White, Petrol. Refiner (March, 1958), p. 173. Copyright Gulf Publishing Company, Houston, Texas (1958).*]

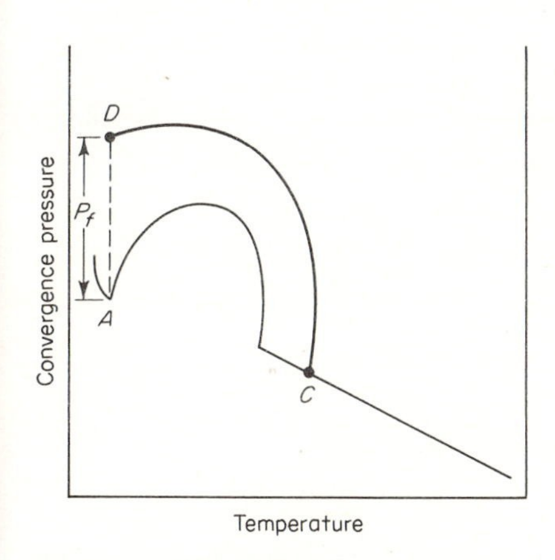

FIGURE 2.31 Construction for nonaliphatic light component, aliphatic heavy component. [*From Lenoir and White, Petrol. Refiner (March, 1958), p. 173. Copyright Gulf Publishing Company, Houston, Texas (1958).*]

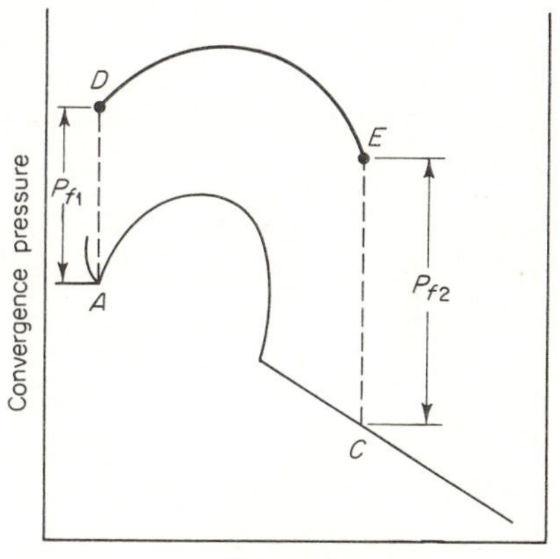

FIGURE 2.32 Construction for both components nonaliphatic. [*From Lenoir and White, Petrol. Refiner (March, 1958), p. 173. Copyright Gulf Publishing Company, Houston, Texas (1958).*]

Mixtures Containing Hydrogen, Methane

Lenoir [31] developed a method of predicting a correction factor for hydrocarbon K values where the mixture contains hydrogen and/or methane. The mole fraction of hydrogen in the liquid phase or methane in the liquid phase is used as a correlating parameter as well as the K value of the component in the mixture containing neither hydrogen nor methane (from the regular K-value charts).

Since the equilibrium vaporization ratios depend on the mixture parameter of convergence pressure, it is necessary to evaluate the convergence pressure for the hydrogen–hydrocarbon mixture. Figure 2.33 was devised by Lenoir to be used in the evaluation of the convergence pressure for hydrogen–aliphatic hydrocarbon binary systems or for mixtures of aliphatic compounds containing relatively large amounts of hydrogen in the liquid. In these cases the $(\text{EBP})_L$ is that of H_2, $-393°$F (not the normal boiling point of $-398°$F), and the chart is used to determine the convergence pressure at the system temperature and the correct $(\text{EBP})_H$.

When the mixture contains an appreciable amount of methane, it is necessary to evaluate the EBP of the pseudolight component,

$$\text{EBP}_L = \frac{\sum_1^n (T_L/T_i)^{5.7} x_i t_i}{\sum_1^n (T_L/T_i)^{5.7} x_i} \tag{2.67}$$

and that for the pseudoheavier component by

$$\text{EBP}_H = \frac{\sum_m^h (T_i/T_H)^2 x_i t_i}{\sum_m^h (T_i/T_H)^2 x_i} \tag{2.68}$$

Components 1 to n have normal boiling points (t_i) less than EBP_H and components m to h have normal boiling points greater than EBP_L. If components are present which have intermediate boiling points, they are included in both EBP_H and EBP_L evaluations. From the correct chart or interpolation for the EBP_L (Figs. 2.33 through 2.36), the convergence pressure for the mixture can be determined and the K values read from the suitable chart corresponding to the correct convergence pressure.

Correction factor charts Figures 2.37 and 2.38 are the Lenoir charts from which the correction factor for aliphatic hydrocarbon K values as a function

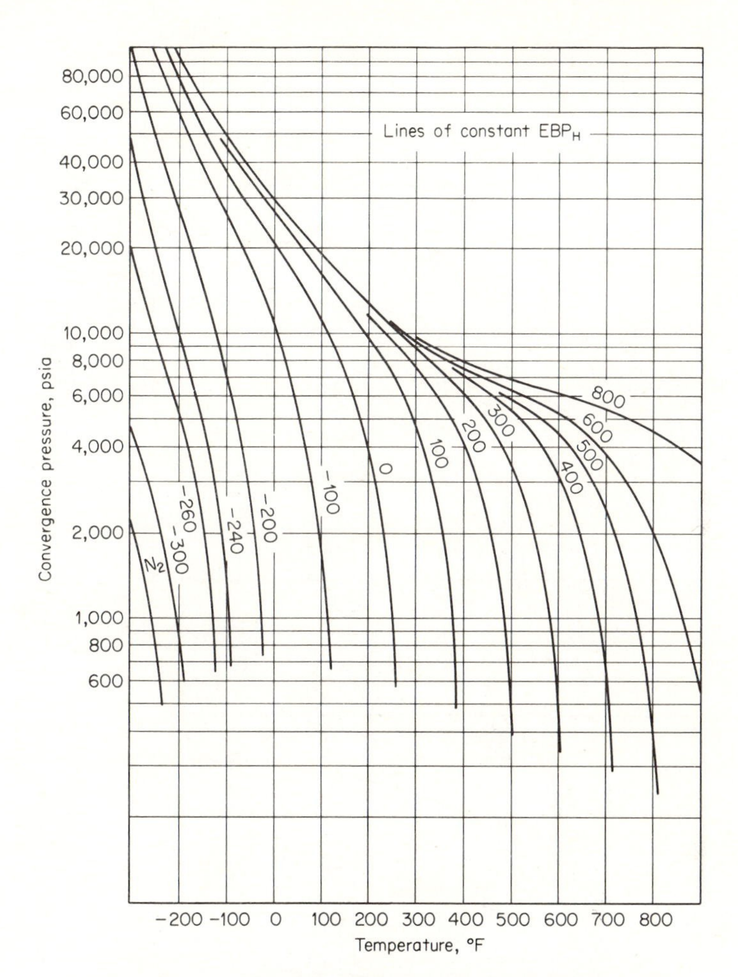

FIGURE 2.33 Convergence pressure as a function of temperature, pressure, and boiling point of the heavier component (lighter component $EBP_L = -393$°F). [*From Lenoir, Hydrocarbon Process. Petrol. Refiner (March, 1965), p. 139. Copyright Gulf Publishing Company, Houston, Texas (1965).*]

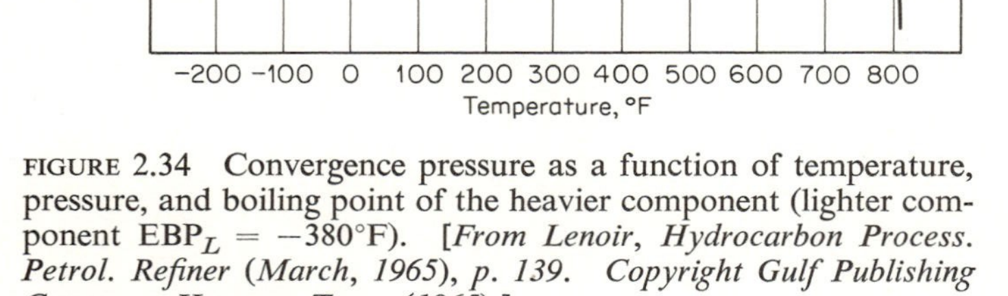

FIGURE 2.34 Convergence pressure as a function of temperature, pressure, and boiling point of the heavier component (lighter component $EBP_L = -380°F$). [*From Lenoir, Hydrocarbon Process. Petrol. Refiner* (*March, 1965*), *p. 139. Copyright Gulf Publishing Company, Houston, Texas* (*1965*).]

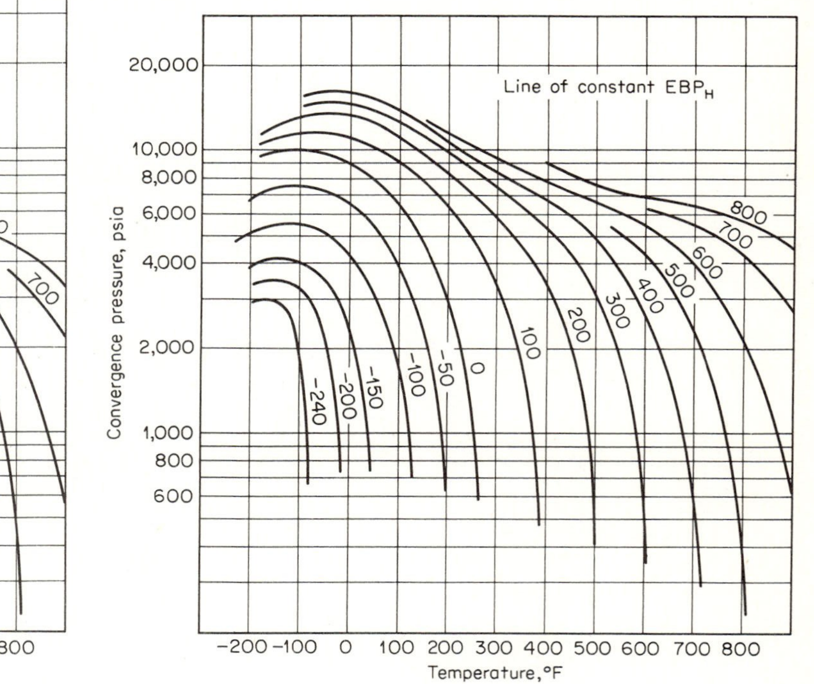

FIGURE 2.35 Convergence pressure as a function of temperature, pressure, and boiling point of the heavier component (lighter component $EBP_L = -370°F$). [*From Lenoir, Hydrocarbon Process. Petrol. Refiner* (*March, 1965*), *p. 139. Copyright Gulf Publishing Co., Houston, Texas* (*1965*).]

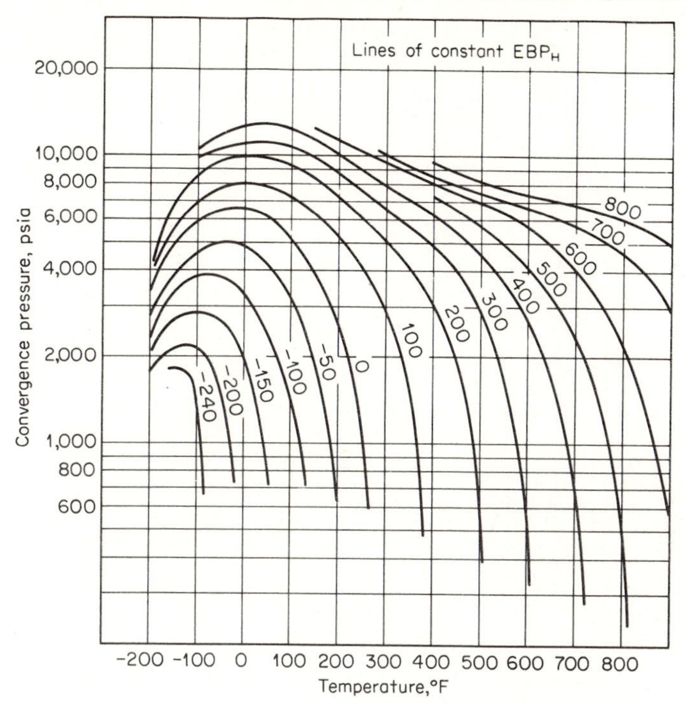

FIGURE 2.36 Convergence pressure as a function of temperature, pressure, and boiling point of the heavier component (lighter component $EBP_L = -360°F$). [*From Lenoir, Hydrocarbon Process. Petrol. Refiner (March, 1965), p. 139. Copyright Gulf Publishing Company, Houston, Texas (1965).*]

of mole fraction of hydrogen in the mixture and the similar correction factor based on methane content can be determined. Figure 2.39 represents a correction factor chart for the methane K value based on hydrogen content of the mixture.

Thus, for mixtures of hydrogen, methane, and aliphatic hydrocarbons, the procedure for determining corrected K values is as follows:

1. Determine the convergence pressure as described above.

2. From Fig. 2.37 determine the correction factor γ_p for the K's of the hydrocarbons heavier than methane, because of the presence of hydrogen in the mixture. From Fig. 2.39 determine the correction factor γ_m for the methane K, because of the presence of hydrogen, and from Fig. 2.38 determine the correction factor γ_h for the K's of the hydrocarbons heavier than methane, because of the presence of methane.

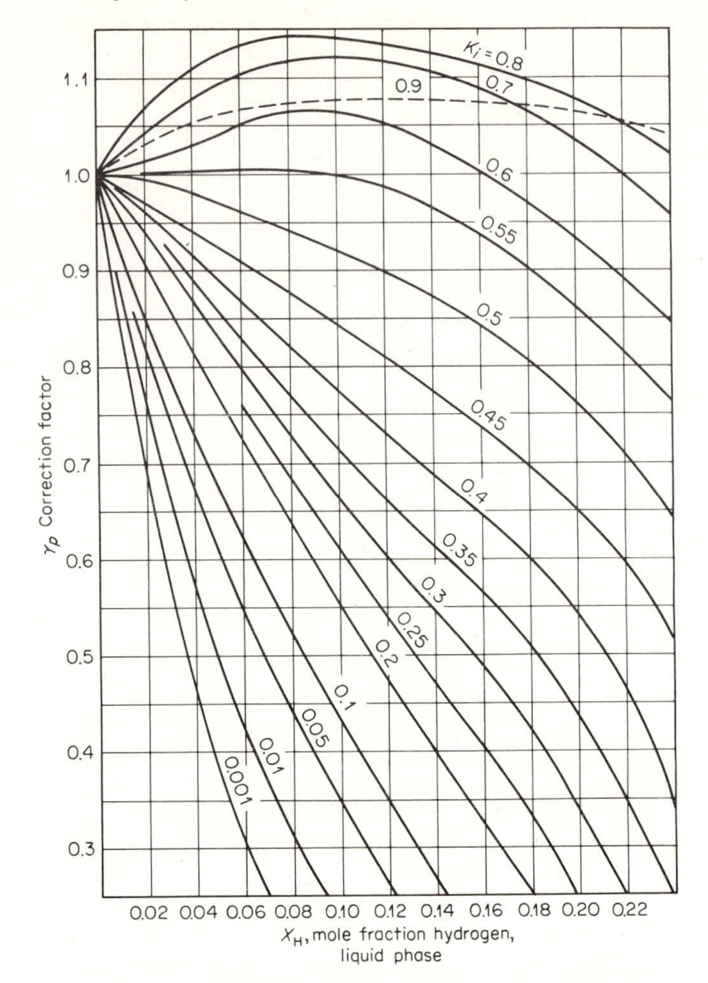

FIGURE 2.37 Aliphatic K correction factors related to hydrogen content. [*From Lenoir, Hydrocarbon Process. Petrol. Refiner (March, 1965), p. 139. Copyright Gulf Publishing Company, Houston, Texas (1965).*]

The K values are then calculated by

$$K_{H_2} = K_{H_2} \qquad \text{(read from Fig. 2.47)} \tag{2.69}$$

$$K_{CH_4} = (K_{CH_4} \text{ from chart})(\gamma_m) \tag{2.70}$$

$$K_i = (K_i \text{ from chart})(\gamma_p)(\gamma_m)(\gamma_h) \tag{2.71}$$

For mixtures of aliphatics and hydrogen without methane,

$$K_i = (K_i \text{ from charts})(\gamma_p) \tag{2.72}$$

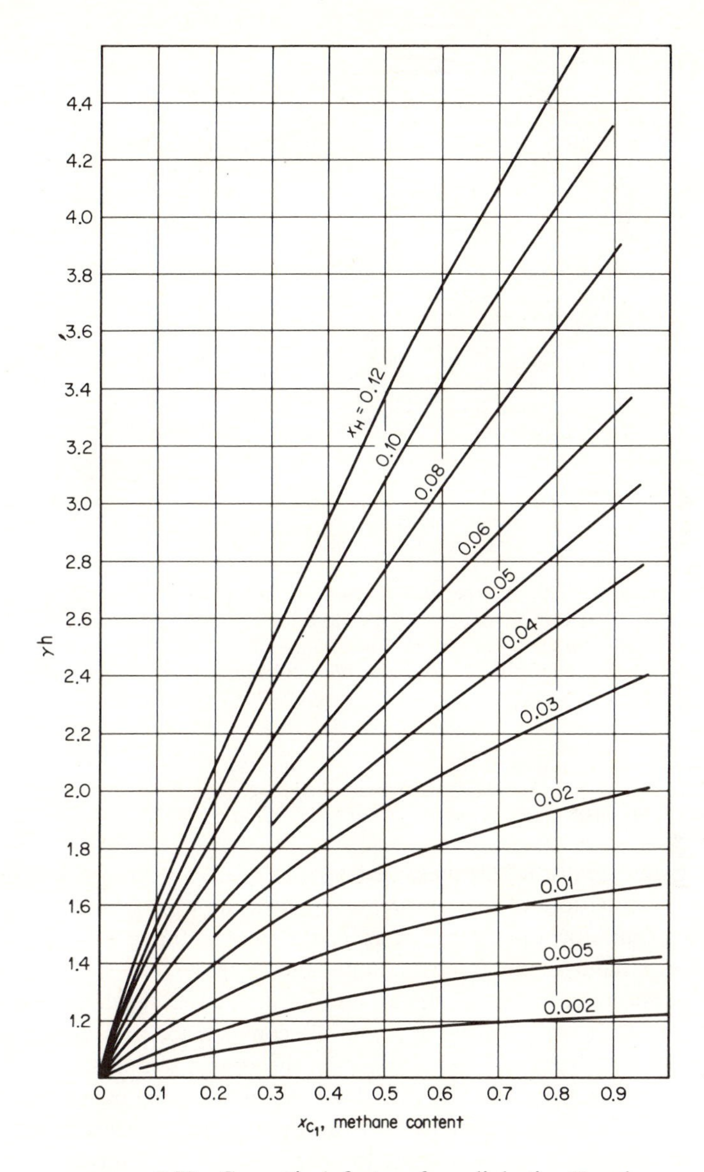

FIGURE 2.38 Correction factor for aliphatic K values related to methane content. [*From Lenoir, Hydrocarbon Process. Petrol. Refiner (March, 1965), p. 139. Copyright Gulf Publishing Company, Houston, Texas (1965).*]

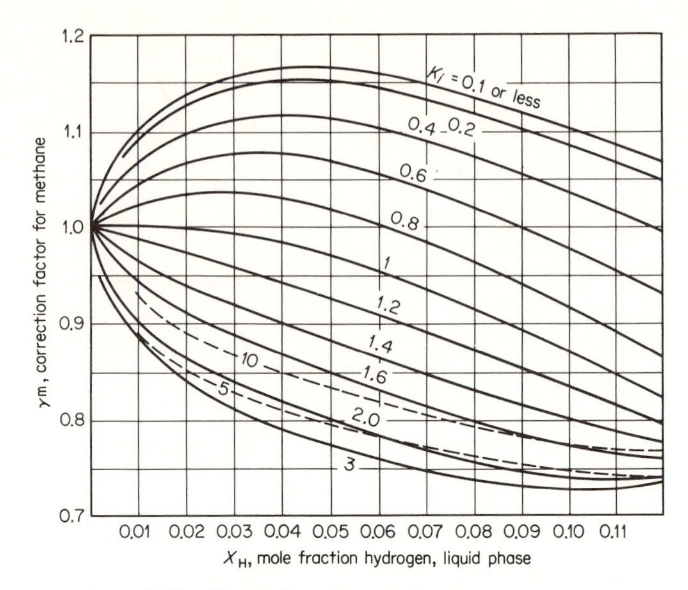

FIGURE 2.39 Correction factors for methane K values for various hydrogen concentrations. [*From Lenoir, Hydrocarbon Process. Petrol. Refiner* (*March*, *1965*), *p. 139. Copyright Gulf Publishing Company, Houston, Texas* (*1965*).]

Hadden-Grayson Convergence Pressure Charts

Hadden and Grayson [23] presented a series of charts along with their K charts for estimation of the convergence pressure for refinery mixtures (Fig. 2.46), and for binary mixtures containing methane as the lightest component, ethane as the lightest component, propane, etc., up to heptane as the lightest component and up to hexadecane as the heavier component in the binary mixtures. On these charts the convergence pressure is related to the identity of the binary mixture and to the system temperature.

For mixtures not included in Fig. 2.46, the binary mixture charts, Figs. 2.40*a* through 2.40*g*, are used to estimate the convergence pressure. The pertinent curve is the one having the low-temperature terminus corresponding to the lightest boiling component and the high-temperature terminus corresponding to the "average heavy component." The average heavy component is determined by the arithmetic average carbon number neglecting the heavy components comprising less than 2% of the mixture. The lightest of the heavy and the heaviest of the heavy components exclusive of the 2% are used in the "average."

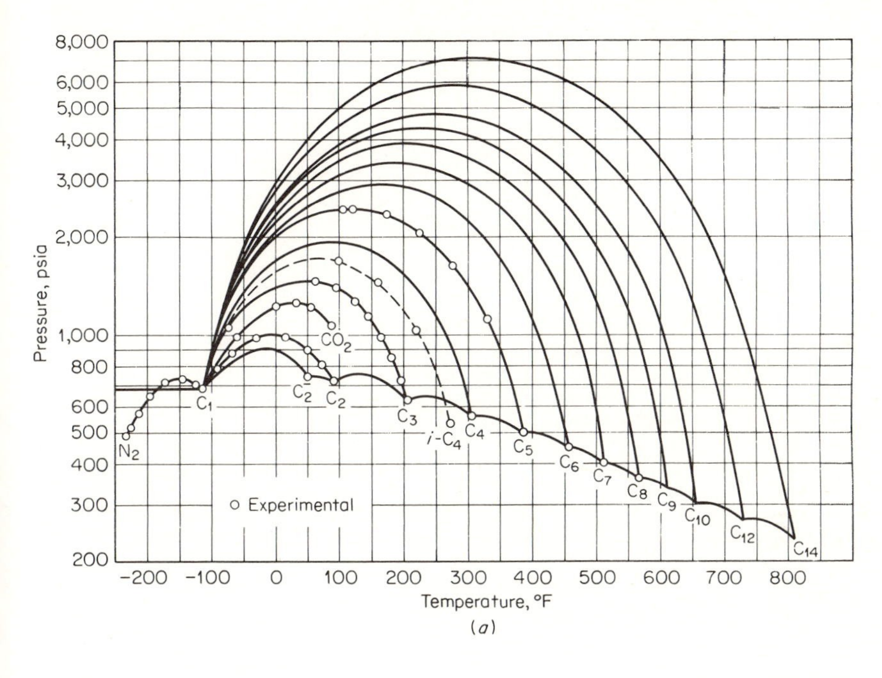

(a)

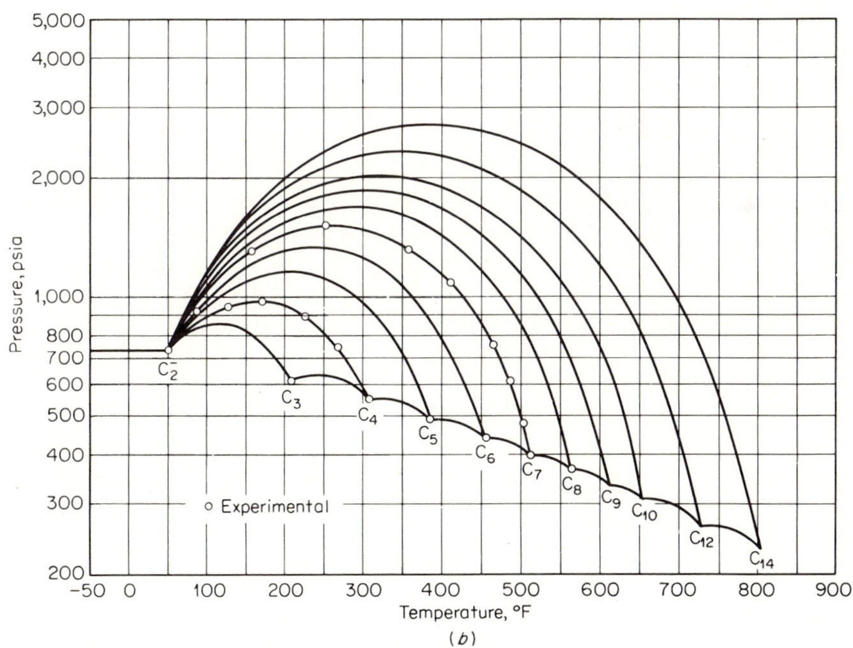

(b)

FIGURE 2.40

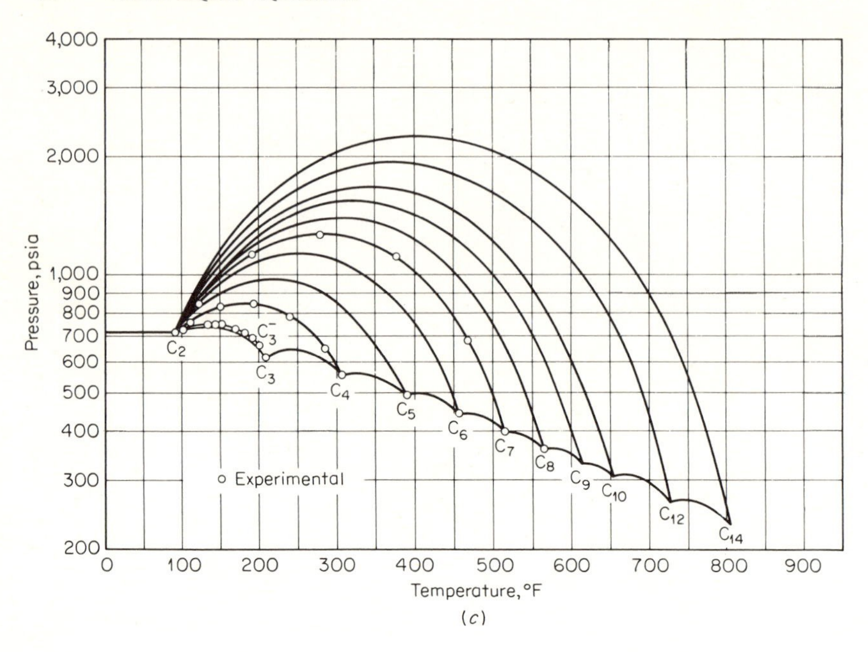

(c)

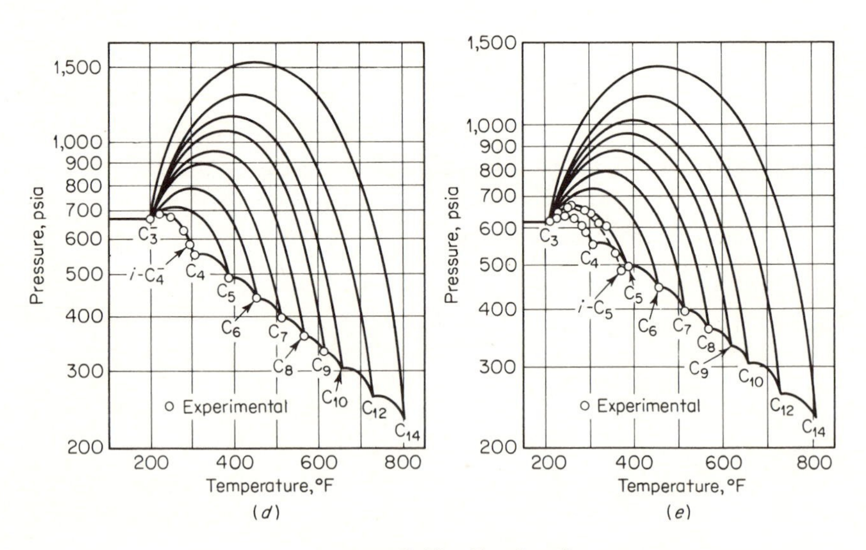

(d) (e)

FIGURE 2.40—*Continued*

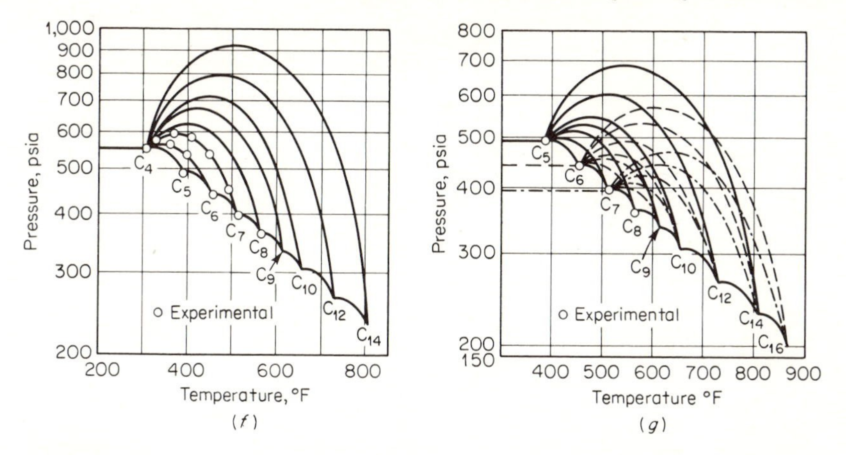

FIGURE 2.40 Convergence pressure data. (*a*) Methane lightest component. (*b*) Ethene lightest component. (*c*) Ethane lightest component. (*d*) Propene lightest component. (*e*) Propane lightest component. (*f*) *n*-Butane lightest component. (*g*) Pentane, hexane, and heptane lightest component(s). [*From Hadden and Grayson, Petrol. Refiner (September, 1961), p. 207. By permission of Socony Mobil Oil Company, Inc.*]

K Values—Selection

In selecting sets of *K* values for use in distillation calculations, a number of considerations must be made. First of all, the *K* values must be applicable to the particular system under consideration and to the conditions of temperature and pressure. They must be readily usable for the type of calculations to be made. If, for example, a digital computer is available and a suitable program is at hand, trial-and-error solution of a nonlinear differential equation to get individual *K* values is not entirely out of reason. On the other hand, if only approximate calculations are to be made with a desk calculator or slide rule, methods for determining *K* involving complex calculations are not desirable.

It is highly desirable to have *K* charts or methods of obtaining *K*'s which involve a minimum of cross-referencing between or among charts. In other words, the simpler the manipulations the better as long as accuracy is not sacrificed.

Finally, it is necessary to match the equilibrium data accuracy and complexity with the accuracy of the calculation methods and correlations used in the distillation design. It is unnecessary to have equilibrium data accurate to within 0.1% and use these data in a correlation which claims only an accuracy of $\pm 10\%$.

Figure 2.41 represents a nomographic-type *K* chart for light hydrocarbons and associated materials based on a convergence pressure of 5000 psia. The

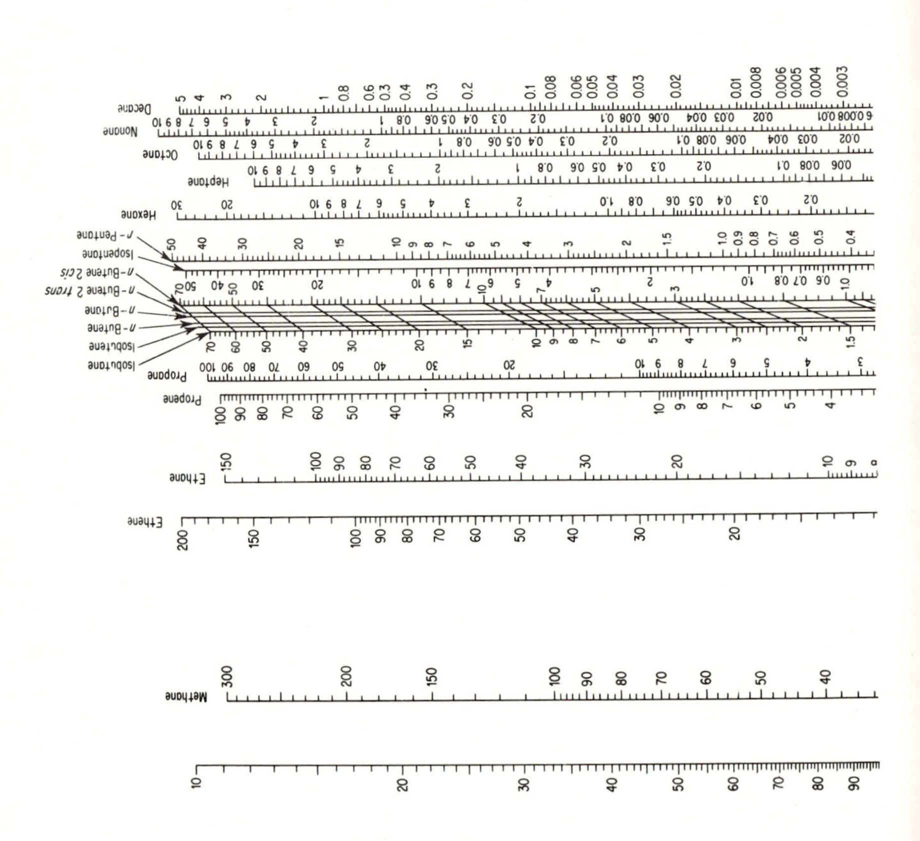
Methane
Ethene
Ethane
Propene
Propane
Isobutane
Isobutene
n-Butene
n-Butane
n-Butene 2 trans
n-Butene 2 cis
Isopentane
n-Pentane
Hexane
Heptane
Octane
Nonane
Decane

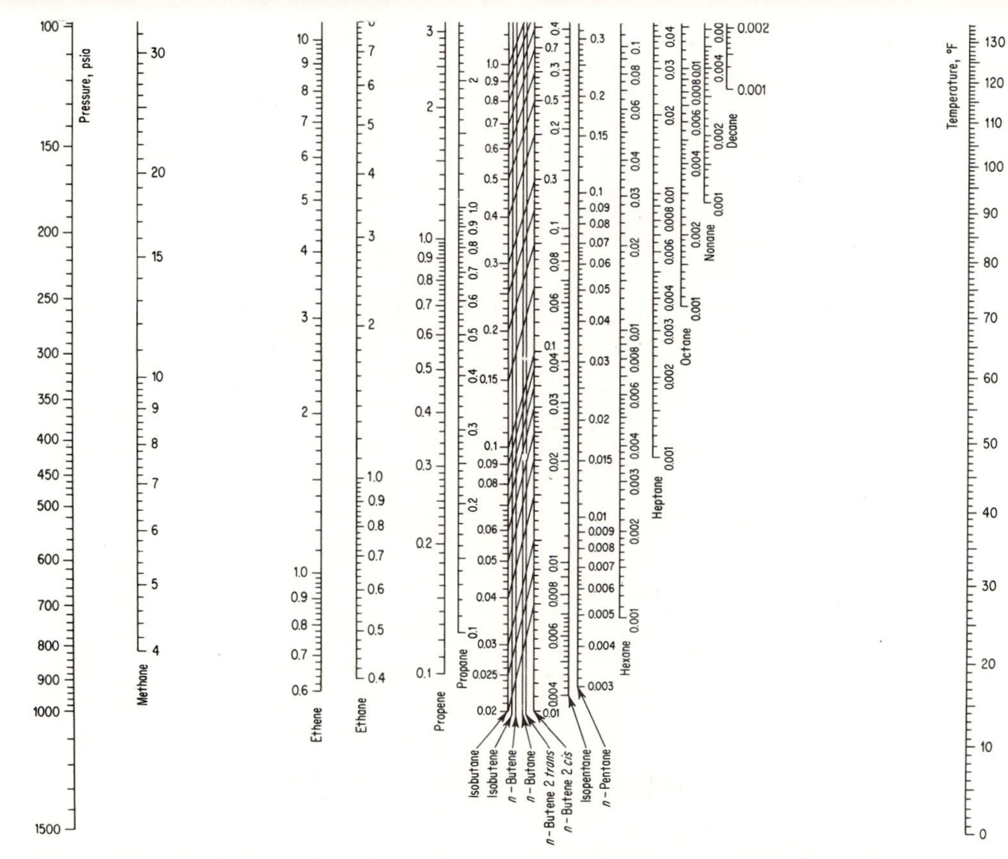

FIGURE 2.41 Vapor-liquid equilibrium constants (light hydrocarbons). [*From Hadden, Chem. Eng. Progr. Symp. Ser., no. 7, **49**:53 (1953). By permission of Socony Mobil Oil Company, Inc.*]

chart was developed by Hadden [21] and is extremely simple to use. For mixtures having a convergence pressure of about 5000 psia it can be used directly. For other convergence pressures the K's can be corrected by the Winn method [54].

Example 2.7

Find the bubble-point temperature of the following mixture at 500 psia. Determine the convergence pressure by the Lenoir-White method and use the Hadden chart corrected for convergence pressure.

Solution

Compound	Mole fraction	T_i, °R
(1) Methane	0.27	201
(2) Ethylene	0.11	305
(3) Ethane	0.35	332
(4) Propylene	0.04	406
(5) Propane	0.13	416
(6) *i*-Butane	0.04	471
(7) Butane	0.06	491
	1.00	

The calculation of EBP_L is as follows:

Compound	T_L/T_i	F_e	$F_e x_i$	$F_L x_i T_i$	EBP_L, °R	EBP_L, °F
1	1.00	1.0	0.27	54.3		
2	0.66	0.09	0.01	3.0		
3	0.605	0.05	0.018	6.0		
4	0.495	0.02	0.001	0.4		
5	0.484	0.015	0.002	0.0		
6	0.427	—	—	—		
			0.301	64.5	214	−246

The calculation of EBP_H is:

Compound	T_i/T_h	F_e	$F_e x_i$	$F_e x_i T_i$	EBP_H, °R	EBP_H, °F
2	0.621	0.39	0.043	13.1		
3	0.677	0.475	0.166	55.2		
4	0.827	0.68	0.027	11.0		
5	0.847	0.72	0.094	39.0		
6	0.960	0.92	0.037	17.3		
7	1.000	1.00	0.060	29.5		
			0.427	165.1	386	−74

By logarithmic interpolation between Figs. 2.19 and 2.20, the convergence pressure $P_c = 1250$ at 0°F. In order to use the nomographs, a "grid" pressure is read from Fig. 2.44 corresponding to $P_c = 1250$. Knowing the grid pressure and temperature, the K values are read from the nomograph.

Component	Mole fraction	K, 0°F	Kx	K, −10°F	Kx
Methane	0.27	3.25	0.877	3.0	0.810
Ethylene	0.11	0.78	0.086	0.70	0.077
Ethane	0.35	0.50	0.175	0.46	0.045
Propylene	0.04	0.15	0.006	0.13	0.005
Propane	0.13	0.13	0.017	0.11	0.014
i-Butane	0.04	0.048	0.002	0.04	0.002
Butane	0.06	0.033	0.002	0.0265	0.002
	1.000		1.165		0.955

The K's at −10°F were computed by using the grid pressure for 0°F because the change is slight in this region. By interpolation,

$$t = -8°\text{F}$$

Hadden and Grayson [23] improved and extended the vapor-liquid equilibrium relationships developed by Hadden [22, 23] and Winn [55] to lower temperatures and to include more compounds. The K charts may be used for mixtures of paraffins and/or olefins, mixtures of aromatic hydrocarbons up to about 150 psia, and special mixtures including hydrogen, methane, HF, CO_2, H_2S methyl mercaptan, and water for certain limited and specified conditions. The K charts are shown in Figs. 2.42 and 2.43. Figure 2.44 relates the grid pressure, system pressure, and convergence pressure. Figure 2.45 is a guide for the method of correcting for convergence pressure and is based on the parameters of estimated convergence pressure and operating pressure.

The location of the pivot point for the evaluation of K's from the nomograms based on the areas in Fig. 2.45 is found as follows:

1. Area A: Use the t_{op} and P_{op} normally used.

2. Area B: Determine grid pressure on Fig. 2.44. Connect t_{op}, P_{op} to $K = 1.0$ on nomogram. Locate on this line the grid pressure as the pivot point.

3. Area C: Determine grid pressure on Fig. 2.44 and use t_{op} and P_g as the pivot point.

4. Area D: P_{cv} must be determined accurately; see [22]. If P_{cv} is above 5000 psia, use area C procedure. If P_{cv} is below 5000 psia, use area B procedure.

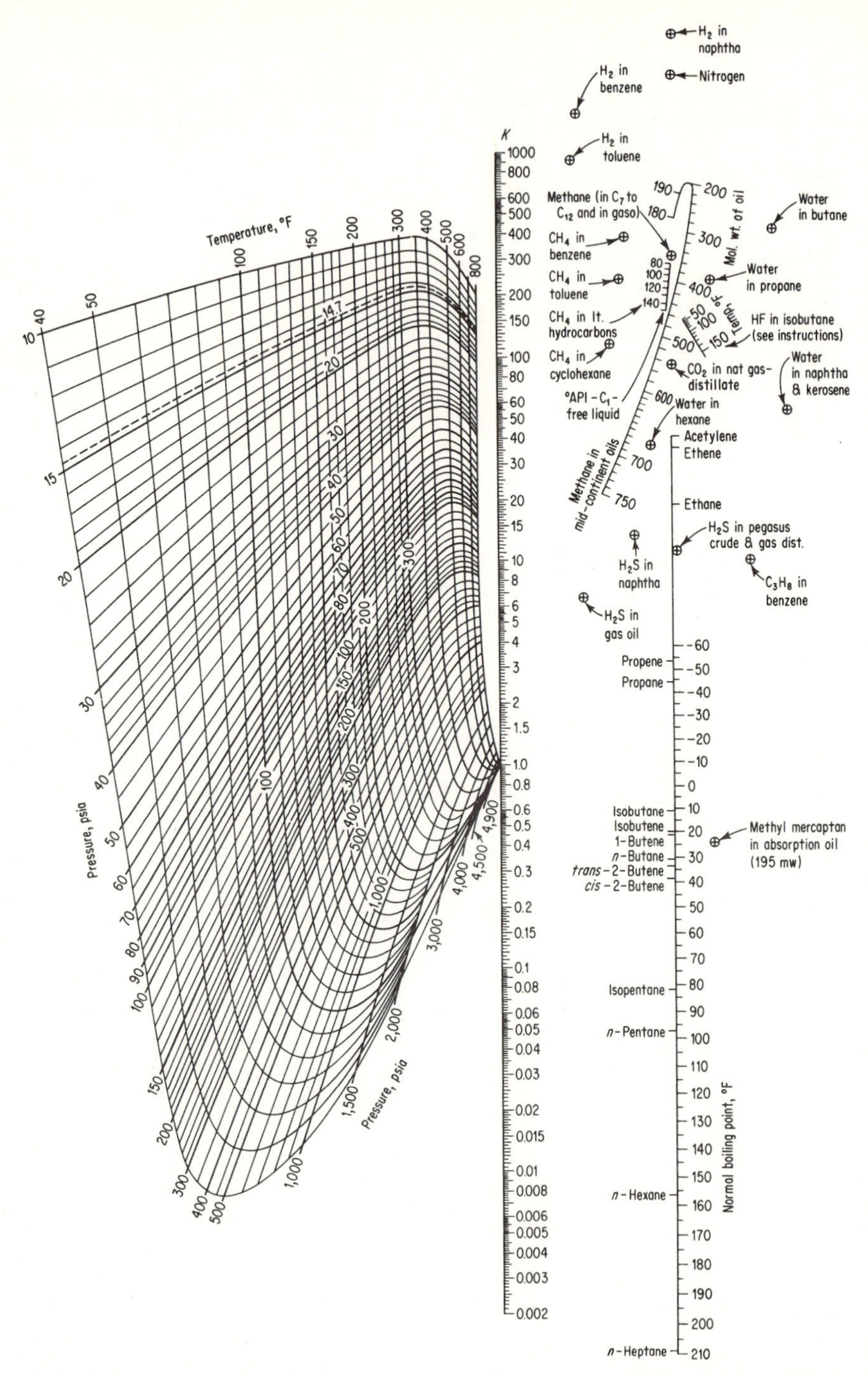

FIGURE 2.42 Vapor-liquid equilibrium constants, 40 to 800°F. [*From Hadden and Grayson, Petrol. Refiner (September, 1961), p. 207. By permission of Socony Mobil Oil Company, Inc.*]

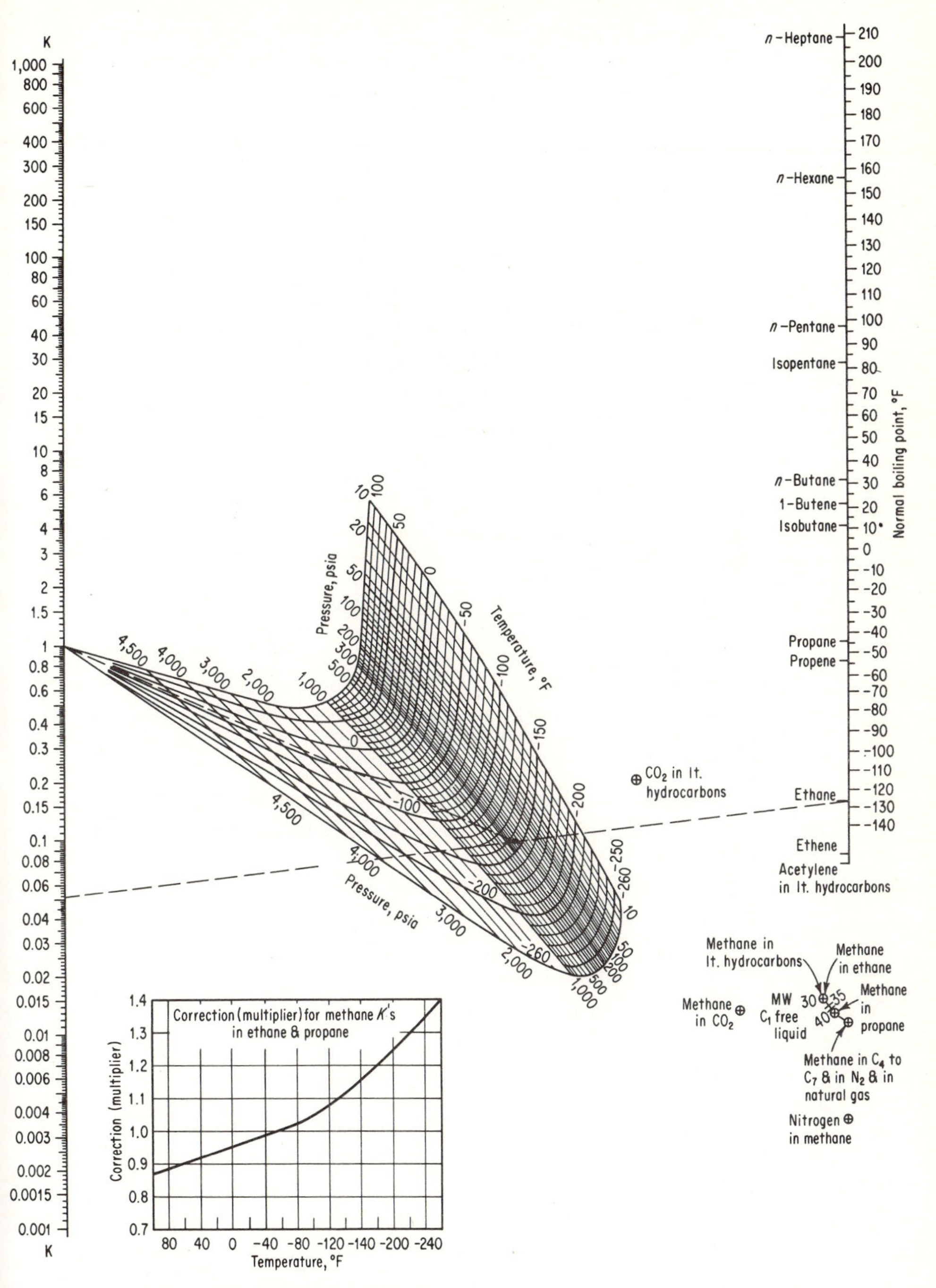

FIGURE 2.43 Vapor-liquid equilibrium constants, −260 to +100°F. [*From Hadden and Grayson, Petrol. Refiner (September, 1961), p. 207. By permission of Socony Mobil Oil Company, Inc.*]

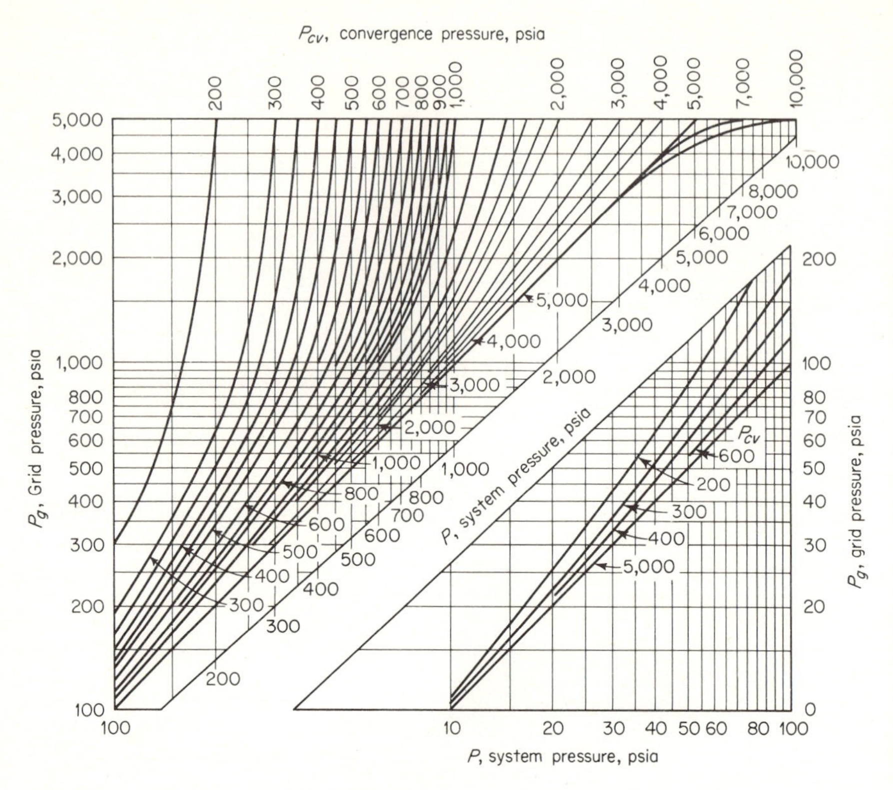

FIGURE 2.44 Grid pressure relationship. [*From Hadden and Grayson, Petrol. Refiner* (*September*, *1961*), *p. 207. By permission of Socony Mobil Oil Company, Inc.*]

Figure 2.46 can be used to estimate approximate P_{cv} values in order to determine grid pressures. The charts in the original reference for P_{cv} (for different mixtures) can be employed to determine the P_{cv}. Also Figs. 2.18 to 2.29 can be used for this purpose.

Table 2.6 lists values of the exponent (volatility exponent) to be used in the equation in the table to determine the value of K for the hydrocarbon fraction having the true boiling point (TBP) 50% temperature which is used to determine the exponent. The equation is based on the K values of ethane and heptane and these are determined in the manner described in the foregoing.

Figure 2.47 is an equilibrium-ratio K chart for hydrogen in hydrocarbon mixtures relating system pressure and K_{H_2} at parameters of convergence pressure ranging up to 10^5 psia. Lenoir and Hipkin [32] developed this chart from binary and multicomponent data. It is interesting to note that

TABLE 2.6 **Volatility Exponent for Pure Hydrocarbons and True Boiling Point 50°F Cuts at 5°F Intervals on Boiling Point [23]**

Temp. interval, °F	Volatility exponent b									
	Pure hydrocarbons, NBP, °F			50°F cuts, TBP mid-boiling-point temperatures, °F						
	200	300	400	100	200	300	400	500	600	700
00	—	0.360	0.834	—	−0.081	0.326	0.792	1.348	2.016	2.801
05	—	0.381	0.861	—	−0.061	0.348	0.817	1.379	2.053	2.843
10	0.003	0.403	0.888	—	−0.041	0.370	0.842	1.410	2.090	2.886
15	0.021	0.425	0.916	—	−0.021	0.392	0.868	1.441	2.126	2.928
20	0.040	0.447	0.944	—	−0.002	0.414	0.894	1.472	2.164	2.972
25	0.060	0.469	—	−0.376	0.018	0.436	0.921	1.504	2.202	3.014
30	0.079	0.491	—	−0.356	0.038	0.458	0.948	1.536	2.240	3.058
35	0.098	0.514	—	−0.336	0.058	0.481	0.974	1.569	2.278	3.102
40	0.117	0.537	—	−0.316	0.078	0.504	1.002	1.602	2.316	3.146
45	0.137	0.560	—	−0.296	0.098	0.526	1.029	1.634	2.355	3.190
50	0.156	0.584	—	−0.276	0.118	0.550	1.057	1.668	2.394	3.234
55	0.176	0.607	—	−0.257	0.138	0.572	1.085	1.701	2.434	3.280
60	0.196	0.631	—	−0.237	0.159	0.596	1.113	1.735	2.474	3.324
65	0.216	0.656	—	−0.218	0.180	0.620	1.142	1.769	2.514	3.370
70	0.236	0.680	—	−0.198	0.200	0.644	1.170	1.804	2.554	3.415
75	0.256	0.705	—	−0.178	0.221	0.668	1.199	1.838	2.594	3.461
80	0.277	0.730	—	−0.158	0.242	0.692	1.228	1.874	2.635	3.507
85	0.297	0.756	—	−0.139	0.262	0.716	1.258	1.908	2.676	3.553
90	0.318	0.781	—	−0.120	0.284	0.742	1.288	1.944	2.718	3.600
95	0.339	0.808	—	−0.100	0.305	0.766	1.318	1.980	2.759	3.646

$$K_H = \frac{K_7}{(K_2/K_7)^b}$$

where K_H = K value of the cut or high-boiling pure hydrocarbon
K_2 = K value of ethane at the temperature, pressure, and convergence pressure of the system
K_7 = K value of n-heptane at the temperature, pressure, and convergence pressure of the system

the correlation is claimed to cover the temperature range of −200 to 303°F, and that the K values are essentially independent of temperature when relatively large amounts of hydrogen are present.

K_{10} *Nomograms for K Evaluation*

Cajander et al. [11] developed nomograms, Figs. 2.48 and 2.49, to enable evaluation of equilibrium vaporization ratios which use as a correlating

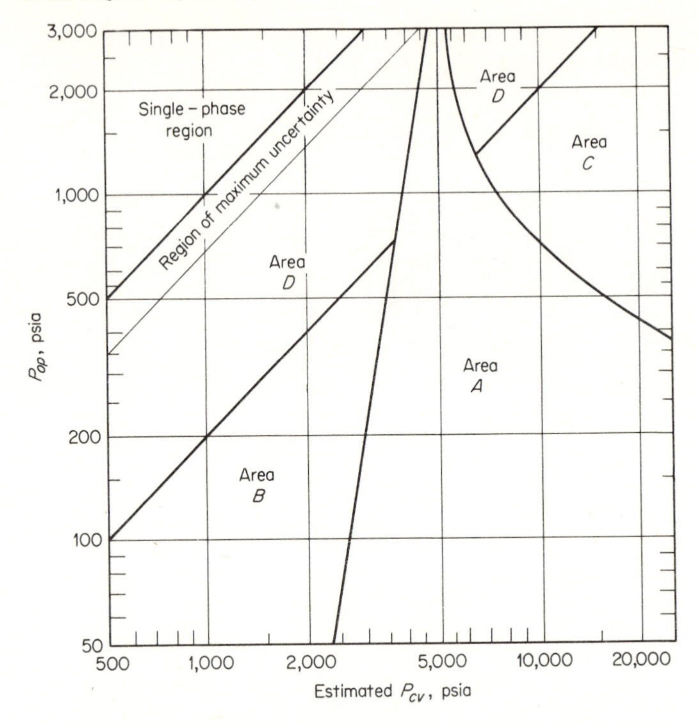

FIGURE 2.45 Requirements for K nomogram corrections based on P_{cv} and P_{op}. [*From Hadden and Grayson, Petrol. Refiner (September, 1961), p. 207. By permission of Socony Mobil Oil Company, Inc.*]

parameter the equilibrium vaporization ratio at 10 psia for many types of hydrocarbons. Figures 2.50*a* through 2.50*l* represent K_{10} values for the individual compounds in the various families of hydrocarbons at 10-psia pressure as a function of temperature. These K_{10} values are related to the K values at the operating pressure and to the convergence pressure of the mixture on Figs. 2.48 and 2.49.

The authors indicate the K_{10} method to be sound for mixtures that behave ideally at low pressures and for mixtures of one molecular type, such as paraffins and olefins. The accuracy is less for mixtures of cycloparaffins, paraffins, and olefins. It is possible that large errors can result from this method if mixtures containing aromatics–paraffins, aromatics–olefins, and aromatics–cycloparaffins are considered. In other words, the method does not allow for activity-coefficient deviations caused by mixing dissimilar molecular types. In such cases the predicted K values must be corrected for liquid-phase nonideality by generalized activity coefficients or some similar modification.

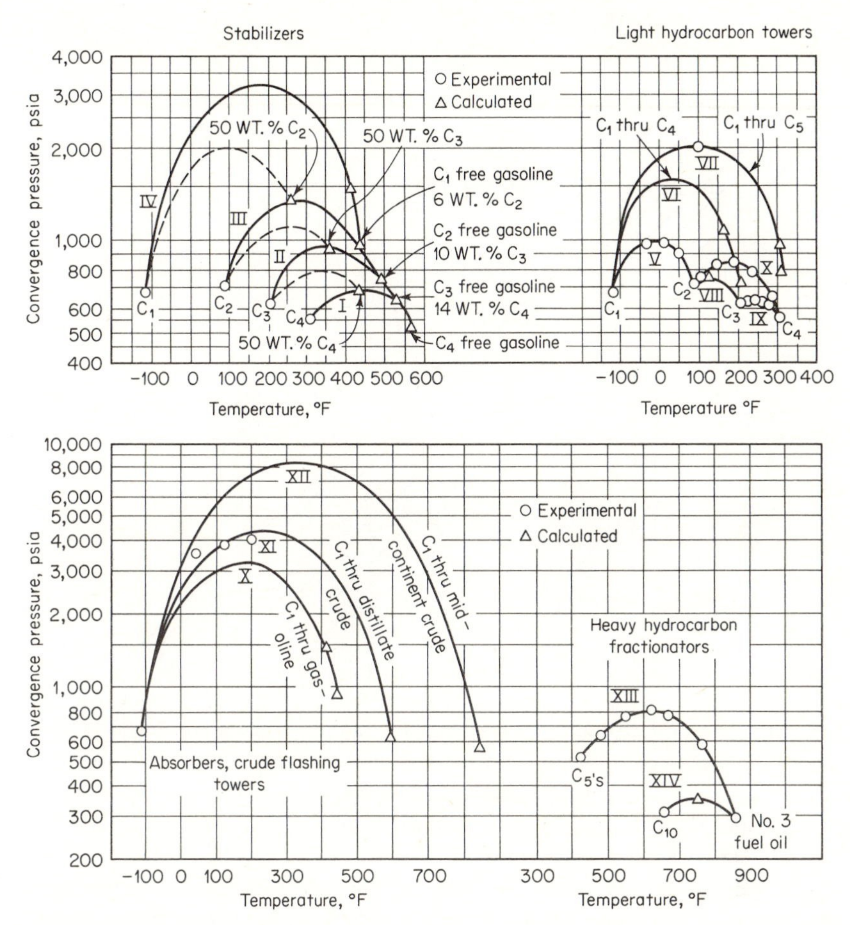

FIGURE 2.46 Convergence pressure data for typical refinery mixtures. [*From Hadden and Grayson, Petrol. Refiner (September, 1961), p. 207. By permission of Socony Mobil Oil Company, Inc.*]

Gamson and Watson [19] developed generalized correlations for relating properties of compounds with temperature and pressure. The K charts of Smith and Smith [48] were based upon these relations modified by Smith and Watson [49]. A generalized K chart based upon reduced conditions for compounds having a critical compressibility factor of 0.27 is given in Hougen et al. [26]. A method is described wherein the K value is modified by a correction equation when the compound's critical compressibility factor varies from 0.27.

For heavy hydrocarbons and petroleum fractions, the K charts developed by Poettman and Mayland [40] may be employed. It is necessary to evaluate the average boiling point and characterization factor in order to use these

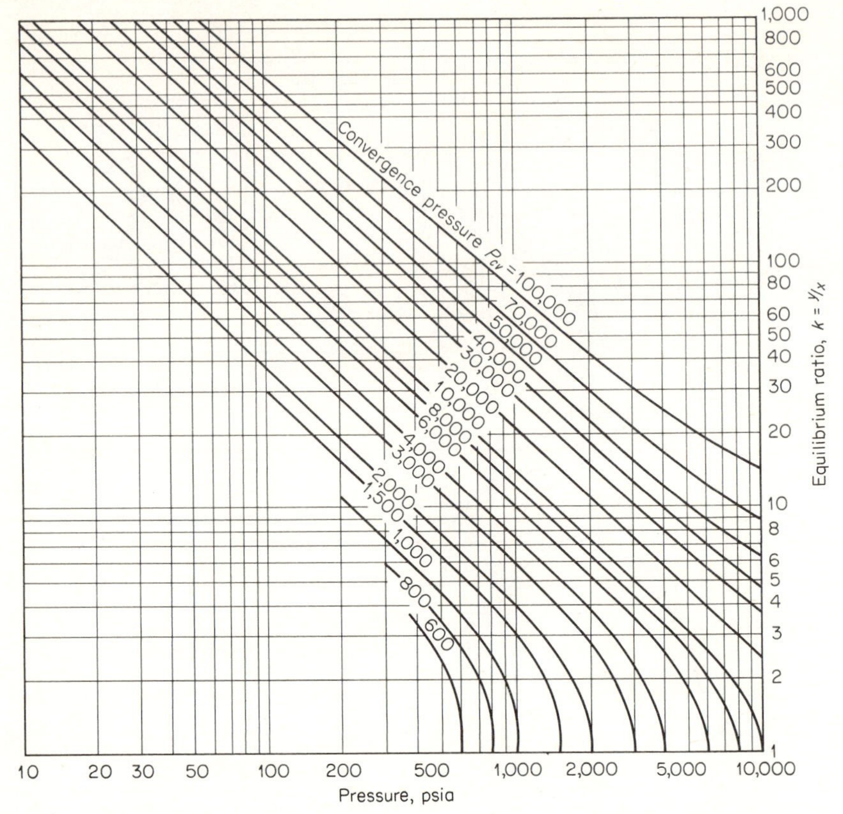

FIGURE 2.47 Equilibrium ratios of hydrogen. [*From Lenoir and Hipkin, A.I.Ch.E. J., 3:318 (1957).*]

charts. Also, the necessity for frequent interpolation between the charts is somewhat of a disadvantage. For most cases the charts of K values by Hadden and Grayson, Figs. 2.42 through 2.46, are satisfactory for petroleum fractions and higher boiling hydrocarbons.

For multicomponent mixtures involving compounds other than hydrocarbons, the activity-coefficient correlations based on a suitable equation of state are probably more satisfactory. Otherwise it is necessary to know the component binary coefficients and use equations of the type described earlier in this chapter to calculate activity coefficients for the components in the mixture.

Approximate Extrapolation and Interpolation of K Values

K's of pseudocompounds or compounds of intermediate molecular weight having the same molecular characteristics as the compounds for which the

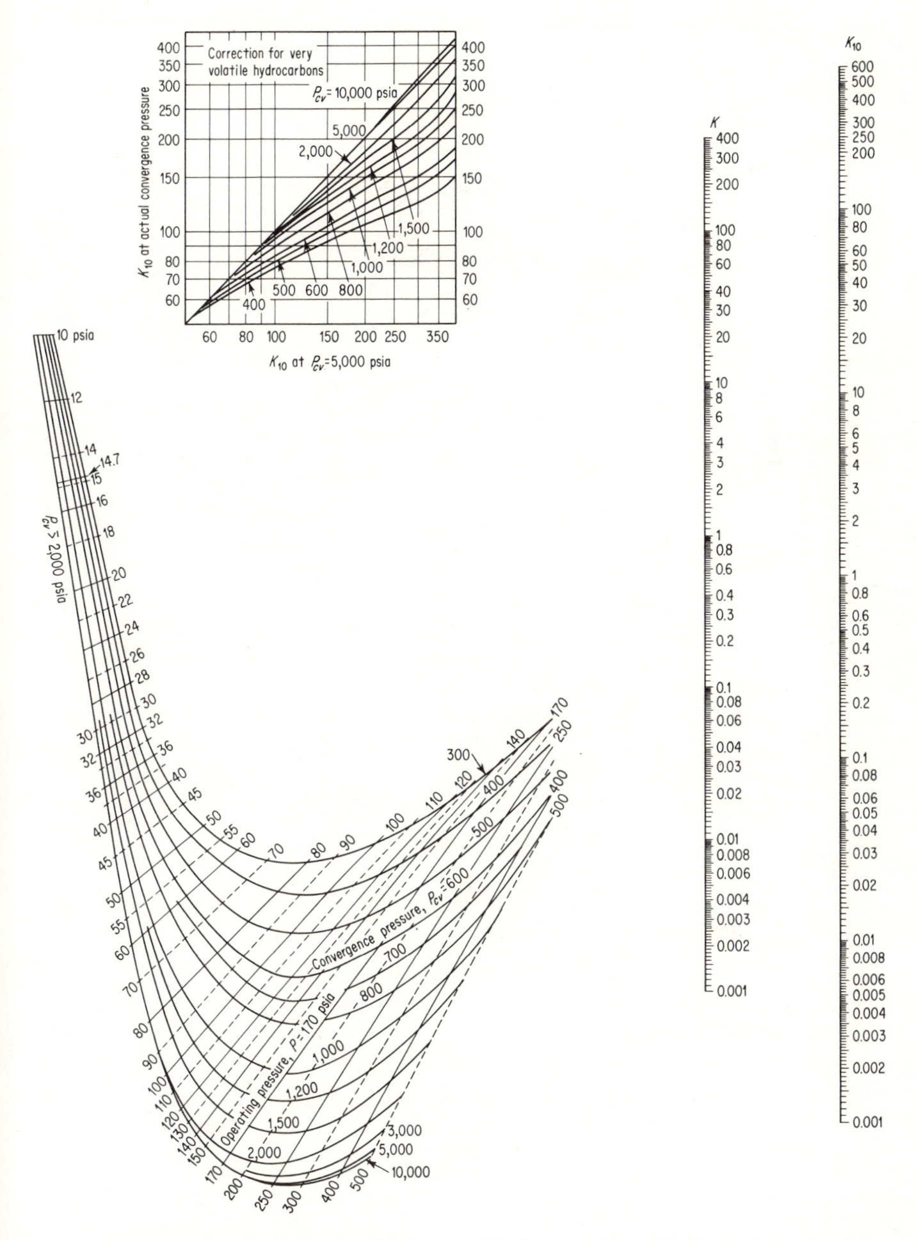

FIGURE 2.48 Nomogram for predicting equilibrium ratios of hydrocarbons in the low-pressure range. [*From Cajander et al., J. Chem. Eng. Data,* **5**:*251 (1960). Reproduced by permission of C. F. Braun and Company.*]

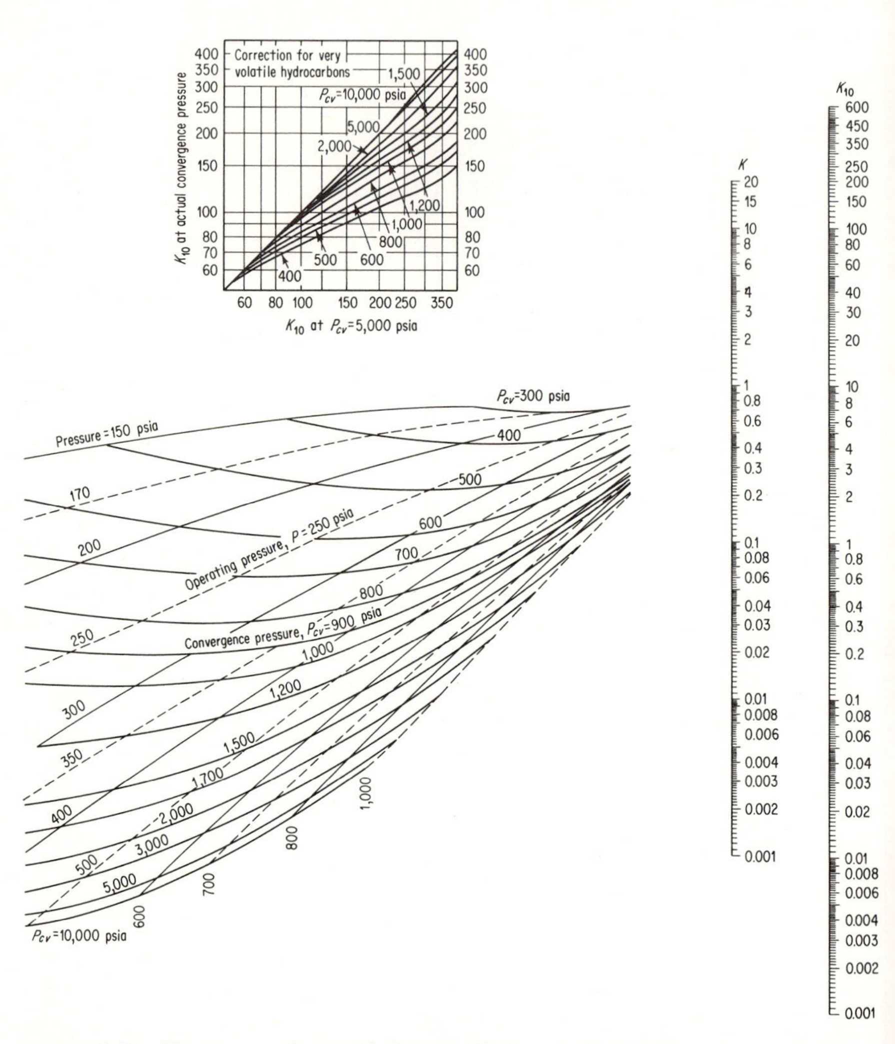

FIGURE 2.49 Nomogram for predicting equilibrium ratios of hydrocarbons in the high-pressure range. [*From Cajander et al., J. Chem. Eng. Data, 5:251 (1960). Reproduced by permission of C. F. Braun and Company.*]

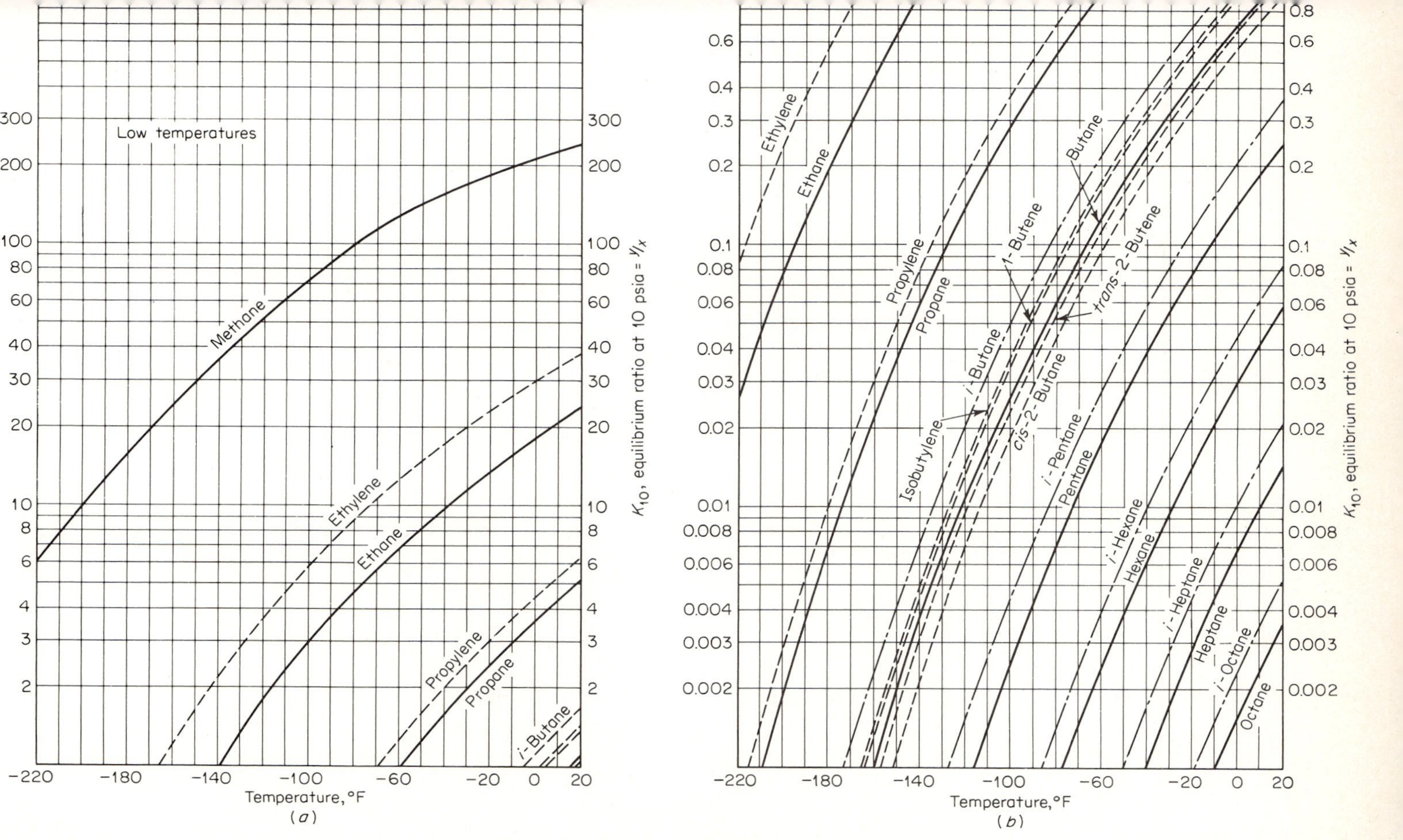

Low temperatures
Methane
Ethylene
Ethane
Propylene
Propane
i-Butane
Temperature, °F
(a)
K_{10}, equilibrium ratio at 10 psia = y/x
Ethylene
Ethane
Propylene
Propane
i-Butane
Isobutylene
1-Butene
Butane
trans-2-Butene
cis-2-Butane
i-Pentane
Pentane
i-Hexane
Hexane
i-Heptane
Heptane
i-Octane
Octane
Temperature, °F
(b)
K_{10}, equilibrium ratio at 10 psia = y/x

FIGURE 2.50

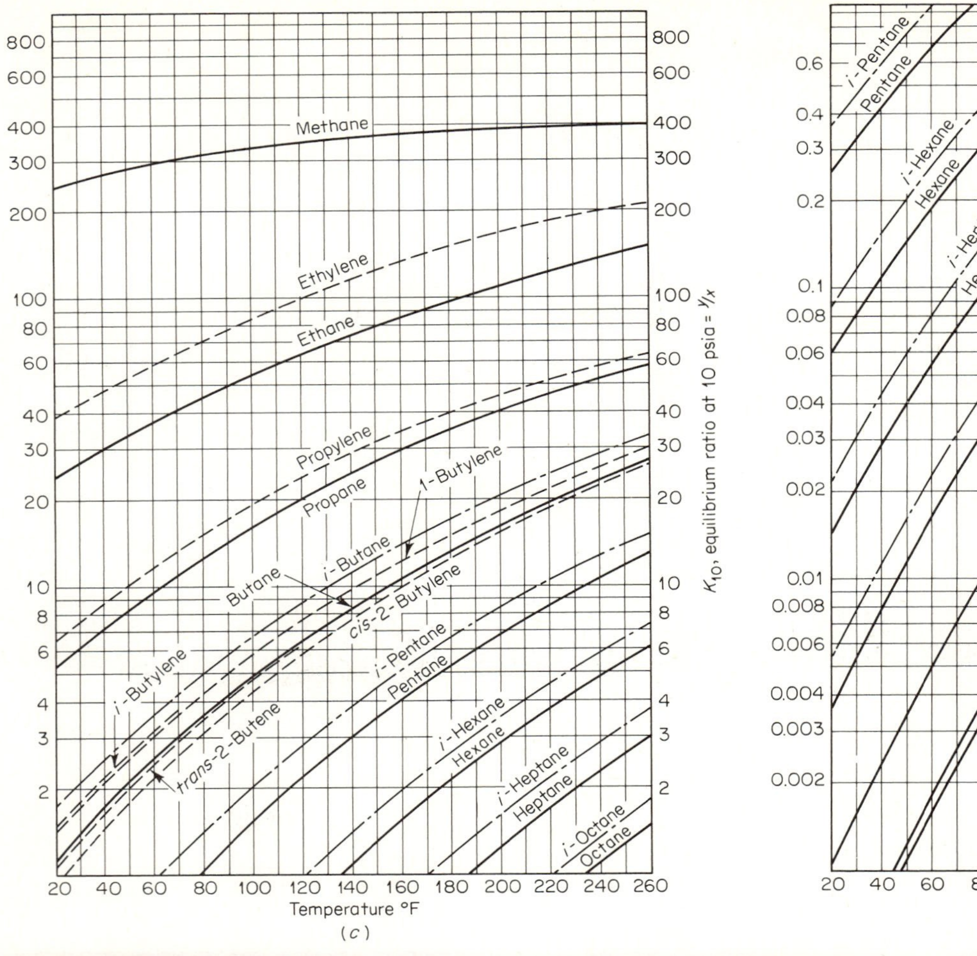
Methane
Ethylene
Ethane
Propylene
Propane
1-Butylene
i-Butane
Butane
cis-2-Butylene
i-Pentane
Pentane
i-Hexane
Hexane
i-Heptane
Heptane
i-Octane
Octane
trans-2-Butene
i-Butylene
K_{10}, equilibrium ratio at 10 psia = y/x
Temperature °F
(c)

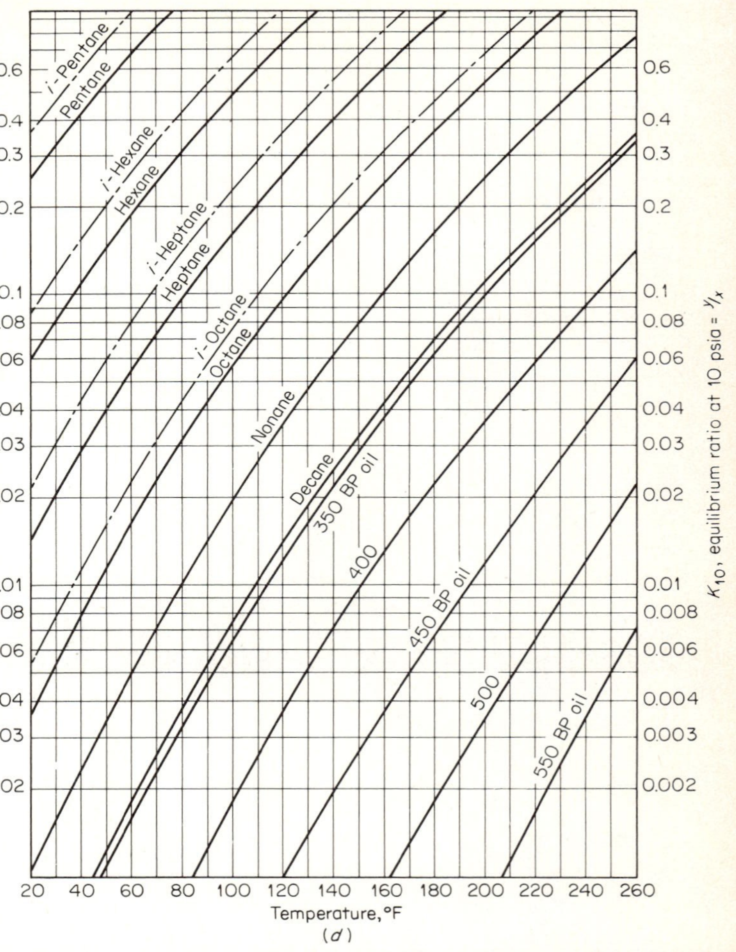
i-Pentane
Pentane
i-Hexane
Hexane
i-Heptane
Heptane
i-Octane
Octane
Nonane
Decane
350 BP oil
400
450 BP oil
500
550 BP oil
K_{10}, equilibrium ratio at 10 psia = y/x
Temperature, °F
(d)

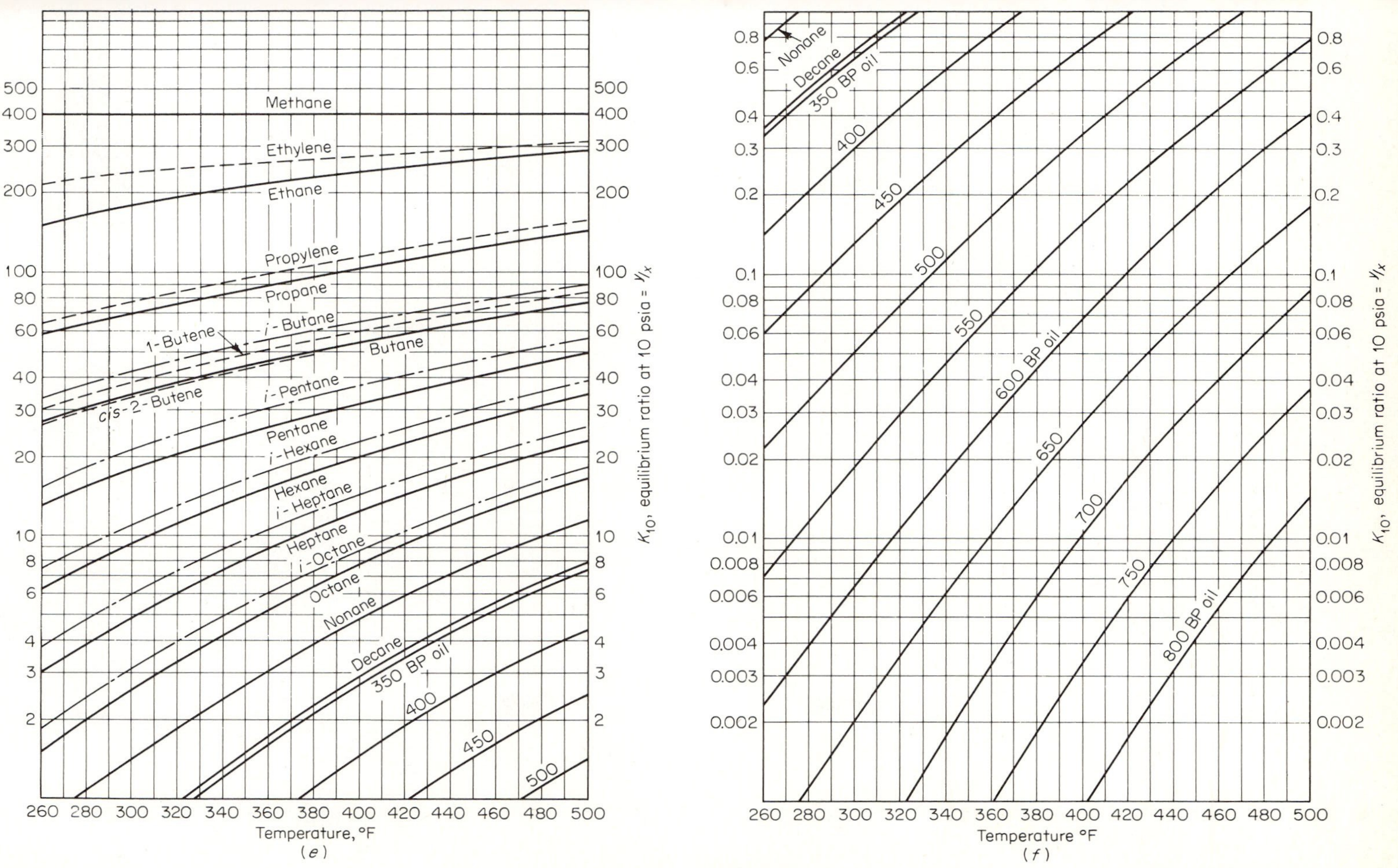

FIGURE 2.50—*Continued*

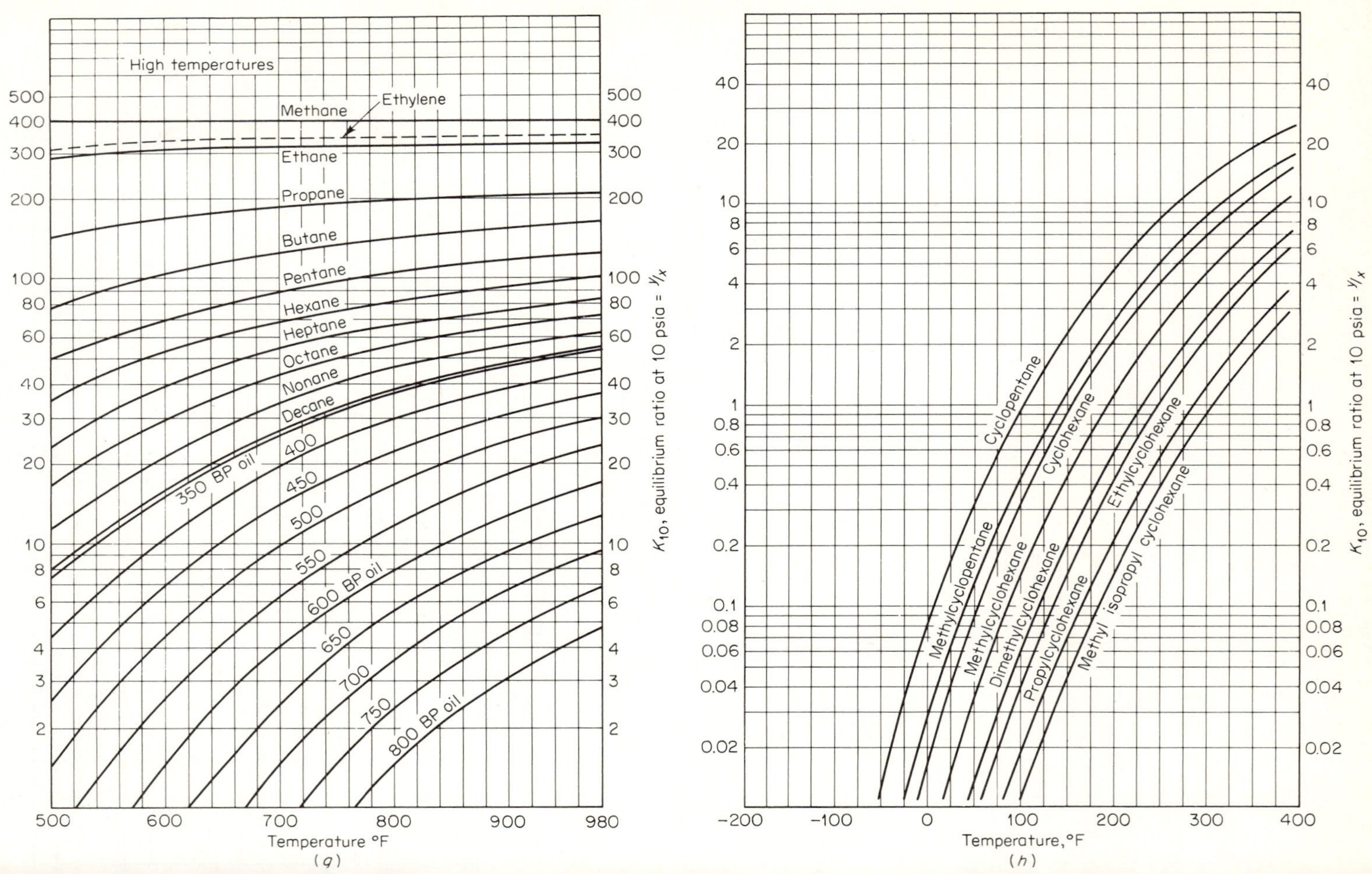

High temperatures
Methane
Ethylene
Ethane
Propane
Butane
Pentane
Hexane
Heptane
Octane
Nonane
Decane
350 BP oil
400
450
500
550
600 BP oil
650
700
750
800 BP oil
K_{10}, equilibrium ratio at 10 psia = y/x
Temperature °F
(g)
Cyclopentane
Cyclohexane
Ethylcyclohexane
Methylcyclopentane
Methylcyclohexane
Dimethylcyclohexane
Propylcyclohexane
Methyl isopropyl cyclohexane
K_{10}, equilibrium ratio at 10 psia = y/x
Temperature, °F
(h)

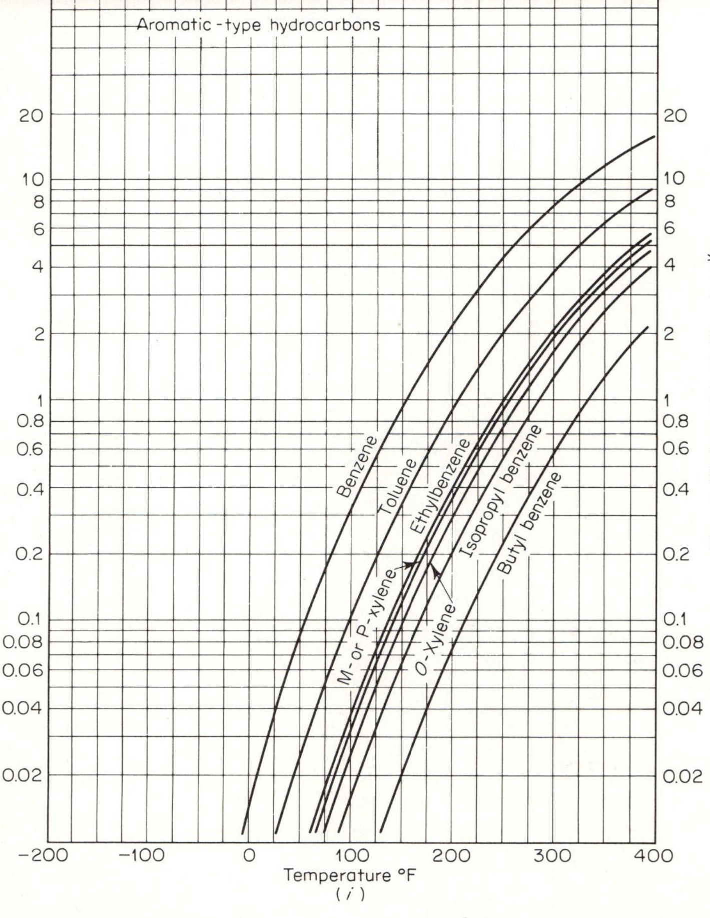

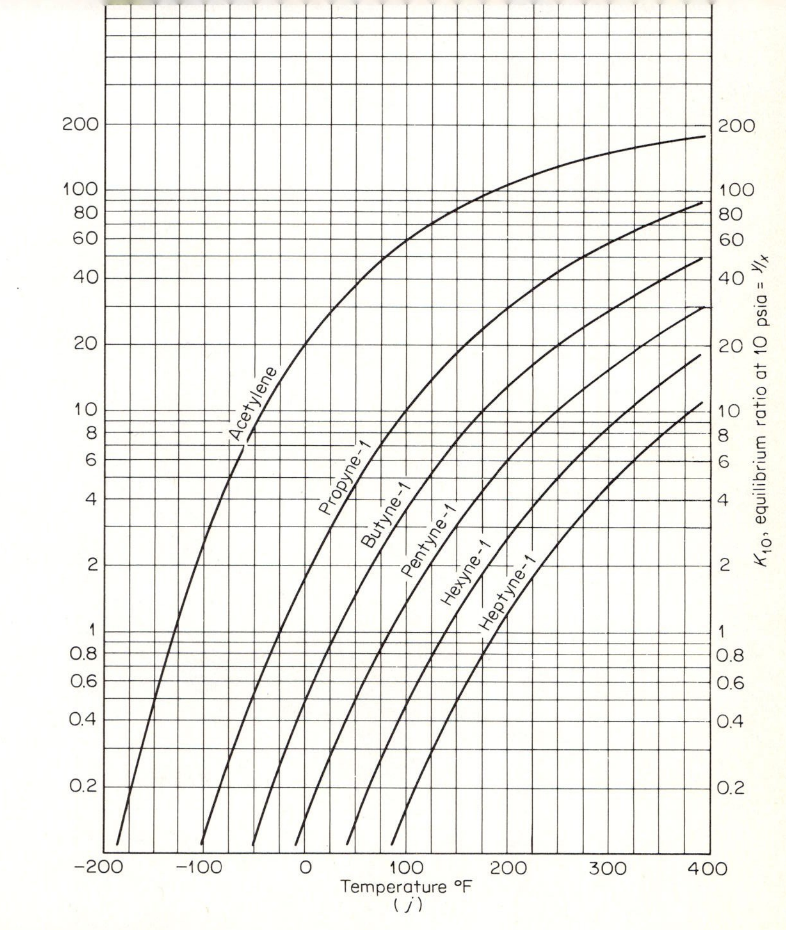

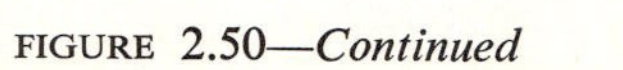
FIGURE 2.50—*Continued*

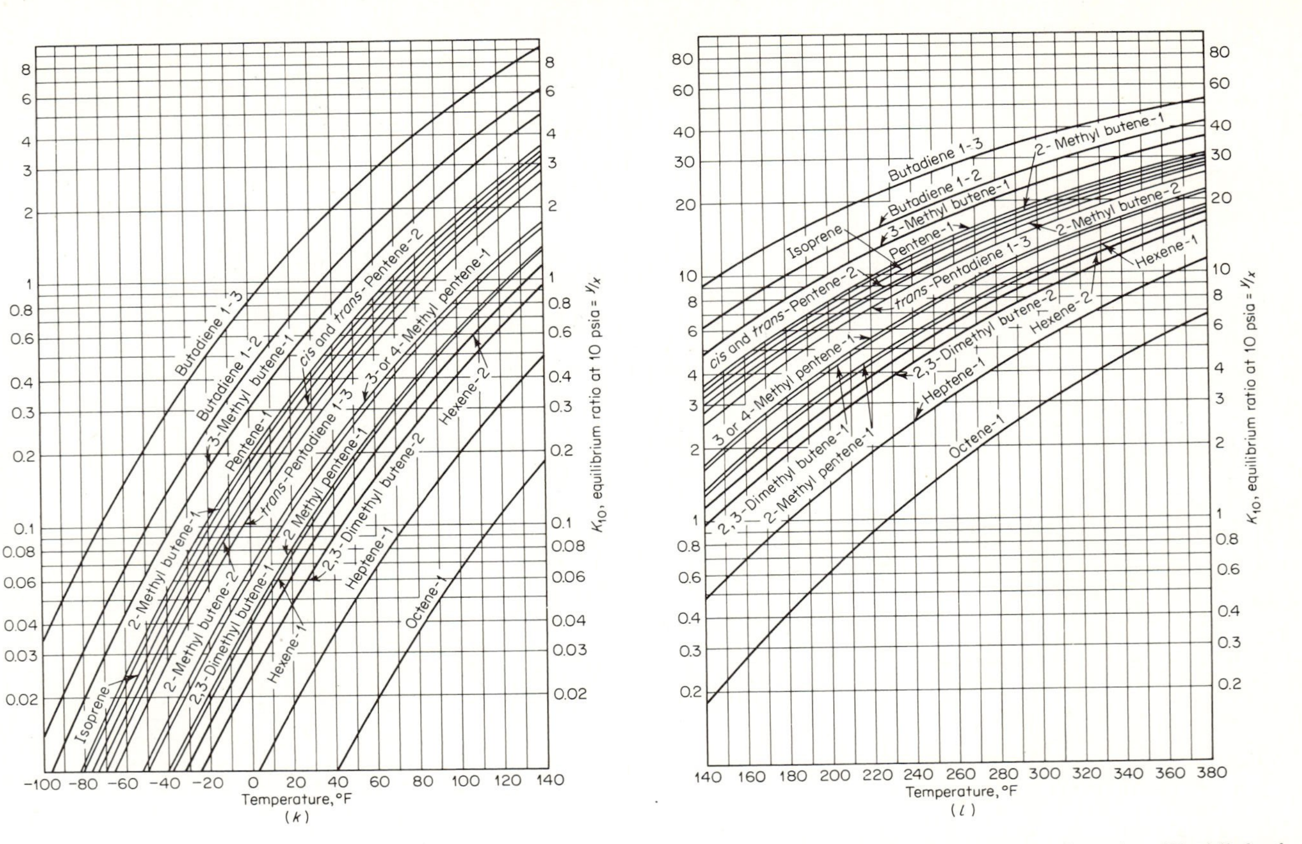

FIGURE 2.50 Equilibrium ratios of hydrocarbons. (*a*) Aliphatic hydrocarbons, −220 to 20°F, K_{10} greater than 1. (*b*) Aliphatic hydrocarbons, −220 to 20°F, K_{10} less than 1. (*c*) Aliphatic hydrocarbons, 20 to 260°F, K_{10} greater than 1. (*d*) Aliphatic hydrocarbons, 20 to 260°F, K_{10} less than 1. (*e*) Aliphatic hydrocarbons, 260 to 500°F, K_{10} greater than 1. (*f*) Aliphatic hydrocarbons, 260 to 500°F, K_{10} less than 1. (*g*) Aliphatic hydrocarbons, 500 to 980°F. (*h*) Naphthenes. (*i*) Aromatics. (*j*) Acetylenes. (*k*) Unsaturated hydrocarbons, 100 to 140°F. (*l*) Unsaturated hydrocarbons, 140 to 380°F. [*From Cajander et al., J. Chem. Eng. Data*, **5**:251 (1960). *Reproduced by*

K's are available may be estimated by plotting log K versus molecular weight at a constant temperature and pressure and interpolating or extrapolating over a narrow range of molecular weights. The relation is a straight line.

Extrapolation of K values with pressure may be done by plotting log K versus P_t at constant temperature and extending the straight line. Extrapolation is not recommended to pressures greater than a reduced pressure of 0.4. Extrapolation with temperature may be done by plotting log K versus T at constant pressure. This is satisfactory to a T_r of about 0.5.

APPENDIX: RELATIONSHIP BETWEEN LI AND COULL AND BONHAM VAPOR-LIQUID EQUILIBRIUM RELATIONS

Binary Equations

The binary equations of the van Laar form developed by Li and Coull are:

$$(T \log \gamma_1)^{-0.5} = \frac{b_1}{b_2}\left(\frac{b_2}{k_{12}}\right)^{0.5}\frac{x_1}{x_2} + \left(\frac{b_2}{k_{12}}\right)^{0.5} \tag{1a}$$

$$(T \log \gamma_2)^{-0.5} = \frac{b_2}{b_1}\left(\frac{b_1}{k_{12}}\right)^{0.5}\frac{x_1}{x_2} + \left(\frac{b_1}{k_{12}}\right)^{0.5} \tag{1b}$$

where $k_{ij} = a_{ij}/2.3R$
a_{ij} = binary interaction constant
b_i = molar volume of component i

Similarly, the Bonham equations of the van Laar type for binary systems are:

$$T \ln \gamma_1 = \left(\frac{x_2 B_{12}^{0.5}}{x_1 A_{12} + x_2}\right)^2 \tag{2a}$$

$$T \ln \gamma_2 = \left(\frac{x_1 A_{12} B_{21}^{0.5}}{x_1 A_{12} + x_2}\right)^2 \tag{2b}$$

where A, B are van Laar binary constants.

Since $T \ln \gamma_1 = 2.3T \log \gamma_1$,

$$(T \log \gamma_1)^{-0.5} = \frac{x_1 A_{12} + x_2}{x_2\left(\dfrac{B_{12}}{2.3}\right)^{0.5}}$$

$$= \left(\frac{x_1}{x_2}\right) A_{12}\left(\frac{2.3}{B_{12}}\right)^{0.5} + \left(\frac{2.3}{B_{12}}\right)^{0.5} \tag{3}$$

Inspection of Eqs. (1*a*) and (3) indicates that

$$\left(\frac{2.3}{B_{12}}\right)^{0.5} = \left(\frac{b_2}{k_{12}}\right)^{0.5} \tag{4}$$

$$\frac{B_{12}}{2.3} = \frac{k_{12}}{b_2} = B_{12}^* \tag{5}$$

By definition,

$$A_{12} = \frac{b_1}{b_2} \tag{6}$$

Starting with Eq. (1*b*) and following the foregoing procedure results in

$$\frac{B_{21}}{2.3} = \frac{k_{12}}{b_1} = B_{21}^* \tag{7}$$

Then

$$A_{12}\left(\frac{B_{21}}{2.3}\right)^{0.5} = \frac{b_1}{b_2}\left(\frac{k_{12}}{b_1}\right)^{0.5} \tag{8}$$

$$A_{12}^2\frac{k_{12}}{b_1} = \left(\frac{b_1}{b_2}\right)^2\frac{k_{12}}{b_1} \tag{9}$$

$$A_{12}B_{21}^* = B_{12}^* \tag{10}$$

In terms of any binary system,

$$A_{ij} = \frac{b_i}{b_j} \tag{11}$$

$$B_{ij}^* = \frac{k_{ij}}{b_j} \tag{12}$$

$$B_{ji}^* = \frac{k_{ji}}{b_i} \tag{13}$$

Note: $k_{ij} = k_{ji}$.

Ternary Equations

The Li and Coull–van Laar type of equation for ternary systems utilizing only binary constants is

$$T\log\gamma_1 = \frac{\dfrac{k_{12}}{b_1}\dfrac{b_2}{b_1}x_2^2 + \dfrac{k_{13}}{b_1}\dfrac{b_3}{b_1}x_3^2 + \left(\dfrac{k_{12}}{b_1}\dfrac{b_3}{b_1} + \dfrac{k_{13}}{b_1}\dfrac{b_2}{b_1} - \dfrac{k_{23}}{b_1}\right)x_2x_3}{[x_1 + (b_2/b_1)x_2 + (b_3/b_1)x_3]^2} \tag{14}$$

Equations for $T \log \gamma_2$ and $T \log \gamma_3$ may be written following the rotation principle

$$\begin{array}{ccc} & 1 & \\ \nearrow & & \searrow \\ 3 & \longleftarrow & 2 \end{array}$$

Considering Eq. (14), multiply both numerator and denominator by $(b_1/b_2)^2$:

$$T \log \gamma_1 = \frac{\dfrac{k_{12}}{b_1}\dfrac{b_1}{b_2}x_2^2 + \dfrac{k_{13}}{b_1}\dfrac{b_3}{b_1}\left(\dfrac{b_1}{b_2}\right)^2 x_3^2 + \left[\dfrac{k_{12}}{b_1}\dfrac{b_3}{b_1}\left(\dfrac{b_1}{b_2}\right)^2 + \dfrac{k_{13}}{b_1}\dfrac{b_1}{b_2} - \dfrac{k_{23}}{b_1}\left(\dfrac{b_1}{b_2}\right)^2\right]x_2x_3}{[x_1(b_1/b_2) + x_2 + (b_3/b_1)(b_1/b_2)x_3]^2} \tag{15}$$

According to the Li and Coull assumption that the partial molal volume of a component in solution is essentially the same in any binary involving the component,

$$\left(\frac{b_1}{b_2}\right)_{12} = \frac{(b_1/b_3)_{13}}{(b_2/b_3)_{23}} \tag{16}$$

then, substituting the equivalent terms,

$$\frac{k_{12}}{b_1}\frac{b_1}{b_2} = B_{21}^*A_{12} = B_{12}^* \tag{17}$$

$$\frac{k_{13}}{b_1}\frac{b_3}{b_1}\left(\frac{b_1}{b_2}\right)^2 = B_{31}^*A_{31}\left(\frac{b_1}{b_2}\right) = B_{31}^*A_{32}A_{12}$$

$$= B_{13}^*A_{31}A_{31}(A_{12})^2 = B_{13}^*A_{32}^2 \tag{18}$$

$$\frac{k_{12}}{b_1}\frac{b_3}{b_1}\left(\frac{b_1}{b_2}\right)^2 = B_{12}^*A_{32} \tag{19}$$

$$\frac{k_{13}}{b_1}\frac{b_1}{b_2} = B_{13}^*A_{32} \tag{20}$$

$$\frac{k_{23}}{b_1}\left(\frac{b_1}{b_2}\right)^2 = B_{23}^*A_{32}A_{12} \tag{21}$$

Substituting the values of Eqs. (17) to (21) in Eq. (15),

$$T \log \gamma_1 = \frac{B_{12}^*x_2^2 + B_{13}^*A_{32}^2x_3^2 + (B_{12}^*A_{32} + B_{13}^*A_{32} - A_{32}A_{12}B_{23}^*)x_2x_3}{(A_{12}x_1 + x_2 + A_{32}x_3)^2} \tag{22}$$

Similarly, the equivalent terms in the $T \log \gamma_2$ and $T \log \gamma_3$ equations can be evaluated.

Bonham Ternary Equations

$$T \ln \gamma_1 = \frac{B_{12}x_2^{\ 2} + A_{32}^{\ 2}B_{13}x_3^{\ 2} + 2A_{32}B_{12}^{0.5}B_{13}^{0.5}x_2x_3}{(A_{12}x_1 + x_2 + A_{32}x_3)^2} \tag{23}$$

Since

$$B_{ij} = 2.3(B_{ij}^*)$$

the equation is changed to the form

$$T \log \gamma_1 = \frac{B_{12}^*x_2^{\ 2} + A_{32}^{\ 2}B_{13}^*x_3^{\ 2} + 2A_{32}(B_{12}^*B_{13}^*)^{0.5}x_2x_3}{(A_{12}x_1 + x_2 + A_{32}x_3)^2} \tag{24}$$

Similarly, the other $T \log \gamma$ equations can be written by using the rotation principle. Equations (22) and (24) are the same equations if

$$2A_{32}(B_{12}^*B_{13}^*)^{0.5} = A_{32}(B_{12}^* + B_{13}^* - B_{23}^*A_{12}) \tag{25}$$

(Also, similar correspondence of terms can be shown in the other $T \log \gamma$ equations.) From the definitions of the van Laar constants,

$$A_{ij} = \frac{b_i}{b_j} \tag{6}$$

$$B_{ij} = \frac{b_i}{R}\left(\frac{a_i^{0.5}}{b_i} - \frac{a_j^{0.5}}{b_j}\right)^2 \tag{26}$$

$$\left(\frac{B_{12}}{A_{12}}\right)^{0.5} + B_{23}^{0.5} + \left(\frac{B_{31}}{A_{32}}\right)^{0.5} = 0 \tag{27}$$

$$\left(\frac{1}{b_1/b_2}\frac{b_1}{R}\right)^{0.5}\left(\frac{a_1^{0.5}}{b_1} - \frac{a_2^{0.5}}{b_2}\right) + \left(\frac{b_2}{R}\right)^{0.5}\left(\frac{a_2^{0.5}}{b_2} - \frac{a_3^{0.5}}{b_3}\right) + \left(\frac{1}{b_3/b_2}\frac{b_3}{R}\right)^{0.5}\left(\frac{a_3^{0.5}}{b_3} - \frac{a_1^{0.5}}{b_1}\right) = 0 \tag{28}$$

$$\left(\frac{b_2a_1}{R}\right)^{0.5}\frac{1}{b_1} - \left(\frac{b_2a_2}{R}\right)^{0.5}\frac{1}{b_2} + \left(\frac{b_2a_2}{R}\right)^{0.5}\frac{1}{b_2} - \left(\frac{b_2a_3}{R}\right)^{0.5}\frac{1}{b_3} + \left(\frac{b_2a_3}{R}\right)^{0.5}\frac{1}{b_3} - \left(\frac{b_2a_1}{R}\right)^{0.5}\frac{1}{b_1} = 0 \tag{29}$$

Also $B_{ij}^{0.5} = -(B_{ji}A_{ij})^{0.5}$ by definition, and $A_{ij} = 1/A_{ji}$.

The van Laar constants in the Bonham equations can be used, i.e.,

$$\left(\frac{B^*_{12}}{A_{12}}\right)^{0.5} + (B^*_{23})^{0.5} + \left(\frac{B^*_{31}}{A_{32}}\right)^{0.5} = 0 \tag{30}$$

$$-(B^*_{23})^{0.5} = \left(\frac{B^*_{12}}{A_{12}}\right)^{0.5} + \left(\frac{B^*_{31}}{A_{32}}\right)^{0.5} = \left(\frac{B^*_{12}}{A_{12}}\right)^{0.5} - \left(\frac{B^*_{13}A_{31}}{A_{32}}\right)^{0.5}$$

$$B^*_{23} = \frac{B^*_{12}}{A_{12}} + \frac{B^*_{13}A_{31}}{A_{32}} - 2\left(\frac{B^*_{12}B^*_{13}A_{31}}{A_{12}A_{32}}\right)^{0.5} \tag{31}$$

Since

$$\frac{A_{31}}{A_{32}} = \frac{b_3/b_1}{b_3/b_2} = \frac{b_2}{b_1} = A_{21} = \frac{1}{A_{12}} \tag{32}$$

$$2(B^*_{12}B^*_{13})^{0.5}\frac{1}{A_{12}} = \frac{B^*_{12}}{A_{12}} + B^*_{13}A_{21} - B^*_{23} \tag{33}$$

Multiply both sides by $A_{12}A_{32}$,

$$2A_{32}(B^*_{12}B^*_{13})^{0.5} = A_{32}(B^*_{12} + B^*_{13} - B^*_{23}A_{12}) \tag{34}$$

This is Eq. (25). Similarly, a correspondence between the terms in the other $T \log \gamma$ equations can be shown in the same manner.

Thus it can be proved that the Li and Coull equations are the same as those of the Bonham-van Laar type. Since experimental data were used to check the validity of the Li and Coull relations, the same data can validate the Bonham equations.

Because of the somewhat greater difficulty in establishing the binary constants for the Li and Coull equations, it would appear that the Bonham equations using the van Laar constants are much easier to use for binary and multicomponent systems.

PROBLEMS

1. Calculate the isothermal vapor and liquid composition lines for $t = 120°C$, $P_t = 760$ mm Hg for the system benzene–toluene–ethylbenzene, assuming ideality.

Components	Antoine constants		
	A	*B*	*C*
Benzene	6.898	1206.35	220.24
Toluene	6.953	1343.94	219.58
Ethylbenzene	6.954	1421.91	212.93

2. Compute the following points of ternary vapor-liquid equilibrium data for the system methanol–ethanol–1-propanol at 760 mm Hg, assuming ideality in both phases.

Mole % methanol in liquid:	25	50	75
Mole % ethanol in liquid:	40	30	10

Determine y and t for each point:

Components	BP, 760 mm Hg, °C	Vapor pressure, mm Hg, 80°C
MeOH	64.7	1335
EtOH	78.4	810
1-PrOH	97.8	375

3. Determine the ternary vapor compositions for the system methyl ethyl ketone–heptane–toluene at 760 mm Hg pressure for the four points below using the following binary γx data to get the binary constants and the Wohl ternary van Laar equations.

MEK–*n*-heptane	MEK–toluene	*n*-C_7–toluene
$x_{MEK} = 0.507$	$x_{MEK} = 0.468$	$x_{n\text{-}C_7} = 0.45$
$y_{MEK} = 0.645$	$y_{MEK} = 0.685$	$y_{n\text{-}C_7} = 0.54$
$\log \gamma_{MEK} = 0.1221$	$\log \gamma_{MEK} = 0.0358$	$\log \gamma_{n\text{-}C_7} = 0.038$
$\log \gamma_{n\text{-}C_7} = 0.1315$	$\log \gamma_{Tol} = 0.045$	$\log \gamma_{Tol} = 0.32$

Components	Points of ternary data			
	1	2	3	4
x_{MEK}	0.198	0.403	0.499	0.704
x_{Tol}	0.690	0.334	0.347	0.090

The vapor pressure is:

For *n*-C_7: $\log P = 6.905 - 1268.59/(216.95 + t)$.

For toluene: $\log P = 6.9533 - 1343.94/(219.38 + t)$.

For MEK: $\log P = 6.9742 - 1209.6/(216 + t)$.

4. Determine the same points of data as for Prob. 3, using pure-component properties to evaluate the binary constants.

Components	P_c, atm	t_c, °C
MEK	39.5	260
n-C_7	27	267
Toluene	40.3	320.8

5. Assume the following mixture to be ideal and determine the K values for each component at $t = 150°F$ and $P_T = 100$ psia.

Components	x	Vapor pressure, mm Hg, 200°F
Neopentane	0.45	7210
n-Pentane	0.15	3750
n-Hexane	0.11	1525
Toluene	0.18	450
Heptane	0.11	660

6. (*a*) Use the convergence pressure method of Lenoir and White to determine the convergence pressure of the mixture.
(*b*) Determine the convergence pressure by the Hadden-Grayson method.

7. Determine the K values by using the convergence pressure calculated from Prob. 6.
(*a*) Use the Hadden-Grayson nomograph, correcting for convergence pressure.
(*b*) Use the K_{10} method.

8. Calculate the K values for the mixture of Prob. 7, including 10% of H_2S (the analysis above is multiplied by 0.90), by the Hadden-Grayson nomograph.

Nomenclature

E_{si} = see Eq. (2.52)
Fe_H = multiplying factor, Eq. (2.65)
Fe_L = multiplying factor, Eq. (2.65)
f = fugacity
K_i = y_i/x_i
i, j, k = component designations
L = liquid
m = class designation
m, n = exponents
n = components
P_c = critical pressure
P_T = total pressure
P_i = vapor pressure for component i
P_{op} = system pressure
P_{cv} = convergence pressure
P_g = grid pressure
p_F = defined factor of pressure, Eq. (2.66)
p_i = partial pressure, component i
R = gas-law proportionality factor
R = class designation
r = component designation in class

S = specific class designation X_m
s = component designation in class X_R
T = absolute temperature
T_c = critical temperature
t = temperature
T_i = boiling point, absolute temperature
t_i = boiling temperature
V = vapor
x = mole fraction liquid
y = mole fraction vapor

Constants

$A, A', a, B, B', b, C, C^*, c, D, d, k$

Greek letters

α = relative volatility
γ = activity coefficient
γ_p = K correction factor for hydrocarbons (H_2)
γ_m = K correction factor for methane
γ_h = K correction factor for hydrocarbons (CH_4)
λ_{ij} = factor proportional to molecular interaction energies
Λ_{ij} = defined factor, Eq. (2.55)
ν_i° = fugacity coefficient of pure liquid i at system conditions
$\sum$ = summation
ϕ_i = fugacity coefficient of component i in vapor mixture

REFERENCES

1. Beattie, J. A., and O. C. Bridgeman: *J. Am. Chem. Soc.*, **49**:1665 (1927).
2. Benedict, M. G., B. Webb, and L. C. Rubin: *Chem. Eng. Progr.*, **47**:419 (1951).
3. Benedict, M., G. B. Webb, and L. C. Rubin: *Chem. Eng. Progr.*, **47**:449 (1951).
4. Benedict, M., G. B. Webb, L. C. Rubin, and L. Friend: *Chem. Eng. Progr.*, **47**:571 (1951).
5. Benedict, M., G. B. Webb, L. C. Rubin, and L. Friend: *Chem. Eng. Prog.*, **47**:609 (1951).
6. Benedict, M., C. A. Johnson, E. Solomon, and L. C. Rubin: *Trans. A.I.Ch.E.*, **41**:371 (1941).
7. Black, Cline: *Ind. Eng. Chem.*, **50**:403 (1958).
8. Black, Cline: *Ind. Eng. Chem.*, **51**:211 (1959).
9. Bonham, M. S.: M.S. Thesis in Chemical Engineering, MIT, 1941.
10. Brown, G. G.: *Petrol. Eng.* (June, 1940), p. 55.
11. Cajander, B. C., H. G. Hipkin, and J. M. Lenoir: *J. Chem. Eng. Data*, **5**:251 (1960).
12. Chao, K. C., and O. A. Hougen: *Chem. Eng. Sci.*, **7**:246 (1958).
13. Chao, K. C., and J. D. Seader: *A.I.Ch.E. J.*, **7**:598 (1961).

14. De Priester, C. L.: *Chem. Eng. Progr. Symp. Ser.*, **49**:(7):1 (1953).
15. Dreisbach, R. R.: Physical Properties of Chemical Compounds I, II, III, *Monograph Ser. nos.* 15, 22, 29 (1955, 1959, 1961).
16. Edmister, W. C., and C. L. Ruby: *Chem. Eng. Progr.*, **51**:95F (1955).
17. Edwards, B. S., Frank Hashmall, Roger Gilmont, and D. F. Othmer: *Ind. Eng. Chem.*, **46**:194 (1954).
18. Flory, P. J.: *J. Chem. Phys.*, **10**:51 (1942).
19. Gamson, B. W., and K. M. Watson: *Natl. Petrol. News, Tech. Sec.*, **36**:R623 (September 6, 1944).
20. Gilmont, Roger, E. A. Weinman, Franklin Kramer, Eugene Miller, Frank Hashmall, and D. F. Othmer: *Ind. Eng. Chem.*, **43**:120 (1950).
21. Hadden, S. T.: *Chem. Eng. Progr. Symp. Ser.*, **49**(7):53 (1953).
22. Hadden, S. T.: *Chem. Eng. Progr.*, **44**:135 (1948).
23. Hadden, S. T., and H. G. Grayson: *Petrol. Refiner* (September, 1961), p. 207.
24. Hala, E., J. Pick, V. Fried, and O. Vilim: "Vapor Liquid Equilibrium" (translated by G. Standart), Pergamon, New York, 1958.
25. Hoffman, P. S., J. R. Walker, R. E. Felt, and J. H. Weber: *A.I.Ch.E. J.*, **8**:508 (1962).
26. Hougen, O. A., K. M. Watson, and R. A. Ragatz: "Chemical Process Principles—Part II: Thermodynamics," 2d ed., Wiley, New York, 1959.
27. Huggins, M. L.: *Ann. N.Y. Acad. Sci.*, **43**:1 (1942).
28. Ibl, N. V., and B. F. Dodge: *Chem. Eng. Sci.*, **2**:120 (1953).
29. Katz, D. L., and Karl Hackmuth: *Ind. Eng. Chem.*, **29**:1072 (1937).
30. Kellogg, M. W., Co.: "Liquid Vapor Equilibrium in Mixtures of Light Hydrocarbons, Equilibrium Constants," New York, 1950, *Chem. Eng. Progr.*, **47**:609 (1951).
31. Lenoir, J. M.: *Hydrocarbon Process. Petrol. Refiner* (March, 1965), p. 139.
32. Lenoir, J. M., and H. G. Hipkin: *A.I.Ch.E. J.*, **3**:318 (1957).
33. Lenoir, J. M., and G. A. White: *Petrol. Refiner* (October, 1953), p. 167.
34. Lenoir, J. M., and G. A. White: *Petrol. Refiner* (March, 1958), p. 173.
35. Li, Y. M., and J. Coull: *J. Inst. Petrol.*, **34**:692 (1948).
36. Myers, H. S., and J. M. Lenoir: *Petrol. Refiner* (February, 1957), p. 167.
37. N.G.P.A.: "Engineering Data Book," Natural Gasoline Suppliers and Processors Association, Tulsa, Oklahoma, 1957.
38. Organick, E. I., and G. G. Brown: *Chem. Eng. Progr. Symp. Ser.*, **48**(2):97 (1952).
39. Orye, R. V., and J. M. Prausnitz: *Ind. Eng. Chem.*, **57**(5):18 (1965).
40. Poettman, F. H., and B. J. Mayland: *Petrol. Refiner* (July, 1949), p. 101.
41. Prausnitz, J. M., and J. H. Duffin, *Petrol. Refiner* (May, 1960), p. 213.
42. Prausnitz, J. M., W. C. Edmister, and K. C. Chao: *A.I.Ch.E. J.*, **6**:215 (1960).
43. Redlich, O., and A. T. Kister: *Ind. Eng. Chem.*, **40**:345 (1948).
44. Reid, R. C., and T. K. Sherwood: "The Properties of Gases and Liquids," McGraw-Hill, New York, 1958.
45. Robinson, C. S., and E. R. Gilliland: "Elements of Fractional Distillation," 4th ed., McGraw-Hill, New York, 1950.
46. Scheibel, E. G., and Daniel Friedland: *Ind. Eng. Chem.*, **39**:1329 (1947).
47. Severns, W. H., A. Sesonske, R. H. Perry, and R. L. Pigford: *A.I.Ch.E. J.*, **1**:401 (1955).
48. Smith, K. A., and R. B. Smith: *Petrol. Process.* (December, 1949), p. 1355.

49. Smith, K. A., and K. M. Watson: *Chem. Eng. Progr.*, **45**:494 (1949).
50. van Laar, J. J.: *Z. Phys. Chem.*, **185**:35 (1929).
51. Weimer, R. F., and J. M. Prausnitz: *Petrol. Refiner* (September, 1965), p. 237.
52. White, R. R.: *Trans. A.I.Ch.E.*, **41**:539 (1945).
53. White, R. R., and G. G. Brown: *Natl. Petrol. News* (October, November, 1942), p. R374.
54. Winn, F. W.: *Chem. Eng. Progr. Symp. Ser.*, **48**(2):121 (1952).
55. Winn, F. W.: *Petrol. Refiner* (June, 1954), p. 131.
56. Wilson, G. M.: *J. Am. Chem. Soc.*, **86**:127 (1964).
57. Wohl, K.: *Trans. A.I.Ch.E.*, **42**:215 (1946).
58. Wohl, K.: *Chem. Eng. Progr.*, **49**:218(1953).

chapter 3

Complex system vapor-liquid equilibria

A complex system is defined as one composed of such a large number of components that it is not feasible to identify them or to determine the composition of the mixture in terms of pure components. This type of system is represented primarily by petroleum mixtures, although some mixtures encountered in the chemical industry may fall into this category. Because these mixtures cannot reasonably be represented by a series of true components having specific compound properties, it is necessary to characterize them in some indirect manner by empirically determined and averaged properties.

3.1. TRUE-BOILING-POINT CURVE

The vaporization properties of petroleum mixtures are characterized by a batch-distillation curve wherein the percentage distilled or recovered is plotted against the temperature at which it is distilled. Theoretically, in a "true-boiling-point" (TBP) distillation, a distillation system capable of making very close separations is utilized so that the compounds present in the mixture will be separated each at its own boiling point and in the quantity present in the original mixture. Figure 3.1 represents a TBP distillation curve derived from a batch fractionation of a mixture of 30 vol % of A and 70 vol % of B. A boils at t_A and B boils at t_B at the total pressure of the distillation. The stepwise plot shows the theoretical TBP relationship

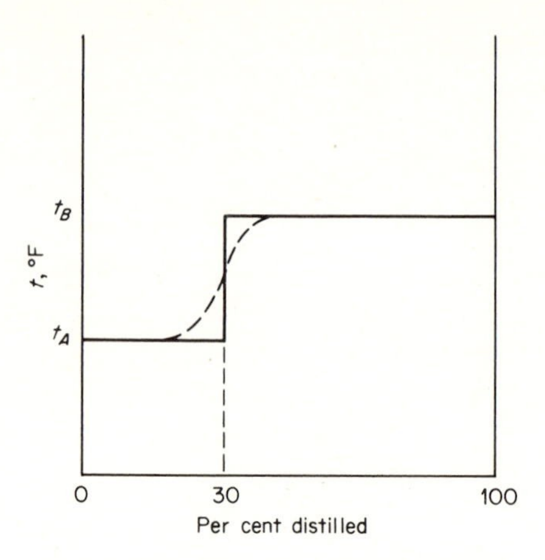

FIGURE 3.1 True-boiling-point curve of a binary mixture.

with perfect fractionation while the smooth curve represents the actual TBP curve with imperfect fractionation, i.e., incomplete separation of the two components. Figure 3.2 shows similar curves for a seven-component mixture. The step curve again indicates perfect fractionation or separation, and the smooth curve shows a somewhat less than perfect separation. The TBP curve of a complex mixture is illustrated in Fig. 3.3 wherein the number of components is large and the volume percent of any one of the components is very small. This TBP curve, if determined under perfect fractionation

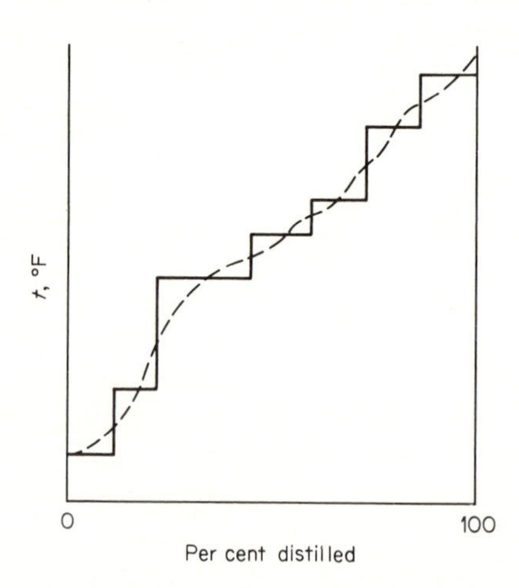

FIGURE 3.2 True-boiling-point curve of a seven-component mixture.

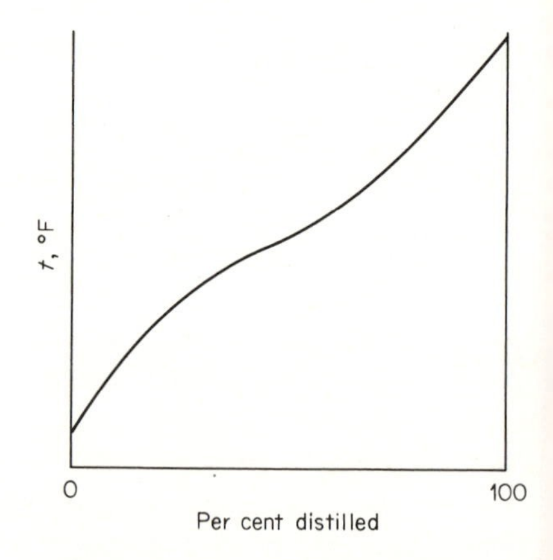

FIGURE 3.3 True-boiling-point curve of a complex mixture.

conditions, describes the boiling point–volume-percent-distilled characteristics of the mixture. The curve actually represents the boiling points of the components at the pressure of the distillation. Because the TBP curve represents the boiling temperature for each component at the total pressure, any point on the curve can be considered a point on the vapor pressure curve of a pseudocompound. Thus, provided the material represented by the point on the curve can be characterized as, for example, paraffinic, aromatic, etc., in the case of hydrocarbons, or as belonging to a family of compounds represented by, for example, a Cox-type vapor pressure chart, the point can be transposed to a new temperature at a different pressure by means of a vapor pressure chart, merely by finding the temperature on the suitable curve corresponding to the desired pressure.

3.2. ASTM OR ENGLER-TYPE DISTILLATION

Vaporization characteristics of petroleum and some complex organic chemical mixtures are determined by means of a simple batch distillation wherein only a very little fractionation is encountered. This is designated as an *Engler-type* distillation. The vaporization properties of petroleum products are usually specified on the basis of the ASTM-D158 distillation [1] in which 100 or 200 cu cm of sample are subject to a batch distillation under closely specified conditions. In general, all batch distillations in which only a small amount of fractionation occurs fall into the Engler-type category.

Figure 3.4 shows a typical Engler-type batch-distillation curve for three components compared with the typical TBP curve for the same three components. The two types of distillation curves for a complex mixture are compared in Fig. 3.5.

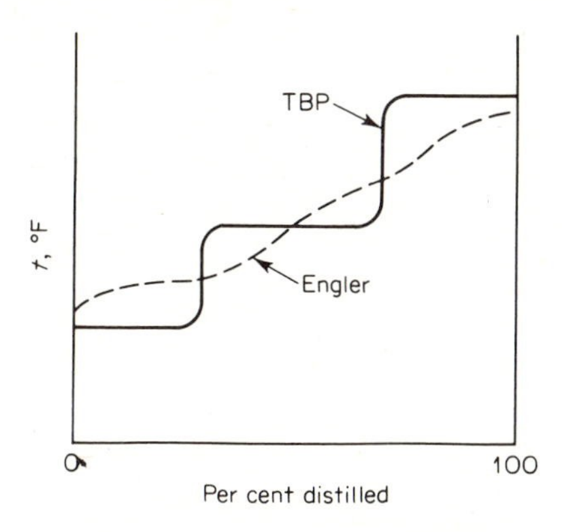

FIGURE 3.4 Comparison of TBP and Engler for a three-component mixture.

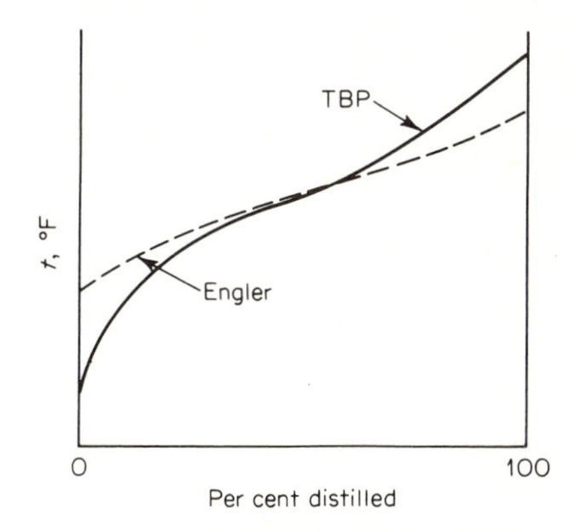

FIGURE 3.5 Comparison of TBP and Engler for a complex mixture.

3.3. EQUILIBRIUM FLASH VAPORIZATION (EFV) CURVES

Equilibrium flash vaporization (EFV) curves are derived from data obtained from operation of either a recirculating-batch still or a continuous-flow vaporizer. The recirculating batch stills, Colburn [11], Othmer [21], or other similar types, maintain the pressure and temperature constant, and the vapor and liquid are kept in intimate contact for a sufficient length of time for the system to attain equilibrium. In a continuous-flow vaporizer [20] steady-state conditions of constant flow rates, temperature, and pressure are maintained with the two phases in intimate contact, so that equilibrium is attained. The term *equilibrium flash* probably originated in the petroleum industry wherein crude oils containing appreciable quantities of light materials under pressure were piped to separators at lower pressure and allowed to vaporize suddenly or "flash" through a pressure-reducing valve into the separator. Since the term *equilibrium* implies that all components in the original material must appear in both vapor and liquid under flash vaporization conditions, and since the original material is a complex mixture whose vaporization characteristics are described analytically by either TBP or Engler curves, the characteristics of the vapor and liquid can be described by the same type of curves.

Figures 3.6 to 3.8 illustrate the TBP curves of the feed material, the vapor, and the liquid resulting from flash vaporization under temperature and pressure conditions such that 50%, 10%, and 80%, respectively, of the feed material was vaporized.

If a number of samples of the same material are flashed at the same

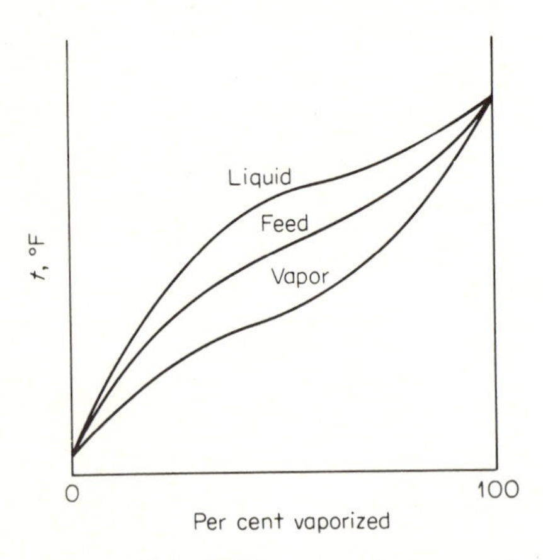

FIGURE 3.6 TBP curves of vapor and liquid products from flash vaporization of 50% of the feed.

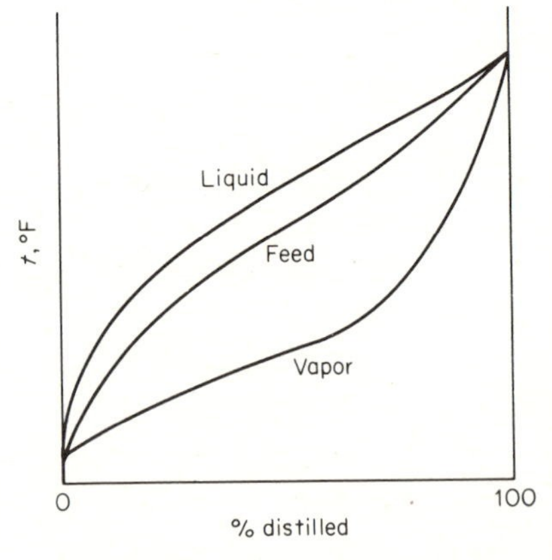

FIGURE 3.7 TBP curves from 10% vaporization.

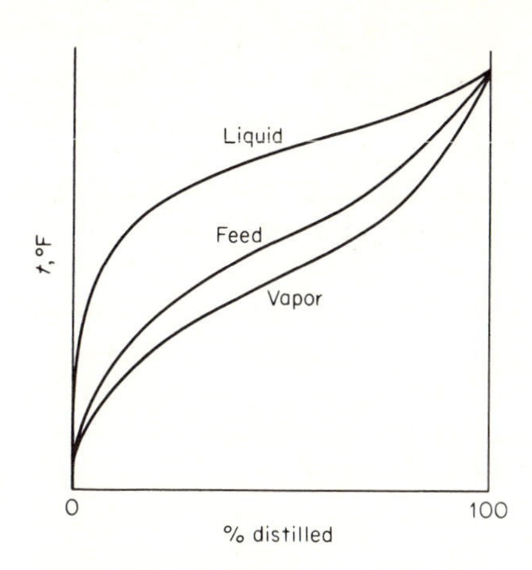

FIGURE 3.8 TBP curves from 80% vaporization.

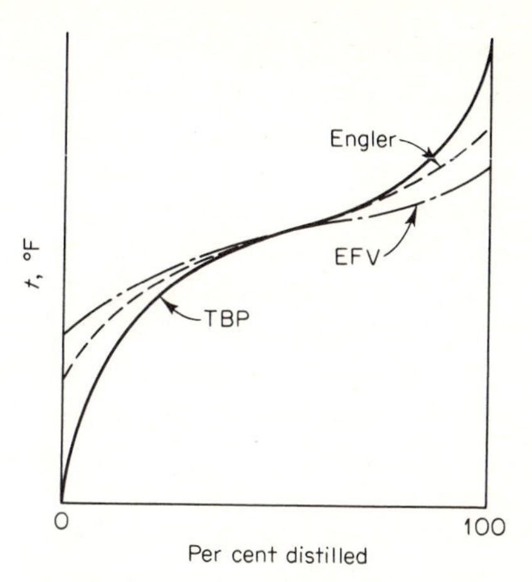

FIGURE 3.9 Equilibrium flash curve compared with Engler-type and TBP curves.

pressure but at different temperatures between the bubble-point and dew-point temperatures of the feed material, an EFV curve for the charge may be obtained such as that shown in Fig. 3.9.

This curve is useful to the engineer in that the relative amount vaporized when a complex mixture is subjected to equilibrium flash distillation at a given temperature and pressure can be read directly from the curve. Figure 3.9 indicates that the 10–70% slope of the EFV curve is less than the slope of the Engler-type curve which in turn, is less than that of the TBP. This follows because the EFV operation involves no fractionation and the TBP involves a reasonably complete fractionation, and it can be expected that a better separation would be obtained by fractionation. The experimental determination of EFV curves is an extremely time-consuming and expensive operation, because with continuous-flow equipment it is difficult to reach equilibrium conditions even though elaborate devices and techniques involving excessive time for each run are used. This has resulted in the development of a number of correlations [2–8, 12, 15–17, 19–25, 28, 30] to relate EFV curves with the analytical TBP and Engler-type curves for use in the petroleum industry.

3.4. INTERRELATION BETWEEN EFV, TBP, AND ENGLER DISTILLATION CURVES

Nelson [14], Nelson and Harvey [15], and Packie [22] developed correlations relating the 10–70% slopes and the 50% vaporized temperatures of the EFV and TBP and EFV and Engler curves. The 10–70% slope is the temperature corresponding to the 70% vaporized temperature minus the 10% vaporized temperature divided by the 60% interval between them, i.e., 70–10%. This slope value has the units of degrees per percent. The relationships, derived essentially from data at atmospheric pressure, enable the approximate evaluation of a straight-line EFV curve. With Nelson's [14] method of correcting for curvature, where the ratio of the slope of any increment of the TBP curve (or ASTM curve) to the slope of the same increment (percentage interval) of the EFV curve is assumed equal to the ratio of the 10–70% slope of the TBP (or ASTM curve) to the 10–70% slope of the EFV curve, reasonably representative EFV curves can be drawn.

$$\left(\frac{\text{Slope TBP}}{\text{Slope EFV}}\right)_{10-70\%} = \left(\frac{\text{slope TBP}}{\text{slope EFV}}\right)_{\text{any increment}}$$

Since many of the higher boiling petroleum oils are characterized by vacuum TBP curves, Okamoto and Van Winkle [19] derived correlations between EFV and TBP curves by calculation of theoretical equilibrium flash data for a series of pseudomulticomponent mixtures and relating the 10–70% slopes and the 50% vaporized temperatures over a pressure range of 10 to 760 mm Hg. They also developed correlations [19] relating the atmospheric TBP with the EFV distillation curves for pressures from 10 to 760 mm Hg pressure. In addition, experimental vacuum EFV data on petroleum oil fractions and atmospheric ASTM data on the same fractions were determined and correlated [20].

Edmister and Pollock [8] developed a method for correlating the atmospheric and superatmospheric EFV curves with experimental atmospheric ASTM data or atmospheric TBP data. The method enables the construction of a phase diagram of temperature versus pressure for the 0% · · · 100% distilled points on the EFV curve for any petroleum oil whose ASTM or TBP curves are known. A typical phase diagram for a petroleum naphtha–kerosene blend derived from this correlation is shown in Fig. 3.10. Filak et al. [9] predicted EFV data using the original Edmister, Nelson, and Okamoto and Van Winkle correlations. They then checked these against EFV data obtained in plant distillation equipment on the same stocks and found the predicted and plant data varied a maximum of ±13%. The latter correlations give better agreement based upon examination of the small amount of comparative data available.

3.5. DETERMINATION OF EFV CURVES—SUPERATMOSPHERIC PRESSURE

The EFV curves for petroleum oils at pressures above atmospheric pressure are determined by Edmister's phase-diagram method from the analytical

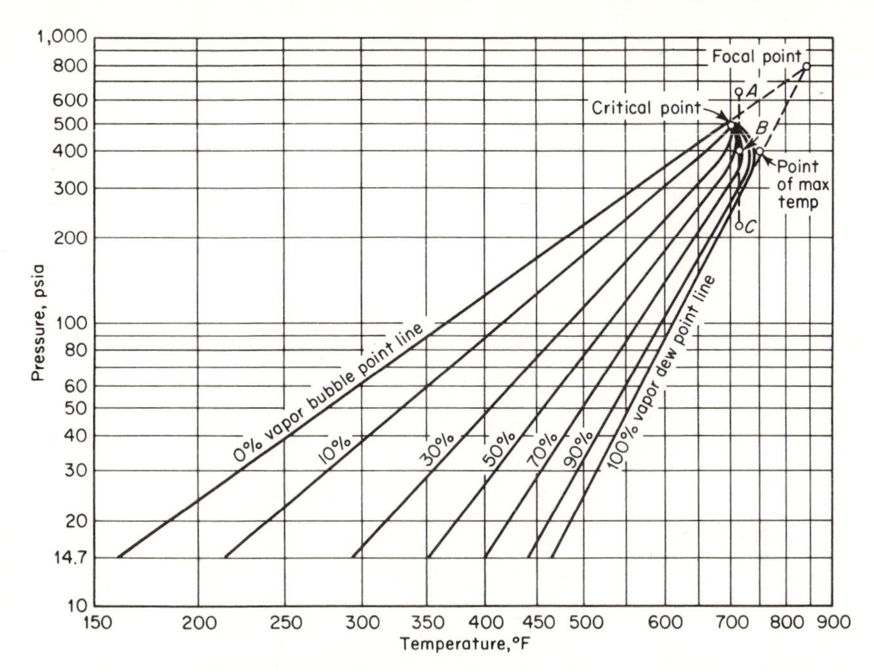

FIGURE 3.10 Phase diagram for naphtha–kerosene blend. [*From Edmister and Pollock, Chem. Eng. Prog.*, **44**:*905 (1948). Reproduced by permission.*]

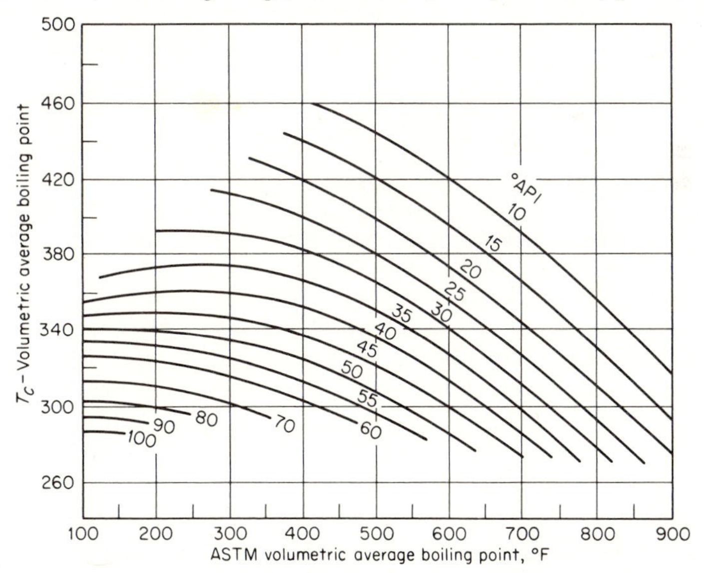

FIGURE 3.11 Critical temperatures of petroleum fractions. [*From Edmister, Applied Hydrocarbon Thermodynamics, Gulf Pub. Company, Houston, Texas (1961). Reproduced by permission.*]

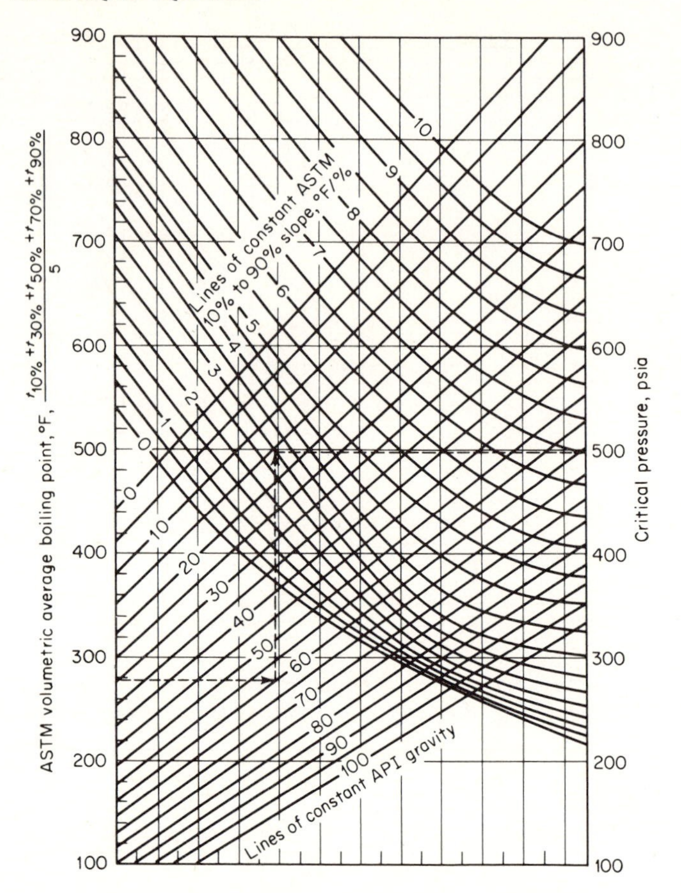

FIGURE 3.12 Critical pressures of petroleum fractions. [*From Edmister, Applied Hydrocarbon Thermodynamics, Gulf Publishing Company, Houston, Texas (1961). Reproduced by permission.*]

atmospheric ASTM data by constructing a phase diagram such as that shown in Fig. 3.10. (The ASTM data may be determined experimentally or predicted from other analytical distillation curves such as the TBP curve.) The procedure is as follows:

1. The volumetric average boiling point (VABP) is calculated by integrating the ASTM distillation curve to find the average boiling temperature or by adding all temperatures (usually 10) determined and dividing by the number of points represented.

2. The 10–90% ASTM slope is evaluated.

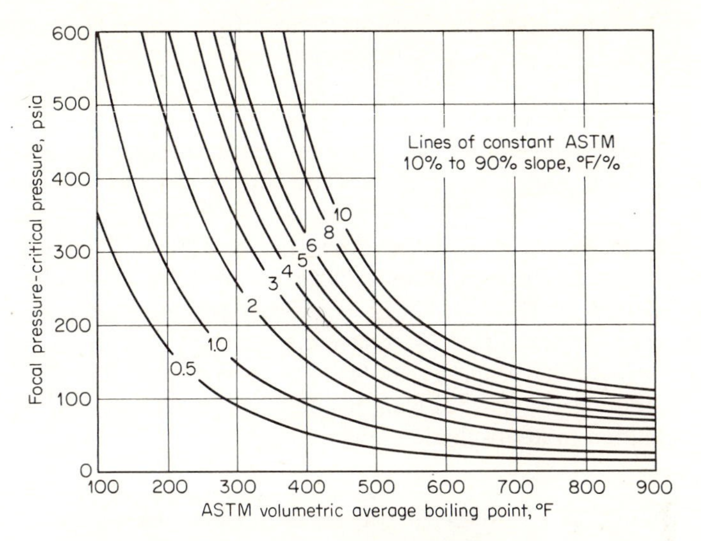

FIGURE 3.13 Phase-diagram focal pressure of petroleum fractions. [*From Edmister, Applied Hydrocarbon Thermodynamics, Gulf Publishing Company, Houston, Texas (1961). Reproduced by permission.*]

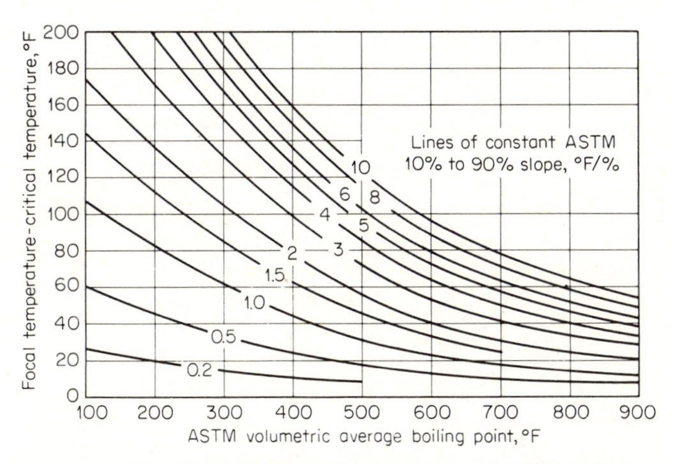

FIGURE 3.14 Phase-diagram focal temperature of petroleum fractions. [*From Edmister, Applied Hydrocarbon Thermodynamics, Gulf Publishing Company, Houston, Texas (1961). Reproduced by permission.*]

3. The critical temperature T_c is determined from Fig. 3.11 by using the VABP calculated in step 1 and API gravity.

4. The critical pressure P_c is determined from Fig. 3.12 by use of the VABP, 10–90% ASTM slope, and API gravity.

5. The focal temperature T_F for the phase diagram is determined from Fig. 3.14, and the focal pressure P_F is determined from Fig. 3.13.

6. The point T_F, P_F is plotted, as shown in Fig. 3.10, on a log scale versus reciprocal temperature scale graph. Also (usually) the atmospheric EFV data are plotted for the 0%, 10%, 20%, 30%, etc., distilled points. The phase diagram is constructed by connecting the focal point and the atmospheric data for 0%, 10%, . . . , 100% distilled with straight lines, and the EFV temperatures can be read from the diagram for any pressure, from the critical condition to atmospheric pressure.

3.6. ATMOSPHERIC AND SUBATMOSPHERIC EFV RELATIONSHIPS FROM ASTM AND TBP DATA

Edmister and Okamoto [7] reworked some of the correlations previously reported, incorporating additional new EFV data and presented methods and

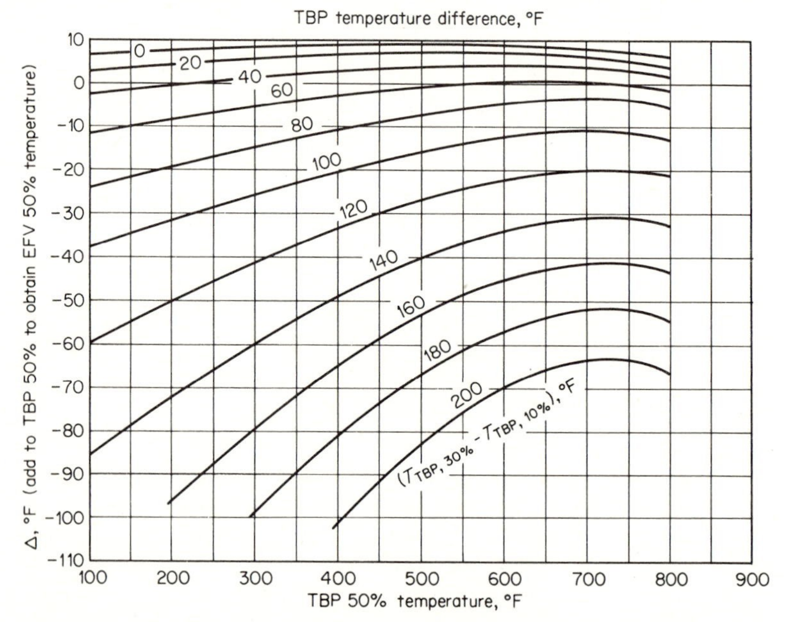

FIGURE 3.15 TBP 50% temperature versus EFV 50% temperature. [*From Edmister, Applied Hydrocarbon Thermodynamics, Gulf Publishing Company, Houston, Texas (1961). Reproduced by permission.*]

charts for relating ASTM, TBP, and EFV distillation curves at atmospheric and subatmospheric pressures. Figures 3.15 through 3.18 are working charts published by Edmister [4] for the evaluation of EFV data from ASTM or TBP data, and Fig. 3.19 interrelates ASTM and TBP atmospheric data.

Where heavy oils are distilled in the laboratory to determine their distillation characteristics, vacuum distillations of the Engler type, ASTM-D1160, 10 mm Hg, and 10 mm Hg TBP type are conducted. In these cases the procedure suggested by Edmister is that of determining the EFV distillation curve for 10 mm Hg and transposing it by means of an empirical correlation to an EFV curve at the subatmospheric pressure desired. Figures 3.20 and 3.21 enable evaluation of the EFV 10 mm Hg curve from the TBP 10 mm Hg data, and Figs. 3.22 and 3.23 relate ASTM-D1160 (10 mm Hg) and EFV 10 mm Hg curves. Transposition of the 10 mm Hg EFV distillation curve to a corresponding distillation curve at other pressures from 10 to 760 mm Hg can be done by using Fig. 3.24. Figure 3.25 interrelates the 10 mm Hg ASTM-D1160 data and the 10 mm Hg TBP distillation data.

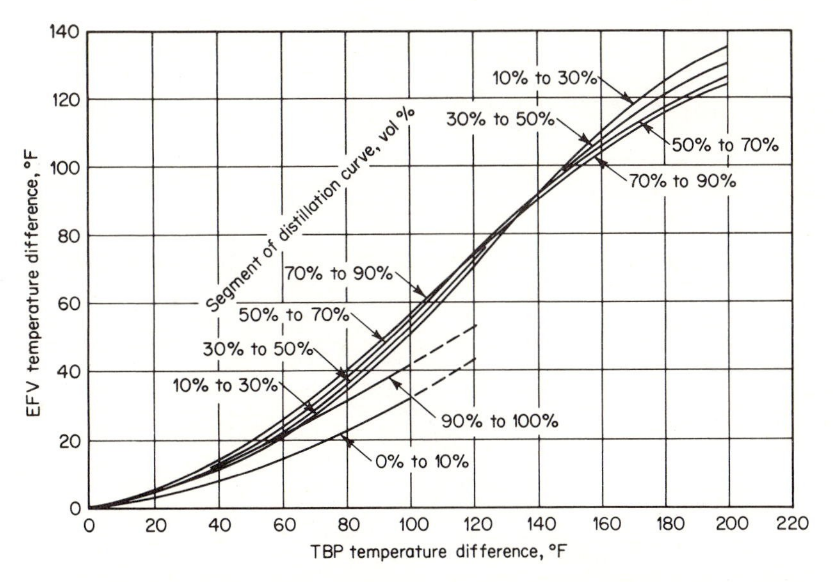

FIGURE 3.16 TBP temperature difference versus EFV temperature difference. [*From Edmister, Applied Hydrocarbon Thermodynamics, Gulf Publishing Company, Houston, Texas (1961). Reproduced by permission.*]

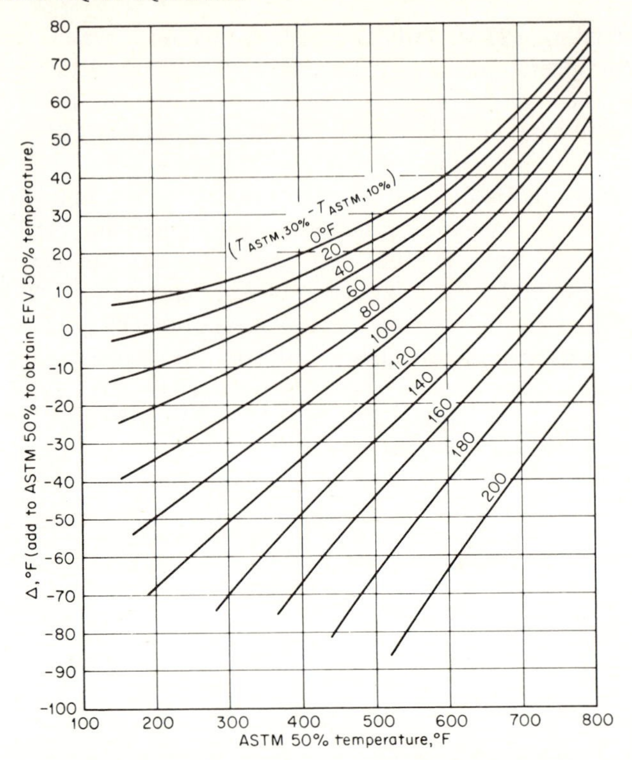

FIGURE 3.17 ASTM 50% temperature versus EFV 50% temperature. [*From Edmister, Applied Hydrocarbon Thermodynamics, Gulf Publishing Company, Houston, Texas (1961). Reproduced by permission.*]

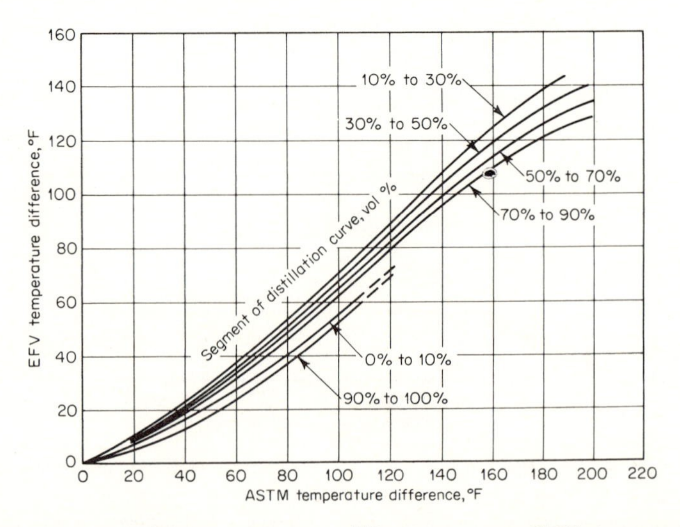

FIGURE 3.18 ASTM temperature difference versus EFV temperature difference. [*From Edmister, Applied Hydrocarbon Thermodynamics, Gulf Publishing Company, Houston, Texas (1961). Reproduced by permission.*]

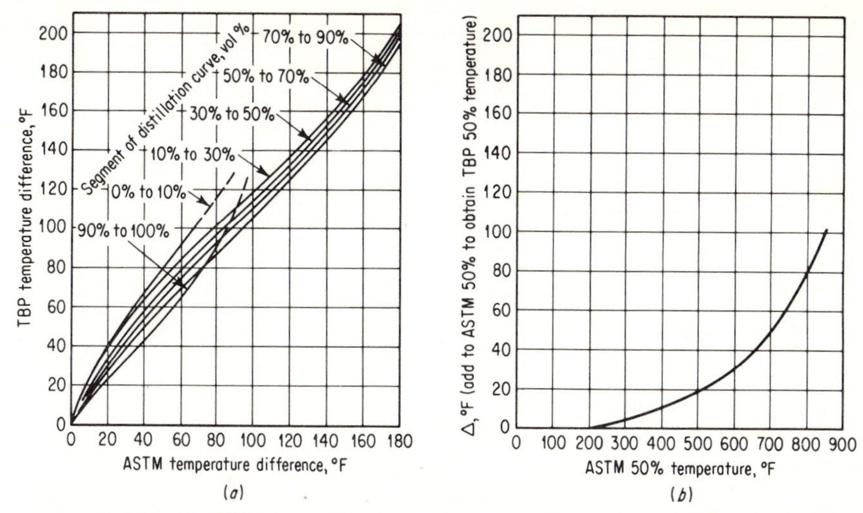

FIGURE 3.19 (*a*) ASTM temperature difference versus TBP 50% temperature difference. (*b*) ASTM 50% temperature versus TBP 50% temperature. [*From Edmister, Applied Hydrocarbon Thermodynamics, Gulf Publishing Company, Houston, Texas (1961). Reproduced by permission.*]

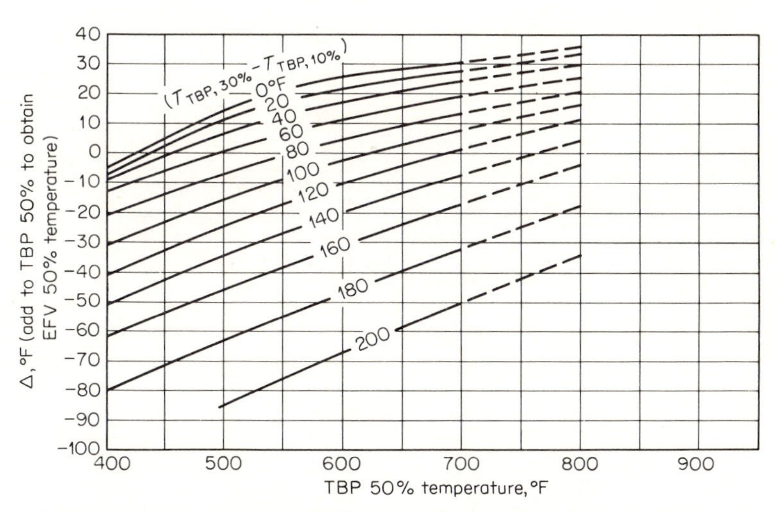

FIGURE 3.20 A 10 mm Hg TBP 50% temperature versus a 10 mm Hg EFV 50% temperature. [*From Edmister, Applied Hydrocarbon Thermodynamics, Gulf Publishing Company, Houston, Texas (1961). Reproduced by permission.*]

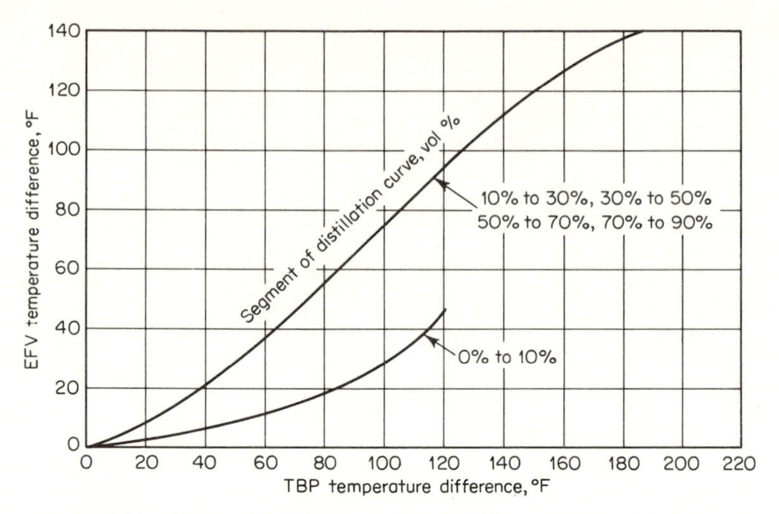

FIGURE 3.21 10 mm Hg TBP temperature difference versus 10 mm Hg EFV temperature differences. [*From Edmister, Applied Hydrocarbon Thermodynamics, Gulf Publishing Company, Houston, Texas (1961). Reproduced by permission.*]

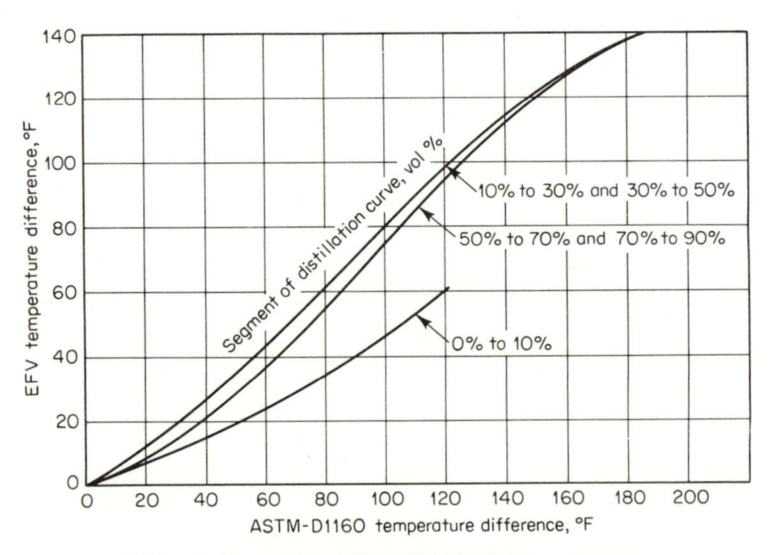

FIGURE 3.22 A 10 mm Hg ASTM-D1160 50% temperature versus a 10 mm Hg EFV 50% temperature. [*From Edmister, Applied Hydrocarbon Thermodynamics, Gulf Publishing Company, Houston, Texas (1961). Reproduced by permission.*]

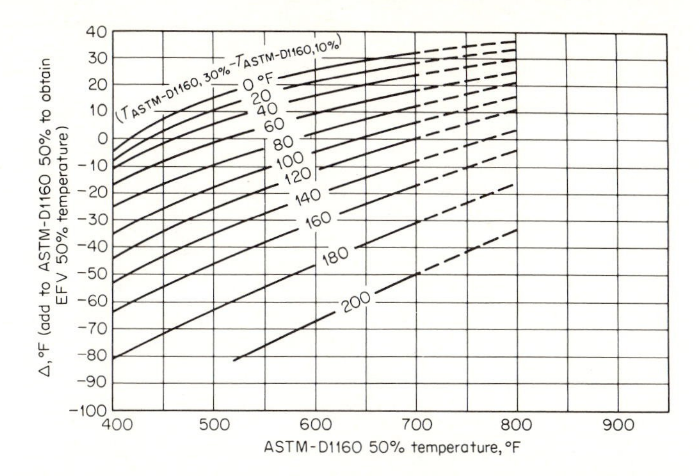

FIGURE 3.23 A 10 mm Hg ASTM-D1160 temperature difference versus a 10 mm Hg EFV temperature difference. [*From Edmister, Applied Hydrocarbon Thermodynamics, Gulf Publishing Company, Houston, Texas (1961). Reproduced by permission.*]

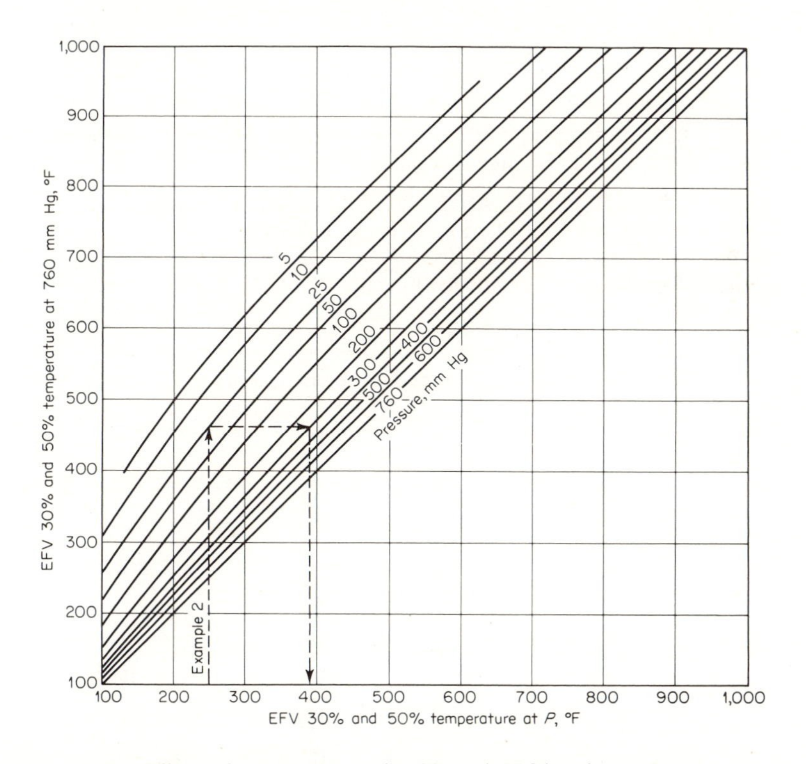

FIGURE 3.24 Effect of pressure on the 30 and 50% point temperatures on the vacuum EFV. [*From Edmister, Applied Hydrocarbon Thermodynamics, Gulf Publishing Company, Houston, Texas (1961). Reproduced by permission.*]

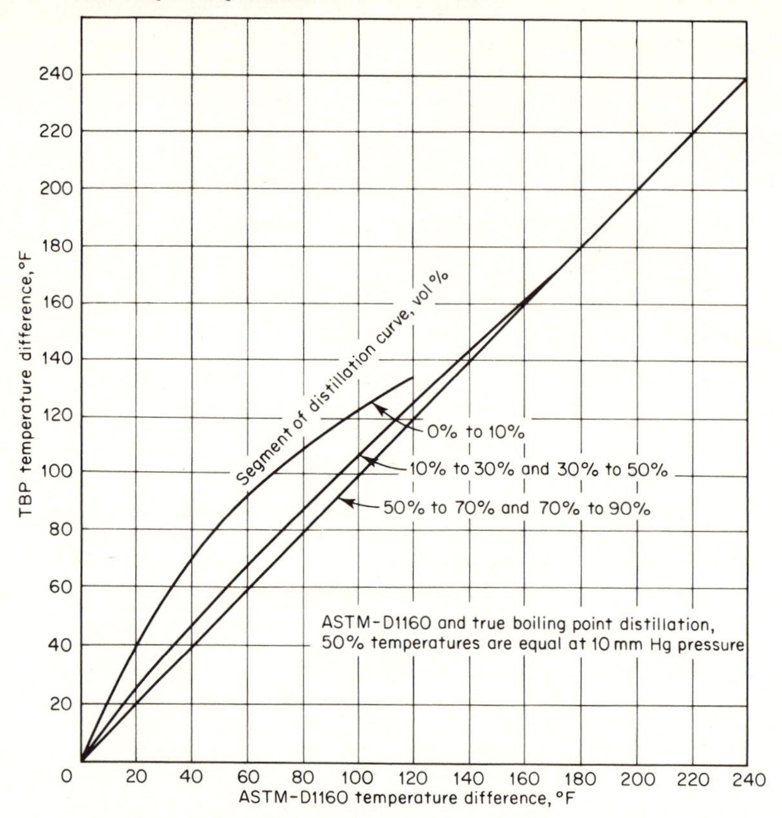

FIGURE 3.25 A 10 mm Hg ASTM-D1160 distillation versus a 10 mm Hg true-boiling-point distillation. [*From Edmister, Applied Hydrocarbon Thermodynamics, Gulf Publishing Company, Houston, Texas (1961). Reproduced by permission.*]

Example 3.1

A petroleum stock (API = 61.0) to be processed for jet fuel has the following ASTM-D158 760 mm Hg distillation range:

%	°F
0	95
5	125
10	150
20	210
30	255
40	290
50	330
60	370
70	410
80	455
90	510
End pt. 98	580

Determine the atmospheric EFV curve by using the Edmister-Pollock method (Figs. 3.17, 3.18).

Solution

The tabulation of Edmister and Okamoto [7] is followed in solving for the atmospheric EFV curve.

Vol % distilled	ASTM			EFV	
	t, °F	Interval, %	Δt	Δt	t, °F
0	95	—	—	—	150
—	—	0–10	55	25	—
10	150	—	—	—	175
—	—	10–30	105	75	—
30	255	—	—	—	250
—	—	30–50	75	47	—
50	330	—	—	—	297
—	—	50–70	80	48	—
70	410	—	—	—	345
—	—	70–90	100	62	—
90	510	—	—	—	407
—	—	10–100	87	42	—
100	597	—	—	—	449

The ASTM 30 − 10% temperature difference is 105°F.

The EFV 50% temperature is read from Fig. 3.17. Temperature differences (functions of percent distilled) are read from Fig. 3.18.

Example 3.2

For the stock in Example 3.1 determine the EFV curve at 100 psia by the Edmister-Pollock phase-diagram method, Figs. 3.11 through 3.14.

Solution

The atmospheric EFV curve has been determined in Example 3.1. With the focal point established, the phase diagram can be drawn and the EFV points determined for 100 psia.

$$\text{VABP} = \frac{150 + 255 + 330 + 410 + 510}{5} = \frac{1655}{5} = 331\text{°F}$$

API = 61, 10–90% slope = 4.5° per percent.

From Fig. 3.12,	$P_c = 475$ psia	
From Fig. 3.13,	$P_F - P_c = 385$	$P_F = 860$ psia
From Fig. 3.11,	$T_c - \text{VABP} = 310$	$T_c = 641$
From Fig. 3.14,	$T_F - T_c = 147$	$T_F = 788$

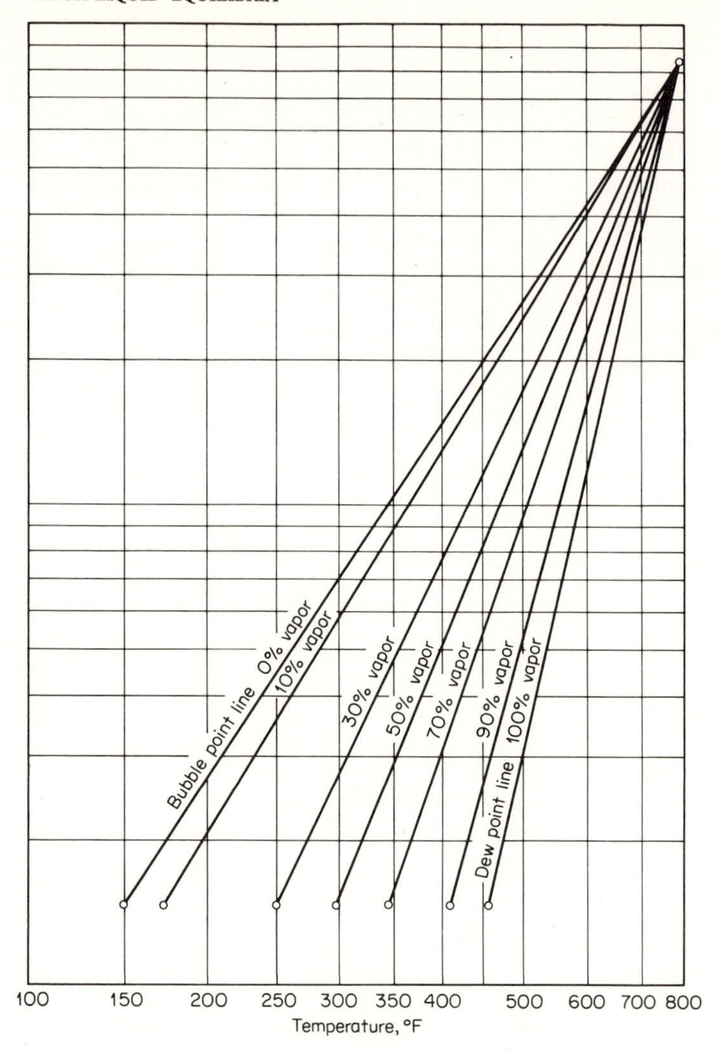

EXAMPLE 3.2 Phase diagram for a jet fuel feed stock.

The phase diagram is plotted by connecting the focal point and the atmospheric EFV data points.

Example 3.3

(*a*) For the stock whose atmospheric ASTM-D158 distillation is given in Example 3.1, determine the EFV curve at 400 mm Hg by the Edmister-Okamoto method (Figs. 3.17, 3.18, and 3.24).

Solution

From Fig. 3.17, $t_{\mathrm{ASTM}\ 30\%} - t_{\mathrm{ASTM}\ 10\%} = 105°$; $t_{\mathrm{EFV}\ 50\%} = 330 - 33 = 297°\mathrm{F}$.

From Fig. 3.18, $P = 760$ mm Hg					From Fig. 3.24, $P = 400$ mm Hg
%	Interval, %	ASTM, Δt	EFV, Δt	t, °F	t, °F
0	—	—	—	150	108
	0–10	55	25	—	
10	—	—	—	175	133
	10–30	105	75	—	
30	—	—	—	250	208
	30–50	75	47	—	
50	—	—	—	297	255
	50–70	80	48	—	
70	—	—	—	345	303
	70–90	100	62	—	
90	—	—	—	407	365
100	90–100	87	42	449	407

Edmister's Method

When the atmospheric-pressure TBP analytical curve is available, the EFV atmospheric distillation curve is obtained as follows:

1. Read from Figs. 3.15 and 3.16 the EFV 50% temperature corresponding to the TBP 50% temperature, and then determine the EFV temperature differences for the 30–50%, 50–70% slopes, etc., from the corresponding TBP temperature differences. The actual EFV temperatures are then calculated from these differences.

2. If atmospheric ASTM data are available, the same procedure is followed as outlined in step 1 except Figs. 3.17 and 3.18 are utilized.

3. When 10 mm Hg TBP data are available, the EFV 10 mm Hg distillation curve is evaluated by using Figs. 3.20 and 3.21. The 50% EFV temperature at 10 mm Hg pressure is determined from the TBP 10 mm Hg distillation temperature by using the correct temperature difference between the 30% TBP and 10% TBP (at 10 mm Hg). The differences in EFV temperatures from the various percent distilled points are determined from Figs. 3.21 and the actual 10 mm Hg pressure EFV temperatures evaluated from the differences.

4. With the ASTM-D1160 10 mm Hg data a similar procedureis followed except Figs. 3.22 and 3.23 are used to evaluate the 10 mm Hg EFV distillation curve.

5. To obtain the EFV distillation curve at pressures other than 10 mm Hg, Fig. 3.24 is utilized. When the 50% EFV temperature at, e.g., 10 mm Hg is available, the 50% temperature for the EFV at the desired pressure is determined by locating the temperature on the abscissa, then rising vertically until the 10 mm Hg pressure line is reached, and following this intersection at constant value of temperature (ordinate) until intersection of the desired pressure line is reached; then read the new EFV 50% temperature on the abscissa. Construct the rest of the curve by assuming that the difference in 50% temperatures is the same as that for all other points on the curves, i.e., they are parallel. If the 30% EFV temperature is available and not the 50% temperature, follow the same procedure, using the 30% temperatures in the same manner as the 50% temperatures were used in the foregoing.

Van Winkle Method

Van Winkle [28] developed correlations relating the ASTM-D158 (atmospheric pressure) and ASTM-D1160 (10 mm Hg pressure) distillations with

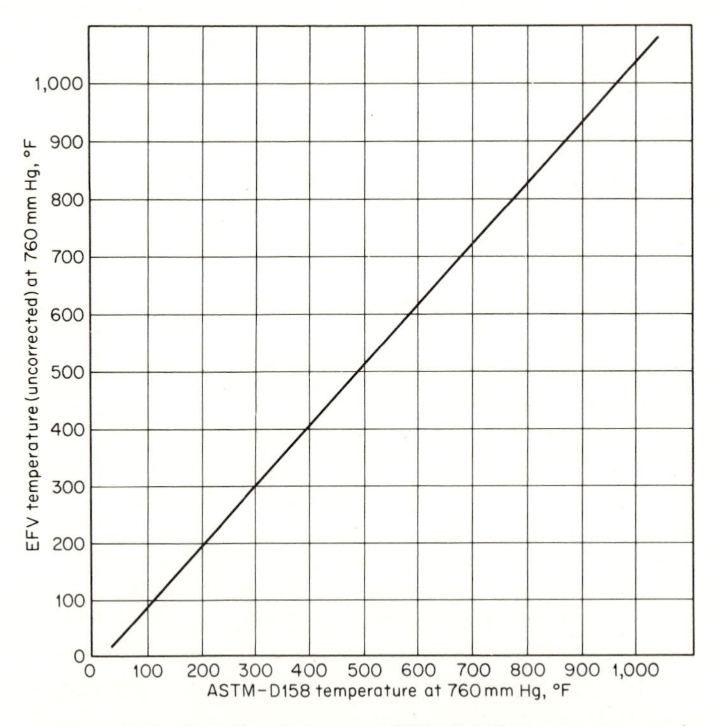

FIGURE 3.26 Relation between ASTM-D158 temperature and EFV uncorrected temperatures at 760 mm Hg. [*From Van Winkle, Hydrocarbon Process. Petrol. Refiner* (*April, 1964*), *p. 139. Copyright Gulf Publishing Company, Houston, Texas* (*1964*).]

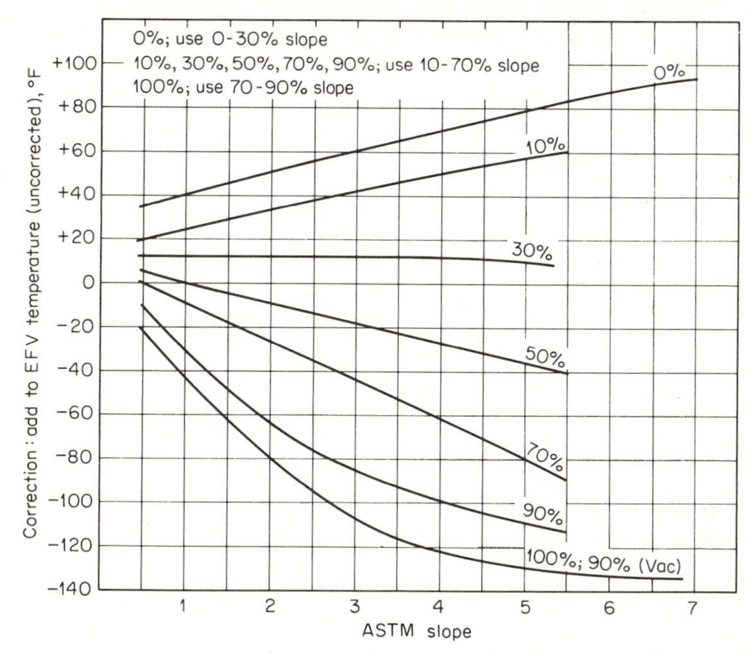

FIGURE 3.27 Correction of EFV temperatures for percent distilled. [*From Van Winkle, Hydrocarbon Process. Petrol. Refiner* (*April, 1964*), *p. 139. Copyright Gulf Publishing Company, Houston, Texas* (*1964*)].

EFV and TBP atmospheric and vacuum distillations. These relationships are somewhat less complicated to use and provide essentially the same accuracy as the more detailed procedures of Edmister.

The "uncorrected" EFV temperatures at 760 mm Hg, determined from the ASTM-D158 (760 mm Hg) distillation data by means of Fig. 3.26, are corrected by using the correct distillation-curve slope and percent distilled parameters of Fig. 3.27. ASTM-D158 (760 mm Hg) data can be converted to subatmospheric EFV data by means of Fig. 3.28. Similarly, ASTM-D1160 (10 mm Hg) data can be converted to EFV data at subatmospheric pressures by means of Fig. 3.29 to get the ASTM-D158 curve; then Figs. 3.26 through 3.28 are used as previously described. The atmospheric 50% TBP temperature is related to the uncorrected atmospheric EFV 50% temperature (and at subatmospheric pressures) through Fig. 3.30. Figure 3.31 is a correction chart to correct the uncorrected EFV 50% temperature to the temperatures corresponding to the various percent distilled points on the basis of atmospheric TBP slope. If atmospheric TBP data are available, the uncorrected EFV 50% temperature at the pressure desired

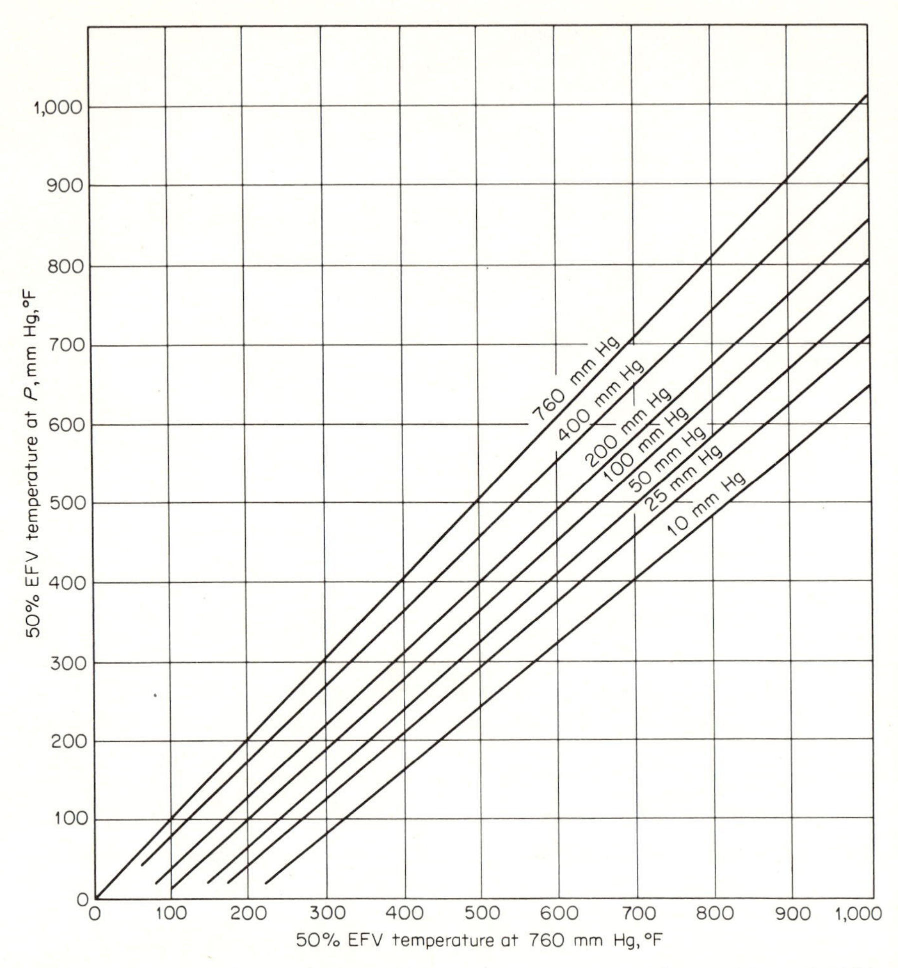

FIGURE 3.28 Relation between 50% EFV temperature at 760 mm Hg and 50% EFV temperature at other pressures. [*From Van Winkle, Hydrocarbon Process. Petrol. Refiner* (*April, 1964*), *p. 139. Copyright Gulf Publishing Company, Houston, Texas* (*1964*).]

can be read from Fig. 3.30 corresponding to the atmospheric 50% TBP temperature. Figure 3.31 can then be used to determine the temperatures for the desired percents distilled by the indicated corrections to the 50% temperature.

If the vacuum TBP data are available, they can be corrected to the 760 mm Hg TBP data by means of a vapor pressure chart. Then the 760 mm Hg TBP data thus obtained are corrected to the EFV data at the desired pressure by the foregoing method.

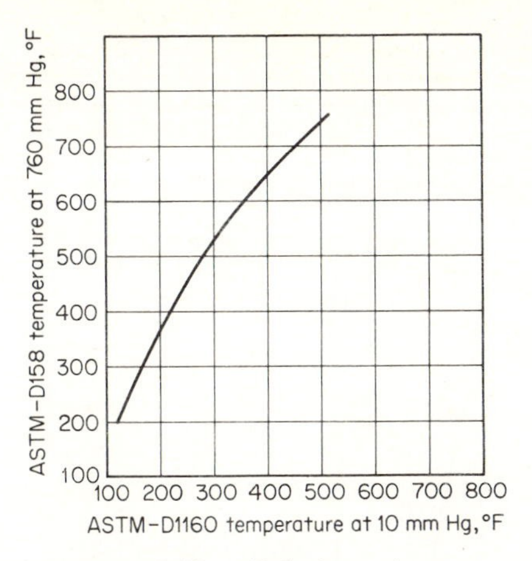

FIGURE 3.29 Relation between ASTM-D1160 temperature at 10 mm Hg and ASTM-D158 temperatures at 760 mm Hg. [*From Van Winkle, Hydrocarbon Process. Petrol. Refiner (April, 1964), p. 139. Copyright Gulf Publishing Company, Houston, Texas (1964).*]

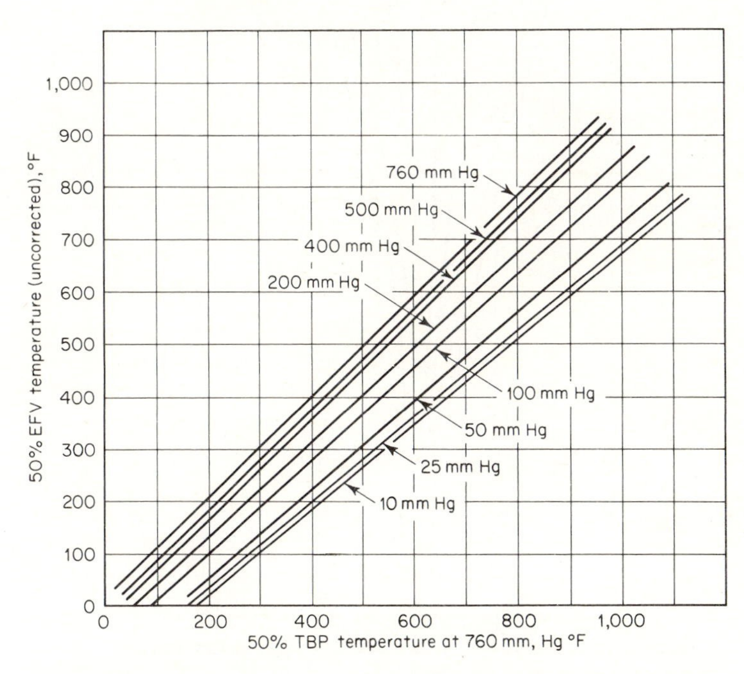

FIGURE 3.30 Relation between 50% TBP temperature to 50% EFV at 760 mm Hg. [*From Van Winkle, Hydrocarbon Process. Petrol. Refiner (April, 1964), p. 139. Copyright Gulf Publishing Company, Houston, Texas (1964).*]

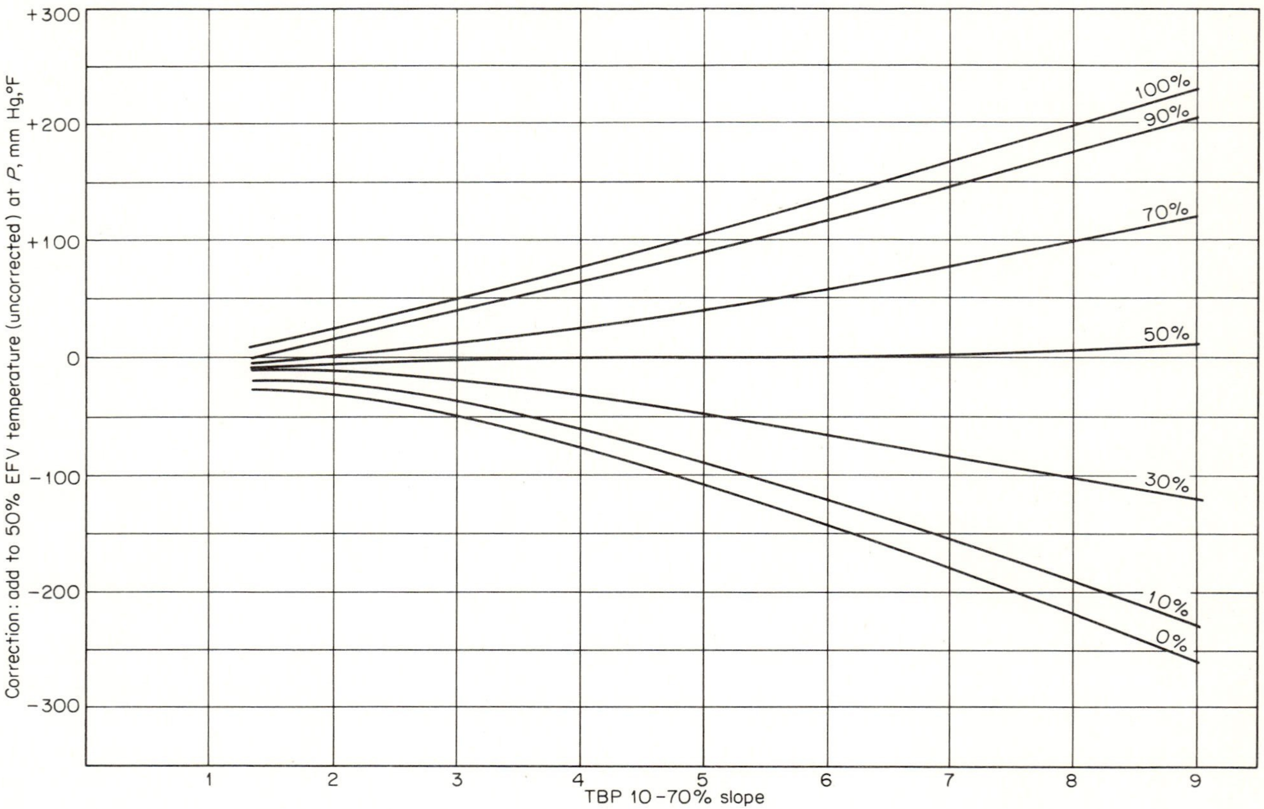

FIGURE 3.31 Correlation between EFV 50% temperature, TBP slope (10–70%) and percent distilled for the EFV curve. [*From Van Winkle, Hydrocarbon Process. Petrol. Refiner* (April, 1964), p. 139. Copyright Gulf Publishing Company, Houston

Example 3.4

It is desired to obtain the atmospheric EFV curve for a petroleum stock having the ASTM-D158 distillation characteristics shown in Table 3.1.

Solution

From Fig. 3.26 read the atmospheric EFV uncorrected temperatures and record in Table 3.1. Correct with the use of Fig. 3.27.

TABLE 3.1

% Distilled	ASTM-D158, t, °F	EFV uncorrected, t, °F	Correction, t, °F	EFV corrected, t, °F
Initial	225	218	+97	315
10	322	322	+58	370
30	445	450	+10	460
50	542	552	−37	515
70	635	650	−83	567
90	740	762	−110	652
EP	815	842	−130	712

The ASTM-D158 slopes are as follows:

$$0\text{–}30\% \text{ slope} = \frac{445 - 225}{30} = 7.33°/\%$$

$$10\text{–}70\% \text{ slope} = \frac{635 - 322}{60} = 5.21°/\%$$

$$70\text{–}90\% \text{ slope} = \frac{740 - 635}{20} = 5.25°/\%$$

Example 3.5

For the material in Example 3.4 determine the EFV at 100 mm Hg pressure.

Solution

Use Fig. 3.28 to determine the EFV 50% temperature at 100 mm Hg from the EFV 50% temperature at 760 mm Hg. Thus,

$$t_{\text{EFV } 50\%,\ 100 \text{ mm Hg}} = 377°\text{F}$$

The difference,

$$t_{50\%,\ 760 \text{ mm Hg}} - t_{50\%,\ 100 \text{ mm Hg}} = 515 - 377 = 138°\text{F}$$

Assuming a constant-temperature difference for the whole curve, the following results:

% Distilled	EFV, 100 mm Hg, t, °F
0	177
10	232
30	322
50	377
70	429
90	514
End pt.	574

Example 3.6

The following data result from an ASTM-D1160 10 mm Hg distillation. Determine the EFV at 100 mm Hg.

% Distilled	t, °F
0	270
10	337
30	393
50	434
70	483
90	542

From Fig. 3.29 the equivalent ASTM-D158 temperatures are read and recorded on Table 3.2.

TABLE 3.2

% Distilled	ASTM-D1160, 10 mm Hg, t, °F	ASTM-D158, 760 mm Hg, t, °F	EFV, 760 mm Hg, uncorrected, t, °F	Correction, t, °F	EFV, 760 mm Hg, corrected, °F	EFV, 100 mm Hg, °F
0	270	484	491	+81	572	410
10	337	552	591	+38	629	467
30	393	641	659	+12	671	509
50	434	686	708	−14	694	532
70	483	733	758	−38	720	558
90	542	786	812	−78	734	563

Figures 3.26 to 3.28 are then utilized as in Examples 3.1 and 3.2.

The ASTM-D158 slopes are:

$$0\text{–}30\% \text{ slope} = \frac{641 - 484}{30} = 5.23°/\%$$

$$10\text{–}70\% \text{ slope} = \frac{733 - 552}{60} = 2.60°/\%$$

$$70\text{–}90\% \text{ slope} = \frac{786 - 733}{20} = 2.65°/\%$$

Example 3.7

Determine the EFV curves at 760 and 100 mm Hg pressure from the atmospheric TBP data shown in Table 3.3.

TABLE 3.3

% Distilled	TBP, 760 mm Hg, t, °F	EFV, 760 mm Hg, t, °F			EFV, 100 mm Hg, t, °F		
		Uncorr.	Δ	Corr.	Uncorr.	Δ	Corr.
Initial	105	—	−212	428		−212	286
10	290	—	−184	456		−184	314
30	510	—	−98	542		−98	400
50	650	640	+5	645	498	+5	503
70	760	—	+93	733		+93	591
90	830	—	+168	808		+168	666
End pt.	950	—	+190	830		+190	688

$$\text{TBP slope} = \frac{760 - 290}{60} = 7.8°/\%$$

Interrelationships between ASTM and TBP distillation curve data are necessary for transposition from one type of distillation curve to the other if complete experimental data are not available. Correlations relating the 50% temperatures and slopes have been reported for atmospheric pressure [18] and subatmospheric pressure [27].

3.7. CHARACTERISTICS OF EQUILIBRIUM VAPOR AND LIQUID

The vaporization characteristics of the vapor and liquid formed when a complex material such as petroleum oil is flashed under equilibrium conditions may be estimated in a number of ways. Essentially the various methods involve the arbitrary breakup of the material into cuts representing pseudo-components. This is accomplished by dividing the volume percent or mole percent temperature curve arbitrarily into a number of sections or cuts by drawing constant-temperature lines intersecting the distillation curve such that the areas above the lines and below each of the lines are equal. (This is a graphical differentiation procedure using relatively large finite-differential lengths of line.) This is shown on Fig. 3.34. Each pseudocomponent or cut represents a definite volume or mole percent interval which has a boiling point at the pressure of the distillation curve at the temperature at which the cut was designated, and its properties are characterized as average properties of the cut or fraction.

Obryadchakoff [17] analyzed a large number of vapor and liquid samples resulting from flashing petroleum oils at different temperatures and pressures by using both Engler and TBP distillations to characterize the equilibrium materials. When a plot was constructed for each oil studied wherein the volume percent distilled in the vapor sample was plotted against the volume percent distilled in the liquid sample at various temperatures, a curve representing a vapor-liquid equilibrium curve of the characteristic form (xy) resulted. Such xy curves derived from both TBP and Engler analytical distillations were found to be essentially independent of temperature.

Katz and Brown [12] found that a family of equilibrium curves, with parameters of the 10–70% slope of the original material from which the vapor and liquid samples resulted on flashing, could be drawn to represent the equilibrium relationships both on the TBP and Engler basis. Such curves are shown in Figs. 3.32 and 3.33. In addition, the authors incorporated a

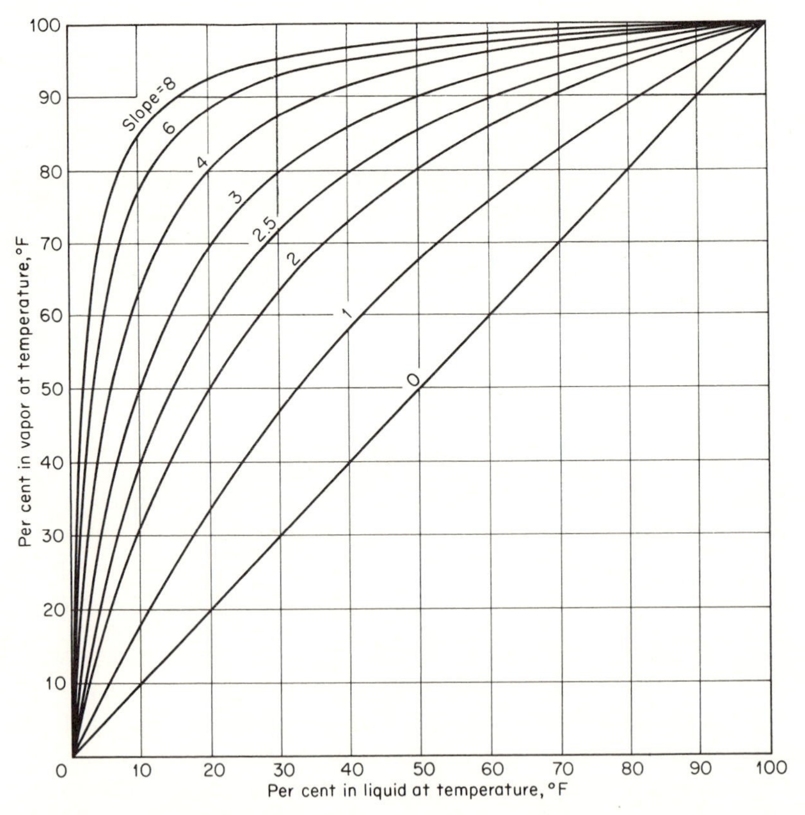

FIGURE 3.32 Equilibrium TBP distillation curves. [*From Katz and Brown, Ind. Eng. Chem.*, **25**:*1373* (*1933*).]

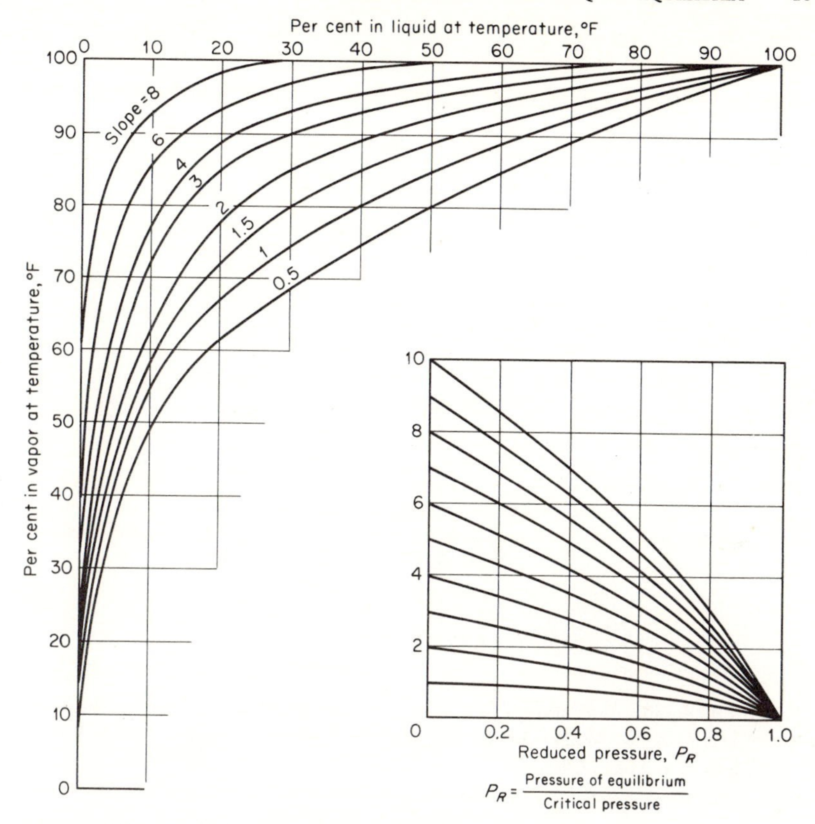

FIGURE 3.33 Equilibrium ASTM distillation curves. [*From Katz and Brown, Ind. Eng. Chem.*, **25**:*1373* (*1933*).]

pressure correction for the slope parameter. The equilibrium curves represent a plot of volume percent of material boiling below t in the liquid versus the volume percent of the material boiling below t in the vapor. Thus, the equilibrium relationship cannot be considered as a true equilibrium relationship for definable pseudocomponents (other than by temperature). However, this type of equilibrium curve has proved useful in estimating the distillation characteristics of equilibrium vapors and liquids.

Example 3.8

Determine the TBP curves of the vapor and liquid formed on flashing 67% of the feed.

Given: TBP slope = (725 − 200)/60 = 8.75°/%; percent vapor on flashing = 67.0.

Summary:

Material	Boiling below t, °F, TBP	TBP curves		
		x_F	y	x
A	200	10	0.125	0.005
B	380	30	0.44	0.015
C	550	50	0.73	0.04
D	725	70	0.95	0.20
E	900	90	0.99	0.73

Sample calculation:

$$Fx_F = Vy + Lx \qquad y = \frac{F}{V}x_F - \frac{L}{V}x$$

where

$$F = 1.0 \qquad V = 0.67 \qquad L = 0.33$$

$$L/V = 0.5 \qquad F/V = 1.5$$

For $x_F = 0.5$,

$$y = 1.5(0.5) - 0.5x \qquad y = \underset{\text{intercept}}{0.75} - \underset{\text{slope}}{0.5x}$$

On the TBP equilibrium curves, Fig. 3.32, draw a line of slope $= -0.5$ from the intercept on the y axis of 0.75. Where this line crosses the equilibrium curve of slope 8.75, read $y = 0.73$ and $x = 0.04$. *Check:*

$$\begin{aligned} 0.5 &= (0.67)(0.73) + (0.33)(0.04) \\ &= 0.487 + 0.013 \\ 0.5 &= 0.5 \qquad \textit{Check} \end{aligned}$$

Vaswani [29] published a number of charts which relate the TBP slope and the percent on the TBP distillation curves of the products at temperatures corresponding to the 10, 20, . . . , 90% distilled of the charge material. These were based on experimental data reported in the literature and represent a slightly different method of plotting the equilibrium curves.

Edmister [5], following the method of Lewis and Wilde [13], proposed a method of calculation of the TBP distillation curves for the vapor and liquid resulting from EFV of a feed material on which TBP data are available. In this method the TBP curve of the feed is broken up into a number of pseudocomponents, as shown in Fig. 3.34, to which boiling temperatures can be assigned and whose gravities, molecular weights, etc., can be estimated. The estimated properties can be used to determine the mole fractions of the pseudocuts in the original feed. K values can be determined for cuts at the temperature and pressure of the equilibrium vaporization, and multicomponent flash calculation methods as outlined in Chap. 4 can be used to

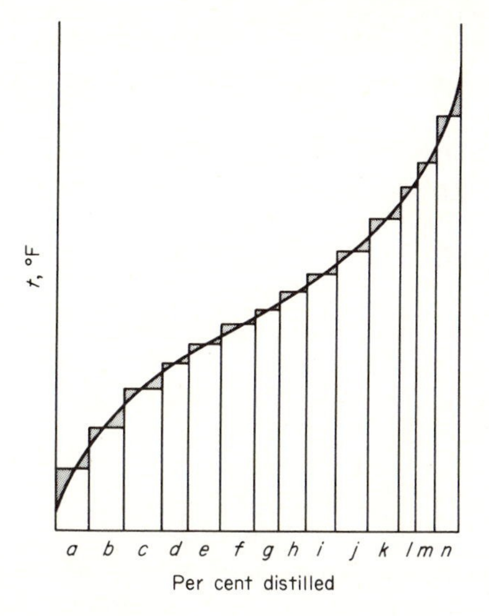

FIGURE 3.34 Pseudocomponents from TBP.

establish the composition of the vapor and liquid. From the estimated properties, the volume percent distilled temperature curves can be constructed for both the vapor and liquid.

K values for petroleum oils and fractions have been presented by Hadden and Grayson [10] (see Chap. 2), Poettman and Mayland [24] (reproduced in the NGPA Engineering Data Book), and others.

Edmister and Bowman [4, 6] developed a method for characterizing the equilibrium vapor and liquid TBP curves from the feed TBP. This method assumes that the TBP curves can be represented by straight lines on a plot for log α versus fraction distilled, and that the relative volatilities of the pseudocomponents are independent of composition of the mixture. This would be true for strictly ideal solutions.

Ten Eyck and Othmer [26] presented charts to transpose EFV data from one pressure to another. In addition, the equilibrium flash curve can be constructed by their method from a knowledge of temperature, pressure, and one value of percent vaporized.

Nomenclature

API = American Petroleum Institute
ASTM = American Society for Testing Materials
EFV = equilibrium flash vaporization
K = equilibrium vaporization ratio

P = pressure, psia
P_c = critical pressure, psia
P_F = focal pressure, psia
t = temperature, °F
t_c = critical temperature, °F
t_F = focal temperature, °F
TBP = true-boiling-point temperature, °F
T_c = critical temperature, °R
T_F = focal temperature, °R
VABP = volumetric average boiling point, °F
x = mole fraction in liquid
y = mole fraction in vapor
Δ = temperature difference, °F

REFERENCES

1. American Society for Testing Materials: Methods of Test for Petroleum Products and Lubricants, *Am. Soc. Testing Materials. Comm.* D-2, Philadelphia, 1958.
2. Chu, J. C., and E. J. Staffel: *J. Inst. Petrol.*, **41**:375 (1955).
3. Edmister, W. C.: "Applied Hydrocarbon Thermodynamics," Gulf Publishing, Houston, 1961.
4. Edmister, W. C.: *Ind. Eng. Chem.*, **47**:1685 (1955).
5. Edmister, W. C.: *Petrol. Refiner* (December, 1949), p. 140.
6. Edmister, W. C., and J. R. Bowman: *Chem. Eng. Progr. Symp. Ser.*, **48**:3:46 (1952).
7. Edmister, W. C., and K. K. Okamoto: *Petrol. Refiner* (August, 1959), p. 38; (September, 1959), p. 271.
8. Edmister, W. C., and D. H. Pollock: *Chem. Eng. Progr.*, **44**:905 (1948).
9. Filak, G. A., H. L. Sandlin, and C. J. Stockholm: *Petrol. Refiner* (April, 1955), p. 153.
10. Hadden, S. T., and H. G. Grayson: *Petrol. Refiner* (September, 1961), p. 207.
11. Jones, C. A., E. M. Schoenborn, and A. P. Colburn: *Ind. Eng. Chem.*, **35**:666 (1943).
12. Katz, D. L., and G. G. Brown: *Ind. Eng. Chem.*, **25**:1373 (1933).
13. Lewis, W. K., and D. L. Wilde: *Trans. A.I.Ch.E.*, **21**:99 (1928).
14. Nelson, W. L.: "Petroleum Refinery Engineering," 4th ed., McGraw-Hill, New York, 1958.
15. Nelson, W. L., and R. J. Harvey: *Oil Gas J.* (June 17, 1948), p. 77.
16. Nelson, W. L., and Mott Souders: *Petrol. Eng.*, **3**(1):131 (1931).
17. Obryadchakoff, S. N.: *Ind. Eng. Chem.*, **24**:1155 (1932).
18. Oehler, H. A., and M. Van Winkle: *Petrol. Eng.* (January, 1955), p. 205.
19. Okamoto, K. K., and M. Van Winkle: *Petrol. Refiner* (August, 1949), p. 113; (January 1950), p. 91.
20. Okamoto, K. K., and M. Van Winkle, *Ind. Eng. Chem.*, **45**:429 (1953).
21. Othmer, D. F.: *Ind. Eng. Chem.*, **20**:743 (1928).
22. Packie, J. W.: *Trans. A.I.Ch.E.*, **37**:51 (1941).
23. Piromoov, R. S., and G. A. Beiswenger: *Proc. Am. Petrol. Inst.*, **10**(2):52 (1929).

24. Poettman, F. H., and B. J. Mayland: *Petrol. Refiner* (July, 1949), p. 101.
25. Ragatz, E. G., E. R. McCartney, and R. E. Haylett: *Ind. Eng. Chem.*, **25**:957 (1935)
26. Ten Eyck, E. H., and D. F. Othmer: *Petrol. Refiner* (September, 1953), p. 229.
27. Van Winkle, M.: *Petrol. Process.* (November, 1954), p. 1738.
28. Van Winkle, M.: *Hydrocarbon Process. Petrol. Refiner* (April, 1964), p. 139.
29. Vaswani, N. R.: *Oil Gas J.* (October 3, 1955), p. 49; (October 17, 1955), p. 141; (October 31, 1955), p. 132; (December 5, 1955), p. 143; (December 12, 1955), p. 117.
30. Watson, K. M., and E. F. Nelson: *Ind. Eng. Chem.*, **25**:880 (1933).

chapter 4

Equilibrium and simple distillation

The process of distillation is used to separate compounds or mixtures of compounds by utilizing the differences in their volatilities or their vapor pressures. The separation may be essentially complete or it may be incomplete, wherein the relative concentration of the components being separated is merely increased in the distillate or in the residual liquid with respect to the relative concentrations in the original mixture. Distillation processes may be classified in accordance with the number of components in the original mixture as *binary*, when there are two components; as *multicomponent*, when there are more than two definable components; and as *complex*, when the number and individual components are unidentifiable as compounds. They may be classed according to the type of separation as *equilibrium* or *equilibrium flash; differential* or *fractionating*. Furthermore, they may be designated as *batch* or *continuous* and as *pressure*, *vacuum*, or *steam* distillations. Almost any combination of these rather arbitrarily designated classifications can be utilized to obtain separations of materials. Some are more suitable than others for a particular situation.

4.1. EQUILIBRIUM DISTILLATION

In equilibrium distillation it is assumed that all components existing in the liquid phase exist in the vapor phase and equilibrium is established. Also the temperature and pressure must be the same in both phases while they are in contact. Although true equilibrium can never actually be reached, because

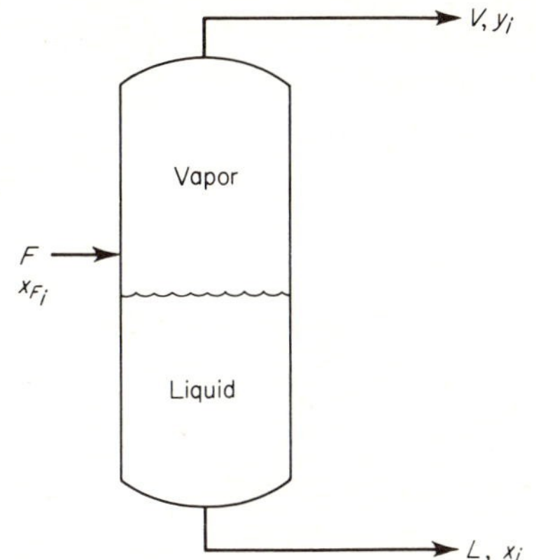

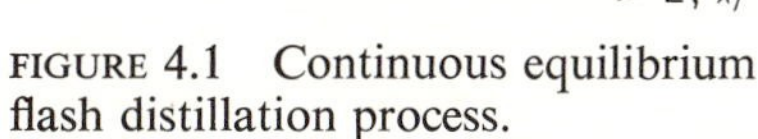
FIGURE 4.1 Continuous equilibrium flash distillation process.

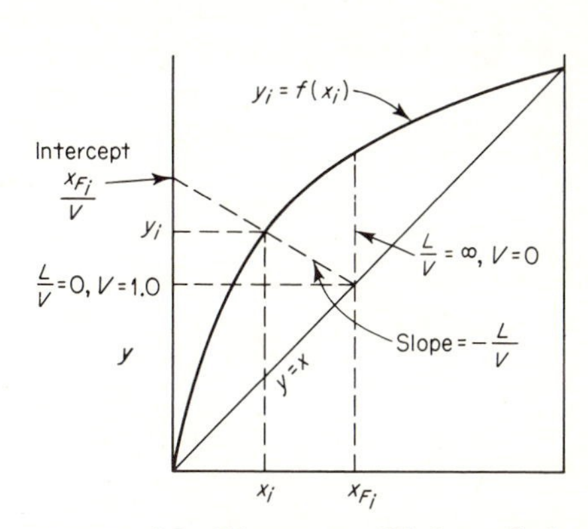

FIGURE 4.2 Binary equilibrium flash diagram.

it would require either infinite contact time or infinite area of contact between the phases, it can be closely approximated by proper adjustment of conditions in suitable equipment.

Figure 4.1 illustrates schematically a continuous EFV operation. The equations describing the process quantitatively are:

Over-all material balance,

$$F = V + L \tag{4.1}$$

Component material balance,

$$Fx_{F_i} = Vy_i + Lx_i \tag{4.2}$$

For 1 mole of F,

$$x_{F_i} = Vy_i + Lx_i \tag{4.3}$$

$$y_i = \frac{x_{F_i}}{V} - \frac{Lx_i}{V} \tag{4.4}$$

Equation (4.4) is an equation of a straight line on a y_i versus x_i plot, where the slope of the line is $-L/V$ and the intercept is x_{F_i}/V (see Fig. 4.2).

The limits of operation of an EFV with respect to fraction vaporized are from $V = 0$ to $V = 1.0$. A diagram of the type shown in Fig. 4.2 is suitable for any component in a mixture where it is possible to evaluate the equilibrium curve for the component in question.

For a given temperature and pressure, the compositions of both vapor and liquid can be obtained from Eqs. (4.5) and (4.6) provided the y/x values can be evaluated or estimated (or the variables upon which the y/x values depend can be evaluated). Where ideal systems may be assumed, the values of the activity coefficient γ and the fugacity coefficient ν are unity, and the y/x or K values become equal to the ratio of the vapor pressure of the component to the total pressure. Thus the equations reduce to

$$x_i = \frac{1}{V}\frac{x_{F_i}}{K_i + L/V} = \frac{1}{V}\frac{x_{F_i}}{P_i/P_t + L/V} \tag{4.5}$$

$$y_i = \frac{1}{V}\frac{K_i x_{F_i}}{K_i + L/V} = \frac{1}{V}\frac{(P_i/P_t)x_{F_i}}{P_i/P_t + L/V} \tag{4.6}$$

The phase compositions can be calculated by using the vapor pressures of the components when the mole fraction vaporized, the temperature, and the total pressure are known.

Binary System Flash Vaporization

In a binary system at equilibrium with two phases present, the phase rule indicates that only two variables need to be fixed to fix the system.

$$V = C + 2 - P = 2 \tag{4.7}$$

If the total pressure and temperature are known, the compositions of the vapor and the liquid as well as the fraction vaporized can be evaluated as follows:

$$y_i = f(x_i)$$

$$t = f(x_i)$$

Example 4.1

A ternary system composed of 20% ethanol, 50% 1-propanol, and 30% 2-propanol can be considered essentially an ideal system. (*a*) Determine the temperature necessary to vaporize 50% of the mixture at 5-atm pressure. (*b*) Determine the bubble-point temperature and dew-point temperature of the mixture at 5 atm.

Solution

(*a*) The composition of the liquid in equilibrium with the vapor is found by Eq. (4.5):

$$x_i = \frac{1}{V}\frac{x_{F_i}}{P_i/P_t + L/V}$$

The temperature at which $x_1 + x_2 + x_3 = 1.0$ is found by trial and error.

At $t = 135°C$:

Component	x_F	P_i, mm Hg	$\frac{P_t}{P_i} + 1$	$x = \frac{2x_F}{(P_i/P_t) + 1}$
1	0.20	5001	2.316	0.173
2	0.50	2760	1.726	0.579
3	0.30	4395	2.157	0.278
				1.030

At $t = 137°C$:

Component	x_F	P_i, mm Hg	$\frac{P_t}{P_i} + 1$	$x = \frac{2x_F}{(P_i/P_t) + 1}$
1	0.20	5287	2.391	0.167
2	0.50	2932	1.772	0.564
3	0.30	4651	2.224	0.270
				1.001

The temperature is found to be 137°C.

(*b*) *Bubble point:* The liquid composition is that of the feed. At the bubble point,

$$\sum_i y_i = \sum_i \frac{x_i P_i}{P_t} = 1.0$$

At $t = 135°C$:

Component	P_i, mm Hg	$y = x_F \frac{P_i}{P_t}$
1	5001	0.264
2	2760	0.363
3	4395	0.347
		0.974

At $t = 136°C$:

Component	P_i, mm Hg	$y = x_F \frac{P_i}{P_t}$
1	5143	0.271
2	2845	0.375
3	4522	0.357
		1.003

By interpolation the bubble-point temperature is found to be 135.9°C.

Dew point: The vapor composition is identical to that of the feed.

$$\Sigma x_i = 1.0 = \Sigma y_i \frac{P_t}{P_i}$$

At $t = 137.5°C$:

Component	P_i, mm Hg	$x_i = y_i \frac{P_t}{P_i}$
1	5361	0.142
2	2976	0.638
3	4717	0.242
		1.022

At $t = 138°C$:

Component	P_i, mm Hg	P_t/P_i	$x_i = y_i \frac{P_t}{P_i}$
1	5435	0.699	0.140
2	3021	1.258	0.629
3	4783	0.794	0.238
			1.007

At $t = 139°C$:

Component	P_i, mm Hg	P_t/P	$x_i = y_i \frac{P_t}{P_i}$
1	5586	0.680	0.136
2	3113	1.221	0.610
3	4919	0.773	0.232
			0.978

By interpolation the dew-point temperature is 138.2°C.

The x and y data are determined from the equilibrium curves, ty and tx, and the fraction vaporized, V, is calculated from Eq. (4.13)

$$y_iV = \frac{K_i x_{F_i}}{K_i + (1 - V)/V} \tag{4.8}$$

$$y_iVK_i + y_i(1 - V) = K_i x_{F_i} \tag{4.9}$$

$$y_iV\frac{y_i}{x_i} + y_i - y_iV = \frac{y_i}{x_i}x_{F_i} \tag{4.10}$$

$$V\left(\frac{y_i^2}{x_i} - y_i\right) = \frac{y_i}{x_i}x_{F_i} - y_i \tag{4.11}$$

$$V = \frac{(y_i/x_i)x_{F_i} - y_i}{y_i^2/x_i - y_i} = \frac{y_i x_{F_i} - y_i x_i}{y_i^2 - y_i x_i} \tag{4.12}$$

$$V = \frac{x_{F_i} - x_i}{y_i - x_i} \tag{4.13}$$

If the total pressure and quantity vaporized (or fraction vaporized) are known, the compositions can be calculated by Eqs. (4.14) and (4.15).

$$y_i = \frac{x_{F_i}}{V} - \frac{1 - V}{V}x_i \tag{4.14}$$

$$x_i = \frac{x_{F_i}/V - y_i}{(1 - V)/V} \tag{4.15}$$

and t can be determined from the tx data or from the K versus P_t and t relations.

Multicomponent EFV

The phase rule indicates that the number of variables to be fixed to fix equilibrium between two phases is equal numerically to the number of components. Thus, for a mixture containing eight components, eight variables must be fixed to fix the system. These variables may be composition variables, pressure and temperature. The compositions may be feed compositions $(C - 1)$, vapor compositions $(C - 1)$ or liquid compositions $(C - 1)$ or $C - 1$ independent compositions from either of the streams F, V, and L. The feed compositions are related to the vapor and liquid compositions for a fixed quantity of feed (e.g., 1 mole) by material balance equations and equilibrium composition relations.

Limiting Conditions

In designing an EFV process it is necessary that conditions be selected such that both vapor and liquid phases exist. The conditions for the two-phase region lie between the bubble-point temperature (pressure) and the dew-point temperature (pressure). The bubble-point temperature is the temperature at which the first bubble of vapor is formed on heating the liquid at constant pressure. The bubble-point pressure is the pressure at which the first bubble of vapor is formed on lowering the pressure on the liquid at constant temperature. The dew-point temperature is the temperature at which the first droplet of liquid is formed as the vapor mixture is cooled at constant pressure, and the dew-point pressure is that at which the first droplet of liquid is formed as the pressure is increased on the vapor at constant temperature.

Mathematically, the bubble point is defined by

$$\sum_1^n y_i = 1.0 = \sum_1^n K_i x_i = \sum_1^n \frac{\gamma_i P_i}{\nu_i P_t} x_i \tag{4.16}$$

and the dew point is defined by

$$\sum_1^n x_i = 1.0 = \sum_1^n \frac{y_i}{K_i} = \sum_1^n \frac{y_i}{\gamma_i P_i / \nu_i P_t} \tag{4.17}$$

The usual condition for a flash calculation is one in which the feed composition is known, and it is desired to determine the quantity of liquid and vapor formed on flashing as a function of temperature and pressure and the compositions of the vapor and liquid. In such cases the limits, bubble point and dew point, are first determined. A temperature and pressure are selected within the two phase conditions and the fraction vaporized is calculated from Eq. (4.20) or (4.21).

$$\sum_1^n x_i = 1.0 = \frac{1}{V}\sum_1^n \frac{x_{F_i}}{K_i + L/V} = \frac{1}{V}\sum_1^n \frac{x_{F_i}}{\gamma_i P_i/\nu_i P_t + L/V} \tag{4.18}$$

or

$$\sum_1^n y_i = 1.0 = \frac{1}{V}\sum_1^n \frac{K_i x_{F_i}}{K_i + L/V} = \frac{1}{V}\sum_1^n \frac{(\gamma_i P_i/\nu_i P_t) x_{F_i}}{\gamma_i P_i/\nu_i P_t + L/V} \tag{4.19}$$

$$V = \sum_1^n \frac{x_{F_i}}{K_i + L/V} = \sum_1^n \frac{x_{F_i}}{\gamma_i P_i/\nu_i P_t + L/V} \tag{4.20}$$

or

$$V = \sum_1^n \frac{K_i x_{F_i}}{K_i + L/V} = \sum_1^n \frac{(\gamma_i P_i/\nu_i P_t) x_{F_i}}{\gamma_i P_i/\nu_i P_t + L/V} \tag{4.21}$$

Example 4.2

A multicomponent hydrocarbon mixture is composed as follows: $CH_4 = 0.02$, $C_2H_6 = 0.08$, $C_3H_8 = 0.25$, $i\text{-}C_4H_{10} = 0.05$, $n\text{-}C_4H_{10} = 0.3$, $n\text{-}C_5H_{12} = 0.30$. The mixture is flashed under such a temperature condition at 100 psia that 50% of the $n\text{-}C_4H_{10}$ is retained in the liquid. Determine (*a*) temperature; (*b*) composition of vapor and liquid.

Solution

The bubble-point temperature and dew-point temperature are first calculated at 100 psia. K's for this example were obtained from Fig. 2.41.

	Bubble point			Dew point		
Component	x	K, 20°F	$y = Kx$	y	K, 157°F	$x = y/K$
C_1	0.02	27.7	0.554	0.02	33.5	—
C_2	0.08	3.08	0.246	0.08	8.8	0.009
C_3	0.25	0.59	0.148	0.25	3.25	0.077
$i\text{-}C_4$	0.05	0.192	0.010	0.05	1.6	0.031
$n\text{-}C_4$	0.30	0.125	0.038	0.30	1.2	0.250
$n\text{-}C_5$	0.30	0.025	0.008	0.30	0.46	0.652
			1.004			1.019

The bubble point $\cong$ 20°F; the dew point $\cong$ 157°F.

The solution is by trial and error. A temperature is chosen, K values are read, and Eq. (4.18) is solved at several assumed values of L/V until a solution is obtained, namely,

$$\sum_i x_i = 1.0$$

Then if $(xL/yV)_{n\text{-}C_4} = 1.0$, the assumed temperature was the correct one. If not, further temperatures are assumed and the procedure is repeated until the correct ratio of n-C_4 in vapor to n-C_4 in liquid is obtained.

Trial 1: (*a*) Assume $t = 120°F$; $L/V = 1.0$.

Component	K	$K + L/V$	$x_F(1/V)$	x	y	xL/yV
C_1	32.4	33.4	0.04	0.001		
C_2	7.2	8.2	0.16	0.020		
C_3	2.35	3.35	0.50	0.149		
$i\text{-}C_4$	1.07	2.07	0.10	0.048		
$n\text{-}C_4$	0.8	1.8	0.60	0.333	0.266	1.25
$n\text{-}C_5$	0.263	1.263	0.60	0.475		
				1.026		

(*b*) Assume $t = 120°F$; $L/V = 1.4$.

Component	K	$K + L/V$	$x_F(1/V)$	x	y	xL/yV
C_1	32.4	33.8	0.48	0.001		
C_2	7.3	8.6	0.192	0.022		
C_3	2.35	3.75	0.600	0.160		
i-C_4	1.07	2.47	0.120	0.049		
n-C_4	0.8	2.2	0.720	0.327	0.262	1.75
C_5	0.263	1.663	0.720	0.433		
				0.922		

Since the ratio is 1.0, a higher temperature is assumed.
Trial 2: Assume $t = 130°F$; $L/V = 0.80$.

Component	K	$K + L/V$	$x_F(1/V)$	x	y	xL/yV
C_1	32.5	33.3	0.036	0.001		
C_2	7.6	8.4	0.144	0.0171		
C_3	2.58	3.38	0.450	0.133		
i-C_4	1.2	2.0	0.090	0.045		
n-C_4	0.9	1.7	0.540	0.318	0.286	0.888
C_5	0.305	1.105	0.540	0.489		
				1.003		

Interpolation between trials 1 and 2 gives 128°F as the next iterant.
Trial 3: (*a*) Assume $t = 128°F$; $L/V = 0.9$.

Component	K	$K + L/V$	$x_F(1/V)$	x	y	xL/yV
C_1	32.5	33.4	0.038	0.001		
C_2	7.55	8.45	0.152	0.018		
C_3	2.53	3.43	0.475	0.138		
i-C_4	1.18	2.08	0.095	0.046		
n-C_4	0.89	1.79	0.570	0.318	0.283	1.01
C_5	0.298	1.198	0.570	0.476		
				0.997		

(*b*) Assume $t = 128°F$; $L/V = 0.885$.

Component	K	$K + L/V$	$x_F(1/V)$	Liquid comp.	Vapor comp.	xL/yV
				x	y	
C_1	32.5	33.385	0.0377	0.001	0.037	
C_2	7.55	8.435	0.151	0.018	0.135	
C_3	2.53	3.415	0.471	0.138	0.349	
i-C_4	1.18	2.065	0.094	0.046	0.054	
n-C_4	0.89	1.775	0.566	0.319	0.284	0.995
C_5	0.298	1.183	0.566	0.478	0.142	
				1.000	1.001	

Trial 3*b* is successful.

The equations are solved by assuming values of V (mole fraction vaporized based on 1 mole of feed) and calculating V by performing the summation indicated by the equation. Although this trial and error is straightforward, the convergence of V calculated and V assumed may be rather indefinite, particularly as the fraction vaporized approaches 1.0 or 0. In order to converge the values more definitely, Holland and Davisson [2], McReynolds [4], Wilson [8], Bachelor [1], Reilly [7], and others have developed mathematical treatment for this purpose. Poettman and Rice [5] have published nomographs for solution of the equations. If a large number of flash calculations are to be made, the equations can be programmed for solution by machine computation.

When the fraction vaporized, temperature, and pressure are known, the composition of the vapor and liquid can be calculated by

$$y_i = \frac{1}{V}\frac{K_i x_{F_i}}{K_i + L/V} = \frac{1}{V}\frac{(\gamma_i P_i/\nu_i P_t)x_{F_i}}{\gamma_i P_i/\nu_i P_t + L/V} \tag{4.22}$$

$$x_i = \frac{1}{V}\frac{x_{F_i}}{K_i + L/V} = \frac{1}{V}\frac{x_{F_i}}{\gamma_i P_i/\nu_i P_t + L/V} \tag{4.23}$$

If the fraction vaporized is known and either the temperature or total pressure, the compositions of the vapor and liquid are determined by trial and error from Eqs. (4.18) and (4.19). Several temperatures or pressures are assumed and several corresponding sets of K values or their equivalents are evaluated. The correct temperature or pressure is found, usually by interpolation, when the summation is equal to unity. Equations (4.18) through (4.23) are generally valid for any system, regardless of the number of components, under EFV conditions.

4.2. DIFFERENTIAL VAPORIZATION–CONDENSATION

Differential vaporization is a batch operation in which the mixture to be distilled is charged to a still pot and the mixture is heated to the bubble point. At the bubble point boiling starts and continues as the distillate is continuously removed as a vapor and condensed (usually) to a liquid product externally from the still. A gradual increase in boiling temperature of the liquid takes place as the lower boiling components are distilled off until the desired quantity of distillate is obtained. This type of distillation is used in laboratory and pilot-plant work to concentrate desirable material, either in the distillate or residue, where some loss of the material can be tolerated.

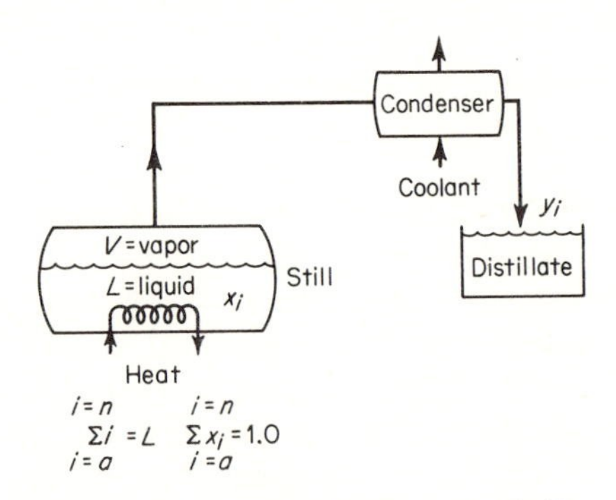

FIGURE 4.3 Batch differential distillation.

It is also used for analytical evaluation of boiling range and vaporization characteristics of mixtures as, for example, in the ASTM or Engler distillation. In differential distillation the vapor evolving at any instant from the boiling liquid mixture is assumed to be in equilibrium with it. Thus, the composition of the liquid changes continuously throughout the distillation process, and the composition of the differential element of the vapor also changes continuously, but it is assumed to be in equilibrium at any instant with the liquid composition remaining in the still. Figure 4.3 schematically represents a batch differential-distillation process to which the following equations are applicable for quantitative evaluation.

If L is the quantity, in moles of liquid in the still at any given time, and dL is the differential amount vaporized but still in contact with the remaining liquid, then by material balance for the component i,

$$Lx_i = (L - dL)(x_i - dx_i) + (y_i + dy_i)\,dL \tag{4.24}$$

Neglecting products of differentials,

$$(x_i - y_i)\,dL = L\,dx_i \tag{4.25}$$

Integrating over the change in quantity of liquid from the initial condition to the final condition and from the initial concentration to the final concentration of the liquid,

$$\int_{L_2}^{L_1} \frac{dL}{L} = \int_{x_{i2}}^{x_{i1}} \frac{dx_i}{y_i - x_i} \tag{4.26}$$

$$\ln \frac{L_1}{L_2} = \int_{x_{i2}}^{x_{i1}} \frac{dx_1}{y_i - x_i} \tag{4.27}$$

where L_1 = initial liquid, moles
L_2 = final (residual) liquid, moles
x_i = mole fraction of component i in liquid
y_i = mole fraction of component i in vapor
dL = differential quantity of liquid distilled in time, $d\theta$
dx_i = differential change in liquid composition in time, $d\theta$
dy_i = differential change in vapor composition in time, $d\theta$

This equation is known as the Rayleigh [6] equation.

For a binary system when experimental $x_i - y_i$ data are available, the integral can be integrated graphically by plotting $1/(y_i - x_i)$ between the limits of x_{i1} and x_{i2}. Where relative volatilities are constant, that is, α = a constant, the integral may be solved directly by

$$\ln \frac{L_1}{L_2} = \frac{1}{\alpha - 1} \left(\ln \frac{x_{i1}}{x_{i2}} + \alpha \ln \frac{1 - x_{i2}}{1 - x_{i1}} \right) \tag{4.28}$$

For nonideal systems where K or activity-coefficient data are available, Eq. (4.29) may be used.

$$y = \frac{\gamma x P}{\nu P_t} = Kx$$

$$\ln \frac{L_1}{L_2} = \int_{x_{i2}}^{x_{i1}} \frac{dx_i}{x_i(K_i - 1)} = \int_{x_{i2}}^{x_{i1}} \frac{dx_i}{x_i(\gamma_i P_i / \nu_i P_t - 1)} \tag{4.29}$$

Differential condensation as a process to increase the concentration of the light component in the vapor and the heavy components in the liquid upon gradually cooling a batch of vapor and continually removing the liquid has little commercial use. In actual condensing equipment in a continuous process, vapor may be partially condensed but the liquid remains in contact with the residual vapor. In such cases equilibrium flash conditions may be approximated, and those equations applicable to equilibrium flash distillation are used.

Example 4.3

A mixture of water and phenol is subjected to a differential batch distillation at 260 mm Hg. The original composition contains 80 mole % water.

(*a*) What amount of an original batch of 100 moles of mixture must be distilled to produce a residual concentration (in the still) of 20 mole % water?

(*b*) What is the concentration of the total distillate?

Phenol-Water at 260 mm Hg*

Wt % water in liquid	Wt % water in vapor
1.54	41.10
4.95	79.72
6.87	82.79
7.73	84.45
19.63	89.91
28.44	91.05
39.73	91.15
82.99	91.86
89.95	92.77
93.38	94.19
95.74	95.64

* From *Ind. Eng. Chem.*, **17**:199 (1925).

The calculations can be carried out on a weight fraction basis, and the results are later converted to moles. The amount of mixture remaining in the still pot is obtained by graphical integration:

$$\ln \frac{L_1}{L_2} = \int_{x_2}^{x_1} \frac{dx}{y - x}$$

where $x_1 = 0.4336$ wt % water (0.80 mole fraction)
$x_2 = 0.0408$ wt % water (0.20 mole fraction)

Integration of the function by Simpson's rule applied to values of the integrand read from the graph give

$$\ln \frac{L_1}{L_2} = 0.602 \qquad L_2 = \frac{L_1}{1.826}$$

Material Balance—Basis: 100 Moles of Charge

	Moles H_2O	lb H_2O	Moles PhOH	lb PhOH	Total lb	Wt. fr. H_2O
Charge	80	1441	20	1882	3323 (L_1)	0.4336
Resid.	4.12	74.2	18.55	1745.8	1820 (L_2)	0.0408
Dist.	75.88	1366.8	1.45	136.2	1503	
	80.00	1441.0	20.0	1882.0	3323	

(*a*) The moles distilled are:

H_2O	75.88
PhOH	1.45
Total	77.33

(*b*) The concentration of the total distillate is

$$x_{H_2O} = 0.981$$

Differential Vaporization—Multicomponent

In differential distillation of multicomponent mixtures the lighter components (lower boiling components) are concentrated in the vapor and the heavier materials are concentrated in the liquid. The process may be conducted at constant pressure, wherein the temperature is gradually increased as the distillation progresses until the desired distillate or residue is obtained; or the process may be a constant-temperature process wherein the pressure is gradually decreased while the liquid and vapor remain at constant temperature throughout the distillation. These processes at one time were widely used in the petroleum and natural gasoline industries for the "weathering" of natural gasoline and liquefied petroleum gas to obtain the desired vapor pressure. The tendency today is to fractionate the material to get better yields as well as better product quality.

Constant-pressure Differential Distillation

The equations used to describe the process are basically those for binary systems. The letters a, b, c, etc., designate the moles of a, b, c, etc., and da, db, dc, etc., designate the moles of a, b, c, etc., vaporized over a differential period of time. Thus, since $L\,dx + x\,dL = y\,dL$,

$$Lx_a = a \qquad Lx_b = b \qquad \text{etc.} \tag{4.30}$$

$$da = L\,dx_a + x_a\,dL \tag{4.31}$$

$$db = L\,dx_b + x_b\,dL \tag{4.32}$$

$$\cdots\cdots\cdots\cdots\cdots$$

$$\frac{da}{db} = \frac{y_a\,dL}{y_b\,dL} = \frac{y_a}{y_b} \tag{4.33}$$

In the case of nonideal vapors and liquids where

$$y = \frac{\gamma x P}{\nu P_t} \tag{4.34}$$

the ratio of vapor compositions of components a and b is expressed by

$$\frac{y_a}{y_b} = \frac{\gamma_a P_a x_a / \nu_a P_t}{\gamma_b P_b x_b / \nu_b P_t} = \frac{x_a}{x_b} \frac{\gamma_a P_a / \nu_a}{\gamma_b P_b / \nu_b} \tag{4.35}$$

Where K values are commonly available as in the case of hydrocarbon mixtures,

$$K_a = \frac{\gamma_a P_a}{\nu_a P_t} \tag{4.36}$$

and

$$\frac{y_a}{y_b} = \frac{x_a K_a}{x_b K_b} \tag{4.37}$$

$$\alpha_{ab} = \frac{K_a}{K_b} \tag{4.38}$$

$$\frac{y_a}{y_b} = \frac{x_a}{x_b} \alpha_{ab} \tag{4.39}$$

For ideal systems

$$\alpha_{ab} = \frac{P_a}{P_b}$$

Example 4.4

A hydrocarbon mixture contains the following: $CH_4 = 0.03$, $C_2H_6 = 0.05$, $C_3H_8 = 0.10$, $i\text{-}C_4H_{10} = 0.20$, and a residual material whose average characteristics are those of decane, $C_{10}H_{22} = 0.62$. This material is to be weathered by differential vaporization at constant temperature until not more than 1% of the residue is $i\text{-}C_4H_{10}$. Determine the final pressure. Temperature is 25°C.

Solution

Basis—1 mole of initial mixture. The relative volatilities are calculated at 25°C, assuming ideal behavior.

Component	Vapor pressure, mm Hg	α_{id}
(*a*) C_1	Above critical	—
(*b*) C_2	31,510	12.07
(*c*) C_3	7139	2.734
(*d*) $i\text{-}C_4$	2611	1.000*
(*e*) C_{10}	1.37	0.000525

* The reference component is $i\text{-}C_4$.

There will be only negligible amounts of C_3 and lighter remaining while most of the C_{10} will be unvaporized. The final moles of i-C_4 remaining is estimated at 1% of 0.62 mole. The final moles are calculated by the Rayleigh equation:

$$\ln \frac{n_{i1}}{n_{i2}} = \alpha_{id} \ln \frac{n_{d1}}{n_{d2}} = \alpha_{id} \ln \frac{0.200}{0.0062}$$

Component	n_{i1}	$\alpha_{id} \ln \frac{0.200}{0.0062}$	n_{i1}/n_{i2}	n_{i2}	x_{i2}	$x_{i2}P_i$, mm Hg
(*a*) C_1	0.03	—	—	—	0	0
(*b*) C_2	0.05	18.21	10	—	0	0
(*c*) C_3	0.10	4.12	13,200	—	0	0
(*d*) i-C_4	0.20	1.509	32.26	0.0062	0.01	26.1
(*e*) C_{10}	0.62	0.00079	1.0018	0.619	0.99	1.4
				0.6252	1.00	27.5

The final pressure $P_{t_2} = 27.5$ mm Hg.

The Rayleigh equation then becomes

$$\ln \frac{a_1}{a_2} = \alpha_{ab} \ln \frac{b_1}{b_2} = \alpha_{ac} \ln \frac{c_1}{c_2} = \cdots \tag{4.40}$$

The usual practice in batch differential-distillation-process calculations is to evaluate the yield and composition of the distillate and residue in producing products of desired characteristics. For a constant-pressure operation it is necessary to evaluate the temperature range of the distillation such that average α's may be used in Eq. (4.40). The method is as follows:

1. The initial temperature will be the bubble-point temperature, and this is calculated in the same manner as indicated before.

$$\Sigma y = \Sigma Kx = 1.0$$

2. The final temperature is assumed and, because the average α's are not extremely sensitive to practical differences in final temperature, the relative volatilities thus determined may be considered to be suitable for the calculation.

The final temperature will be the bubble-point temperature of the residual liquid. If the vapor pressure at some temperature is specified for the final product (or some composition specification is fixed so that the vapor pressure

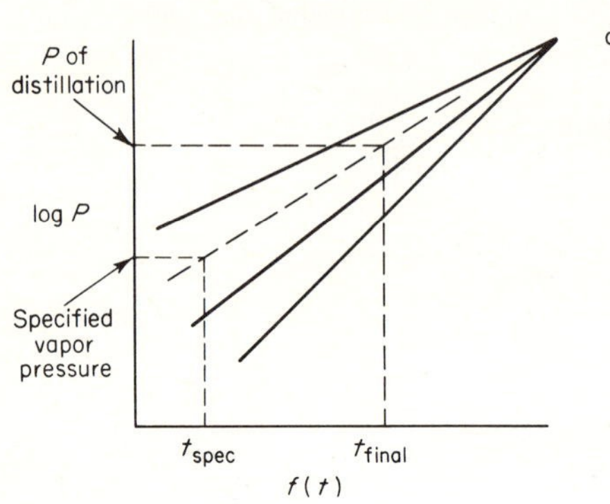

FIGURE 4.4 Evaluation of temperature from Cox chart.

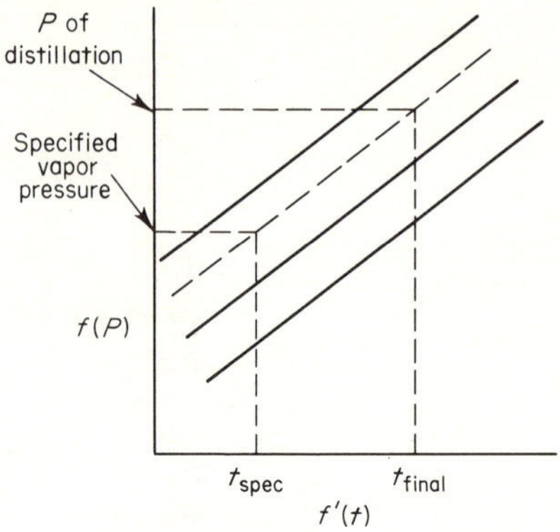

FIGURE 4.5 Evaluation of temperature from special modified vapor pressure chart.

at some temperature may be calculated), the final temperature of the distillation may be estimated from a vapor pressure chart such as a Cox chart for the compounds involved. This is accomplished by locating the point on the chart corresponding to the specified vapor pressure (at the specified temperature) and drawing a line either through the focal point (in the case of the Cox chart) or parallel to the lines on charts whose scales are adjusted to make them parallel, through the specification point, and intersecting the pressure line corresponding to the pressure of distillation. The temperature may be read from this intersection.

3. Determine average α's for all components. Usually the heaviest component is used as a reference.

$$\alpha_{in,av} = \frac{\alpha_{in,t_1} + \alpha_{in,t_2}}{2} \tag{4.41}$$

4. The final moles of some component remaining in the liquid (usually heaviest component, but it may be some component of intermediate volatility) are assumed.

5. The moles of the other components remaining in the liquid are calculated by

$$\log \frac{a_1}{a_2} = \log \frac{n_1}{n_2} \alpha_{an}$$

$$\cdots\cdots\cdots\cdots\cdots \tag{4.42}$$

$$\log \frac{i_1}{i_2} = \log \frac{n_1}{n_2} \alpha_{in}$$

6. The final composition is calculated by

$$x_{i2} = \frac{i_2}{\sum_{1}^{n} i_2} \tag{4.43}$$

7. The bubble-point pressure of mixture is then calculated at the final temperature. If the pressure thus calculated is not equal to the distillation pressure, a new assumption is made in step 4 and the process is repeated until a solution is obtained.

The solution thus obtained is unique only if the relative volatilities of the hydrocarbons are relatively insensitive to temperature, and the calculated results are reasonably accurate (calculated—not approximated—e.g., by rounding off compositions to the third place).

Constant Temperature—Multicomponent Differential Distillation

In constant-temperature differential distillation the relative volatility values are considered constant because the temperature is constant and can be evaluated directly by

$$\alpha_{an} = \frac{K_a}{K_n} \quad \text{or} \quad \alpha_{an} = \frac{P_a}{P_n}$$

This assumes that the pressure level is not sufficiently high and the change in pressure from start to finish of the process is not great enough to affect the α values significantly. Average values should be used if there is much variation. The method is as follows:

1. The bubble-point pressure of the mixture to be distilled is determined for the temperature at which the distillation is to be conducted. This pressure is the initial pressure of the distillation.

2. Determine the final pressure of the distillation from the specification vapor pressure at the specified temperature (or the specified composition from which the vapor pressure can be determined) on a vapor pressure plot.

3. Assume the final moles of the reference component as before.

4. Calculate the final moles of *a*, *b*, *c*, etc., from Eq. (4.42).

5. Calculate the composition of the residue at the end of the distillation by using Eq. (4.43).

6. Determine the bubble-point pressure of the residue.

7. If the pressure calculated is equal to the final pressure of the distillation, the solution is complete. If it is not, repeat steps 3 through 6 until agreement is reached.

Steam Distillation

Steam distillation is a term applied to a distillation process with open steam, i.e., wherein the steam is in direct contact with the distilling system in either a batch or a continuous operation. More generally, a distillation conducted in the presence of any inert added component such as nitrogen, carbon dioxide, flue gas, etc., is governed by the same fundamental relationship as steam distillation. Steam is widely used because of its energy level, cheapness, and availability. Steam or inert distillation is commonly used in the following situations:

1. To separate relatively small amounts of volatile impurity from a large amount of material.

2. To separate appreciable quantities of higher boiling materials.

3. To recover high-boiling materials from small amounts of impurity which have a higher boiling point.

4. Where the material to be subjected to distillation is thermally unstable or will react with other components associated with it at the boiling temperature.

5. Where the material cannot be distilled by indirect heating even under low pressure because of the high boiling temperature.

6. Where direct-fired heaters cannot be used because of danger.

Batch Steam Distillation—Binary System plus Steam

If a system composed of *a* (volatile) and *b* (nonvolatile), completely miscible in the liquid phase, is to be separated by batch steam distillation and both *a* and *b* are immiscible with water, *a* and steam would appear in the vapor. The total pressure would be

$$P_t = p_a + p_s = x_a P_a + P_s \tag{4.44}$$

This assumes that *a* and *b* form an ideal liquid mixture. If the total pressure were fixed at any point in the distillation process (at a definite value of x_a), the temperature would be fixed to satisfy Eq. (4.44). Since the actual partial pressure of *a* in the vapor is somewhat less than the theoretical because of mass- and heat-transfer resistance in the actual process, a vaporization efficiency *E* is commonly used.

$$E = \frac{p_a}{x_a P_a} = \frac{p_a}{\left(\dfrac{a}{a+b}\right) P_a} \tag{4.45}$$

The value of *E* varies widely with the system, rate of distillation, type of equipment, etc., but will usually range from about 0.6 to 0.95.

Using E in Eq. (4.44) gives

$$P_t = Ex_aP_a + P_s \tag{4.46}$$

In considering a batch steam distillation process, there are three items of importance usually evaluated: (*a*) the quantity of steam required; (*b*) the temperature profile of the distillation; and (*c*) the time to complete the process. Generally, the following equation describes the process and can be used for evaluation of the items of interest.

$$\frac{ds/d\theta}{-da/d\theta} = \frac{s}{a} = \frac{p_s}{p_a} \tag{4.47}$$

$$\frac{ds}{-da} = \frac{P_t - p_a}{p_a} = \frac{P_t - Ex_aP_a}{Ex_aP_a} \tag{4.48}$$

$$\frac{ds}{-da} = \frac{P_t}{EP_a\dfrac{a}{a+b}} - 1 \tag{4.49}$$

$$ds = -\frac{P_t}{EP_a\dfrac{a}{a+b}}\,da + da \tag{4.50}$$

$$= -\frac{P_ta + P_tb}{EP_aa}\,da + da \tag{4.51}$$

$$= -\frac{P_ta}{EP_aa}\,da + da - \frac{P_tb}{EP_aa}\,da \tag{4.52}$$

$$= -\left(\frac{P_t}{EP_a} - 1\right)da - \frac{P_tb}{EP_aa}\,da \tag{4.53}$$

Integrating Eq. (4.53) between initial and final conditions gives

$$s = \left(\frac{P_t}{EP_a} - 1\right)(a_1 - a_2) + \frac{P_tb}{EP_a}\ln\frac{a_1}{a_2} \tag{4.54}$$

At constant pressure and temperature the quantity of steam to carry out the distillation to distill $a_1 - a_2$ moles of a is calculated from Eq. (4.54). The time required can be calculated by dividing the total moles of steam needed by the steam rate in moles per unit time.

If all or almost all of the volatile component a is to be distilled, the partial pressure of the component p_a will become very small and the total pressure becomes essentially equal to the partial pressure of the steam, $P_t \cong p_s = P_s$.

In order to prevent condensation of the steam, the temperature should be selected to be higher than the saturation temperature of the steam at the pressure of the distillation.

If the system to be steam distilled were composed of two volatile components, both a and b,

$$P_t = p_a + p_b + p_s \tag{4.55}$$

$$P_t - p_s = P_t - y_s P_t = \gamma_a x_a P_a + \gamma_b x_b P_b \tag{4.56}$$

When the composition with respect to steam in the vapor, that is, y_s is maintained constant and the pressure is fixed, the vapor composition y_a and y_b is fixed for either a fixed liquid composition or fixed temperature, and the ab system distills at an "effective" pressure equal to $P_t - p_s$. This is a very useful process because open steam allows miscible materials to be distilled at temperatures below their normal boiling temperature without the use of vacuum.

Batch Steam Distillation—Rates

The general rate equation for batch steam distillation may be written as

$$-d(bX_a) = R_s Y_a \, d\theta \tag{4.57}$$

where b = moles of component b at any time in the liquid
R_s = steam rate = moles of steam in the vapor per unit time
X_a = moles of a per mole of b in liquid
Y_a = moles of a per mole of steam in vapor

$$Y_a = \frac{p_a}{P_t - p_a} = \frac{Ep_a^*}{P_t - Ep_a^*} \tag{4.58}$$

where

$$E = \frac{p_a}{p_a^*} = \frac{\text{Actual partial pressure}}{\text{Equilibrium partial pressure}} = \frac{p_a}{x_a \gamma_a P_a}$$

$$p_a^* = y_a P_t = x_a \gamma_a P_a$$

$$y_a = \frac{Y_a}{1 + Y_a} \tag{4.59}$$

$$p_a^* = \frac{Y_a}{1 + Y_a} P_t = x_a \gamma_a P_a \tag{4.60}$$

There are two general methods of operating a batch steam distillation with regard to the rate of steam flow—the constant steam rate and variable steam rate. These are considered in conjunction with the type of distilling system as follows:

Binary system—*a* volatile, *b* nonvolatile, steam rate constant, temperature constant Equation (4.57) becomes

$$-X_a\,db - b\,dX_a = R_s Y_a\,d\theta$$

$$db = 0$$

$$\frac{-b}{R_s}\int_{X_{a2}}^{X_{a1}} \frac{dX_a}{Y_a} = \int_0^\theta d\theta \tag{4.61}$$

$$\frac{-b}{R_s}\int_{X_{a1}}^{X_{a2}} \left(\frac{P_T}{Ep_a^*} - 1\right) dX_a = \theta \tag{4.62}$$

$$\frac{-b}{R_s}\int_{X_{a1}}^{X_{a2}} \left(\frac{P_t}{Ex_a\gamma_a P_a} - 1\right) dX_a = \theta \tag{4.63}$$

$$\frac{-b}{R_s}\int_{X_{a1}}^{X_{a2}} \left(\frac{P_t}{E\dfrac{X_a}{1+X_a}\gamma_a P_a} - 1\right) dX_a = \theta \tag{4.64}$$

$$\frac{-b}{R_s}\left(\frac{P_t}{E\gamma_a P_a}\int_{X_{a1}}^{X_{a2}} \frac{1+X_a}{X_a}\,dX_a - \int_{X_{a1}}^{X_{a2}} dX_a\right) = \theta \tag{4.65}$$

$$\frac{-b}{R_s}\left[\frac{P_t}{E\gamma_a P_a}\left(\int_{X_{a1}}^{X_{a2}} \frac{dX_a}{X_a} + \int_{X_{a1}}^{X_{a2}} dX_a\right)\int_{X_{a1}}^{X_{a2}} dX_a\right] = \theta \tag{4.66}$$

$$\frac{-b}{R_s}\left\{\frac{P_t}{E\gamma_a P_a}\left[\ln\frac{X_{a2}}{X_{a1}} + (X_{a2} - X_{a1})\right] - (X_{a2} - X_{a1})\right\} = \theta \tag{4.67}$$

This equation applies essentially to ideal systems so that γ is relatively constant and approximates unity. Where this is not true, the activity-coefficient variation with composition must be known analytically or by means of experimental data. Where experimental data are available, the integral may be evaluated graphically.

Binary system—*a* volatile, *b* nonvolatile, steam rate varies, temperature constant

$$-b\,dX_a = R_s Y_a\,d\theta$$

$$-b\int_{X_{a1}}^{X_{a2}} \frac{dX_a}{Y_a} = \int_0^\theta R_s\,d\theta \tag{4.68}$$

$$-b\left\{\frac{P_t}{E\gamma_a P_a}\left[\ln\frac{X_{a2}}{X_{a1}} + (X_{a2} - X_{a1})\right] - (X_{a2} - X_{a1})\right\} = \int_0^\theta R_s\,d\theta \tag{4.69}$$

In this case for the time required for the distillation, R_s, the steam rate, must be known as a function of time. If $R_s = a + b\theta$, the right-hand side of Eq. (4.68) becomes

$$\int_0^\theta (a + b\theta)\, d\theta = b\theta^2 \tag{4.70}$$

If the steam rate is a function of the composition remaining in the still, that is, $R_s = f(X)$, this function is included in the integral on the left-hand side of the equation.

Binary system—both *a* and *b* volatile, steam rate constant It has been noted that the distillation of both components in the presence of steam with no liquid water phase present is the same as a vacuum distillation of the binary system at an effective pressure which equals P_t minus the partial pressure of the steam. If the steam rate is constant but the amount vaporized with time decreases as the distillation progresses, the effective pressure will decrease and the temperature will increase, decrease, or remain constant depending upon the type of system and the conditions. For example, if the system has a relatively narrow boiling range, the temperature will usually decrease toward the end of the distillation because the boiling temperature of *b* (which is near that of *a*) is being reduced owing to the lowering of the effective distillation pressure as the amount of the system vaporized decreases. On the other hand, if the boiling point of *b* is much greater than that of *a*, the reduction of the effective distillation pressure can be insufficient to compensate for the greater boiling temperature of *b*, and the temperature will increase. If the change in effective pressure is just sufficient to compensate for the boiling-point change with pressure, the temperature can remain constant.

To calculate the composition of the liquid and vapor and the time necessary for the distillation requires a complex trial-and-error solution unless one of the conditions previously described can be assumed. If constant temperature can be assumed, the Rayleigh equation (constant temperature, varying pressure) can be used to determine the composition and quantity of material distilled, as described in the first part of this chapter.

Equation (4.70) can be used to determine the time required for the distillation if some relation between composition of the liquid or vapor and quantity of liquid or vapor with time can be assumed or established.

Binary system—both *a* and *b* volatile, steam rate varying This situation offers the same difficulty of quantitative evaluation as the previous case. Again it is necessary to assume that the steam rate can be changed with time in such a manner that, for example, the effective pressure remains constant. If this can be done, the amount distilled and composition of liquid and vapor can be calculated by the Rayleigh equation. The time required for the distillation can be determined if the change in composition of the vapor and liquid with respect to time or with respect to steam rate can be assumed or fixed.

Continuous Steam Distillation

In many cases steam distillation or steam stripping is conducted as a continuous steady-state operation. The flow of steam may be countercurrent to the flow of the feed or cocurrent with it, as shown in Figs. 4.6 and 4.7.

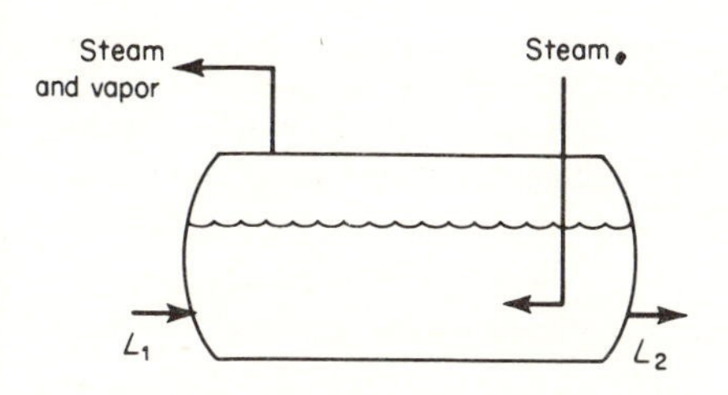

FIGURE 4.6 Continuous countercurrent steam distillation.

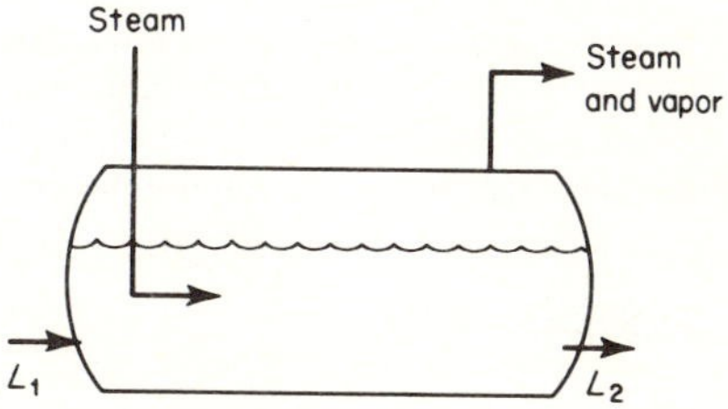

FIGURE 4.7 Continuous cocurrent steam distillation.

The quantitative evaluation of the pertinent variables for continuous steady-state steam distillation is relatively simple compared with that involved in the batch differential processes. The equations for continuous steam distillation for a binary system with a volatile and b nonvolatile are:

$$b(X_{a1} - X_{a2}) = sY_a \tag{4.71}$$

where b = moles of b per unit time

X_{a1} = moles of a per mole b entering

X_{a2} = moles a per mole b leaving

s = moles of steam per unit time

Y_a = moles of a per mole of steam

$$s = \frac{b(X_{a1} - X_{a2})}{Y_a} \tag{4.72}$$

$$s = b(X_{a1} - X_{a2})\frac{P_t - Ep^*}{Ep^*} \tag{4.73}$$

For countercurrent flow the moles of steam are given by

$$s = b\left[\frac{P_t}{EP_a}\frac{X_{a1} - X_{a2}}{X_{a1}} + \left(\frac{P_t}{EP_a} - 1\right)(X_{a1} - X_{a2})\right] \tag{4.74}$$

For cocurrent flow,

$$s = b\left[\frac{P_t}{EP_a}\frac{X_{a1} - X_{a2}}{X_{a2}} + \left(\frac{P_t}{EP_a} - 1\right)(X_{a1} - X_{a2})\right] \quad (4.75)$$

Example 4.5

An amount of 100,000 lb of a food extract—approximate molecular weight is 450—containing 1.3 wt % hexane is to be subjected to steam distillation to reduce the hexane content to not more than 0.01 wt %. Assume $E = 0.7$, total pressure 1000 mm Hg, $t = 100°C$. Determine the steam consumption needed for the continuous countercurrent process.

Solution

$$s = \left[b\frac{P_t}{EP_a}\frac{X_1 - X_2}{X_1} + \left(\frac{P_t}{EP_a} - 1\right)(X_1 - X_2)\right]$$

Assume a feed rate with respect to the feed of 1000 lb/hr.

$$b = \frac{1000 - 1000(0.013)}{450} = \frac{987}{450} = 2.19 \text{ moles/hr}$$

$$X_1 = \frac{(1000)(0.013)/86}{2.19} = 0.151 \text{ mole hexane/2.19 moles feed} = 0.069 \text{ mole/mole}$$

$$X_2 = \frac{0.0986/86}{2.19} = 0.00053 \text{ mole/mole}$$

$$\frac{P_t}{EP_a} = \frac{P_t}{(0.7)\ P_a} = \frac{1000}{(0.7)\ (1860)} = 0.77$$

$$s = 2.19\left[(0.77)\frac{0.069 - 0.00053}{0.069} + (0.77 - 1)(0.069 - 0.00053)\right]$$

$$= 1.69(0.99) + [(-0.23)(0.0685)]2.19$$

$$= 1.69 - 0.0344 = 1.64 \text{ moles steam/hr}$$

or

$$(1.64)(18) = 29.5 \text{ lb steam/1000 lb feed}$$

Multicomponent Batch Steam Distillation

The most common situation encountered in multicomponent batch steam distillation is that wherein n components are present, some of which are volatile and some nonvolatile, and wherein all components of the mixture to be distilled are miscible in the liquid phase. Holland and Welch [3] derived an equation to describe the process:

$$\sum_{i \neq s,r} \frac{L_c^\circ}{\beta_i}\left[1 - \left(\frac{L_b}{L_b^\circ}\right)^{\beta_c^\circ}\right] + L_r^\circ \ln \frac{L_b^\circ}{L_b} = \frac{E_b P_b}{P_t}(L^\circ - L + S) \tag{4.76}$$

Where the nonvolatile components do not alter the mole fractions of the volatile components in the still, Eq. (4.76) reduces to

$$\sum_{i \neq s,r} \frac{L_i^\circ}{\beta_i}\left[1 - \left(\frac{L_b}{L_b^\circ}\right)^{\beta_i}\right] = \frac{E_b P_b}{P_t}(L^\circ - L + S) \tag{4.77}$$

where $E_i = p_i/P_i x_i$ = vaporization efficiency of i (4.78)

L = moles of volatile components in still at any time

$L = L^\circ$ at start of distillation

L_r° = moles of nonvolatile material in still

P_i = vapor pressure of i at still temperature

p_i = partial pressure of i in vapor

$\beta_i = (E_i/E_b)\alpha_{ib}$ (4.79)

α_{ib} = relative volatility of i referred to $b = P_i/P_b$

$\sum_{i \neq s,r}$ = summation over all components except steam and nonvolatiles

b = reference component

r = nonvolatile component

Example 4.6

It is desired to separate three high boiling materials from a nonvolatile material by batch steam distillation, and to remove 95% of the least volatile material. Determine the moles of steam required when the distillation is carried out at 100°C and 200 mm Hg pressure. The data are:

Initial mixture	100 moles	P, 200 mm Hg, 100°C	E
Component 1	30	20	0.9
Component 2	25	14	0.9
Component 3	25	8	0.9
Nonvolatile	20	—	—

Solution

Assume component 3 as base component.

$$\beta_3 = \beta_b = 1.0 \qquad \beta_1 = \frac{E_1}{E_b}\frac{P_1}{P_b} = \frac{0.9}{0.9}\frac{20}{8} = 2.5$$

$$\beta_2 = \frac{E_2}{E_b}\frac{P_2}{P_b} = \frac{0.9}{0.9}\frac{14}{8} = 1.75 \qquad \beta_r = \frac{E_r}{E_b}\frac{0}{8} = 0$$

$$L_T = \text{total moles in still at end of distillation}$$

$$= L_1^\circ\left(\frac{L_b}{L_b^\circ}\right)^{\beta_1} + L_2^\circ\left(\frac{L_b}{L_b^\circ}\right)^{\beta_2} + L_3^\circ\left(\frac{L_b}{L_b^\circ}\right)^{\beta_3} + L_r^\circ\left(\frac{L_b}{L_b^\circ}\right)^{\beta_r}$$

$$= 30\left[\frac{0.05(25)}{25}\right]^{2.5} + 25(0.05)^{1.75} + 25(0.05)^{1.0} + 20$$

$$= 30(0.00055) + 25(0.00525) + 25(0.05) + 20$$

$$= 0.0165 + 0.131 + 1.25 + 20 = 21.398$$

$$L = L_T - L_r^\circ = 21.398 - 20 = 1.398$$

$$\sum_{i\neq s,r} \frac{L_1^\circ}{\beta_1}\left[1 - \left(\frac{L_b}{L_b^\circ}\right)^{\beta_1}\right] = \frac{30}{2.5}[1 - (0.05)^{2.5}] + \frac{25}{1.75}[1 - (0.05)^{1.75}]$$

$$+ \frac{25}{1.0}[1 - (0.05)^{1.2}]$$

$$= 12(1 - 0.00055) + 14.3(1 - 0.00525) + 25(1 - 0.05)$$

$$= 12 + 14.3 + 23.75 = 50.05$$

$$L_r^\circ \ln \frac{L_b^\circ}{L_b} = 20 \ln \frac{25}{(0.05)25} = 20(1.301)(2.303) = 60.1$$

$$50.5 + 60.1 = \frac{E_b P_b}{P_t}(L^\circ - L + s) = \frac{0.9(8)}{200}(80 - 1.398 + s)$$

$$110.15 - 0.036(78.602) = 0.036s$$

$$\frac{110.15 - 2.83}{0.036s} = s$$

$$\frac{107.32}{0.036} = 2980 \text{ moles steam}$$

PROBLEMS

1. The benzene–cyclopentane vapor-liquid equilibrium data at 760 mm Hg pressure are as follows:

t, °C	x_C	y_C
76.65	0.05	0.147
69.9	0.169	0.395
64.1	0.309	0.568
59.3	0.473	0.712
55.5	0.653	0.823
51.1	0.888	0.944

Components	Antoine constants		
	A	B	C
Cyclopentane	6.887	1124.16	231.36
Benzene	6.898	1206.35	220.24

(*a*) To what temperature must a mixture of 40 mole % cyclopentane be heated at 760 mm Hg pressure to vaporize 50% of the feed?
(*b*) What will be the composition of the vapor?
(*c*) Calculate the bubble-point and dew-point temperatures for a mixture containing 50 mole % cyclopentane.

2. (*a*) Using the K values of Fig. 2.42, determine the dew point and bubble point for the mixture at 100 psia.
(*b*) Determine the volume of liquid resulting from flashing this mixture 30°F above its bubble-point temperature at 100 psia.

Component	x
C_2H_6	0.22
C_3H_6	0.16
i-C_4H_8	0.04
C_4H_8	0.21
C_5H_{12}	0.33
CO_2	0.04

3. A feed liquid consisting of 1200 moles of mixture containing 30 mole % naphthalene and 70 mole % dipropylene glycol is differentially distilled at 100 mm Hg until the final total distillate contains 55 mole % naphthalene.
(*a*) Determine the amount of distillate.
(*b*) Determine the concentration of naphthalene in the residue still liquid.

Vapor-Liquid Equilibrium Data at 100 mm Hg, Mole %

x_n	y_n
5.4	22.3
11.6	41.1
28.0	62.9
50.6	74.8
68.7	80.2
80.6	84.4
84.8	86.4
88.0	88.0

4. The following demethanized light hydrocarbon mixture is to be batch-distilled at its bubble-point pressure at 100°F by raising the temperature as the distillation proceeds until 30% of the original butane content remains in the liquid.
(*a*) Determine the composition of the residue liquid.
(*b*) Determine the fraction of the original feed remaining as liquid at the end of the distillation.

Component	x
Ethane	0.02
Propane	0.10
i-Butane	0.08
n-Butane	0.14
i-Pentane	0.06
n-Pentane	0.20
n-Hexane	0.40

5. An amount of 4000 moles of nonvolatile paint oil contains 5 mole % of hexane after the initial flashing operation.
(*a*) If a constant-steam-rate batch steam distillation were used at a total pressure of 800 mm Hg and 210°F, how much steam would be needed to reduce the hexane mole fraction of the liquid of 0.0011? Assume $E = 0.6$, $P_{\text{Hex}} = 1920$ mm Hg.
(*b*) If a countercurrent continuous process is used with the same conditions of temperature and pressure but with 100 moles/hr of feed flow, determine the moles of steam needed.

Nomenclature

C = number of components
D = moles of distillate
d = differential operator
E = vaporization efficiency
F = moles of feed
K = equilibrium vaporization ratio
L = moles of liquid
p_i = partial pressure, component *i*
P_i = vapor pressure, component *i*

P = number of equilibrium phases
P_t = total pressure
R = rate, moles per unit time
V = vapor
V = degrees of freedom
X = mole ratio in liquid
Y = mole ratio in vapor
a, b, c, i, n = moles of a component
s = moles of steam
t = temperature
x = mole fraction of a component in liquid
y = mole fraction of a component in vapor

Greek letters

α = relative volatility
β = defined quantity, Eq. (4.79)
γ = activity coefficient
ν = fugacity coefficient
θ = time
Σ = summation

Subscripts

B = residue
D = distillate
F = feed
1, 2, 3 = component designation
i, j = component designation
a = volatile
b = reference component
r = nonvolatile

REFERENCES

1. Bachelor, J. B.: *Petrol. Refiner* (October, 1957), p. 113.
2. Holland, C. D., and R. R. Davisson: *Petrol. Refiner* (March, 1957).
3. Holland, C. D., and N. E. Welch: *Petrol. Refiner* (May, 1957), p. 251.
4. McReynolds, E. E.: *Petrol. Refiner* (March, 1945), p. 84.
5. Poettman, F. H., and R. B. Rice: *Petrol. Refiner* (December, 1947), p. 128.
6. Rayleigh: *Phil. Mag.*, **4**(6):521 (1902).
7. Reilly, P. M.: *Petrol. Refiner* (July, 1951), p. 132.
8. Wilson, C. L.: *Petrol. Refiner* (June, 1952), p. 131.

part

2

Fractional distillation

chapter

5

General considerations

The ultimate application of distillation is for the purpose of separating two or more components occurring in a mixture to produce products which meet certain specifications. These specifications may be sales specifications which require certain purity or characteristics based on boiling range, or process specifications which require purity or concentration with respect to one or more components for use in subsequent processes. The specified purity required may range from a wide-boiling-range material, such as a gasoline fuel or lubricating oil, to a composition of 99.999% with impurities limited to parts per million. Although it is possible to obtain specified materials boiling over a temperature range by means of, e.g., differential distillation or even equilibrium flash distillation, it is impossible to obtain the maximum yield of such material by these methods. The economic considerations of yield of material of specified characteristics has resulted in the development of the process of fractional distillation which may be of the *stage* type, which is the most widely used, or of the *differential* type. The latter is discussed in Chap. 15.

From a simplified point of view, stage fractional distillation may be considered to be a process in which a series of flash vaporization stages are arranged in series in such a manner that the products from each stage are fed to adjacent stages, as shown in Fig. 5.1. The vapor produced in one stage is conducted to the stage above and the liquid to the stage below. In turn, this stage receives the liquid from the stage above and the vapor from the stage below as its feed material. In this arrangement the concentration of the lower boiling component or components is being increased in the vapor from each stage in the direction of vapor flow and decreased in the liquid

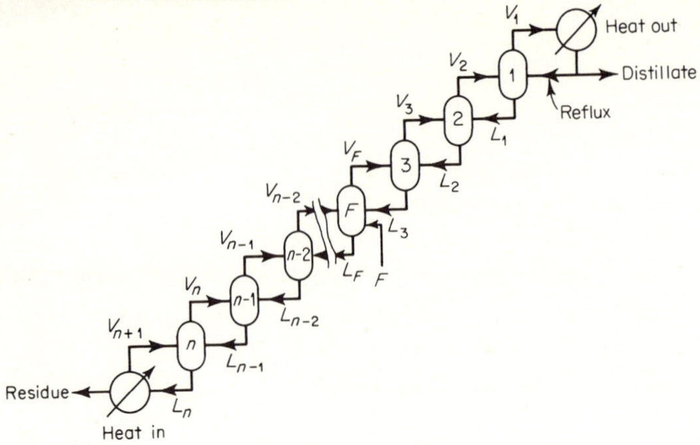

FIGURE 5.1 Schematic fractional distillation process.

in the direction of the liquid flow. Because the lighter boiling constituents are concentrating in the vapor from each successive stage, the temperature decreases from stage to stage and reaches the minimum as the final vapor is produced from the process. Similarly, the temperature increases along the direction of flow of the liquid, and the maximum temperature is reached at the point where the liquid product is withdrawn from the process. Since temperature is a measure of the level of heat energy, it is obvious that heat energy is necessary to the distillation process.

In addition to the heat energy involved in maintaining a temperature differential, an amount of heat energy roughly equivalent to the latent heat of the vapor evolved from the last stage (with respect to vapor flow) must be supplied. This heat energy may be supplied in the feed, in the last stage from which the liquid product is withdrawn, or in both places.

5.1. REFLUX

If the reflux were eliminated in the schematic fractional distillation process shown in Fig. 5.1, there would be no liquid returned to stage 1 and, therefore, there could be no condensation of V_2 to supply liquid leaving stage 1. The vapor leaving stage 1 then would be the same quantity and composition of the vapor leaving stage 2, and the vapor leaving stage 2 would be of the same quantity and composition of the vapor leaving stage 3, etc. In addition, the liquid reflux, as it passes from stage to stage, serves as an absorbing liquid for the heavier components in the vapor, thereby aiding the fractionating process by concentrating the light components in the vapor and the heavier components in the liquid. Thus, it becomes apparent that in order for fractional distillation to function, it is necessary to supply a liquid stream to the last stage from which the vapor product is produced to remove all or part of the latent heat contained in the vapor in the last vapor stage. This liquid stream is called *reflux*. It is produced conveniently by condensing all or part of the vapor leaving stage 1 and returning some of the liquid to stage 1, although an intercooler in the column shell could produce the reflux internally. The ratio of the mass (in pounds or moles) of liquid returned to the process to the mass (in pounds or moles) of the liquid or vapor product is called the *operating reflux ratio* L/D. The ratio of the mass (in pounds or moles) of liquid flowing from any stage to the next lower stage to the mass (in pounds or moles) of vapor rising to the stage is called the *internal reflux ratio* L_1/V_2, L_2/V_3, etc.

5.2. THE FRACTIONATING COLUMN

Because the relative densities of the vapors and liquids naturally cause the liquids to flow downward and the vapors to flow upward, and because in practice a pressure differential across the column is induced to increase the relative counterflow rate of liquid and vapor, the requirements shown schematically in Fig. 5.1 have been met by the device called a *fractionating column*. The essential arrangement is shown schematically in Fig. 5.2.

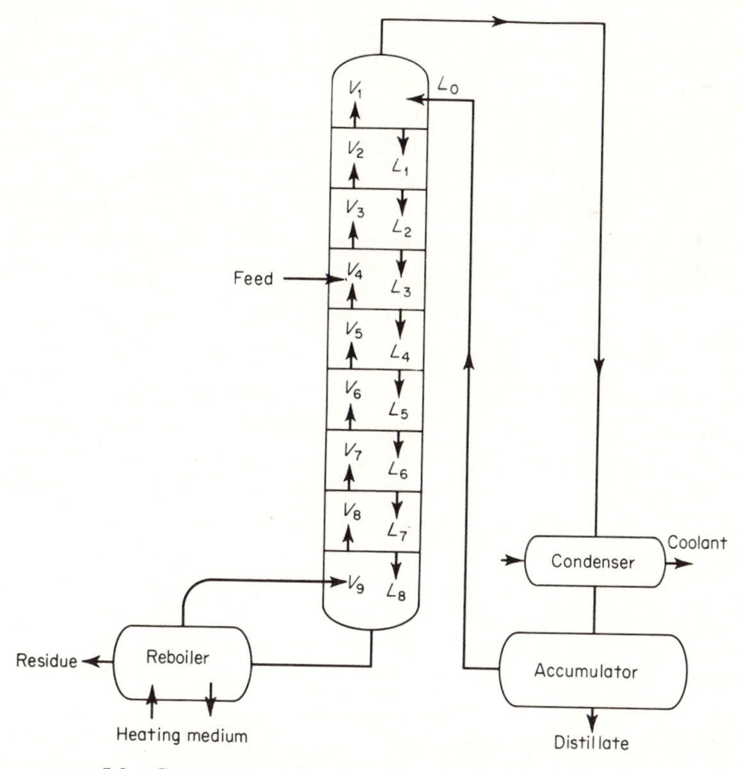

FIGURE 5.2 Conventional fractionating column.

Comparison of the figures shows that each stage in Fig. 5.1 corresponds to a contact stage commonly called a *plate* or *tray* in Fig. 5.2.

The minimum essentials for a fractionation process are:

1. Feed stream
2. Distillate product stream
3. Residual liquid product stream
4. Contacting stages which serve as flash or equilibrium vaporization stages
5. A means of introducing heat energy, such as a reboiler
6. A means of removing heat energy (to product reflux), such as a reflux condenser
7. Suitable liquid and vapor flow paths

5.3. THE CONTACT STAGE—PLATE OR TRAY

The contact stage is a device for intimately contacting the vapor and liquid introduced to the stage so that, ideally, the vapor and liquid leaving the stage are in equilibrium. A schematic representation of a plate is shown in Fig. 5.3. The vapor V_{n+1} rising to the plate n from plate $n + 1$ and the liquid L_{n-1} falling from plate $n - 1$ to plate n are intimately contacted by mixing so that the vapor V_n and the liquid L_n approach a state of equilibrium in both composition and temperature. If equilibrium were reached, the efficiency of the contact stage would be 100%, and it could be truly classed as an *equilibrium* stage or an *equilibrium flash* stage. However, since the closeness to approach to equilibrium is a function of the rate operations of mass and heat transfer, it is theoretically impossible for true equilibrium to be reached. This would require infinite area, or infinite time of contact of the phases, or

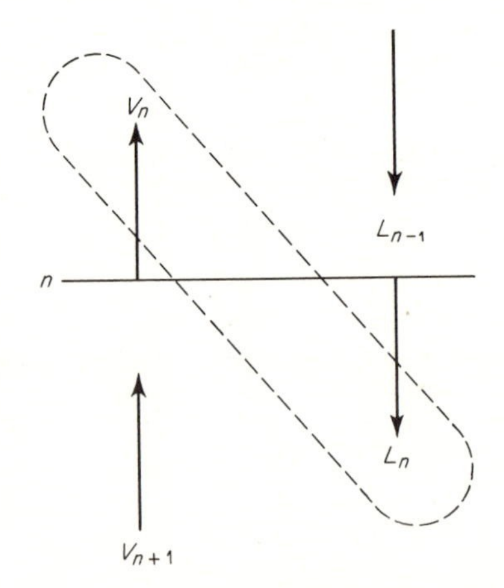

FIGURE 5.3 The plate or contact stage.

both. In addition, as equilibrium is approached, the rate of transfer decreases because the driving forces, both temperature difference and concentration difference, approach zero. In spite of the fact that true equilibrium contacting devices are not possible, such are assumed in the conventional methods of calculation for the number of stages in the design of fractionating columns. Indeed, under optimum conditions near approach to equilibrium can be reached.

5.4. DESIGN CONSIDERATIONS IN FRACTIONATING PROCESSES

There are two main factors to be considered in designing fractionating columns, both of which are controlled by economic considerations. (*a*) The fractionation process must produce the desired quality of product and (*b*) it must be capable of handling the quantity of material to be processed with provision for fluctuation in the quantity without changing the quality of the product within specification limits. Almost all steps in the process design affect, in one way or another, the two aforementioned considerations.

The general design procedure is to fix or calculate the following:

1. The pressure of the distillation
2. The number of equilibrium stages necessary for the specified separation
3. The efficiency of the stages and the actual number of stages required
4. The plate design
5. The column design
6. The column accessories

At this point it is necessary to clarify the terms *separation* and *quality of product*. Fractionation of a multicomponent system in a single fractionating tower will enable the separation only between two components. If a seven-component mixture were to be fractionated, a separation could be made between components 1 and 2, or 2 and 3, or 3 and 4, etc. The components separated are designated the *light key and lighter than light key components* and *heavy key and heavier than heavy key components*. If there were intermediate components boiling between the designated light and heavy keys, these would be designated as *intermediate* or *distributed* keys. Thus, if component 3 were the light key and component 5 the heavy key, component 4 would be the distributed key. Components 1 and 2 would be the lighter than light keys and components 6 and 7 the heavier than heavy keys. The light materials would appear in the distillate, the heavier materials in the bottoms, and the distributed key in both places. In a binary system there are only the two keys, light and heavy. It is evident from the discussion that C components require $C - 1$ columns for complete separation.

Separation and *quality of product* are related in that the composition of the stream—distillate or bottoms—determines the quality of the product. In the case of a ternary mixture, components 1 and 3 could be separated almost completely, but both streams could be of such a concentration with respect to the light key in the distillate and the heavy key in the bottoms, because of the presence of the distributed key, that their quality would be unsatisfactory. On the other hand, the distillate could be composed of high-purity component 1, but with, e.g., 80% of it contained in the bottoms. In this case the quality of the distillate would be high but the separation poor.

Pressure

The pressure at which the distillation is conducted may be fixed by one or more of several considerations.

In general, at higher pressures, the equilibrium concentrations of vapor and liquid approach one another; at the critical pressure, $y = x$, and only one phase exists. Because of this, the separation becomes "more difficult" as pressure increases or, for a given separation and comparable efficiency, more equilibrium stages are required to make the separation. Conversely, as pressure is decreased, the separation becomes "easier," or fewer stages are required because the difference between the equilibrium vapor and liquid concentrations increases. However, the deciding factor in selection of an operating pressure (based on pressure effect on separability) is one of economics.

Vapor volume increases as pressure decreases, and larger diameter columns are required to handle the increase in vapor volume. This increase in size of column increases the investment cost. Increasing pressure increases wall thickness, number of stages, and height, thus also increasing the cost. Therefore, it is necessary to establish the optimum for a particular situation.

Another factor controlling the pressure selection is the bubble-point pressure of the reflux and the type of coolant used in the condensers. For example, if water were the cooling medium used in the condenser and the cooling-tower return temperature at the worst condition were 110°F, the reflux temperature could be expected to be at least 125°F. If the bubble-point pressure for the reflux were, e.g., 150 psia at 125°F, the column would have to be designed to be operated at 150 psia. If the coolant were some refrigerant which could cool the reflux to a temperature of 30°F, the bubble-point pressure for the reflux at that temperature might be 50 psia, and the distillation column could then be designed to be operated at that lower pressure.

Another factor influencing the selection of column pressure in design is the thermal stability of the components in the distilling mixture. Many compounds decompose, polymerize, condense, or interact when the temperature reaches some critical value. In such cases it is necessary to reduce the design column pressure so that this critical reaction temperature will not be reached at any place in the column. Usually the reboiler temperature is the controlling one since it is maintained at the highest temperature in the normally constituted distillation column.

Other factors influencing the distillation pressure must often be considered. In some cases a homogeneous azeotropic system may be pressure sensitive with respect to azeotropic composition. Thus, it may be possible by judicious selection of pressures, for two columns working together, to separate the essentially pure components from an otherwise inseparable azeotropic mixture. Also, column pressure selection may be dictated by the pressure at which the feed is supplied or by the pressure at which the column products must be supplied to subsequent processing steps. In general, this requirement is the least important of the considerations discussed.

Number of Equilibrium Stages

Once the pressure at which the distillation is to be conducted is established, the number of theoretical or equilibrium stages is determined. Equilibrium stages are evaluated because this number of stages represents the minimum number necessary to make the separation desired under the assumption of 100% efficiency for each stage. Also, the methods of computation are not complicated at this point by modifications necessary to determine over-all efficiency or individual stage efficiency.

In analyzing the operation of a fractionating column with respect to the variables of operation and design related to the number of plates, the various factors of pressure, temperature, material balance, component material balances, enthalpy balances, and heat additions and removals from the column, condenser, and reboiler are considered. In determining the number of variables to be fixed, the number of variables are listed and the number of independent equations involving the variables are written, and the difference between them determines the degrees of freedom or the number of variables to be fixed to fix the operation of the column. This analysis leads to $C(6 + 2n) + 5n + 7$ total variables and $C(5 + 2n) + 5n + 1$ independent equations which can be written describing the process, where C is the number of components in the distilling system and n represents the number of plates. Thus, the difference is $C + 6$, which equals the degrees of freedom or the number of variables to be fixed to fix the column operation.

The variables usually fixed in column design are as follows:

Feed rate	1
Feed composition	$C - 1$
Feed enthalpy	1
Distillate and bottoms composition through key distribution	2
Reflux temperature or enthalpy	1
Pressure on column	1
Reflux ratio L/D (or L/V) or no. of plates	1
Total	$C + 6$

This assumes that for design purposes the feed is introduced on the correct tray or stage. This analysis agrees with that of Gilliland and Reed [5] when the differences in assumptions and variables considered are accounted for.

Thus, for a binary system the number of variables to be fixed would be 8 and for a ten-component system the number would be 16, when a *total condenser* is used. Where a *partial condenser* is used, the degrees of freedom reduce to $C + 5$ because the use of a partial condenser fixes the *composition* of the *reflux liquid* and the enthalpy or temperature of the reflux liquid. Thus, the pressure P_t, the feed quantity, feed composition, and reflux ratio are established to fix the system. For a binary $C + 5 = 7$ variables need to be established.

Actual Stages—Efficiency

The actual number of stages or plates to be specified for the fractionating column is related to the number of theoretical stages or equilibrium stages by the individual plate efficiency or the average plate efficiency (or the so-called *column efficiency*), and the actual number of stages is calculated by dividing the number of equilibrium stages by the efficiency factor.

The approach to equilibrium is a rate factor intimately tied into the rate of mass and rate of heat transfer. In examining all the possible variables which contribute to these rate processes, it is found that they can be generally put into three classes:

1. *Operating variables*

(*a*) Temperature
(*b*) Pressure
(*c*) Liquid flow rate
(*d*) Vapor flow rate

2. *Design variables*

(*a*) Diameter of column
(*b*) Plate spacing
(*c*) Total area of vapor passage through plate
(*d*) Area of individual vapor passages through plate
(*e*) Plate thickness
(*f*) Spacing of vapor passages through plate
(*g*) Height of liquid through which vapor passes
(*h*) Height of foam through which the vapor passes
(*i*) Length of liquid path
(*j*) Resistance to liquid flow (liquid gradient)
(*k*) Area of liquid-vapor contact per unit plate area
(*l*) Special design features—inlet baffle, foam baffle, etc.

3. *System variables*
   (*a*) Surface tension
   (*b*) Density of liquid
   (*c*) Viscosity of liquid
   (*d*) Density of vapor
   (*e*) Viscosity of vapor
   (*f*) Relative volatility of components in system
   (*g*) Diffusivity

There are 23 variables included in the list, and there are others which possibly contribute to the mass-transfer mechanism which are not definable in simple terms. In addition, there are undoubtedly interactions between the variables which must be taken into account in evaluating the efficiency of the stage or plate and of the column. Thus, the accurate evaluation of efficiency for the subsequent determination of actual stages is extremely difficult. Approximate estimation of efficiency may be done by a number of methods which are described fully in Chap. 13.

Plate Design or Stage Design

The plate contact device is designed to accomplish two purposes. One is to contact the vapor and liquid in such a manner that area and time of contact are at a maximum in order to obtain the highest rate of mass transfer. The other is to handle the required quantities of vapor and liquid without excessive pressure drop and with stable operation. The design of a particular type of plate is based upon a number of interrelated factors to produce the best possible efficiency and operating characteristics. Design methods for plates of various types are given in Chap. 14.

Column Design

The column design consists of designing the shell and plate arrangement with the necessary liquid and vapor flow paths, the vapor-disengaging space above the top plate, the liquid sump in the bottom of the column, the feed manifold and inlet nozzles, the reflux return, and the reboiler vapor distributor below the bottom plate.

Accessory Design

The column accessories consist of the condenser or condensers, the reflux accumulator, the reflux and product pumps, the reboiler, bottoms product pump, superstructure, including walkways and ladders, insulation, and instrumentation.

5.5. EQUILIBRIUM STAGES—GENERAL CONSIDERATIONS

Limits of Operability

A fractionating column by its inherent nature has two limits of operation based upon reflux ratio. These limits are fixed by (*a*) minimum reflux, on one hand, and (*b*) total reflux, on the other.

Under the conditions of minimum reflux there is insufficient liquid returned to the column to enable progressive enrichment of the vapor passing up the column with respect to the lighter components. The condition is a function of the relative volatility of the components being distilled and the equilibrium characteristics of each component. It is represented mathematically by the relation

$$y_{n+1} = y_n = y_{n-1} = y_{n-2} = \cdots \tag{5.1}$$

and

$$x_{n+1} = x_n = x_{n-1} = x_{n-2} = \cdots \tag{5.2}$$

for each and all components, and an infinite number of plates would be necessary for a separation, each plate causing an infinitesimal change in liquid and vapor composition. Actual operation of a column below or at minimum reflux is impossible, so that the actual minimum reflux representing one limit of operation must be determined mathematically by calculation or graphical solution, and the condition is incapable of experimental proof. Mathematical analysis of this limit seems to indicate one zone of constant composition in a fractionating column operating at minimum reflux on a binary system, two zones of constant composition, one above and one below the feed, in the case of a multicomponent system, and either one or two in the case of a ternary system. This is shown schematically in Fig. 5.4.

The other operating limit based on reflux ratio is that at total reflux. This requires the least number of stages for the desired separation, but practically no overhead product or bottom product can be made and no feed is introduced. Figure 5.5 schematically illustrates a column operating at total reflux. It is possible to operate experimentally a fractionating column under essentially the conditions imposed by the restriction of total reflux where the system inventory is large, and only very small samples of distillate and bottoms are removed for analysis so that the inventory is not noticeably disturbed. In practice this operation would never be performed because it would result in a negligible amount of product made.

It is evident from the foregoing discussion that any fractionating column must be operated with respect to reflux ratio between the limiting conditions of total reflux and minimum reflux. Although the minimum number of equilibrium stages is required at total reflux, neither feed, distillate, nor bottoms quantities are produced. On the other hand, at minimum reflux

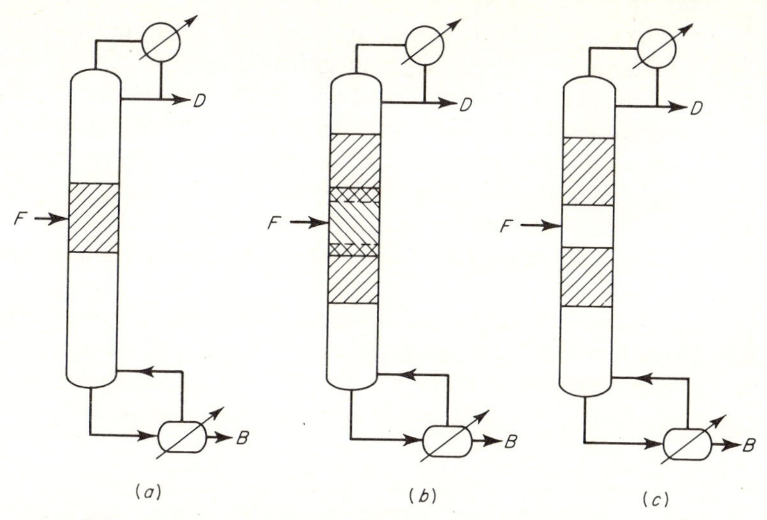

FIGURE 5.4 Minimum reflux—zones of constant composition. (*a*) Binary. (*b*) Ternary. (*c*) Multicomponent.

the column cannot produce the desired product because an infinite number of stages would be required.

Thus, the actual operating reflux ratio which must be used is controlled by economic considerations. These are primarily evaluated in terms of the operating and investment costs. A large reflux ratio involves greater operating costs because provision must be made for recirculating larger quantities of reflux liquid, more coolant is required for the condenser system,

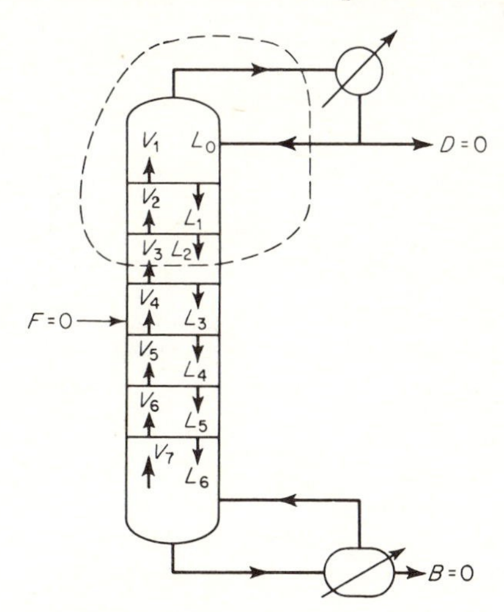

FIGURE 5.5 Schematic representation of total reflux operation.

and larger quantities of heating medium—fuel or steam—are needed to handle the vaporization of the larger quantities of liquid. On the other hand, a low reflux ratio requires the column to have a greater number of plates and greater height, thus increasing the investment cost. Occasionally a factor which is generally not considered becomes important in selection of reflux ratio. This is the increased investment cost necessary to provide adequate condenser area, reboiler area, or heater size, as well as plate area to provide space for downcomers which are capable of handling the increased liquid load.

The engineer arrives at the decision as to the proper (meaning economically feasible) reflux ratio on the basis of total cost of the distillation operation per unit of product produced. As a first step in the procedure this may be estimated by approximate designs and cost figures, and the column design

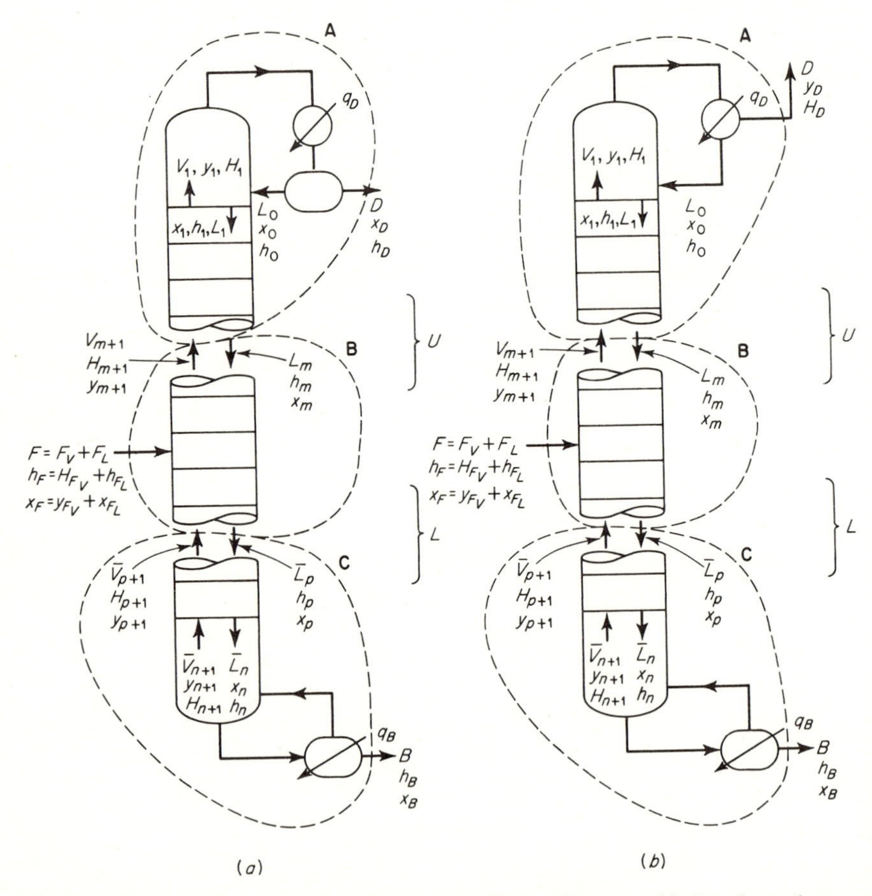

FIGURE 5.6 Schematic diagrams of fractionating columns. (*a*) Total condenser. (*b*) Partial condenser.

is based on this preliminary estimate. However, before it is finalized in complete engineering form, a careful economic analysis may be necessary. This is particularly true if the fractionation cost is a significant proportion of the total product cost and if the process is marginal from the over-all viewpoint.

Quantitative Relationships

General Fractionation Equations

If we assume steady-state operation, which is usual for most fractionating column operations, a series of quantitative relationships may be written relating over-all material balances, component material balances, and enthalpy balances which apply to all fractionating columns and to any type of system. The schematic diagram of a fractionating column is shown in Fig. 5.6, and the sections referred to in the following discussion are designated by the lettered envelopes about which the equations are written. The rectifying section, above the feed, is in the A envelope. The stripping section, below the feed, is in the C envelope, and the section around the feed is in the B envelope. The equations in final form relate internal reflux ratio L/V as well as external or operating ratio L_0/D in terms of composition, enthalpy, and total stream quantities.

Writing material and enthalpy balances about the condenser for each column results in the following:

Balances	Column *a* Total condenser		Column *b* Partial condenser	
Over-all	$V_1 = L_0 + D$	(5.3*a*)	$V_1 = L_0 + D$	(5.3*b*)
Component i	$V_1 y_{i_1} = L_0 x_{i_0} + D x_{i_D}$	(5.4*a*)	$V_1 y_{i_1} = L_0 x_{i_0} + D y_{i_D}$	(5.4*b*)
	$y_{i_1} = x_{i_0} = x_{i_D}$	(5.5*a*)	$y_{i_1} \neq x_{i_0} \neq y_{i_D};\ K_{i_D} = \dfrac{y_{i_D}}{x_{i_0}}$	(5.5*b*)
			$V_1 y_{i_1} = L_0 x_{i_0} + D x_{i_0} K_{i_D}$	(5.6*b*)
	$\dfrac{L_0}{V_1} = 1 - \dfrac{D}{V_1}$	(5.7*a*)	$\dfrac{L_0}{V_1} = \dfrac{y_{i_1}}{x_{i_0}} - \dfrac{D K_{i_D}}{V_1}$	(5.7*b*)
			$\dfrac{L_0}{V_1} = \dfrac{y_{i_1} - x_{i_0} K_{i_D}}{x_{i_0}(1 - K_{i_D})}$	(5.8*b*)
	$\dfrac{L_0}{D} = \dfrac{V_1}{D} - 1$	(5.9*a*)	$\dfrac{L_0}{D} = \dfrac{V_1}{D}\dfrac{y_{i_1}}{x_{i_0}} - K_{i_D}$	(5.9*b*)
			$\dfrac{L_0}{D} = \dfrac{x_{i_0} K_{i_D} - y_{i_1}}{y_{i_1} - x_{i_0}}$	(5.10*b*)
Enthalpy	$V_1 H_1 + q_D = L_0 h_0 + D h_D$	(5.11*a*)	$V_1 H_1 + q_D = L_0 h_0 + D H_D$	(5.11*b*)

The total heat removed in the condenser can be expressed in terms of the heat per unit mass of distillate stream times the mass of stream.

Column a Total condenser	Column b Partial condenser
$q_D = DQ_D$ (5.12a)	$q_D = DQ_D$ (5.12b)
$V_1H_1 + DQ_D = L_0h_0 + Dh_D$ (5.13a)	$V_1H_1 + DQ_D = L_0h_0 + DH_D$ (5.13b)
$V_1H_1 + (V_1 - L_0)Q_D = L_0h_0 + (V_1 - L_0)h_D$ (5.14a)	$V_1H_1 + (V_1 - L_0)Q_D = L_0h_0 + (V_1 - L_0)H_D$ (5.14b)
$H_1 + \left(1 - \frac{L_0}{V_1}\right)Q_D = \frac{L_0h_0}{V_1} + \left(1 - \frac{L_0}{V_1}\right)h_D$ (5.15a)	$H_1 + \left(1 - \frac{L_0}{V_1}\right)Q_D = \frac{L_0}{V_1}h_0 + \left(1 - \frac{L_0}{V_1}\right)H_D$ (5.15b)
$\frac{L_0}{V_1} = \frac{H_1 - h_D + Q_D}{h_0 - h_D + Q_D}$ (5.16a)	$\frac{L_0}{V_1} = \frac{H_1 - H_D + Q_D}{h_0 - H_D + Q_D}$ (5.16b)
$\frac{L_0}{D} = \frac{h_D - Q_D - H_1}{H_1 - h_0}$ (5.17a)	$\frac{L_0}{D} = \frac{H_D - Q_D - H_1}{H_1 - h_0}$ (5.17b)

The equations relating the streams in envelope A are given in the table on page 207.

In writing equations around the streams in envelope C, it is noted that they are the same for both columns because the distillate stream quantity, composition, and enthalpy do not enter into the equations. Thus, one set of equations will suffice for both columns [(a) total condenser and (b) partial condenser]:

Over-all:

$$\bar{L}_p = \bar{V}_{p+1} + B \tag{5.28}$$

Component i:

$$\bar{L}_p x_{i_p} = \bar{V}_{p+1}\, y_{p+1} + Bx_{i_B} \tag{5.29}$$

$$\bar{L}_p x_{i_p} = \bar{V}_{p+1} y_{i_{p+1}} + \bar{L}_p x_{i_B} - \bar{V}_{p+1} x_{i_B} \tag{5.30}$$

$$\frac{\bar{L}_p}{\bar{V}_{p+1}} = \frac{y_{i_{p+1}} - x_{i_B}}{x_{i_p} - x_{i_B}} \tag{5.31}$$

Balances	Column a Total condenser	Column b Partial condenser
Over-all	$V_{m+1} = L_m + D$ (5.18a)	$V_{m+1} = L_m + D$ (5.18b)
Component i	$V_{m+1}y_{i_{m+1}} = L_m x_{i_m} + Dx_{i_D}$ (5.19a)	$V_{m+1}y_{i_{m+1}} = L_m x_{i_m} + Dy_{i_D}$ (5.19b)
	$V_{m+1}y_{i_{m+1}} = L_m x_{i_m} + V_{m+1}x_{i_D} - L_m x_{i_D}$ (5.20a)	$V_{m+1}y_{i_{m+1}} = L_m x_{i_m} + V_{m+1}y_{i_D} - L_m y_{i_D}$ (5.20b)
	$\dfrac{L_m}{V_{m+1}} = \dfrac{y_{i_{m+1}} - x_{i_D}}{x_{i_m} - x_{i_D}}$ (5.21a)	$\dfrac{L_m}{V_{m+1}} = \dfrac{y_{i_{m+1}} - y_{i_D}}{x_{i_m} - y_{i_D}}$ (5.21b)
In terms of L_m/D	$\dfrac{L_m}{D} = \dfrac{x_{i_D} - y_{i_{m+1}}}{y_{i_{m+1}} - x_{i_m}}$ (5.22a)	$\dfrac{L_m}{D} = \dfrac{y_{i_D} - y_{i_{m+1}}}{y_{i_{m+1}} - x_{i_m}}$ (5.22b)
Enthalpy	$V_{m+1}H_{m+1} + DQ_D = L_m h_m + Dh_D$ (5.23a)	$V_{m+1}H_{m+1} + DQ_D = L_m h_m + DH_D$ (5.23b)
	$V_{m+1}H_{m+1} + V_{m+1}Q_D - L_m Q_D = L_m h_m + V_{m+1}h_D - L_m h_D$ (5.24a)	$V_{m+1}H_{m+1} + V_{m+1}Q_D - L_m Q_D = L_m h_m + V_{m+1}H_D - L_m H_D$ (5.24b)
	$H_{m+1} + Q_D - \dfrac{L_m}{V_{m+1}} Q_D = \dfrac{L_m}{V_{m+1}} h_m + h_D - \dfrac{L_m}{V_{m+1}} h_D$ (5.25a)	$H_{m+1} + Q_D - \dfrac{L_m}{V_{m+1}} Q_D = \dfrac{L_m}{V_{m+1}} h_m + H_D - \dfrac{L_m}{V_{m+1}} H_D$ (5.25b)
	$\dfrac{L_m}{V_{m+1}} = \dfrac{H_{m+1} - h_D + Q_D}{h_m - h_D + Q_D}$ (5.26a)	$\dfrac{L_m}{V_{m+1}} = \dfrac{H_{m+1} - H_D + Q_D}{h_m - H_D + Q_D}$ (5.26b)
In terms of L_m/D	$\dfrac{L_m}{D} = \dfrac{h_D - H_{m+1} - Q_D}{H_{m+1} - h_m}$ (5.27a)	$\dfrac{L_m}{D} = \dfrac{H_D - H_{m+1} - Q_D}{H_{m+1} - h_m}$ (5.27b)

Enthalpy:

$$\bar{L}_p h_p + q_B = \bar{V}_{p+1}H_{p+1} + Bh_B \tag{5.32}$$

$$q_B = BQ_B \tag{5.33}$$

$$\bar{L}_p h_p + \bar{L}_p Q_B - \bar{V}_{p+1}Q_B = \bar{V}_{p+1}H_{p+1} + \bar{L}h_B - \bar{V}_{p+1}h_B \tag{5.34}$$

$$\bar{L}_p(h_p + Q_B - h_B) = \bar{V}_{p+1}(H_{p+1} - h_B + Q_B) \tag{5.35}$$

$$\frac{\bar{L}_p}{\bar{V}_{p+1}} = \frac{H_{p+1} - h_B + Q_B}{h_p - h_B + Q_B} \tag{5.36}$$

Material and enthalpy balances written for the streams entering and leaving the reboiler give the same expressions as Eqs. (5.31) and (5.36) except they are in terms of the liquid stream leaving the column $\bar{L}_n$ and the vapor stream $\bar{V}_{n+1}$ entering from the reboiler. Thus,

$$\frac{\bar{L}_n}{\bar{V}_{n+1}} = \frac{y_{i_{n+1}} - x_{i_B}}{x_{i_n} - x_{i_B}} \tag{5.37}$$

$$\frac{\bar{L}_n}{\bar{V}_{n+1}} = \frac{H_{n+1} - h_B + Q_B}{h_n - h_B + Q_B} \tag{5.38}$$

Equations relating the streams in envelope *B*, around the feed plate, are as follows (again, one set of equations describes the relationships for both columns):

Over-all:

$$F_V + F_L + \bar{V}_{p+1} + L_m = \bar{L}_p + V_{m+1} \tag{5.39}$$

Component *i*:

$$F_V y_{i,F_V} + F_L x_{i,F_L} + \bar{V}_{p+1} y_{i_{p+1}} + L_m x_{i_m} = \bar{L}_p x_{i_p} + V_{m+1} y_{i_{m+1}} \tag{5.40}$$

$$F_V + \bar{V}_{p+1} = V_{m+1} \tag{5.41}$$

$$F_L + L_m = \bar{L}_p \tag{5.42}$$

Equations (5.41) and (5.42) are based on the assumption that the distilling system is essentially ideal in its behavior, and the feed is introduced at the correct plate or point in the column so there is no condensation or vaporization caused by addition of the feed. Although this situation is rarely encountered, calculations based on such assumptions are sufficiently accurate for engineering purposes. Correction for nonideality thermal effects requires data which are unavailable except for special (usually the simplest) cases.

Eliminating the streams below the feed by substitution, the following steps express L_m/V_{m+1} in terms of the remaining variables for the general case of mixed vapor and liquid feed.

$$F_V y_{i,F_V} + F_L x_{i,F_L} + V_{m+1} y_{i_{p+1}} - F_V y_{i_{p+1}} + L_m x_{i_m} = F_L x_{i_p} + L_m x_{i_p} + V_{m+1} y_{i_{m+1}} \tag{5.43}$$

$$L_m x_{i_m} - L_m x_{i_p} = V_{m+1} y_{i_{m+1}} - V_{m+1} y_{i_{p+1}} + F_L x_{i_p} - F_L x_{i,F_L} - F_V y_{i,F_V} + F_V y_{i_{p+1}} \tag{5.44}$$

$$\frac{L_m}{V_{m+1}} = \frac{y_{i_{m+1}} - y_{i_{p+1}}}{x_{i_m} - x_{i_p}} + \frac{F_L(x_{i_p} - x_{i,F_L})}{V_{m+1}(x_{i_m} - x_{i_p})} + \frac{F_V(y_{i_{p+1}} - y_{i,F_V})}{V_{m+1}(x_{i_m} - x_{i_p})} \tag{5.45}$$

If the feed is a saturated liquid, the last term in Eq. (5.45) drops out. If the feed is a saturated vapor, the middle term on the right side of Eq. (5.45) drops out. If necessary, similar equations can be written in terms of the stripping section internal reflux ratio $\bar{L}_p/\bar{V}_{p+1}$ by following the above procedure. The internal reflux ratio in the rectifying section may be written in terms of enthalpy as well as composition around the feed plate (envelope B).

$$F_V H_{F_V} + F_L h_{F_L} + \bar{V}_{p+1} H_{p+1} + L_m h_m = V_{m+1} H_{m+1} + \bar{L}_p h_p \quad (5.46)$$

$$F_V H_{F_V} + F_L h_{F_L} + V_{m+1} H_{p+1} - F_V H_{p+1} + L_m h_m = V_{m+1} H_{m+1} + F_L h_p + L_m h_p \quad (5.47)$$

$$L_m(h_m - h_p) = V_{m+1}(H_{m+1} - H_{p+1}) + F_L(h_p - h_{F_L}) + F_V(H_{p+1} - H_{F_V}) \quad (5.48)$$

$$\frac{L_m}{V_{m+1}} = \frac{H_{m+1} - H_{p+1}}{h_m - h_p} + \frac{F_L}{V_{m+1}} \frac{h_p - h_{F_L}}{h_m - h_p} + \frac{F_V}{V_{m+1}} \frac{H_{p+1} - H_{F_V}}{h_m - h_p} \quad (5.49)$$

Again, for a saturated liquid feed, the last term drops out, and for a saturated vapor feed the second term on the right-hand side is eliminated.

Enthalpies of Liquid and Vapor Streams

Enthalpies of liquid mixtures and of vapor mixtures constantly appear in equations used for distillation calculations. In a liquid mixture or solution the molal enthalpy of the mixture at a given temperature and pressure is the sum of the partial molal enthalpies of the components composing the mixture.

$$h_m = \sum_1^n \bar{h}_i \quad (5.50)$$

In "regular" or ideal mixtures the partial molal enthalpy of the component is the product of its mole fraction in the mixture and its enthalpy if it existed as a pure component at the same temperature and pressure. In nonideal solutions there are heat effects resulting from physical molecular interaction, orientation, induction, and dispersion forces and chemical molecular association or dissociation (hydrogen bonding or hydrogen bond breaking). These are usually classed as *heats of mixing*. Unfortunately, it is extremely difficult to predict mixture enthalpies with accuracy and relatively few experimental data have been determined and reported. Therefore, unless accurate data are available or can be predicted, it is customary to assume ideal liquid mixtures to approximate liquid mixture enthalpy.

For gaseous or vapor mixtures at normal temperatures and pressures the assumption of ideality is reasonably correct and errors introduced into the calculations are relatively slight. The vapor mixture enthalpy may be estimated by

$$H_m = \sum_1^n h_i y_i + \sum_1^n y_i \lambda_{i,t} \quad (5.51)$$

where data are not available. This approximation is based on the assumption that the individual components are vaporized at the reference temperature t_r, the vapors are heated to t, and then mixed.

Reflux and Number of Plates

Reflux liquid is the liquid usually produced by partially or completely condensing the distillate vapor from the fractionating column. It may be a bubble-point liquid resulting from the partial condensation of the vapor and whose composition is essentially the equilibrium composition with that of the vapor, or it may be the liquid condensate cooled below the bubble point of the liquid. It is common practice in some distillation processes, usually in the petroleum industry, to produce additional reflux liquid by removing liquid from a tray or plate at an intermediate location in the column, cooling it by passing it through an external exchanger, and returning it to the column at a temperature below its bubble point. A reflux ratio must be established before any quantitative fractionation design calculations can be made.

Earlier it was pointed out that a fractionating column can only produce the desired products between the limits of *minimum* reflux and *total* reflux. Therefore, the designer must be able to evaluate both limits quantitatively for the system which he is considering to ensure selection of a suitable range of reflux ratios for economic evaluation of the process. Figure 5.7 illustrates the relationship between reflux ratio and number of plates.

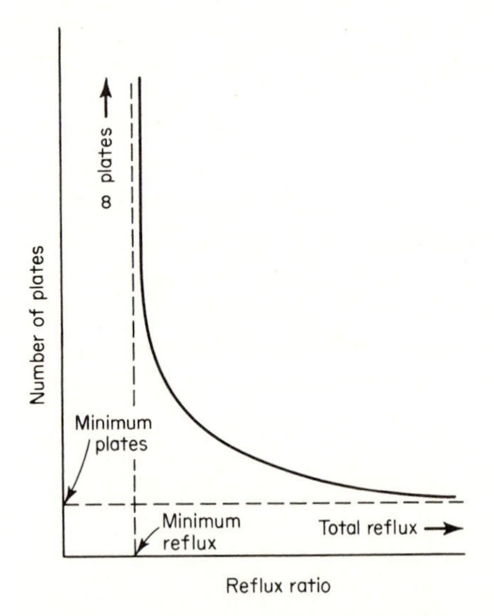

FIGURE 5.7 Schematic relationship between reflux ratio and number of plates or stages.

Minimum Reflux

Minimum reflux for a particular set of conditions may be estimated or calculated by the following methods:

1. Graphical methods (strictly applicable only to binary systems).

2. Short-cut-calculation methods involving the use of one or more simplifying assumptions regarding the relationship between key-component composition ratios in the feed and in the upper zone of constant composition and in the lower zone of constant composition, relative volatility of the key components, specified relationships between the compositions in the zones of constant composition and in the feed or in the products, and others.

3. Plate-to-plate calculations involving stepwise material and enthalpy-balance calculations until the liquid or vapor compositions obtained by calculation from successive stages become identical. This is a tedious trial-and-error method, particularly where multicomponent systems are being considered and the use of a digital computer is a necessity for those systems. There are several methods suitable for making this type of calculation by means of a computer. Holland's book [6] serves as a good reference for discussion of some of the calculation methods used and the pertinent equation forms.

General fractionation equations—minimum reflux Equations (5.3) to (5.49) can be modified to fit the condition of minimum reflux ratio, internal and external, by assuming that there are two zones of constant composition at minimum reflux conditions for a multicomponent system, one above the feed designated as U on Fig. 5.6 and one below the feed designated as L on the same figure. In zone U,

$$y_{m-1} = y_m = y_{m+1} = y_{m+2} = \cdots \tag{5.52}$$

and

$$x_{m-1} = x_m = x_{m+1} = x_{m+2} = \cdots \tag{5.53}$$

Thus, the composition of neither the liquid stream nor the vapor stream changes in the zone of constant composition from plate to plate. Similarly in the stripping section L

$$y_{p-1} = y_p = y_{p+1} = y_{p+2} = \cdots \tag{5.54}$$

and

$$x_{p-1} = x_p = x_{p+1} = x_{p+2} = \cdots \tag{5.55}$$

and the same situation exists. Since the composition of the liquid falling to the plate is the same as that falling from the plate, the vapor and liquid streams passing one another are in equilibrium with one another (when equilibrium stages are assumed).

$$K_{i_U} = \frac{y_{i_{m+1}}}{x_{i_m}} = \frac{y_{i_m}}{x_{i_{m+1}}} = \cdots \tag{5.56}$$

Also with no change in composition, there is no change in temperature and throughout each zone of constant composition the temperature is constant.

In the stripping zone L,

$$K_i = \frac{y_{i_{p+1}}}{x_{i_p}} = \frac{y_{i_p}}{x_{i_{p-1}}} = \cdots \tag{5.57}$$

Therefore, at minimum reflux conditions, Eqs. (5.21*a*), (5.21*b*), (5.22*a*), and (5.22*b*) may be modified to read as follows:

Total condenser	Partial condenser
$\left(\frac{L_m}{V_{m+1}}\right)_{\min} = \frac{K_{i_m}x_{i_m} - x_{i_D}}{x_{i_m} - x_{i_D}}$ (5.58*a*)	$\left(\frac{L_m}{V_{m+1}}\right)_{\min} = \frac{y_{i_m} - y_{i_D}}{y_{i_m}/K_{i_m} - y_{i_D}}$ (5.58*b*)
$\left(\frac{L_m}{D}\right)_{\min} = \frac{x_{i_D} - K_{i_m}x_{i_m}}{K_{i_m}x_{i_m} - x_{i_m}}$ (5.59*a*)	$\left(\frac{L_m}{D}\right)_{\min} = \frac{y_{i_D} - y_{i_m}}{y_{i_m} - y_{i_m}/K_{i_m}}$ (5.59*b*)

The enthalpy Eqs. (5.26*a*), (5.26*b*), (5.27*a*), and (5.27*b*) are:

Total condenser	Partial condenser
$\left(\frac{L_m}{V_{m+1}}\right)_{\min} = \frac{H_m - h_D + Q_D}{h_m - h_D + Q_D}$ (5.60*a*)	$\left(\frac{L_m}{V_{m+1}}\right)_{\min} = \frac{H_m - H_D + Q_D}{h_m - H_D + Q_D}$ (5.60*b*)
$\left(\frac{L_m}{D}\right)_{\min} = \frac{h_D - H_m - Q_D}{H_m - h_m}$ (5.61*a*)	$\left(\frac{L_m}{D}\right)_{\min} = \frac{H_D - H_m - Q_D}{H_m - h_m}$ (5.61*b*)

Similarly, Eqs. (5.31) and (5.36) become

$$\left(\frac{\bar{L}_p}{\bar{V}_{p+1}}\right)_{\min} = \frac{K_{i_p}x_{i_p} - x_{i_B}}{x_{i_p} - x_{i_B}} \tag{5.62}$$

and

$$\left(\frac{\bar{L}_p}{\bar{V}_{p+1}}\right)_{\min} = \frac{H_p - h_B + Q_B}{h_p - h_B + Q_B} \tag{5.63}$$

For the section around the feed, Eq. (5.45) becomes

$$\left(\frac{L_m}{V_{m+1}}\right)_{\min} = \frac{y_{i_m} - y_{i_p}}{y_{i_m}/K_{i_m} - y_{i_p}/K_{i_p}} + \frac{F_L}{V_m}\frac{x_{i_p} - x_{i,F_L}}{x_{i_m} - x_{i_p}} + \frac{F_V}{V_m}\frac{y_{i_p} - y_{i,F_V}}{y_{i_m}/K_{i_m} - y_{i_p}/K_{i_p}} \tag{5.64}$$

In the case of a saturated liquid feed the last term drops out, and for a saturated vapor feed the second term on the right side is eliminated.

The enthalpy equations change only in that the subscripts are simplified and no differentiation is made between one plate and the ones adjacent to it in the zone of constant composition.

$$\left(\frac{L_m}{V_{m+1}}\right)_{\min} = \frac{H_m - H_p}{h_m - h_p} + \frac{F_L}{V_{m+1}}\frac{h_p - h_{F_L}}{h_m - h_p} + \frac{F_V}{V_{m+1}}\frac{H_p - H_{F_V}}{h_m - h_p} \tag{5.65}$$

In the case of a binary system where it is commonly assumed that the upper and lower zones of constant composition merge at the feed plate, Eqs. (5.52) and (5.54) become

$$y_{m-1} = y_m = y_{m+1} = y_{p-1} = y_p = y_{p+1} = \cdots \tag{5.66}$$

and Eqs. (5.53) and (5.55) become

$$x_{m-1} = x_m = x_{m+1} = \cdots = x_{p-1} = x_p = x_{p+1} \tag{5.67}$$

Also, Eqs. (5.56) and (5.57) are the same:

$$K_{i_U} = K_{i_L} = \frac{y_{i_{m+1}}}{x_{i_m}} = \frac{y_{i_m}}{x_{i_{m+1}}} = \frac{y_{i_{p+1}}}{x_{i_p}} = \frac{y_{i_p}}{x_{i_{p+1}}} \tag{5.68}$$

Similar modifications of the Eqs. (5.58) through (5.63) can be made for their application to binary systems by eliminating the subscripts, m, $m + 1$, p, $p + 1$.

Equations (5.64) and (5.65) become Eqs. (5.69) and (5.70) when the composition subscripts and enthalpy subscripts are dropped out because of constant composition and temperature.

$$\left(\frac{\bar{L} - L}{V - \bar{V}}\right)_{\min} = \frac{y_{i,F_V} - y_i}{x_i - x_{i,F_L}} \tag{5.69}$$

$$\left(\frac{\bar{L} - L}{V - \bar{V}}\right)_{\min} = \frac{H_{F_V} - H_V}{h_L - h_{F_L}} \tag{5.70}$$

Total Reflux

Under the condition of total reflux there is no distillate, feed, or bottoms stream. When these restrictions are applied to the general fractionation equations, L_0/D has no meaning because $D = 0$ and $L_0 = V_1$, and the total reflux ratio internally is $L_0/V_1 = 1.0$. The difference between any two streams passing in the column at total reflux is zero and, thus,

$$L_0 = V_1 \qquad V_2 = L_1 \qquad V_3 = L_2 \qquad \text{etc.}$$

$$x_0 = y_1 \qquad x_1 = y_2 \qquad x_2 = y_3 \qquad \text{etc.}$$

The enthalpy balance around the condenser (both total and partial) is

$$V_1 H_1 + q_D = L_0 h_0 \tag{5.71}$$

Expressing

$$q_D = V_1 Q_1 \tag{5.72}$$

$$\left(\frac{L_0}{V_1}\right)_T = \frac{H_1 + Q_1}{h_0} = 1.0 \tag{5.73}$$

Similarly, around the reboiler,

$$\left(\frac{\bar{L}_n}{\bar{V}_{n+1}}\right)_T = \frac{H_{n+1} - Q_{n+1}}{h_n} = 1.0 \tag{5.74}$$

and

$$q_B = V_{n+1} Q_{n+1} \tag{5.75}$$

It must be remembered that q_D and q_B represent thermodynamic values where q is positive when it represents heat *added* to the system from the surroundings. Q_1 is then negative and Q_{n+1} is positive in value.

Variables in the evaluation of minimum reflux From the discussion and equations in the foregoing section, it is found that the same factors must be arbitrarily fixed or calculated to determine quantitatively the minimum reflux as those that are fixed in other fractionation calculations involving reflux and number of stages.

1. Feed quantity, composition, and enthalpy
2. Distillate quantity, composition, and enthalpy or temperature
3. Bottoms quantity, composition, and enthalpy or temperature
4. Composition and temperature in the upper zone of constant composition, and composition and temperature in the lower zone of constant composition
5. Pressure at which the distillation is to be carried out
6. Type of condenser—total or partial

Items 1, 2, 3, 5, and 6 are evaluated in all fractionation process designs because they are related to the feed availability and characteristics and the product specifications. Pressure is selected on the basis of a number of engineering considerations, and the type of condenser is related closely to the selection of the pressure. Therefore, the temperature and composition of the liquid and/or vapor in the zones of constant composition become the key factors in determining the minimum reflux ratio.

"Rigorous" minimum reflux calculations Rigorous calculation of minimum reflux involves the plate-to-plate calculation from the top of the column toward the feed plate and from the bottom of the column toward the feed plate until no change in composition of the liquid from plate to plate is encountered. This represents the pinch zone or zone of constant composition; that in the rectifying section above the feed plate is the upper pinch zone and that below is the lower pinch zone.

A necessary requirement for this type of calculation procedure is a known composition of the distillate and the bottoms, and this necessitates an assumption of the distribution of the components between the distillate and bottoms products. Although this distribution can be estimated, it is rarely possible to establish it exactly for a given feed composition and for an assumed separation between the key components. However, for an assumed or calculated component distribution between the distillate and bottoms product, the following method can be used.

1. Assume a minimum operating reflux ratio $(L_0/D)_{\min}$ and calculate the corresponding $(L/V)_{\min}$ and $(\bar{L}/\bar{V})_{\min}$ for the rectifying and stripping sections, respectively, accounting for the thermal condition of the feed.

2. Calculate the dew-point temperature of the vapor, V_1, from the top plate.

3. Determine the composition of the liquid falling from plate 1, L_1, from equilibrium data.

4. By material balance calculate the composition of the vapor rising from plate 2 by using Eqs. (5.18*a*) and (5.19*a*).

5. Calculate the dew-point temperature of V_2 and the equilibrium composition of L_2.

6. Repeat these stepwise calculations until there is no change in composition (or temperature) from plate to plate.

7. Calculate the bubble-point temperature of the liquid in the reboiler and the composition of the vapor in equilibrium with it.

8. By material balance calculate the liquid falling from the bottom plate by using Eqs. (5.28) and (5.29).

9. Continue the stepwise calculations up the column until the compositions do not change from plate to plate.

10. Add traces of lighter than light key components to the compositions in the lower pinch zone and continue upward to the feed plate (where the feed compositions essentially match those from the upward plate-to-plate calculation), and add the feed into the material on the tray.

11. Continue until the heavier than heavy key components have essentially disappeared.

12. Compare the compositions calculated from the stepwise procedure from the top of the column with those from step 11. If they are essentially the same, the right minimum reflux has been assumed. If not, a new minimum is assumed and the calculations are repeated.

It may be more convenient to add traces of heavier than heavy key to the upper-pinch-zone compositions and continue calculations downward, adding the feed at the correct place, to the point where the lighter than light key materials essentially disappear and compare with the lower-pinch-zone compositions calculated stepwise from the bottom.

This technique is a tedious one even if machine calculation is used, and the results are still approximate. More accurate results could be obtained by incorporating the enthalpy balances into the stepwise procedure. However, since there is some uncertainty with regard to prediction of the enthalpies of multicomponent mixtures, the increase in accuracy might be little or none.

The above procedure is based on the observation that in mixtures containing nothing lighter than the light key, the lower pinch zone occurs at the feed and the upper zone is above the feed. Where the mixture contains nothing heavier than the heavy key, the upper pinch zone occurs at the feed and the lower zone is below the feed. When the mixture contains both lighter than light and heavier than heavy key components, the two zones are located above and below the feed, respectively.

Short-cut or approximate minimum reflux calculation methods The interrelationship of the composition and temperature in the zones of constant composition to the feed and product compositions forms the basis for the approximate or short-cut minimum reflux calculations.

Many short-cut minimum reflux calculation methods have been devised. Colburn's [2] method essentially enables the calculation of the minimum reflux ratio for the separation of the key components as if they composed a binary system; the value thus obtained is then corrected to account for the volatility effects of the heavier than heavy key components in the upper zone of constant composition and of the lighter than light key components in the lower zone. He assumes constant molal overflow and relative volatilities at the temperatures in each of the zones. Underwood's [13] method consists of solving an equation relating feed composition, thermal condition of the feed, and relative volatilities, α's, of the components at the average temperature of the column for a factor θ which lies numerically between the relative volatilities of the keys, and using this factor in a second equation relating

$(L/D)_{min}$, α of the components and the distillate compositions. He assumes constant molal volatility for each component.

Brown and Martin [7] based their method for minimum reflux on the observation that, at the point of calculated minimum reflux, using plate-to-plate calculations, the ratio of the key components in the liquid in the zones of constant composition was essentially equal to that in the feed liquid. They related the evaluation of the temperatures in zones of constant composition and the reflux ratio to the relative volatility of the keys and other components so that this ratio was established. Gilliland [4] assumed two cases for minimum reflux: (1) the concentrations of all components are the same for a number of plates above and below the feed plate and (2) the ratio of the key components in both zones of constant composition is the same at minimum reflux.

Mayfield and May [9] modified Underwood's method and presented equations which apply to complete separations only. Scheibel and Montrose [10] developed empirical equations to apply to systems whose relative volatilities are constant. Essentially, they compute the minimum reflux ratio for separation of the binary composed of the light key and the heavy key, assuming all components lighter than the light key have infinite volatility and all heavier than the heavy key have zero volatility. The value thus established is corrected empirically for the actual volatility of the light and heavy materials neglected in the first step. Shiras et al. [11] presented a series of equations which are based upon assumptions of constant molal overflow and constant relative volatility, and which were limited to the assumed location of the pinch zones in the column. Bachelor [1] presented a method for calculation of minimum reflux by using a mathematical relaxation technique wherein successive calculation steps are followed to converge with any degree of accuracy required to the true minimum reflux. May [8] devised a method based on calculation of minimum reflux for an equivalent series of binaries which resolve to the multicomponent mixture. Equimolal overflow and constant relative volatility are assumed and the method does not apply to split keys.

Recommended short-cut methods Three methods are described here in detail.

1. The Brown-Martin method [7], which is the least complicated, requiring fewer calculations and giving calculated values of minimum reflux on the safe side, i.e., the $(L/D)_{min}$ calculated is greater than the $(L/D)_{min}$ actual—is recommended for primarily multicomponent hydrocarbon distillation situations where great accuracy is not required and where comparative values for different pressures of distillation will suffice.

2. The Underwood [13] method, which requires somewhat more extensive calculations and assumes that relative volatilities are constant, is recommended as an intermediate method giving reasonable engineering accuracy for systems which can be classed as approaching ideality under the conditions of the distillation.

3. Colburn's [2] method, which assumes constant relative volatility in each zone of constant composition and corrects for lighter than light key and heavier than heavy key volatilities, is recommended as probably the most accurate short-cut means of evaluating minimum reflux ratio. However, it must be emphasized that none of these methods is highly accurate.

Brown and Martin Method [7]

In multicomponent systems if there are assumed to be two zones of constant composition requiring infinite plates for separation of the components, one above the feed plate and one below the feed plate, it can be postulated that there are sections above and below each zone of constant composition wherein some separation between the components is possible. Under the conditions of minimum reflux, even in multicomponent systems, *if the separation takes place between the two least volatile components in the feed in the upper part of the column*, the ratio of the mole fractions of the key components in the liquid in the zone of constant composition (*upper zone*) would be the same as that in the liquid portion of the feed. In this particular case, calculation of the minimum reflux ratio necessary to produce the *ratio* of the mole fractions of the cut components in the upper zone to that of the cut components in the feed establishes the correct value.

In almost all practical situations, however, the separation takes place between components of intermediate volatility, and some of the less volatile components are rectified from the vapor *above* the feed plate and between the feed and the upper zone of constant composition. This rectification changes the ratio of the cut components to a somewhat higher value in the upper zone of constant composition. Therefore, the *actual minimum reflux ratio* is *slightly less* than that calculated on the basis of matching the ratios of the cut components in the upper zone and in the feed. However, calculation of the minimum reflux ratio on this basis gives a value on the safe side.

Method of Calculation (*Rectifying Section*) From a material balance for the heavy key component around the top of the column and between the plates in the upper pinch zone (see Fig. 5.8),

$$VK_j x_j = Dx_{j_D} + Lx_j \tag{5.76}$$

If none of the heavy key appeared in the distillate but did appear in the upper zone of constant composition,

$$VK_j x_j = Lx_j \tag{5.77}$$

$$K_j = \frac{L}{V} \tag{5.78}$$

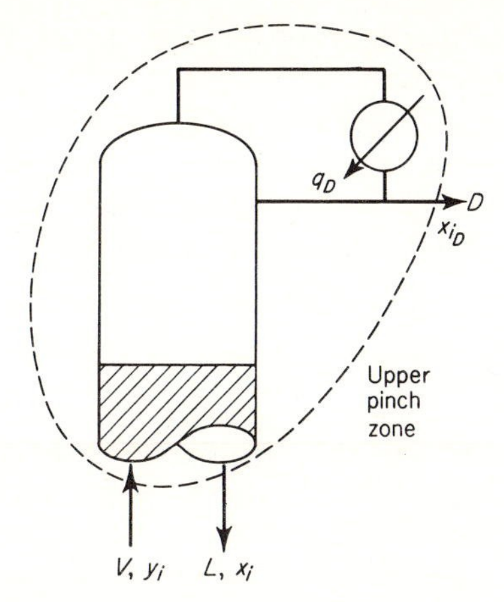

FIGURE 5.8 Upper zone material balance.

Thus, the temperature in the pinch zone would be that at which the $K_{HK} = (L/V)_{\min}$.

The stepwise method is as follows:

1. Assume a value for $(L/V)_{\min}$. This is equal to K_{HK}.

2. At the column pressure determine the temperature from suitable K charts. Since the distillate actually contains some of the heavy key, the temperature will be slightly higher in the pinch zone and the K values, and, thus, the $(L/V)_{\min}$ will be slightly higher than the value indicated previously. Therefore, assume a slightly higher value and evaluate the temperature.

The component material balance is calculated as follows:

$$y_i V = Dx_{i_D} + Lx_i$$

$$V = D + L$$

$$y_i = K_i x_i$$

$$(D + L)K_i x_i = Dx_{i_D} + Lx_i$$

$$K_i Dx_i + K_i Lx_i = Dx_{i_D} + Lx_i$$

$$K_i Dx_i + K_i Lx_i - Lx_i = Dx_{i_D}$$

$$x_i = \frac{Dx_{i_D}}{K_i D + K_i L - L}$$

$$x_i = \frac{Dx_{i_D}}{K_i D + L(K_i - 1)}$$

$$\Sigma x = 1.0 = \sum \frac{Dx_{i_D}}{K_i D + L(K_i - 1)}$$

3. Evaluate the K values for all components in the distillate including the heavy key at the temperature and pressure in the upper zone of constant composition.

4. From the foregoing equations calculate the liquid composition in the pinch zone by calculating the compositions down to and including that of the light key and obtaining the composition of the heavy key by difference.

5. Assume two more minimum reflux values and repeat the calculations.

6. Plot the ratio of x_{LK}/x_{HK} versus $L_{\min}$ as shown in Fig. 5.9.

7. At the point where the ratio of the liquid compositions of the light and heavy key equals the ratio in the liquid portion of the feed, read the $L_{\min}$ and calculate $(L/V)_{\min}$.

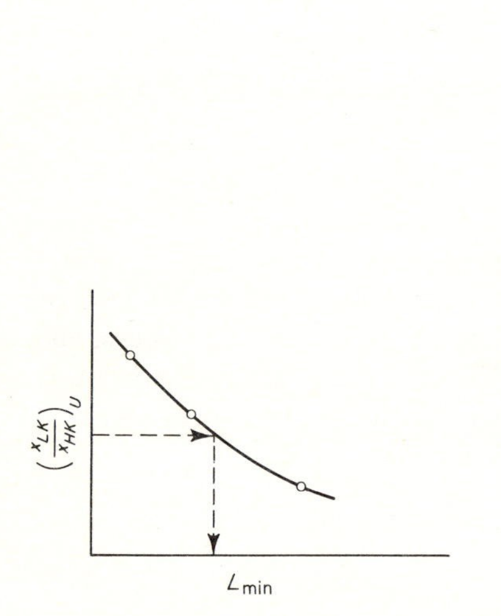

FIGURE 5.9 Minimum L.

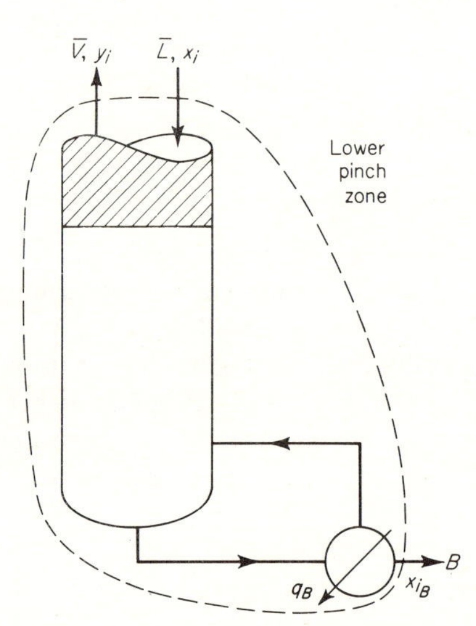

FIGURE 5.10 Lower zone material balance.

This value is a bit higher than the actual minimum and, unless great accuracy is needed, may be used as the actual. The error is on the safe side.

Method of Calculation (Stripping Section) In considering the stripping section involving the lower zone of constant composition, equations (see Fig. 5.10) similar to those for the rectifying section can be used to evaluate minimum reflux.

The component balance equations are:

$$\bar{L}x_i = \bar{V}y_i + Bx_{i_B}$$
$$\bar{L} = \bar{V} + B$$
$$y_i = K_i x_i$$
$$\bar{L}x_i = \bar{V}K_i x_i + Bx_{i_B}$$
$$\bar{L}x_i = (\bar{L} - B)x_i K_i + Bx_{i_B}$$
$$\bar{L}x_i = \bar{L}x_i K_i - Bx_i K_i + Bx_{i_B}$$
$$\bar{L}x_i - \bar{L}x_i K_i + Bx_i K_i = Bx_{i_B}$$
$$x_i(\bar{L} - \bar{L}K_i + BK_i) = Bx_{i_B}$$
$$x_i = \frac{Bx_{i_B}}{BK_i + L(1 - K_i)}$$
$$\Sigma x_i = 1.0 = \sum \frac{Bx_{i_B}}{BK_i + \bar{L}(1 - K_i)}$$

If none of the light key component *i* appears in the bottom product,

$$\bar{L}x_i = \bar{V}y_i + Bx_{i_B} \qquad x_{i_B} = 0 \tag{5.79}$$
$$\bar{L}x_i = \bar{V}K_i x_i \tag{5.80}$$
$$\frac{\bar{L}}{\bar{V}} = K_{LK} \tag{5.81}$$

Three values of $\bar{L}/\bar{V}$ are selected and the K's of the components are evaluated at temperatures corresponding to values somewhat *less* than $K_{LK} = \bar{L}/\bar{V}$. For several values of $\bar{L}/\bar{V}$ compute the x values and the ratios of x_{LK}/x_{HK}. Plot as shown in Fig. 5.11 and read the approximate minimum $\bar{L}$, from which the minimum reflux ratio is calculated.

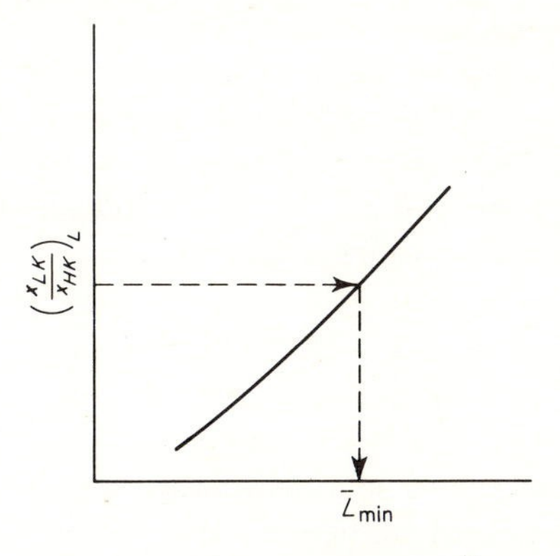

FIGURE 5.11 Minimum $\bar{L}$.

The $(\bar{L}/\bar{V})_{min}$ is the approximate minimum reflux ratio when the separation is based on the stripping section of the column. This value is somewhat *more* than the actual $(\bar{L}/\bar{V})_{min}$ and is therefore a safe design value.

Because in this method of calculation the values of $(\bar{L}/\bar{V})_{min}$ and $(L/V)_{min}$ are not consistent (because the same components do not appear in both zones in the calculations), the design minimum reflux ratio is that giving the higher value of $(L/V)_{min}$. This is attributed to "more difficult" separation. Either both $(L/V)_{min}$ and $(\bar{L}/\bar{V})_{min}$ are calculated and the equivalent highest $(L/V)_{min}$ is selected or the number of plates in the rectifying section and the stripping section are determined for total reflux, and the one requiring the most plates (feed to distillate or bottoms) is used to designate the "more difficult" separation section for which the higher value of $(L/V)_{min}$ is determined.

Underwood's Method [13]

Underwood assumed constant molal overflow and constant relative volatility at the mean column temperature. His method is applicable to close separations and to rough separations. Where the light key and heavy key are adjacent, i.e., there are no intermediate distributed keys, the following stepwise method is used:

1. The value of θ, the correct value of which must lie numerically between the α values of the keys, is found by trial and error from

$$\sum_1^n \frac{x_F}{(\alpha - \theta)/\alpha} = 1 - q = \frac{(\alpha_a x_a)_F}{\alpha_a - \theta} + \frac{(\alpha_b x_b)_F}{\alpha_b - \theta} + \cdots + \frac{(\alpha_n x_n)_F}{\alpha_n - \theta} \tag{5.82}$$

where x_{a_F} = mole fraction total of a in the feed
α_a = relative volatility of a referred to the heaviest component or to the heavy key
q = the number of moles of saturated liquid formed on the feed plate by the introduction of one mole of feed. For a bubble-point feed $q = 1.0$; for a dew-point feed $q = 0$

2. Substitute the value of θ into Eq. (5.83) and with the use of the distillate compositions calculate the value of $(L/D)_{min}$.

$$\left(\frac{L_0}{D}\right)_{min} + 1 = \sum_1^n \frac{x_D}{(\alpha - \theta)/\alpha} = \frac{(\alpha_a x_a)_D}{\alpha_a - \theta} + \frac{(\alpha_b x_b)_D}{\alpha_b - \theta} + \cdots + \frac{(\alpha_n x_n)_D}{\alpha_n - \theta} \tag{5.83}$$

In the situation where there is a distributed key or a component having a relative volatility intermediate between that of the light key and that of the heavy key, there are two values of θ to be solved for. One value lies between α_{LK} and α_{DK} (distributed key) and one value lies between α_{DK} and α_{HK}.

The distillate will contain the amounts of the light and heavy key desired for the split. It will also contain the distributed key in an unknown (as yet) amount, and all components lighter than the light key will be present along with some amount of the distributed key.

(*a*) Assume that the ratios of the lighter than light key compositions to that of the light key are the same as in the feed. This means also that no heavier than heavy key is present in the distillate. Set up Eqs. (5.84) to (5.89) to relate x_{d_D} to the feed compositions and the composition of the light key in the distillate x_{c_D}. Thus if a, b, c, d, e, f ($c = LK$, $d = DK$, $e = HK$) make up the feed mixture,

$$\left(\frac{x_a}{x_c}\right)_D = \left(\frac{x_a}{x_c}\right)_F \qquad \left(\frac{x_b}{x_c}\right)_D = \left(\frac{x_b}{x_c}\right)_F \tag{5.84}$$

Since

$$x_{a_D} + x_{b_D} + x_{c_D} + x_{d_D} = 1.0 \tag{5.85}$$

$$x_{a_D} = x_{c_D}\frac{x_{a_F}}{x_{c_F}} \qquad x_{b_D} = x_{c_D}\frac{x_{b_F}}{x_{c_F}} \tag{5.86}$$

$$x_{c_D}\frac{x_{a_F}}{x_{c_F}} + x_{c_D}\frac{x_{b_F}}{x_{c_F}} + x_{c_D} + x_{d_D} = 1.0 \tag{5.87}$$

or

$$x_{c_D}\left(\frac{x_{a_F}}{x_{c_F}} + \frac{x_{b_F}}{x_{c_F}} + 1\right) + x_{d_D} = 1.0 \tag{5.88}$$

If x_{c_D} is known,

$$x_{d_D} = 1.0 - x_{c_D}\left(\frac{x_{a_F}}{x_{c_F}} + \frac{x_{b_F}}{x_{c_F}} + 1\right) \tag{5.89}$$

where x_d is composition of distributed key component.

(*b*) Solve Eq. (5.82) including components a through n,

$$\sum_1^n \frac{x_F}{(a - \theta)/\alpha} = 1 - q$$

getting two values of θ, one lying between α of the light key and α of the distributed key, and one between α of the distributed key and α of the heavy key.

(*c*) Substitute the θ values in Eq. (5.90):

$$\left(\frac{L}{D}\right)_{\min} + 1 = \sum \frac{x_D}{(\alpha - \theta)/\alpha} + \frac{x_{d_D}}{(\alpha - \theta)/\alpha} \tag{5.90}$$

This gives two equations with two unknowns, $(L/D)_{\min}$ and x_{d_D}, which can be solved simultaneously for the two unknowns.

The Colburn Method [2]

1. In the stepwise procedure a value of $(L/V)_{\min}$ is assumed and the estimation of the temperatures in both the upper and lower zones of constant composition is made by (*a*) assuming in the upper zone of constant composition,

$$t = t_{\text{top}} + \frac{t_{\text{bottom}} - t_{\text{top}}}{3}$$

or (*b*) using the Brown-Martin method. Assume $x_{D,HK} = 0$.

Therefore in the upper zone of constant composition,

$$\frac{y_{n+1,HK}}{x_{n+1,HK}} = K_{HK} = \left(\frac{L}{V}\right)_{\min}$$

Assume $(L/V)_{\min} = K_{HK}$. At the column pressure the temperature in the upper zone is that which will give $K_{HK} = (L/V)$ assumed. Similarly, in the lower zone, the temperature is that which will give $(\bar{L}/\bar{V})_{\min}$ corresponding to K_{LK}.

2. The approximate compositions in the upper zone of constant composition are calculated by the following equations, assuming $LK = c$; $HK = d$; $m =$ upper zone; $p =$ lower zone.

$$(Lx_a)_m = \frac{(Dx_a)_D}{\left(\frac{K_a}{K_d}\right)_m - 1 + \left(\frac{K_a}{K_d}\right)_m \frac{(Dx_d)_D}{(Lx_d)_m}} \quad \text{or} \quad \frac{(Dx_a)_D}{(\alpha_{ad})_m - 1 + (\alpha_{ad})_m \frac{(Dx_d)_D}{(Lx_d)_m}} \tag{5.91}$$

$$(Lx_b)_m = \frac{(Dx_b)_D}{(\alpha_{bd})_m - 1 + (\alpha_{bd})_m \frac{(Dx_d)_D}{(Lx_d)_m}} \tag{5.92}$$

$$(Lx_c)_m = \frac{(Dx_c)_D}{(\alpha_{cd})_m - 1 + (\alpha_{cd})_m \frac{(Dx_d)_D}{(Lx_d)_m}} \tag{5.93}$$

$(Lx_d)_m$ is determined later. $(Lx_e)_m, (Lx_f)_m, \ldots = 0$.

[*Note:* On the basis of the assumed temperature and material balances, all terms on the right-hand side of the equations are known except $(Lx_d)_m$. When K's are used, errors in assumed temperatures are relatively negligible in the effect on the solution.]

For sharp separations, i.e., $(Dx_a)_D$ is a small number (x_a is 5% or less), the equations reduce to

$$(Lx_a)_m = \frac{(Dx_a)_D}{(\alpha_{ad})_m - 1} \cdots \tag{5.94}$$

For poor separations $(Dx_d)_D$ is not small and $(Lx_d)_m$ may be approximated (in the above equations only) by

$$(Lx_d)_m = \left(\frac{x_d}{x_c}\right)_F \frac{(Dx_c)_D}{\alpha_{cd} - 1} \tag{5.95}$$

3. The approximate compositions in the lower zone of constant composition are calculated by

$$(Lx_a)_p, (Lx_b)_p = 0$$

$(Lx_c)_p$ is determined later.

$$(Lx_d)_p = \frac{(Bx_d)_B}{1 - (\alpha_{dc})_p + (\alpha_{dc})_p \dfrac{(Bx_c)_B}{(Lx_c)_p}} \tag{5.96}$$

$$(Lx_e)_p = \frac{(Bx_e)_B}{1 - (\alpha_{ec})_p + (\alpha_{ec})_p \dfrac{(Bx_c)_B}{(Lx_c)_p}} \tag{5.97}$$

Note: If the separation is sharp, $(Bx_c)_B$ is small, and the equations reduce to

$$(Lx_d)_p = \frac{(Bx_d)_B}{1 - (\alpha_{dc})_p} \tag{5.98}$$

If the separation is not sharp and $(Bx_c)_B$ is appreciable, $(Lx_c)_p$ may be approximated (for the above equations only) by

$$(Lx_c)_p = \left(\frac{x_c}{x_d}\right)_F \frac{(Bx_d)_B}{1 - (\alpha_{dc})_p} \tag{5.99}$$

4. (*a*) Calculate for the upper zone,

$$\frac{(\alpha_{cd} - 1)\alpha_{cd}}{\alpha_{ad}} = \frac{(K_c/K_d - 1)(K_c/K_d)}{K_a/K_d} \tag{5.100}$$

for all components lighter than the light key (in this case for a and b).

(*b*) Using Fig. 5.12 evaluate C_m factors for all components lighter than the light key (C_a, C_b, in this case).

(*c*) Evaluate the summation

$$\Sigma\,(CL_x)_m = (C_aLx_a)_m + (C_bLx_b)_m \tag{5.101}$$

5. (*a*) Calculate for the lower zone of constant composition

$$(\alpha_{ed} - 1)\alpha_{ed} = \left(\frac{K_e}{K_d} - 1\right)\frac{K_e}{K_d} \tag{5.102}$$

for all components heavier than the heavy key (i.e., for *e* and *f*).

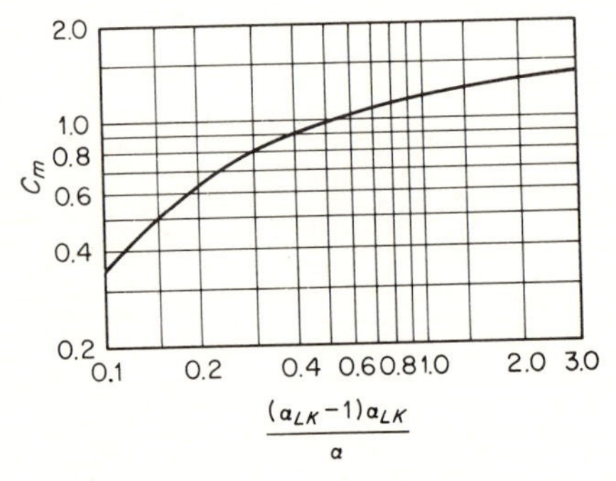

FIGURE 5.12 Correction factor C_m. [*From Colburn, Trans. A.I.Ch.E.*, **37:** 805 (1941).]

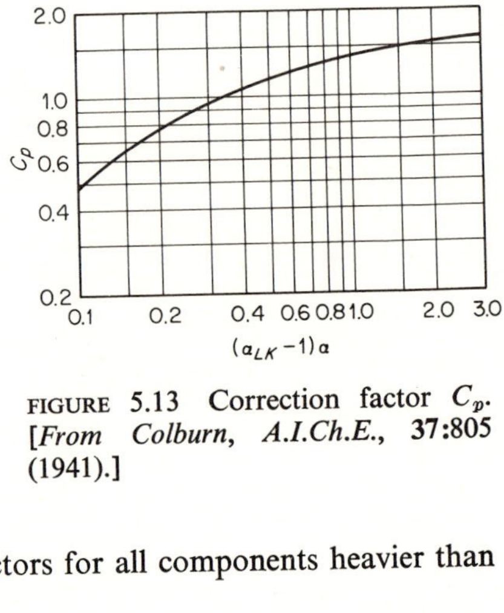

FIGURE 5.13 Correction factor C_p. [*From Colburn, A.I.Ch.E.*, **37:**805 (1941).]

(*b*) Using Fig. 5.13, evaluate C_p factors for all components heavier than the heavy key (C_e,C_f).

(*c*) Evaluate the summation

$$\Sigma\left(C\frac{K}{K_d}Lx\right)_p = \left(C_e\frac{K_e}{K_d}Lx_e\right)_p + \left(C_f\frac{K_f}{K_d}Lx_f\right)_p \tag{5.103}$$

6. Solve the following for $(Lx_d)_p$:

$$\frac{(Lx_c)_p(Lx_d)_m}{(Lx_d)_p(Lx_c)_m} = \frac{L_pL_m}{\left[L_p - \Sigma\left(C\frac{K}{K_d}L_x\right)_p\right][L_m - \Sigma\,(CL_x)_m]} \tag{5.104}$$

as follows:

(*a*)
$$(Lx_c)_m = \frac{(Dx_c)_D}{\left(\frac{K_c}{K_d}\right)_m - 1 + \left(\frac{K_c}{K_d}\right)_m \frac{(Dx_d)_D}{(Lx_d)_m}} \tag{5.105}$$

(*b*)
$$(Lx_d)_p = \frac{(Bx_d)_B}{1 - \left(\frac{K_d}{K_c}\right)_p + \left(\frac{K_d}{K_c}\right)_p \frac{(Bx_c)_B}{(Lx_c)_p}} \tag{5.106}$$

(*c*)
$$L_p = L_m + F_L \tag{5.107}$$

where F_L = liquid portion of feed.

$$L_p = (Lx_c)_p + (Lx_d + Lx_e + Lx_f)_p \tag{5.108}$$

$$L_m = (Lx_d)_m + (Lx_a + Lx_b + Lx_c)_m \tag{5.109}$$

$$(Lx_c)_p + (Lx_d + Lx_e + Lx_f)_p = (Lx_d)_m + (Lx_a + Lx_b + Lx_c)_m + F_L \tag{5.110}$$

$$(Lx_c)_p = (Lx_d)_m + (Lx_a + Lx_b + Lx_c)_m + F_L - (Lx_d + Lx_e + Lx_f)_p \tag{5.111}$$

(*d*) With all the terms in the above equation known except $(Lx_d)_m$, solve it for $(Lx_d)_m$ by trial and error.

7. Determine $(Lx_c)_p$, L_p, and L_m.
8. Calculate L/V knowing L_m, D, and L_p, B.
9. If the L/V calculated does not equal the L/V assumed, repeat.
10. *Note:* The following equation must be satisfied.

$$\frac{r_p}{r_m} = \frac{(x_c/x_d)_p}{(x_c/x_d)_m} = \frac{1}{\left[1 - \sum\left(C_p \frac{K}{K_d} x\right)_p\right][1 - \sum (C_m x)_m]} = \psi \tag{5.112}$$

where $\Sigma\,[C_p(K/K_d)x]_p$ is for all components heavier than the heavy key and $\Sigma\,(C_m x)_m$ is for all components lighter than the light key.

Colburn Minimum Reflux Ratio Where Intermediate Distributed Components Are Present The method is as follows:

1. Designate the intermediate component as j and assume the distribution of j in the distillate and in the bottoms. Under step 2 above,

$$(Lx_j)_m = \frac{(Dx_j)_D}{\left(\frac{K_j}{K_d}\right)_m - 1 + \left(\frac{K_j}{K_d}\right)_m \frac{(Dx_d)_D}{(Lx_d)_m}} \tag{5.113}$$

2. Under step 3 above,

$$(Lx_j)_p = \frac{(Bx_j)_B}{1 - \left(\frac{K_j}{K_c}\right)_p + \left(\frac{K_j}{K_c}\right)_p \frac{(Bx_c)_B}{(Lx_c)_p}} \tag{5.114}$$

3. Under step 4 above, no correction—same as before.
4. Under step 6 above, no correction—same as before.
5. Under step 6 above,

$$L_p = (Lx_c)_p + (Lx_j + Lx_d + Lx_e + Lx_f)_p \tag{5.115}$$

$$L_m = (Lx_d)_m + (Lx_a + Lx_b + Lx_c + Lx_j)_m \tag{5.116}$$

6. Trial-and-error solution gives an approximate solution with the distributed key.

7. Check by comparing with the assumed distribution of the intermediate components.

$$\frac{(Lx_c)_m}{(Lx_c)_p} < \frac{(Lx_j)_m}{(Lx_j)_p} < \frac{(Lx_d)_m}{(Lx_d)_p} \tag{5.117}$$

If this relation is not satisfied, juggle the distribution of the intermediate component and solve for $(Lx_j)_m$ and $(Lx_j)_p$ until the relation is satisfied.

Knowing the composition of both zones of constant composition, the bubble-point calculations can be made to check the assumed temperatures in these zones. Using K values, however, the sensitivity of the calculations to temperature is slight.

Example 5.1

A 60 mole % methanol, 20 mole % ethanol, 20 mole % 1-propanol mixture is to be fractionated to produce a distillate containing 99% of the methanol in the feed and having a purity of 99 mole % methanol. Assume ideal behavior of the mixture. The data are:

Data	MeOH	EtOH	1-PrOH
Antoine constants			
A	7.87863	8.04494	7.99733
B	1473.11	1554.3	1569.70
C	230.0	222.65	209.5
λ, cal/g mole, BP	8491.0	9255.0	10,030.0
$C_{p,\text{molal}}$, cal/(°C)(g mole)	19.11	26.26	34.74

Calculate the minimum reflux ratio for the following situations by (*a*) the Colburn method and (*b*) the Brown-Martin method.

1. Bubble-point feed, bubble-point reflux, total condenser
2. Bubble-point feed, reflux 20°C below the bubble point, total condenser
3. Dew-point vapor feed, bubble-point reflux, total condenser
4. Dew-point vapor feed, reflux 20°C below the bubble point, total condenser

Solution

(*a*) COLBURN METHOD. For (1):

Material Balance Basis—1.0 Mole of Feed

Component	Feed	Distillate		Bottoms	
		Moles	Mole fr.	Moles	Mole fr.
(1) MeOH (*LK*)	0.60	0.594	0.99	0.006	0.015
(2) EtOH (*HK*)	0.20	0.006	0.01	0.194	0.485
(3) 1-PrOH	0.20	—	—	0.200	0.500
	1.00	0.600	1.00	0.400	1.000

Column temperatures are calculated assuming that Raoult's law holds for the mixture:

Temperature at top plate = 65°C (by interpolation)
Temperature at feed plate = 71.5°C
Temperature, upper pinch ≃ 68°C
Temperature at reboiler = 86°C
Temperature, lower pinch ≃ 79°C

An approximate value of $(L/D)_{\text{min}}$ is calculated following the suggestion of Colburn:

$$\left(\frac{L}{D}\right)_{\text{min}} = \frac{1}{(\alpha_{12}-1)}\left[\frac{(x_1)_D}{(x_1)_m} - \alpha_{12}\frac{(x_2)_D}{(x_2)_m}\right]$$

where

$$(x_1)_m = \frac{r_f}{(1+r_f)(1+\alpha_{32}(x_3)_F} \qquad r_f = \left(\frac{x_1}{x_2}\right)_F$$

m = upper pinch $\quad p$ = lower pinch

$$(x_1)_m \cong \frac{3}{4[1+0.446(0.2)]} = 0.689$$

$$\left(\frac{L}{D}\right)_{\text{min}} = \frac{1}{1.72-1}\left(\frac{0.99}{0.689} - \text{neg.}\right) = 2.0$$

Having assumed a value of $(L/D)_{\text{min}}$, values of x_p and x_m are calculated with Eqs. (5.95) and (5.105). For this particular ternary, Eq. (5.105) reduces to

$$(x_1)_m = \frac{(x_1)_D}{(\alpha_{12} - 1)\dfrac{L}{D} + \dfrac{\alpha_{12}(x_2)_D}{(x_2)_m}}$$

Also, $x_1 + x_2 = 1.0$ in the upper pinch. The equation can be solved explicitly for $(x_1)_m$. An iterative procedure is used, however, since convergence is rapid. In the lower pinch Eq. (5.95) reduces to

$$(x)_p = \frac{\alpha_{12}x_B}{(\alpha_{12} - \alpha)\dfrac{L}{B} + \alpha\dfrac{(x_1)_B}{(x_1)_p}}$$

In the lower pinch all components are present and $x_1 + x_2 + x_3 = 1.0$. Starting values of x_2 and x_3 are obtained by assuming that the second term in the denominator is negligible and then calculating x_1 by difference. Convergence to final values is usually obtained with one iteration. Abbreviated results of the trials are given in the following table:

Trial	Upper pinch					Lower pinch							r_p/r_m	ψ
	L/D	t_m, °C	α_{12}	x_1	x_2	L/B	t_p,°C	α_{12}	α_3	x_1	x_2	x_3		
1	2.0	68	1.72	0.662	0.338	5.5	79	1.662	0.463	0.654	0.22	0.126	1.52	1.06
2	1.8	69	1.72	0.723	0.277	5.2	70	1.716	0.445	0.649	0.222	0.129	1.00	1.06
3	1.81	68	1.72	0.726	0.274	5.215	70	1.716	0.445	0.650	0.221	0.129	1.11	1.06

The pinch-zone temperatures are corrected by computing the bubble-point temperature at pinch composition. The value of r_p/r_m is very sensitive to changes in L/D while ψ is not so sensitive. The final interpolated value of $(L/D)_{\min}$ is 1.805.

For (2): Heat and material balances written around the rectifying section result in

$$V_{m+1} - L_m = D$$

$$V_{m+1}H_{m+1} - L_m h_m = D(h_D - Q_d)$$

In order for the solution to (1) to apply, the internal reflux ratios must be held constant, necessitating a revision in the external reflux ratio. That is, for a balance below the mth plate and above the first plate,

$$V' - L_0' = V_{m+1} - L_m = D$$

$$V'H_1' - L_0'(h_D - \Delta) = V_{m+1}H_{m+1} - L_m h_m = D(h_D - Q_d)$$

where V_1', L_0' = the altered vapor and liquid rates, respectively
Δ = amount of supercooling, Btu/lb mole

The new external reflux ratio is

$$\left(\frac{L_0'}{D}\right)_{\min} \cong \frac{(h_D - Q_d) - H_1}{H_1 - (h_D - \Delta)}$$

Using the data of (1),

$$1.805 = \frac{(h_D - Q_d) - H_1}{8491}$$

approximately,

$$\Delta = (20)(19.9) = 382$$

$$\left(\frac{L_0'}{D}\right)_{min} = \frac{(891)(1.805)}{(8491 + 382)} = 1.73$$

For (3): The composition of the liquid in equilibrium with the vapor feed is determined to provide a value of r_f. Then an initial value of $(L/D)_{min}$ is calculated as before.

Dew-point Temperature and Composition of Liquid in Equilibrium with Feed

Component	y_F	K, 78°C	$x_F = y_F/K$
(1) MeOH (*LK*)	0.60	1.645	0.365
(2) EtOH (*HK*)	0.20	0.988	0.202
(3) 1-PrOH	0.20	0.454	0.441
			1.008

$$r_f \cong 1.81$$

$$(x_1)_m \cong 0.536 \qquad \left(\frac{L}{D}\right)_{min} \cong 2.8$$

The calculations for minimum reflux are carried on as in the case of liquid feed except that the internal reflux is calculated differently. A summary of the trials is given:

Trial	Upper pinch					Lower pinch							r_p/r_m	ψ
	L/D	t_m, °C	α_{12}	x_1	x_2	L/B	t_p, °C	α_{12}	α_{32}	x_1	x_2	x_3		
1	2.8	71.5	1.704	0.495	0.505	4.2	82	1.637	0.467	0.54	0.294	0.166	1.87	0.179
2	2.6	71	1.701	0.535	0.465	3.7	72	1.700	0.446	0.528	0.299	0.173	1.535	1.079
3	2.35	70.5	1.707	0.581	0.419	3.52	72.5	1.696	0.448	0.478	0.330	0.193	1.043	1.089
4	2.40	70.5	1.707	0.565	0.435	3.60	72.5	1.696	0.448	0.486	0.326	0.188	1.148	1.087

The final value of $(L/D)_{min}$ is taken to be 2.37.

For (4):

$$\left(\frac{L_0'}{D}\right)_{min} = \frac{8491(2.37)}{8491 + 382} = 2.26$$

(*b*) BROWN-MARTIN METHOD. For (1):

Upper pinch zone. The ratio of light key to heavy key in the feed is

$$\left(\frac{x_1}{x_2}\right)_F = \frac{0.6}{0.2} = 3$$

Component 3 is excluded from the upper pinch and, consequently,

$$\left(\frac{x_1}{x_2}\right)_m = 3 \qquad x_1 = 0.75 \qquad x_2 = 0.25$$

$$L = \frac{D[(x_D/x)_i - K_i]}{K_i - 1} \qquad i = 1, 2$$

Several temperature trials are made to find the L value which satisfies the equations for components 1 and 2 simultaneously.

Trial 1:

$$t = 70°C \qquad K_1 = 1.224 \qquad K_2 = 0.714$$

$$L(1) = \frac{0.6\left(\dfrac{0.99}{0.75} - 1.224\right)}{1.224 - 1} = 0.257$$

$$L(2) = \frac{0.6\left(\dfrac{0.01}{0.25} - 0.714\right)}{0.714 - 1} = 1.41$$

Trial 2:

$$t = 67°C \qquad K_1 = 1.091 \qquad K_2 = 0.630$$

$$L(1) = 1.510 \qquad L(2) = 0.957$$

Trial 3:

$$t = 68°C \qquad K_1 = 1.134 \qquad K_2 = 0.658$$

$$L(1) = 0.833 \qquad L(2) = 1.084$$

By interpolation, $L = 1.04$,

$$(L/D)_{\min} = 1.733$$

Lower pinch zone. All components are present in the pinch zone and no simplification is possible. Several temperature trials are made to find the internal reflux ratio at which $(x_1/x_2)_m = 3.0$.

Trial 1: Assume $(\bar{L}/\bar{V})_{\min} = 1.180$, $K_1 = 1.180$, $t = 69°C$, $K_2 = 0.687$, $K_3 = 0.302$.

$$\bar{L} = \bar{V} + 0.4 = 1.18\bar{V}$$

$$\bar{V} = \frac{0.4}{0.18} = 2.222 \qquad \bar{L} = 2.622$$

$$x_2 = \frac{B(x_2)_B}{BK_2 + \bar{L}(1 - K_2)} = \frac{(0.4)(0.485)}{(0.4)(0.687) + 2.662(1 - 0.687)} = 0.177$$

$$x_3 = 0.103$$

$$x_1 = 1 - 0.177 - 0.103 = 0.720$$

$$\left(\frac{x_1}{x_2}\right)_m = 4.07$$

Trial 2: Assume $(\bar{L}/\bar{V})_{\min} = 1.224$, $K_1 = 1.224$, $t = 70°C$, $K_2 = 0.714$, $K_3 = 0.316$, $\bar{V} = 1.786$, $\bar{L} = 2.186$.

$$x_2 = 0.213 \qquad x_3 = 0.123$$

$$x_1 = 1 - 0.213 - 0.123 = 0.664$$

$$\left(\frac{x_1}{x_2}\right)_m = 3.07$$

Trial 3: Assume $(\bar{L}/\bar{V})_{\min} = 1.252$, $K_1 = 1.252$, $t = 70.6°C$, $K_2 = 0.733$, $K_3 = 0.326$, $\bar{V} = 1.587$, $\bar{L} = 1.987$.

$$x_2 = 0.235 \qquad x_3 = 0.136$$

$$x_1 = 1 - 0.235 - 0.136 = 0.629$$

$$\left(\frac{x_1}{x_2}\right)_m = 2.677$$

By interpolation, $\bar{L} = 2.155$ and $(L/D)_{\min} = 1.925$.

The values of minimum external reflux ratio calculated at upper and lower pinch do not agree. The larger value is chosen as a factor of safety.

For (2): *Cold reflux.* Using the reflux ratio calculated in (1), i.e., 1.925,

$$\left(\frac{L_0'}{D}\right)_{\min} = \frac{8491}{8491 + 382}\, 1.925 = 1.84$$

For (3): *Dew-point vapor feed.* The liquid at the feed plate is taken as that in equilibrium with the feed. From the solution of (1), under the Colburn method,

$$\left(\frac{x_1}{x_2}\right)_F = \frac{0.365}{0.202} = 1.81$$

For the *upper pinch zone* the trials are summarized:

Trial	t, °C	$L(1)$	$L(2)$
1	70	0.833	1.439
2	69	1.19	1.26
3	68.5	1.46	1.18

By interpolation, $L = 1.24$, $(L/D)_{\min} = 2.067$.

For the *lower pinch zone*:

Trial	t, °C	$\bar{L}$	$(x_1/x_2)_m$
1	70.6	1.987	2.677
2	72	1.65	1.913
3	72.5	1.553	1.713

By interpolation, $\bar{L} = L = 1.6$, $(L/D)_{\min} = 2.67$. The larger value, $(L/D)_{\min} = 2.67$ is chosen.

For (4): Using cold reflux and a dew-point vapor feed,

$$\left(\frac{L_0'}{D}\right) = (0.957)(2.67) = 2.55$$

Example 5.2

A ternary mixture at its bubble point composed of components a, b, and c is to be fractionated to produce the products indicated. The system is considered to be essentially ideal. Calculate the minimum reflux by the Underwood method. The reflux is a bubble-point liquid.

Component	x_F	α	x_D	x_B
a	0.047	4.20	0.1263	0.000
(LK) b	0.072	1.57	0.1913	0.001
(HK) c	0.881	1.00	0.6824	0.999

$$q = 1.0$$

Solution

$$1 - q = 0 = \sum \frac{x_F}{(\alpha - \theta)/\alpha}$$

Assume $\theta = 1.5028$.

Component	α	$\dfrac{\alpha - \theta}{\alpha}$	x_F	$\dfrac{x_F}{(\alpha - \theta)/\alpha}$
a	4.20	0.6425	0.047	0.0731
b	1.57	0.0428	0.072	1.680
c	1.00	−0.5028	0.881	−1.753
				0.00

Thus, the assumed value of θ is correct. Solve for $(L/D)_{\min}$:

Component	$\dfrac{\alpha - \theta}{\alpha}$	x_D	$\dfrac{x_D}{(\alpha - \theta)/\alpha}$
a	0.6425	0.1263	0.1965
b	0.0428	0.1913	4.4470
c	−0.5028	0.6824	−1.3570
			3.310

$$(L/D)_{\min} = \frac{x_D}{(\alpha - \theta)/\alpha} - 1 = 3.310 - 1.0 = 2.310.$$

Example 5.3

A multicomponent mixture of the composition below and consisting of 66% vapor is to be fractionated to produce the products shown. Determine the minimum reflux by the Underwood method. The reflux is a bubble-point liquid.

Solution

Component	x_F	x_D	x_B	α
a	0.26	0.434	—	100
b	0.09	0.150	—	24.6
(*LK*) *c*	0.25	0.411	0.010	10.0
(*HK*) *d*	0.17	0.005	0.417	4.85
e	0.11	—	0.274	2.08
f	0.12	—	0.299	1.00

Base on the *heaviest* component (not the heavy key). Since the feed consists of 66% vapor, $1 - q = 0.66$.

$$\sum \frac{x_F}{(\alpha - \theta)/\alpha} = 1 - q = 0.66$$

Trial 1*:* Assume $\theta = 6.73$.

Component	α	$\dfrac{\alpha - \theta}{\alpha}$	x_F	$\dfrac{x_F}{(\alpha - \theta)/\alpha}$
a	100	0.9327	0.26	0.279
b	24.6	0.726	0.09	0.124
c	10.0	0.327	0.25	0.765
d	4.85	0.388	0.17	−0.438
e	2.08	−2.235	0.11	−0.049
f	1.00	−6.73	0.12	−0.018
				0.661

Assumed $0.66 = 1 - q$. Calculated $0.661 = 1 - q$. This is a satisfactory solution. The $(L/D)_m$ calculation is:

Component	$\alpha - \theta$	x_D	$\dfrac{x_D}{(\alpha - \theta)/\alpha}$	$(L/D)_m + 1 = 1.911$
a	0.938	0.434	0.462	
b	0.726	0.150	0.207	$(L/D)_m = 0.911$
c	0.327	0.411	1.255	
d	−0.388	0.005	−0.013	

Determination of the minimum plates at total reflux The same general classification of calculation procedures for evaluation of minimum plates at total reflux can be used as was used for evaluation of minimum reflux.

1. Graphical methods applicable only to binary systems
2. Short-cut methods
3. Plate-to-plate methods

The graphical methods are discussed in Chap. 6 under Binary Fractionation, and the other methods are described in the following.

Short-cut or approximate methods for total reflux In an ideal mixture the ratio of vapor pressures or the ratio of the equilibrium vaporization ratios, K's, of the key components is constant over the range of temperatures considered, i.e., the relative volatilities are constant. If this can be assumed without the introduction of excessive error, the number of equilibrium stages at total reflux may be calculated by the Fenske equation or the Underwood equation.

The Fenske Method [3] *for Total Reflux*

The Fenske equation is

$$N_m = \frac{\log\,[(x_{LK}/x_{HK})_D(x_{HK}/x_{LK})_B]}{\log\,(\alpha_{LK/HK})_{\text{av}}} \tag{5.118}$$

where N_m are the minimum theoretical stages at total reflux.

Normally, the reboiler is considered one theoretical stage, and also a partial condenser, if one is used, is considered one theoretical stage. Thus, the number of plates (theoretical) in the column utilizing a reboiler and a partial condenser is $N - 2$.

The relative volatility α used in Eq. (5.118) may be evaluated as the arithmetic average α between the top and bottom temperatures.

Thus,

$$\alpha_{\text{av}} = \frac{\alpha_{t_0} + \alpha_{t_B}}{2} \tag{5.119}$$

The Underwood Method for Total Reflux [12]

The Underwood method is essentially the same as that of Fenske. The minimum plates are calculated by

$$N'_m + 1 = \frac{\log\,[(x_{LK}/x_{HK})_0(x_{HK}/x_{LK})_{n+1}]}{\log \dfrac{K_{LK,\text{av}}}{K_{HK,\text{av}}}} \tag{5.120}$$

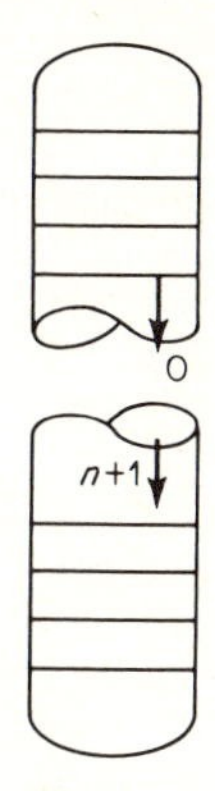

FIGURE 5.14 Location of plate 0 and plate $n + 1$.

where N'_m = number of plates between 0 and $n + 1$ (see Fig. 5.14)
K_{av} = arithmetic average K between t_0 and t_{n+1}

Underwood recommends the calculation by plate-to-plate (see pp. 215 and 239 and Chap. 7) methods down three or four plates from the top and up three or four plates from the bottom, and applying the equation to the section between. However, many engineers use Underwood's method applied over the whole column (without making the few plate-to-plate calculations). The recommendation is based on the observation that the greatest temperature gradients are usually encountered in the top and bottom few plates.

Example 5.4

The following feed is available at its bubble point at 100 psia and it is to be fractionated to produce a residue containing 95% of the isobutane in the feed and the composition with respect to propane not more than 0.1 mole %. Determine the minimum plates at total reflux by (*a*) the Fenske method and (*b*) the Underwood method. Use K values from Fig. 2.41.

Feed	x_F
(1) Ethane	0.06
(2) Propene	0.09
(3) Propane	0.18
(4) Isobutane	0.18
(5) Butene	0.25
(6) Butane	0.24

Solution

The material balance is fixed by the specifications on the bottoms (constant vapor and liquid enthalpics are assumed in this example). Basis—1 mole of feed.

Component	Distillate		Residue	
	Mole	Mole fr.	Mole	Mole fr.
(1)	0.060	0.177	—	—
(2)	0.090	0.266	—	—
(3) *LK*	0.179	0.530	0.00066	0.001
(4) *HK*	0.009	0.027	0.171	0.258
(5)	—	—	0.250	0.378
(6)	—	—	0.240	0.363
	0.338	1.00	0.662	1.000

(*a*) FENSKE METHOD. The distillate and bottoms temperatures are calculated.

Dew point of distillate				Bubble point of bottoms			
Component	y_1	K, 36°F	$x_1 = y_1/K$	Component	x_n	K, 127°F	$y_n = Kx_n$
1	0.177	3.7	0.048	3	0.001	2.5	0.0025
2	0.266	0.98	0.271	4	0.258	1.15	0.297
3	0.530	0.79	0.671	5	0.378	1.0	0.378
4	0.027	0.275	0.010	6	0.363	0.88	0.319
			1.000				0.9965

$$\alpha_{\text{top}} = 2.87 \qquad \alpha_{\text{bottom}} = 2.17 \qquad \bar{\alpha} = (2.87 \times 2.17)^{1/2} = 2.51$$

$$N_m = \frac{\log \left(\dfrac{0.530/0.027}{0.001/0.258} \right)}{\log 2.51} = 9.3$$

(*b*) UNDERWOOD METHOD. Plate-to-plate calculations are carried out to the third plate and Eq. (5.120) is applied to those remaining. At total reflux, $L_1 = V_2$, and hence the vapor rising from plate 2 is determined by the dew-point calculation of part (*a*). L_2 is determined by an equilibrium calculation.

Plate 2				Plate 3		
Component	y_2	K, 46°F	$x_2 = y_3$	Component	K, 53°F	x_3
1	0.048	4.1	0.012	1	4.4	0.003
2	0.271	1.15	0.236	2	1.27	0.186
3 (*LK*)	0.671	0.94	0.712	3	1.04	0.684
4 (*HK*)	0.010	0.34	0.0294	4	0.39	0.076
5	—	0.275	0.011	5	0.32	0.034
			1.000	6	—	0.014
						0.997

The amount of component 5 is estimated with the use of Eq. (5.118) applied to the nine-plate column of part (*a*).

$$\left(\frac{x_4}{x_5}\right)_2 \cong (\bar{\alpha}_{45})^7 \left(\frac{x_4}{x_5}\right)_9 \cong (1.21)^7 \frac{0.258}{0.378} \cong 2.6$$

Plate $p-1$			Plate $p-2$		
Component	x_{p-1}	K, 126°F	$y_{p-1} (= x_{p-2})$	K, 125°F	y_{p-2}
2	—	—	—	2.83	0.002
3 (*LK*)	0.0025	2.48	0.0062	2.45	0.015
4 (*HK*)	0.299	1.13	0.339	1.12	0.380
5	0.380	0.99	0.371	0.97	0.366
6	0.319	0.87	0.228	0.84	0.236
	1.000		1.000		0.999

$$(K_{LK})_{av} = \frac{1.04 + 2.45}{2} = 1.75$$

$$(K_{HK})_{av} = \frac{0.39 + 1.12}{2} = 0.75$$

$$N' + 1 = \frac{\log \dfrac{0.684/0.076}{0.0062/0.339}}{\log (1.75/0.75)} = 7.3$$

Total plates = 12.3.

Plate-to-plate evaluation of number of stages of total reflux At total reflux the quantity of liquid falling from a plate and the quantity of vapor rising

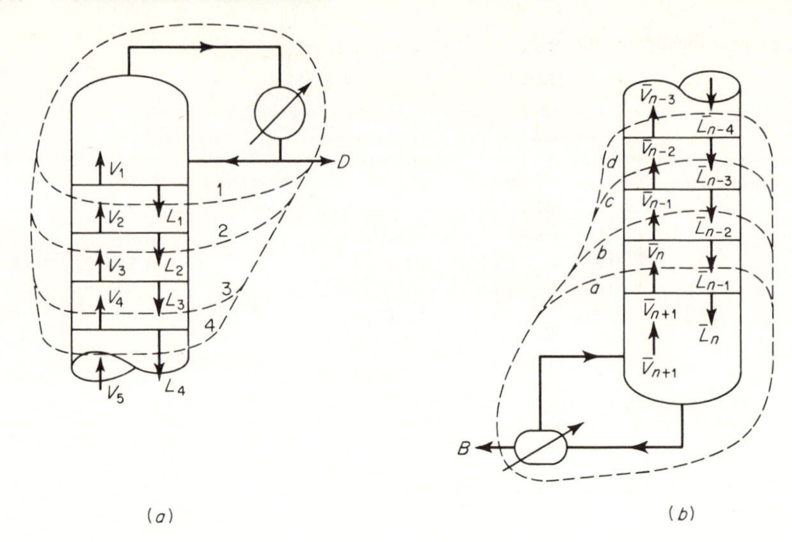

FIGURE 5.15 Material balances. (*a*) Rectifying section. (*b*) Stripping section.

to it are equal (Fig. 5.15). Similarly, the composition of the liquid and the composition of the vapor passing between plates is the same, i.e.,

$$L_1 = V_2 \qquad L_2 = V_3 \qquad L_3 = V_4 \qquad \cdots$$

$$x_{i1} = y_{i2} \qquad x_{i2} = y_{i3} \qquad x_{i3} = y_{i4} \qquad \cdots$$

The plate-to-plate calculation methods utilizing these relationships are as follows:

1. Starting with the quantity and composition of V_1 (which has the same composition as D), by equilibrium relationship determine the composition of L_1 by assuming, e.g., three trial temperatures for plate 1 and using the corresponding K values or α values. The defining dew-point equation enables calculation of the summation of the composition values for each temperature.

$$\sum_1^n x_{i1} = 1.0 = \sum_1^n \frac{y_{i1}}{K_{i1}}$$

By plotting the $\sum_1^n x_i$ versus assumed temperature, the correct temperature is found from the curve at the point at which the summation equals unity.

2. The vapor composition from plate 2 equals the liquid composition from plate 1, i.e., $x_{i1} = y_{i2}$. Therefore, calculate the temperature of plate 2 as above and continue down the column until the bottoms composition and temperature is reached. To calculate progressively plate to plate from top to bottom of the column, all components must be present in both distillate (V_1) and bottoms (L_n) streams. It is difficult to estimate minute quantities of the heavier than heavy keys in the distillate and lighter than light keys in the bottoms for sharp separations. Therefore, it is common practice to compute up from the bottom, calculating the bubble-point temperature of the liquid on each plate as well as down from the top, calculating the dew-point temperature on each plate, and plotting the compositions of the streams

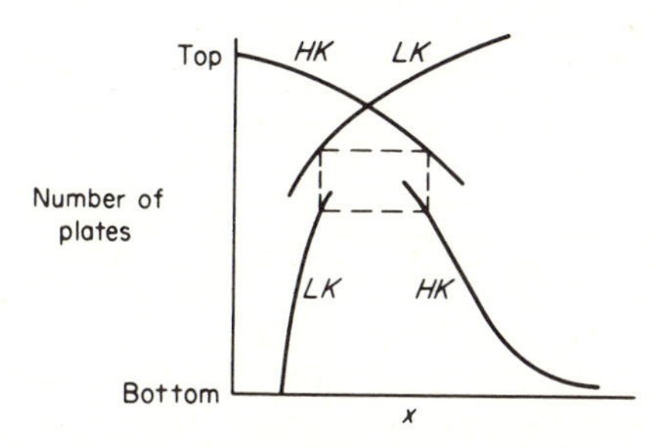

FIGURE 5.16 Matching key compositions.

versus plate number. At the point where the calculated compositions of the keys in the liquid or vapor agree as the calculations proceed from the two directions, the plates down from the top and up from the bottom are added to get the minimum plates (see Fig. 5.16).

Empirical Evaluation of Number of Equilibrium Stages as a Function of the Limits of Total-reflux Minimum Stages and Minimum-reflux Infinite Stages

The number of theoretical or equilibrium stages required for a given separation at a given reflux ratio may be determined by use of empirical correlations such as those of Brown and Martin [7] or Gilliland [4]. The Brown-Martin correlation, more commonly used for hydrocarbon systems, relates the ratio

$$\frac{[(\bar{L}V/\bar{V}L) - 1]}{[(\bar{L}V/\bar{V}L) - 1]_{\text{min reflux}}}$$

and the ratio of the number of equilibrium stages N at reflux ratio L/V to the number of equilibrium stages $N_{\min}$ at total reflux with parameters of

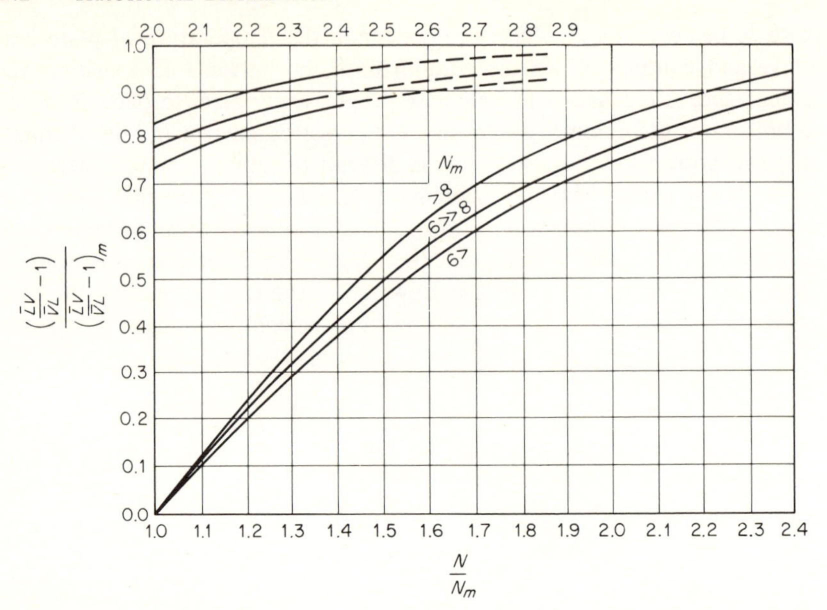

FIGURE 5.17 Brown-Martin correlation of reflux ratio, theoretical stages, minimum reflux ratio, and minimum stages. [*From Martin and Brown, Trans. A.I.Ch.E.*, **35**:679 (1939).]

$N_{\min} = 6$, $N_{\min} = 6$ to 8, and $N_{\min} = 8$. This correlation is shown in Fig. 5.17. The Gilliland correlation relates the terms

$$\frac{N - N_m}{N + 1} \quad \text{and} \quad \frac{L/D - (L/D)_{\min}}{L/D + 1}$$

where N = the number of equilibrium stages at reflux ratio L/D
N_m = the number of equilibrium stages at total reflux
$(L/D)_{\min}$ = the minimum reflux

This correlation is shown in Fig. 5.18.

The method of determining the number of equilibrium stages at a desired reflux ratio is as follows:

1. Determine the minimum reflux ratio by one of the short-cut methods given in this chapter.

2. Determine the minimum number of stages at total reflux.

3. Compute the values of the ordinate, in the event the Brown-Martin relationship is used, or the abscissae, where the Gilliland method is used, and evaluate the corresponding abscissae or ordinate.

4. From this value with the minimum number of equilibrium stages known, the number of equilibrium stages at the required reflux is calculated.

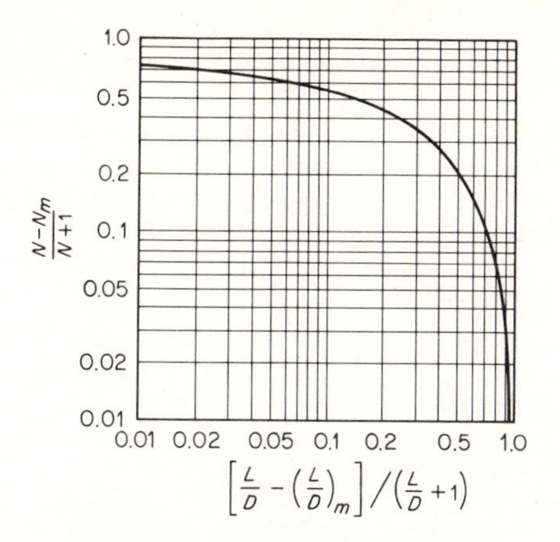

FIGURE 5.18 Gilliland correlation of reflux ratio, theoretical stages, minimum reflux ratio, and minimum stages. [*From Gilliland, Ind. Eng. Chem.*, **32:**1101 (1940).]

Generally, the Brown-Martin relation has been used more for hydrocarbons, whereas the Gilliland relation has been applied more to the nonhydrocarbon systems. Again it must be emphasized that these correlations are empirical and generalized with respect to system and conditions. Therefore, it cannot be expected that highly accurate results can be obtained.

PROBLEMS

1. A benzene–toluene cut at its bubble point, 62 mole % benzene, 38 mole % toluene (assume ideal behavior), is to be fractionated at 120 psia to recover toluene of 99.9% purity with a yield of 95% of the toluene in the feed.

Components	Antoine constants		
	A	*B*	*C*
Benzene	6.9057	1211.03	220.79
Toluene	6.9533	1343.94	219.38

Determine the minimum reflux by the following methods: (*a*) Colburn; (*b*) Underwood; (*c*) Brown-Martin.

2. Determine the minimum plates at total reflux by (*a*) Fenske method; (*b*) Underwood method (plate to plate for three plates down and three plates up).

3. For an operating reflux, $L/D = 1.5(L/D)_{min}$, calculate the internal reflux ratios L/V and $\bar{L}/\bar{V}$ for the case where the reflux liquid is returned to the column at its bubble point.
4. It is desired to fractionate at 1000 mm Hg pressure the ternary system—hexane, 60 mole %, 2-methyl pentane, 12 mole %, and heptane, 28 mole %—to obtain a distillate containing 99.9% of the hexane and residue containing 99.9% of the heptane. The 2-methyl pentane will distribute between the distillate and bottoms. The feed and reflux are at their bubble points. Assume ideal behavior.

Component	Antoine constants		
	A	*B*	*C*
Hexane	6.8778	1171.53	224.37
2-Methyl pentane	6.8391	1135.41	226.57
Heptane	6.9024	1268.12	216.9

(*a*) Calculate the minimum reflux by the Colburn, the Underwood, and the Brown-Martin methods.

(*b*) Determine the minimum plates by the Fenske method.

5. A natural gasoline of the following composition is available as a saturated vapor feed at 50 psia from the absorption-stripping operation. It is to be fractionated to remove the light ends (stabilized) to obtain a residue which will have an absolute vapor pressure of 24 psia at 100°F. Using the *K* chart, Fig. 2.41, for both a partial condenser and a total condenser, determine (*a*) minimum reflux by Underwood and by the Brown-Martin method; (*b*) minimum plates by the Fenske method.

Feed Material

Component	*x*
Ethane	0.06
Propane	0.12
i-Butane	0.11
n-Butane	0.22
i-Pentane	0.06
n-Pentane	0.18
Hexanes + (use *K* for heptane)	0.25

6. The following hydrocarbon mixture (at its bubble-point temperature) is to be fractionated at a pressure of 200 mm Hg to produce a distillate containing 95% of the dodecene and 1% of tridecane contained in the feed. Determine the minimum reflux by the Brown-Martin method and the minimum plates by the Fenske method. Assume ideality.

Olefin Feed

Component	x	ρ, 20°C, g/cu cm	MABP, 760 mm Hg, °C
1-Nonene	0.03	0.7292	146.9
1-Decene	0.05	0.7408	170.6
1-Undecene	0.09	0.7503	192.7
1-Dodecene	0.36	0.7584	213.4
1-Tridecene	0.08	0.7653	232.8
1-Tetradecene	0.10	0.7713	251.1
1-Pentadecene	0.22	0.7764	268.2
Polymer (triacontene, C_{30})	0.07	0.8127	448

Nomenclature

a, b, c, d, etc. = components

B = moles residue/hr

C_m = correction factor, Fig. 5.12

C_p = correction factor, Fig. 5.13

D = moles distillate/hr

F = moles feed/hr

F_L = moles of liquid feed/hr

F_V = moles of vapor feed/hr

H = enthalpy of vapor, Btu/mole

h = enthalpy of component in liquid, Btu/mole

$\bar{H}$ = partial molal enthalpy of component in vapor, Btu/mole

HK = heavy key component

K = equilibrium vaporization ratio

L = lower pinch zone

L = moles liquid/hr, rectifying section

$\bar{L}$ = moles liquid/hr, stripping section

L_F = moles liquid from feed plate/hr

LK = light key component

L/D = operating or external reflux ratio

$(L_0/D)_{\min}$ = minimum external reflux ratio

L/V = internal reflux ratio, rectifying section

$\bar{L}/\bar{V}$ = internal reflux ratio, stripping section

$(L/V)_{\min}$ = minimum internal reflux ratio (rectifying)

$(\bar{L}/\bar{V})_{min}$ = minimum internal reflux ratio (stripping)
N = number of theoretical plates
N_m = minimum theoretical plates
N'_m = minimum theoretical plates (between reference plates—Underwood)
P = vapor pressure
P_t = total pressure
Q = heat input rate, Btu/(mole of stream)(hr)
q = total heat input, Btu/hr
q = moles of liquid introduced at the feed plate by the addition of 1 mole of feed
t = temperature
U = upper pinch zone
V = moles vapor/hr, rectifying section
$\bar{V}$ = moles vapor/hr, stripping section
x, y = mole fraction of component in liquid, vapor

Greek letters

α = relative volatility
γ = activity coefficient
ν = fugacity coefficient
θ = defined quantity, Eq. (5.82)
ψ = defined quantity, Eq. (5.112)
Σ = summation
λ = molal latent heat

Subscripts

1, 2, 3 = plate numbers
B = residue
c = light key component or component
D = distillate (or condenser)
d = heavy key component or "distributed" component
F = feed
i, j = component designations
n, $n-1$, $n+1$, etc. = nth plate, plate above nth, plate below nth
L = liquid in rectifying section; lower pinch section
$\bar{L}$ = liquid in stripping section
m = mixture
m, $m+1$, etc. = refers to rectifying section

$\left.\begin{array}{l} p \\ p+1, \\ \text{etc.} \end{array}\right\}$ = refers to stripping section

min = minimum

U = upper pinch section

REFERENCES

1. Bachelor, J. B.: *Petrol. Refiner* (June, 1957), p. 161.
2. Colburn, A. P.: *Trans. A.I.Ch.E.*, **37**:805 (1941).
3. Fenske, M. R.: *Ind. Eng. Chem.*, **24**:482 (1932).
4. Gilliland, E. R.: *Ind. Eng. Chem.*, **32**:1101 (1940).
5. Gilliland, E. R., and C. E. Reed: *Ind. Eng. Chem.*, **34**:551 (1942).
6. Holland, C. D.: "Multicomponent Distillation," Prentice-Hall, Englewood Cliffs, N.J., 1963.
7. Martin, H. Z., and G. G. Brown: *Trans. A.I.Ch.E.*, **35**:679 (1939).
8. May, J. A.: *Ind. Eng. Chem.*, **41**:2775 (1949).
9. Mayfield, F. D., and J. A. May: *Petrol. Refiner* (April, 1946), p. 101.
10. Scheibel, E. G., and C. F. Montrose: *Ind. Eng. Chem.*, **38**:614 (1946).
11. Shiras, R. N., D. N. Hanson, and C. H. Gibson: *Ind. Eng. Chem.*, **42**:871 (1950).
12. Underwood, A. J. V.: *Trans. Inst. Chem. Engrs. (London)*, **10**:112 (1932).
13. Underwood, A. J. V.: *Chem. Eng. Progr.*, **44**:603 (1948).

chapter

Binary fractionation—number of equilibrium stages

In Chap. 5 it was emphasized that the number of equilibrium stages was an important factor in all fractionating column design because the economics of column size and cost are directly related to this factor. Because the calculations for evaluating the number of equilibrium stages for a binary fractionation are less complicated than for multicomponent and complex mixtures, there have been a number of systems devised for this evaluation, both analytical and graphical. Some of these are discussed in this chapter.

The number of variables to be fixed to fix the operation of a binary fractionating column is $C + 6$ or $2 + 6 = 8$.

The usual procedure is to fix the following:

Feed rate	1
Feed composition	1
Feed enthalpy	1
Distillate composition	1
Bottoms composition	1
Reflux temperature	1
Column pressure	1
Reflux ratio	1
Total	8

The number of equilibrium stages may then be calculated analytically by using material balance and enthalpy balance equations, or graphically by using graphical representations of the equations.

6.1. ANALYTICAL DETERMINATION OF NUMBER OF EQUILIBRIUM STAGES

Sorel Method [7]

The Sorel method uses plate-to-plate calculations involving alternately material and enthalpy balances and equilibrium calculations to determine the quantities of liquid and vapor, the temperature of each plate, and the compositions of each stream from each plate. It is the basis for one of the types of computer programs used in plate-to-plate calculations. The method assumes the pressure of the distillation, the operating reflux ratio, L_0/D, the temperature or enthalpy of the reflux stream, and the use of a total condenser.

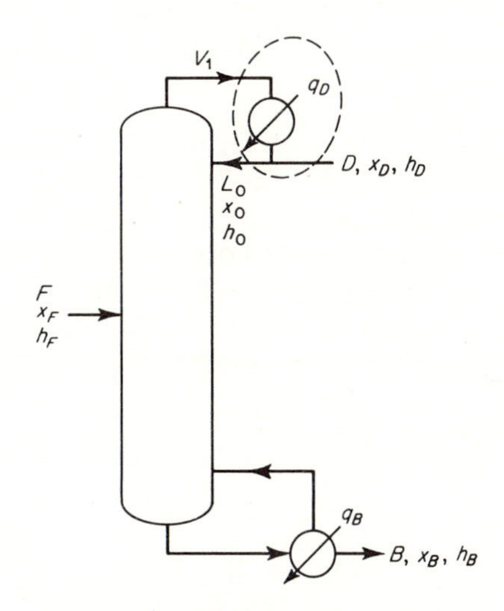

FIGURE 6.1 Material and enthalpy balances around the condenser.

From over-all material balance (see Fig. 6.1),

$$F = D + B \tag{6.1}$$

and from the component material balance,

$$Fx_F = Dx_D + Bx_B \tag{6.2}$$

Since F, x_F, x_D, and x_B are known from process requirements and specifications on the products,

$$Fx_F = Dx_D + Fx_B - Dx_B \tag{6.3}$$

$$D = \frac{F(x_F - x_B)}{x_D - x_B} \tag{6.4}$$

The operating reflux ratio L_0/D is fixed, thus L_0 is known. Also for a total condenser $y_1 = x_0 = x_D$. Material and enthalpy balances around the condenser give

$$V_1y_1 = L_0x_0 + Dx_D \tag{6.5}$$

and

$$V_1H_1 + q_D = L_0h_0 + Dh_D \tag{6.6}$$

Since plate 1 is considered an equilibrium stage, the temperature on plate 1 is the dew-point temperature of V_1 at the pressure of the distillation. Therefore $x_1 + x_2 = 1.0$ and

$$\frac{y_1\nu_1P_t}{\gamma_1P_1} + \frac{(1 - y_1)\nu_2P_t}{\gamma_2P_2} = 1.0 \tag{6.7}$$

The vapor composition y_1, fugacity coefficients, activity coefficients, and total pressure are known, therefore, temperatures are selected to obtain values of P_1 and P_2 which satisfy Eq. (6.7). The value of q_D may be determined from Eq. (6.6) once the temperature of V_1 and, therefore, the enthalpy H_1 is known. The unknown quantities L_1 and V_2, t_2 (or H_2), and y_2 must be solved for by trial and error with the following equations:

$$V_2 = L_1 + D \tag{6.8}$$

$$V_2y_2 = L_1x_1 + Dx_D \tag{6.9}$$

$$V_2H_2 + q_D = L_1h_1 + Dh_D \tag{6.10}$$

and

$$H_2 = f(y_2,t_2) \tag{6.11}$$

The enthalpy is evaluated from an enthalpy-concentration chart, or from data, or from ty data from which the enthalpy may be computed if heat capacities and latent heats are available.

Equation (6.10) may be written as

$$V_2f(y_2,t_2) - L_1h_1 + q_D = Dh_D \tag{6.12}$$

From

$$V_2 - L_1 = D$$

and

$$V_2 y_2 - L_1 x_1 = D x_D$$

$$L_1 = \frac{D(x_D - y_2)}{y_2 - x_1} \tag{6.13}$$

$$V_2 = \frac{D(x_D - x_1)}{y_2 - x_1} \tag{6.14}$$

Then

$$D \frac{x_D - x_1}{y_2 - x_1} f(y_2,t_2) - D \frac{x_D - y_2}{y_2 - x_1} h_1 + q_D = D h_D \tag{6.15}$$

All variables in the above equations are known except y_2, so that values of y_2 are selected and the corresponding values of $f(y_2,t_2)$ are obtained and substituted in the equation. When the correct values of y_2 are selected, the equation is satisfied.

From the *txy* or *Hxy* data the temperature on plate 2 is determined as well as the composition and enthalpy of L_2. The quantities of L_2 and V_3 and H_3, y_3 are calculated by use of equations similar to Eqs. (6.12) through (6.15).

$$H_3 = f(y_3,t_3) \tag{6.16}$$

$$V_3 - L_2 = D \tag{6.17}$$

$$V_3 y_3 - L_2 x_2 = D x_D \tag{6.18}$$

$$V_3 H_3 - L_2 h_2 = D h_D - q_D \tag{6.19}$$

$$D \frac{x_D - x_2}{y_3 - x_2} f(y_3,t_3) - D \frac{x_D - y_3}{y_3 - x_2} h_2 = D h_D - q_D \tag{6.20}$$

Values of y_3 are selected, $f(y_3,t_3)$ are determined, and substituted in Eq. (6.20). The equation is satisfied when the correct values of y_3 are selected. Then L_2 and V_3 are calculated and these stepwise calculations are continued down the column, plate by plate, until the liquid composition equals or becomes less than the average composition of the feed. Material and enthalpy balances are then written about the feed plate, as indicated in Fig. 6.2.

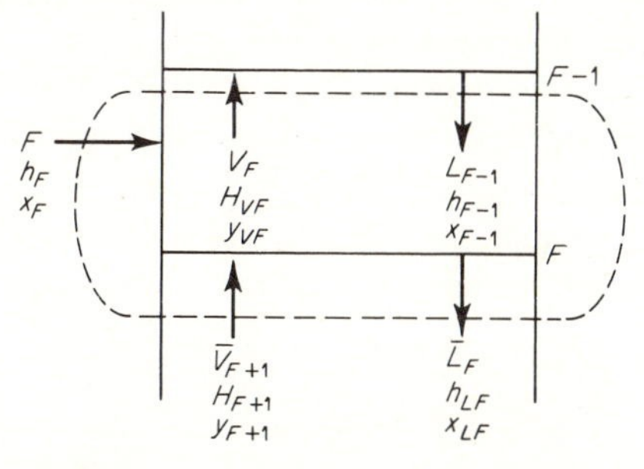

FIGURE 6.2 Material and enthalpy balances around the feed plate.

$$F + \bar{V}_{F+1} + L_{F-1} = V_F + \bar{L}_F \tag{6.21}$$

$$Fx_F + \bar{V}_{F+1}y_{F+1} + L_{F-1}x_{F-1} = V_F y_{VF} + \bar{L}_F x_{LF} \tag{6.22}$$

$$Fh_F + \bar{V}_{F+1}H_{F+1} + L_{F-1}h_{F-1} = V_F H_{VF} + \bar{L}_F h_{LF} \tag{6.23}$$

Since V_F, H_{VF}, and y_{VF} are calculated by stepwise calculations down the column, the unknowns are $\bar{V}_{F+1}$, H_{F+1}, y_{F+1}, $\bar{L}_F$, h_{LF}, and x_{LF}. The composition of $\bar{L}_F$ is calculated by the equilibrium relationship at the dew-point temperature of the vapor V_F.

$$x_{1_{LF}} + x_{2_{LF}} = \frac{y_{1_{VF}}\nu_{1_{VF}}P_t}{\gamma_{1_{LF}}P_{1_{LF}}} + \frac{(1 - y_{1_{VF}})\nu_{2_{VF}}P_t}{\gamma_{2_{LF}}P_{2_{LF}}} = 1.0 \tag{6.24}$$

In a binary system the vapor-liquid equilibrium data relate $x = f(y)$, $x,y = f'(t)$. With the composition of $\bar{L}_F$ known as well as the temperature, the enthalpy of $\bar{L}_F$ can be calculated or determined from Hx data. This leaves the quantity of $\bar{L}_F$, and the quantity, composition, and enthalpy of $\bar{V}_{F+1}$ unknown. Rearrangement of Eqs. (6.22) and (6.23) results in

$$\bar{V}_{F+1} = \frac{V_F y_{VF} - Fx_F - L_{F-1}x_{F-1} + (F + L_{F-1} - V_F)x_{LF}}{y_{F+1} - x_{LF}} \tag{6.22a}$$

and

$$\bar{V}_{F+1} = \frac{V_F H_F - Fh_F - L_{F-1}H_{F-1} + (F + L_{F-1} - V_F)h_{LF}}{H_{F+1} - h_{LF}} \tag{6.23a}$$

In a binary mixture the *txy* data or *Hxy* data relate the composition and enthalpy of the stream thus:

$$H_{F+1} = f(y_{F+1}, t_{F+1})$$

Trial-and-error calculations establish the values of V_{F+1}, y_{F+1}, H_{F+1}, and the temperature on the plate below the feed plate; then stepwise calculations are carried out in the same manner as those above the feed plate until the composition of the liquid from the stage is equal to or less than the composition of the bottoms product.

If it is desirable to make the plate-to-plate calculations starting at the bottom of the column and proceeding upward to the feed plate, the following procedure is used.

The quantity of heat in the reboiler, q_B, may be calculated from Eq. (6.25), derived from an over-all material balance around the whole column (see Fig. 6.3).

$$q_B = Dh_D + Bh_B - Fh_F - q_D \tag{6.25}$$

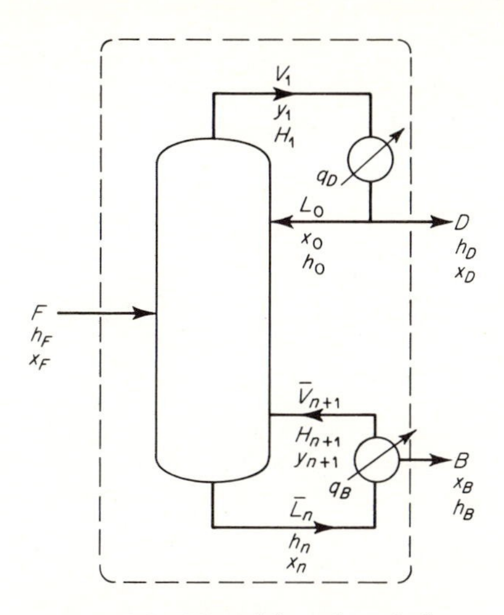

FIGURE 6.3 Material and enthalpy balance envelope for entire column.

Since the reboiler is generally assumed to be an equilibrium stage, V_{n+1} and B are in equilibrium and

$$y_{n+1} = \frac{x_B \gamma_B P_B}{\nu_{n+1} P_t} \tag{6.26}$$

The temperature in the reboiler is determined as the bubble-point temperature by trial-and-error calculation using

$$\Sigma y_{n+1} = \sum \frac{x_B \gamma_B P_B}{\nu_{n+1} P_t} = 1.0 \tag{6.26a}$$

With q_B, known, the following equations are used to evaluate the composition, quantity, and temperature of L_n (see Fig. 6.4),

$$\bar{L}_n = \bar{V}_{n+1} + B \tag{6.27}$$

$$\bar{L}_n x_n = \bar{V}_{n+1} y_{n+1} + B x_B \tag{6.28}$$

$$\bar{L}_n h_n = \bar{V}_{n+1} H_{n+1} + B(h_B - Q_B) \tag{6.29}$$

since

$$q_B = B Q_B$$

$$h_n = f(x_n, t_n) \tag{6.30}$$

Equation (6.29) may be written as

$$\bar{L}_n f(x_n, t_n) = \bar{V}_{n+1} H_{n+1} + B(h_B - Q_B) \tag{6.31}$$

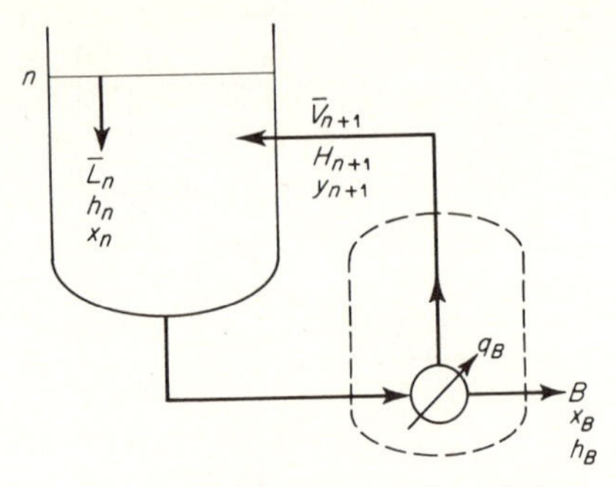

FIGURE 6.4 Material and enthalpy balances around the reboiler.

From Eqs. (6.27) and (6.28),

$$\bar{L}_n = \frac{B(x_B - y_{n+1})}{x_n - y_{n+1}} \tag{6.27a}$$

$$\bar{V}_{n+1} = \frac{B(x_B - x_n)}{x_n - y_{n+1}} \tag{6.28a}$$

then

$$B\frac{x_B - y_{n+1}}{x_n - y_{n+1}} f(x_n,t_n) = B\frac{x_B - x_n}{x_n - y_{n+1}} H_{n+1} + B(h_B - Q_B) \tag{6.32}$$

All variables in Eq. (6.32) are known except x_n. Values of x_n are selected and corresponding values of $f(x_n,t_n)$ are calculated or obtained from *hx* data and substituted in the equation. When the correct values of x_n are selected, the equation is satisfied and the solution is obtained. The temperature on the plate *n* is then obtained from *txy* or *hxy* data, and the vapor composition y_n is computed from equilibrium relationships at the temperature and pressure of plate *n*. The quantities L_{n-1}, V_n, h_{n-1}, and x_{n-1} are solved for by Eqs. (6.33) through (6.36).

$$h_{n-1} = f(x_{n-1},t_{n-1}) \tag{6.33}$$

$$\bar{L}_{n-1} = \bar{V}_n + B \tag{6.34}$$

$$\bar{L}_{n-1}x_{n-1} = \bar{V}_n y_n + Bx_B \tag{6.35}$$

$$\bar{L}_{n-1}h_{n-1} = \bar{V}_n H_n + B(h_B - Q_B) \tag{6.36}$$

Rearrangement and substitution results in

$$B\frac{x_B - y_n}{x_{n-1} - y_n} f(x_{n-1},t_{n-1}) = B\frac{x_B - x_{n-1}}{x_{n-1} - y_n} H_n + B(h_B - Q_B) \tag{6.37}$$

Values of x_{n-1} are selected and $f(x_{n-1})$ is evaluated and substituted in the equation until the solution is reached. $\bar{L}_{n-1}$ and $\bar{V}_n$ are then calculated by means of material balance.

The plate-to-plate calculations are continued until the approximate feed composition is reached. Equations (6.21) through (6.23) are used to evaluate the quantities, compositions, and enthalpies of the streams above the feed plate.

Simplified Calculations—Lewis Modification

For many binary fractionation situations the calculations may be simplified without greatly affecting the accuracy of the results. Lewis [1] modified Sorel's method by assuming equimolal overflow, which is equivalent to assuming equimolal latent heats and heat capacities and no heats of mixing. Furthermore, he assumed L_0 to be a saturated liquid. Equimolal overflow assumes that the moles of vapor flow are constant in the rectifying section above the feed, i.e., $V_1 = V_2 = V_3 = \cdots$, etc., $\bar{V}_{F+1} = \bar{V}_{F+2} = \bar{V}_{F+3} = \cdots$ in the stripping section, $L_1 = L_2 = L_3 = \cdots$, and $\bar{L}_F = \bar{L}_{F+1} = \bar{L}_{F+2} = \cdots$.

The composition of L_1 is calculated or obtained from xy or txy data. (Because of the foregoing assumptions, it is not necessary to use the enthalpy balance equations until a point in the column is reached where material or energy is added or removed.) Then, since $L_1 = L_0$ and $V_2 = V_1$, the composition of V_2 is calculated by

$$V_2 y_2 = L_1 x_1 + D x_D \tag{6.38}$$

The plate-to-plate computation is continued down the column until the composition of the liquid becomes equal to or less than the composition of the feed x_F. At this point the feed stream represents both an addition of material and enthalpy, and the evaluation of the quantity and composition of the streams $\bar{V}_{F+1}$ and $\bar{L}_F$ must be determined by means of Eqs. (6.21), (6.22), and (6.23). Once these quantities are established, the plate-to-plate calculations are continued until the composition of the liquid equals or is less than the composition of the bottom product. In the section below the feed (stripping section) $\bar{L}_F = \bar{L}_{F+1} = \bar{L}_{F+2} = \cdots$ and $\bar{V}_{F+1} = \bar{V}_{F+2} = \bar{V}_{F+3} = \cdots$.

Because of the tedious calculations involving trial-and-error solutions, graphical methods have been devised in which the equilibrium relations, material balance relations, and enthalpy balance relations can be expressed as plots or curves and the calculations can be carried out graphically. The two principal methods are those of McCabe-Thiele [2] and Ponchon-Savarit [3, 5].

6.2. GRAPHICAL METHODS

The McCabe-Thiele Method

The McCabe-Thiele method is based on the Lewis modification of the Sorel method. It assumes equimolal overflow in the rectifying section, in the stripping section, and equimolal latent heats, L_0 is a saturated liquid, and the column pressure and reflux ratio.

A material balance written for the envelope *A*, Fig. 6.5, cutting below any plate in the rectifying section and the top of the column gives the

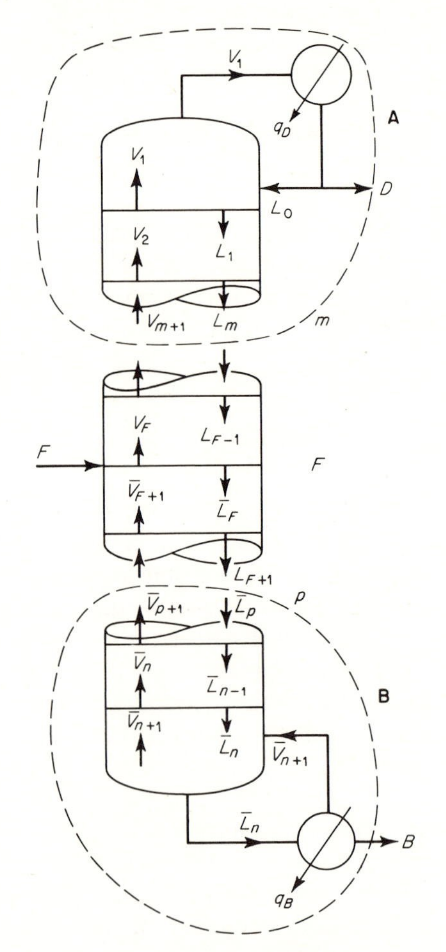

FIGURE 6.5 Material balance envelopes for equation development.

following:

$$V_{m+1} = L_m + D \tag{6.39}$$

$$V_{m+1}y_{m+1} = L_m x_m + Dx_D \tag{6.40}$$

$$y_{m+1} = \frac{L_m}{V_{m+1}} x_m + \frac{D}{V_{m+1}} x_D \tag{6.41}$$

This is an equation of a straight line on a plot of vapor composition versus liquid composition, where L_m/V_{m+1} is the slope and D/V_{m+1} is the intercept which passes through the point $x_D = y_D$ and x_m, y_{m+1}. Since all L values are equal and all V values are equal, the subscript can be dropped to give Eq. (6.42), which is the equation of the "operating line" or material balance line for the rectifying section.

$$y_{m+1} = \frac{L}{V} x_m + \frac{D}{V} x_D \tag{6.42}$$

Writing corresponding equations around the stripping section, envelope B, Fig. 6.5, of the column below the feed,

$$y_{p+1} = \frac{\bar{L}}{\bar{V}} x_p - \frac{B}{\bar{V}} x_B \tag{6.43}$$

Again, these equations represent the equation of a straight line with slope $\bar{L}/\bar{V}$ and intercept B/V passing through $x_B = y_B$ and x_p, y_{p-1}. This is the equation of the operating line or material balance line in the stripping section.

A plot of the equilibrium data of a "normal" (nonazeotropic) system is shown in Fig. 6.6.

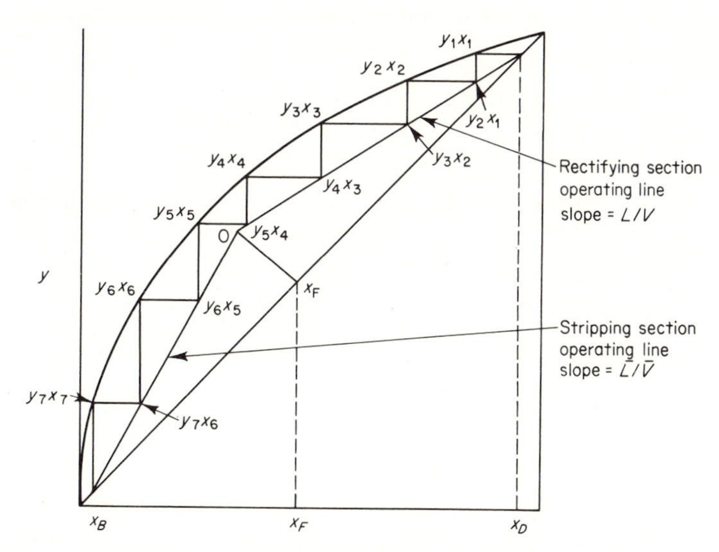

FIGURE 6.6 The McCabe-Thiele graphical method.

The determination of the number of equilibrium stages by use of the diagram is as follows:

Starting with the distillate composition $x_D = y_D$, draw a horizontal line to intersect the equilibrium curve at the intersection y_1x_1. Draw a vertical line to intersect the operating line at y_2x_1 and continue this operation until the liquid composition is reached equal to or less than x_B. The number of intersections of the equilibrium curve represent the number of equilibrium stages necessary for the separation. After the point O, where the operating lines meet, the construction intersections are on the stripping-section operating line, which indicates these stages are in the stripping section below the feed. This construction is for the case of optimum feed-plate location.

The intersection of the operating lines takes place at a point which is a function of the feed composition and the material, and enthalpy balance relationships around the feed plate. This is conveniently expressed as

$$\bar{L} = L + qF \tag{6.44}$$

$$q = \frac{\bar{L} - L}{F} \tag{6.45}$$

or

$$V = \bar{V} + (1 - q)F \tag{6.46}$$

where q is the number of moles of saturated liquid formed on the feed plate by the introduction of 1 mole of feed. A saturated liquid feed of the same composition as the liquid on the feed plate introduces 1 mole of liquid/mole of feed and $q = 1.0$. If the feed is a saturated vapor of the same composition as the vapor from the feed plate, no additional liquid is formed and $q = 0$. With a cold liquid feed, $q > 1$; with a superheated vapor feed, $q < 0$.

Since $\bar{V} = \bar{L} - B$,

$$y_{p+1} = \frac{L + qF}{L + qF - B} x_p - \frac{Bx_B}{L + qF - B} \tag{6.47}$$

This equation gives the slope of the operating line in the stripping section as $(L + qF)/(L + qF - B)$. Drawing this line through x with the indicated slope until it intersects with the rectifying operating line locates one end of the q line. The compositions of the distillate, bottoms, and feed are all plotted on the 45° line representing $y = x$, and the other end of the q line lies at the location of the feed composition on the $x = y$ line. The slope of this q line is related to the thermal condition of the feed. Using the Eqs. (6.44) and (6.45),

$$q = \frac{\bar{L} - L}{F} \tag{6.48}$$

$$q - 1 = \frac{\bar{V} - V}{F} \tag{6.49}$$

Designating y_i and x_i as the intersection compositions of the two operating lines, the material balance equations around the top of the column and bottom of the column, and the bottom of the column and around the feed result in

$$(q - 1)y_i = qx_i - x_F \tag{6.50}$$

$$y_i = \frac{q}{q-1}x_i - \frac{x_F}{q-1} \tag{6.51}$$

Equation (6.51) is the equation of the q line having a slope of $q/(q - 1)$ and terminating at x_F on the 45° line and at x_i, y_i. This equation locates the intersection of the operating lines for all values of q. Thus for:

1. Saturated liquid feed, $q = 1$, slope $= \infty$, $x_F = x_i$
2. Saturated vapor feed, $q = 0$, slope $= 0$, $x_F = y_i$
3. Cold liquid feed, $q > 1.0$, slope $= +$, $x_F < x_i$
4. Superheated vapor, $q < 0$, slope $= +$, $x_F > x_i$
5. Two-phase, q ranges from 1.0 to 0.0, slope $= -$, $x_F \geqslant x_i$

The actual intersection points may be calculated by

$$x_i = \frac{(L/D + 1)x_F + (q - 1)x_D}{L/D + q} \tag{6.52}$$

$$y_i = \frac{(L/D)x_F + qx_D}{L/D + q} \tag{6.53}$$

Figure 6.7 illustrates the q-line relations.

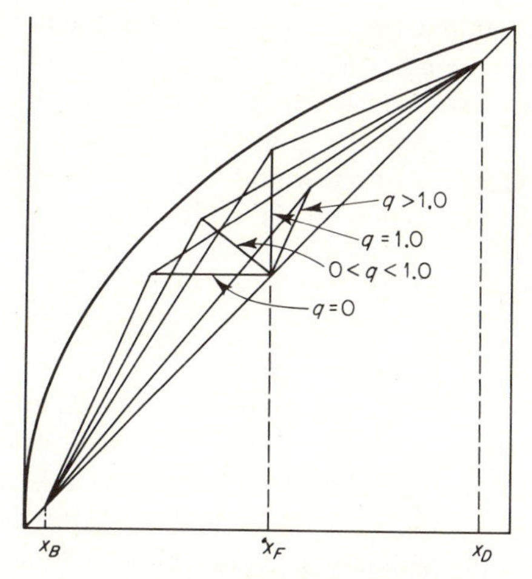

FIGURE 6.7 The q-line location.

Minimum Reflux—McCabe-Thiele Method

Minimum reflux conditions in the McCabe method are found by drawing a straight line from the composition of the distillate located on the $x = y$ line to the point of intersection with the q line drawn through the feed composition on the $x = y$ line and the equilibrium curve, as shown in Fig. 6.8. Where the system forms an azeotrope or shows azeotropic tendencies (the equilibrium curve crosses the $x = y$ line or approaches it), a line is drawn through the distillate composition but tangent to the equilibrium curve.

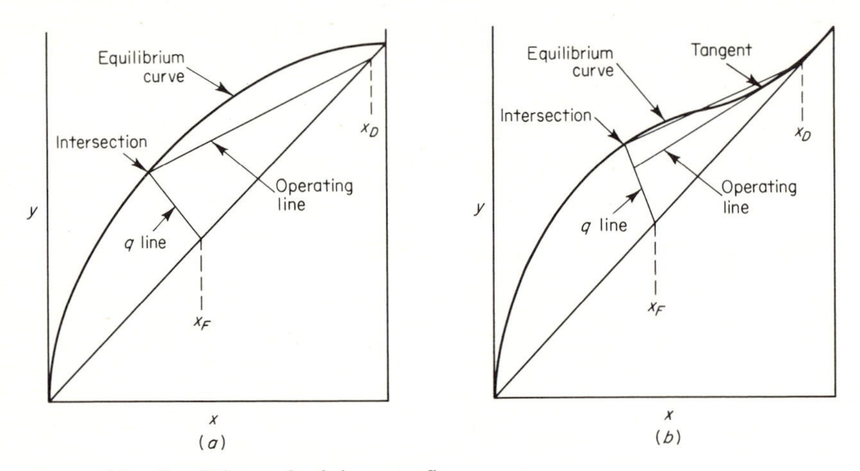

FIGURE 6.8 Conditions of minimum reflux.

The operating line having the greater slope fixes the minimum reflux. It is obvious that under these conditions an infinite number of stages would be required because the stepwise construction could not pass the point of intersection or tangency to complete the graphical construction to evaluate the number of plates. This condition would require an infinite number of plates—the condition of minimum reflux. Illustrations of such conditions are shown in Fig. 6.8.

Total Reflux

The condition of total reflux is encountered when L/V, the slope of the operating line in the rectifying section, is equal to 1.0, or the slope of the operating line in the stripping section, $\bar{L}/\bar{V}$, is equal to 1.0. Since they must intersect, they intersect on the $x = y$ line and, therefore, coincide with the $x = y$ line. Also the composition of the liquid and vapor passing one another between plates is the same, i.e., $x_m = y_{m+1}$. The number of theoretical stages is at a minimum at total reflux, as illustrated in Fig. 6.9.

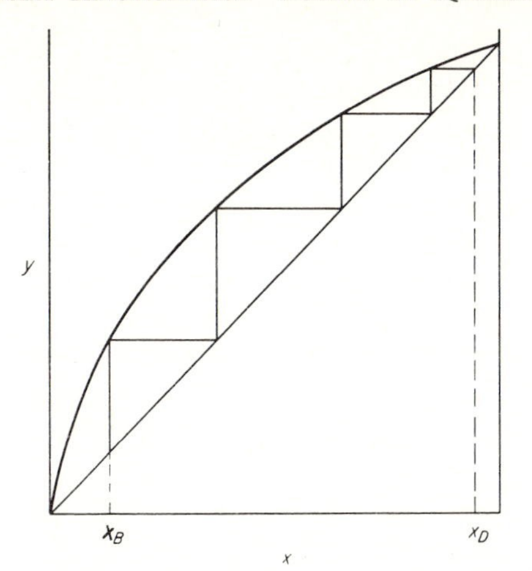

FIGURE 6.9 Equilibrium stages and total reflux.

Number of Stages at Low Concentrations

At very low concentrations it becomes difficult to carry out the graphical construction for the number of stages even when the scales are enlarged. In such cases the equilibrium data and operating line in the low-concentration region may be plotted on log-log paper, and the construction is carried out with relatively high accuracy because the operating line and equilibrium lines do not converge rapidly. In the low-concentration region the equilibrium relation follows Henry's law and $y = kx$, and this relationship is used to plot the equilibrium line if data are not available. Smoker [6] developed equations for handling these operations analytically. The Smoker equation is

$$n = \frac{\log \dfrac{x'_D\left[1 - \dfrac{mc(\alpha - 1)}{-mc^2}x'_n\right]}{x'_n\left[1 - \dfrac{mc(\alpha - 1)}{-mc^2}x'_D\right]}}{\log(\alpha/mc^2)} \tag{6.54}$$

where n is the number of theoretical plates required to produce a separation between x'_D and x'_n, $x'_D = x_D - k$, and $x'_n = x_n - k$; k is the composition of the liquid where the operating line intersects the equilibrium line. The operating line equation is

$$y = mx + b \tag{6.55}$$

$$c = 1 + (\alpha - 1)k \tag{6.56}$$

$$\alpha = \text{average relative volatility (assumed constant)}$$

$$= \frac{\alpha_{\text{top}} + \alpha_{\text{bottom}}}{2}$$

It should be noted that the Smoker equation is not limited to only the range of small concentrations but can be used for stage calculations over the entire range. In such a case the equation is written twice, once for each section of the tower, and average α's for the rectifying section and for the stripping section are determined.

Example 6.1

Pertinent data on the binary system heptane–ethyl benzene are as follows.

Vapor-Liquid Equilibria at 760 mm Hg

t, °C	x_H	y_H	γ_H	γ_{EB}
136.2	0	0	—	1.00
129.5	0.08	0.233	1.23	1.00
122.9	0.185	0.428	1.19	1.02
119.7	0.251	0.514	1.14	1.03
116.0	0.335	0.608	1.12	1.05
110.8	0.487	0.729	1.06	1.09
106.2	0.651	0.834	1.03	1.15
103.0	0.788	0.904	1.00	1.22
100.2	0.914	0.963	1.00	1.27
98.5	1.00	1.00	1.00	—

Data	Heptane	Ethyl benzene
Antoine constants		
A	6.9024	6.9537
B	1268.12	1421.91
C	216.9	212.9
ρ_{av}, g/cu cm	0.611	0.795
λ_{av}, cal/g mole	7575	8600
cal/g	75.75	81.0
$(C_{p,L})_{va}$, cal/(g mole) (°C)	51.90	43.4
cal/(g)(°C)	0.519	0.407
$(C_{p,V})_{av}$, cal/(g mole) (°C)	39.8	30.9
cal/(g)(°C)	0.398	0.291

A feed mixture composed of 42 mole % heptane, 58 mole % ethyl benzene is to be fractionated at 760 mm Hg to produce a distillate containing 97 mole % heptane and a residue containing 99 mole % ethyl benzene.

(*a*) Using a reflux ratio $(L/D) = 2.5$, determine the number of equilibrium stages needed for a saturated liquid feed and bubble-point reflux by the McCabe-Thiele graphical method.

(*b*) Determine the minimum reflux ratio.

(*c*) Determine the minimum number of equilibrium stages at total reflux.

Solution

(*a*)
$$\frac{L}{V} = \frac{L}{L+D} \qquad L = 2.5D \; \frac{L}{V} = \frac{2.5}{2.5+1} = \frac{2.5}{3.5} = 0.715$$

The slope of the operating line in the rectifying section is 0.715. Theoretical stages = 10.

(*b*)
$$\left(\frac{L}{V}\right)_{\min} = \frac{0.97-0.68}{0.97-0.42} = \frac{0.29}{0.55} = 0.527$$

$$\left(\frac{L_0}{D}\right)_{\min} = 1.115$$

(*c*) Approximately seven stages.

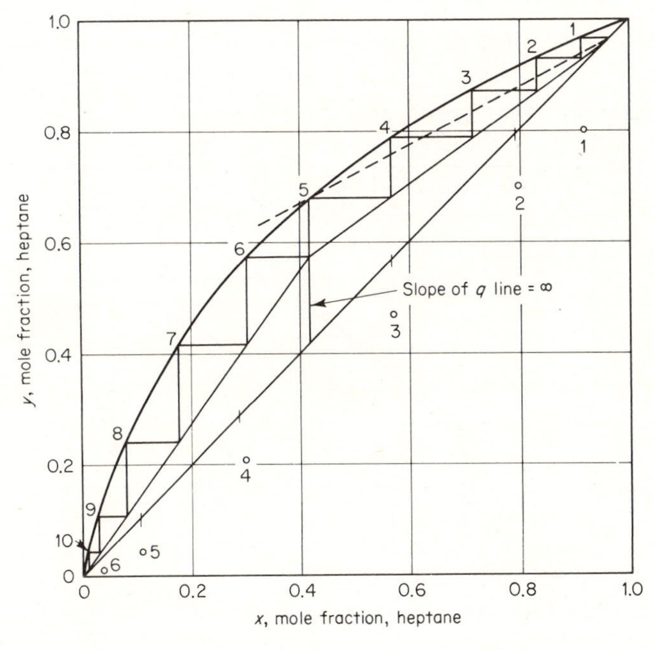

EXAMPLE 6.1 Graphical solution. (*a*) Theoretical stages, 10. (*b*) Minimum reflux,

$$(L/V)_{\min} = (0.97 - 0.68)/(0.97 - 0.42) = 0.29/0.55 = 0.527$$

$$(L_0/D)_{\min} = 1.115$$

(*c*) Approximately seven stages at total reflux (O = total-reflux intersections).

Example 6.2

For the system in Example 6.1 but with a feed composed of 40% vapor and 60% liquid, determine by the McCabe-Thiele method (*a*) the number of equilibrium stages for an $L/D = 2.5$; (*b*) the minimum reflux; (*c*) the minimum number of equilibrium stages at total reflux.

Solution

The slope of the operating line in the rectifying section is again 0.715. The slope of the q line is

$$\frac{q}{q-1} = \frac{0.6}{0.4} = -1.5$$

See figure for Example 6.2.

(*a*) Eleven stages.

(*b*)
$$\left(\frac{L}{V}\right)_{\min} = \frac{0.97 - 0.58}{0.97 - 0.31} = \frac{0.39}{0.66} = 0.592$$

$$\left(\frac{L_0}{D}\right)_{\min} = 1.45$$

(*c*) Approximately seven stages (the same as for Example 6.1).

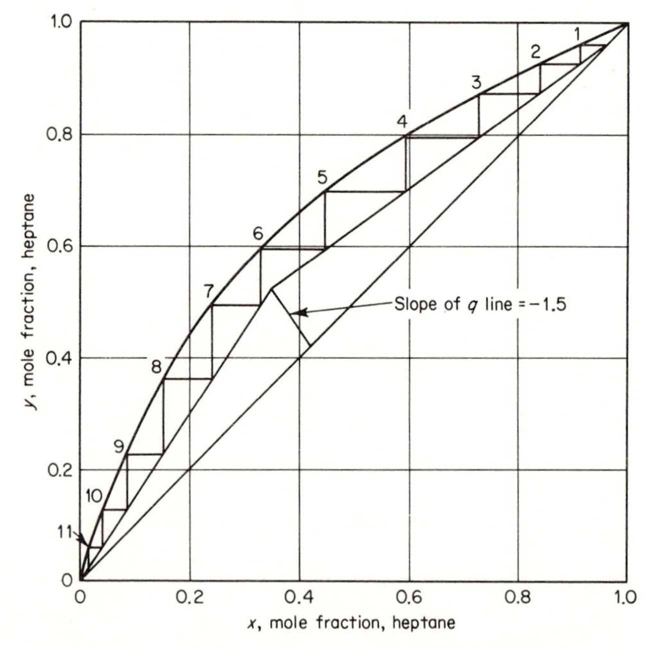

EXAMPLE 6.2 Graphical solution. (*a*) Theoretical stages, 11. (*b*) $(L/V)_{\min} = 0.592$, $(L_0/D)_{\min} = 1.45$. (*c*) Approximately seven stages.

Example 6.3

For the system in Example 6.1 but with a saturated vapor feed, repeat parts (*a*), (*b*), and (*c*) of Example 6.2.

Solution

The slope of the operating line in the rectifying section is 0.715. The slope of the q line is 0. See figure for Example 6.3.

(*a*) Approximately 20 stages.

(*b*)

$$\left(\frac{L}{V}\right)_{\min} = \frac{0.97 - 0.42}{0.97 - 0.19} = \frac{0.55}{0.78} = 0.705$$

$$\left(\frac{L_0}{D}\right)_{\min} = 2.39$$

(*c*) Approximately seven stages (the same as for Example 6.1).

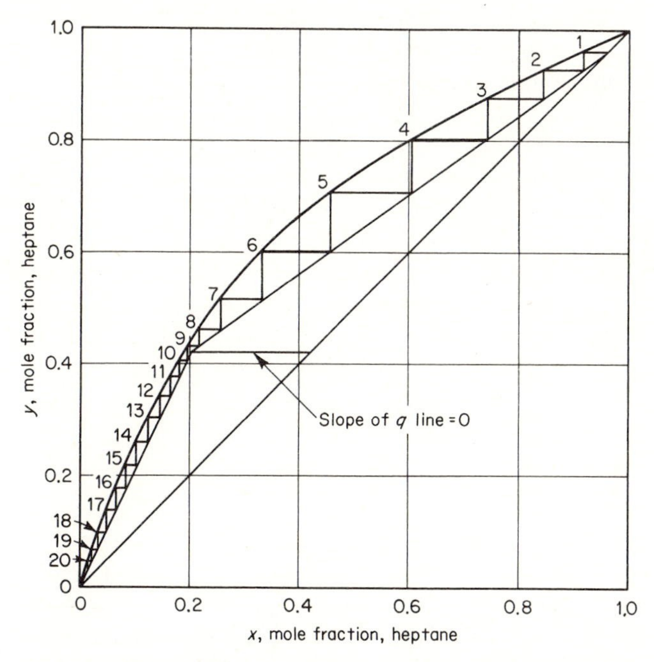

EXAMPLE 6.3 Graphical solution. (*a*) Theoretical stages, 20. (*b*) $(L/V)_{\min} = 0.705$, $(L_0/D)_{\min} = 2.39$. (*c*) Approximately seven stages.

Ponchon-Savarit Graphical Method

The Ponchon [3]-Savarit [5] graphical method [4] for the interrelation of the variables of binary fractionation calculations involves no simplifying assumptions such as those involved in the McCabe-Thiele method, and variations in molal enthalpy of the vapors and liquids as a function of composition are incorporated. In order to utilize this method, it is necessary to construct an enthalpy-concentration diagram for the particular binary system over a temperature range covering the two-phase vapor-liquid region at the pressure of the distillation.

The construction of such an enthalpy-concentration diagram requires the following data: (1) heat capacity as a function of temperature, composition, and pressure; (2) heat of mixing and dilution as a function of temperature and composition; (3) latent heats of vaporization as a function of composition and pressure or temperature; and (4) bubble-point temperature as a function of composition and pressure. The diagram is usually based upon a given reference state, such as liquid, at a reference temperature of, e.g., 32°F, at a given pressure. The saturated-liquid line may be established by first calculating the enthalpy of each of the pure-liquid components A and B at their boiling points at the pressure desired (these are the terminal points at $x = 0$ and $x = 1.0$), and then establishing the enthalpy of intermediate compositions between the pure components by

$$h_{\text{mix}} = \bar{h}_A + \bar{h}_B + \Delta H_{\text{sol}} \tag{6.57}$$

$$h_{\text{mix}} = x_A C_A (t - t_{\text{ref}}) + x_B C_B (t - t_{\text{ref}}) + \Delta H_{\text{sol}} \tag{6.58}$$

The saturated-vapor line may be established by calculating the enthalpy of the pure-component vapors by adding the latent heats of vaporization to the pure-component saturated-liquid enthalpies, and by adding the latent heat of vaporization at various compositions to the saturated-liquid-mixture enthalpies. For practical purposes the latent heats of vaporization may be calculated as

$$\lambda_{\text{mix}} = x_A \lambda_A + x_B \lambda_B \tag{6.59}$$

When only pure-component data are available, the saturated-vapor and saturated-liquid enthalpy lines may be drawn as straight lines between saturated-liquid enthalpies of the pure components and saturated-vapor enthalpies of the pure components. This is an approximation neglecting heat of solution when the enthalpies are plotted against mole fraction or against mass or weight fraction. However, it is used frequently for estimations when no mixture data are available.

Diagrams of the type described above are shown in Figs. 6.10 and 6.11. Figure 6.10 is an enthalpy-concentration diagram for a binary system when the concentration is expressed in mole fraction and Fig. 6.11 is a diagram in which mass (or weight) fraction is the concentration variable.

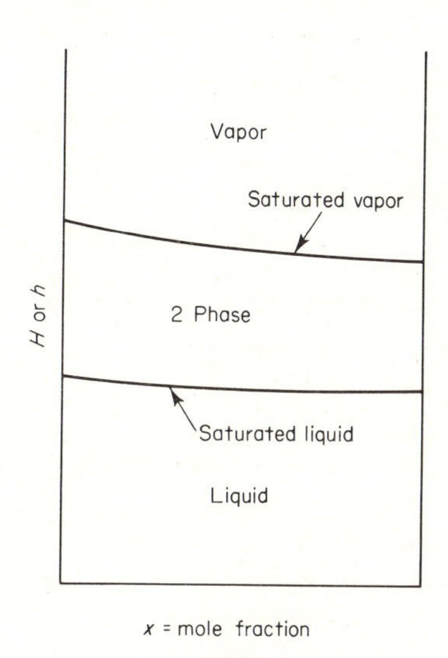

FIGURE 6.10 Enthalpy-concentration plot—mole fraction basis.

FIGURE 6.11 Enthalpy-concentration plot—mass fraction basis.

Example 6.4

Devise an enthalpy-concentration diagram for the heptane–ethyl benzene system at 760 mm Hg, using the pure liquid at 0°C as the reference state and assuming zero heat of mixing, i.e., $\bar{h}_i = x_i h_i$. Tabulate the data as well as plot it on a weight fraction basis and a mole fraction basis.

Solution

t, °C	x_H	h_L, cal/mole	H_V, cal/mole	g H / mole mix.	g EH / mole mix.	w_H, weight fraction	h_L, cal/g	H_V, cal/g
136.2	0	5920	14,520	0	106.2	0	55.7	137.0
129.5	0.08	5697	14,203	8	97.6	0.076	53.7	134.4
122.9	0.185	5520	13,920	18.5	86.5	0.176	52.3	132.4
119.7	0.251	5440	13,870	25.1	79.5	0.240	51.8	131.5
116.0	0.335	5360	13,610	33.5	70.6	0.321	51.3	130.5
110.8	0.487	5260	13,450	48.7	54.5	0.471	50.8	128.5
106.2	0.651	5170	13,110	65.1	37.0	0.636	50.3	127.9
103.0	0.788	5160	12,940	78.8	22.5	0.776	50.9	127.8
100.2	0.914	5124	12,784	91.4	9.1	0.91	51.1	129.0
98.5	1.000	5100	12,675	100.0	0	1.00	51.0	132.0

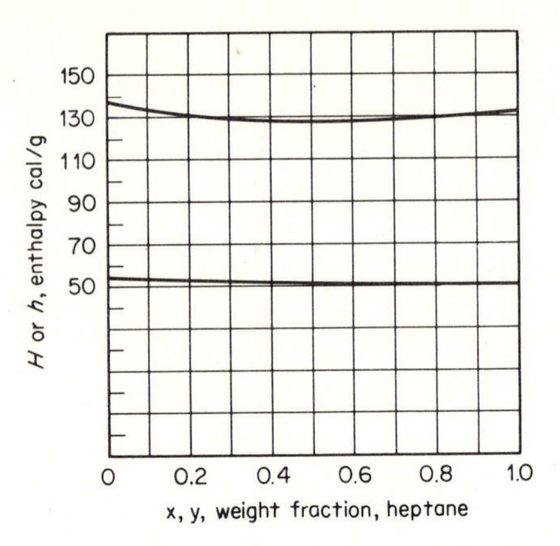

EXAMPLE 6.4 Enthalpy-concentration diagram—weight fraction basis.

The example calculations are:

$$t = 136.2°\text{C}$$

$$x_{\text{H}} = 0.0 \qquad x_{\text{EB}} = 1.0$$

$$h = x[C_p(t - t_0)]$$

$$h_L = 1.0[43.4(136.2 - 0)] = 5920 \text{ cal/mole}$$

$$H_V = h_L + \lambda_{L,t} = 5920 + 8600 = 14{,}520 \text{ cal/mole}$$

$$t = 129.5°\text{C}$$

$$x_{\text{H}} = 0.08 \qquad x_{\text{EB}} = 0.92$$

$$h_L = x_{\text{H}}C_{p,\text{H}}\,(129.5) + x_{\text{EB}}C_{p,\text{EB}}\,(129.5)$$

$$h_L = 0.08(51.9)(129.5) + 0.92(43.4)(129.5) = 537.0 + 5160 = 5697 \text{ cal/g mole}$$

$$H_V = h_{L,\text{H}} + h_{L,\text{EB}} + x_{\text{H}}\lambda_{\text{H}} + x_{\text{EB}}\lambda_{\text{EB}}$$

$$= 5697 + 0.08(7575) + 0.92(8600)$$

$$= 5697 + 606 + 7900 = 14{,}203 \text{ cal/g mole}$$

On a weight basis (Fig. Ex. 6.4),

$$t = 136.2°\text{C}$$

$$x_{\text{H}} = 0.0 \qquad x_{\text{EB}} = 1.0 \qquad w_{\text{EB}} = 1.0$$

$$h_L = 5920/106.2 = 55.7 \text{ cal/g}$$

$$H_V = 14{,}520/106.2 = 137.0 \text{ cal/g}$$

$$t = 129.0°\text{C}$$

$$x_{\text{H}} = 0.08 \qquad x_{\text{EB}} = 0.92 \qquad w_{\text{H}} = 0.076 \qquad w_{\text{EB}} = 0.924$$

$$h_L = 0.076(0.519)(129.5) + 0.924(0.407)(129.5)$$

$$= 5.1 + 48.6 = 53.7 \text{ cal/g}$$

$$H_V = h_L + (0.076)(75.75) + (0.924)(81)$$

$$= 53.7 + 5.7 + 75 = 134.4 \text{ cal/g}$$

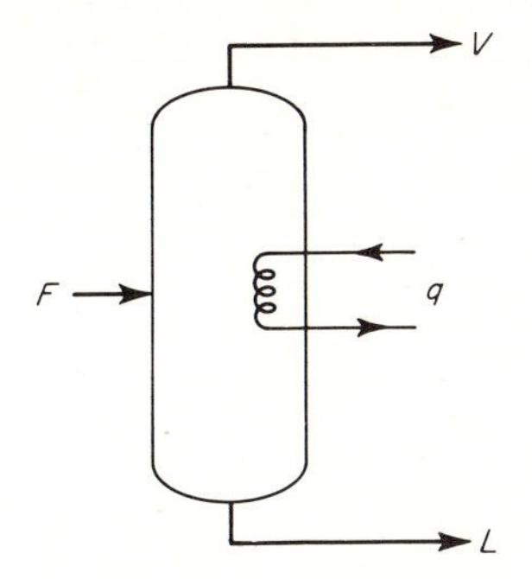

FIGURE 6.12 Steady-state flow system with phase separation and heat added.

The enthalpy-concentration diagram may be used to evaluate graphically the enthalpy and composition of streams added or separated. Referring to Fig. 6.12, a process is schematically described in which a feed stream F, having the composition x_F and enthalpy h_F, is introduced into a separator wherein it is separated into two streams V, enthalpy H, composition y, and L, enthalpy h, and composition x. Heat is added to the system as q. The over-all material balance, component material balance, and enthalpy balance equations are as follows:

$$F = V + L$$

$$Fx_F = Vy + Lx$$

$$Fh_F + q = VH + Lh$$

From these equations, by substitution and rearrangement, Eqs. (6.60) and (6.61) result.

For the adiabatic case where $q = 0$,

$$V(H - h_F) = L(h_F - h) \tag{6.60}$$

$$V(y - x_F) = L(x_F - x) \tag{6.61}$$

$$\frac{L}{V} = \frac{H - h_F}{h_F - h} \tag{6.62}$$

$$\frac{L}{V} = \frac{y - x_F}{x_F - x} \tag{6.63}$$

$$\frac{H - h_F}{y - x_F} = \frac{h_F - h}{x_F - x} \tag{6.64}$$

Consider Fig. 6.13. The slope of the line $\overline{VL}$ is $(H - h)/(y - x)$; also the slope of the line $\overline{FL}$ is $(h_F - h)/(x_F - x)$. Since the slopes are the same and since the lines go through the same point, the points lie on the same straight line. This is shown both by the construction and by the material and enthalpy balance equations.

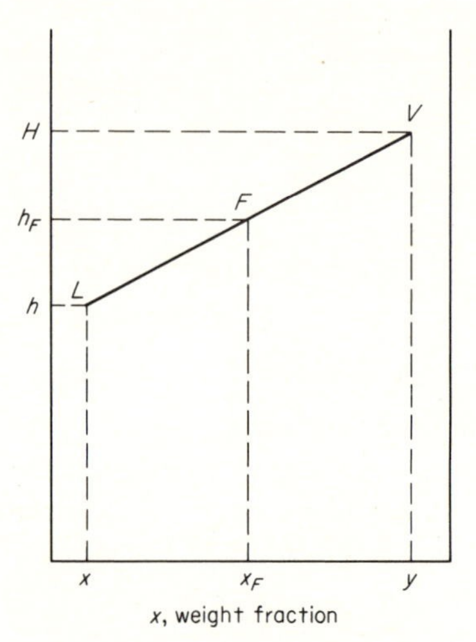

FIGURE 6.13 Enthalpy-concentration lines—adiabatic, $q = 0$.

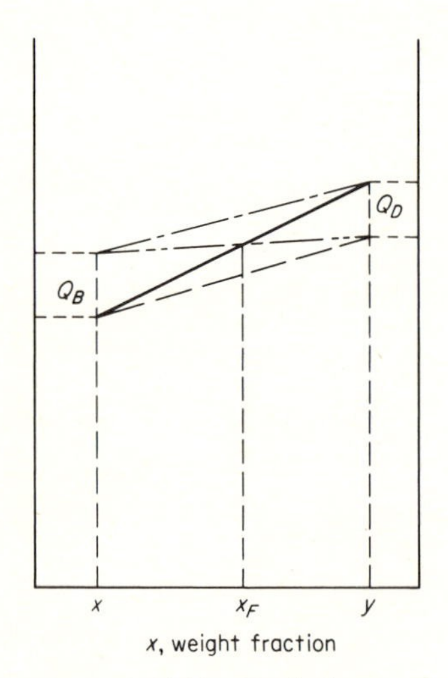

FIGURE 6.14 Enthalpy-concentration lines—nonadiabatic, $q \neq 0$.

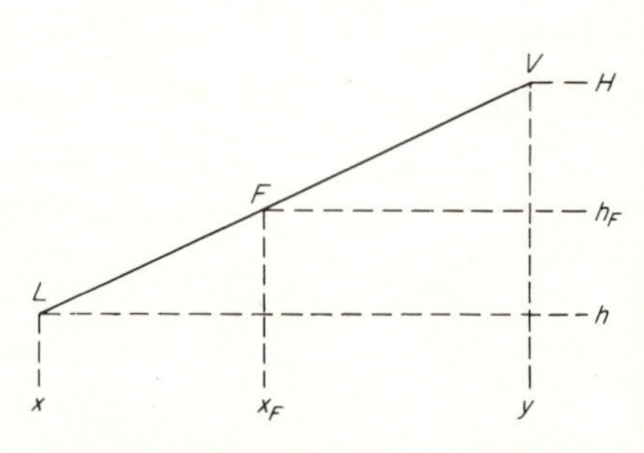

FIGURE 6.15 Lever-arm principle—enthalpy-concentration diagram.

Similarly in the nonadiabatic case, if the value of q is allocated to a particular stream (for example, $q = Q_F F = Q_L L = Q_V V$, where Q_F, Q_L, and Q_V, respectively, are in terms of Btu per pound mole or Btu per pound if the concentrations are in terms of weight fraction of stream F, L, and V), the same straight-line relation is true, as indicated in Fig. 6.14.

By the lever-arm principle (Fig. 6.15), the quantities of streams and their compositions and enthalpies can be established graphically by measuring the distances along the lines and ratioing them as indicated.

$$\frac{\overline{LF}}{\overline{LV}} = \frac{V}{F} \qquad \frac{\overline{FV}}{\overline{LV}} = \frac{L}{F} \qquad \frac{\overline{FV}}{\overline{LF}} = \frac{L}{V} \cdots$$

The enthalpy-concentration diagram, when used for a quantitative study of a binary-system fractionation process, has certain necessary characteristics, some of which are shown in Fig. 6.16. The saturated-vapor and saturated-liquid lines must be shown in terms of the enthalpy—Btu per pound or Btu per mole—and concentration, weight or mass fraction, or mole fraction. It is also helpful to plot a conjugate line on the same diagram which relates the equilibrium values of x and y.

Similarly the ratios of the absolute values of the concentrations

$$\frac{\overline{xx_F}}{\overline{xy}} = \frac{y}{x_F} \qquad \frac{\overline{x_F y}}{\overline{xx_F}} = \frac{x}{y} \cdots$$

along the abscissa may be related to the ratios of the distances, and this is

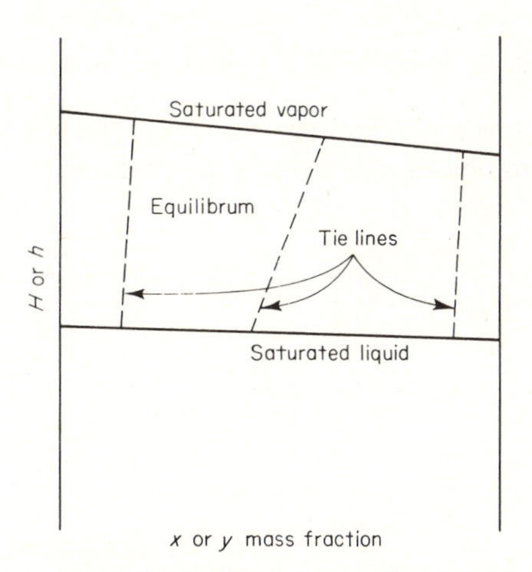

FIGURE 6.16 Two-phase-region boundaries and tie lines.

also true for the ordinate-value ratios. Because

$$\frac{\overline{Hh}}{\overline{\overline{h_F h}}} = \frac{h_F}{H} \qquad \frac{\overline{Hh_F}}{\overline{\overline{h_F h}}} = \frac{h}{H} \cdots$$

similar triangles are involved on rectilinear coordinates, and the ratios of the quantities of the streams are also equivalent to the corresponding distances along the abscissa or ordinate, e.g.,

$$\frac{F}{V} = \frac{\overline{LV}}{\overline{\overline{LF}}} = \frac{\overline{xy}}{\overline{\overline{xx_F}}} = \frac{\overline{Hh}}{\overline{\overline{h_F h}}}$$

Application to Binary Fractionation

For the conditions of a total condenser and partial condenser shown in Fig. 6.17, the graphical representations on the Ponchon diagram are shown in Fig. 6.18. The quantitative relations are developed in the following equations:

Total condenser	*Partial condenser*	
$V_1 = L_0 + D$	$V_1 = L_0 + D_V$	(6.65)
$V_1 y = L_0 x + D x_D$	$V_1 y = L_0 x + D_V y_D$	(6.66)
$q_D + V_1 H_1 = L_0 h_0 + D h_D$	$q_D + V_1 H_1 = L_0 h_0 + D_V H_D$	(6.67)

Designating $Q_D = \dfrac{q_D}{D}$,

$$V_1 H_1 = L_0 h_0 + D(h_D - Q_D) \qquad V_1 H_1 = L_0 h_0 + D_V(H_D - Q_D) \quad (6.68)$$

Since material balances, enthalpy balances, and composition relations can be determined graphically, the reflux ratios can be indicated on the enthalpy axis, as shown in Fig. 6.18. These are

$$\frac{L_0}{D} = \frac{h_D - Q_D - H_1}{H_1 - h_D} \quad \text{and} \quad \frac{L_0}{D_V} = \frac{H_D - Q_D - H_1}{H_1 - h_0} \quad (6.69)$$

Internal reflux is shown as

$$\frac{L_0}{V_1} = \frac{h_D - Q_D - H_1}{h_D - Q_D - h_0} \quad \text{and} \quad \frac{L_0}{V_1} = \frac{H_D - Q_D - H_1}{H_D - Q_D - h_0} \quad (6.70)$$

Similar relations can be illustrated in terms of compositions, although graphically this is less accurate than the result using the enthalpy scale.

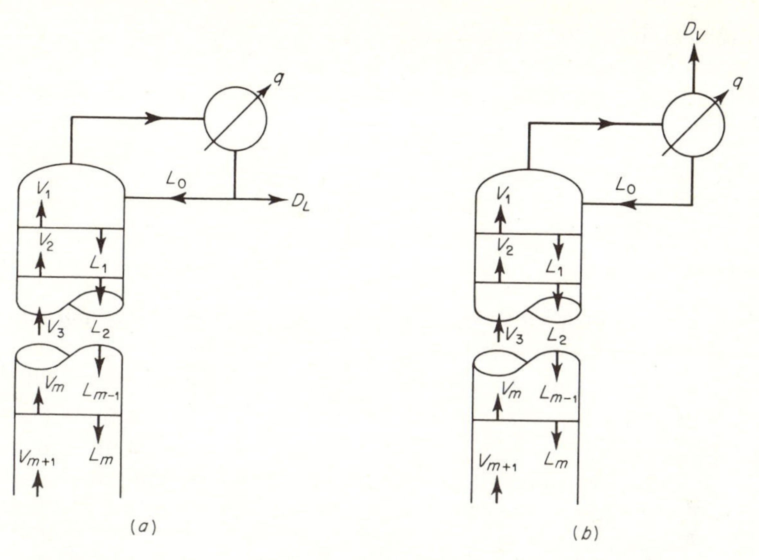

FIGURE 6.17 Schematic illustration of columns. (*a*) Rectifying section, total condenser. (*b*) Rectifying section, partial condenser.

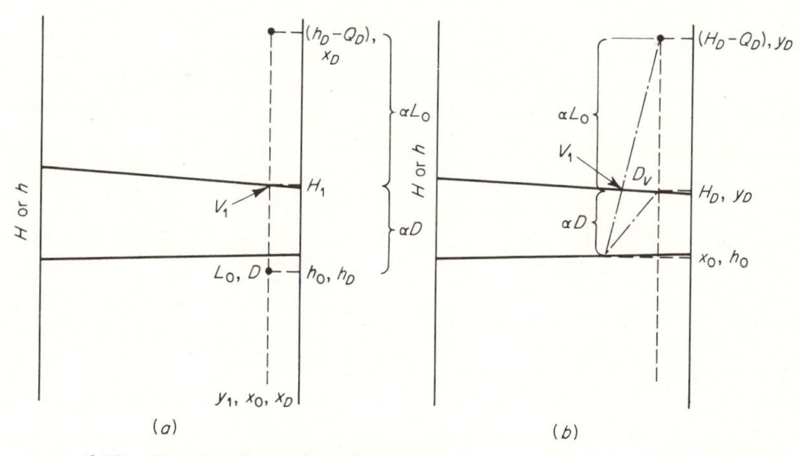

FIGURE 6.18 Graphical relationship for reflux ratio on the Ponchon diagram. (*a*) Total condenser. (*b*) Partial condenser.

The internal reflux between each plate, until a point in the column is reached where a stream is added or removed, can be shown as

$$\frac{L_m}{V_{m+1}} = \frac{h_D - Q_D - H_{m+1}}{h_D - Q_D - h_m} \qquad \text{and} \qquad \frac{L_m}{V_{m+1}} = \frac{H_D - Q_D - H_{m+1}}{h_D - Q_D - h_m} \tag{6.71}$$

Rectifying section The material and enthalpy balance equations may be rearranged in the form of differences such as follows:

$$V_1 - L_0 = V_2 - L_1 = V_3 - L_2 = \cdots = V_{m+1} - L_m = D \quad (6.72)$$

$$V_1 y_1 - L_0 x_0 = V_2 y_2 - L_1 x_1 = \cdots = V_{m+1} y_{m+1} - L_m x_m = D x_D \quad (6.73)$$

$$V_1 H_1 - L_0 h_0 = V_2 H_2 - L_1 h_1 = \cdots = V_{m+1} H_{m+1} - L_m h_m = D(h_D - Q_D) \quad (6.74)$$

These three independent equations (commonly called *companion* equations) can be written for each section of the column between each plate. They show that on the enthalpy scale and on the composition scale the differences in enthalpy and in composition always pass through the same point, x_D, $h_D - Q_D$ or y_D, $H_D - Q_D$. This is designated as d, the difference point, and all lines corresponding to the combined material and enthalpy balance equations (operating line equations) for the rectifying section of the column pass through this intersection.

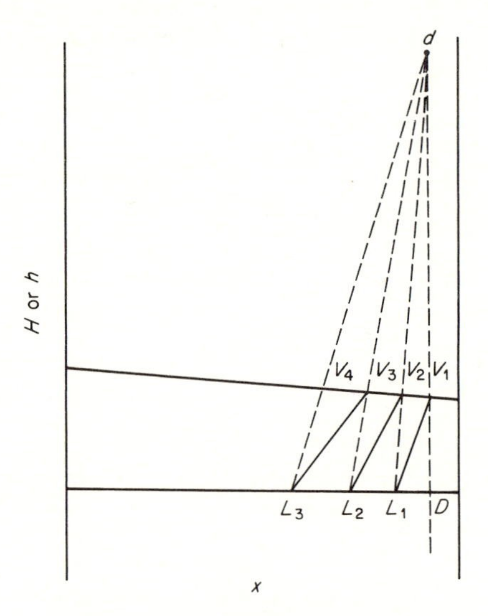

FIGURE 6.19 Rectifying section—number of stages.

The plate-to-plate graphical procedure for determining the number of equilibrium stages is as follows. On Fig. 6.19 the reflux ratio L_0/D, x_D, and enthalpy of the distillate, h_D, are used to establish the location of the difference point d. Equilibrium data alone establish the point, L_1 at x_1, h_1. Since L_1 is assumed to be a saturated liquid, x_1 must lie on the saturated-liquid line.

The operating line is drawn between L_1 and d, and it intersects the saturated-vapor line at V_2, y_2, H_2. Again equilibrium data establish L_2, x_2, h_2, and the procedure is repeated until the feed plate is reached (or until a side stream is removed).

Stripping section In Fig. 6.20, the material and enthalpy balance equations written in the form of difference-point equations around reboiler I and bottom stream B, and between any two plates in section II are:

$$\bar{L}_n - \bar{V}_{n+1} = B = \bar{L}_{n-1} - \bar{V}_n = \bar{L}_{n-2} - \bar{V}_{n-1} = \cdots = \bar{L}_p - \bar{V}_{p+1} \quad (6.75)$$

$$\bar{L}_n h_n - \bar{V}_{n+1}H_{n+1} = B(h_B - Q_B) = \bar{L}_{n-1}h_{n-1} - \bar{V}_n H_n = \cdots$$
$$= \bar{L}_p h_p - \bar{V}_{p+1}H_{p+1} \quad (6.76)$$

$$\bar{L}_n x_n - \bar{V}_{n+1}y_{n+1} = Bx_B = \bar{L}_{n-1}x_{n-1} - \bar{V}_n y_n = \cdots$$
$$= \bar{L}_p x_p - \bar{V}_{p+1}y_{p+1} \quad (6.77)$$

In Fig. 6.21 the reflux ratio $\bar{L}/\bar{V}$, x_B, and the heat in the reboiler q_B fix the location of the difference point d', x_B, $h_B - Q_B$. All operating lines in the stripping section pass through this point. The graphical construction to determine the number of equilibrium stages in the stripping section is as follows:

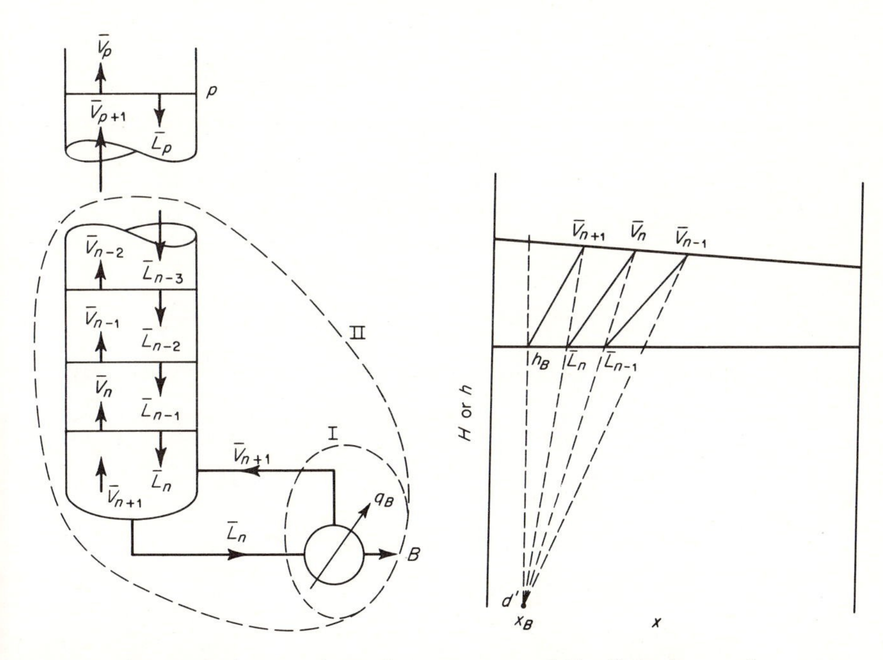

FIGURE 6.20 Stripping section of a fractionating column.

FIGURE 6.21 Stripping section—number of stages.

Assuming the reboiler to be an equilibrium stage, the vapor $\bar{V}_{n+1}$ is in equilibrium with the bottoms stream. Therefore equilibrium data establish the value of y_{n+1} on the saturated-vapor line. This intersection also establishes H_{n+1}. A line is drawn through d' and y_{n+1}, H_{n+1}. The intersection of this line with the saturated-liquid line establishes x_n, h_n. Equilibrium data establish the value of y_n, H_n, and a line through this point and d' establishes x_{n-1}, h_{n-1} at the intersection with the saturated-liquid line. This procedure is repeated until the feed plate is reached in terms of composition or until a side stream is withdrawn.

Feed-plate location Equations (6.72) through (6.74) show that the difference point d in the rectifying section lies at x_D, $h_D - Q_D$, and the difference point d' in the stripping section is located at x_B, $h_B - Q_B$, by Eqs. (6.75) through (6.77).

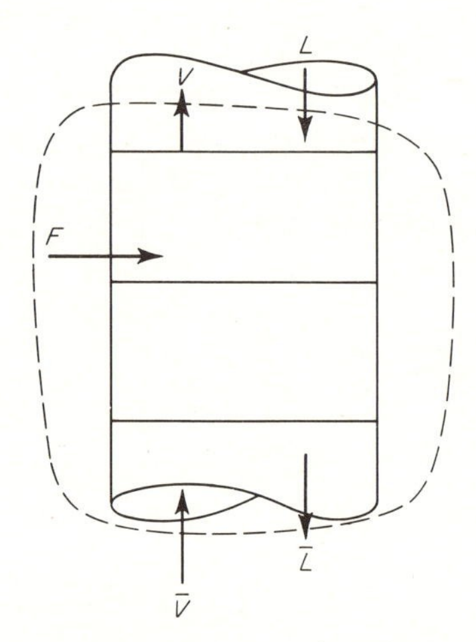

FIGURE 6.22 Fractionating column—feed section.

The equations for the section around the feed plate (shown schematically in Fig. 6.22) are

$$F = V - L + \bar{L} - \bar{V} \tag{6.78}$$

$$Fx_F = Vy - Lx + \overline{Lx} - \overline{Vy} \tag{6.79}$$

$$Fh_F = VH - Lh + \overline{Lh} - \overline{VH} \tag{6.80}$$

and in terms of the difference points,

$$Fx_F = d_x + d'_x \tag{6.81}$$

$$Fh_F = d_h + d'_h \tag{6.82}$$

Thus the point x_F, h_F lies on a straight line connecting the points x_B, $h_B - Q_B$, and x_D, $h_D - Q_D$.

The graphical construction for the number of theoretical stages in the whole column is shown in Fig. 6.23.

The construction may start from either side of the diagram, indicating either the conditions at the top or the bottom of the column, and proceed as indicated for the rectifying and stripping sections of the column. In either case, when an equilibrium tie line crosses the line connecting the difference points through the feed condition, the other difference point is used to complete the construction. This is the condition where the feed is introduced at the optimum location. In Fig. 6.23 the feed plate is the fourth plate from the top.

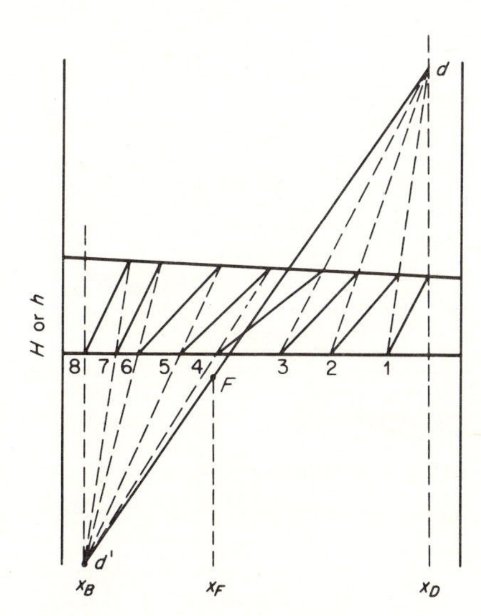

FIGURE 6.23 Complete fractionating column—graphical determination of number of stages.

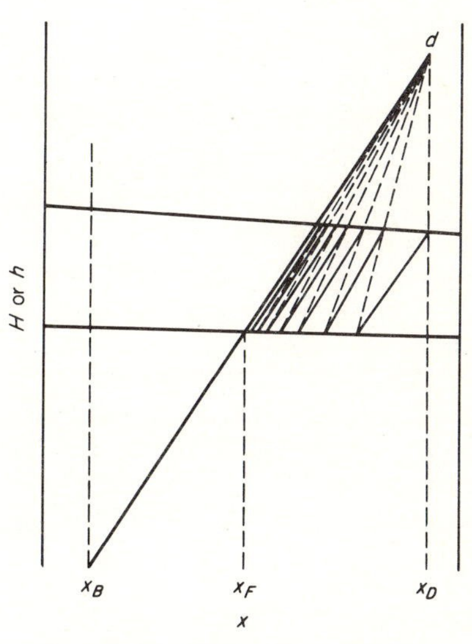

FIGURE 6.24 Minimum reflux on an enthalpy-concentration diagram.

Minimum reflux—Ponchon diagram The condition of minimum reflux is encountered when any equilibrium tie line coincides with any operating line, as shown in Fig. 6.24. Usually this coincidence takes place on the operating line through the feed condition, and the location of d for the condition of minimum reflux is found by extending the equilibrium tie line through the feed condition to intersect the composition line through x_D or y_D.

Total reflux The condition of total reflux is shown in Fig. 6.25. For the condition of total reflux the operating lines are parallel and vertical, converging at infinity. This results from the condition of $V = L$ since $D = 0$, $B = 0$, $Q_D = q_D/D = \infty$, and $Q = q_B/B = \infty$. Thus the enthalpy coordinate $h_D - Q_D = h_D - (-\infty)$ and $h_B - Q_B = h_B - \infty$. Obviously the minimum number of plates are encountered at the condition of total reflux.

Side streams and multiple feeds In some situations, where feeds of different composition are available, it is possible to introduce each at its proper location in the column, and thereby reduce the total number of plates necessary by

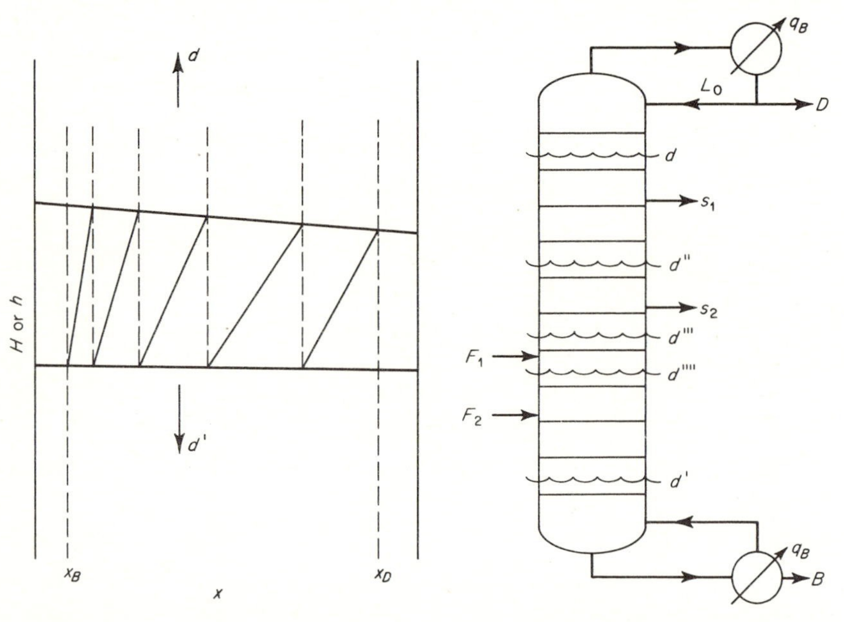

FIGURE 6.25 Total reflux on an enthalpy-concentration diagram.

FIGURE 6.26 Multifeed and multiside stream column.

one or two, compared with the number required when the feeds are premixed. Similarly, process requirements may require two or more distillate products of different composition, and these may be removed from the column as distillate and side streams. The Ponchon method may be applied to these situations by writing the difference-point equations for each section of the column between streams added or removed. The number of difference points will always be equal to the number of streams N (including D and B) minus 1 (see Fig. 6.26).

Although this appears to be a desirable practice, it is the opinion of the author that for most binary-system and multicomponent-system distillations this is very unsatisfactory. Engineering design must include not only consideration of the cost of the additional plates but all other factors—not the least of which is control of the process. The more fluctuations introduced into a fractionation system because of stream quantity, temperature, and composition variation, the more difficult becomes the control of the process to maintain steady-state operation. Therefore this practice is not recommended where component composition specifications are to be met. In the case of complex systems, such as petroleum oils and fractions, and others, the practice has some advantages and is followed. However, the component purity specifications are not the criteria of satisfactory operation.

Example 6.5

Using the enthalpy-concentration diagram based on molal enthalpy and mole fraction from Example 6.4, determine the following for the conditions in Example 6.1, assuming a saturated liquid feed. (*a*) The number of theoretical stages for an operating reflux ratio of $L_0/D = 2.5$; (*b*) minimum reflux ratio L_0/D; (*c*) minimum equilibrium stages at total reflux; (*d*) condenser duty feeding 10,000 lb of feed/hr, Btu/hr; (*e*) reboiler duty, Btu/hr.

Solution

See figure for Example 6.5.

(*a*) Ten stages.

(*b*)

$$\left(\frac{L_0}{D}\right)_{\min} = 1.18$$

(*c*) Seven stages.

(*d*)

$$h_D - Q_D = 31{,}700 \text{ cal/g mole}$$

$$h_D = 5100 \text{ cal/g mole}$$

$$Q_D = 26{,}600 \text{ cal/g mole}$$

$$Q_D = 26{,}600 \text{ cal/g mole} \; \frac{1.8 \text{ Btu/lb mole}}{\text{cal/g mole}} \left(0.426 \frac{\text{mole } D}{\text{mole } F}\right) \frac{10{,}000 \text{ lb } F\text{/hr}}{103 \text{ lb/mole } F}$$

$$= 2{,}045{,}000 \text{ Btu/hr}$$

(*e*)

$$h_B - Q_B = 5800 - (-14{,}400) = 20{,}200 \text{ cal/g mole}$$

$$Q_B = 20{,}200 \text{ cal/g mole} \; \frac{1.8 \text{ Btu/lb mole}}{\text{cal/g mole}} \; \frac{0.574 \text{ lb mole } B}{\text{lb mole } F} \; \frac{10{,}000 \text{ lb } F\text{/hr}}{103 \text{ lb } F\text{/mole } F}$$

$$= 2{,}020{,}000 \text{ Btu/hr}$$

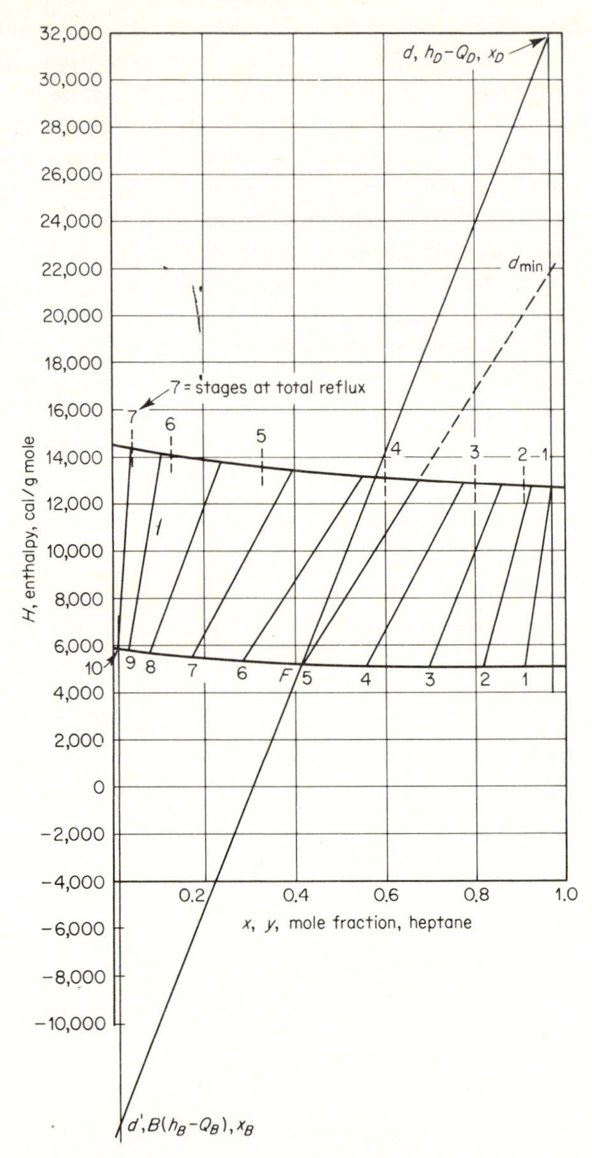

EXAMPLE 6.5 Enthalpy-concentration diagram—mole fraction basis.

PROBLEMS

1. A mixture of 30 mole % benzene and 70 mole % toluene are to be fractionated to produce 99 mole % benzene and 99 mole % toluene. Assume ideal behavior, and 1-atm total pressure.
   (*a*) Determine the minimum reflux ratio L/D required for this separation if the feed

were introduced (i) at its bubble point; (ii) at its dew point; (iii) as a liquid at 100°F.

(*b*) Determine the number of stages required for each of the cases in Prob. 1 for a reflux ratio equal to 1.8 $(L/D)_{min}$.

(*c*) Determine the minimum stages at total reflux for each of the cases in Prob. 1.

2. A fractionating column operates at 760 mm Hg pressure and is equipped with a total condenser. The distillate (and reflux) is produced at 100°F and the bottoms product at its bubble point. The feed enters at 150°F.

Data	Feed	Distillate	Bottoms
Ethanol, wt %	30.0	92.0	0.5
Water, wt %	70.0	8.0	99.5
T, °F	150	100	BP
Enthalpy, Btu/lb	95.5	40.3	

(*a*) Determine the minimum reflux $(L/D)_{min}$.

(*b*) Determine the number of equilibrium stages needed for a reflux ratio $L/D = 3.0$.

(*c*) For a feed of 10,000 lb/hr determine the hourly heat load in the reboiler.

(*d*) Determine the gallons of water per hour needed in the condenser. Inlet water temperature is 80°F; outlet water temperature is 125°F.

(*e*) Repeat (*a*) and (*b*) with the McCabe-Thiele method.

The equilibrium and enthalpy data are shown in the following tables. Ethanol–water; 1 atm.

Equilibrium Data

T, °F	Weight fraction, ethanol	
	Liquid	Vapor
212.0	0.00	0.000
210.1	0.01	0.103
208.5	0.02	0.192
206.9	0.03	0.263
204.8	0.04	0.325
203.4	0.05	0.377
197.2	0.10	0.527
189.2	0.20	0.656
184.5	0.30	0.713
181.7	0.40	0.746
179.6	0.50	0.771
177.8	0.60	0.794
176.2	0.70	0.822
174.3	0.80	0.858
173.0	0.90	0.912
172.8	0.94	0.942
172.7	0.96	0.959
172.8	0.98	0.978
173.0	1.00	1.000

Saturated Enthalpy Data

Wt. fr. ethanol	h_L, Btu/lb	h_V, Btu/lb
0	180.1	1150
10	159.8	1082
20	144.3	1012.5
30	135.0	943
40	128.2	873
50	122.9	804
60	117.5	734
70	111.1	664
80	103.8	596
90	96.6	526
100	89	457.5

3. A continuous bubble-cap fractionating column operating at 1-atm pressure is to be used for fractionating 18,000 lb/hr of an ethanol–water mixture containing 47.5% ethanol. The column is equipped with a *partial* condenser producing a vapor product at its dew point and the reflux at its bubble point. Assume the partial condenser to be an equilibrium stage.

Data	Feed	Distillate (vapor)	Bottoms
Ethanol, wt %	47.5	91.2	1.0
Water, wt %	52.5	8.8	99.0
Enthalpy, Btu/lb	600	—	—

(*a*) Determine the minimum reflux $(L/D)_{\min}$.
(*b*) Determine the equilibrium stages required at an operating reflux ratio $= 3.5(L/D)_{\min}$.
(*c*) Repeat (*a*) and (*b*) and use the McCabe-Thiele method.

4. A fractionating column is to be designed to fractionate two feeds *A* and *B* to produce a distillate product containing 90 wt % ethanol and a bottom product containing 1 wt % ethanol. $L/D = 1.55$.

Data	Feed *A*	Feed *B*	Distillate	Bottoms
Quantity, lb/hr	2035	7965	—	—
Ethanol, wt %	80	50	90	1
Enthalpy, Btu/lb	170	650	50	—

Determine (*a*) number of theoretical plates (feeds introduced on correct plates); (*b*) location of feed plates; (*c*) heat input required per hour in reboiler; (*d*) pounds per hour of distillate product; (*e*) minimum reflux ratio $(L/D)_{\min}$.

5. 13,700 lb of 99.5 mole % acetone per day are to be produced by continuous fractionation of a feed containing 35 mole % acetic acid and 65 mole % acetone at 760 mm Hg. The feed enters the column at 80°F and the distillate leaves at its bubble point from the total condenser. Draw the *Hx* diagram for this system at atmospheric pressure by plotting the terminal saturated-vapor and saturated-liquid enthalpies (pure components) and connect them by straight lines. Use reference temperature of 32°F.

Data	HOAC	Acetone
Molecular weight	60.05	58.8
Bubble point, °C	118.2	51.13
Melting point, °C	16.68	−94.3
Latent ht. of vapor at bubble point, cal/g	96.75	126
Latent ht. of fusion at melting point, cal/g	46.68	—
Spec. ht., cal/(g) (°C)	0.56	0.52

Equilibrium Data at 760 mm Hg

t, °C	Acetone, mole %	
	x	y
112.1	4.2	10.8
107.4	8.2	22.5
104.6	15.8	35.6
94.3	22.6	56.4
90.4	27.1	63.0
86.3	30.7	70.9
78.6	43.3	84.4
70.8	55.0	91.8
65.6	66.8	96.6
60.7	93.5	99.7

Determine (*a*) minimum reflux ratio $(L/D)_{\min}$; (*b*) number of theoretical stages for a reflux ratio of 2.5; (*c*) minimum reflux ratio for saturated-liquid feed; (*d*) minimum reflux ratio for saturated-vapor feed.

6. *Separation of air by Linde double column.* It is desired to separate 3,500,000 std cu ft (60°F and 1 atm) of air into nitrogen and oxygen (day = 24 hr). Both products are to be 99.0% pure. Assume the separation to be a binary separation. The separation is to be accomplished in a Linde double column. Specify temperatures, pressures, reflux ratios, the heat transfer in all stills and condensers, and the number of equilibrium plates required in a typical operation. Include a plot of composition as a function of plate number for the column and specify the condition of the air before it enters the first reboiler.

For a description of the Linde double column, see Dodge [8], Ruheman [9], and Williams [10].

Enthalpy and Vapor-Liquid Equilibrium Data for an Oxygen–Nitrogen System

T, °R	Composition (liquid)	Mole fr. nitrogen (vapor)	Enthalpy sat. liquid	Btu/lb mole sat. vapor
		Pressure, 5 atm, 760 mm Hg		
196.0	0.0000	0.0000	1432	4107
193.9	0.0500	0.1225	—	—
191.8	0.1000	0.2275	1470	4093
188.0	0.2000	0.3985	1512	4082
184.5	0.3000	0.3319	1551	4067
181.7	0.4000	0.6385	1587	4052
179.0	0.5000	0.7250	1623	4040
176.9	0.6000	0.7970	1662	4023
174.7	0.7000	0.8585	1704	4007
172.7	0.8000	0.9119	1749	3987
171.0	0.9000	0.9583	1792	3962
169.7	1.0000	1.0000	1833	3937
		Pressure, 1 atm, 760 mm Hg		
162.3	0.0000	0.0000	1047	3987
160.0	0.0500	0.1735	—	—
158.0	0.1000	0.3100	1065	3972
154.1	0.2000	0.4919	1090	3967
151.1	0.3000	0.6405	1124	3955
148.4	0.4000	0.7350	1160	3942
146.3	0.5000	0.8046	1202	3927
144.7	0.6000	0.8591	1245	3912
143.0	0.7000	0.9031	1290	3897
141.7	0.8000	0.9399	1333	3872
140.3	0.9000	0.9717	1375	3847
139.1	1.0000	1.0000	1415	3822

Enthalphy-composition data calculated by Hansen from [9], p. 89.

Equilibrium data by Dodge and Dunbar, *J. Am. Chem. Soc.*, **49**:591 (1927), and Dodge, *Chem. Met. Eng.*, **35**:622 (1928).

Nomenclature

B = residue, moles/unit time

D = distillate, moles/unit time

F = feed, moles/unit time

f = function of

H = enthalpy of vapor, Btu/lb mole

h = enthalpy of liquid, Btu/lb mole

K = equilibrium vaporization ratio

L = liquid, moles/unit time, rectifying section

$\bar{L}$ = liquid, moles/unit time, stripping section
P_1 = vapor pressure, component 1
P_t = total pressure
Q_B = Btu/lb mole of residue (heat in reboiler)
q_B = total heat in reboiler, Btu/unit time
Q_D = Btu/lb mole of distillate (heat in condenser)
q_D = total heat in condenser, Btu/unit time
t = temperature
V = Vapor, moles/unit time in rectifying section
$\bar{V}$ = Vapor, moles/unit time in stripping section

Greek letters

γ = activity coefficient
ν = fugacity coefficient
λ = latent heat of vaporization
Σ = summation

Subscripts

B = residue
D = distillate
F = feed
1, 2 = component designation
0 = reflux
m, n, p = plate number

REFERENCES

1. Lewis, W. K.: *Ind. Eng. Chem.*, **14**:492 (1922).
2. McCabe, W. L., and E. W. Thiele: *Ind. Eng. Chem.*, **17**:605 (1925).
3. Ponchon, M.: *Tech. Moderne*, **13**:20 (1921).
4. Randall, M., and B. Longtin: *Ind. Eng. Chem.*, **30**:1063, 1188, 1311 (1938); **31**:908, 1295 (1939).
5. Savarit, R.: *Arts et métiers* (1922), pp. 65, 142, 178, 241, 266, 307.
6. Smoker, E. H.: *Trans. A.I.Ch.E.*, **34**:165 (1938).
7. Sorel, M.: "La Rectification de l'alcool," Paris, 1893.
8. Dodge, B. F.: "Chemical Engineering Thermodynamics," McGraw-Hill, New York, 1944.
9. Ruheman, M.: "The Separation of Gases," Clarendon Press, Oxford, 1940.
10. Williams, V. C.: Thermodynamic Properties of Air, *Trans. A.I.Ch.E.*, **39**:93 (1943).

chapter

Ternary and multicomponent system fractionation—number of stages

The term *multicomponent* system can be used to describe all systems containing more than two components. In industrial practice, true binary systems are almost never encountered and true ternary systems are found only infrequently. Thus the quantitative calculation techniques for multicomponent fractionation are the most generally applicable to all fractionations, although certain simplifications can be made if actual or pseudobinary and ternary systems can be assumed.

It is axiomatic that for close separation of N components $N - 1$ fractionating columns are required. Thus one column suffices to separate a binary system, two for a ternary, and eight for a nine-component system. Where only rough separations are necessary, such as the separation of a multicomponent mixture into a "light" boiling fraction and a "heavy" boiling fraction, the fractions can be considered as pseudocomponents and only one column would be required. In this case if, e.g., a definite boiling range, vapor pressure, or composition split were specified, the design of the fractionation column would be based on the separation between two "key" components.

Thus the product composition, distillate, bottoms, or both, can be fixed by specification, and the engineer designs the fractionation process to make the separation desired.

The number of variables V to be fixed in fixing the operation of the fractionation system for design purposes is

$$V = C + 6$$

The usual variables conveniently or necessarily fixed are given for a ternary and for an eight-component mixture:

Number of Variables

Variable	Ternary mixture	Eight-component mixture
Feed rate	1	1
Feed compositions	2	7
Feed enthalpy	1	1
Reflux temperature	1	1
Column pressure	1	1
Reflux ratio	1	1
Distillate and/or bottoms compositions*	2	2
	9	14

* The distillate and bottoms compositions are usually specified by fixing the recovery of one key component in the distillate as a percentage of that component in the feed and the recovery of the heavy key in the bottoms as a percentage of that component in the feed. Only when the light key is the lowest boiling material or the heavy key the highest boiling material in the mixture can both the purity (composition) and percent recovery be specified for one or the other of the keys.

7.1. PRELIMINARY CALCULATIONS

Selection of Key Components

In the case where "pure" components are to be produced, the key components are the compounds boiling adjacent to one another on the temperature scale. The material having the lower boiling point is designated as the *light key* component and the next heavier as the *heavy key* component. Where only rough separations are required, the keys are selected to give the best yield of product of the desired characteristics. In some instances the keys selected are not adjacent but have an intermediate boiling component between them. They are then designated as *light key*, *heavy key*, and *intermediate* (boiling) or *distributed key*.

Feed Condition

Because the condition of the feed as a single- or two-phase mixture and its heat content are important factors in fractionating column design, the limits of the two-phase region for the feed mixture with respect to temperature and pressure as well as the enthalpy of the feed must be evaluated. The two-phase region limits are determined by calculation of the bubble-point and dew-point temperatures for a number of pressures which encompass the desired pressure range of operation. The enthalpy of the feed may be calculated or estimated depending upon the availability of data for the multicomponent system considered. If it is necessary or desirable to introduce a feed material at its process temperature and pressure, its dew-point and bubble-point temperatures at that pressure are determined first, and if it is a two-phase mixture, the quantities of vapor and liquid are calculated by Eq. (7.1) or (7.2) as described in Chap. 4.

$$V = \sum_{1}^{n} \frac{x_{F_i}}{K_i + L/V} = \sum_{1}^{n} \frac{x_{F_i}}{\gamma_i P_i/\nu_i P_t + L/V} \tag{7.1}$$

$$V = \sum_{1}^{n} \frac{K_i x_{F_i}}{K_i + L/V} = \sum_{1}^{n} \frac{(\gamma_i P_i/\nu_i P_t) x_{F_i}}{\gamma_i P_i/\nu_i P_t + L/V} \tag{7.2}$$

Selection of Column Pressure

Generally at higher pressures the number of stages required for a given separation is greater than the number required at lower pressures and, based on this consideration alone, all distillations should be run at the lowest possible pressure. Because vapor volume increases as pressure decreases, the diameter of the column would have to be increased to handle the increased vapor volume. Also as the pressure is decreased, the boiling point decreases, and in many normally gaseous mixtures such as light hydrocarbons, refrigeration would be necessary to attain temperatures low enough to condense the overhead vapor so that liquid reflux could be supplied to the column. All of these things must be considered in selecting an operating pressure for a fractionating column.

Under normal conditions the temperature of available cooling water determines the condensing temperature and therefore the fractionating column operating pressure. For example, the highest temperature of the cooling water recirculated from the cooling tower is 105°F (the design is always based on the "worst" condition, not the "average") and a light hydrocarbon mixture is to be distilled. The vapor pressure-temperature and dew-point-temperature–pressure relationship of the mixture are as shown in the table.

t, °F	Bubble-point pressure, psia	Dew-point pressure, psia
80	250	180
90	260	200
100	275	220
110	308	260
120	350	295
130	400	325

Then following considerations would be taken into account:

1. What is the lowest temperature to which the vapor could be cooled? Or, how closely will the reflux temperature approach that of the cooling water? With well-designed condensers and clean liquids the approach could be expected to be around 15°F. Thus the temperature would be 105 + 15 = 120°F. At 120°F a total condenser would require a pressure of 350 psia and a partial condenser a pressure of 295 psia for the system shown.

2. What type of condenser will be most suitable for this column? A number of factors are to be considered under this question.

(*a*) *Condition of distillate product.* If the distillate product can be satisfactorily produced as a vapor and if the other criteria can be satisfied, a partial condenser could be selected. By "satisfactorily" is meant a salable or usable product. If, for example, the distillate product were to be sold as a liquid, a product from the distillation column in vapor form would have to be either cooled or compressed, or both, to produce it in a liquid form. It might be more economical to do this during the fractionation procedure than subsequently.

(*b*) *Control of internal reflux.* Normally, when a partial condenser is utilized, the temperature of the reflux liquid is the bubble-point temperature of the reflux or dew-point temperature of the vapor product. This would require the circulation of more reflux liquid to increase the internal reflux. If a total condenser were used, the reflux liquid could be returned to the column as a supercooled liquid, and the internal reflux could be controlled by the temperature of the reflux liquid returned to the column.

(*c*) Since the type of condenser fixes in general the operating pressure on the column, the economic consideration of investment and operating costs for the fractionation process utilizing the two types of condensers would have to be made. A guide to the factors and their relationship to the type of condenser is as follows:

Variable	Type of condenser	
	Partial	Total
Product	Vapor	Liquid
Pressure	Lower	Higher
Temperature	Same (cooling water)	Same
No. of plates	Less	More
Column-shell thickness	Less	More
Capacity based on vapor velocity	Less	More

Before a reliable economic comparison can be made, the column design must be made for each type of condenser for a number of reflux ratios, and in many cases for a number of pressures. In general, all variables affecting the economics of the fractionation process should be considered. Whether or not all of these are studied depends upon the particular set of conditions at the time the design is made; whether it is a feasibility study for a future process, a pilot-plant design, or a final plant design.

7.2. DESIGN PROCEDURE

The discussion in Chap. 5 indicated that the limits of operation of a fractionating column are *minimum reflux* and *total reflux*. The condition of actual operation with regard to reflux is selected on the basis of the relationship between the number of plates for a number of reflux ratios lying between the minimum and total reflux and the evaluation of operating and investment costs resulting from the various combinations. Therefore, the minimum reflux ratio and the minimum number of theoretical plates is determined by one of the methods outlined in Chap. 5.

Once the limits are established, the number of theoretical stages at selected reflux ratios may be determined by one of several methods: (*a*) analytically, using algebraic plate-to-plate calculations involving equilibrium-, material-, and enthalpy-balance relationships; (*b*) empirical short-cut methods in which the minimum reflux, actual reflux, minimum plates, and actual theoretical plates are related; and (*c*) in the case of ternary systems, graphical methods using ternary distillation diagrams.

A common stepwise procedure is as follows:

1. Decide on the key components. This is determined by product requirements and the processing scheme selected.

2. Decide on the distribution of the key components. This is determined also by the product requirements with regard to purity specifications and yield.

3. Make a material balance to establish the approximate composition of the distillate and its rate, and the composition of the bottoms product and its rate. This is done as follows: The recovery of the light key component as a fraction of that component contained in the feed establishes the moles of that component in the distillate and bottoms. Similarly, the recovery of the heavy key component as a fraction of that component contained in the feed establishes the number of moles of the heavy key in the distillate and bottoms product. The distribution of the key components in the distillate and bottoms is thus established. Since the feed contains all components, theoretically at least and practically to some extent, both distillate and bottoms will have to contain at least some of each component. The distribution of these components into the two streams must be established before detailed calculations can be made to determine the number of stages. One method for estimating distribution was devised by Geddes [13]. He related the distribution of the components in the distillate and bottoms with their relative volatilities by using the type of plot introduced by Hengstebeck [19]. This correlation is represented by

$$\log \frac{i_D}{i_B} = C \log \alpha_{ik} \tag{7.3}$$

Figure 7.6 shows a plot of this equation wherein the logarithm of the ratio of the number of moles of any component in the distillate to the moles of the same component in the bottoms product is plotted against the logarithm of the relative volatility of the component referred to some other component in the mixture at the average column temperature. The relative volatilities are determined for all components present in the feed, and their distribution is read from the straight line plotted as shown in Fig. 7.6 through the values for the keys.

The average column temperature may either be taken as the feed temperature or as the arithmetic average between the dew-point temperature of the vapor rising from the top plate and the bubble-point temperature of the bottoms product. These are determined as follows: Assume that all of the components lighter than the light key are recovered completely in the distillate. This fixes the moles of all lighter than light key components in the distillate and, with the specified moles of light and heavy keys, establishes the approximate total moles of distillate. The approximate composition can then be computed. Similarly, assume that all of the heavier than heavy key components are completely recovered in the bottoms product to determine the approximate total moles and approximate composition of the bottoms as in the foregoing, including the portion of the light key contained.

(*a*) If the distillate consists of low-boiling components such that the cooling-water temperature is the controlling factor, the bubble-point temperature of the distillate (where a total condenser is used, or dew-point temperature at the vapor distillate product from a partial condenser) is fixed at 15 to 20°F above the cooling-water temperature. The top column pressure is calculated as the bubble-point or the dew-point pressure. The pressure drop across the column is estimated and added to the top column pressure. The bubble-point temperature is then calculated for the bottoms product, based on the assumed composition and at the estimated pressure.

(*b*) If the cooling-water temperature is not a controlling factor, a column pressure is fixed and the dew-point temperature of the vapor rising from the top plate is computed, based upon the distillate composition calculated from the distribution correlation Eq. (7.3), using the feed temperature or the average of the estimated top plate and bottoms temperatures to get the α's. At this point all of the variables to fix the column operation have been fixed or determined except the operating reflux ratio and the number of theoretical stages.

4. Establish a reflux ratio (usually as some factor times the minimum) and calculate the number of theoretical stages. The minimum reflux is calculated by one of the methods in Chap. 5.

Rigorous Calculation of the Number of Equilibrium Stages

The rigorous plate-to-plate calculation of the number of equilibrium stages consists generally of fixing the pressure, reflux ratio, and component distribution in the distillate and bottoms products, and conducting trial-and-error calculations of the temperatures, compositions, and quantities on each plate. Calculations are started with the distillate and continued down the column to the feed plate and, starting with the bottoms compositions, quantity, and temperature, are continued up the column, plate by plate, to the feed plate. Such a procedure involves the use of material balance, enthalpy balance, and equilibrium relations to obtain the most accurate results.

The correct number of stages is reached when the feed temperature, composition, and condition are matched from the two sets of calculations originating from the top and from the bottom conditions. The exact matching can rarely, if ever, be accomplished, even with a large number of trials, because of the difficulty in determining the correct distribution of the components in the distillate and bottoms products. Therefore it has been common practice to establish matching criteria ranging from somewhat gross approximations of the correct match to reasonably close, although not exact, approximations of the match.

The usual approximations consist of matching the ratios of the compositions or quantities of the light to heavy key components in the liquid or vapor—calculated from the top and from the bottom of the column—to the ratios of the compositions or quantities of the light to heavy key components in the corresponding liquid or vapor portion of the feed. Other methods require the exact match of all of the component quantities or compositions at the feed plate resulting from plate-to-plate calculations from the top and bottom of the column, without attempting to match the feed compositions or even the temperature of the feed and its thermal condition.

This introduces the problem of proper feed plate location, which is relatively simple in binary fractionation but somewhat more difficult in multicomponent system fractionation.

Estimation of Feed Plate Location

Where the plate-to-plate calculation method does not require the feed plate to be designated initially, Gilliland's [16] method, based on binary-system studies, relates the ratios of the compositions of the light and heavy key components to their ratio in the feed based upon the thermal condition of the feed vapor, liquid, or mixed phase.

Thus, for a vapor feed the approximate feed plate match is considered reached when the ratio of the key compositions in equilibrium with the total feed vapor compositions lies between the ratio of the key compositions in the liquid at the plate above the feed plate and the next plate above that.

$$\left(\frac{x_{LK}}{x_{HK}}\right)_{F-1} < \left(\frac{x^*_{LK}}{x^*_{HK}}\right)_{F_T} < \left(\frac{x_{LK}}{x_{HK}}\right)_{F-2} \tag{7.4}$$

For a mixed vapor-liquid feed the match is considered to be accomplished when the ratio of the composition of the keys in the liquid portion of the mixed feed lies between the calculated ratios of the keys in the liquid on the plate above and on the feed tray.

$$\left(\frac{x_{LK}}{x_{HK}}\right)_{F} < \left(\frac{x_{LK}}{x_{HK}}\right)_{F_L} < \left(\frac{x_{LK}}{x_{HK}}\right)_{F-1} \tag{7.5}$$

For a liquid feed the approximate match is considered to be reached when the composition of the key components in the liquid feed lies between the calculated values in the liquid on the feed tray and in the liquid on the tray above.

$$\left(\frac{x_{LK}}{x_{HK}}\right)_{F} < \left(\frac{x_{LK}}{x_{HK}}\right)_{F_T} < \left(\frac{x_{LK}}{x_{HK}}\right)_{F-1} \tag{7.6}$$

Although these are only approximate relations, they serve as a guide in determining the feed plate location and, thus, enables the designer to judge when a reasonable solution to the number of equilibrium plates has been reached.

Kirkbride [23] devised an equation for estimation of the ratio of the number of plates above and below the feed which enables the estimation of the feed tray location. This relationship is shown in

$$\log\frac{m}{p} = 0.206\log\left\{\frac{B}{D}\left(\frac{x_{HK}}{x_{LK}}\right)_F\left[\frac{(x_{LK})_B}{(x_{HK})_D}\right]^2\right\} \tag{7.7}$$

where m = number of theoretical stages above the feed plate
p = number of theoretical stages below the feed plate

Sorel Method [35]

The "rigorous" plate-to-plate calculation procedure using mole fractions is outlined in the following steps:

1. (*a*) For a total *condenser* (Fig. 7.1) the temperature of plate 1 is calculated as the dew-point temperature of the vapor having the same composition as the distillate D. Thus a trial-and-error calculation is carried out until the relation

$$\Sigma x_i = 1.0 = \sum \frac{y_i}{K_i} = \sum \frac{y_i \nu_i P_t}{\gamma_i P_i} \tag{7.8}$$

is satisfied. This is done by selecting a number of temperatures (usually a minimum of three), obtaining K or equivalent values at the column pressure

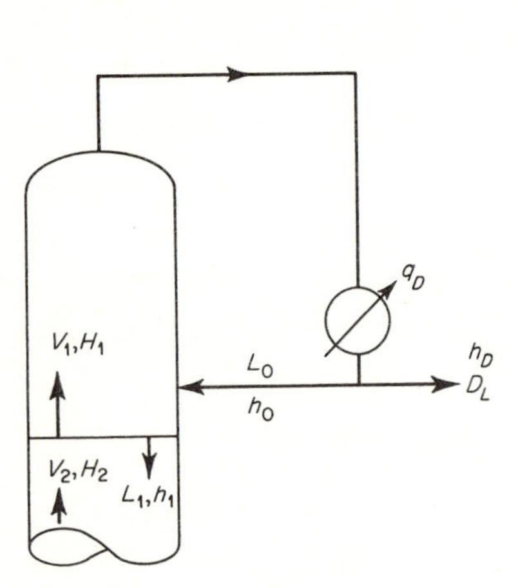

FIGURE 7.1 Total condenser.

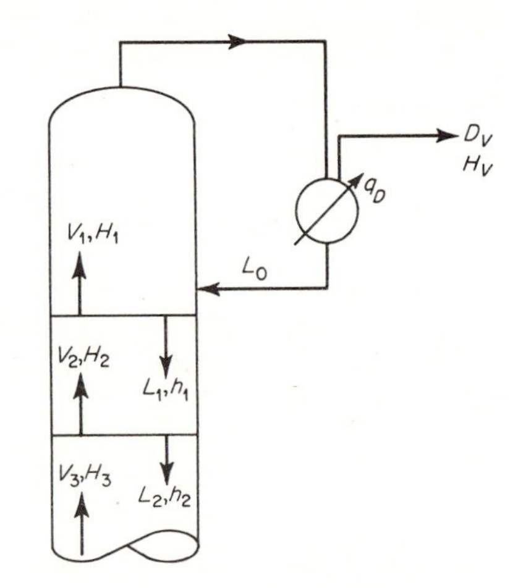

FIGURE 7.2 Partial condenser.

and the selected temperature for each component, and summing the y/K values. Plot the summation versus temperature and read the temperature where the summation equals 1.0. The liquid composition in equilibrium with the vapor from plate 1 is calculated by obtaining the K values at the temperature and pressure on plate 1 and by evaluating y/K for each component.

(*b*) Where a *partial condenser* (Fig. 7.2) is used, the temperature of the vapor product and reflux liquid must be determined. This temperature is the dew-point temperature of the vapor and is calculated in the manner previously described—by selecting temperatures, obtaining the K values, and summing the y/K values for each temperature. A plot of the summation versus temperature enables the correct temperature to be read corresponding to the point where the sum equals unity (see Fig. 7.3). The composition of the reflux liquid L_0 is determined by obtaining K values for each component at the computed temperature and pressure and evaluating $x_i = y_i/K_i$.

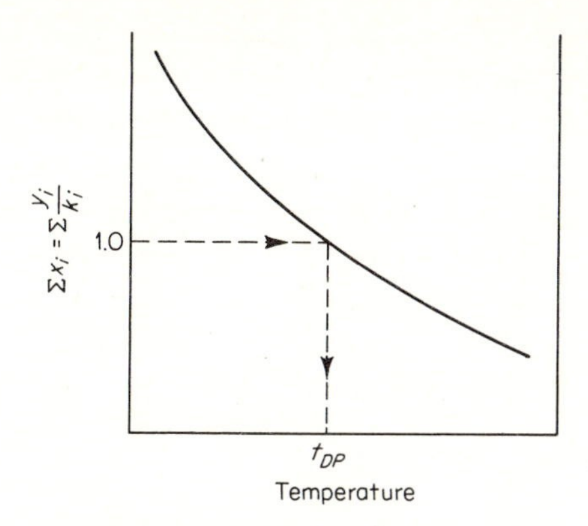

FIGURE 7.3 Determination of dew point.

2. The quantity of vapor rising from plate 1 is determined by material balance. Since L_0/D and D are known or fixed,

$$V_1 = L_0 + D \tag{7.9}$$

or

$$V_1 = \left(\frac{L_0}{D} + 1\right)D \tag{7.10}$$

3. The condenser duty or heat removed in the condenser is then calculated by enthalpy balance because all temperatures and pressures are known.

$$q_D = L_0 h_0 + D h_D - V_1 H_1 \tag{7.11}$$

4. The composition of the vapor V_2 rising to plate 1 (or V_1 in the case of the partial condenser) is calculated by material and enthalpy balance. The equations for each case are as follows:

Total condenser	*Partial condenser*	
$V_2 = L_1 + D$	$V_1 = L_0 + D$	(7.12)
$V_2 y_{i2} = L_1 x_{i1} + D x_{i_D}$	$V_1 y_{i1} = L_0 x_{i0} + D x_{i_D}$	(7.13)
$V_2 H_2 + qD = L_1 h_1 + D h_D$	$V_1 H_1 + qD = L_0 h_0 + D h_D$	(7.14)
$H_2 = f(y_{i2}, y_{j2}, y_{k2}, \ldots, t_2, P_t)$	$H_1 = f(y_{i1}, y_{j1}, y_{k1}, \ldots, t_1, P_t)$	(7.15)

With a total condenser and reflux liquid cooled to a temperature below its

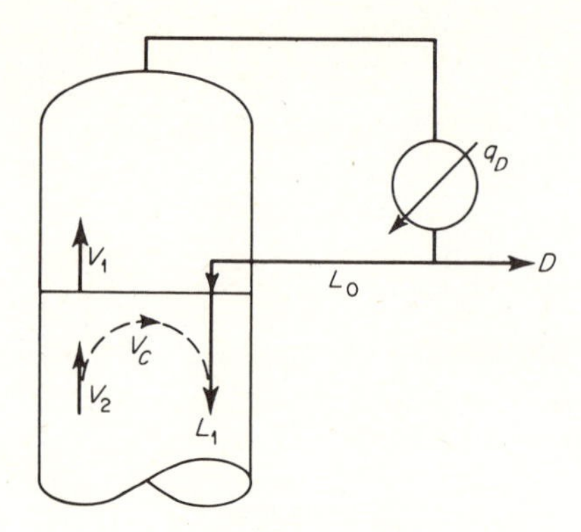

FIGURE 7.4 Reflux caused by supercooled liquid reflux from condenser.

bubble point (see Fig. 7.4), there are insufficient equations to evaluate the composition of V_2, quantity of V_2 and L_1, and the temperature t_2. However, if it is assumed that the molal latent heat of the mixture is essentially constant and varies but little with temperature and composition, and that the molal heat capacities of the components in the system are essentially constant and vary but slightly with temperature, the foregoing variables may be evaluated. Where the assumptions can be considered applicable, Eqs. (7.18) through (7.20) are used to obtain a solution.

$$V_C\lambda_1 = L_0 C_{p_L}(t_0 - t_1) \tag{7.18}$$

$$V_C + V_1 = V_2 \qquad L_0 + V_C = L_1 \tag{7.19}$$

$$t_2 = \frac{DC_{p_L}(t_0 - t_1) - qD - V_2\lambda_1}{V_2 C_{p_{V1}}} + t_1 \tag{7.20}$$

5. With the quantity of V_2 known, the composition can be calculated by Eq. (7.13) since the quantities L_1 and D and the compositions x_{i1} and x_{i_D} are known.

Note: Enthalpies of the streams may be calculated approximately by the following:

$$h_0 = \sum_1^n \bar{h}_0 = x_{i0}C_{p_i}(t_0 - t_r) + x_{j0}C_{p_j}(t_0 - t_r) + \cdots + x_{n0}C_{p_n}(t_0 - t_r) \tag{7.16}$$

$h_0 = h_D$ with a total condenser
For a partial condenser,

$$H_D = \sum_1^n \bar{H}_D = \sum_1^n x_i[\lambda_{i_r} + C_{p_v}(t_D - t_r)] \tag{7.17}$$

6. Determine the composition of L_2 in equilibrium with V_2 at t_2 and the column pressure by calculating the equilibrium concentrations x_{i2} by

$$x_{i2} = \frac{y_{i2}}{K_{i2}}$$

7. The vapor V_3 rising to plate 2 can be calculated if the partial molal enthalpies of the components in the mixture are known by equations comparable to Eqs. (7.12) through (7.15). If the Lewis-Matheson assumptions of equimolal overflow and boilup are made, V_3 equals V_2.

8. Evaluate the composition of V_3 by component material balance.

9. Determine the temperature of V_3 by the equation comparable to Eq. (7.20).

10. Repeat the calculations plate by plate until a vapor composition is reached which approximates the composition of the vapor portion of the feed, or until a liquid composition is reached which approximates the composition of the liquid portion of the feed. With several components in the mixture and by using the approximate relations, it is unrealistic to expect to duplicate the compositions in the feed exactly. As a guide, if the calculated ratio of the compositions of the key components approximately matches that of the ratio of the keys in the feed, and if the calculated temperature essentially matches the temperature of the feed, the number of equilibrium stages is approximately correct.

11. Based on the foregoing assumptions, the quantity of liquid and vapor below the feed plate can be calculated as

$$\bar{L}_F = L_{F-1} + F_L \tag{7.21}$$

$$\bar{V}_{F+1} = V_F - F_V \tag{7.22}$$

(see Fig. 7.5).

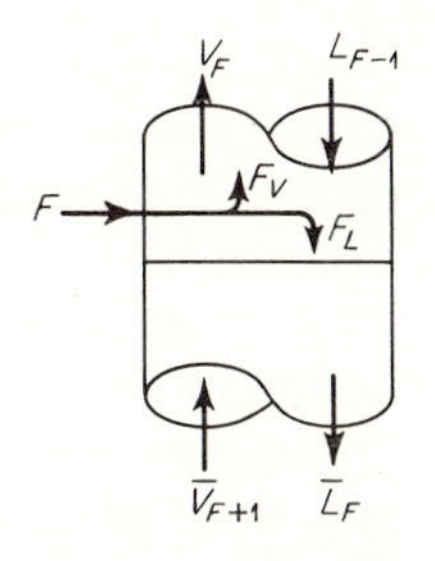

FIGURE 7.5 Condition at the feed plate.

12. Calculate the bubble-point temperature of the bottoms product B, if not determined above.

13. The composition of $\bar{V}_{n+1}$ is the equilibrium composition with B and is calculated by $y_{i_{n+1}} = (K_i x_i)_{n+1}$.

14. Since the quantities of $\bar{L}_n$ and $\bar{V}_{n+1}$ are known, the composition of $\bar{L}_n$ can be determined by material balance.

$$x_{i_n} = \frac{\bar{V}_{n+1} y_{i_{n+1}} + B x_{i_B}}{\bar{L}_n} \tag{7.23}$$

15. The temperature of plate n is the bubble-point temperature of L_n and is calculated in the same manner as the reboiler temperature.

16. The computations are then continued, plate by plate, up the column, establishing liquid compositions by material balance, temperature by bubble-point calculation, and vapor composition by the equilibrium calculations until the vapor and liquid compositions approximate the compositions of the vapor and liquid portions of the feed or until the key component ratio matches approximately that of the feed. The same observations apply here as to the calculations proceeding down from the top of the column. If the key component composition ratio approximately matches that of the feed and the plate temperature is essentially the same as that of the feed, this usually is sufficient.

Enthalpy balances should be used in the case of systems wherein equimolal latent heats and heat capacities cannot be assumed without appreciable error. In such cases the liquid and vapor quantities are not constant in each section of the column. To include enthalpy balances as part of the calculation procedure requires accurate enthalpy data on the mixture of components or an accurate means of calculating these data. In many cases such data are not available, and the resulting calculations are little more accurate than those based on the assumption of equimolal overflow. However, if reasonably accurate data are available, a more realistic evaluation of the number of equilibrium stages can be made by inclusion of enthalpy balances.

Throughout the discussion of plate-to-plate design it has been assumed that the pressure throughout the column is essentially constant. Actually, of course, the pressure is not constant, or there would be no vapor flow from the bottom of the column to the top and the column would be inoperable. Thus, because the pressure does change from plate to plate, consideration must be given to account for the effect of changing pressure on the vapor-liquid equilibrium relationships. The average maximum pressure drop across a commercial fractionating plate operating under high-load conditions can be estimated as equivalent to 6 in. of liquid flowing. With water at room temperature this would represent about a 0.22-psi pressure drop. Usually with other systems at higher temperatures the pressure drop is much less, usually around 0.1 to 0.2 psia. Thus, in a column operating at, e.g., 300 psia at the condenser, the pressure at the bottom of a 100-plate column would be 310 to 320 psia. Although the difference in K values between pressures of 300 and 310 psia would be slight, a significant difference can be noted between 300 and 320 psia. This difference would have to be accounted for in the calculations. If a column containing 100 plates were operated with the condenser maintained at a pressure of 30 psia, the bottom pressure would be 40 to 50 psia. The K values are considerably different for pressures of 30 and 40 psia and much more for pressures of 30 and 50 psia.

Under vacuum conditions the plate is designed to give the lowest possible pressure drop. Even with the best designs and at low liquid heads, the pressure drop will range from 0.01 to 0.05 psi, or approximately 1 to 3 mm Hg. Thus, with the top pressure at 50 mm Hg, the pressure at the bottom of a 25-plate tower would be 75 to 125 mm Hg pressure, and the equilibrium vaporization ratio or K values are much different at these pressures.

The problem of accounting for the effect of changing pressure throughout the column on the vapor-liquid equilibrium data cannot be completely and satisfactorily solved, but the following approximate solutions have been used.

1. Decide on a plate design and calculate an average pressure drop through the plate under expected operating conditions. Estimate plate efficiency and divide into the average pressure drop per plate to get an equivalent theoretical plate pressure drop. At each successive plate in the plate-to-plate calculation, add pressure corresponding to the drop through one more (theoretical) plate in determining the equilibrium vaporization ratios of K values. Continue the calculations to the feed plate. Assume the same number of theoretical plates between the reboiler and the feed plate as calculated between the condenser and feed plate, and calculate the pressure at the reboiler as the feed plate pressure plus the number of theoretical plates between the feed and reboiler times the pressure drop per theoretical plate. The bubble-point temperature is calculated at this pressure. The equilibrium vaporization ratios are then determined for each successive plate up the column at pressures equivalent to one less theoretical plate pressure drop. If the number of plates thus computed differs greatly from the assumed number, use the calculated number, recalculate the reboiler pressure, and repeat the stepwise calculation.

2. Estimate the number of theoretical plates needed for the separation by short-cut methods, and compute the pressure in the reboiler by adding to the top tower pressure the number of theoretical plates times the pressure drop per plate. Average the pressure at the top and bottom and evaluate the equilibrium vaporization ratios or K values for all plates at this average pressure.

3. Evaluate the feed liquid bubble-point pressure if the feed is a saturated liquid, and use this pressure in obtaining K values for all plates throughout the column.

All of these methods are inexact but serve to aid in correcting the calculated results in the right direction. In the case of the higher pressure columns, accounting for the pressure effect on equilibrium is not as important as in the case of vacuum columns where accounting for pressure effect is mandatory.

Lewis-Matheson Method [25]

In the Sorel method the amounts of vapor and liquid and, thus, the internal reflux and vapor flow are calculated for each plate. The Lewis-Matheson method [25] simplifies the calculations by assuming a constant L/V in the rectifying section and a constant L/V in the stripping section, and uses moles of each component in the various streams rather than mole fraction. This is a common practice when there exist little or no reliable mixture enthalpy data for the system or when the system is relatively ideal in its behavior.

The Lewis-Matheson method first fixes the pressure, the distribution of the components in the distillate and bottoms products, the feed plate location, and the reflux ratio. Then plate-to-plate computation is carried out from the top of the column toward the feed plate location. The temperature is assumed for each plate, and trial-and-error equilibrium calculations are made to check the assumed temperature. The same kind of stepwise, trial-and-error calculations are made, starting from the bottom plate of the column or the reboiler and proceeding upward toward the feed plate location. The approximately correct solution results when the approximate relations between the ratio of the light and heavy key component quantities or compositions calculated in the plate liquids and the ratio of the key component quantities in the feed shown by one of the Eqs. (7.4) to (7.6) are satisfied. Once the key component ratio is matched, adjustment of the nonkey compositions in the distillate and bottoms is made and the calculations are repeated until the match is satisfactory.

The exact match of the feed plate temperature, component quantities, compositions, or component quantity ratios represents a highly improbable situation because of the approximation involved in establishing the distribution of all components in the distillate and bottoms products. However, for most purposes the guides given in the foregoing are sufficient for a reasonable evaluation of the number of theoretical stages.

The following equations are the principal ones used in this method:

$$v_{n+1} = l_n + d \tag{7.24}$$

$$v_n = K_n \frac{V}{L} l_n \tag{7.25}$$

$$\bar{l}_n = \bar{v}_{n+1} + b \tag{7.26}$$

$$\bar{v}_n = K_n \frac{\bar{V}}{\bar{L}} \bar{l}_n \tag{7.27}$$

$$\Sigma v = V \qquad \Sigma l = L \qquad \Sigma \bar{v} = \bar{V} \qquad \Sigma \bar{l} = \bar{L} \tag{7.28}$$

A more exact solution can be obtained by comparing the liquid compositions or moles of each component at the designated feed plate which result from computation from the distillate compositions and from the bottoms compositions to the feed plate. If the difference in the composition or moles is greater than that considered satisfactory for a suitable match, e.g., $\Delta x_i = 0.0000 \pm 0.0005$, the correction method of Bonner [8] can be used to establish a new value of D, d_i, and x_{i_D},

$$\Delta d_i = -\frac{(x_F)_T - (x_F)_B}{(x_F)_T/d_i + (x_F)_B/b_i} \tag{7.29}$$

$$d_{i_N} = d_i + \Delta d_i \tag{7.30}$$

$$b_{i_N} = b_i - \Delta d_i \tag{7.31}$$

where $(x_F)_T$ and $(x_F)_B$ = liquid compositions calculated from the top and bottom of the column, respectively

d_i, b_i = the former moles of individual component in the distillate and bottoms

d_{i_N}, b_{i_N} = the new values

Repetitive calculations are carried out by readjusting the moles of components in the distillate and bottoms until the agreement at the feed plate is met. The recovery of the components is checked, and if the specifications are not satisfied, the number of plates, reflux ratio, or feed plate location can be varied until the satisfactory solution is obtained.

Thiele-Geddes Method [36]

In the Thiele-Geddes [36] method and in the Hummel [21] modification, the temperature profile throughout the column is assumed, and it is the independent variable which is modified by plate-to-plate calculation until the assumed and calculated values agree. The distillate and bottoms compositions and quantities are dependent variables which are determined when the temperature profile is fixed. Other assumptions in this method are: column pressure, number of plates, feed plate location, reflux ratio L_0/D and L_m/V_{m+1} (and $\bar{L}_p/\bar{V}_{p+1}$), and the feed condition.

The compositions in the stripping section are expressed as ratios to the bottoms compositions, x_{i_p}/x_{i_B} and y_{i_p}/x_{i_B}, and the compositions in the rectifying section are expressed as ratios to the distillate compositions x_{i_m}/x_{i_D} and y_{i_m}/x_{i_D}. The numerical values of the compositions are not established until the calculation procedure is completed. In some modifications of the method the ratios of the moles of components are determined in the stripping and rectifying sections of the column, i.e.,

$$\frac{\bar{l}_{i_p}}{\bar{l}_{i_B}} \qquad \frac{\bar{v}_{i_p}}{\bar{l}_{i_B}} \qquad \frac{l_{i_m}}{l_{i_D}} \qquad \frac{v_{i_m}}{l_{i_D}}$$

The calculation procedure usually starts with the bottoms composition ratio $x_{i_p}/x_{i_B} = 1.0$ for all components (or $\bar{l}_{i_p}/\bar{l}_{i_B} = 1.0$ for all components). Using the assumed temperature, the composition ratios or mole ratios are calculated by the equilibrium relations

$$\frac{y_{i_p}}{x_{i_B}} = K_{i_p} \frac{x_{i_p}}{x_{i_B}} \tag{7.32}$$

$$\frac{\bar{v}_{i_p}}{b_i} = \frac{\bar{V}}{\bar{L}} K_{i_p} \frac{\bar{l}_{i_p}}{b_i} \tag{7.33}$$

The composition ratio for the liquid is calculated by material balance using the assumed $\bar{L}/\bar{V}$,

$$\frac{x_{i_{p-1}}}{x_{i_B}} = \frac{\bar{V}}{\bar{L}} \left(\frac{y_{i_p}}{x_{i_B}} - 1 \right) + 1 \tag{7.34}$$

$$\frac{\bar{l}_{i_{p-1}}}{b_i} = \frac{\bar{v}_{i_p}}{b_i} + 1 \tag{7.35}$$

Equilibrium and material balance calculations are continued toward the feed plate until the values of y_{i_F}/x_{i_B} or v_{i_F}/b_i are obtained.

Starting with the distillate and setting $y_{i_D}/x_{i_D} = 1.0$ or $v_{i_1}/d_i = 1.0$,

$$\frac{x_{i_m}}{x_{i_D}} = \frac{y_{i_m}/K_{i_m}}{x_{i_D}} \tag{7.36}$$

$$\frac{l_{i_m}}{d_i} = \frac{L}{VK_{i_m}} \frac{v_{i_m}}{d_i} \tag{7.37}$$

The composition ratios or the mole ratios of the liquids in the rectifying section are calculated by material balance using

$$\frac{y_{i_{m+1}}}{x_{i_D}} = \frac{L}{V}\left(\frac{x_{i_m}}{x_{i_D}} - 1\right) + 1 \tag{7.38}$$

$$\frac{v_{i_{m+1}}}{d_i} = \frac{l_{i_{m+1}}}{d_i} + 1 \tag{7.39}$$

Equilibrium and material balance calculations are continued to the feed plate until the y_{i_p}/x_{i_D} or v_{i_F}/d_i values are obtained. At this point the values of

$$\frac{x_{i_D}}{x_{i_B}} = \frac{y_{i_F}/x_{i_B}}{y_{i_F}/x_{i_D}} \tag{7.40}$$

or

$$\frac{d_i}{b_i} = \frac{v_{i_F}/b_i}{v_{i_F}/d_i} \tag{7.41}$$

are calculated.

The total moles of distillate and bottoms can be obtained from

$$1.0 = \frac{x_{1_F}}{(D/F)(1 - x_{1_B}/x_{1_D}) + x_{1_B}/x_{1_D}} + \frac{x_{2_F}}{(D/F)(1 - x_{2_B}/x_{2_D}) + x_{2_B}/x_{2_D}} + \cdots + \frac{x_{n_F}}{(D/F)(1 - x_{n_B}/x_{n_D}) + x_{n_B}/x_{n_D}} \tag{7.42}$$

$$D = \Sigma d_i = \sum \frac{Fx_{i_F}}{1 + b_i/d_i} \tag{7.43}$$

The total moles of bottoms and the moles of the individual component are evaluated by material balance. The compositions can then be calculated and the liquid compositions on each tray are determined. If they do not sum to unity, new temperatures are assumed and the calculations are repeated. After the summations of liquid compositions on each plate become equal to unity, the final distillate and bottoms compositions and quantities are fixed and the recoveries of each component can be calculated.

7.3. SHORT-CUT METHODS

The short-cut methods which allow determination of the number of theoretical plates as a function of reflux ratio, minimum plates, and minimum reflux are commonly used to study the effect of reflux ratio on investment and operating costs with a minimum of tedious and extensive calculations. These methods are also useful in making comparative mathematical studies of other variables related to reflux ratio and number of stages. The Colburn

[12] and Underwood [37] minimum reflux methods are used for more accurate calculation of minimum reflux while the Brown-Martin [10] method can be used for safe approximations. The Underwood [38] or Fenske methods are useful for evaluation of minimum plates, and the Brown-Martin [10] and Gilliland [15] correlations relate minimum reflux, minimum plates, operating reflux, and number of theoretical plates. These methods are described thoroughly in Chap. 5.

The steps to be followed in the use of short-cut methods are the same as those for the "exact" or modified plate-to-plate calculations insofar as calculating the overhead composition and quantity as well as the top-plate and reboiler temperatures. From that point on the simple steps are:

1. Calculate the minimum reflux ratio.

2. Calculate the minimum number of plates.

3. Using the results obtained in steps 1 and 2 and the Brown-Martin or Gilliland correlation, determine the number of theoretical stages for the selected reflux ratio.

7.4. COMPUTER RIGOROUS CALCULATION OF NUMBER OF THEORETICAL PLATES

The advent of high-speed machine computation has made it possible for the tedious trial-and-error plate-to-plate design calculations described here to be made with facility. Unfortunately, the basic methods of calculation have not been improved to take adequate advantage of the rapidity and accuracy of the computing devices, and the same assumptions and approximations are included in the computer programs that were inherent in the older, certainly slower, and, for the most part, less accurate calculation procedures. Therefore, the principal advantage lies in the ability of the computer (when correctly programmed) to solve complex trial-and-error iterative calculations with ease in a short time. Prior to the general availability of computers, few such calculations were attempted.

The computer programs developed for use in distillation calculations are essentially for the purpose of relating reflux ratio and number of theoretical or equilibrium stages for a given feed with the component distribution ratios in the distillate and bottoms products for a specific set of operating conditions, feed rate, temperature, and pressure. Generally these computer methods fall into two categories which utilize the equations and approach previously described. One method involves the assumption of the component distribution in the distillate and bottoms products, the feed plate location, and the reflux ratio, and computation is carried out plate to plate from the top of the column toward the feed plate and from the bottom of the column toward the feed plate until the composition ratios or composition calculated from both directions match within designated limits in the vicinity of the feed location. The other method involves the assumption of the reflux ratio, number of plates, and the temperature profile and, starting from the feed compositions, the calculation is carried out plate to plate for the number of stages specified in the rectifying section. By repeatedly modifying the assumed temperature on each plate and repeating the calculations, the bubble- or dew-point temperature for the liquid or vapor at each plate is matched. The calculation proceeds in a similar manner for the number of stages specified in the stripping section to obtain the bubble point or dew point of the vapor and liquid at each plate. If the first trial does not converge, a different number of plates, reflux ratio, or temperature profile is assumed and the calculation is repeated.

The first method utilizes the Lewis-Matheson [25] method of calculation described before, which assumes composition of the distillate and of the bottoms, requires fixing the component distribution between the two products, and includes stepwise calculation from the top toward the feed tray and from the bottom toward the feed tray. The reflux ratio, column pressure, feed composition and condition, feed tray location, and the feed and product rates are fixed, and the number of theoretical plates are calculated for the rectifying and stripping sections. Since an exact match cannot be expected unless the component distribution between the overhead and bottoms is correctly assumed, the principal difficulty is in selecting the initial distribution values and varying the successive trial values to get a solution with a reasonable number of trials—in other words, to get a rapid convergence to the solution.

Bonner [7, 8] suggested procedures in which for the first trial (*a*) essentially no heavier than heavy key materials appear in the distillate, essentially no lighter than light key components appear in the bottoms product, and then the nonkeys are introduced in minute quantities into the calculations; (*b*) a linear temperature profile exists; and (*c*) the vapor rates are constant. The component distribution for each trial is adjusted according to the difference between the calculated and assumed values. The suggested program calculates for each of the trials based on one of the foregoing assumptions, assuming the other two remain constant. Shelton and McIntyre [34] proposed a similar procedure. Newman [29] proposed a method for correcting the temperatures assumed in the Thiele-Geddes method, using a successive approximation scheme.

In the programs using the basic Lewis-Matheson method there is usually no provision made for the effect of enthalpy change on the quantities of vapor and liquid flowing through the column from plate to plate and for the effects of solution nonideality.

Holland et al. [20, 27] developed a computer procedure based upon the Thiele-Geddes [36] method which converges rapidly to the true solution. Trial values of the ratio of moles of component in the bottoms to moles of component in the distillate are compiled for each component. A multiplier correction θ, the same for every component, is used to obtain values which satisfy the over-all balances and the stream rates. Lyster [26] recommends "forcing" procedures which force the computer program to converge on the correct answer at a more rapid rate. In addition to the material balances, enthalpy balances are included in the program.

Mills [28] devised a program for multicomponent mixtures, up to 10 components, which includes effects of solution nonideality. Greenstadt et al. [17] describe a computer program using essentially the Lewis-Matheson method for multicomponent systems in which the distillate and bottoms compositions and conditions are fixed, and tray-to-tray calculations to the feed are carried out to obtain convergence at the feed condition. Differences in material and heat balances are used to correct for new trials.

Rose et al. [33] have proposed a computer routine for multicomponent-system calculations to relate feed distillate and bottoms conditions and compositions with reflux ratio and number of stages. This method is based upon a relaxation technique and uses batch-distillation equations for the condition of appreciable liquid holdup on the trays. In effect, the gradual change in compositions on the plates and in the products is calculated until steady-state conditions are achieved. This method gives an exact answer for each set of conditions assumed and does not require successive trials.

Amundson et al. [1, 4, 5] utilize a matrix solution from wherein the heat and material balance equations for each tray in the column are solved simultaneously for the whole column for each component. A linear temperature gradient is assumed, and the correct solution results when the compositions on each plate sum to unity. If the gradient is incorrect, a new one is assumed and the calculation is repeated until convergence is obtained.

The American Institute of Chemical Engineers has printed and made available computer programs in the form of manuals [2, 3]. Computer Program Manual No. 8 [2] is concerned primarily with the multicomponent-distillation computation method and No. 4 [3] presents a method for complex-system distillation calculations.

Rose et al. [32] describe a program for computing tray requirements for the ternary methanol, ethanol, and water.

Baer et al. [6] have a program for computation of minimum reflux.

O'Brien and Franks [30] suggest the use of an analog computer for plate-to-plate multicomponent calculations.

Waterman and Frazier [39] developed a program utilizing the Chao-Seader [11] and Redlich-Kwong [31] correlations to compute equilibria and enthalpies for rigorous plate-to-plate calculations based on the modified Lewis-Matheson method.

See Gerster [14] for a summary discussion of computer methods applied to distillation and Hanson et al. [18] on computer methods applied to multistage processes.

Example 7.1

The feed to a butane-pentane splitter of the following composition is to be fractionated into a distillate product containing 95% of the *n*-butane contained in the feed and a bottoms product containing 95% of the isopentane in the feed. The reflux ratio for the fractionation will be $1.3(L_0/D)_{\text{min}}$, and the column pressure will be 100 psia at the top plate. The reflux and feed are at their bubble-point temperatures. The conditions estimated for the column are: distillate and reflux bubble-point temperature at 100 psia = 145°F; bottoms bubble-point temperature at 102 psia = 215°F; feed plate pressure = 101 psia. (This assumes a ΔP/plate = 0.08 psi or 4.5 in. of fluid.) Assume the K values of Fig. 2.41 apply. The feed composition is:

Component	x_F
i-C_4	0.06
n-C_4	0.17
i-C_5	0.32
n-C_5	0.45

Determine the number of equilibrium stages needed by four calculation methods.

A. SHORT-CUT METHOD. Use (1) Underwood's method for minimum reflux, (2) Fenske's method for minimum plates, (3) Gilliland's correlation for determination of number of theoretical plates at the operating reflux of $1.3(L_0/D)_{\text{min}}$, and (4) Kirkbride's approximate method for feed plate location.

B. SOREL METHOD. Use plate-to-plate calculations including enthalpy balances and the *same* L/V as that determined in (A). Assume the molal enthalpies of the mixtures are additive on a mole fraction basis.

C. LEWIS-MATHESON METHOD. Use plate-to-plate calculations, assuming L/V and $\bar{L}/\bar{V}$ are constant and have the same values as those determined in (A).

D. THIELE-GEDDES METHOD. Use plate-to-plate calculations, assuming L/V and $\bar{L}/\bar{V}$ to be the same values as those determined in (A), the number of theoretical stages are the same as those determined in (A), and a linear temperature profile between the top plate and reboiler, for the first approximation.

Solution

A. SHORT-CUT METHOD. Evaluation of component distribution is done as follows:

(a) Evaluate the bubble-point temperature of the feed at 101 psia by trial and error. Assume t = 188°F.

Component	x_{F_i}	K_i	$x_{F_i}K_i$
i-C_4	0.06	2.15	0.129
n-C_4 (LK)	0.17	1.70	0.289
i-C_5 (HK)	0.32	0.835	0.268
n-C_5	0.45	0.700	0.315
			1.002

The bubble-point temperature of the feed is $\simeq$ 188°F.

(b) Determine the approximate distribution of the components in the distillate and bottom product using the average α's at 188°F and 101 psia and Hengstebecks' method.

Reference component is $i\text{-}C_5$.

$$\alpha_{i\text{-}C_4/i\text{-}C_5} = \frac{K_{i\text{-}C_4}}{K_{i\text{-}C_5}} = 2.58 \qquad \log \alpha = 0.411$$

$$\alpha_{C_4/i\text{-}C_5} = \frac{K_{C_4}}{K_{i\text{-}C_5}} = 2.04 \qquad \log \alpha = 0.309$$

$$\alpha_{i\text{-}C_5/i\text{-}C_5} = \frac{K_{i\text{-}C_5}}{K_{i\text{-}C_5}} = 1.00 \qquad \log \alpha = 0.0$$

$$\alpha_{C_5/i\text{-}C_5} = \frac{K_{C_5}}{K_{i\text{-}C_5}} = 0.839 \qquad \log \alpha = -0.077$$

For a 95% split between the keys,

$$\log \frac{(n\text{-}C_4)_D}{(n\text{-}C_4)_B} = \log \frac{0.1615}{0.0085} = \log 19.0 = 1.279$$

$$\log \frac{(i\text{-}C_5)_D}{(i\text{-}C_5)_B} = \log \frac{0.016}{0.304} = \log 0.0526 = -1.279$$

From Fig. 7.6,

$$\log \frac{(i\text{-}C_4)_D}{(i\text{-}C_4)_B} = 2.1$$

$$\frac{(i\text{-}C_4)_D}{(i\text{-}C_4)_B} = 125.9$$

$$\text{moles } (i\text{-}C_4)_B/\text{mole feed} = 4.77 \times 10^{-4}$$

$$\text{moles } (i\text{-}C_4)_D/\text{mole feed} = 5.9523 \times 10^{-2}$$

$$\log \frac{(C_5)_D}{(C_5)_B} = -1.89 \qquad \frac{(C_5)_D}{(C_5)_B} = 1.289 \times 10^{-2}$$

$$\text{moles } (C_5)_B/\text{mole feed} = 0.4441$$

$$\text{moles } (C_5)_D/\text{mole feed} = 0.0059$$

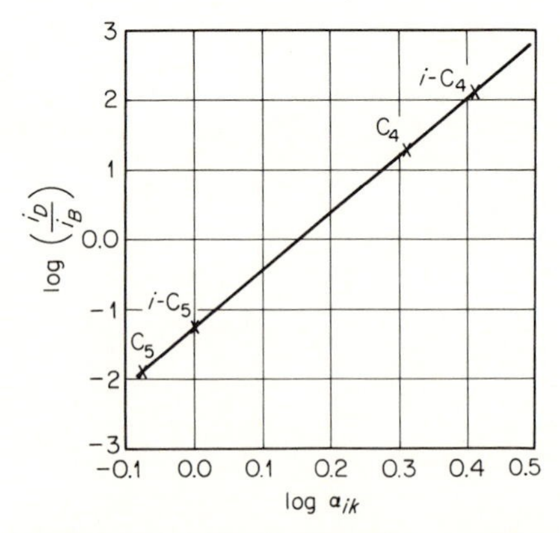

FIGURE 7.6 Distribution of components in distillate and bottoms.

Summary:

Component	Feed		Distillate		Bottoms	
	Moles	x_F	Moles	x_D	Moles	x_B
i-C_4	0.06	0.06	0.0595	0.2450	0.00048	0.0006
n-C_4	0.17	0.17	0.1615	0.6650	0.00850	0.0112
i-C_5	0.32	0.32	0.0160	0.0659	0.30400	0.4022
n-C_5	0.45	0.45	0.0059	0.0242	0.44411	0.5860
	1.00	1.00	0.2429	1.0001	0.75709	1.0000

(*c*) Calculate the bubble-point temperature of the distillate based on the estimated distribution of the components at 100 psia. Assume $t = 135°F$.

Component	x_{Di}	K_i	$K_i x_i$
i-C_4	0.2450	1.28	0.3145
n-C_4	0.6650	0.98	0.6510
i-C_5	0.0659	0.419	0.0276
n-C_5	0.0242	0.339	0.0082
			1.0013

The bubble-point temperature of the distillate is $t \cong 135°F$.

(*d*) Calculate the dew-point temperature of the distillate. Assume $t = 145°F$.

Component	y_1	K	$x = y/K$
i-C_4	0.2450	1.41	0.1740
n-C_4	0.6650	1.06	0.6280
i-C_5	0.0659	0.475	0.1390
n-C_5	0.0242	0.390	0.0620
			1.0030

The dew-point temperature of the distillate is $t \cong 145°F$.

(*e*) Calculate the bubble-point temperature of the bottoms product at 102 psia. Assume $t = 209°F$.

Component	x_B	K	Kx_B
i-C_4	0.0006	2.7	0.0016
n-C_4	0.0112	2.18	0.0244
i-C_5	0.4022	1.09	0.4380
n-C_5	0.5860	0.93	0.5450
			1.0090

The bubble-point temperature of the bottoms product is $t \simeq 209°F$.

1. Minimum plates—Fenske's method:

$$N = \frac{\log\,[(x_{LK}/x_{HK})_D(x_{HK}/x_{LK})_B]}{\log \alpha_{\text{av}}}$$

$$= \frac{\log\,[(0.665/0.0659)(0.4022/0.0112)]}{\log\,(1.7/0.835)}$$

$$= 8.26 \text{ theoretical stages}$$

2. Minimum reflux—Underwood's method:

$$1 - q = \sum_1^n \frac{x_{F_i}}{(\alpha_i - \theta)/\alpha_i} = 0 \qquad q = 1.0$$

$$(L/D)_m + 1 = \sum_1^n \frac{x_{iD}}{(\alpha_i - \theta)/\alpha_i}$$

Evaluated at $P_t = 101$ psia, $t = 188°F$ (trial and error):

Component	x_F	α	θ	$\alpha - \theta$	$\dfrac{\alpha - \theta}{\alpha}$	$\dfrac{x_F}{(\alpha - \theta)/\alpha}$	x_D	$\dfrac{x_D}{(\alpha - \theta)/\alpha}$
i-C_4	0.06	3.06	1.935	1.125	0.368	0.163	0.2450	0.665
n-C_4	0.17	2.43	1.935	0.495	0.204	0.834	0.6650	3.260
i-C_5	0.32	1.19	1.935	−0.745	−0.626	−0.512	0.0590	−0.0942
n-C_5	0.45	1.00	1.935	−0.935	−0.935	−0.491	0.0242	−0.0258
						+0.006 (OK)		+3.805

$$(L/D)_m = 3.805 - 1.0 = 2.805$$

$$(L/D)_{op} = 1.3\,(L/D)_m = 3.65$$

3. Number of theoretical stages—Gilliland's method:

$$\frac{L/D - (L/D)_m}{L/D + 1} = \frac{3.65 - 2.805}{3.65 + 1} = 0.182$$

From Fig. 5.18,

$$\frac{N - N_m}{N + 1} = 0.46$$

$$N = 16.5 \text{ theoretical stages}$$

4. Feed plate location—Kirkbride's approximate method:

$$\log \frac{m}{p} = 0.206 \log \left\{\frac{(x_{HK})_F}{(x_{LK})_F} \frac{B}{D} \cdot \left[\frac{(x_{LK})_B}{(x_{HK})_D}\right]^2\right\}$$

$$= 0.206 \log \left[\frac{0.32}{0.17} \frac{0.7571}{0.2429} \left(\frac{0.0112}{0.0659}\right)^2\right]$$

$$\log \frac{m}{p} = 0.206 \log 0.172 = -0.157$$

$$\frac{m}{p} = 0.70 \qquad m + p = 16.5$$

$$0.700p + p = 16.5$$

$$p = 9.7 \qquad m = 6.9$$

The feed is on the sixth or seventh plate.

B. SOREL METHOD

Assumptions:

1. The feed is a saturated liquid; basis—1 mole. $x_{i\text{-}C_4} = 0.0600$, $x_{C_4} = 0.1700$, $x_{i\text{-}C_5} = 0.3200$, $x_{C_5} = 0.4500$.
2. Reflux ratio $L_0/D = 3.70$; $L_0/V_1 = 0.910/1.158 = 0.786$.
3. $P_t = 100$ psia at top, $= 102$ psia at bottom.
4. There is no heat of mixing.
5. Enthalpies of pure components can be considered as linear functions of temperature.

$$H_i = H_i^\circ + (t - 135)S$$

where H_i° = enthalpy of component i at 135°F

S = average slope between 210°F and 120°F

Absolute error is less than 1%.

6. $K = A_1 + A_2T + A_3T^2$ (T = °R).
7. Number of plates, 19.

8. Feed plate location, 6.

9. Starting distillate composition is based on specified recovery of key components (from the short-cut method, 95% i-C_5 in bottoms, 95% C_4 in distillate); $x_{i\text{-}C_4} = 0.2450$, $x_{C_4} = 0.6650$, $x_{i\text{-}C_5} = 0.0658$, $x_{C_5} = 0.0242$.

10. Starting distillate rate per mole of feed $= 0.2460$; $B = 0.7540$.

Procedure:

1. Calculate t_1 and x_{i1} by calculating the dew-point temperature of V_1, where $\Sigma(y_i/K_i) = x_i = 1.000 \pm 0.005$ ($y_{i1} = x_{i_D}$).

2. Assume t_2 and calculate L_1 from

$$L_m = \frac{-q_D + D(\Sigma x_{i_D} h_{i_D} - \Sigma y_{i_{m+1}} H_{i_{m+1}})}{\Sigma y_{i_{m+1}} H_{i_{m+1}} - \Sigma x_{i_m} h_{i_m}}$$

3. Calculate y_{i2} from

$$y_{i_{m+1}} = \frac{L_m x_{i_m} + D x_{i_D}}{V_{m+1}}$$

4. Calculate K_i values at t_2 and calculate the dew-point temperature $\Sigma x_{i2} = \Sigma(y_{i2}/K_{i2}) = 1.0000 \pm 0.005$.

5. If the summation does not meet the specified limits, assume a different t_2 and repeat step 5 until the requirement is met.

6. Calculate, repeating steps 2 through 5, until the feed plate is reached.

7. Calculate x_{i_B} and B from

$$F x_{i_F} = B x_{i_B} + D x_{i_D}$$

8. Calculate the bubble-point temperature of the bottoms product, t_B, and the $y_{i_{n+1}}$ values:

$$\Sigma y_{i_B} = \Sigma K_i x_{i_B}$$

9. Calculate L_{p-1}:

$$L_{p-1} = \frac{-q_B + B(\Sigma x_{i_B} h_{i_B} - \Sigma y_{i_p} H_{i_p})}{\Sigma x_{i_{p-1}} h_{i_{p-1}} - \Sigma y_{i_p} H_{i_p}}$$

10. Calculate $x_{i_{p-1}}$:

$$x_{i_{p-1}} = \frac{B x_{i_B} + (L_{p-1} - B) y_{i_p}}{L_{p-1}}$$

11. Assume t_p and calculate K_{i_p}.

12. Calculate the bubble point,

$$\Sigma y_{i_p} = \Sigma K_{i_p} x_{i_p} = 1.0000 \pm 0.005$$

13. If the summation does not satisfy the requirement, assume a new t_p and repeat step 12 until the requirement is met.

14. Calculate stepwise, repeating steps 9 through 13 until the feed plate is reached.

15. Calculate $(x_{i_F})_T$, the compositions of the feed plate from the top down.

16. Calculate $(x_{i_F})_B$, the compositions of the feed plate from the bottom up.

17. If the $\Delta x_i = (x_{i_F})_T - (x_{i_F})_B \neq 0.0000 \pm 0.005$, a correction to d_i (and x_{i_D}) is made as follows by Bonner's method [8]:

$$d_{i_{\text{corr}}} = d_i - \frac{\Delta x_i}{(x_{i_F})_T/d_i + (x_{i_F})_B/b_i}$$

18. The calculation procedure is repeated until the required convergence is achieved.

19. If convergence is achieved with the adjusted moles of distillate but the recovery specifications are not satisfied, the following procedure is used:

(*a*) The number of plates is varied for the same specified feed plate location.

(*b*) The specified feed plate location can be varied for the same number of plates within rather narrow limits (two to four plates).

One or both of these steps can be taken until the readjustment of the distillate quantity and composition gives the specified recovery. A number of combinations of total plate and feed plate location may give the desired recovery and the optimum combination can be found.

Enthalpy Data: Molal Enthalpy of Components, 100 psia, Btu/Mole

Component	Vapor			Liquid		
	130, °F	175, °F	220, °F	130, °F	175, °F	220, °F
i-C_4	17,580	18,540	19,830	9,850	11,600	13,320
n-C_4	18,550	19,700	20,880	10,420	12,050	13,900
i-C_5	21,950	23,400	24,800	11,890	14,020	16,200
n-C_5	23,000	24,480	25,770	12,230	14,400	16,550

Condenser duty $q_D = 9934$ Btu/mole feed
Reboiler duty $q_B = 10{,}151$ Btu/mole feed

Sorel Method—Tabulated Results

Component	Final distillate		Final feed		Final bottoms	
	Moles	x_D	Moles	x_F	Moles	x_B
i-C_4	0.0597	0.2451	0.0600	0.0600	0.0001	0.0002
C_4	0.1620	0.6625	0.1700	0.1700	0.0075	0.0109
i-C_5	0.0156	0.0637	0.3200	0.3200	0.3043	0.4027
C_5	0.0068	0.0284	0.4500	0.4500	0.4440	0.5861
	0.2441	0.9997	1.0000	1.0000	0.7559	0.9999

Recovery of $(LK)_D = 95.13\%$, $(HK)_B = 95.14\%$; $L_0/V_1 = 0.786$, $V_1 = 1.158$, $L_0 = 0.910$; $t_D = 136.6$°F, $t_F = 188$°F, $t_B = 210.6$°F.

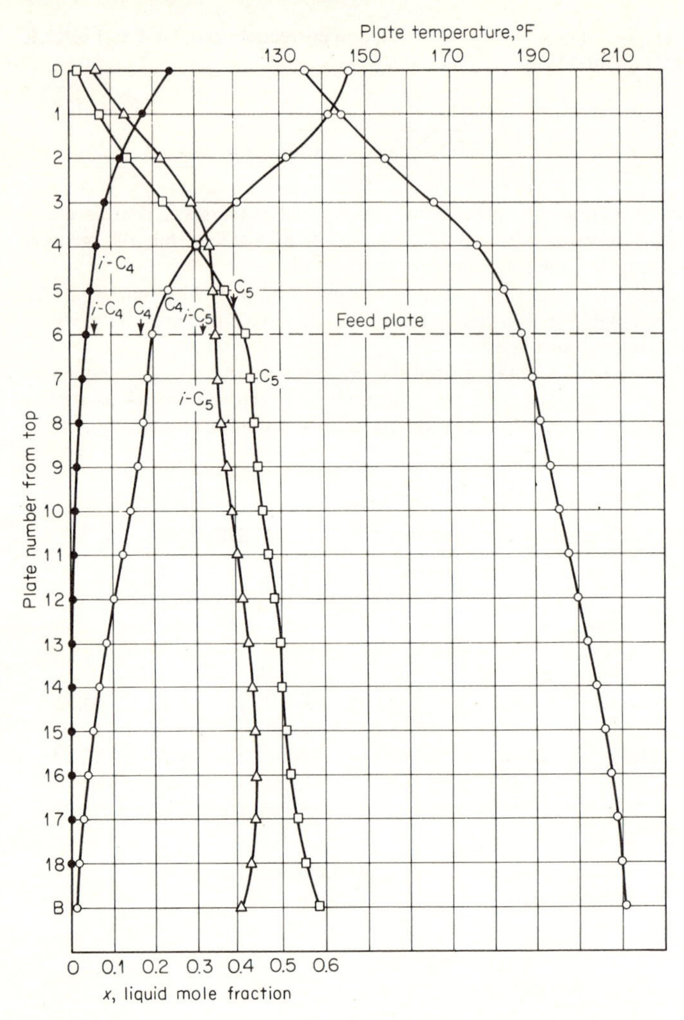

Plate-to-plate calculation—Sorel method.

Effect of Variation of Variables on Recovery ($L_0/D = 3.7$)

No. of plates	Feed plate	D, moles	Recovery	
			Light key	Heavy key
17	6	0.2432	0.9367	0.9475
18	6	0.2437	0.9435	0.9493
19	6	0.2441	0.9514	0.9513
20	6	0.2443	0.9562	0.9527
20	6	0.2439	0.9546	0.9531

Computed Plate Data—Direct Iteration

Plate no. from top	t, °F	L	V	L/V	$x_{i\text{-}C_4}$	x_{C_4}	$x_{i\text{-}C_5}$	x_{C_5}
1	145.0	0.8704	1.1473	0.76	0.1739	0.6158	0.1354	0.0742
2	155.4	0.8401	1.1145	0.754	0.1206	0.5165	0.2197	0.1437
3	166.5	0.8225	1.0842	0.759	0.0845	0.4008	0.2909	0.2250
4	176.1	0.8194	1.0666	0.769	0.0627	0.3036	0.3322	0.3017
5	182.7	0.8167	1.0636	0.775	0.0508	0.2376	0.3462	0.3670
6	187.2	0.8182	1.0608	0.771	0.0444	0.1974	0.3464	0.4190
6	187.8	1.8218	1.0659	1.71	0.0432	0.1926	0.3423	0.4225
7	189.7	1.8268	1.0709	1.71	0.0335	0.1846	0.3519	0.4306
8	191.5	1.8250	1.0691	1.71	0.0255	0.1733	0.3633	0.4405
9	193.9	1.8362	1.0804	1.70	0.0190	0.1588	0.3746	0.4497
10	196.0	1.8432	1.0874	1.70	0.0138	0.1415	0.3866	0.4598
11	198.1	1.8493	1.0935	1.70	0.0098	0.1229	0.3989	0.4704
12	200.3	1.8552	1.0993	1.69	0.0068	0.1038	0.4108	0.4812
13	202.4	1.8644	1.1085	1.685	0.0046	0.0852	0.4212	0.4915
14	204.2	1.8700	1.1141	1.68	0.0030	0.0678	0.4299	0.5020
15	206.0	1.8824	1.1265	1.67	0.0019	0.0523	0.4353	0.5120
16	207.2	1.8826	1.1267	1.67	0.0012	0.0388	0.4378	0.5244
17	208.8	1.8926	1.1367	1.67	0.0070	0.0276	0.4344	0.5383
18	210.0	1.9044	1.1485	1.66	0.0004	0.0183	0.4231	0.5566
19	210.6	0.7559	1.1485	—	0.0002	0.0109	0.4027	0.5861

C. LEWIS-MATHESON METHOD

Assumptions:

1. Constant molal overflow.

2. Feed is a saturated liquid; basis—1 mole. $x_{i\text{-}C_4} = 0.0600$, $x_{C_4} = 0.1700$, $x_{i\text{-}C_5} = 0.3200$, $x_{C_5} = 0.4500$.

3. Reflux ratio = 3.70.

4. Total plates, 17; with reboiler, 18.

5. Feed plate, 6.

6. Constant pressure, 101 psia.

7. $K = A_1 + A_2T + A_3T^2$.

8. *Starting* distillate composition is based on specified recovery of key components (from the short-cut method, 95% C_4 in distillate and 95% i-C_5 in bottoms).

$$
\begin{aligned}
x_{i\text{-}C_4} &= 0.2450 & d_{i\text{-}C_4} &= 0.0595 \\
x_{C_4} &= 0.6650 & d_{C_4} &= 0.1650 \\
x_{i\text{-}C_5} &= 0.0658 & d_{i\text{-}C_5} &= 0.0160 \\
x_{C_5} &= 0.0242 & d_{C_5} &= 0.0059 \\
& & & \overline{0.2429}
\end{aligned}
$$

Procedure:

1. Using D and L_0/D, calculate V and L in the section above the feed plate.

2. Calculate the dew-point temperature of V_1 to determine composition of L_1; $\Sigma y_1/K_1 = \Sigma x_1 = 1.0$.

(*a*) If after a specified number of iterations the calculations do not give $\Sigma x_1 = 1.0 \pm 0.0005$, print out reason for terminating calculations.

(*b*) Stop.

3. If $\Sigma x_1 = 1.0 \pm 0.0005$, by material balance determine the composition of V_2.

4. Continue alternating dew-point calculations to obtain $\Sigma y_m/K_m = 1.0 \pm 0.005 = \Sigma x_m$ and material balances to obtain y_{m+1} until the liquid composition on the designated feed plate (not feed composition) is reached.

5. Calculate from over-all material balance calculation, $b_i = Bx_{i_B}$, for all components.

6. Calculate the bubble-point temperature for the bottom product, $\Sigma K_i x_{i_B} = 1.0 \pm 0.0005 = \Sigma y_{n+1}$, and the vapor compositions rising to the bottom plate. If the calculations do not converge to give $\Sigma K_i x_{i_B} = 1.0 \pm 0.0005$ after a predetermined number of iterations, repeat steps 2*a* and *b*.

7. If $\Sigma y_{n+1} = 1.0 \pm 0.0005$, calculate $\bar{V}$ and $\bar{L}$ below the feed plate.

8. Continue alternating bubble-point and material balance calculations until the composition of the liquid on the plate designated as the feed plate is determined.

9. Determine the difference in the calculated feed composition (resulting from computations from the top of the column to the feed plate and from the bottom of the feed plate) for each component. Do each of the composition differences meet the specified convergence requirement (which is $\Delta x = 0.0 \pm 0.0005$)?

(*a*) If not, use the Δx to calculate an improved value of x_d and D (or x_d) by the Bonner [8] method:

$$\Delta d_i = -\frac{(x_F)_T - (x_F)_B}{(x_F)_T/d_i + (x_F)_B/b_i}$$

$$d_{i_N} = d_i + \Delta d_i$$

$$b_{i_N} = b_i - \Delta b_i$$

$$x_{D,i_N} = \frac{d_{i_N}}{\Sigma d_{i_N}}$$

Keep original D or let $D_N = \Sigma\, d_{i_N}$.

(*b*) Repeat starting with step 1.

10. If the convergence is satisfied, are the specified recoveries of the key components satisfied?

(*a*) If not satisfied, change number of plates, or feed plate location, or reflux ratio.

(*b*) Repeat starting with step 1.

11. If the convergence is satisfied, the program is successfully completed. D, x_{i_D}, x_{i_B}, L_0/D, plate temperatures and liquid and vapor compositions on each plate are printed.

12. Stop.

Lewis-Matheson Method—Tabulated Results

Compound	Final distillate, x_D	Final feed, x_F	Final bottoms, x_B
i-C_4	0.2461	0.0600	0.00022
C_4	0.6677	0.1700	0.01014
i-C_5	0.0607	0.3200	0.40348
C_5	0.0260	0.4500	0.58616

Recovery of $(LK)_D = 0.955$, $(HK)_B = 0.955$; $t_D = 136.5°F$, $t_F = 188°F$, $t_B = 210.9°F$.

Plate Data Computed

Plate no.	t, °F	$x_{i\text{-}C_4}$	x_{C_4}	$x_{i\text{-}C_5}$	x_{C_5}
1	144.2	0.1761	0.6263	0.1290	0.0686
2	154.5	0.1228	0.5296	0.2125	0.1350
3	165.7	0.0854	0.4115	0.2870	0.2161
4	175.3	0.0621	0.3084	0.3336	0.2959
5	183.1	0.0490	0.2360	0.3506	0.3644
6	188.0	0.0420	0.1912	0.3466	0.4202
7	190.4	0.0315	0.1774	0.3596	0.4315
8	193.0	0.0230	0.1602	0.3733	0.4435
9	195.5	0.0165	0.1409	0.3870	0.4557
10	198.0	0.0149	0.1205	0.4003	0.4677
11	200.3	0.0078	0.1003	0.4125	0.4793
12	202.5	0.0052	0.0813	0.4231	0.4904
13	204.4	0.0034	0.0640	0.4314	0.5012
14	206.0	0.0021	0.0490	0.4369	0.5120
15	207.4	0.0013	0.0362	0.4386	0.5239
16	208.7	0.0008	0.0256	0.4353	0.5383
17	209.8	0.0004	0.0170	0.4247	0.5578

Effect of Variation of Variables on Recovery

No. of plates*	Feed plate from top	$\left(\frac{L_0}{D}\right)_{op}$	D, based on $F = 1.0$	Recovery of light key in distillate	Recovery of heavy key in bottoms
17	6	3.70	0.2431	0.955	0.954
17	8	3.70	0.2455	0.958	0.946
20	16	3.70	0.2316	0.875	0.929
18	7	3.70	0.2323	0.939	0.970
18	12	3.70	0.2336	0.914	0.949
16	6	3.70	0.2421	0.946	0.953
16	8	3.70	0.2445	0.948	0.944
16	10	3.70	0.2450	0.929	0.930
30	15	3.70	0.2454	0.990	0.952
15	5	3.70	0.2427	0.930	0.949
17	10	3.70	0.2469	0.945	0.933
16	6	3.20	0.2370	0.898	0.947
16	6	4.70	0.2410	0.974	0.965
16	6	6.20	0.2400	0.989	0.972
13	5	3.48	0.2449	0.905	0.935
13	5	3.48	0.2466	0.910	0.934
17	7	3.48	0.2499	0.958	0.938

* Exclusive of reboiler.

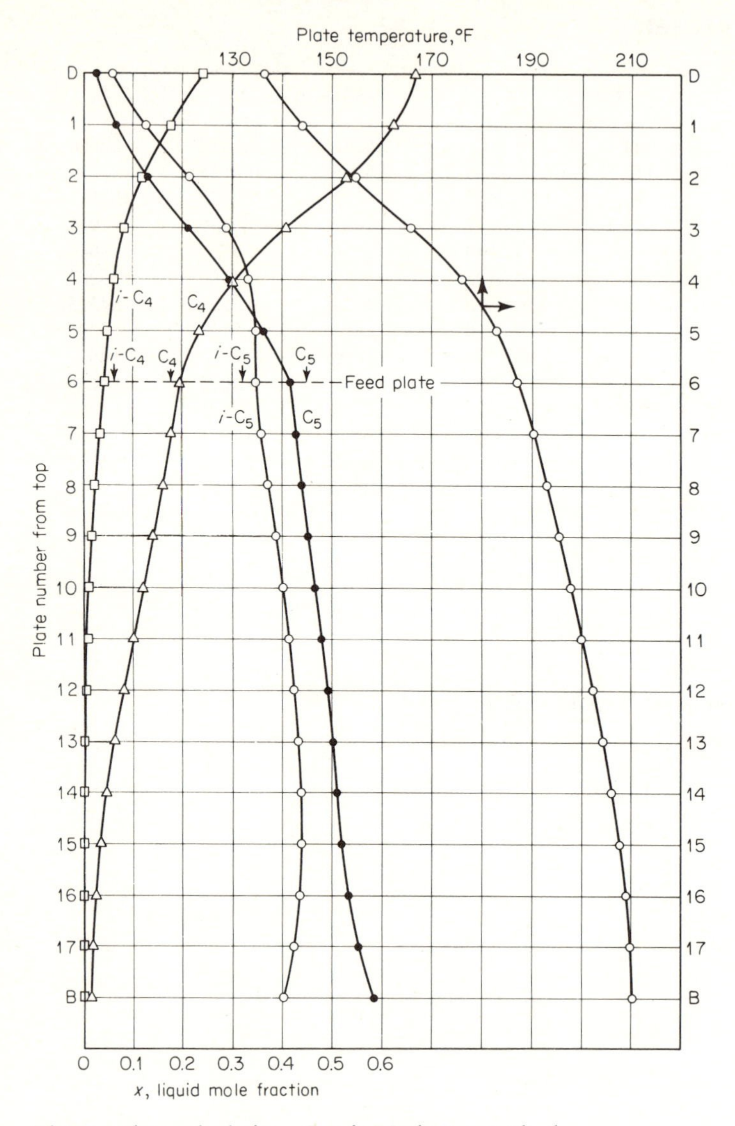

Plate-to-plate calculation—Lewis-Matheson method.

D. THIELE-GEDDES METHOD

Assumptions: Assumptions 1 to 7 are the same as for part (*C*).

8. The quantity of D is assumed to be 0.2431 mole/mole of feed.

9. The initial temperature profile is assumed to be linear between the 136 and 210.8°F.

Procedure

1. Calculate internal reflux ratios, $L/V = 3.70/(3.70 + 1.0)$, $\bar{V}/\bar{L} = 1 - B/(3.7D + F)$.

2. Using the initial assumed temperature profile, calculate moles of each component in bottoms product by $x_{i1}/x_{i_B} = 1.0$.

3. Calculate mole ratios of components in the liquid for each plate from the reboiler to the assigned feed plate:

$$R_p y = \frac{y_{i_p}}{x_{i_B}} = \frac{x_{i_p} K_{i_p}}{x_{i_B}}$$

$$R_p x = \frac{y_{i_{p+1}}}{x_{i_B}} = \frac{1 - B}{3.7D + F}\left[\left(\frac{y_{i_p}}{x_{i_B}} - 1\right) + 1\right]$$

(Numbering from the reboiler up the column.)

4. Set $y_{i,\mathrm{top}}/x_{i_D} = 1.0$.

5. Calculate the mole ratios of the components from the top plate to the feed plate:

$$R_m x = \frac{x_{i_m}}{x_{i_D}} = \frac{y_{i_m}}{K_{i_m} x_{i_D}}$$

$$R_m y = \frac{y_{i_{m-1}}}{x_{i_D}} = \frac{3.7}{3.7 + 1.0}\left[\left(\frac{x_{i_m}}{x_{i_B}} - 1\right) + 1\right]$$

6. Calculate the ratio of x_{B_i}/x_{Di}:

$$R = \frac{(y_{i_F})_m / x_{i_D}}{(y_{i_F})_p / x_{i_B}} = \frac{x_{i_B}}{x_{i_D}}$$

7. Calculate the over-all material balance:

$$x_{i_D} = \frac{x_{i_F}}{D + B(R)}$$

$$x_{i_B} = x_{i_D}(R)$$

8. Calculate the mole fractions on each plate:

$$x_{i_p} = x_{i_B}(R_p x) \qquad x_{i_m} = x_{i_D}(R_m x)$$

$$y_{i_p} = x_{i_B}(R_p y) \qquad y_{i_m} = x_{i_D}(R_m y)$$

9. Calculate Σx_i on each plate. If not equal to 1.0000 ± 0.0005, normalize by dividing $x_i/\Sigma x_i$ and calculate a bubble point to get a new estimate for the temperature on the plate.

10. Does the $\Sigma x_i = 1.0$ for each plate? If not, start with step 2 and repeat if number of iterations specified has not been exceeded. If the number has been exceeded, print results.

11. If $\Sigma x_i = 1.0$ on all plates, print results.

Thiele-Geddes Method—Tabulated Results

Component	Final distillate		Final feed		Final bottoms	
	Moles	x_D	Moles	x_F	Moles	x_B
i-C_4	0.0600	0.2461	0.0600	0.0600	0.0001	0.0002
n-C_4	0.1622	0.6673	0.1700	0.1700	0.0076	0.0103
i-C_5	0.0145	0.0594	0.3200	0.3200	0.3052	0.4037
n-C_5	0.0063	0.0257	0.4500	0.4500	0.4440	0.5863
	0.2431	0.9985	1.0000	1.0000	0.7569	1.0005

Recovery of $(LK)_D = 0.9542$, $(HK)_B = 0.9549$; $t_D = 136.5°\mathrm{F}$, $t_F = 188°\mathrm{F}$, $t_B = 210.8°\mathrm{F}$.

Plate Data Computed

Plate no. from top	t, °F	$x_{i\text{-}C_4}$	x_{C_4}	$x_{i\text{-}C_5}$	x_{C_5}
1	144.1	0.1766	0.6276	0.1279	0.0680
2	154.3	0.1232	0.5315	0.2111	0.1341
3	165.5	0.0858	0.4133	0.2857	0.2152
4	175.6	0.0624	0.3098	0.3327	0.2952
5	183.0	0.0492	0.2369	0.3500	0.3639
6	187.9	0.0422	0.1921	0.3461	0.4196
7	190.3	0.0317	0.1784	0.3590	0.4309
8	192.8	0.0232	0.1614	0.3726	0.4428
9	195.4	0.0166	0.1421	0.3863	0.4550
10	197.8	0.0116	0.1217	0.3997	0.4671
11	200.2	0.0079	0.1014	0.4119	0.4787
12	202.3	0.0053	0.0822	0.4226	0.4899
13	204.2	0.0034	0.0648	0.4310	0.5007
14	205.9	0.0022	0.0496	0.4366	0.5116
15	207.4	0.0014	0.0367	0.4384	0.5235
16	208.6	0.0008	0.0260	0.4352	0.5380
17	209.8	0.0005	0.0173	0.4247	0.5576
18	210.8	0.0002	0.0103	0.4035	0.5860

Comparison of Results—Various Methods

Data	Short-cut	Sorel	Lewis-Matheson	Thiele-Geddes
D	0.2429	0.2441	0.2431	0.2431
$x_D = i\text{-}C_4$	0.2450	0.2451	0.2461	0.2461
$x_D = C_4$	0.6650	0.6625	0.6677	0.6673
$x_D = i\text{-}C_5$	0.0659	0.0637	0.0607	0.0594
$x_D = C_5$	0.0242	0.0284	0.0260	0.0257
B	0.7571	0.7559	0.7569	0.7569
$x_B = i\text{-}C_4$	0.0006	0.0002	0.0002	0.0002
$x_B = C_4$	0.0112	0.0109	0.0101	0.0103
$x_B = i\text{-}C_5$	0.4022	0.4027	0.4035	0.4037
$x_B = C_5$	0.5860	0.5861	0.5862	0.5863
L_0/D	3.65	3.70	3.70	3.70
No. of plates*	16.5	19	18	18
Feed plate†	6.9	6	6	6
$x_F = i\text{-}C_4$	0.060	0.0432	0.0420	0.0422
$x_F = C_4$	0.17	0.1926	0.1912	0.1921
$x_F = i\text{-}C_5$	0.32	0.3423	0.3466	0.3461
$x_F = C_5$	0.45	0.4225	0.4202	0.4196
t_D, °F	135	136.6	136.5	136.5
t_1, °F	145	145.0	144.2	144.1
t_F, °F	188	187.8	188.0	187.9
t_B, °F	209	210.6	209.8	210.8
% Rec., LK	95.0	95.13	95.50	95.42
% Rec., HK	95.0	95.14	95.40	95.49

* Including reboiler. † From top.

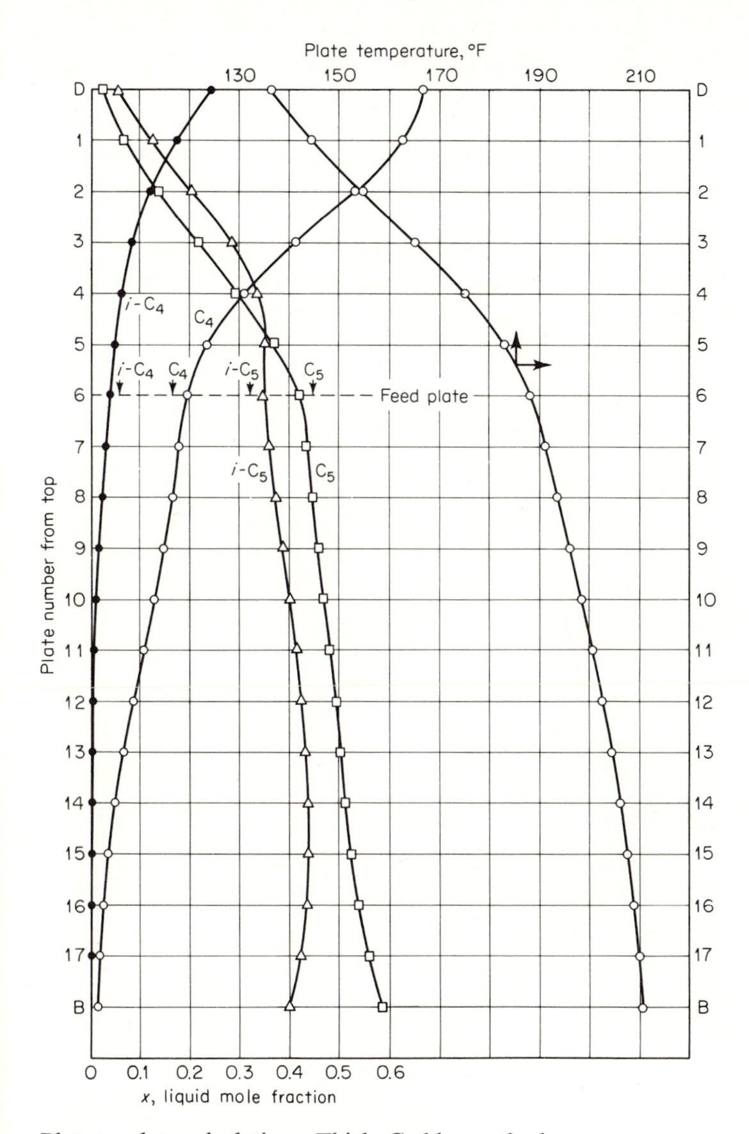

Plate-to-plate calculation—Thiele-Geddes method.

7.5. GRAPHICAL EVALUATION OF THE NUMBER OF THEORETICAL STAGES FOR TERNARY SYSTEMS

Ternary systems represent a special multicomponent situation. Not only can the analytical multicomponent calculation methods described in this chapter be used for the evaluation of the number of theoretical stages, but, if certain simplifying assumptions are made, graphical methods can also be used for this purpose. Such methods, developed by Bonilla [9], Hunter and Nash [22], and Kirschbaum [24], are useful in visualization of the distillation process and are reasonably accurate for use in making a number of comparative evaluations of the number of stages. Accuracy is somewhat subjective because of the personal factor in making the constructions and because of the simplifying assumptions on which the methods are based.

It is assumed that there is maintained (*a*) equimolal overflow and constant internal reflux ratio in the rectifying section and also in the stripping section of the column; (*b*) the feed is introduced at the correct plate; and (*c*) usually the systems are assumed to be ideal.

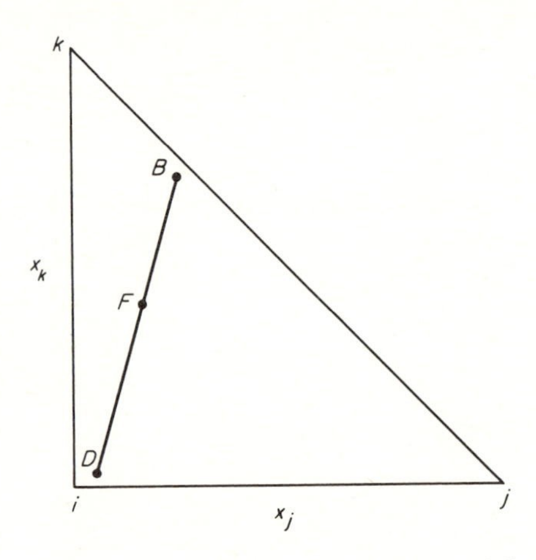

FIGURE 7.7 Ternary graphical material balance.

A rectangular coordinate plot such as that shown in Fig. 7.7 may be used to represent a ternary system, and material balances may be determined in the same manner as shown in Chap. 6. Thus, the quantity of B of composition x_{Bi}, x_{Bj}, x_{Bk} is proportional to the distance $\overline{FD}$, the quantity of D of composition x_{Di}, x_{Dj}, x_{Dk}, is proportional to the distance $\overline{FB}$, and the quantity of A, composition x_{Fi}, x_{Fj}, x_{Fk}, is proportional to the distance $\overline{BD}$. This is true when any two or more streams are added or subtracted to form one or more new streams. If i is the light component and k the heavy component, with j being the intermediate boiling component, and assuming i and j being the light and heavy keys, respectively, the distillate, bottoms, and feed might be located as shown on the figure. With the compositions of the feed, distillate, and bottoms fixed, the quantities of distillate and bottoms are fixed. It must be noted, however, that compositionwise the three points D, F, and B must lie on a straight line and, since the ratios of the stream quantities are proportional to distances along the line, the compositions cannot be fixed independently.

Total Reflux

The construction for total reflux is shown on Fig. 7.8. Because there is no distillate or bottoms—the streams $L_1 = V_2$, $L_2 = V_3$, etc., and the compositions $x_1 = y_2$, $x_2 = y_3$, etc.—V_1 is the distillate composition (it is totally condensed and returned to the column) and L_1 at the composition x_{i_1} is located by means of equilibrium data. This is also the composition of V_2. The composition of L_2 is found from equilibrium data and the procedure is continued down the column until the approximate composition of the bottoms product L_n is obtained. The minimum number of stages are thus determined.

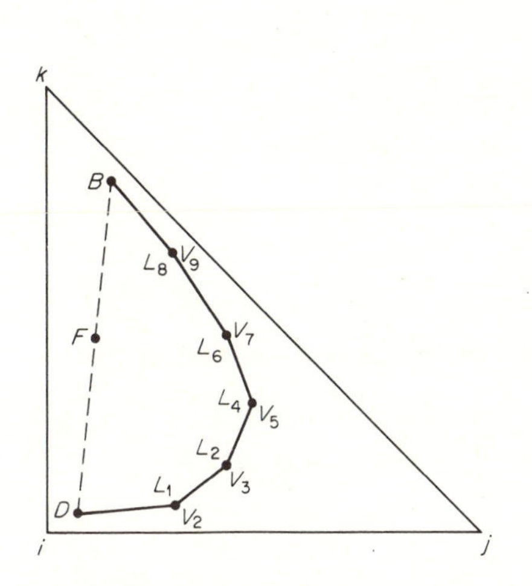

FIGURE 7.8 Condition of total reflux.

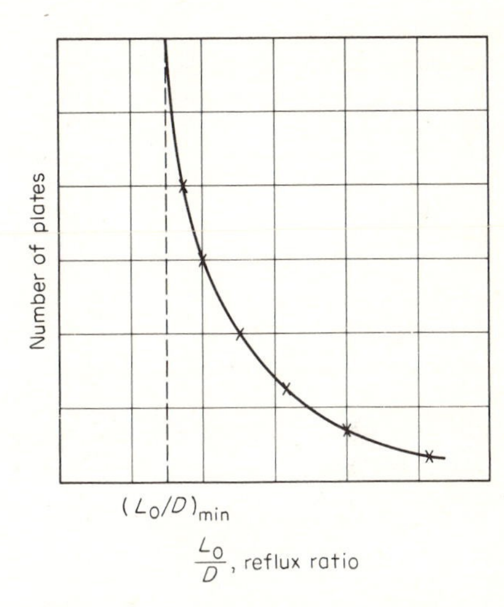

FIGURE 7.9 Evaluation of minimum reflux ratio.

Minimum Reflux

The evaluation of minimum reflux using the ternary graphical method is not very satisfactory. When it is done graphically, a reflux ratio is selected and the construction described in the next section is followed—down from the top in the rectifying section and up from the bottom in the stripping section—for workable reflux ratios until a pinch zone, where the equilibrium tie lines and the operating lines essentially coincide, is reached. The number of steps are counted down from the top and up from the bottom and are plotted against the reflux ratio selected. A number of these evaluations are made and plotted. The curve should become asymptotic to the minimum reflux ratio, as shown in Fig. 7.9, where the number of plates becomes infinite.

Graphical Evaluation of Number of Stages at an Operating Reflux Ratio

The operating reflux ratio is selected as some value between the minimum value and the total reflux and is based on economic considerations representing the optimum relation between investment and operating costs. After the reflux ratio is selected, the graphical construction to determine the number of equilibrium stages is carried out as indicated in the following: Referring to Fig. 7.10 it can be seen there are two difference points. Above the feed, D

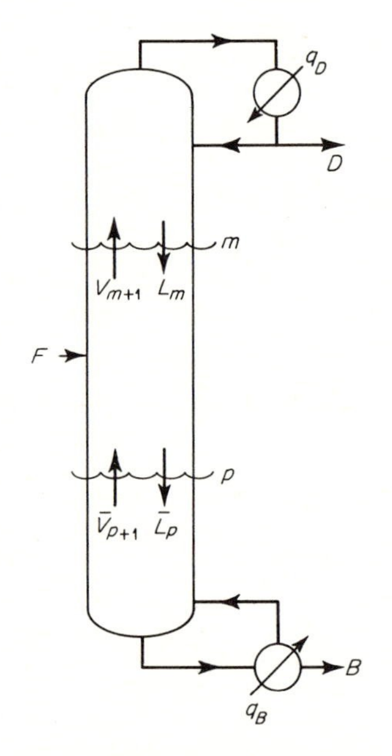

FIGURE 7.10 Regular ternary fractionation difference points.

is the difference point, and D, the vapor rising to the plate, V_{m+1}, and the liquid leaving the plate, L_m, lie on a straight line. Below the feed, B is the difference point, and B, the vapor rising from the plate, $\bar{V}_{p+1}$ and the liquid leaving the plate, $\bar{L}_p$, all lie on a straight line. (*Note:* Any two of the

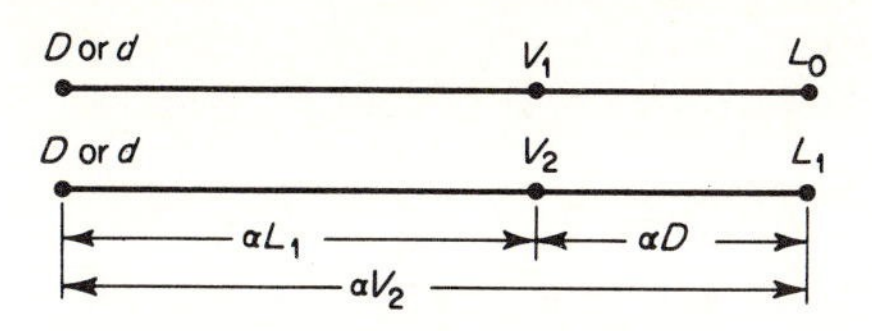

FIGURE 7.11 Reflux ratio from the lever-arm principle.

three-component-composition balances will give the same difference points.) For the rectifying section, the graphical relationships shown in Fig. 7.11 are applicable.

The assumed or determined value of external or operating reflux ratio L_0/D is used in material balance calculations to get the internal reflux ratio $L_0/V_1 = L_1/V_2$, etc. Assuming $L_0/D = 3.0$, then $L_0/V_1 = 3/(3+1) = 0.75$. Below the top plate $L_1/V_2 = \overline{dV_2}/\overline{V_2L_1}$ since $V_1 - L_1 = d$.

The stepwise method is outlined as follows:

1. Starting with the composition V_1 (L_0, d, or D), find L_1 from the equilibrium data.

2. Draw a line from the point V_1 to L_1 and measure the distance between d and L_1.

3. For the case where L_0/D was assumed to be 3.0, multiply the length of line $\overline{dL_1}$ by 0.75 to locate V_2.

4. Locate L_2 from the equilibrium data.

5. Repeat until the equilibrium tie line crosses the isothermal vapor or liquid line passing through the feed (see below).

For the stripping section the internal reflux ratio $\bar{L}/\bar{V}$ is determined graphically in accordance with the condition of the feed. If the feed is a *saturated liquid*,

$$\frac{\bar{L}}{\bar{V}} = \frac{\bar{L}}{\bar{L} + B}$$

Since it was assumed previously that $D = 1.0$,

$$F = B + D = B + 1$$

For the compositions, based on the light key,

$$x_F = 0.1 \qquad x_D = 0.9 \qquad x_B = 0.01$$

the values of

$$F = 9.9 \qquad B = 8.9 \qquad D = 1.0$$

$$\bar{L} = L + F = 3 + 9.9 = 12.9$$

$$\bar{V} = \bar{L} - B = 12.9 - 8.9 = 4.0$$

$$\frac{\bar{L}}{\bar{V}} = \frac{12.9}{4.0} = 3.2$$

$$\frac{\bar{L}_B}{\bar{V}_{B+1}} = \frac{\bar{L}_{B-1}}{\bar{V}_B} = \cdots$$

See Fig. 7.12 for proportionalities.

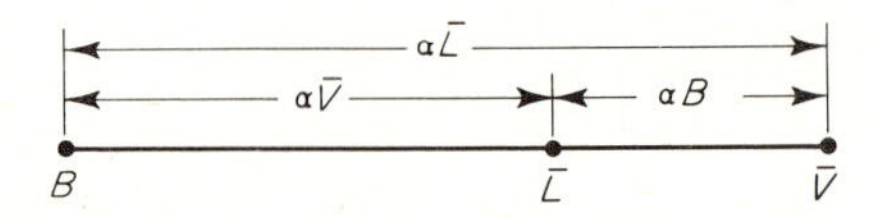

FIGURE 7.12 Stripping section reflux ratio.

The stepwise method for the number of equilibrium stages in the stripping section is as follows:

1. Determine $\bar{L}/\bar{V}$ in the stripping section of the column.
2. Locate x_B, (d').
3. Locate V_B from equilibrium data.
4. Draw the line $\overline{V_B d'}$ and measure its length.
5. Divide the value of $\bar{L}/\bar{V}$ to locate $\bar{L}_{B-1}$.
6. Locate L_{B-1} and from equilibrium data get V_{B+1}.
7. Connect d' and V_{B-1}.

Repeat these steps until the equilibrium tie line crosses the isothermal vapor or liquid line through the feed.

To prevent thermal upset in the column, the feed should be introduced on the plate whose temperature is more nearly that of the feed. If the feed is a saturated liquid, the bubble-point temperature should be calculated; if it is a saturated vapor, the dew-point temperature at the feed plate pressure should be calculated; if the feed is a two-phase feed, the quantity and compositions of the two phases should be evaluated and the temperature. Because the feed composition usually cannot be matched in the column, it is introduced at the correct temperature rather than at the correct composition. Thus the isothermal liquid line through the feed, the isothermal vapor line through the feed, or both, are calculated for the corresponding feed condition. Equations (7.44) and (7.45) are used for this purpose:

$$x_j = \frac{P_t - P_i}{P_j - P_i} - x_k \frac{P_k - P_i}{P_j - P_i} \tag{7.44}$$

$$y_j = \frac{P_j}{P_t}\frac{P_t - P_i}{P_j - P_i} - y_k \frac{P_j}{P_k}\frac{P_k - P_i}{P_j - P_i} \tag{7.45}$$

These are shown schematically in Fig. 7.13.

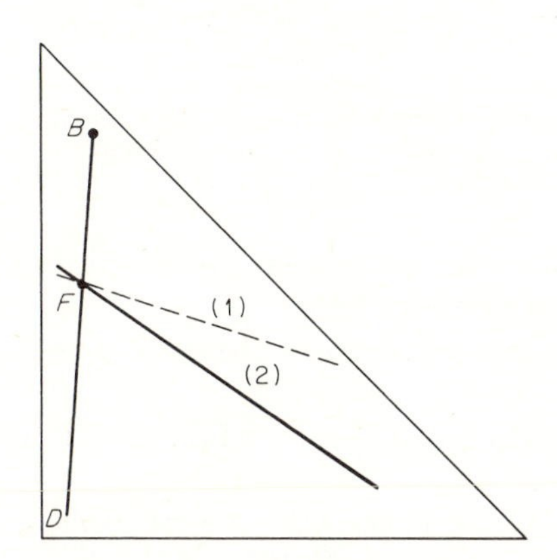

FIGURE 7.13 Isothermal vapor and liquid lines through the feed. (1) Isothermal liquid line. (2) Isothermal vapor line.

Referring to Fig. 7.14, the construction recommended for the cases of saturated liquid feed, saturated vapor feed, and two-phase is as follows:

1. *Saturated liquid feed.* Start the construction from the bottom of the column, as previously noted, until the composition of the liquid falls closest to the isothermal liquid line through the feed. At this point change to the top difference point, after adding the $\bar{L}$ graphically to the feed to locate the value of L; $F_L + L = \bar{L}$ or $L = \bar{L} - F_L$.

2. *Saturated vapor feed.* Starting from the top of the column, continue the construction until the vapor composition falls nearest the isothermal vapor line through the feed. Calculate the new $\bar{V}$ by $\bar{V} = V - F_V$ and change to the bottom difference point to continue the plate-to-plate construction.

3. *Mixed feed.*

(*a*) Calculate the quantity and composition for both the vapor and liquid in the feed.

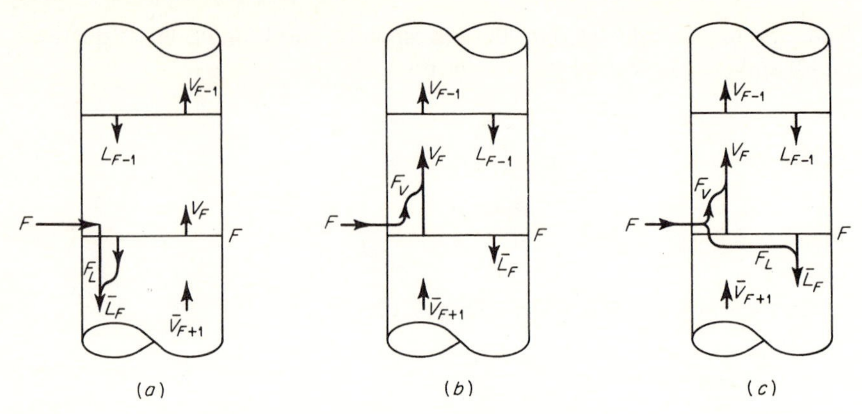

FIGURE 7.14 Feed addition to vapor and liquid streams. (*a*) Saturated liquid feed. (*b*) Saturated vapor feed. (*c*) Two-phase feed.

(*b*) Construct the isothermal vapor and liquid lines through each composition, respectively (see Fig. 7.15).

(*c*) Carry on the construction from the bottom until $\bar{L}$ comes close to the isothermal liquid line.

(*d*) Add $-F_L$ to $\bar{L}$ to get L; $L = \bar{L} - F_L$. L is the liquid from plate above feed.

(*e*) $\bar{V}$ should be added on the plate closest to the isothermal vapor line, and V_F should be determined from $V_F = F_V + \bar{V}$.

(*f*) $V_F - L_{F-1}$ should terminate at D. For the case where they do not, connect L_{F-1} with D and from L/V find V_F. Also connect V_F (from $F_V + \bar{V}$) with D and find L_{F-1}.

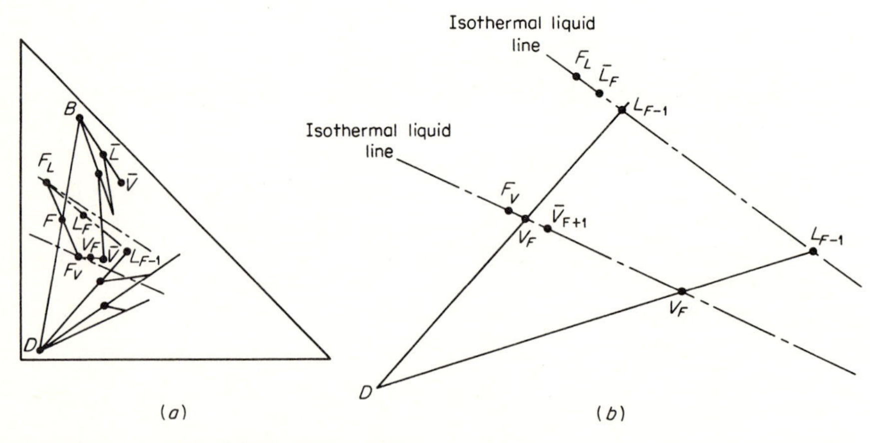

FIGURE 7.15 Mixed feed construction.

(*g*) Select values of V_F and L_{F-1} wherein the L/V ratio is satisfied and continue construction to D to get the number of theoretical stages.

PROBLEMS

1. A hydrocarbon mixture of the following composition is to be fractionated into the products indicated.

Components	Feed	Distillate	Bottoms
Isobutane	0.12	—	—
Butane	0.31	98% of C_4 in feed	—
Pentane	0.17	—	99.5% of C_5 in feed
Hexane	0.40	—	—

Use K values from Fig. 2.41. Assume an operating pressure of 50 psia, a saturated liquid feed, a total condenser, and bubble-point reflux. Calculate (*a*) composition of distillate and bottoms; (*b*) temperature of V_1 and reboiler (bottoms) liquid.

2. (*a*) Calculate the minimum reflux ratio by the Brown-Martin method and by the Colburn method.
(*b*) Calculate the minimum stages at total reflux by the Underwood method.
(*c*) Calculate the theoretical stages required for the following reflux ratios by the Brown-Martin correlation and the plot stages versus reflux ratio:

$$(L_0/D)_{min} \times 1.5$$
$$(L_0/D)_{min} \times 2.0$$
$$(L_0/D)_{min} \times 3.0$$
$$(L_0/D)_{min} \times 4.0$$

3. Assuming a dew-point vapor feed and a partial condenser, repeat calculations listed in Prob. 2*a* to *c*.

4. An alcohol mixture is to be fractionated at 20 psia to separate the ethanol from the remaining homologs. For this problem, assume the system to be ideal.

Data	EtOH	2-PrOH	1-PrOH	1-BuOH
	x, Mole fraction in liquid			
Feed	0.62	0.12	0.17	0.09
Distillate	0.995	—	—	—
Bottoms (nmt)	0.01	—	—	—
Antoine constants				
A	8.045	6.6604	7.997	8.275
B	1554.3	813.06	1569.7	1873.9
C	222.7	132.9	209.5	230.0

The feed will be introduced at its bubble-point temperature at 30-psia pressure. A total condenser with bubble-point reflux will be used.

(*a*) Determine the minimum reflux by means of the Underwood method.

(*b*) Determine the minimum stages at total reflux by the Underwood method.

(*c*) Determine the theoretical stages required by the Gilliland correlation using a reflux ratio $(L_0/D) = 2(L_0/D)_{\min}$.

5. Determine the minimum plates for the fractionation in Prob. 4 by plate-to-plate calculation from the top of the column and from the bottom, matching the compositions of the key components at the feed. (*Note:* Make a plot of K versus t at 30 psia for each component to simplify equilibrium evaluation.)

6. The ternary system, butane–2-methyl butane (isopentane)–*n*-pentane is to be fractionated at its bubble-point pressure at 100°F to recover 99% of the 2-methyl butane in the residue, and the residue is to contain a maximum of 1% butane.

Data	Butane	2-Methyl butane	Pentane
Feed, x_F	0.34	0.15	0.51
Distillate, x_D	—	—	—
Residue, x_B	0.01 (max)	0.99	

Assume the equilibrium ratios (Fig. 2.41) are applicable.

(*a*) Calculate the minimum reflux ratio by the Brown-Martin method.

(*b*) For a reflux ratio of $2.2(L_0/D)_{\min}$ and bubble-point reflux, determine the number of equilibrium stages needed by the conventional ternary graphical method.

7. For the conditions in Prob. 6 calculate the number of equilibrium stages needed, analytically, assuming the molal latent heats and heat capacities to be constant.

If the distillation were to be carried out with a dew-point vapor feed using a partial condenser and operated at the dew-point pressure of the distillate at 100°F, determine the number of theoretical stages graphically.

8. Assume conditions of Prob. 6:

(*a*) Calculate the minimum stages at total reflux by the Fenske method.

(*b*) Using the values of minimum reflux and minimum stages, determine the number of theoretical stages needed by both the Gilliland and Brown-Martin correlations.

9. Repeat the calculations in Prob. 8, but under the conditions of a dew-point vapor feed, and utilize a partial condenser as in Prob. 7.

10. Graphically determine the number of stages if the fractionation were operated at 100 psia, with bubble-point liquid feed, total condenser, bubble-point reflux, and $L_0/D = 2.5(L_0/D)_{\min}$.

Nomenclature

A_1, A_2, A_3 = constants

b = moles of component in the bottoms

B = moles residue per unit time

C = phase-rule components
C = constant
C_p = heat capacity, Btu/(lb mole)(°F)
D = moles distillate per unit time
d = moles component in the distillate
d = difference point, rectifying section
d' = difference point, stripping section
d'' = difference point, intermediate section
D_L = moles liquid distillate per unit time
D_V = moles vapor distillate per unit time
F = moles feed per unit time
F_L = moles liquid in feed per unit time
F_V = moles vapor in feed per unit time
H = enthalpy of vapor, Btu/lb mole
H° = reference enthalpy of vapor
h = enthalpy liquid, Btu/lb mole
i = moles of component i
K = equilibrium vaporization ratio
l = moles of component in the liquid
L = moles liquid per unit time, rectifying section
$\bar{L}$ = moles liquid per unit time, stripping section
L_F = moles liquid from feed plate per unit time
L_0 = moles reflux liquid returned to plate 1 per unit time
m = number of equilibrium stages above feed plate
N = number of stages at operative reflux
N = number of columns
N_m = minimum number of stages (total reflux)
p = number of equilibrium stages below the feed plate
P = vapor pressure
P_t = total pressure
q = heat added to system per unit time
q_D = condenser duty
q_B = reboiler duty
T = temperature, °R
t = temperature, °F
v = moles of component in the vapor
V = degrees of freedom
V = moles vapor per unit time, rectifying section
$\bar{V}$ = moles vapor per unit time, stripping section
V_C = moles of vapor condensed necessary to heat reflux
= liquid to t
V_F = moles vapor per unit time rising from feed plate
x, y = mole fraction in the liquid and vapor, respectively

Greek letters

α	= proportional to or relative volatility
ν	= fugacity coefficient
γ	= activity coefficient
λ	= latent heat of vaporization, Btu/lb mole
Σ	= summation
Δ	= difference

Subscripts

0, 1, 2, etc.	= plate numbers
B	= residue
$(\ )_B$	= calculated from the bottom of the column
D	= distillate
F	= feed
F_T	= total feed
HK, LK	= heavy key and light key, respectively
i, j, k	= components
p	= any plate, stripping section
m	= any plate, rectifying section
m	= minimum
r	= reference temperature
$(\ )_T$	= calculated from the top of the column

REFERENCES

1. Acrivos, A., and N. R. Amundson: *Ind. Eng. Chem.*, **47**:1533 (1955).
2. American Institute of Chemical Engineers: "Multicomponent Distillation," New York, A.I.Ch.E. Computer Program Manual No. 8.
3. American Institute of Chemical Engineers: "Complex Tower Distillation," New York, A.I.Ch.E. Computer Program Manual No. 4.
4. Amundson, N. R., A. J. Pontinen, and J. W. Tierney: *J. Am. Inst. Chem. Engrs.*, **5**:295 (1959).
5. Amundson, N. R., and A. J. Pontinen: *Ind. Eng. Chem.*, **50**:730 (1958).
6. Baer, R. M., J. D. Seader, and R. D. Crozier: *Chem. Eng. Progr.*, **55**:88 (1959).
7. Bonner, J. S.: *Chem. Eng. Progr. Symp. Ser.*, **55**(2):87 (1959).
8. Bonner, J. S.: *Petrol. Process.*, **11**:6 (June, 1956).
9. Bonilla, C. F.: *Trans. A.I.Ch.E.*, **37**:669 (1941).
10. Brown G. G., and H. Z. Martin: *Trans. A.I.Ch.E.*, **35**:679 (1939).
11. Chao, K. C., and J. D. Seader: *A.I.Ch.E. J.*, **7**:598 (1961).
12. Colburn, A. P. *Trans. A.I.Ch.E.*, **37**:805 (1941).
13. Geddes, R. L.: *A.I.Ch.E. J.*, **4**:389 (1958).
14. Gerster, J. A.: *Ind. Eng. Chem.*, **52**:645 (1960).
15. Gilliland, E. R.: *Ind. Eng. Chem.*, **32**:1101 (1940).
16. Gilliland, E. R.: *Ind. Eng. Chem.*, **32**:918 (1940).

17. Greenstadt, J., Y. Bard, and B. Morse: *Ind. Eng. Chem.*, **50**:1644 (1958).
18. Hanson, D. M., J. H. Duffin, and G. F. Somerville: "Computation of Multistage Separation Processes," Reinhold, New York (1962).
19. Hengstebeck, R. J.: *Trans. A.I.Ch.E.*, **42**:309 (1946).
20. Holland, C. D.: "Multicomponent Distillation," Prentice-Hall, Englewood Cliffs, N.J., 1963.
21. Hummel, H. H.: *Trans. A.I.Ch.E.*, **40**:445 (1944).
22. Hunter, T. G., and A. W. Nash: *Trans. Chem. Eng. World Power Conf. London*, **11**:409 (1936).
23. Kirkbride, C. G.: *Petrol. Refiner*, **23**:32 (1944).
24. Kirschbaum, I.: "Distillation and Rectification" (translated by M. Wulfingholf), Chemical Publishing (1948).
25. Lewis, W. K., and G. L. Matheson: *Ind. Eng. Chem.*, **24**:494 (1932).
26. Lyster, W. N.: *Chem. Eng. Progr.*, **55**:90 (July, 1959).
27. Lyster, W. N., S. L. Sullivan, D. S. Billingsly, and C. D. Holland: *Petrol. Refiner*, **38**:221 (June, 1959); **38**:151 (July, 1959); **38**:139 (October, 1959).
28. Mills, A. K.: *Chem. Eng. Progr.*, **55**:93 (July, 1959).
29. Newman, J. S.: *Hydrocarbon Process.*, **42**(4):141 (1963).
30. O'Brien, N. G., and R. G. E. Franks: *Chem. Eng. Progr. Petrol. Refiner Symp. Ser.*, **55**(21):25 (1959).
31. Redlich, O., and J. N. Kwong: *Chem. Rev.*, **44**:233 (1949).
32. Rose, A., R. E. Stillman, T. J. Williams, and H. C. Carlson: *Chem. Eng. Progr. Symp. Ser.*, **55**(21):79 (1959).
33. Rose, A., R. F. Sweeny, and V. N. Schrodt: *Ind. Eng. Chem.*, **50**:737 (1958).
34. Shelton, R. O., and R. L. McIntyre: *Chem. Eng. Progr. Symp. Ser.*, **55**(21):69 (1959).
35. Sorel, M.: "La Rectification de l'alcool," Paris, 1893.
36. Thiele, E. W., and R. L. Geddes: *Ind. Eng. Chem.*, **25**:289 (1933).
37. Underwood, A. J. V.: *Chem. Eng. Progr.*, **44**:603 (1948).
38. Underwood, A. J. V.: *Trans. Inst. Chem. Engrs. (London)*, **10**:112 (1932).
39. Waterman, W. W., and J. P. Frazier: *Hydrocarbon Process. Petrol. Refiner*, **44**(9):155 (1965).

chapter

Complex-system fractionation—estimation of number of equilibrium stages

A *complex* system may be described as one for which the number of individual compounds and their identity is unknown and, therefore, individual "pure" component properties cannot be used in quantitative evaluation of the properties of the system or in distillation calculations pertaining to the system. The most common industrially important complex systems are those represented by petroleum oils. Crude oil, natural gas, natural gasoline, and associated products represent an estimated 80% of all liquid materials which are subjected to distillation in the course of their processing and purification. Other examples are food and allied material resulting from refining animal and vegetable products, organic fuels, paint-blending agents, and others. One important class of materials falling into this classification are the residual mixtures or heavier component mixtures resulting from reaction processes which permit polymerization, alkylation, and condensation of some of the lighter products into complex mixtures of higher boiling materials.

In the case of complex systems, the technique of quantitative fractionation design calculations is completely empirical and, generally, the principal method used is based upon the pseudomulticomponent concept. In this method the mixture is considered to be composed of a number of cuts or portions, each of which can be characterized with an average boiling point (and, therefore, each can have a vapor pressure curve), an average density, an average molecular weight, average critical properties, and an average molecular characterization. With such properties estimated for each of the cuts or portions and with the availability of some further correlation of vapor-liquid equilibrium concentrations as a function of temperature and pressure, the multicomponent fractionation methods discussed in Chap. 7 can be applied to the complex mixture for fractionator design. References [1, 9] should be consulted for computer methods for complex system fractionation calculations.

No method, including the pseudomulticomponent method, is exact because of the assumptions necessary to assign average properties to each of the pseudocomponents and to reduce the complexity of the calculations, and because of the generalized nature of the correlations. Nevertheless, in spite of the inaccuracies of the methods, it is possible to design fractionators that are adequate both as to size and number of stages or plates which produce satisfactory products. It must be emphasized that the designs resulting from use of the available methods are approximate, because no quantitatively exact and satisfactory method has been developed to handle even multicomponent fractionation, much less complex-mixture fractionation. The reasons are numerous and the more important ones are as follows:

1. The product specifications are based on empirical tests, the results of which are difficult to correlate accurately with properties commonly used in fractional distillation calculations.

2. Product analytical distillations are of the batch type whereas the plant distillations are continuous.

3. In many cases one or more side-draw products are removed from the column between the feed and the overhead distillate streams.

4. In many cases, such as in petroleum refinery topping columns, all of the heat energy supplied to the column is supplied in the feed stream, i.e., there is no reboiler.

5. Products derived from adjacent draw-off points in the column overlap or gap with respect to distillation range, depending on the type of analytical distillation used to characterize the product. There is no exact separation at any point along the distillation curve.

6. Where side streams are withdrawn from the column, side-stream strippers are used to strip the light material from the product. The addition of stripping medium which returns to the column as vapor and the withdrawal of liquid in the side streams cause varying vapor and liquid loads throughout the rectifying section of the column.

7. Efficiencies are difficult if not impossible to define and, thus, to evaluate.

8. Unconventional refluxing of the column is resorted to in many cases.

9. Precise equilibrium data are very difficult if not impossible to obtain.

10. The equilibrium data are defined in a wide variety of terms, many of which are not consistent with conventional concepts.

Because somewhat different approaches are used in the design of conventional-type columns for complex-system fractionation, i.e., where the feed to the column is separated into two products, distillate and bottoms, and columns from which side streams are withdrawn (petroleum topping columns), each of the situations will be discussed separately with regard to quantitative design.

8.1. PSEUDOCOMPONENT DESIGN METHOD

Conventional Fractionation

A conventional fractionation schematic arrangement is shown on Fig. 8.1. The stepwise calculations involved in the evaluation of number of equilibrium stages are outlined as follows:

1. Decide upon the preciseness of separation desired because this will determine the number of pseudocomponents to be postulated in step 2.

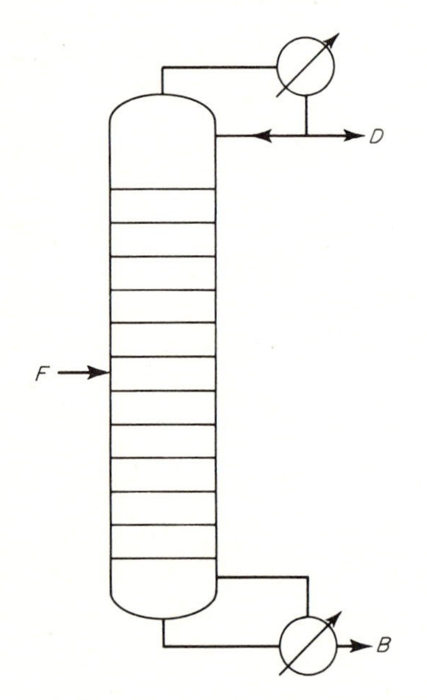

FIGURE 8.1 Conventional distillation.

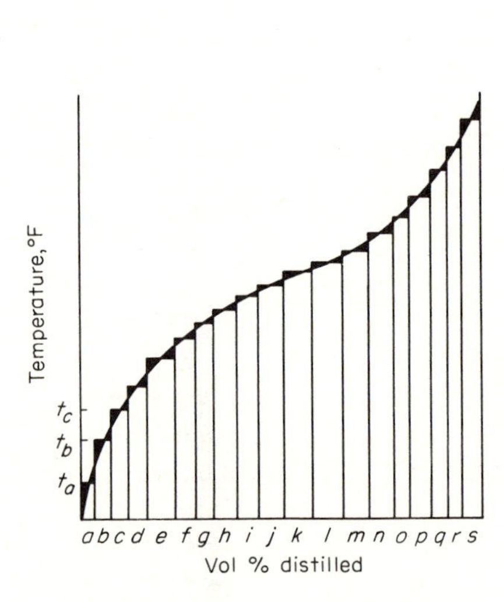

FIGURE 8.2 Breakup of complex system into pseudocomponents.

2. If a precise separation is required, use at least 50 pseudocomponents. For less precise separations, 10 to 15 pseudocomponents may suffice. Step off, as if a graphical integration were to be performed on the TBP curve, as shown in Fig. 8.2, a number of constant-temperature cuts so that the areas above and below the line between it and the TBP curve are equal as shown. The *a* component is designated as boiling at t_a at 760 mm Hg, the *b* component at t_b, etc.

3. The density midpercent curve is drawn from the evaluation data on the cuts, as shown in Fig. 8.3 [or determined for each cut by empirical relationships such as those of Watson et al. [21, 22] in which API gravity, K (characterization factor), and average boiling point are correlated].

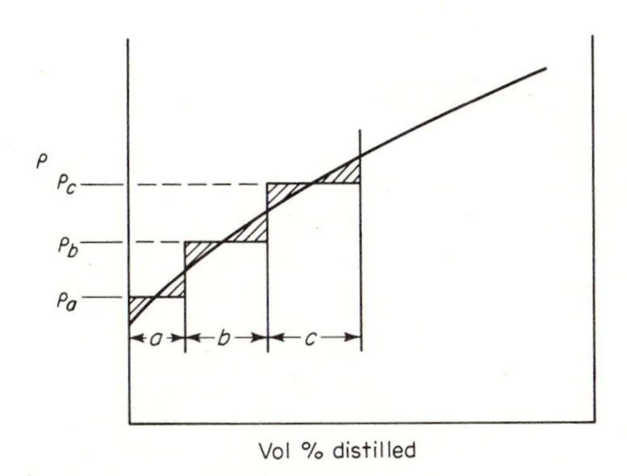

FIGURE 8.3 Density midpercent curve showing density of pseudocomponents.

4. The molecular weight of the cuts *a*, *b*, *c*, etc., are determined by either laboratory evaluation or by a series of empirical steps outlined as follows.

(*a*) From the slope and average-boiling-point correlations for petroleum oils [19] determine the ASTM 50% temperature for each cut (assuming straight lines) and the slope of the ASTM temperature-volume percent curve for each cut. This is the volumetric average slope (10–90% slope or any segment slope because it is a straight line) and 50% temperature that is designated as the volumetric average boiling point on Fig. 8.4.

(*b*) Use Fig. 8.4 to determine the cubic average and mean average boiling points; then use Fig. 8.5 to determine the molecular weight. Also record the characterization factor K_c because it will be used in selecting the proper equilibrium vaporization ratios ($K = y/x$) and vapor pressure curves.

5. The mole fraction represented by each cut can be determined by assuming 100 volumes of mixture, usually gallons or cubic feet, and converting to pounds for each cut. Divide the pounds for each cut by the molecular weight to get the moles for each cut and add the moles for all cuts to get the total number of moles. Then divide the moles of each cut by the total moles to get mole fractions for each of the cuts or pseudocomponents.

6. From the specifications on the distillate and/or bottoms products decide upon the key components and determine whether or not distributed key components should be considered.

If a rather precise separation is desired, the separation based upon the key distribution is sufficient with almost all (in quantity) of the lighter than light key components appearing in the distillate with the light key and almost all

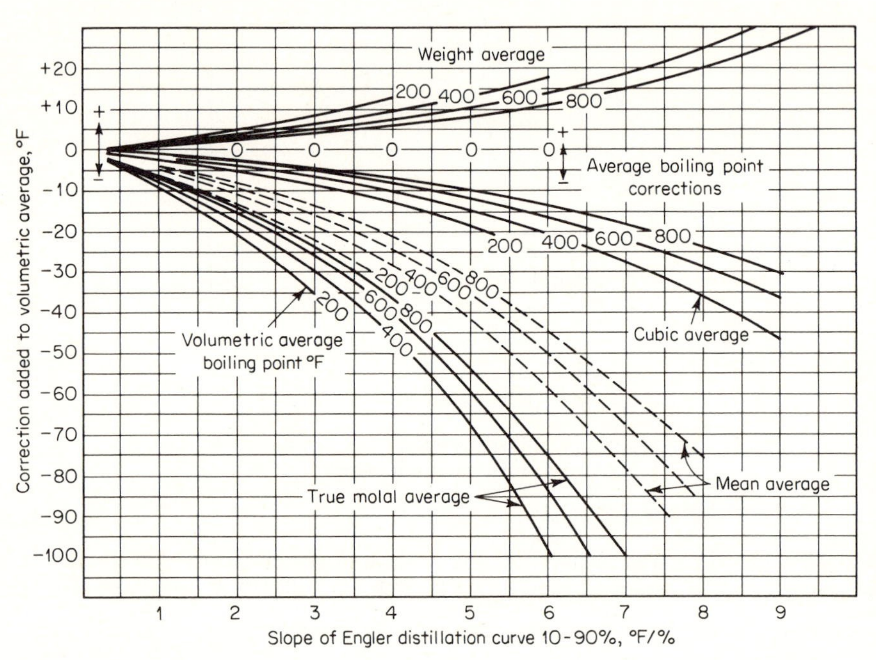

FIGURE 8.4 Relation between average boiling points. [*From Smith and Watson, Ind. Eng. Chem.*, ***29**:1408* (*1937*).]

(in quantity) of the heavier than heavy key components appearing in the bottoms with the heavy key. Where less precise separations are required and some quantity of all of the pseudocomponents appears in both the distillate and bottoms products, a means of determining the distribution of all components is necessary. Geddes [6] shows that the distribution of components into the distillate and bottoms streams can be represented by plotting the logarithm of the vapor pressure of each component (or relative volatility of each component) against the distribution of the component as moles in the distillate divided by moles in the bottoms at the average column temperature. These points fall on a straight line, and when the distributions of any two components are assumed, the distribution of all the others is fixed (see Chap. 7).

By constructing TBP curves of the distillate and bottoms products from specification ASTM (or TBP curves) and comparing the TBP curves for a number of assumed distributions, that distribution giving the TBP curves closest to those specified can be determined.

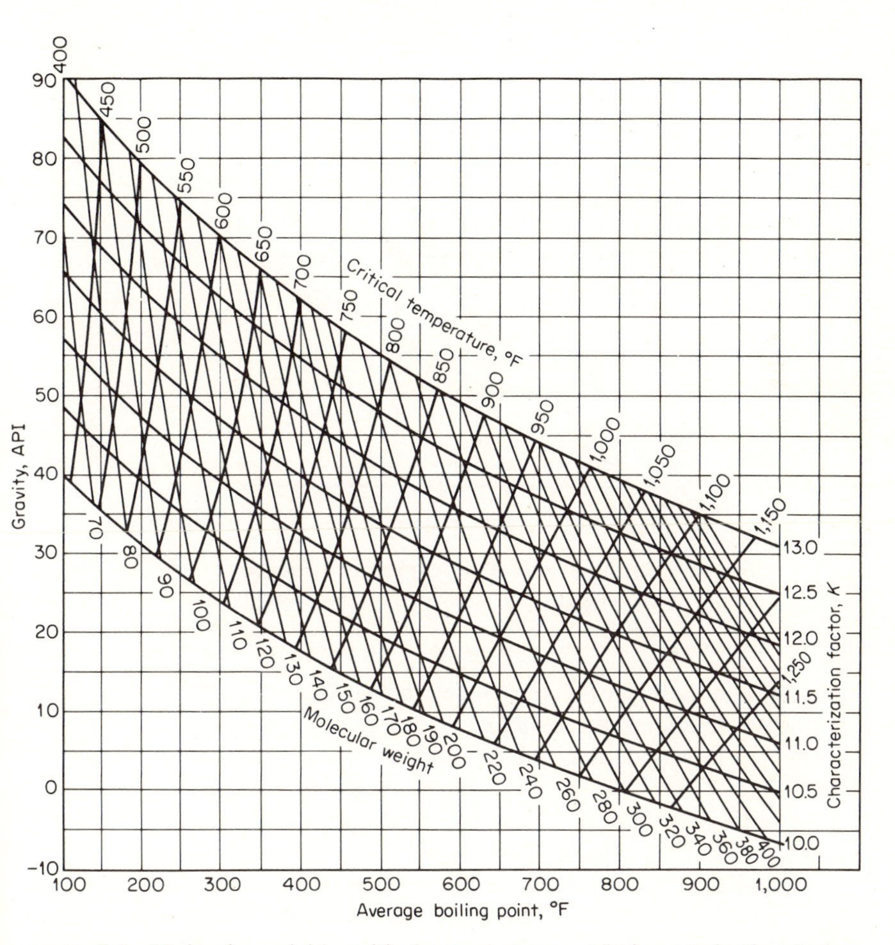

FIGURE 8.5 Molecular weights, critical temperatures, and characterization factors of petroleum fractions. Characterization factor K—use mean average; molecular weight—use mean average. [*From Watson et al., Ind. Eng. Chem.*, ***25****:880 (1933)*; ***27****:1460 (1935)*.]

It was shown in Chap. 7 that when the split between two keys is specified along with the reflux ratio, the distribution is fixed for all components appearing in the feed, in the distillate, and in the bottoms. Thus, theoretically the fractionation calculation may be based on any two key components.

Practically, however, it should be based on the key components representing the split between the distillate and bottoms products if perfect fractionation is desired. Figure 8.6 illustrates schematically the TBP curve of the feed, the TBP curves of the distillate and bottoms, and the TBP curves of the two streams if perfect fractionation were obtained.

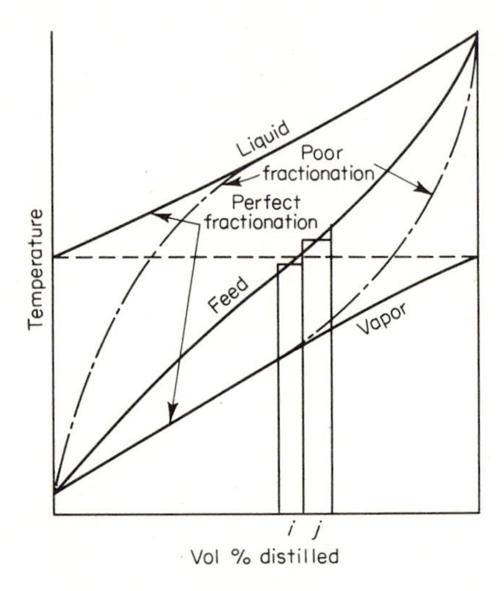

FIGURE 8.6 TBP curves of products obtained by precise or by rough fractionation.

7. Use the Colburn [4], the Underwood [17], or the Brown-Martin [3] method to determine the minimum reflux for the separation, in the same manner as it was determined for multicomponent mixtures discussed in Chap. 7.

8. Use the Underwood [17] or Fenske [5] method to determine the minimum plates at total reflux (Chap. 7).

9. Use either the Brown-Martin [3] or Gilliland [7] correlation for determining the number of theoretical stages corresponding to the selected reflux ratio.

10. Design the column in the manner shown in Part 4, Chaps. 12 to 14. Plate-to-plate calculations may be made by the method outlined in references [1] and [9] for machine computation.

Example 8.1

A crude oil having the following characteristics is to be topped to produce a jet fuel in a fractionating tower with no side streams, but with a reboiler. Steam is introduced at the bottom of the column so that the effective distillation pressure can be considered as 760 mm Hg.

Crude Oil TBP Distillation at 760 mm Hg

% Distilled	t, °F	ρ, g/cu cm	API gravity	Products, Example Fig. 8.1	Products, Example Fig. 8.2
0	105				
5	230	0.725	63.5		Light naphtha
10	300	0.796	46.7		
20	392	0.830	39.0	Jet fuel	
30	458	0.852	34.5		
40	505	0.865	32.0		Diesel fuel base stock
50	542	0.875	30.8		
60	585	0.890	27.5		
70	640	0.912	23.5	Residue	
80	720	0.930	20.4		Residue
90	880	0.980	13.1		
99	1090				

From the specifications the 0 to 55 vol % portion of the crude oil will satisfy the boiling range requirement of jet fuel. Determine the number of equilibrium stages needed for the desired distillate.

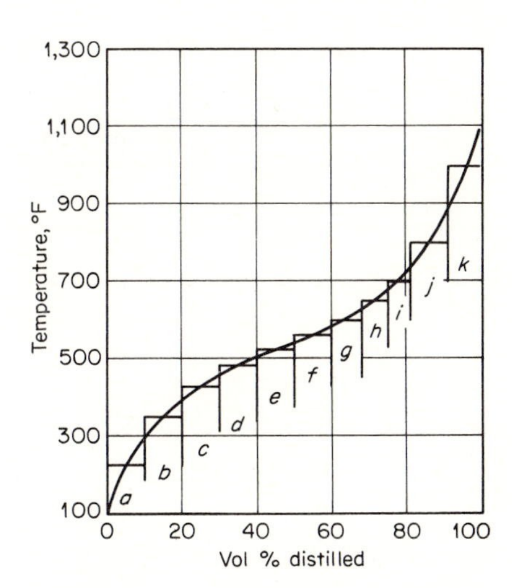

EXAMPLE 8.1 TBP curve of feed and component designation.

Solution

1. Separation: The separation between cuts having a TBP initial endpoint of 600°F can safely be specified as 90% of the heavy key in the residue and as 90% of the light key in the distillate.

2. Selection of pseudocomponents: Refer to the TBP plot. Eleven cuts are selected for pseudocomponents. Their properties are as follows:

Component	Vol %	TBP 50%, t, °F	ρ, g/cu cm	°API	ρ, lb/cu ft	lb/100 cu ft mixture	Mol. weight	Moles/100 cu ft mixture	Mol. frac.
a	10	225	0.745	58.5	46.50	465.0	102	4.55	0.1730
b	10	350	0.815	42.0	50.91	509.1	141	3.61	0.1370
c	10	430	0.842	36.5	52.58	525.8	165	3.19	0.1215
d	10	485	0.860	33.0	53.70	537.0	192	2.60	0.0990
e	10	528	0.870	31.0	54.36	543.6	210	2.59	0.0987
f	10	565	0.880	29.3	54.94	549.4	227	2.42	0.0920
g	8	600	0.896	26.5	55.91	447.3	242	1.85	0.0705
h	7	650	0.913	23.5	57.00	349.0	270	1.48	0.0562
i	6	700	0.930	22.3	57.45	344.3	300	1.15	0.0434
j	10	800	0.955	16.8	59.60	596.0	353	1.69	0.0641
k	9	1000	1.030	6.0	64.25	578.3	485	1.19	0.0450
					54.45	5444.8		26.32	1.0000

Component	TBP, 10–70% slope	ASTM, 10–70% slope	TBP 50%, t, °F	ASTM 50%, t, °F	ASTM, 10–90% slope	CABP	MABP	K	P, psia, 500°F	$\alpha_{LK/HK}$
a	2.17	1.3	225	215	1.3	213	209	11.9	380	
b	1.00	0.6	350	332	0.6	330	325	11.5	135	
c	0.67	0.4	430	410	0.67	409	404	11.4	41	
d	0.58	0.35	485	465	0.58	464	460	11.4	21	
e	0.58	0.35	528	510	0.58	509	506	11.5	13	
f	0.42	0.30	565	540	0.42	539	537	11.4	8.8	
g	0.50	0.35	600	575	0.50	574	573	11.4	6.3	1.31
h	0.67	0.4	650	620	0.67	619	617	11.4	4.8	1.00
i	0.50	0.35	700	670	0.50	669	667	11.5	2.7	
j	1.50	0.9	800	765	1.50	763	761	11.5	0.8	
k	2.34	1.5	1000	950	2.34	950	948	11.6	0.003	

3. The light key g and the heavy key h are selected on the basis of the specification boiling point ASTM 95% temperature at 590°F.

4. The composition of the distillate and bottoms is as follows:

Component	Feed, mole/mole		Distillate		Bottoms	
	Moles	x_F	Moles	x_D	Moles	x_B
a	0.1730	0.1730	0.1730	0.2200		
b	0.1370	0.1370	0.1370	0.1740		
c	0.1215	0.1215	0.1215	0.1540		
d	0.0990	0.0990	0.0990	0.1255		
e	0.0987	0.0987	0.0987	0.1251		
f	0.0920	0.0920	0.0900	0.1140	0.0020	0.0095
$(LK)\,g$	0.0705	0.0705	0.0635	0.0806	0.0070	0.0330
$(HK)\,h$	0.0562	0.0562	0.0057	0.0072	0.0505	0.2380
i	0.0434	0.0434			0.0434	0.2045
j	0.0641	0.0641			0.0641	0.3025
k	0.0450	0.0450			0.0450	0.2120
			0.7884	1.0004	0.2120	0.9995

5. Distribution and relative volatility:

$$\alpha_{LK/HK} = \frac{6.3}{4.8} = 1.31 \qquad \frac{(\text{moles } LK)_D}{(\text{moles } LK)_B} = \frac{0.0635}{0.007} = 9.09$$

$$\log \alpha = 0.1175 \qquad \log 9.09 = 0.958$$

$$\alpha_{HK/HK} = \frac{4.8}{4.8} = 1.00 \qquad \frac{(\text{moles } HK)_D}{(\text{moles } HK)_B} = 0.1105$$

$$\log \alpha = 0.0 \qquad \log 0.1105 = -0.956$$

$$\alpha_{a/h} = \frac{380}{4.8} = 79 \qquad \log \alpha = 1.898$$

$$\alpha_{b/h} = \frac{135}{4.8} = 28.1 \qquad \log \alpha = 1.449$$

$$\alpha_{c/h} = \frac{41}{4.8} = 8.55 \qquad \log \alpha = 0.932$$

$$\alpha_{d/h} = \frac{21}{4.8} = 4.37 \qquad \log \alpha = 0.640$$

$$\alpha_{e/h} = \frac{13}{4.8} = 2.71 \qquad \log \alpha = 0.432$$

$$\alpha_{f/h} = \frac{8.8}{4.8} = 1.83 \qquad \log \alpha = 0.263$$

$$\alpha_{g/h} = \frac{6.3}{4.8} = 1.31 \qquad \log \alpha = 0.1175$$

$$\alpha_{h/h} = \frac{4.8}{4.8} = 1.00 \qquad \log \alpha = 0.00$$

$$\alpha_{i/h} = \frac{2.7}{4.8} = 0.563 \qquad \log \alpha = -0.25$$

$$\alpha_{j/h} = \frac{0.8}{4.8} = 0.167 \qquad \log \alpha = -0.779$$

$$\alpha_{k/h} = \frac{0.003}{4.8} = 0.00063 \qquad \log \alpha = -3.204$$

Since a straight line results when the log α of a component versus log (x_D/x_B) is plotted, the relationship may be expressed as

$$\log \alpha = \text{m} \log \frac{x_D}{x_B} + b$$

$$m = \frac{0.1175 - 0.0}{0.958 - (-0.956)} = 0.0615 \qquad b = 0.0586$$

$$\log \alpha = 0.0615 \log \frac{x_D}{x_B} + 0.0586$$

$$\log \frac{x_D}{x_B} = \frac{\log \alpha - 0.0586}{0.0615}$$

$$\log \frac{x_{a_D}}{x_{a_B}} = \frac{1.898 - 0.0586}{0.0615} = \frac{1.394}{0.0615} = 30.0$$

$$\log \frac{x_{b_D}}{x_{b_B}} = 22.7 \qquad \log \frac{x_{g_B}}{x_{g_D}} = 0.958$$

$$\log \frac{x_{c_D}}{x_{c_B}} = 14.2 \qquad \log \frac{x_{h_D}}{x_{h_B}} = -0.956$$

$$\log \frac{x_{d_D}}{x_{d_B}} = 9.45 \qquad \log \frac{x_{i_D}}{x_{i_B}} = -5.01$$

$$\log \frac{x_{e_D}}{x_{e_B}} = 6.06 \qquad \frac{x_{e_D}}{x_{e_B}} = 1.15 \times 10^6 \qquad \log \frac{x_{j_D}}{x_{j_B}} = -13.6$$

$$\log \frac{x_{f_D}}{x_{f_B}} = 3.33 \qquad \frac{x_{f_D}}{x_{f_B}} = 2.14 \times 10^3 \qquad \log \frac{x_{k_D}}{x_{k_B}} = -53$$

For all practical purposes the foregoing calculation indicates negligible amounts of lighter than light key in the bottoms and negligible amounts of heavier than heavy key in the distillate.

6. Minimum plates—Fenske's method:

$$N_{\min} = \log \frac{(x_{LK}/x_{HK})_D(x_{HK}/x_{LK})_B}{\log (\alpha_{LK/HK})_{\text{av}}}$$

$$= \log \frac{(11.2)(7.2)}{\log 1.31} = \frac{1.908}{0.116} = 16.4 \text{ plates}$$

7. Minimum reflux—Underwood's method: For saturated liquid feed $q = 1.0$,

$$\sum_1^h \frac{x_F}{(\alpha - \theta)/\alpha} = 0 \qquad \alpha_{LK/HK} = 1.31$$

Assume $\theta = 1.045$. Then, for the components,

$$a: \frac{0.1730}{(79 - 1.045)/79} = 0.176 \qquad b: \frac{0.1370}{(28.1 - 1.045)/28.1} = 0.142$$

$$c: \frac{0.1215}{(8.55 - 1.045)/8.55} = 0.138 \qquad d: \frac{0.0990}{(4.37 - 1.045)/4.37} = 0.130$$

$$e: \frac{0.0987}{(2.71 - 1.045)/2.71} = 0.161 \qquad f: \frac{0.0920}{(1.83 - 1.045)/1.83} = 0.214$$

$$g: \frac{0.0705}{(1.31 - 1.045)/1.31} = 0.349 \qquad h: \frac{0.0562}{(1.0 - 1.045)/1.0} = -1.250$$

$$i: \frac{0.0434}{(0.563 - 1.045)/0.563} = -0.048 \qquad j: \frac{0.0641}{(0.167 - 1.045)/0.167} = -0.0123$$

$$k: \frac{0.0450}{(0.00063 - 1.045)/0.00063} = \text{neg.}$$

Check: 1.311 − 1.310 = 0.001.

$$\left(\frac{L}{D}\right)_{\min} + 1 = \sum_1^n \frac{x_D}{(\alpha - \theta)/\alpha}$$

$$a: \frac{0.22}{0.986} = 0.223 \qquad b: \frac{0.174}{0.962} = 0.180$$

$$c: \frac{0.154}{0.879} = 0.165 \qquad d: \frac{0.126}{0.765} = 0.165$$

$$e: \frac{0.125}{0.615} = 0.204 \qquad f: \frac{0.114}{0.432} = 0.264$$

$$g: \frac{0.0806}{0.145} = 0.555 \qquad h: \frac{0.007}{-0.045} = -0.156$$

$$\sum_1^n \frac{x_D}{(\alpha - \theta)/\alpha} = 1.61 = \left(\frac{L}{D}\right)_{\min} + 1 \qquad \left(\frac{L}{D}\right)_{\min} = 0.61$$

8. Theoretical plates for an operating reflux ratio (use the Brown-Martin chart, Fig. 5.17):

$$\left(\frac{L}{V}\right)_{\min} = \frac{L}{L + D} = \frac{0.61}{0.61 + 1.00} = 0.38$$

$$\frac{L}{D} = 3\left(\frac{L}{D}\right)_{\min} = 1.83 \qquad \frac{L}{V} = \frac{1.83}{1.83 + 1.0} = 0.65$$

$$V_{\min} = (L + D)_{\min} = (0.788)(0.38) + 0.788 = 0.30 + 0.788 = 1.088$$

$$L_{\min} = (0.38)(0.788) = 0.3$$

$$V = L + D = (1.83)(0.788) + 0.788 = 1.44 + 0.788 = 2.228$$

$$L = 1.44$$

$$\bar{L}_{\min} = F_L + L_{\min} = 0.212 + 0.3 = 0.512$$

$$\bar{V}_{\min} = V_{\min} - F_V = 1.088 - 0.788 = 0.3$$

$$\bar{L} = F_L + L = 0.212 + 1.44 = 1.652$$

$$\bar{V} = V - F_V = 2.228 - 0.788 = 1.44$$

$$\frac{\dfrac{\bar{L}}{\bar{V}}\dfrac{V}{L} - 1}{\left(\dfrac{\bar{L}}{\bar{V}}\dfrac{V}{L} - 1\right)_{\min}} = \frac{(1.15)(1.55) - 1}{(1.7)(3.62) - 1}$$

$$= \frac{1.78 - 1}{6.15 - 1} = \frac{0.78}{6.15} = 0.127$$

$$\frac{N}{N_{\min}} = 1.1$$

$$N = (16.4)(1.1) = 18 \text{ theoretical plates}$$

Fractionation with Side Streams

The procedure of determining the number of stages for a side-stream column, shown schematically in Fig. 8.7, using the pseudomulticomponent method is outlined briefly as follows:

1. Draw the TBP curve for the feed material and select the components as done under "Conventional Fractionation." At least 30 to 50 pseudocomponents should be used.

2. From the density midpercent curve determine the density of each cut or pseudocomponent.

3. Evaluate the molecular weight of each component and the mole fraction represented by the component.

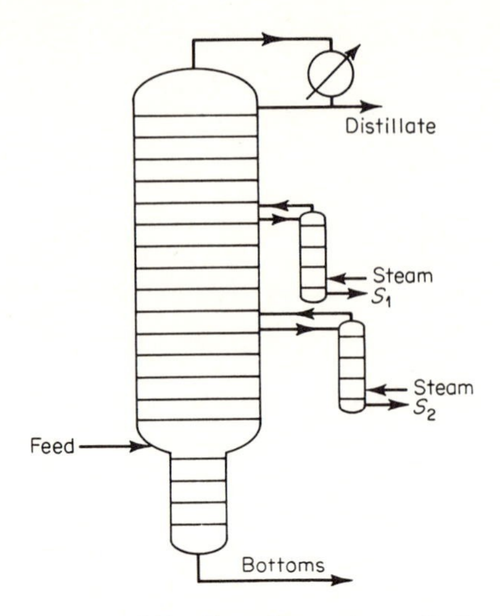

FIGURE 8.7 Complex-system distillation with side-stream products.

4. From the specifications for the distillate and those for each side stream related to the TBP endpoint of the product, determine the volume percent yield of each product. This is the theoretical yield of each product. The total volume percent of the distillate plus all side-stream products equal the theoretical percent of the feed to be vaporized at the feed plate.

5. Usually the total pressure at which the distillation will be carried on is chosen, and a maximum temperature is selected based (in petroleum distillation as well as in many others where high-boiling materials are encountered) on a safe value with regard to thermal degradation or decomposition. In such cases the maximum percent vaporized in an EFV is fixed. To get around this invariance, a stripping medium, introduced as a vapor, is added to the bottoms, which is immiscible in the liquid state with the distilling system or which has such a high volatility compared with the distilling system that it may be considered essentially as a fixed gas. This medium, which is usually steam but may be composed of inert gases, lowers the effective distilling pressure of the distilling system, because the effective pressure is the total pressure minus the partial pressure of the stripping medium.

$$P_{\text{system}} = P_t - p_{\text{stripping medium}} \tag{8.1}$$

6. With the total pressure selected and the maximum flash or feed plate temperature fixed, the amount of stripping medium for a selected temperature at a value somewhat under the maximum or the temperature for a selected quantity of stripping medium is determined by trial and error. For example, the total pressure, quantity of stripping medium, and desired percent vaporized are selected. This includes the vapor necessary to supply the distillate and side-stream products plus the material stripped from the residue in the stripping section of the column. If this represents 2% or more of the total vapor, it must be evaluated and included. If it is smaller than 2%, it may be neglected in the calculation of the partial pressure. The properties of the vapor are determined through the density and molecular-weight correlations so that the number of moles of vapor can be calculated. The number of moles of vapor plus the moles of stripping medium are added, and the mole fraction of the vapor is determined by dividing by the total number of moles. The effective partial pressure or distilling pressure is the mole fraction of the vapor times the total pressure.

7. The EFV curve is determined at the pressure calculated previously by use of the distillation curve correlations in Chap. 3.

8. The temperature read from the EFV curve corresponding to the percent vaporized is compared with the maximum allowable temperature. If it is equal to or lower than the maximum allowable temperature, it is satisfactory. Otherwise, a larger quantity of stripping medium is selected and the calculations are repeated.

9. The composition of the vapor formed at the feed plate is computed by the conventional multicomponent flash equations, i.e.,

$$y_i = 1.0 = \frac{1}{V} \sum \frac{K_i x_{F_i}}{L/V + K_i} \tag{8.2}$$

or

$$V = \sum \frac{K_i x_{F_i}}{L/V + K_i} \tag{8.3}$$

Thus,

$$y_i = \frac{K_i}{V} \sum \frac{x_{F_i}}{K_i + L/V} \tag{8.4}$$

Since the temperature, pressure, L/V, and V are known, the values of y_i can be calculated. The recommended K values are those of Hadden and Grayson [8] or those of Poettman [13] and Smith [15] when petroleum oils are being distilled. Where other systems are encountered, equilibrium data must be available or predictable from available data. The approximate TBP curves for the vapor and liquid at the feed plate can be determined by the pseudocomponent method using the Brown-Katz [2] modified Obryadchacoff [11] equilibrium curves.

10. The quantity and composition of the residue entering the stripping section below the feed is evaluated as follows: The TBP curve of the residue is determined by subtracting the moles of each component vaporized (i.e., y_iV, assuming 1 mole of feed) from the moles originally in the feed and by evaluating the mole fraction of each component.

11. The identity and quantity of material to be stripped from the residuum may be determined in a number of ways:

(*a*) From the residue specification flash point. This requires a knowledge of the amount of material removed from the residue versus the residue flash point. Thus, experimental data might indicate that the removal of, e.g., 15% would give the correct flash temperature for the residue. On such a basis assume all lighter than the key component to be stripped out plus the amount of the key to satisfy the requirement of the total amount stripped.

(*b*) Because of the steam causing a lower effective partial pressure, assume that the component having a vapor pressure equal to that pressure is the key component. Thus all components lighter than this key are removed completely in addition to the amount of the key component necessary to give the correct boiling point at the partial pressure desired. The exact amount of material in the vapor is calculated by trial and error.

12. The new flash plate temperature corresponding to the readjusted vapor quantity and steam quantity is evaluated by first determining the new partial pressure by

$$p = \frac{\text{moles of vapor}}{\text{moles of vapor} + \text{moles of steam}} P_t \qquad (8.5)$$

The new equilibrium flash curve is drawn and the temperature necessary to produce the required number of moles of vapor is determined.

13. Calculate by heat balance the temperature of the residue or bottoms product leaving the column (see Fig. 8.8). Known: V_F, t_F, F, S_B/V above feed plate, and S_B, t_{S_B}, B.

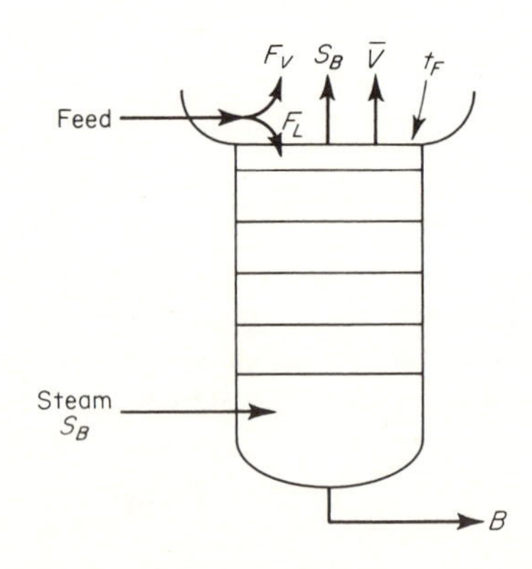

FIGURE 8.8 Heat balance around bottom section and feed.

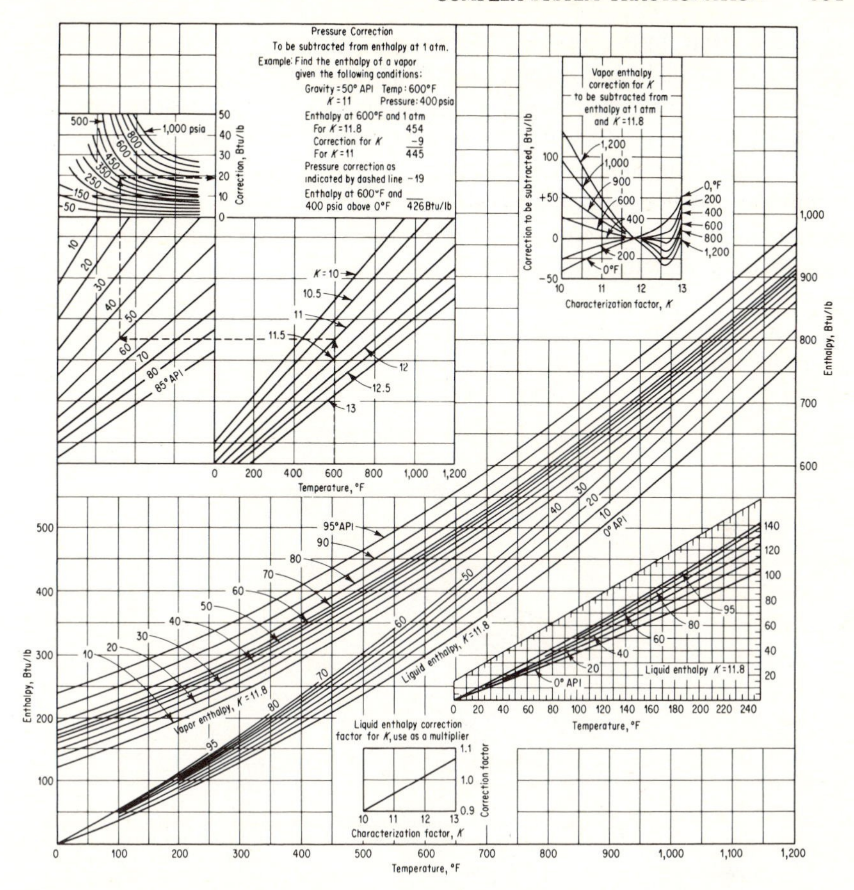

FIGURE 8.9 Enthalpy of petroleum fractions. [*From Bauer and Middleton, Petrol. Refiner (January, 1953), p. 111. Copyright Gulf Publishing Company, Houston, Texas, 1953.*]

Heat balance:

$$FH_F = F_V H_F + F_L h_F \tag{8.6}$$

$$FH_F + S_B H_{SB_{\text{in}}} = F_V H_V + S_B H_{SB_{\text{out}}} + \bar{V} H_{\bar{V}} + B h_B \tag{8.7}$$

$$Bh_B = BC_{p_B}(t - t_{\text{ref}}) = FH_F + S_B(H_{SB_{\text{in}}} - H_{SB_{\text{out}}}) - F_V H_F + \bar{V} H_{\bar{V}} \tag{8.8}$$

From the enthalpy chart, Fig. 8.9, determine t_B corresponding to h_B.

14. Unless perfect fractionation is obtained in the various sections of the column (which implies total reflux and a large number of stages), the effective fractionation between adjacent products does not separate the products as shown on the TBP curve by the cut points indicating theoretical yields. The limits of the fractionation are shown in Fig. 8.10. From Fig. 8.10 it can be seen that the TBP curves of the adjacent products must overlap since the limits represented are 0% for perfect fractionation and some plus value for no fractionation.

This being the case, some of the heavy material which should be rectified from stock 1 will be present in stock 2, and some of the light material which should be completely contained in stock 1 will be present in stock 2. If

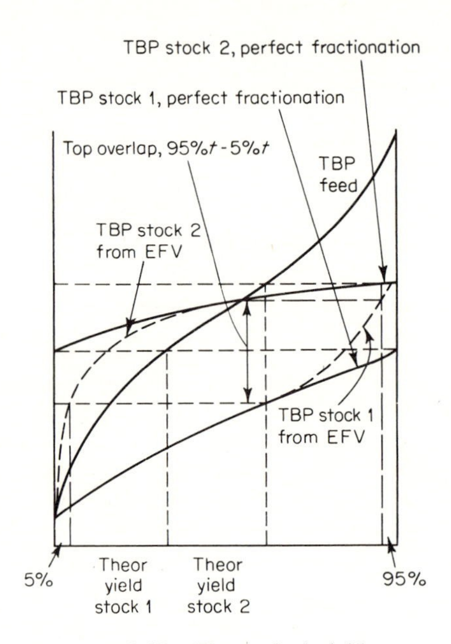

FIGURE 8.10 Theoretical yields.

the distribution could be considered equal (or symmetrical), the actual yield of stocks would be very close to the yields predicted from perfect fractionation. However, this is usually not quite the case in side-draw towers.

The distribution is a function of relative volatility or the slopes of the EFV curves (which are related to the slopes of the TBP curves) or boiling ranges. For example, if two adjacent stocks to be separated have boiling ranges of 200 to 300°F and 300 to 400°F, it would be expected that the contamination of the stocks with each other would be greater than in the case of stocks boiling over the ranges of 100 to 300°F and 300 to 550°F.

Furthermore, since all of the lighter stock must contact the heavier stock on each plate and decreasing amounts of the heavier stock (or components of the heavier stock) contact the lighter material, it would be expected that the lighter material must give up more than it retains with respect to theoretical yield.

Next, as the stocks become heavier (higher boiling components), the relative volatilities become less and separation is less for the same conditions. Therefore the overlap of the curves becomes more symmetrical and the yields approach the theoretical, even though there is more contamination of adjacent stocks with one another.

The effects of these observations are that the light distillate has less yield than indicated by the theoretical-yield calculations. The side-draw-stream yields approach theoretical values as the boiling points increase.

Since each feed stock or crude oil represents a different case, it is impossible to formulate generalized correlations which will give accurate and dependable results. Therefore, approximations useful in estimating roughly the percent of theoretical yield are used unless specific data on a particular topping feed and column are available.

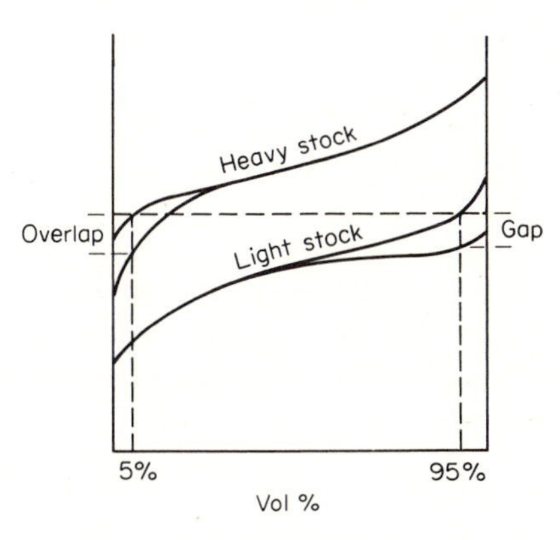

FIGURE 8.11 TBP overlap and gap for adjacent boiling stocks.

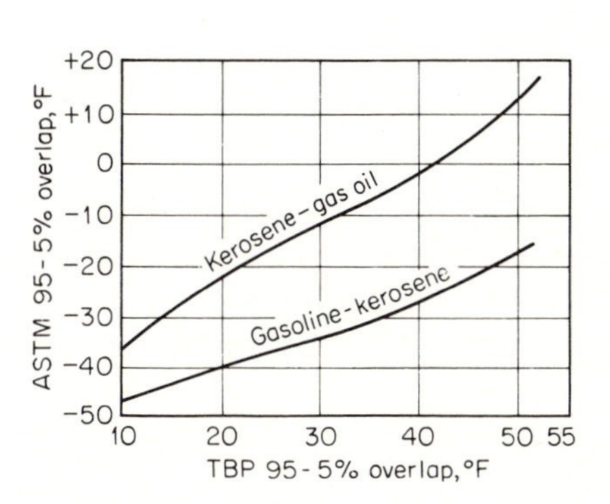

FIGURE 8.12 ASTM 95–5% gap or overlap versus TBP 95–5% gap or overlap. [*From Prater and Boyd, Oil Gas J. (May 2, 1955), p. 71.*]

The usual methods relate the allowable overlap or gap, i.e., the difference in the 95% temperature of the lighter stock and the 5% temperature of the heavier stock adjacent to it, on the ASTM atmospheric distillation curve. Nelson [10] and Packie [12], among others, have related the gap (the 5 to 95% temperature difference is positive) or overlap (the 5 to 95% temperature difference is negative) to the gap or overlap on the atmospheric TBP distillation curves of the separated materials. The TBP gap or overlap is then related to the percent of theoretical yield. Figures 8.12 to 8.14 are taken from Prater and Boyd [14].

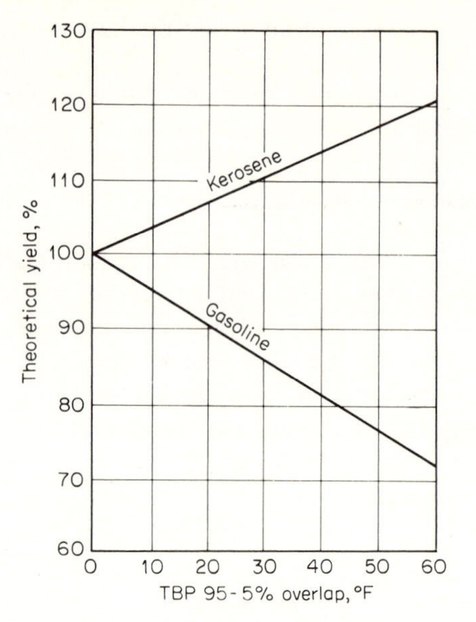

FIGURE 8.13 Percent of theoretical yield versus TBP 95–5% overlap. [*From Prater and Boyd, Oil Gas J.* (*May 2, 1955*), *p. 71.*]

Packie [12], Nelson [10], and others, have also related the ASTM gap or overlap with the product of L/D (number of theoretical plates) for differences between the 50% ASTM temperatures of the stocks being separated with no steam present and with "maximum" steam used in stripping.

A modification of this type of relation which can be used for estimation is

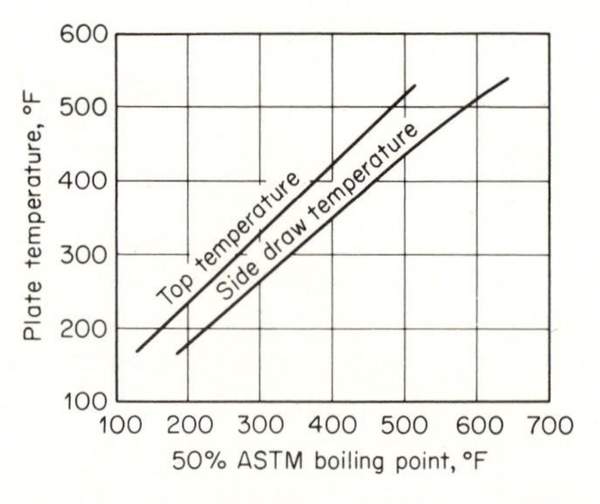

FIGURE 8.14 Plate temperatures related to ASTM 50% boiling point. [*From Prater and Boyd, Oil Gas J.* (*May 2, 1955*), *p. 71.*]

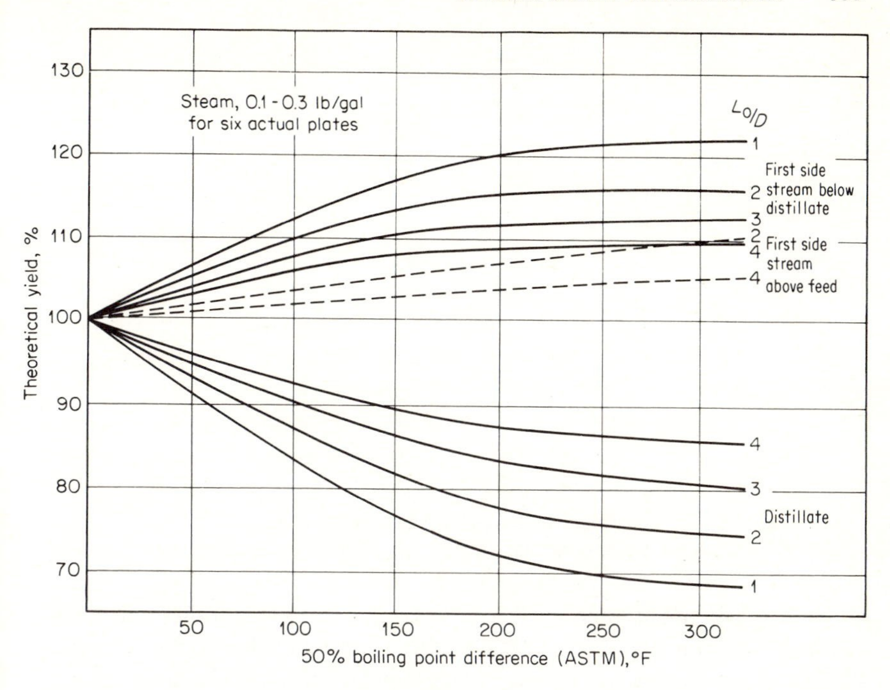

FIGURE 8.15 Percent of theoretical yield for various stocks related to 50% ASTM boiling-point difference.

shown in Fig. 8.15, where for six actual plates the percent theoretical yield is plotted against the difference in 50% ASTM temperatures of adjacent stocks for parameters of reflux ratio L_0/D. These curves are for steam quantities in the range of 0.1 to 0.3 lb/gal. If the actual steam quantities used are greater, the curves representing the constant L_0/D are wider apart. If the steam quantity is less, the curves are closer. Obviously the relationship is highly empirical and good only for approximation. The relations above can be used in several ways:

(*a*) The allowable gap or overlap specified for the various products by customer—or use—demands can be used to determine the percent theoretical yield and the actual yields from which the actual TBP curves of the overhead and side-draw products can be estimated.

(*b*) The actual yields and reflux ratios for assumed quantities of steam and 50% ASTM differences can be determined as a function of reflux ratio L_0/D (for six actual plates).

(*c*) The reflux ratio L_0/D can be determined approximately for different yields of products.

15. Determine the actual yield of distillate by determining the 50% ASTM temperature of the first side-draw product minus the 50% ASTM temperature

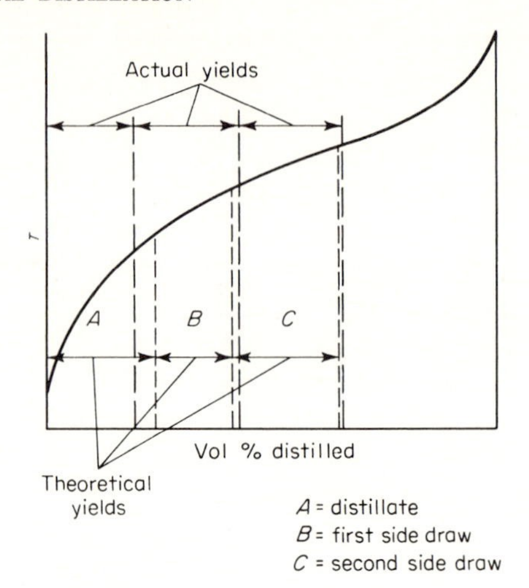

FIGURE 8.16 Relation between theoretical and actual yields.

of the distillate and assume a reflux ratio L_0/D. The percent theoretical yield can be read from Fig. 8.15. This is done for each of the stocks—distillate, first side-draw product, and second side-draw product. The actual yield is then the percent theoretical yield times the theoretical yield (see Fig. 8.16).

16. The composition of the distillate is determined next. The TBP curve can be estimated empirically by assuming that all components in the feed appear in the distillate up to 70 vol % of the distillate. The portion of the TBP curve from the initial boiling point to the 70% point can be drawn through the points representing the midboiling points of the components. The TBP endpoint is assumed to be the temperature on the original feed TBP curve corresponding to the endpoint of the cut whose midpercent temperature corresponds to the TBP cut point for the distillate product. This cut point is derived from distillation specifications. The curve is smoothed in from the 70% temperature to the TBP endpoint. The quantity of each of the cuts between the 70% point and endpoint is determined by the volume percent of each of the cuts between the temperatures defined by the boiling ranges (initial to endpoint) of the cuts.

In the case of a topping column where gasoline is produced as distillate from a crude-oil feed, the actual yield and theoretical yield are evaluated as in the foregoing.

Since the ASTM endpoint of the gasoline is specified, the pseudocomponent having that boiling point (TBP) is the heaviest to be included in the gasoline fraction. The assumption is made that 70% of the gasoline (0 to 70%) will have the same TBP curve as that of the TBP based on the theoretical yield. That is, all the amounts of each of the components present in the first 70% of the gasoline cut will appear in the actual gasoline. The TBP endpoint of the gasoline fraction will be the same as the endpoint of the heaviest component. For example, the highest boiling pseudocomponent might have a TBP boiling range of, say, 390 to 410°F. Thus the TBP endpoint will be 410°F. The remainder of the curve is sketched in between the 70% and the 410°F limit.

The volume percent of each component included in the gasoline fraction from the 70 to 100% is determined from the boiling point of the cut in relation to the sketched-in curve (see Fig. 8.17).

17. The TBP curve of the side-draw products is evaluated by determining the amounts of the cuts appearing in the side-draw stream as follows: The amounts of the cuts appearing in the first side-draw stream, which are distributed between the distillate and the side stream, are found by subtracting the amount of each cut appearing in the distillate from the original amount in the feed. Next it is assumed that all of the intermediate cuts between the last distributed cut and that cut representing 70% of the theoretical yield of the side-draw product S_1 appear only in the side-draw stream. The endpoint of the TBP curve is fixed as the endpoint of the last cut contained in the theoretical yield of the product, as in the case of the distillate. A smooth curve is drawn between the endpoint temperature and the 70 vol % temperature of the product (see Fig. 8.18). The quantities of the cuts between the 70 vol % of the product and the endpoint are read from the smooth curve by determining the volume percent of each cut falling between its initial and endpoint.

18. The TBP curve and volume percent yield of the cuts contained in the next side-draw product, S_2, are determined in the same manner.

19. The volume percent yields of the distillate and side-draw products are

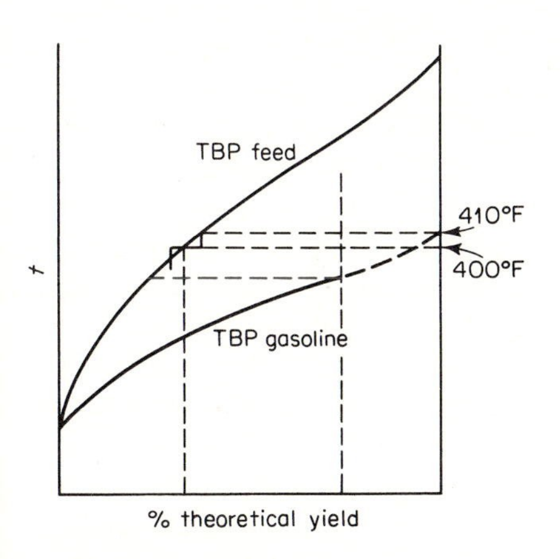

FIGURE 8.17 Selection of TBP endpoint as the endpoint of the last cut included.

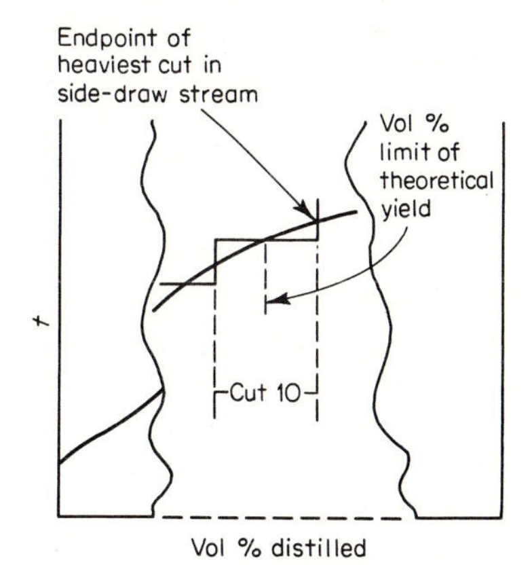

FIGURE 8.18 Relationship between cut TBP endpoint and theoretical yield based on cut midpoint temperature.

added together and subtracted from the amount of vapor formed on flashing plus the vapor stripped from the residue. This difference is the amount of overflash.

The composition of the overflash is determined by subtracting the amount of each cut, which appears in the residue and which is distributed between the overflash and bottoms, from the difference in volume between the amount of each cut in the feed and in the vapor (not including overflash).

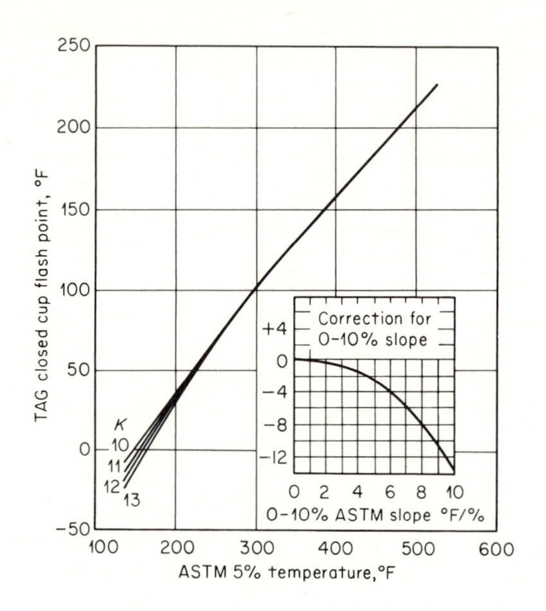

FIGURE 8.19 Correlation of ASTM (0% temperature + 10% temperature)/2 versus TAG closed cup flash point. [*From Van Winkle, Petrol. Refiner (November, 1954), p. 171. Copyright Gulf Publishing Company, Houston, Texas, 1954.*]

20. Determine the amounts of stripping steam needed in the side-stream strippers: From the specifications for the finished side-stream product flash point, distillation range, or volatility, determine the TBP curve for the finished material. Figure 8.19 (see Ref. [19]) may be used to estimate the 5% ASTM distillation temperature. From the TBP curves of the actual side-draw stream and the specified side-draw product, determine the volume percent to be removed by stripping. This can be either estimated from the curves or by determining the amount of each cut distributed between the finished product and the side stream.

The quantity of steam needed for stripping is then evaluated by use of empirical curves, such as those shown in Fig. 8.20, where the steam required

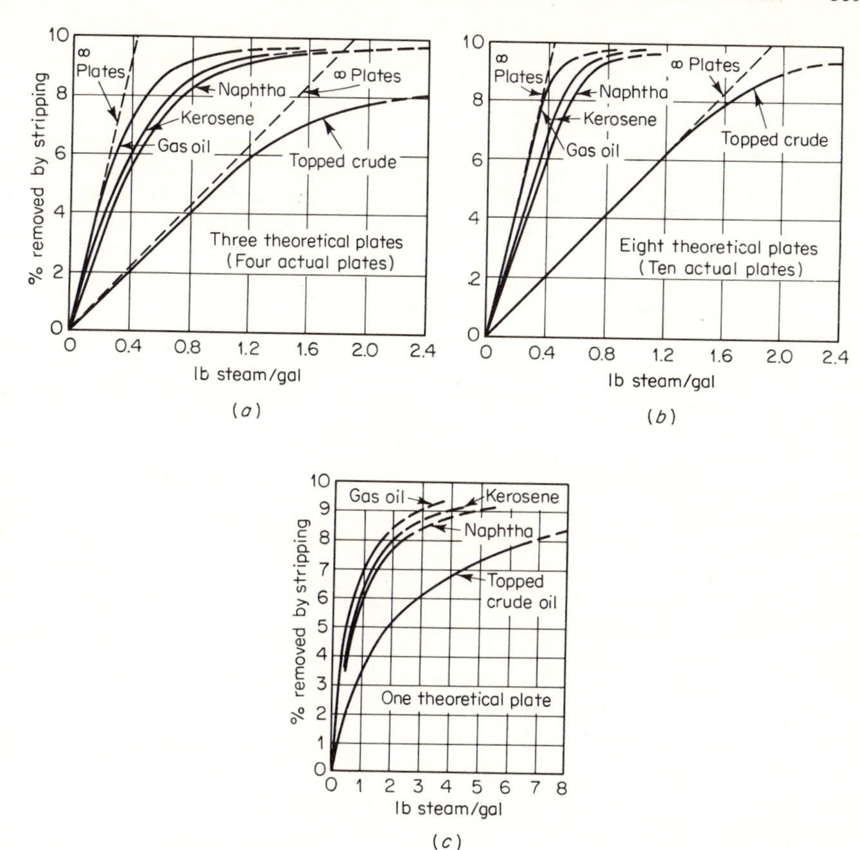

FIGURE 8.20 Approximate steam required for stripping. (*a*) With four plates. (*b*) With ten plates. (*c*) When using only a bath of liquid (one theoretical plate). (*From Nelson, Oil Gas J.* (*March 2, 1944*), *p. 72;* (*July 21, 1945*), *p. 128;* (*May 12, 1945*), *p. 51.*]

is related to the percent stripped for various stocks for a fixed number of plates in the stripper.

For petroleum-oil strippers the usual practice is shown in the following table.

Stock	lb steam/gal
Naphtha	0.2–0.5
Kerosene (or diesel)	0.2–0.5
Gas oil	0.1–0.5
Neutral oils	0.4–0.9
Topped crude	0.4–1.2
Residual cylinder stock	1.0–5.0

Example 8.2

The crude oil whose characteristics are given in Example 8.1 is to be fractionated into (*a*) light naphtha distillate having an ASTM endpoint of 385°F; (*b*) diesel fuel base having an ASTM endpoint of 585°F and a residual oil. The diesel fuel base will be withdrawn as a side stream and the column will be operated as a topping column with one side stream with all of the vapor produced at the feed. The bottom section of the column is merely a stripping section for the residue. The column operates at 18 psia at the top, 21 psia at the feed plate, and 0.4 lb steam/gal of residue is introduced at the bottom. Determine the number of plates needed in the column and in the side-stream stripper by the pseudo-multicomponent method. The feed rate is 500 bbl/hr.

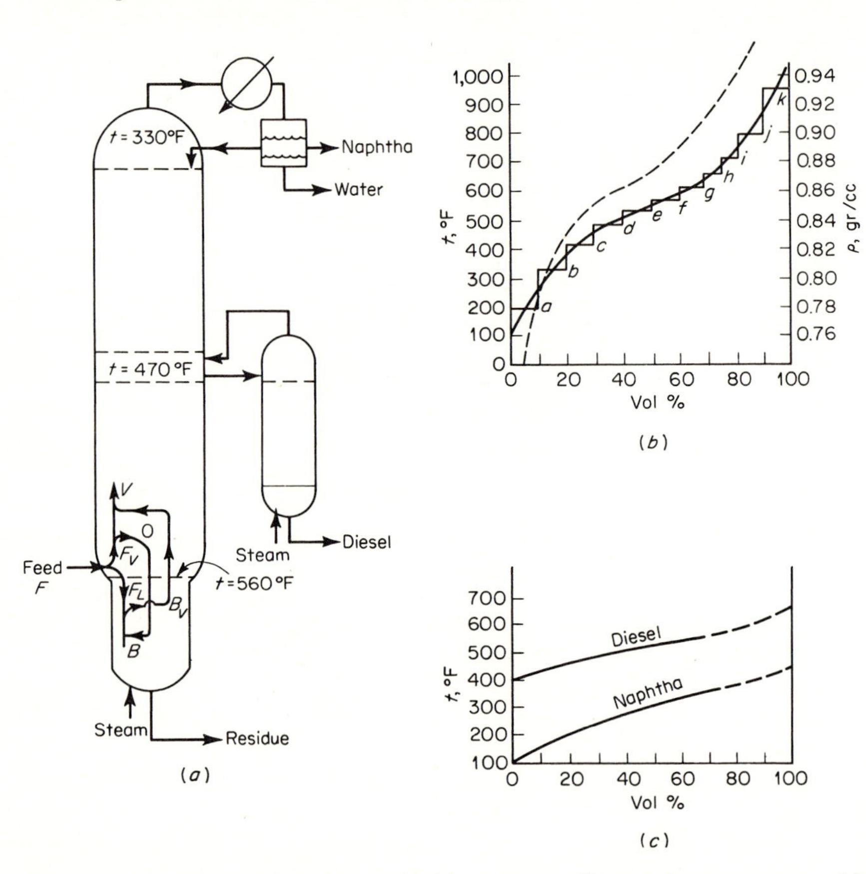

EXAMPLE 8.2 (*a*) Fractionation with side stream. (*b*) Pseudocomponents. (*c*) TBP of products.

Solution

1. Components: The components selected are those having the properties and amounts shown in Example 8.1.

2. Theoretical yield: The TBP cut points can be assumed to differ from the ASTM end-point by 25°F. Thus the naphtha TBP cut point is 405°F and the diesel TBP cut point is 610°F. The theoretical yield of each is read from the TBP curve.

Naphtha = 23.2 vol %
Diesel fuel = 65.0 − 23.2 = 41.8 vol %
Residue = 100 − 65.0 = 35.0 vol %

3. Flash temperature: From the TBP-EFV slope and 50% temperature correlations in Chap. 3, the straight-line flash curve is drawn at 15 psia (corresponding to the total pressure minus the partial pressure of the steam).

$$0.4 \text{ lb steam/gal residue} = \frac{0.4 \text{ lb steam}}{\text{gal residue}} = \frac{0.4 \text{ mole steam}}{18 \text{ gal residue}}$$

Gallons vapor = $63/37$ = 1.70/gal/residue; molecular weight = 180.

$$y_{\text{oil}} = \frac{[(1.70)(52.5)]/[(7.48)(180)]}{\dfrac{0.4}{18} + \dfrac{(1.70)(52.5)}{(7.48)(180)}} = \frac{0.0664}{0.0222 + 0.0664} = 0.75$$

$$P_t \text{ at feed} = 18 + 3 = 21$$

$$p_{\text{oil}} = (21)(0.75) = 15.70 \text{ psia}$$

$$\text{TPB}_{10-70\% \text{ slope}} = \frac{640 - 300}{60} = 5.67°/\%$$

$$\text{TBP}_{50\%,t} = 545°\text{F}$$

$$\text{EFV}_{10-70\% \text{ slope}} = 2.8°/\%$$

$$\text{EFV}_{50\%,t} = 525°\text{F}$$

$$\text{Flash}_{63\%,t} = 525 + (13)(2.8) = 560$$

4. Composition of vapor at feed plate: K at 560°F and 15 psia; $V = 0.80$.

Component	Moles/100 cu ft	Mole fr.	K, 560°F, 15 psia	$L/V = 0.25$	$L/V = 0.26$	Moles/100 cu ft
				$\dfrac{x_F}{K + 0.25}$	$\dfrac{x_F}{K + 0.26}$	
a	4.55	0.173	111	0.0016	0.0016	4.54
b	3.61	0.137	38	0.0036	0.0036	3.59
c	3.19	0.1215	16.7	0.0072	0.0072	3.14
d	2.60	0.099	5.1	0.0184	0.0185	2.48
e	2.59	0.0987	3.2	0.0285	0.0285	2.40
f	2.42	0.0920	2.8	0.0300	0.0300	2.21
g	1.85	0.0705	0.88	0.0624	0.0619	1.43
h	1.48	0.0562	0.38	0.0890	0.0878	0.88
i	1.15	0.0434	0.064	0.1380	0.1338	0.228
j	1.69	0.0641	0.004	0.2560	0.2460	0.025
k	1.19	0.0450	0.002	0.1800	0.1730	0.000
	26.32			0.8147	0.7919	20.923

5. Material to be stripped from bottoms includes 50% of the light key and everything lighter.

Component	Moles/100 cu ft in feed liquid	Moles/100 cu ft removed by stripping	Moles residue/100 cu ft
a	0.01	0.01	
b	0.02	0.02	
c	0.05	0.05	
d	0.12	0.12	
e	0.19	0.19	
f	0.21	0.21	
g	0.42	0.21	0.21
h	0.60		0.60
i	0.922		0.922
j	1.665		1.665
k	1.190		1.190
	5.397	0.81	4.587

Moles residue/100 cu ft feed = 4.587, ρ = 60 lb/cu ft, molecular weight = 350 lb/mole.

6. Effective pressure:

Vapor in flash zone at feed = 20.923 + 0.81 = 21.733 moles/100 cu ft feed

$$\text{Moles steam} = \frac{0.4 \text{ lb}/18}{\text{gal residue}} = 0.0222 \text{ mole steam/gal residue}$$

$$\text{Gallons residue} = \frac{(4.587)(350)(7.48)}{60} = 204 \text{ gal/100 cu ft feed}$$

$$\text{Moles steam} = (204)(0.0222) = 4.54 \text{ moles/100 cu ft}$$

$$y_{\text{oil}} = \frac{21.733}{21.733 + 4.54} = \frac{21.733}{26.273} = 0.825$$

$$p_{\text{oil}} = (0.825)(21) = 17.3 \text{ psia}$$

Because the K values are very little different from those used in step 3, where the calculated pressure was 15.7 psia, and because of the approximations made in the calculations, this is satisfactory.

7. Number of plates in stripping section: With infinite plates and 50% removal of component g, the light key, the minimum moles of steam are calculated as

$$S_{\min} = \frac{L_R E_S(1 + \Sigma X)}{K_{LK}(1 + \Sigma Y)}$$

where $S_{\min}$ = minimum steam requirement, moles/100 cu ft of feed
L_R = moles of liquid residue/100 cu ft of feed
E_S = fractional stripping efficiency
K_{LK} = equilibrium constant, light key
ΣX = moles of components in feed minus moles of residue per mole of residue (per 100 cu ft of feed)

ΣY = moles of components in vapor (excluding steam) stripped from liquid per mole of steam (per 100 cu ft of feed)

$$S_{\min} = \frac{(4.6)(0.5)(1 + 0.81/4.6)}{(0.88)(1 + 0.81/S_{\min})} = \frac{(2.61)(1.176)}{1 + 0.81/S_{\min}}$$

$$S_{\min} = \frac{3.08}{1 + 0.81/S_{\min}} \qquad S_{\min} = 2.27 \text{ moles steam/100 cu ft of feed}$$

Practically, approximately twice the minimum steam requirement is used. Therefore, the assumed value of 0.4 lb/gal residue or 4.54 moles/100 cu ft of feed is satisfactory.

8. Yield and composition of 410° TBP cut point naphtha: A 40°F gap of the ASTM 95–5%, naphtha-diesel boiling curves is assumed. The TBP 95–5% overlap is 20°F (Fig. 8.12). Thus the yield (Fig. 8.13) is 90% of the theoretical or

$$\text{Yield naphtha} = (0.90)(23.2) = 20.9\%$$

The theoretical yield of naphtha includes components *a* and *b* and part of *c*. The TBP curve is drawn for the naphtha, assuming the curve to be the same as in the feed for the first 70% of the naphtha.

Component	Moles/100 cu ft feed	Volume, cu ft/100 cu ft feed	Vapor, vol %	Vol % of feed	Vol % of naphtha
a	4.55	10	14.9	10	50
b	3.61	10	14.9	10	40
c	3.19	10	14.9	1	10
d	2.60	10	14.9		
e	2.59	10	14.9		
f	2.42	10	14.9		
g	1.64	7.1	10.6		
		67.1	100.0	21	

Component	Cumulative vol %	Vol % of component	Vol % of feed	Vol % of deisel	Cumulative vol %	Vol % of component
a	50	100	—			
b	90	100	—			
c	100	10	9	16	16	90
d			10	20	36	100
e			10	17	53	100
f			10	24	77	100
g			7.1	23	100	89
			46.1			

9. Yield and composition of diesel stock: Assume a 10° ASTM 95–5% gap for the diesel stock and the residue.

A 32°F TBP overlap is determined from Fig. 8.12 and 110% of theoretical yield is indicated by Fig. 8.13. Diesel stock yield = (1.08)(41.8) = 46.1.

10. Total yields, approximate:

Naphtha = 21 vol %
Diesel = 46.1 vol %
Residue = 32.9 vol %

11. In summary:

Product	Vol %	°API	lb/bbl	bbl/hr	lb/hr	BP 50%, t, °F	Mol. weight	Latent heat, Btu/lb	$C_{p,\text{av}}$, Btu/(lb)(°F)
Crude oil	100	30.7	305.3	500	152,700	545	220	102	—
Naphtha	21	49.0	274.1	105	28,800	302	121	128	0.74
Diesel	46.1	31.7	303.2	230.5	70,000	540	206	102	0.71
Residue	32.9	19.5	327.8	164.6	53,900	720	335	92	—

12. Temperatures: At diesel side draw (from Fig. 8.14), 470°F; overhead temperature assume 330°F.

13. Heat balance to diesel side draw:

	lb/hr	t_1	t_2	$C_{p,\text{av}}$	λ	Total heat
Cool naphtha	28,800	470	560	0.74	—	1,920,000
Cool diesel	70,000	470	560	0.71	—	4,460,000
Cool overflash	8,858	500	560	0.72	—	392,000
Cool steam	2,770	470	560	0.5	—	125,000
Condense overflash	8,858	—	—	—	94	835,000
Condense diesel	70,000	—	—	—	102	665,000
Total reflux heat at side draw						8,397,000

$$\text{Moles internal reflux} = \frac{8{,}397{,}000}{(206)(102)} = 400 \text{ moles}$$

14. Partial pressure at diesel side-draw plate. $P_t = 19.0$. Use 0.1 lb steam/gal stripped side steam:

$$p_{\text{oil}} = P_t \frac{\text{moles reflux}}{\text{moles reflux} + \text{moles steam}} = \frac{400}{400 + 207.1} 19$$

$$= 12.5 \text{ psia}$$

15. Check by bubble-point pressure of diesel at 470°F:

Component	Moles/hr	x, Mole fr.	K, 12.5 psia, 470°F	Kx
c	19.05	0.240	2.4	0.576
d	17.30	0.218	1.09	0.237
e	17.23	0.217	0.60	0.131
f	16.10	0.204	0.34	0.069
g	9.70	0.122	0.17	0.021
	79.38	1.000		1.034

This is close enough.

Material Balances: Basis—1 hr

Component	Feed, F		Feed vapor, F_V		Stripped from bottom, B_V		Total vapor, V		Feed liquid, F_L		Overflash, O		Total bottoms B	
	Moles	bbl	Moles	bbl	Moles	bbl	Moles	bbl	Moles	bbl	Moles	bbl	Moles	bbl
a	127.8	50	127.5	49.8	0.300	0.2	127.8	50	0.30	0.20	—	—	—	
b	101.4	50	100.8	49.2	0.600	0.8	101.4	50	0.60	0.80	—	—	—	—
c	89.7	50	88.3	49.1	1.405	0.9	89.7	50	1.405	0.90	—	—	—	—
d	73.0	50	69.7	44.2	3.350	5.8	73.0	50	3.35	5.80	—	—	—	—
e	72.8	50	67.5	46.4	5.34	3.6	72.8	50	5.34	3.60	—	—	—	—
f	68.0	50	62.1	45.8	5.90	4.2	68.0	50	5.90	4.20	—	—	—	—
g	52.0	40	40.1	31.0	5.95	4.58	46.05	35.5	11.90	9.15	—	—	5.95	4.6
h	41.6	35	24.7	20.8	—	—	24.7	20.8	16.90	14.20	24.70	30.80	41.6	35.0
i	32.3	30	6.4	5.93	—	—	6.4	5.93	25.90	24.10	6.40	5.93	30.3	30.0
j	47.5	50	0.71	0.75	—	—	0.71	0.75	46.80	49.25	0.71	0.75	47.5	50.0
k	33.4	45	0.00	0.00	—	—	0.00	0.00	33.40	45.00	0.00	0.00	33.4	45.0
	739.5	500	587.81	343.0	22.85	20.3	610.6	363.0	151.7	157.20	31.81	27.48	160.75	164.6

Material Balances: Basis—1 hr

Component	Diesel side stream						Naphtha distillate, D	
	Stripper feed, S		Stripper bottoms, B_S		Vapor from stripper, V_S			
	Moles	bbls	Moles	bbls	Moles	bbls	Moles	bbls
a	0.1	0.04	—	—	0.1	0.04	127.8	50
b	1.0	0.5	—	—	1.0	0.5	101.4	50
c	83.0	46.1	80.7	45	2.3	1.2	9.0	5
d	73.0	50.0	73	50	—	—	—	—
e	72.8	50.0	68.0	50	—	—	—	—
f	68.0	50.0	68.0	50	—	—	—	—
g	46.05	35.5	46.05	35.5	—	—	—	—
h	—	—	—	—	—	—	—	—
i	—	—	—	—	—	—	—	—
j	—	—	—	—	—	—	—	—
k	—	—	—	—	—	—	—	—
	343.95	232.14	340.55	230.5	3.4	1.74	238.2	105
Steam, lb	—	—	53.6	—				

16. Heat balance to top of tower:

	lb/hr	t_2	t_1	C_p	λ	Btu/hr
Cool naphtha vapor	28,800	560	330	0.74	—	4,910,000
Cool diesel vapor	70,000	560	470	0.71	—	4,460,000
Cool overflash	8,858	560	490	0.72	—	446,000
Cool vapor stripped from diesel	510	470	330	0.73	—	52,100
Cool steam (bottoms)	2,770	560	330	0.5	—	319,000
Cool steam (diesel)	965	470	330	0.5	—	67,500
Condense diesel	70,000	—	—	—	102	716,000
Condense overflash	8,858	—	—	—	94	832,000
Condense vapor stripped from diesel	510	—	—	—	106	54,000
						11,802,600

$$\text{Moles internal reflux} = \frac{11{,}802{,}600}{(128)(121)} = 761 \text{ moles}$$

Assume arithmetic average reflux from diesel side-draw plate to top plate.

$$\text{Internal reflux}_{(\text{av})} = \frac{761 + 400}{2} = 580 \text{ moles}$$

17. Check the top tower temperature by the dew point of the vapor from the top plate.

$$\text{Pressure at the top of the tower} = 18 \text{ psia}$$

$$\text{Partial pressure} = \frac{\text{moles vapor}}{\text{moles vapor} + \text{moles stream}} P_t$$

$$p = \frac{V}{V + 207.1} P_t$$

A reflux ratio of $L_0/D = 2.7$ is assumed and the reflux temperature of 200°F:

$$D = 105 \text{ bbl/hr} = 28{,}800 \text{ lb/hr}$$

$$L = (2.7)(28{,}800) = 77{,}800 \text{ lb/hr}$$

Reflux at 330°F,

$$V = L + D = 28{,}800 + 77{,}800 = 106{,}600 \text{ lb/hr}$$

$$= \frac{106{,}600}{121} = 88 \text{ moles/hr}$$

Reflux at 200°F,

$$V = 88 + \frac{(77{,}800)(0.74)(330 - 200)}{(128)(121)} = 88 + 484 = 572$$

$$p = \left(\frac{572}{572 + 207.1}\right) 18 = 13.2 \text{ psia}$$

Component	y	K at 330°F	y/K
a	0.537	3.35	0.160
b	0.425	0.65	0.655
c	0.038	0.21	0.182
	1.000		0.997

This is satisfactory.

18. Theoretical plates in diesel side stripper (Fig. 8.20): For 0.75% of material stripped and 0.1 lb steam/barrel, three theoretical plates are satisfactory.

19. Theoretical plates in stripping section of column (Fig. 8.20): For 0.4 lb steam/gal of residue and 12% of material stripped, approximately 10 theoretical plates are needed.

20. Number of plates in section between feed and diesel side draw: Calculate minimum reflux, minimum plates, and use the Gilliland correlation to get the number of plates for $L/V = 0.73$. LK = component g; HK = component h.

$$\alpha_{g/h} = \frac{K_g}{K_h}$$

At 560°F,

$$\alpha_{g/h} = \frac{0.88}{0.38} = 2.32$$

At 470°F,

$$\alpha_{g/h} = \frac{0.17}{0.08} = 2.12$$

$$\alpha_{\text{av}} = 2.22$$

Using the composition of the LK and HK at the feed and at the diesel side draw as x_{LK}, x_{HK} in "bottoms" and "distillate," respectively, $N_{\min}$ is approximately 3.0 plates. The minimum reflux ratio is determined to be:

$$\left(\frac{L}{D}\right)_{\min} \simeq 1.1 \qquad \text{or} \qquad \frac{L}{V} \simeq 0.525$$

$$\frac{L/D(L/D)_{\min}}{L/D + 1} = \frac{2.7 - 1.1}{2.7 + 1} = \frac{1.6}{3.7} = 0.422$$

$$\frac{N - 3}{N + 1} = 0.3 \qquad \text{(from Fig. 5.17)}$$

$$N = 4.7 \text{ theoretical plates}$$

21. Number of plates in section between diesel side draw and top of column:
At 330°F,

$$\alpha = \frac{K_c}{K_d} = \frac{0.21}{0.078} = 2.7$$

At 470°F,

$$\alpha = \frac{K_c}{K_d} = \frac{2.4}{1.09} = 2.2$$

$$\alpha_{\text{av}} = 2.45$$

Calculating $(L/D)_{\min}$ as before, the value is approximately 0.8. $N_{\min}$ is approximately 3.0.

$$\frac{L/D - (L/D)_{\min}}{L/D + 1} = \frac{2.7 - 0.8}{2.7 + 1} = 0.515$$

$$\frac{N - N_{\min}}{N + 1} = \frac{N - 3}{N + 1} = 0.25 \qquad \text{(from Fig. 5.17)}$$

$$N = 4.35 \text{ theoretical plates}$$

21. Decide on the type of reflux at the top of the column (see Fig. 8.21): In general, cold reflux is better since it is easily controlled, gives greater internal reflux, and fractionation is enhanced. Where column liquid capacity is forced, it may be necessary to use hot reflux to reduce liquid load on plates. However, fractionation is sacrificed.

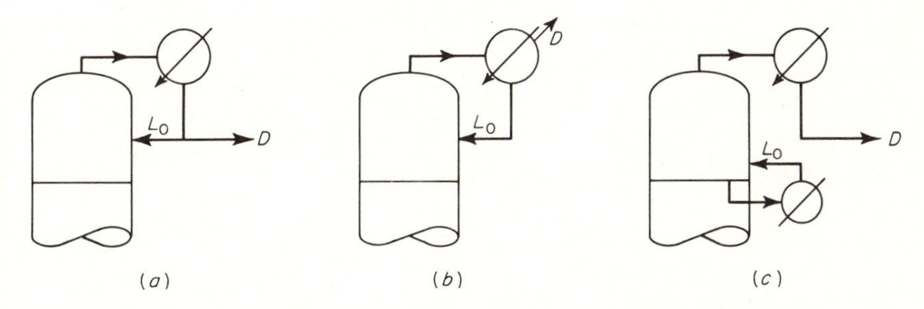

FIGURE 8.21 Types of reflux. (*a*) Cold. (*b*) Hot. (*c*) Circulating.

22. Calculate $V_1 = L_0 + D$ [D is based on corrected yield: (% theoretical yield) (theoretical yield)]. L_0/D was assumed in step 15.

23. Calculate the top tower temperature as the dew-point temperature of the distillate or the 100% temperature on the EFV curve of the distillate product at the partial pressure of the distillate material. Where the steam quantities in the bottoms and side-stream strippers are fixed,

$$S_T = S_B + S_{S1} + S_{S2}$$

and P_t at the top of tower, V_1, and the TBP curve of the distillate product are fixed.

(*a*) Determine the molecular weight of the distillate product.

(*b*) Knowing the volume V_1, determine moles of V_1.

(*c*) Calculate

$$p_{\text{dist}} = y_{\text{dist}} P_t = \frac{\text{moles distillate}}{\text{moles distillate} + \text{moles steam}} P_t$$

24. Assume S_1 side-draw temperature (to be checked later) by using Fig. 8.14.

25. Assume a linear temperature drop between the S_1 side draw and the top plate.

26. Determine the number of plates in the tower assuming the use of six actual plates between product streams and four plates between the feed and first side draw. (Nelson [10] recommends the number of plates shown in the following.)

Recommended actual plates	No. of plates
Light gasoline-naphtha	4–5
Naphtha-kerosene	3–5
Gasoline-kerosene	5–6
Kerosene-gas oil	4–5
Gas oil-lube distillate	4–5
Feed-gas oil	2–4
Stripping	4–5
Side strippers	3–5

27. Calculate the temperature of V_2.

$$t_2 = \frac{t_{S1} - t_1}{6} + t_1 \tag{8.10}$$

28. Calculate $(L_1/V_2)_{\text{act}}$. See Fig. 8.22.

$$V_{2,\text{act}} = V_1 + V_i \qquad L_{1,\text{act}} = L_0 + L_i \qquad V_i = L_i \tag{8.11}$$

$$L_0h_0 + V_2H_2 + V_iH_i + S_TH_S = V_1H_1 + S_TH_{S1} + L_1h_1 + L_ih_i \tag{8.12}$$

$$V_2(H_2 - H_1) + S_T(H_{S2} - H_{S1}) + V_i(H_i - h_i) = L_1(h_1 - h_0) \tag{8.13}$$

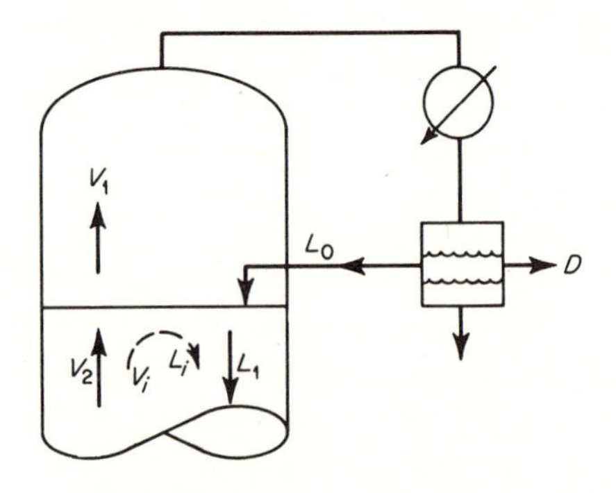

FIGURE 8.22 Effect of cold reflux on internal reflux.

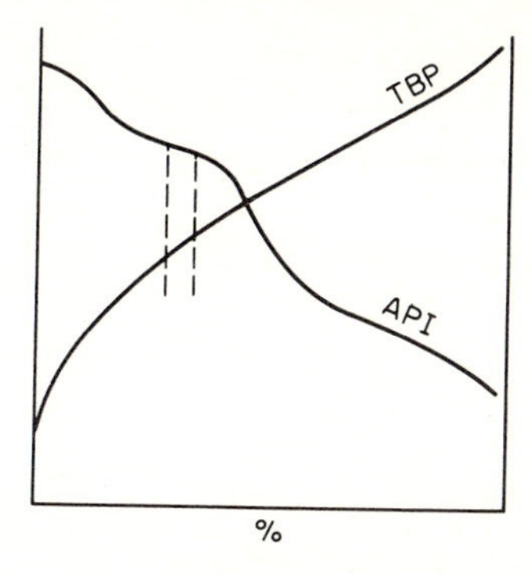

FIGURE 8.23 Estimation of properties of condensed vapor from plate 2.

The molecular weight and other properties of stream $V_i = L_i$ can be estimated as the properties of the next 5% cut following the distillate product. See Fig. 8.23.

The internal reflux ratio is then

$$\frac{L_0 + L_i}{V_1 + V_i} = \frac{L_1 + L_i}{V_2 + L_i} = \left(\frac{L_1}{V_2}\right)_{\text{act}} \tag{8.14}$$

This reflux ratio will hold reasonably constant to the first side-draw plate where a portion of L is removed.

29. Using Fig. 8.14, get an approximate side-draw temperature for each side stream.

30. Check the side-draw temperatures by writing enthalpy balances around the feed plate and around the side-draw plate as shown. The amount of vapor above the side-draw plate is then determined, the p_{oil} is calculated, and the bubble-point temperature is determined as the initial temperature on the EFV curve of the side stream at p_{oil}. It may also be determined as the temperature found by the pseudocomponent method when the $\Sigma y = 1.0 = \Sigma Kx$ at p_{oil}. These must agree in each case (within 1 or 2 degrees).

Latent heat and enthalpy data for petroleum oils are given by Watson et al. [21, 22] and others (see Figs. 8.9 and 8.24).

A number of assumptions pertinent to these calculations are made:

(*a*) In determining p_{oil}, the material leaving the column at the second side draw above the product stream in question acts as inert gas since it is far above its bubble-point temperature (much as the steam is).

(*b*) The material at the next side draw has no effect.

(*c*) The overflash condenses approximately 20°F above the first side-draw temperature.

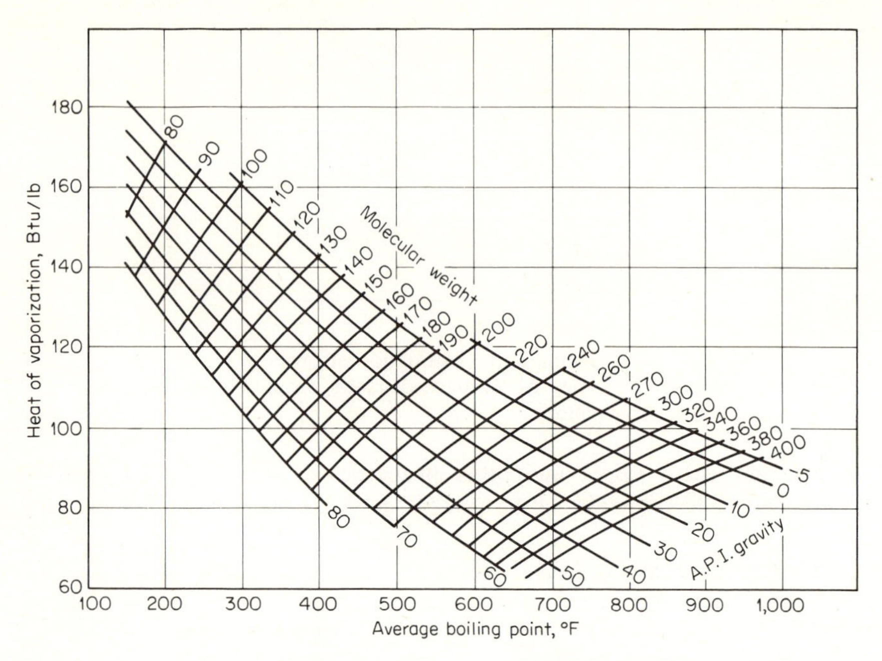

FIGURE 8.24 Latent heat and boiling-point correlation. [*From Watson et al., Ind. Eng. Chem.*, **25**:*880* (*1933*); **27**, *1460* (*1935*).]

(*d*) The vapor to provide reflux above the side-draw plate has the characteristics of the side-draw material.

The diagrams for calculating the two heat balances are shown in Figs. 8.25 and 8.26.

Enthalpy balance for Fig. 8.25 In, t_F Out, $t_{\text{Gas oil}}$	Enthalpy balance for Fig. 8.26 In, t_F Out, t_{Kero}
(lb Gaso)($H_{t,\text{ Gas oil}_V} - H_{t,F_V}$)	(lb Gaso)($H_{t,\text{ Kero}_V} - H_{t,F_V}$)
(lb Kero)($H_{t,\text{ Gas oil}_V} - H_{t,F_V}$)	(lb Kero)($H_{t,\text{ Kero}_L} - H_{t,F_V}$)
(lb Gas oil)($H_{t,\text{ Gas oil}_L} - H_{t,F_V}$)	(lb Gas oil)($H_{t,\text{ Gas oil}_L} - H_{t,F_V}$)
(lb Overflash)($H_{t,\text{ Gas oil}+20°\text{F}_L} - H_{t,F_V}$)	(lb Overflash)($H_{t,\text{ Gas oil}+20°\text{F}_L} - H_{t,F_V}$)
(lb S_B)($H_{t,\text{ Gas oil}} - H_{t,F}$)	(lb $S_{t,\text{ Gas oil}}$)($H_{t,\text{ Kero}_L} - H_{t,\text{ Gas oil}_V}$)
(lb L_V)($H_{t,\text{ Gas oil}_V} - H_{t,\text{ Gas oil}_L}$) = lb $L_L\lambda_{L_V}$	(lb S_B)($H_{t,\text{ Kero}_V} - H_{t,\ F_V}$)
	(lb $S_{\text{Gas oil}}$)($H_{t,\text{ Kero}_V} - H_{t,\text{ Gas oil}_V}$)
	(lb L_V)($H_{t,\text{ Kero}_V} - H_{t,\text{ Kero}_L}$) = lb $L_V\lambda_{L_V}$

For Fig. 8.25,

$$\text{lb } L_V = \frac{H_{\text{Gaso}} + H_{\text{Kero}} + H_{\text{Gas oil}} + H_{\text{Overflash}} + H_{S_B}}{\lambda L_V} \tag{8.15}$$

For Fig. 8.26,

$$\text{lb } L_V = \frac{H_{\text{Gaso}} + H_{\text{Kero}} + H_{\text{Overflash}} + H_{S_B} + H_{\text{Strip}} + H_{S_{\text{Gas oil}}}}{\lambda L_V} \tag{8.16}$$

$$\text{Moles } L_V = \frac{\text{lb } L_V}{(\text{mol wt})L_V} \tag{8.17}$$

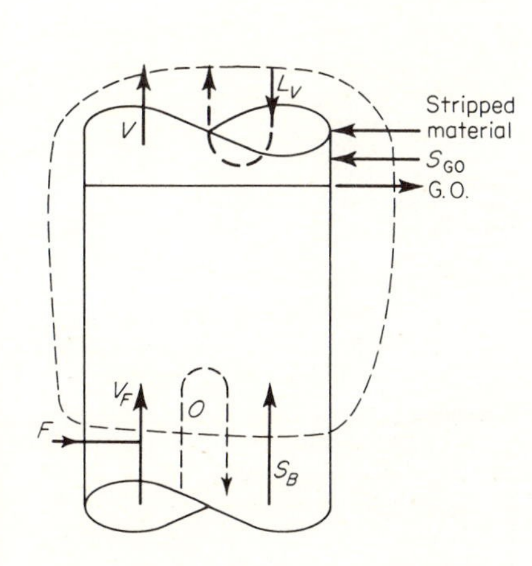

FIGURE 8.25 Streams in the section from the feed to the first side draw.

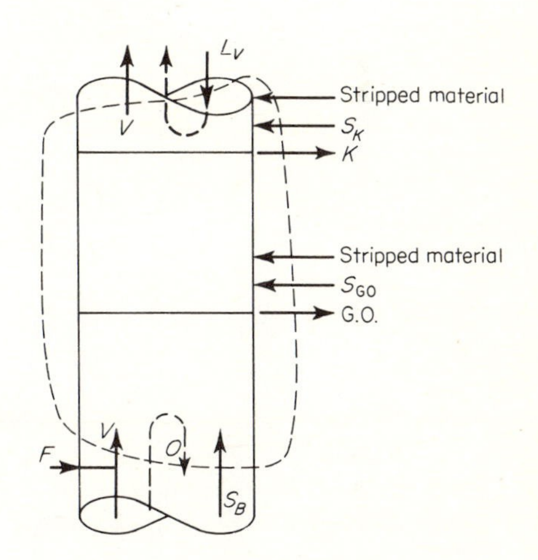

FIGURE 8.26 Streams in the section from the feed to the second side draw.

31. Calculate the bubble-point temperatures for the two stocks at the empirically defined p_{oil} based on the above assumptions. The bubble-point temperature at p_{oil} corresponds to the condition $\Sigma y = 1.0 = Kx$.

The p_{oil} is calculated in each case by:

Inert gases (Fig. 8.25)	Inert gases (Fig. 8.26)
Moles distillate	Moles S_B
Moles S_B	Moles S_2
Moles S_2	Moles S_1
Moles vapor $S_2 = L_V$	Moles vapor $S_1 = L_V$

For Fig. 8.25,

$$p_{\text{oil}} = P_t y_{\text{oil}} = \frac{\text{moles } L_V}{\text{moles (distillate} + S_B + S_2 + L_V)} P_i \tag{8.18}$$

For Fig. 8.26,

$$p_{\text{oil}} = \frac{\text{moles } L_V}{\text{moles } (S_B + S_2 + S_1 + L_V)} P_t \tag{8.19}$$

The temperatures calculated by the heat or enthalpy balance method should check the original temperatures within 1 or 2 degrees. If they do not, a new temperature must be assumed and the calculations are repeated.

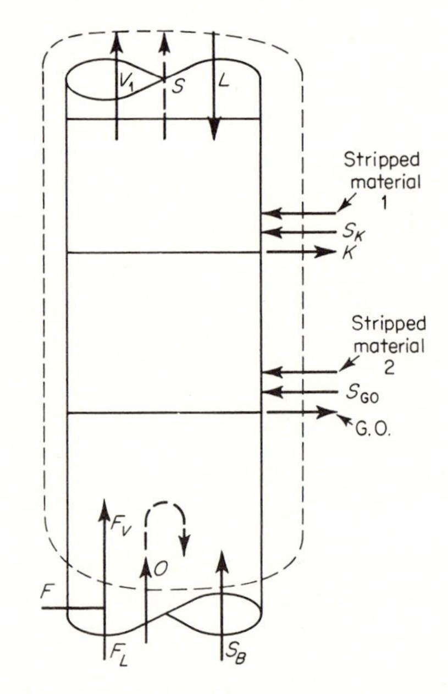

FIGURE 8.27 Streams from the feed to the top plate.

32. Check the assumed reflux ratios L_0/D and L_1/V_2 determined in step 28 by continuing the enthalpy balance calculations to above the top plate (see Fig. 8.27).

The enthalpy balance for Fig. 8.27 is:

(lb distillate)$(H_{t,V_1} - H_{t,F_V})$

(lb S_1)$(H_{t,\text{Kero}_L} - H_{t,F_V})$

$(\text{lb } S_2)(H_{t,\text{Gas oil}_L} - H_{t,F_V})$

$(\text{lb Overflash})(H_{t,\text{Gas oil}+20^\circ F_L} - H_{t,F_V})$

$(\text{lb } SM_{S_1})(H_{t,V_1} - H_{t,\text{Kero},V})$

$(\text{lb } SM_{S_2})(H_{t,\text{Kero}_L} - H_{t,\text{Gas oil}_V})$

$(\text{lb } S_{S_1})(H_{t,V_1} - H_{t,\text{Kero}_V})$

$(\text{lb } S_{S_2})/(H_{t,V_1} - H_{t,\text{Gas oil}_V})$

$(\text{lb } S_B)(H_{t,V_1} - H_{t,F_V})$

$(\text{lb } L_V)(H_{t,L_1} - H_{t,V_1})$

$$\text{lb } L_V = \frac{H_{\text{Dist}} + H_{S_1} + H_{S_2} + H_{\text{Overflash}} + H_{S_B} + H_{S_{S1}} + H_{S_{S2}} + H_{Sm_{S1}} + H_{Sm_{S2}}}{\lambda L_V} \tag{8.20}$$

$$\text{Moles } L_V = \frac{\text{lb } L_V}{(\text{mol wt})_{L_V}} \tag{8.21}$$

The values of L_1/L_2, L_0/V_1, and the dew-point temperature can be checked as the result of this calculation.

33. The condenser duty can be calculated by heat balance.

34. If desired, the number of plates may be checked by using short-cut minimum-reflux and minimum-plate calculations for each section. Determine the theoretical plates used for the actual reflux ratio determined for each section. If the efficiency is known, the actual plates can be determined.

35. The minimum plates at total reflux and the minimum reflux at infinite plates are determined for the section between the feed and the side-draw plate S_2.

(*a*) Selection of key components: Usually the heaviest two components in the *vapor* feed are considered the key components. The α_m to be used in Fenske's [5] equation (or Underwood's [18]) is the geometric mean

$$\alpha_{\min} = (\alpha_{S_2}\alpha_F)^{1/2} = \left[\left(\frac{K_{LK}}{K_{HK}}\right)_{S_2}\left(\frac{K_{LK}}{K_{HK}}\right)_F\right]^{1/2} \tag{8.22}$$

and

$$N_{\min} = \log\frac{(x_{LK}/x_{HK})_{S_2}(x_{LK}/x_{HK})_F}{\log \alpha_{\min}} \tag{8.23}$$

(*b*) Minimum reflux can be determined by any of a number of methods: Colburn [4], Underwood [17], Brown-Martin [3], etc.

36. The number of theoretical plates needed is determined by one of the empirical correlations relating operating reflux ratio and number of theoretical plates with minimum reflux and minimum plates, such as with the Gilliland [7] method and the Brown-Martin method [3]. The problem here is the evaluation of the reflux in the section between the feed and the S_2 plate. The reflux at plate S_2 is very much greater than that at the feed plate because only the liquid overflash returns to the feed plate. The average is used as the reflux ratio for the correlation

$$\frac{L_{S_2} + O_0}{2} = L_{av} \tag{8.24}$$

$$\frac{L_{av}}{D} = \text{reflux ratio} \tag{8.25}$$

37. The number of theoretical plates between S_2 and S_1 and between S_1 and the distillate are determined in the same manner, using the reflux ratios calculated for each of the sections.

38. The efficiency of the plates in a complex-system fractionation is difficult if not impossible to define meaningfully. Therefore, quantitative evaluation of efficiency is not common, and experience factors are used. Nelson [10] suggests the safe design value to be 80%.

PROBLEMS

1. A topping column distillate is to be fractionated into a light naphtha and a heavy naphtha. The characteristics are as follows:

Feed			Light naphtha		Heavy naphtha	
TBP, %	°F	API	TBP, %	°F	TBP, %	°F
Initial	92	—	Initial	92	Initial	250
5	126	75	50	187	50	326
10	140	72	90	242	90	428
20	170	65	100	255	100	440
30	185	61				
40	210	58				
50	230	55				
60	250	53				
70	280	50				
80	325	49				
90	380	48				
100	440	47				

Determine the number of theoretical stages needed by the pseudocomponent method using an operating reflux ratio $L_0/D = 2.5(L_0/D)_{min}$.

2. Using the product boiling-range method determine the number of stages needed for the same situation as in Prob. 1.

3. A crude oil having the following properties is to be fractionated into a gasoline distillate, fuel oil side stream, and residue.

Crude oil			Gasoline		Kerosene dist.		Residue	
TBP, %	°F	API	TBP, %	°F	TBP, %	°F	TBP, %	°F
Initial	105	78.0	Initial	105	Initial	385	Initial	580
5	155	63.5	50	278	50	450		
10	200	59.0	Endpt.	400	Endpt.	590		
20	278	53.5						
30	325	50.0						
40	390	44.0						
50	430	40.0						
60	580	36.6						
70	670	33.0						
80	790	30.0						
90	1000	27.5						

Determine the number of theoretical stages needed for the column and the side-draw stripper by the pseudocomponent method. Use $L_0/D = 1.5(L_0/D)_{min}$.

Nomenclature

B = moles residue per unit time
C_p = heat capacity, Btu/(lb mole)(°F), or Btu/(lb)(°F)
D = moles distillate per unit time
F = moles feed per unit time
H = enthalpy of vapor, Btu/lb mole, or Btu/lb
K = equilibrium vaporization ratio
K = characterization factor
L = moles liquid per unit time
O = moles overflash per unit time
p = partial pressure
P_t = total pressure
S = moles side stream per unit time
S_t = steam, moles per unit time
V = moles vapor per unit time
a, b, c = component designations
h = enthalpy of liquid, Btu/lb mole
t = temperature, °F
x, y = mole fraction liquid and vapor

Greek letters

α = relative volatility
Σ = summation

Subscripts

1, 2 = plate numbers
B = residue
D = distillate
F = feed
HK = heavy key
i, j = component designations
LK = light key
L = liquid
S = side stream

REFERENCES

1. American Institute of Chemical Engineers: "Complex Tower Distillation," New York, A.I.Ch.E. Computer Manual No. 4.
2. Brown, G. G., and D. L. Katz: *Ind. Eng. Chem.*, **25**:1373 (1933).
3. Brown G. G., and H. Z. Martin: *Trans. A.I.Ch.E.*, **35**:679 (1939).
4. Colburn, A. P.: *Trans. A.I.Ch.E.*, **37**:805 (1941).
5. Fenske, M. R.: *Ind. Eng. Chem.*, **24**:482 (1932).
6. Geddes, R. L.: *A.I.Ch.E. J.*, **4**:389 (1958).
7. Gilliland, E. R.: *Ind. Eng. Chem.*, **32**:1220 (1940).
8. Hadden, S. T., and H. G. Grayson: *Petrol. Refiner* (September, 1961), p. 207.
9. Holland, D. D.: "Multicomponent Distillation," Prentice-Hall, Englewood Cliffs, N.J., 1963.
10. Nelson, W. L.: "Petroleum Refinery Engineering," 4th ed., McGraw-Hill, New York, 1958.
11. Obryadchacoff, S. N.: *Ind. Eng. Chem.*, **24**:1155 (1932).
12. Packie, J. W.: *Trans. A.I.Ch.E.*, **37**:51 (1941).
13. Poettman, F., and B. J. Mayland: *Petrol. Refiner* (July, 1949), p. 101.
14. Prater, N. H., and C. W. Boyd: *Oil Gas J.* (May 2, 1955), p. 71.
15. Smith, K. A., and R. B. Smith: *Petrol. Process.* (December, 1949), p. 1355.
16. Smith, R. L., and K. M. Watson: *Ind. Eng. Chem.*, **29**:1708 (1937).
17. Underwood, A. J. V.: *Chem. Eng. Progr.*, **44**:603 (1948).
18. Underwood, A. J. V.: *Trans. Inst. Chem. Engrs. (London)*, **10**:112 (1932).
19. Van Winkle, M.: *Petrol. Refiner* (April, 1964), pp. 139–142.
20. Van Winkle, M.: *Petrol. Refiner* (November, 1954), pp. 171–173.
21. Watson, K. M., and E. F. Nelson: *Ind. Eng. Chem.*, **25**:880 (1933).
22. Watson, K. M., E. F. Nelson, and G. B. Murphy: *Ind. Eng. Chem.*, **27**:1460 (1935).

part

Azeotropic and extractive fractional distillation

chapter

Techniques of separation of azeotropes and other difficult to separate mixtures by fractional distillation

Fractional distillation is a very useful tool for separation of mixtures of components comprising certain types of systems. In general, the simple fractionation technique will apply when:

1. The components have appreciable differences in volatility or the relative volatility of the components to be separated is 1.05 or greater.
2. There is no azeotrope formation.
3. There is no reaction between components.
4. There is no decomposition or polymerization of one or more of the components.
5. The components are capable of vaporization at practical temperatures and pressures.

In addition to the suitability of simple fractionation to separation of components in systems falling within the categories above, by changing the conditions of the distillation process and by using a two-column distillation, certain special types of mixtures may be separated economically by straight fractional distillation. Those most commonly encountered which fall into this category are:

(*a*) Close boiling compounds whose vapor pressure curves cross or which have somewhat different slopes.

(*b*) Homogeneous azeotropes whose compositions are highly pressure sensitive, i.e., the composition of the azeotrope between two key components changes as the pressure on the system changes.

(*c*) Heterogeneous azeotropes between components in the mixture which form two liquid phases and one vapor phase at the azeotropic temperature.

9.1. TECHNIQUES FOR SEPARATION

Close Boiling Components

Vapor pressure curves of systems composed of close boiling components which could be separated by fractionation under selected conditions are shown in Figs. 9.1 and 9.2.

In Fig. 9.1 a binary system is shown where at a total pressure corresponding to (1), the components have essentially the same boiling point and, therefore,

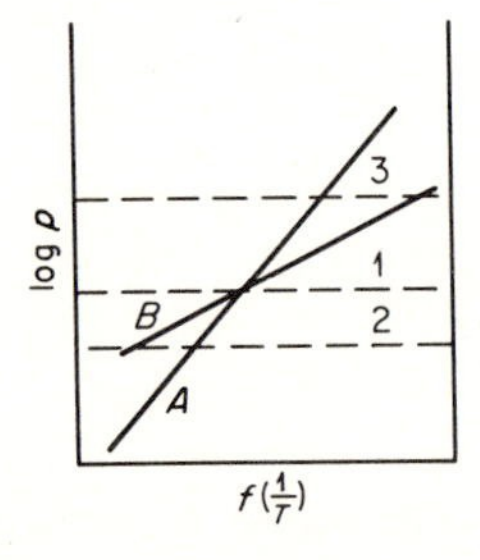

FIGURE 9.1 Binary system whose relative volatility changes from less than 1.0 to more than 1.0 with temperature change.

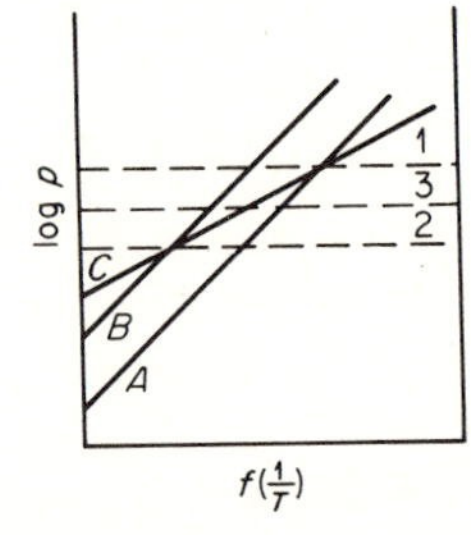

FIGURE 9.2 Ternary system having one component whose relative volatility changes from less than 1.0 to more than 1.0 with temperature change.

at that pressure are inseparable by straight fractionation. At a pressure corresponding to either (2) or (3) the boiling temperatures of the compounds are considerably different and, therefore, probably could be separated. *Tx* diagrams for these situations are indicated in Fig. 9.3. Separation in this case would be obtained by raising or lowering the pressure of the distillation to such a degree that the boiling-point spread would enable separation with a reasonable number of stages and reflux ratio.

In Fig. 9.2 a ternary system is represented wherein compounds *A* and *B* can be separated by straight fractional distillation at pressures corresponding to (2) and (3) but not at (1), and components *B* and *C* can be separated at pressures corresponding to (1) and (3) but not at (2). All three components can be separated at a pressure corresponding to (3) and, therefore, a two-column distillation could be utilized to make close separations between the three components.

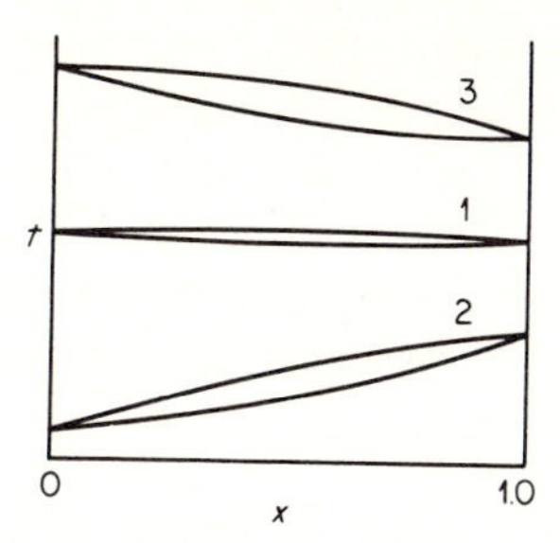

FIGURE 9.3 Reversal of volatility with pressure (see Fig. 9.1). (1) Intermediate pressure. (2) Low pressure. (3) High pressure.

Separation of Homogeneous Azeotropes by Fractionation

Systems containing compounds which form homogeneous azeotropes, either minimum boiling or maximum boiling or both, cannot be separated by simple fractional distillation unless the azeotrope composition is "pressure sensitive" or varies by at least 4 to 5% over a nominal change of total pressure. If the azeotrope composition varies sufficiently with change in pressure, it is possible to use two-column fractionation schemes, such as shown in Figs. 9.4 and 9.5.

Figure 9.4 shows the *tx* diagrams for a minimum boiling azeotropic system at two pressures P_{t1} and P_{t2}. The azeotrope composition is 0.6 at P_{t1} and 0.7 at P_{t2}. Figure 9.5 shows that *tx* diagrams for a maximum boiling azeotropic system having an azeotrope composition of 0.50 at P_{t1} and 0.35 at P_{t2}.

The minimum boiling azeotrope may be separated by feeding the mixture of composition x_F to a column at P_{t2} and recovering component *j* in a relatively high concentration as a bottoms product from the column and, essentially, the azeotrope composition of $x_{D_i} = 0.7$ as a distillate product. The distillate

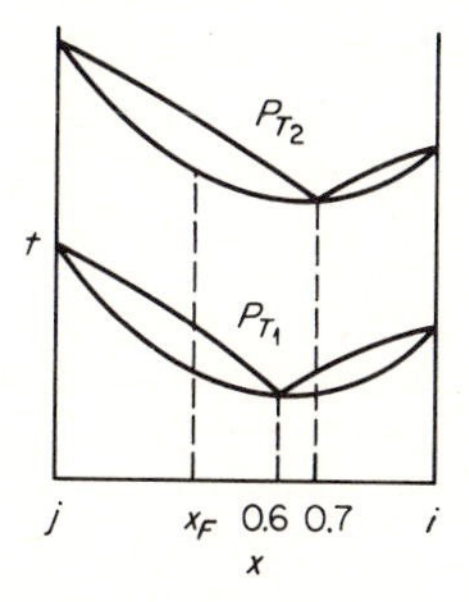

FIGURE 9.4 Pressure-sensitive minimum boiling azeotrope.

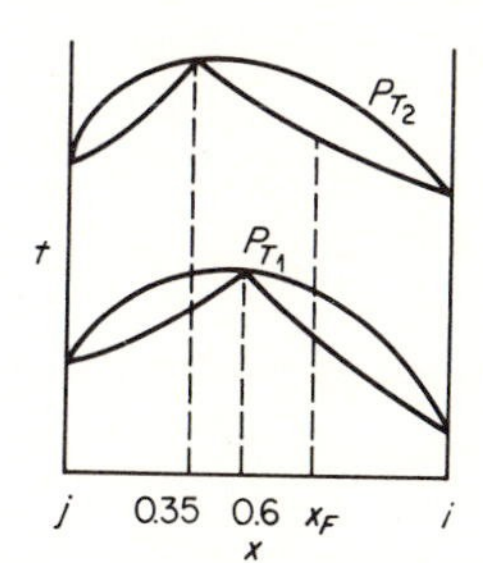

FIGURE 9.5 Pressure-sensitive maximum boiling azeotrope.

product $x_{D_i} = 0.7$ is then fed to the other column maintained at a total pressure P_{t1} and essentially pure i is obtained as a bottoms product. The distillate product, essentially the azeotrope composition $x_{D_i} = 0.6$, is then added to the feed to the column maintained at the pressure P_{t2}.

A somewhat similar treatment may be given the system shown in Fig. 9.5. The column arrangements are shown schematically in Figs. 9.6 and 9.7.

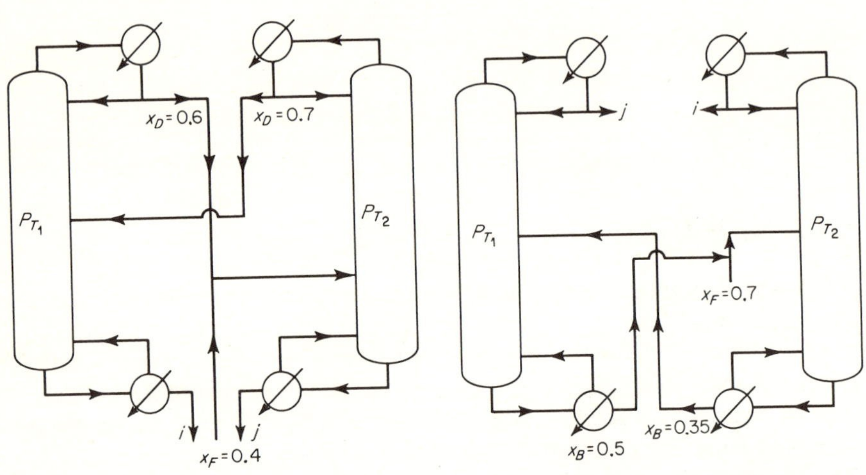

FIGURE 9.6 Separation scheme for minimum boiling azeotropes.

FIGURE 9.7 Separation scheme for maximum boiling azeotropes.

In this manner essentially pure i and j can be produced continuously.

The maximum boiling azeotrope whose tx diagram is shown in Fig. 9.5 may be separated by two-column distillation by feeding the column operated at P_{t1} ($x_{F_i} = 0.4$) and removing pure j as distillate product and essentially the azeotrope composition, $x_{B_i} = 0.50$, as bottoms product. The bottoms product is then fed to the column operated at P_{t2}, and essentially pure i is obtained as distillate and essentially the azeotropic composition $x_{B_i} = 0.4$ is obtained as bottoms product. This material is returned to the system by combining it with the feed to the P_{t1} column.

Separation of Heterogeneous Azeotropes by Fractionation

Heterogeneous azeotropic systems such as furfural–water and aniline–water of the type shown in Fig. 9.8 may be separated by straight fractional distillation utilizing two columns and a suitable liquid-liquid separator (as shown schematically in Fig. 9.9).

The scheme consists of introducing the feed into the separator if it is in the composition range such that it forms two liquid phases at the temperature

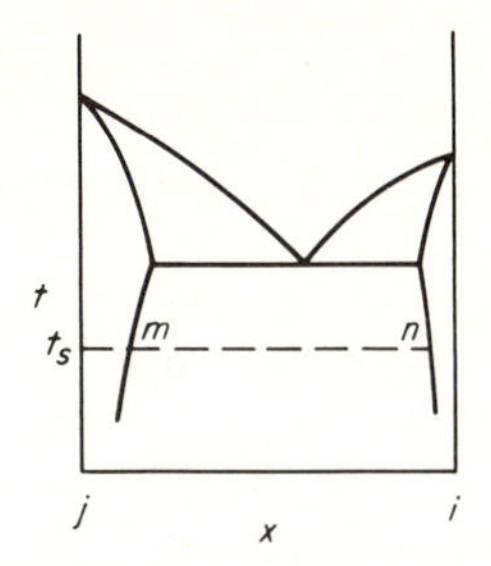

FIGURE 9.8 Heterogeneous minimum boiling azeotrope.

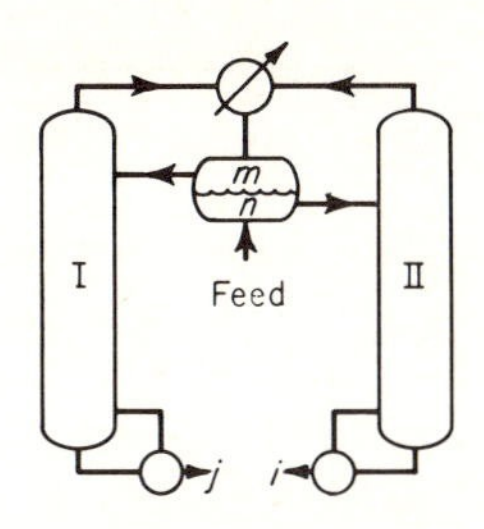

FIGURE 9.9 Separation scheme for a heterogeneous azeotropic mixture (feed—two-phase liquid).

of the separator. If the feed is of such a composition that it consists of a single phase, it is fed to the column distilling the material nearest the composition of the feed. The first situation is illustrated in Fig. 9.9.

The separator is maintained at a temperature level t_s so that two liquid phases are produced, one rich in i, phase n, one rich in j, phase m. The phases are separated and the one rich in i, phase n, is returned as reflux to the top plate of the column which produces i as a bottom product and the phase rich in j, phase m, is returned as reflux to the top plate of the column producing j as a bottom product.

The distillate products from both columns are essentially the azeotropic compositions, one being slightly richer than the azeotrope composition with respect to component i and the other being slightly less. The condensed distillate streams are cooled to the separator temperature t_s and are conducted to the separator wherein the phases of composition m and n are returned to the columns as reflux liquids.

Quantitative Treatment of Separation of Binary Heterogeneous Azeotropes by the Ponchon Method

Case I. The feed consists of two liquid phases (see Fig. 9.10). The over-all material and enthalpy balances around the top of column I are:

$$R_1h_1 + V_{m+1}H_{m+1} = V_1H_1 + L_mh_m \tag{9.1}$$

or

$$R_1h_1 - V_1H_1 = L_mh_m - V_{m+1}H_{m+1} = d_1 \tag{9.2}$$

Around the bottom of column I,

$$L_mh_m = B_1(h_{B1} - Q_{B1}) + V_{m+1}H_{m+1} \tag{9.3}$$

$$L_mh_m - V_{m+1}H_{m+1} = B_1(h_{B1} - Q_{B1}) = d_1 \tag{9.4}$$

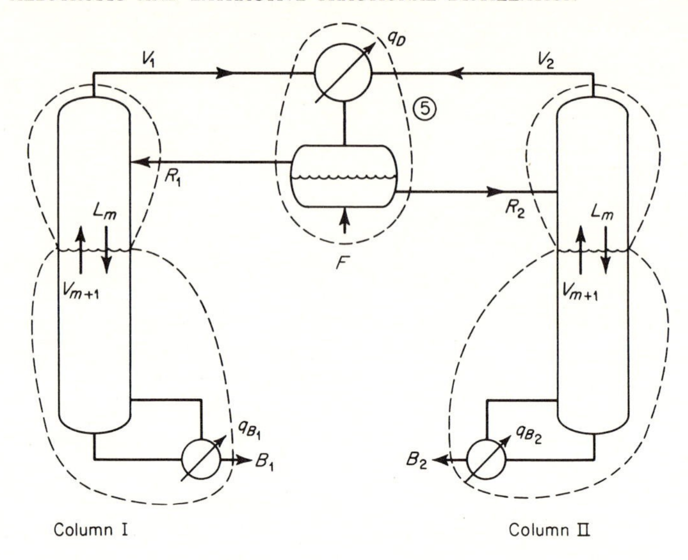

FIGURE 9.10 Material and enthalpy balances (feed—two-phase liquid).

Therefore $B_1(h_{B1} - Q_{B1})$, R_1h_1, and V_1H_1 lie on the same straight line through d_1.

Using the same development for column II, d_2, $B_2(h_{B2} - Q_{B2})$, R_2h_2, and V_2H_2 lie on the same straight line, envelope 5. An over-all material balance and enthalpy balance around the separator and condenser give

$$V_1H_1 + V_2H_2 + Fh_F + q_D = R_1h_1 + R_2h_2 \tag{9.5}$$

$$R_1H_1 - V_1H_1 + R_2h_2 - V_2H_2 = Fh_F + q_D \tag{9.6}$$

or

$$d_1 + d_2 = F(h_F + Q_F) \tag{9.7}$$

On the enthalpy composition diagram the basic construction is shown in Fig. 9.11.

Case II. The feed consists of a single liquid phase (see Fig. 9.12). The material and enthalpy balances around the top of column I between top and feed are:

$$R_1h_1 + V_{m+1}H_{m+1} = L_mh_m + V_1H_1 \tag{9.8}$$

or

$$R_1h_1 - V_1H_1 = L_mh_m - V_{m+1} = d' \tag{9.9}$$

Balances around the bottom of column I and above the feed are:

$$Fh_F + L_mh_m = V_{m+1}H_{m+1} + B_1(h_{B1} - Q_{B1}) = V_{m+1}H_{m+1} + d_1 \tag{9.10}$$

or

$$B_1(h_{B1} - Q_{B1}) - Fh_F = L_mh_m - V_{m+1}H_{m+1} = d' \tag{9.11}$$

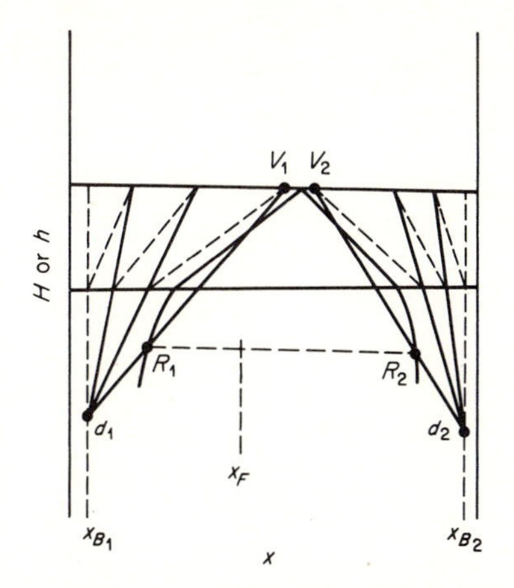

FIGURE 9.11 Composition-enthalpy diagram. Feed separates into two liquid phases at the separator temperature.

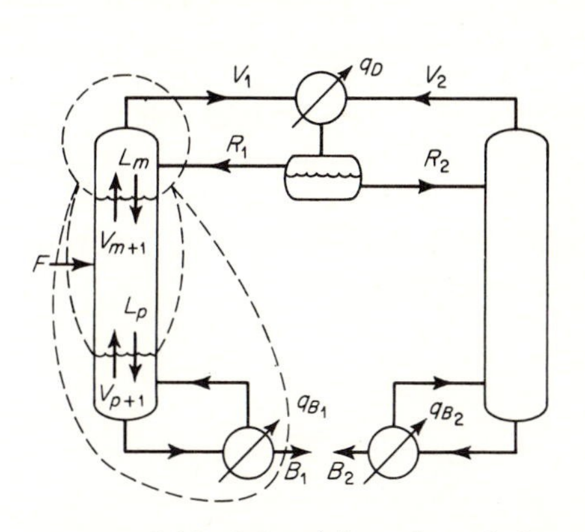

FIGURE 9.12 Material and enthalpy balance (feed—single-phase liquid).

From Eqs. (9.9) and (9.11) it follows that R_1h_1, V_1H_1, and d' lie on a straight line and $B_1(h_{B1} - Q_{B1})$, d_1, Fh_F, and d' lie on a straight line.

Similarly balances are written around the top of column I and below the feed, and around the bottom of column I and below the feed, with Eqs. (9.12) and (9.13) resulting in

$$Fh_F + R_1h_1 - V_1H_1 = L_ph_p - V_{p+1}H_{p+1} = d_1 \qquad (9.12)$$

$$B_1(h_{B1} - Q_{B1}) = L_ph_p - V_{p+1}H_{p+1} = d_1 \qquad (9.13)$$

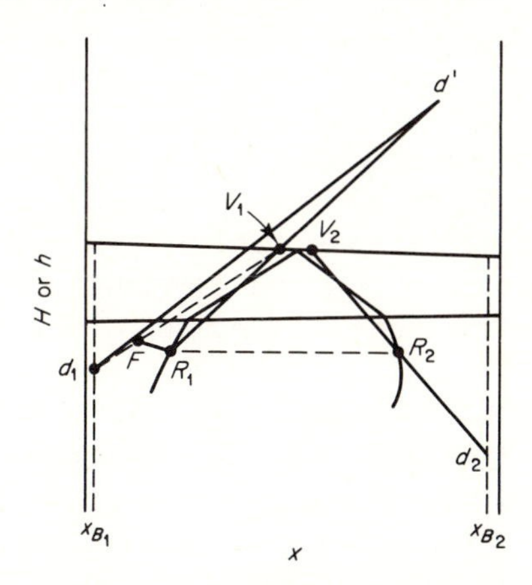

FIGURE 9.13 Enthalpy concentration diagram (feed—single-phase liquid).

The difference-point development for column II is the same as that for case I. The construction on the enthalpy concentration diagram is shown on Fig. 9.13.

Where homogeneous minimum boiling azeotropes occur between two components, it may be feasible to lower the pressure with the corresponding reduction in temperature to such a level that liquid-liquid phase separation occurs. If this can be caused, a simple two-column distillation may be utilized to make essentially a complete separation. This is shown schematically in Fig. 9.14.

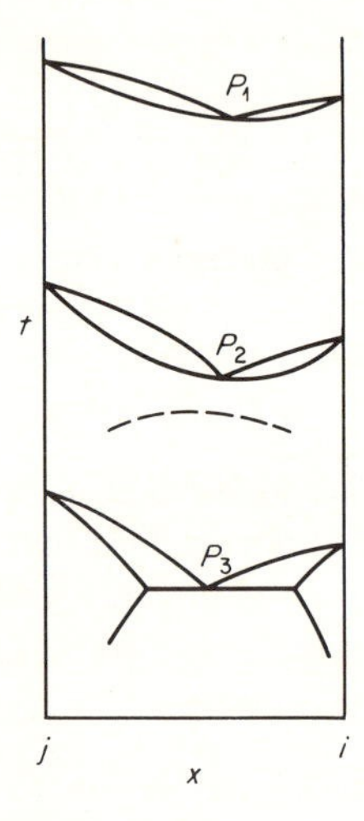

FIGURE 9.14 Possible conversion of homogeneous azeotrope to a heterogeneous azeotrope by lowering system pressure from P_1 to P_3.

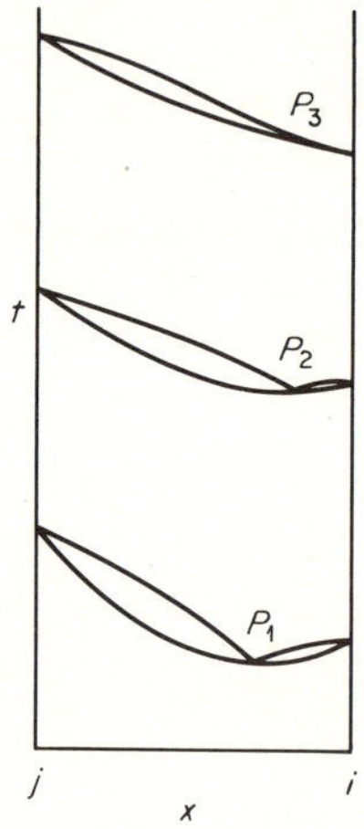

FIGURE 9.15 Shift of azeotrope composition with change in system pressure.

If the azeotrope composition shifts with change in pressure, it might be possible to shift the composition far enough to recover material of satisfactory purity or perhaps, in some cases, even get essentially pure products by using only a single-column distillation. This possibility is shown schematically in Fig. 9.15.

Where components react, as in the case of ethanol and acetic acid, it is possible to lower the pressure to a point where the reaction rate is negligible because of the lowered temperature level. Similarly, where polymerization or decomposition takes place because of the high temperature level, it can be prevented by lowering the pressure in some instances.

Where components have a very high volatility, such as methane and ethane, low-temperature fractionation can be utilized.

Other Special Techniques

In some cases other techniques, alone or in combination with some of the foregoing practices, can make a separation possible through simple fractionation. For example, when a reaction or polymerization takes place, in some cases, some types of inhibitors can be added to prevent the reactions. These are added in small quantity and, for all practical purposes, do not change the character of the simple fractionation technique.

Although the foregoing discussion uses binary and ternary systems for the purpose of ease of visualization, the same separation techniques will apply to multicomponent systems where the key components form characteristic systems falling into the categories discussed. The principal difficulty lies in determining the characteristic behavior of the multicomponent system so that conditions can be selected for their separation.

Nomenclature

B = moles residue per unit time
D = moles distillate per unit time
F = moles feed per unit time
H = enthalpy of vapor, Btu/lb mole
L = moles liquid per unit time
P = vapor pressure
P_t = total pressure
Q = heat added per mole of stream, Btu/mole
R = moles reflux per unit time
V = moles vapor per unit time
d = difference point
h = enthalpy of liquid, Btu/lb mole
t = temperature

Subscripts

B = residue
D = distillate
F = feed
i, j = component designations
p, m, n; $1, 2, 3$ = plate designations

chapter 10

Selection of addition agents for azeotropic and extractive distillation

It was pointed out in Chap. 9 that mixtures of components requiring distillation techniques for their separation other than straight or simple fractional distillation are those in which (*a*) two or more of the components to be separated have but a slight difference in volatility (relative volatility approximately equal to unity) and their vapor pressure curves are essentially of the same slope; (*b*) two or more of the components form homogeneous azeotropes which are not pressure sensitive or which are pressure sensitive but will not provide sufficient relative volatility change near $x_i = 1.0$ or $x_i = 0$; (*c*) one or more of the compounds will decompose or change chemically at temperatures and pressures within economic distillation ranges.

10.1. RELATIVE VOLATILITY AS AN ECONOMIC FACTOR

Relative volatility is a measure of the effective vapor pressure ratio of the two materials which are designated as key components for a given separation, and the cost of the separation is closely related to this property. Although the actual limiting economic values of α_{ij} vary with the value of the products and numerous other conditions, generally separations become uneconomical when $0.95 < \alpha_{ij} < 1.05$. The reasons for this are that in this range a large number of stages are required and the reflux ratio, L_0/D, is large, i.e., above 10. Colburn and Schoenborn [8] offered the approximate generalization that the number of theoretical plates required for a separation of products, each of 99% + purity, varied as the ratio $4/(\alpha - 1)$. They assumed twice the number of theoretical stages at total reflux for this approximation. Thus 80 plates are required for a separation where the relative volatility of the components to be separated, $\alpha = 1.05$, and 40 plates for the situation where $\alpha = 1.1$. Increasing the number of stages increases the capital investment cost and increasing the reflux ratio increases the operating cost. Therefore, it is obvious that changing α, which greatly affects the two factors above, to a value greatly different from unity is highly desirable.

$$\alpha_{ij} = \frac{y_i/x_i}{y_j/x_j} = \frac{\gamma_i P_i \nu_j}{\gamma_j P_j \nu_i} \tag{10.1}$$

Examination of Eq. (10.1) indicates that α may be changed by altering any of the terms in the numerator or denominator in such a manner that the value of the ratio is changed. Under low- and medium-pressure conditions the fugacity coefficients are essentially unity and it is usually not feasible to try to improve the value of relative volatility by changing their ratio. Although the vapor pressures are a function of temperature, the value of their ratio is rarely changed to an appreciable degree by changing the temperature through altering the pressure. Thus the activity coefficient terms or the solution nonideality terms provide the greatest possibilities for modification of relative volatility characteristics of the components. Relative volatility may be changed by increasing or decreasing γ_i and maintaining γ_j at the same value, increasing or decreasing γ_j while maintaining γ_i at the same value, and increasing or decreasing the value of one activity coefficient and changing the other in the opposite direction.

Although changing activity coefficient ratio—and by so doing, changing α by purely physical means (pressure and temperature changes)—is a definite possibility, it generally is uneconomical. On the other hand, however, physical-chemical solution changes offer a number of reasonable possibilities:

1. Add a component to the mixture to be separated which will form a complex or hydrogen bond with component *i* and not with component *j*. In this case the effective vapor pressure of component *i* will decrease and α_{ij} will decrease. If such an addition will reduce α_{ij} to, e.g., 0.8 (or α_{ji} to 1.2 or greater), economical separation is greatly enhanced. This is the technique used extensively in extractive distillation processes.

2. Add a component which will break a complex formed between component *i* and component *j*, or one which will break complexes between molecules of *i*, that is, one which will break hydrogen bonds, thus increasing the value of the effective vapor pressure of *i* to a greater extent than it increases the effective vapor pressure of *j* so that α_{ij} increases. This is essentially the basis for azeotropic distillation processes.

3. Add a component which will cause a minimum boiling azeotropic system to be formed so that component i will have a greater volatility such that it and the volatile added material will be distilled and the component j, having no change in volatility, will concentrate in the liquid and be removed in the bottoms product. This may be done in many cases without reference to hydrogen bonding or bond breaking. Such a method will generally apply to azeotropic distillation.

10.2. PREDICTION OF VAPOR-LIQUID EQUILIBRIUM DATA

Since it is desirable to separate two components by the modification or the changing of their relative volatility, a knowledge of the variables affecting the relative volatility of the two components mixed alone, existing together in a multicomponent mixture, and in the presence of the added component is helpful. Equation (10.1) indicates that α_{ij} varies with concentration, nonideality of the liquid phase expressed in terms of activity coefficient, nonideality of the vapor phase expressed as fugacity coefficient and with temperature, and it is inherently tied to the identity of the components in the mixture and the mixture properties. In an ideal mixture the variance of relative volatility with temperature is not encountered since the ratio of vapor pressures does not change appreciably with temperature and the activity coefficients and fugacity coefficients are unity. The relative volatility α_{ij} depends only on the ratio of vapor pressures which is essentially a constant over the temperature range between the boiling points of the components. Therefore, the ratios of the concentrations of the two components in the two phases will be constant over the temperature range.

In the general practical case the ratio of vapor pressures is not constant, the activity coefficients are other than unity and vary with liquid phase concentration, and the fugacity coefficients may be other than unity. Thus the experimental data from which α is calculated must be available in rather complete form or they must be capable of prediction from relations adequately describing the vapor-liquid equilibrium relationships. Although a rough approximation of the mixture behavior may be predicted on the basis of pure-component characteristics, experimental points derived from an experimental study of the actual mixture in question must be used for accurate design calculations. Also where a solvent is introduced into the mixture to change the relative volatility of the components to be separated, the vapor-liquid equilibria must be determined experimentally or predicted from binary data or other mixture data to determine the new relative volatility so that the effectiveness of the added agent can be evaluated. If binary data and ternary data are available or can be calculated or predicted with some degree of accuracy, the ratios of the effective vapor pressures can be determined and the change in α_{ij} can be evaluated.

Because rarely will sufficient experimental data be available to enable this evaluation to be made, and because time and expense frequently prohibit complete experimental studies, it is highly desirable to be able to resort to reliable prediction methods to calculate the data. All the systems are nonideal and, therefore, the correlations involving nonideality as a function of composition, pressure, and temperature are used.

Nonideal Vapor-Liquid Equilibria

The equations relating the single-component and binary-system properties to the multicomponent-system properties are found in Chaps. 1 and 2 and are referenced in Table 10.1.

TABLE 10.1 **Index of Equations Relating Vapor-Liquid Equilibria Mixture Properties**

Type of equation	Variables	Equation number			
		Single	Binary	Ternary	Multi-component
Van Laar [29]	γ, x, A, B, T	(1.75)–(1.78)	(1.37) (1.38)		
Van Laar (Carlson-Colburn modification [5])	γ, A, B, x		(1.57) (1.58)		
Wohl [32, 33]	γ, n, q, A, B		(1.54)		
[26]	γ, z, q, A, B		(1.55) (1.56)	Margules (2.42)–(2.43) Van Laar (2.28)	
Margules [17]	γ, x, A, B		(1.73) (1.74)		
Redlich-Kister [25]	γ, x, constants			(2.52)	
Benedict [2]	γ, x, A, B, T			(2.46)–(2.48)	
Chao-Hougen [6]	γ, x, constants			(2.49)	
Li-Coull [15]	γ, b, k, T, x		(2.34) (2.35)	(2.31)–(2.33)	
Bonham [4]	γ, T, A, B, x			(2.36)–(2.38)	(2.39)
Black et al. [3,20]	γ^{∞}, T, n, x, P_t, component classes, class variables				(2.51)–(2.53)
Orye-Prausnitz [18] (Wilson equations)	γ, x, P_t, T		(2.57) (2.58)	(2.54)–(2.56)	(2.54)–(2.56)
Kyle-Leng [14]	γ, V_L, x, ϕ, λ			(10.51)	
Weimer-Prausnitz [30]	γ^{∞}, V_L, x, ϕ, λ, τ, T		(10.56)	(10.56)	

The relative volatility of the two components selected as the key components can be computed by selecting the proper equations consistent with the data at hand and evaluating the activity coefficients for the key components in the mixture, using the binary constants and whatever other data are available on the whole mixture or binary, ternary, etc., mixtures comprised of the components in the mixture under study.

Selectivity

Selectivity or the ability of a compound to affect the behavior of other components in solution to the extent that their relative volatilities are changed is the result of molecular interaction. The work of Hildebrand [12], van Arkel [28], London [16], and others has resulted in the recognition of two broad forms of molecular interaction, namely, physical and chemical.

The physical forces causing molecular interactions in which energy effects are thermodynamically positive in sign (endothermic) are classified by Hildebrand [12] as:

1. *Dispersion* forces which tend to cause a perturbation in the electronic motion of one molecule as the result of its being within the field of influence of another. Since this electronic motion is perturbed by light also, there is a very definite relation between light dispersion and frequency; thus the term *dispersion*. This is considered a nonpolar effect.

2. *Induction* forces which are exerted by one molecule on another, the first having a permanent dipole moment which makes it capable of inducing a polarization or induced dipole in the other. This is an attractive force.

3. *Orientation* forces which are exerted by the action of one permanent dipole on another permanent dipole causing molecules to orient with respect to one another. It has been shown [12, 16, 28] that molecules which are nonpolar in makeup or electroneutral—such as the saturated hydrocarbons—when forming a nonideal solution with other nonpolar molecules evidence only endothermic energy effect or positive heats of mixing, since only dispersion forces are involved.

The chemical forces are usually attributed to hydrogen bonding or complexing of the molecules in a solution. These forces cause molecular interactions in which the energy effects are thermodynamically negative in sign or exothermic.

Example 10.1

The system normal octane–ethylcyclohexane at 400 mm Hg has a relative volatility in the range of 1.1 to 1.16 ($\alpha_{n\text{-O}/\text{ECH}}$). Extractive fractionation using either butyl Cellosolve or 2-propanol as a solvent is to be considered. Data derived from adding 1.0, 2.0, and 4.0

moles of solvent to 1.0 mole of 50-50 mixture of octane–ethylcyclohexane are as follows (compositions are on a solvent-free basis):

$\frac{\text{Moles } S}{\text{mole HC mix.}}$	t, °C	x_O	y_O	x_{ECH}	y_{ECH}
	Butyl Cellosolve				
0	106.61	0.482	0.512	0.518	0.488
1	116.3	0.432	0.487	0.568	0.513
2	122.2	0.435	0.502	0.565	0.498
4	132.6	0.435	0.508	0.565	0.492
	2-Propanol				
1	70.0	0.482	0.512	0.518	0.488
2	68.9	0.488	0.524	0.512	0.476
4	67.8	0.496	0.540	0.504	0.460

Determine the relative volatility (O/ECH) versus composition of solvent for the two solvents.

Solution

Butyl Cellosolve as a solvent,

$$\frac{\text{Moles } S}{\text{Total moles mix.}} = 0 \qquad \alpha = \frac{0.512/0.482}{0.488/0.518} = 1.13$$

$$\frac{\text{Moles } S}{\text{Total moles mix.}} = 0.5 \qquad \alpha = \frac{0.487/0.432}{0.513/0.568} = 1.25$$

The remaining results are tabulated as follows:

Components	$\frac{\text{Moles } S}{\text{total moles mix.}}$	α
Butyl Cellosolve	0.0	1.13
	0.5	1.25
	0.67	1.31
	0.8	1.35
2-Propanol	0.0	1.13
	0.5	1.15
	0.8	1.19

Where nonideal mixtures of nonpolar and polar molecules are formed, both dispersion and induction forces are involved with the mixture formation accompanied by an endothermic heat of mixing. When polar-polar mixtures are formed, all three physical effects of dispersion, induction, and orientation are evidenced to contribute to a positive or endothermic heat of mixing. On the other hand, complexing can take place resulting in a negative or exothermic heat of mixing. Thus, in this case, the net heat of mixing may be positive, negative, or zero depending upon the extent and type of the interactions between the components in the mixture.

Prausnitz et al. [1, 22] qualitatively considered evidence of physical and chemical interaction of compounds or complexing when mixed as liquids, and they discussed three criteria—heat of mixing, volume change on mixing, and change in ultraviolet spectra of the compounds alone and in solution.

In examining the activity coefficient of the hydrocarbon at infinite dilution in hydrocarbon–selective agent mixtures, they noted that the activity coefficient of the hydrocarbon increased linearly with the carbon number of the hydrocarbon (Fig. 10.1).

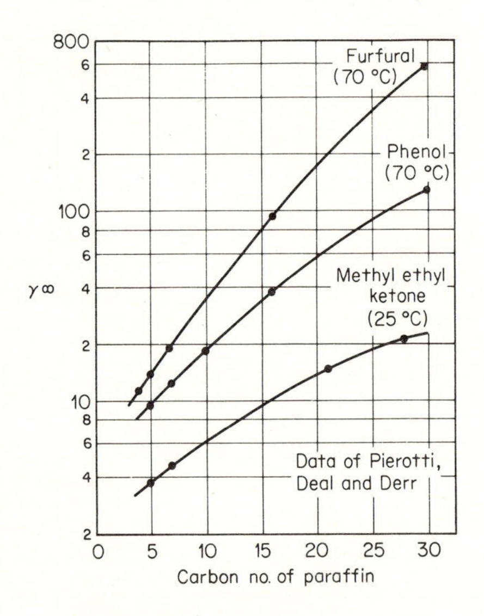

FIGURE 10.1 Variation of infinite-dilution activity coefficient with molecular size of hydrocarbon. [*From Anderson et al., A.I.Ch.E. J.*, **8**:66 (1962).]

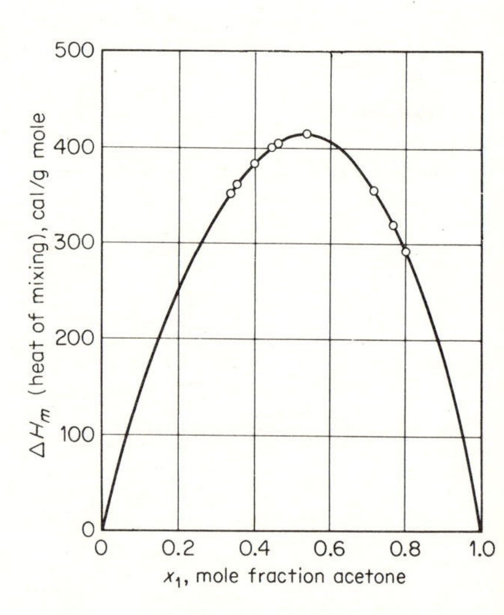

FIGURE 10.2 Variation of ΔH_m, acetone–cyclohexane mixtures at 45°C. [*From Anderson et al., A.I.Ch.E. J.*, **8**:66 (1962).]

Thus in the absence of chemical effects, the larger molecule has the greater activity coefficient. It was also observed that, where chemical effects were evident, complexing tended to increase with increasing unsaturation of the selected molecule for the solvents examined (see Figs. 10.2, 10.3). As

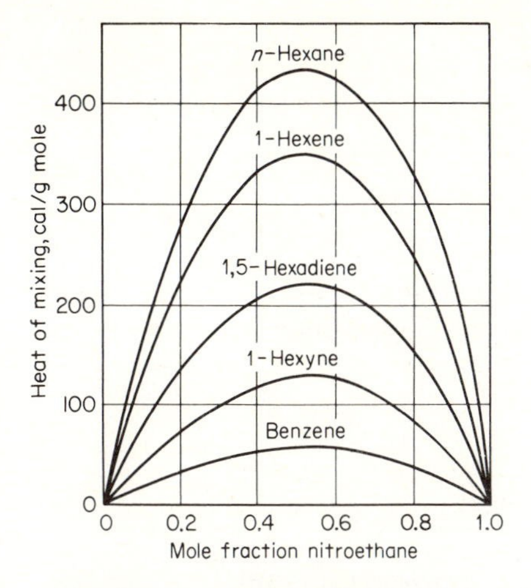

FIGURE 10.3 Variation of ΔH_m with unsaturation, nitroethane–hydrocarbon mixtures at 45°C. [*From Anderson et al., A.I.Ch.E. J.*, **8**:66 (1962).]

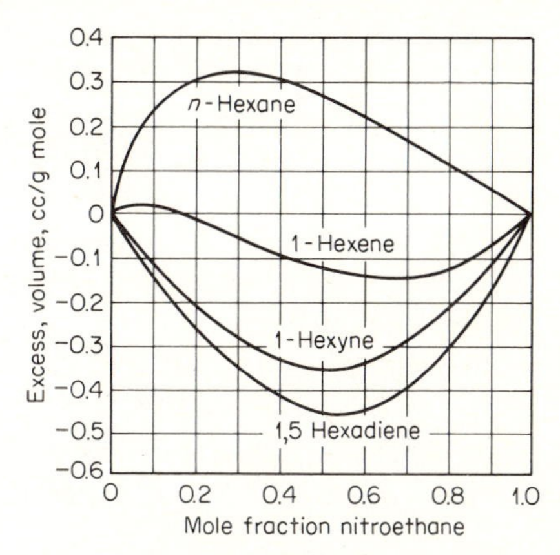

FIGURE 10.4 Mole fraction nitroethane with hydrocarbons at 25°C. Excess volume change on mixing—effect of unsaturation. [*From Anderson et al., A.I.Ch.E. J.*, **8**:66 (1962).]

unsaturation increased, the excess volume tended to decrease as indicated on Fig. 10.4.

Example 10.2

Binary van Laar data derived from experimental data in Example 10.1 systems are as follows:

System	A	A'	B	B'	Average temperature °C	
					Binary	Ternary
n-C_8–ECH	1.00	0.011	9.6	0.011	106	122°C for O–ECH–BC
n-C_8–2-PrOH	0.99	0.84	691	0.85	84	
ECH–2-PrOH	0.99	0.845	699	0.855	86	
n-C_8–BC	0.76	0.56	517	0.74	128	
ECH–BC	0.99	0.585	543	0.59	130	69°C for O–ECH–2-PrOH

Using the Bonham equations, determine the relative volatility of the octane–ethylcyclohexane for the mixture $x_O = 0.17$, $x_{ECH} = 0.17$, $x_{solv} = 0.66$ for each solvent.

Solution

System containing butyl Cellosolve: Use Eqs. (2.36) to (2.38) and designate n-octane, ethylcyclohexane, and solvent with subscripts 1, 2, and 3.

$$T \ln \gamma_1 = \frac{[x_2(B_{12})^{0.5} + x_3A_{32}(B_{13})^{0.5}]^2}{(x_1A_{12} + x_2 + x_3A_{32})^2}$$

$$= \frac{[(0.17)(3.1) + (0.66)(1.01)(22.74)]^2}{[(0.17)(1.0) + (0.17) + (0.66)(1.01)]^2}$$

$$= \frac{(0.526 + 14.5)^2}{(0.17 + 0.17 + 0.667)^2} = \frac{223}{1.012} = 220$$

$$\log \gamma_1 = \frac{220}{(395)(2.303)} = 0.242 \qquad \gamma_1 = 1.745$$

$$T \ln \gamma_2 = \frac{[(0.17)(1.0)(0.321) + (0.66)(1.01)(23.3)]^2}{[(0.17)(1.0) + 0.17 + 0.66(1.01)]^2}$$

$$= \frac{258}{1.012} = 254$$

$$\log \gamma_2 = \frac{254}{(395)(2.303)} = 0.28 \qquad \gamma_2 = 1.92$$

Binary system octane–ethylcyclohexane:

$$T \ln \gamma_1 = \frac{B}{\left(1 + A\dfrac{x_1}{x_2}\right)^2}$$

$$= \frac{9.6}{(1 + 0.17/0.17)^2} = \frac{9.6}{4} = 2.4$$

$$\ln \gamma_1 = \frac{2.4}{379} = 0.00632$$

$$\log \gamma_1 = 0.00274 \qquad \gamma_1 = 1.006$$

$$T \ln \gamma_2 = \frac{AB}{\left(A + \dfrac{x_2}{x_1}\right)^2}$$

$$= \frac{9.6}{A} = 2.4$$

$$\log \gamma_2 = 0.00274 \qquad \gamma_2 = 1.006$$

$$\frac{\alpha_T}{\alpha_B} = \frac{(\gamma_1P_1/\gamma_2P_2)_T}{(\gamma_1P_1/\gamma_2P_2)_B} = \frac{(1.745)(710)/(1.92)(618)}{(1.006)(445)/(1.006)(372)} = \frac{1.042}{1.198} = 0.87$$

System containing 2-propanol:

$$T \ln \gamma_1 = \frac{[(0.17)(3.1) + (0.66)(1.01)(26.28)]^2}{[(0.17)(1.00) + (0.17) + (0.66)(1.01)]^2}$$

$$\log \gamma_1 = \frac{326}{(1.012)(342)(2.303)} = 0.406 \qquad \gamma_1 = 2.54$$

$$\ln \gamma_2 = \frac{[(0.17)(1.00)(0.321) + (0.66)(1.01)(26.43)]^2}{(1.012)(342)} = 0.90$$

$$\log \gamma_2 = 0.39 \qquad \gamma_2 = 2.46$$

$$\alpha_T = \left(\frac{\gamma_1 P_1}{\gamma_2 P_2}\right)_T = \frac{(2.54)(111)}{(2.46)(100)} = 1.149$$

$$\frac{\alpha_T}{\alpha_B} = \frac{1.149}{1.198} = 0.959$$

In summary:

	Added agent			
	2-Propanol		Butyl Cellosolve	
	γ_1, 2.54	γ_2, 2.46	γ_1, 1.745	γ_2, 1.92
Selectivity	0.959		0.87	
Experimental	1.03		1.16	

Figure 10.5 is a plot of the excess volume of benzene and polar solvent mixtures and it indicates that excess volume decreases as the polarity of the solvent increases.

Complexing was also found to cause a change in the ultraviolet spectrum as well as in the infrared spectrum of the mixture of components as compared to those of the individual components.

Commonly, selectivity of an added component or solvent for one of the components to be separated from a mixture is illustrated graphically by plotting the variation of the relative volatility of two components to be separated with parameters of constant composition ratio versus the composition of the added component. This method of plotting, unfortunately, does not illustrate directly the variation of relative volatility with temperature. The bubble-point temperature of the mixture will increase, decrease, or remain essentially constant depending upon the relative boiling points of the key components and the boiling point of the added selective component. Complete studies of relative volatility modification caused by the addition of a selective component should include constant-temperature data or

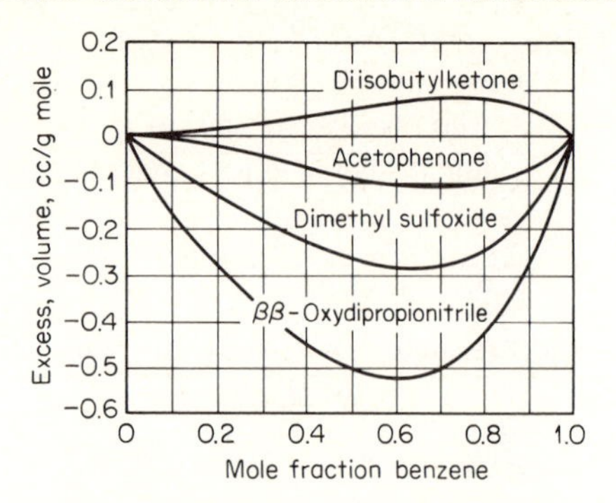

FIGURE 10.5 Mole fraction benzene with solvents at 25°C. Excess volume change on mixing—effect of solvent polarity. [*From Anderson et al., A.I.Ch.E. J.*, **8**:*66* (*1962*).]

sufficient isobaric data run at a number of pressures to enable the effect of temperature to be evaluated.

Types of selectivity There appears to be a number of different types of systems based upon the characteristic behavior of the curves relating the relative volatility and added component concentration.

Normal System

The normal system is one which shows a gradual increase in relative volatility as the composition of the selective agent is increased until the relative volatility becomes constant at some high solvent concentration. Figure 10.6 shows the system toluene–heptane–furfural [11] (and others) which behaves in this manner.

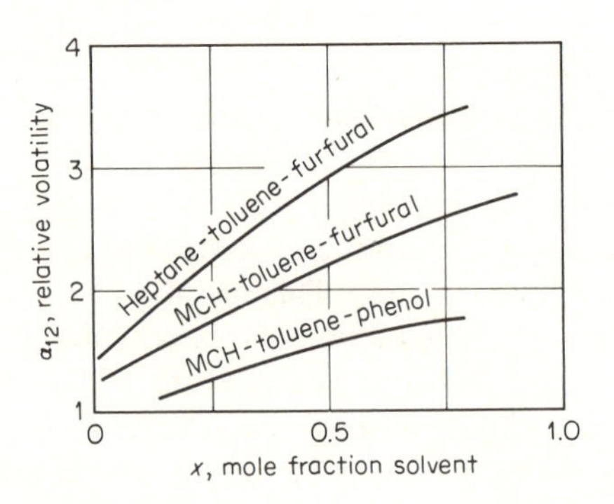

FIGURE 10.6 Normal α modification.

Example 10.3

From the binary experimental A and B values (modified van Laar) and the Wohl two-suffix equations evaluate the relative volatilities for the same composition as in Example 10.2.

Solution

Use Eq. (2.28) for the system containing propanol:

$$\log \gamma_1 = \frac{x_2^2 A_{12}\left(\frac{A_{21}}{A_{12}}\right)^2 + x_3^2 A_{13}\left(\frac{A_{31}}{A_{13}}\right)^2}{\left(x_1 + x_2 \frac{A_{21}}{A_{12}} + x_3 \frac{A_{31}}{A_{13}}\right)^2} + \frac{x_2 x_3 \frac{A_{21}}{A_{12}} \frac{A_{31}}{A_{13}}\left(A_{12} + A_{13} - A_{32}\frac{A_{13}}{A_{31}}\right)}{\left(x_1 + x_2 \frac{A_{21}}{A_{12}} + x_3 \frac{A_{31}}{A_{13}}\right)^2}$$

$$= \frac{(0.17)^2(0.011)\left(\frac{0.011}{0.011}\right)^2 + (0.66)^2(0.84)\left(\frac{0.85}{0.84}\right)^2}{\left[0.17 + 0.17\left(\frac{0.011}{0.011}\right) + 0.66\left(\frac{0.85}{0.84}\right)\right]^2}$$

$$+ \frac{(0.17)(0.66)\left(\frac{0.011}{0.011}\right)\left(\frac{0.85}{0.84}\right)\left[0.011 + 0.84 - 0.855\left(\frac{0.84}{0.85}\right)\right]}{\left[0.17 + 0.17\left(\frac{0.011}{0.011}\right) + 0.66\left(\frac{0.85}{0.84}\right)\right]^2}$$

$$= \frac{0.000318 + 0.375 + 0.1132(0.016)}{(1.008)^2} = \frac{0.377}{1.015} = 0.371$$

$$\gamma_1 = 2.35$$

$$\log \gamma_2 = \frac{x_3^2 A_{23}\left(\frac{A_{32}}{A_{23}}\right)^2 + x_1^2 A_{21}\left(\frac{A_{12}}{A_{21}}\right)^2}{\left(x_2 + x_3 \frac{A_{32}}{A_{23}} + x_1 \frac{A_{12}}{A_{21}}\right)^2} + \frac{x_3 x_1 \frac{A_{32}A_{12}}{A_{23}A_{21}}\left(A_{23} + A_{21} - A_{13}\frac{A_{21}}{A_{12}}\right)}{\left(x_2 + x_3 \frac{A_{32}}{A_{23}} + x_1 \frac{A_{12}}{A_{21}}\right)^2}$$

$$= \frac{(0.66)^2(0.845)\left(\frac{0.855}{0.845}\right)^2 + (0.17)^2(0.11)\left(\frac{0.011}{0.011}\right)^2}{\left[0.17 + 0.66\left(\frac{0.855}{0.845}\right) + 0.17\frac{0.011}{0.011}\right]^2}$$

$$+ \frac{(0.66)(0.17)\left(\frac{0.855}{0.845}\right)\left(\frac{0.011}{0.011}\right)\left[0.845 + 0.001 - 0.84\left(\frac{0.85}{0.84}\right)\right]}{\left[0.17 + 0.66\left(\frac{0.855}{0.845}\right) + 0.17\left(\frac{0.011}{0.011}\right)\right]^2}$$

$$= \frac{0.376 + 0.00289 + 0.1132(0.011)}{(1.008)^2} = \frac{0.380}{1.015} = 0.374$$

$$\gamma_2 = 2.37$$

Binary activity coefficients:

$$\log \gamma_1 = \frac{0.011}{\left[1 + \left(\frac{0.17}{0.17}\right)\left(\frac{0.011}{0.011}\right)\right]^2} = \frac{0.011}{(1+1)^2} = \frac{0.011}{4} = 0.00275$$

$$\gamma_1 = 1.006$$

$$\log \gamma_2 = \frac{0.011}{\left[1 + \left(\frac{0.17}{0.17}\right)\left(\frac{0.011}{0.011}\right)\right]^2} = 0.00275$$

$$\gamma_2 = 1.006$$

Selectivity:

$$\frac{\alpha_T}{\alpha_B} = \frac{(2.35)(111)/(2.37)(100)}{(1.006)(445)/(1.006)(372)} = \frac{1.10}{1.198} = 0.917$$

For the system containing butyl Cellosolve:

$$\log \gamma_1 = \frac{0.000318 + (0.66)^2(0.56)\left(\frac{0.74}{0.56}\right)^2 + (0.17)(0.66)\left(\frac{0.011}{0.011}\right)\left(\frac{0.74}{0.56}\right)\left[0.011 + 0.56 - 0.59\left(\frac{0.56}{0.74}\right)\right]}{\left[0.17 + 0.17\left(\frac{0.011}{0.011}\right) + 0.66\left(\frac{0.74}{0.56}\right)\right]^2}$$

$$= \frac{0.00318 + 0.425 + (0.148)(0.125)}{(1.21)^2} = \frac{0.4438}{1.468} = 0.302$$

$$\gamma_1 = 2.05$$

$$\log \gamma_2 = \frac{(0.66)^2 0.585\left(\frac{0.590}{0.585}\right)^2 + (0.11)^2 0.011\left(\frac{0.011}{0.011}\right)^2}{\left[0.017 + 0.66\left(\frac{0.59}{0.585}\right) + 0.17\left(\frac{0.011}{0.011}\right)\right]^2}$$

$$+ \frac{(0.66)(0.11)\left(\frac{0.59}{0.585}\right)\left(\frac{0.011}{0.011}\right)\left[0.585 + 0.011 - 0.56\left(\frac{0.011}{0.011}\right)\right]}{\left[0.017 + 0.66\left(\frac{0.59}{0.585}\right) + 0.17\left(\frac{0.011}{0.011}\right)\right]^2}$$

$$= \frac{0.2585 + 0.000318 + (0.1131)(0.036)}{(1.006)^2} = \frac{0.2629}{1.01} = 0.2627$$

$$\gamma_2 = 1.83$$

Selectivity:

$$\frac{\alpha_T}{\alpha_B} = \frac{(2.05)(111)/(1.83)(100)}{1.198} = \frac{1.25}{1.198} = 1.04$$

In summary:

	Added agent			
	2-Propanol		Butyl Cellosolve	
	γ_1, 2.35	γ_2, 2.37	γ_1, 2.05	γ_2, 1.83
Selectivity	0.917		1.04	
Experimental	1.03		1.16	

Example 10.4

Calculate the selectivity of phenol for ethylbenzene in the following mixture when 3 moles of solvent per mole of mixture are utilized under atmospheric pressure. Calculate the selectivity as the ratio of relative volatility of ethylcyclohexane to ethylbenzene in the solvent-containing mixture to that in the solvent-free mixture. Use pure-component data to predict the binary constants and the Bonham-type van Laar equations for the activity coefficients.

Component	Mole fraction
(1) *n*-Octane	0.20
(2) Ethylcyclohexane	0.40
(3) Ethylbenzene	0.40

Solution

$$A_{ij} = \frac{T_{c_i}}{T_{c_j}} \frac{p_{c_j}}{p_{c_i}}$$

$$B_{ij} = 4.476 T_{c_i} \left[1 - \left(\frac{p_{c_j}}{p_{c_i}} \right)^{0.5} \right]^2$$

Data	*n*-Octane	Ethylcyclohexane	Ethylbenzene	Phenol
Component	1	2	3	4
Ternary, x	0.20	0.40	0.40	—
With solvent, x	0.05	0.10	0.10	0.75
n, BP, °C	125.7	131.8	136.2	181.8
T_c, °K	569.4	594.3	619.7	692
p_c, atm	24.64	27.45	37.0	60.5
P, mm Hg, at 133°C	—	785	694	—
P, mm Hg, at 169°C	—	1,900	1,700	—
a	37,300	36,400	29,500	22,500
b	238	223	173	118

$A_{13} = 1.38 \quad B_{21} = 74.5 \quad (B_{21})^{0.5} = 8.3 \quad B_{31} = 71.5 \quad (B_{31})^{0.5} = 8.2$

$A_{23} = 1.29 \quad B_{22} = 0 \quad B_{32} = 39.8 \quad (B_{32})^{0.5} = 6.35$

$A_{33} = 1.00 \quad B_{23} = 52.6 \quad (B_{23})^{0.5} = -7.3 \quad B_{33} = 0$

$A_{43} = 0.685 \quad B_{24} = 468 \quad (B_{24})^{0.5} = -21.7 \quad B_{34} = 161 \quad (B_{34})^{0.5} = -12.6$

$$T \ln \gamma_2 = \frac{[x_1A_{13}(B_{21})^{0.5} + x_2A_{23}(B_{22})^{0.5} + x_3A_{33}(B_{23})^{0.5}]^2}{(x_1A_{13} + x_2A_{23} + x_3A_{33})^2}$$

$$= \frac{[(0.2)(1.38)(8.3) + (0.4)(1.29)(0) + (0.4)(1.0)(-7.3)]^2}{[(0.2)(1.38) + (0.4)(1.29) + (0.4)(1.0)]^2}$$

$$= \frac{(2.3 - 2.97)^2}{(1.192)^2} = \frac{0.371}{1.42} = 0.262$$

$$\log \gamma_2 = \frac{0.262}{406(2.303)} = 0.00028 \qquad \gamma_2 \cong 1.00$$

$$T \ln \gamma_3 = \frac{[x_1A_{13}(B_{31})^{0.5} + x_2A_{23}(B_{32})^{0.5} + x_3A_{33}(B_{33})^{0.5}]^2}{(x_1A_{13} + x_2A_{23} + x_3A_{33})^2}$$

$$= \frac{[(0.2)(1.38)(8.2) + (0.4)(1.29)(6.35)]^2}{1.42}$$

$$= \frac{2.26 + 3.27}{1.42} = \frac{30.6}{1.42} = 21.6$$

$$\log \gamma_3 = \frac{21.6}{(406)(2.303)} = 0.023 \qquad \gamma_3 = 1.025$$

$$\alpha_{23} = \frac{1.00}{1.025}\,\frac{785}{694} = 1.10$$

With phenol added:

$$T \ln \gamma_2 = \frac{[x_1A_{13}(B_{21})^{0.5} + x_3A_{33}(B_{33})^{0.5} + x_4A_{43}(B_{24})^{0.5}]^2}{(x_1A_{13} + x_2A_{23} + x_3A_{33} + x_4A_{43})^2}$$

$$= \frac{[(0.05)(1.38)(8.3) + (0.10)(1.0)(-7.3) + (0.75)(0.685)(-21.7)]^2}{[(0.05)(1.38) + (0.10)(1.29) + (0.10)(1.0) + (0.75)(0.685)]^2}$$

$$= \frac{(0.575 - 0.73 - 11.15)^2}{(0.069 + 0.129 + 0.10 + 0.513)^2} = \frac{(-11.305)^2}{(0.811)^2} = \frac{128}{0.658} = 195$$

$$\log \gamma_2 = \frac{195}{(442)(2.303)} = 0.192 \qquad \gamma_2 = 1.56$$

$$T \ln \gamma_3 = \frac{[x_1A_{13}(B_{31})^{0.5} + x_2A_{23}(B_{32})^{0.5} + x_4A_{43}(B_{34})^{0.5}]^2}{0.658}$$

$$= \frac{[(0.05)(1.38)(8.2) + (0.10)(1.29)(6.35) + (0.75)(0.685)(-12.6)]^2}{0.658}$$

$$\log \gamma_3 = \frac{47}{(442)(2.303)} = 0.046 \qquad \gamma_3 = 1.11$$

$$\alpha_{23} = \frac{1.56}{1.11}\,\frac{1900}{1700} = 1.56$$

Selectivity:

$$\frac{(\gamma_{23})_T}{(\gamma_{23})_B} = \frac{1.56}{1.10} = 1.42$$

Maximum

Qozati and Van Winkle [23], (Fig. 10.7) and Chears and Makin [7] (Fig. 10.8) observed in some cases that additions of a selective component produce a maximum in the relative volatility-solvent concentration curves

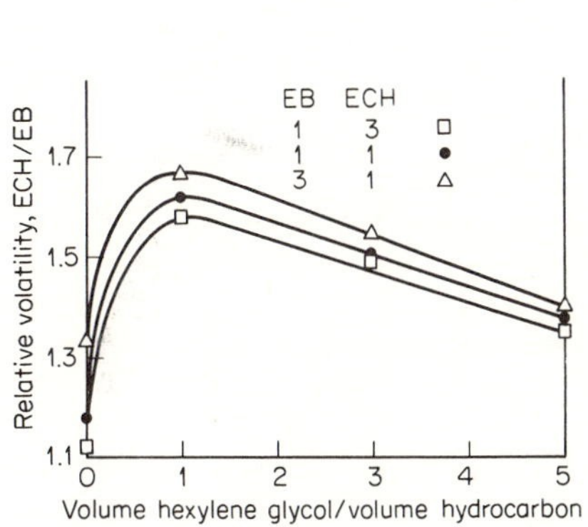

FIGURE 10.7 Maximum relative volatility system—ethylbenzene, ethylcyclohexane, and hexylene glycol. [*From Qozati and Van Winkle, J. Chem. Eng. Data, 5:269 (1960).*]

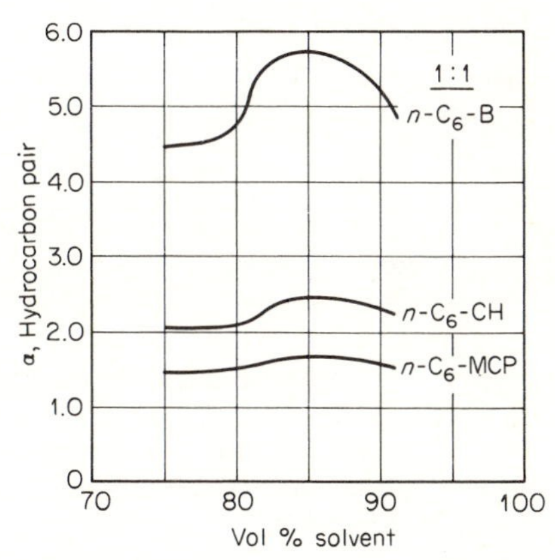

FIGURE 10.8 Maximum relative volatility system with hexane to cycloparaffin or hexane to aromatic; aniline solvent. [*From Chears and Makin, Am. Chem. Soc. Paper (December, 1960).*]

and further additions of solvent beyond this maximum cause the relative volatility to decrease. Either one or combinations of the temperature, dilution, and decomplexing effects may cause this characteristic behavior.

Increasing Rate of Change of Relative Volatility with Increase in Composition of Selective Component

In the mixtures of 2,4-dimethyl pentane–benzene–aniline, 2,4-dimethyl pentane–benzene–furfural, and 2,4-dimethyl pentane–benzene–hexylene glycol, Stephenson and Van Winkle [27] observed that the rate of increase of α with composition increases with increase in composition of the added component. This can be explained by the approach of the system to the point of immiscibility. As the concentration of the solvent is increased, the

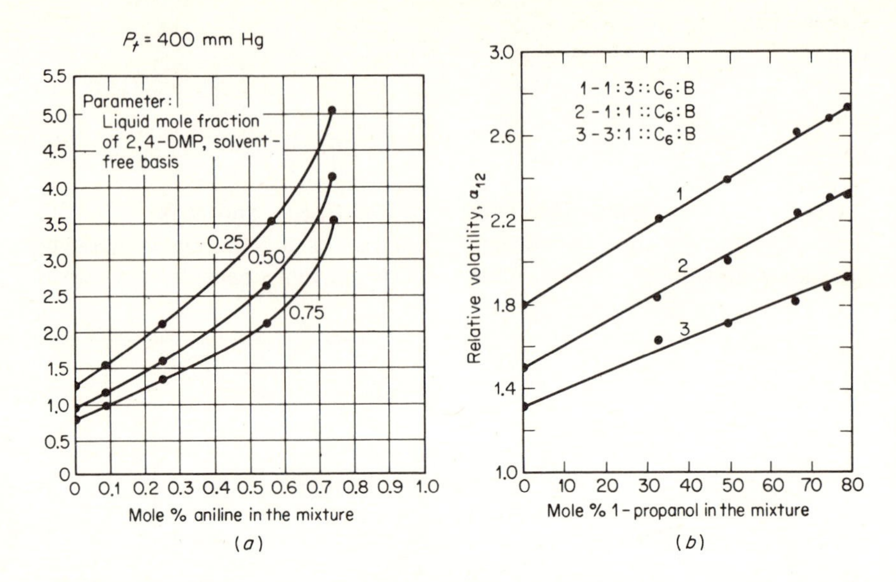

FIGURE 10.9 Relative volatility modification—increasing with added component. (a) Aniline is the solvent. [*From Stephenson and Van Winkle, J. Chem. Eng. Data, 7:510 (1962).* (*b*) Effect of 1-propanol on relative volatility of binary system hexane–benzene at 700 mm Hg. [*From Prabhu and Van Winkle, J. Chem. Eng. Data, 7:210 (1962).*]

2,4-dimethyl pentane tends to become insoluble, its activity coefficient increases, and the ratio of $\gamma_{2,4\text{-DMP}}/\gamma_B$ becomes increasingly larger. If this is indeed true, the composition of the vapor will change from $y_i = x_i\gamma_i P_i/P_t$ to $y_i = \gamma_i P_i/P_t$ at the point of immiscibility. This type of behavior is shown in Fig. 10.9.

Rate of Change of Alpha as a Function of Both Relative Composition of the Key Components and the Composition of the Solvent

A somewhat different alpha-concentration relationship has been observed [21] in which the slope and the change of slope of the curves relating α with solvent composition depend on the relative concentration of the key components. This variation is shown in Fig. 10.10, using data for the normal octane–ethylcyclohexane–butyl Cellosolve system.

This behavior may be explained by the added agent being selective for the component present in the larger amount, and nonselective for the other. As the agent's concentration is increased, the less soluble component becomes still less soluble and its activity coefficient increases, thereby increasing α. Where the less selective component is present in large amounts, the effect of the added agent is much less and the relative volatility increases little, if any.

Factors affecting selectivity The variables affecting selectivity of one compound for another are numerous and the quantitative extent, and in some instances even the qualitative extent, and direction of the effects are little understood. Experimental study of the effects of some of the variables has given some insight to the problem for some systems, but in the study of other systems the experimental results are not readily explained by accepted theory.

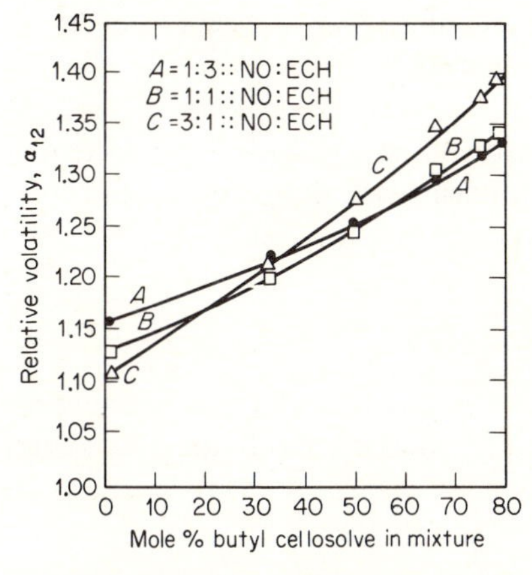

FIGURE 10.10 Effect of butyl Cellosolve on the relative volatility of binary system *n*-octane–ethylcyclohexane at 400 mm Hg absolute. [*From Prabhu and Van Winkle, J. Chem. Eng. Data*, **8**:14 (*1963*).]

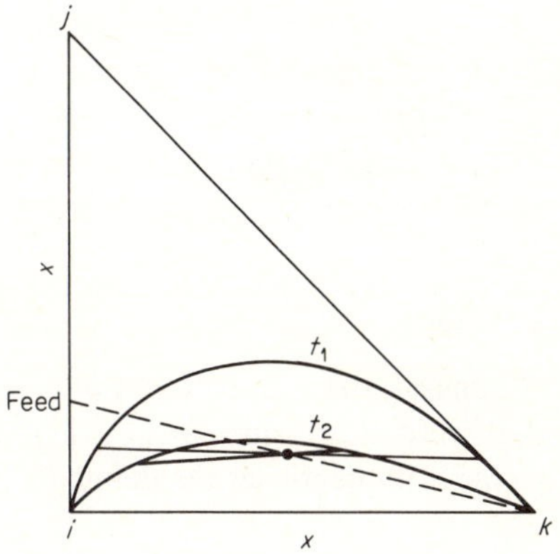

FIGURE 10.11 Selectivity of k for j as a function of temperature; $t_1 < t_2$.

Temperature

Temperature is believed to affect selectivity (Fig. 10.11) in that an increase in temperature tends to increase mutual solubility of compounds in a liquid mixture and thus decreases the selectivity of one component for another. This may be referred to as a *physical* effect as contrasted to a *chemical* effect. The decrease in selectivity may be illustrated by using a liquid-liquid system which may be considered analogous to a liquid-vapor system.

If selectivity S is defined by Eq. (10.2),

$$S = \frac{B'_{ij}}{B''_{ij}} = \frac{x'_i/x'_j}{x''_i/x''_j} \tag{10.2}$$

the selectivities at the two temperatures for the system shown graphically on

Fig. 10.11 are as follows:

$$S_{t1} = \frac{0.92/0.16}{0.02/0.14} = 40$$

$$S_{t2} = \frac{0.80/012}{0.33/0.15} = 3.1$$

The selectivity at the lower temperature t_1 is much greater than at t_2.

In addition to the physical effect of solubility, the chemical effect of complexing is generally considered to be affected by temperature. Prausnitz [22] and others observe that complex stability decreases with increase in temperature and, therefore, the selectivity attributed to complexing is decreased by increase in temperature. This is consistent with the generalization that exothermic reactions are favored by lower temperature levels.

Pressure

In general, the specific effect of pressure on activity coefficients is negligible and, therefore, pressure can be said to have no effect on selectivity. However, if the activity coefficient is defined by

$$\gamma_i = \frac{\nu_i P_t y_i}{x_i P_i}$$

at high pressures, the fugacity coefficient is appreciably different from unity and there is a specific effect of pressure. Also, since increasing the system pressure increases the bubble-point temperature in a boiling system, the temperature increase will generally tend to decrease the selectivity, as previously observed.

Volume Fraction of Solvent

The quantity of solvent or selective agent relative to the quantity of original mixture (as volume fraction, mole fraction, or weight fraction) can exert a strong effect on the selectivity. For example, in an extractive distillation process, the solvent necessarily has a boiling point greater than that of the components originally present and, therefore, additions of solvent will increase the bubble-point temperature of the boiling system. On the other hand, the selective material added in an azeotropic distillation may have a boiling point lower than that of the system, or nearly the same. In this latter instance the temperature effect on selectivity would be favorable or exert no appreciable influence, whereas the higher boiling solvent would probably decrease the selectivity because of the higher temperature distillation.

Also, in both extractive and azeotropic distillation, using higher- and lower-boiling-point selective agents, respectively, it is possible for the dilution effect of further additions of solvent to break complexes formed in the more concentrated solutions, to reduce the absolute values of $(y_i/x_i)/(y_j/x_j)$ to insignificance, and to reduce the solubility of the less soluble component to the point of immiscibility.

Relative Size of Molecule

Prausnitz [1] pointed out that the logarithm of the activity coefficient for individual paraffin hydrocarbons mixed with a polar solvent increases approximately linearly with the number of carbon atoms in the paraffin molecule where there is no hydrogen bonding or chemical effect. Thus, the logarithm of the selectivity is proportional to the difference in size of the paraffin molecules. In addition, the larger molecule will have the greater activity coefficient of that of two differently sized paraffin molecules in the same solvent.

Chemical Effect of H *Bonding*

Many peculiar properties of certain compounds, NH_3, H_2O, phenols, alcohols, etc., have been observed experimentally which indicate that the compounds consist of something other than that shown by the formulas. The generally accepted explanation for this is *hydrogen bonding*. This is evidenced in the ability of the hydrogen atom to act as a link between the electronegative atoms. The hydrogen-bonding theory accounts for molecular association between like and also unlike molecules, usually designated as the *chemical effect* in nonideal behavior of liquids.

According to the Pauli principle [19] the hydrogen atom can be associated with no more than two electrons. Therefore it is impossible for both atoms which are linked in this manner to be attached to the hydrogen by an ordinary covalence. The strength of the hydrogen bond increases with the electronegativity of the bonds linked by the hydrogen. The atoms may be the same or different, but are limited to oxygen, fluorine, and nitrogen. Carbon and oxygen bonds and carbon and nitrogen bonds are formed sometimes, but only when the carbon is attached to a strong electronegative group or atom, e.g., chlorine.

H-bond energies vary from 2 to 8 kcal/mole compared to regular bond strengths of 87 kcal for C—H bonds and 84 for N—H bonds. This accounts for easy breaking of H bonds (e.g., as temperature is raised).

Some complexes are postulated to be in chain form, and in the case of carboxylic acids and N-substituted amides, definite dimers are claimed to be formed. Also both intramolecular and intermolecular hydrogen-bond formations are known.

Ewell et al. [9] made a study of the relationship between hydrogen bonding and azeotrope formation for the purpose of selecting azeotrope-forming combinations to separate close boiling compounds and minimum boiling azeotropes. The following possibilities for the separations were classified.

Close Boiling Compounds

1. The entrainer forms a binary minimum boiling azeotrope with one compound.

2. The entrainer forms binary minimum boiling azeotropes with both compounds, but one boils at a much lower temperature than the other.

3. The entrainer forms a ternary minimum boiling azeotrope whose boiling temperature is appreciably lower than that of any binary azeotrope formed. Also, the ratio of the compositions of the original compounds in the ternary minimum azeotrope must be different from the original composition ratio.

Minimum Boiling Azeotropes

1. The entrainer forms a binary minimum azeotrope boiling at a lower temperature than the original.

2. The entrainer forms a ternary minimum azeotrope boiling at a lower temperature than the binary azeotrope temperature, and in which ratio of components is different than in the original azeotrope.

Ewell et al. [9] concluded that hydrogen can coordinate between two molecules of O_2, N_2, F, and can coordinate between O_2, N_2, and F and C if a number of negative atoms are attached to the carbon atom. They suggested the following classifications of hydrogen bonds as "strong" or "weak," and classify all liquid materials into five classes.

Strong	Weak
O—HO	N—HN
N—HO	O, N { HCCl
	O, N { $HCCl_2$—CCl
O—HN	O, N { HCNO
	O, N { $HCCN_2$

Classification of Liquids

Class I. Liquids capable of forming three-dimensional networks of H bonds: water, glycol, glycerol, amino acids, hydroxylamine, hydroxy acids, polyphenols, amides, etc. These are "abnormal" or "associated" liquids having high dielectric constants and are water soluble.

Class II. Other liquids containing active hydrogen atoms and other donor atoms: acids, phenols, alcohols, primary and secondary amines, oximes, nitro compounds and nitriles with H atoms, NH_3, hydrazine, HF, HCN, etc. The characteristics of class II liquids are the same as class I.

Class III. Liquids composed of molecules containing donor (O, N, F) atoms but no active H atoms: ethers, ketones, aldehydes, esters, tertiary amines. These liquids are water soluble also.

Class IV. Liquids composed of molecules containing active H atoms but no donor atoms: $CHCl_3$, CH_2Cl_2, CH_3CHCl_2, $CH_2Cl—CHCl_2$, etc., which show only slight water solubility.

Class V. All other liquids, i.e., compounds having no hydrogen-bond-forming capabilities: hydrocarbons, CS_2, RSH, nonmetallic elements, etc. This class of materials shows essentially no solubility in water.

TABLE 10.2 **Deviations of Class Combinations from Raoult's Law**

Classes	Deviations	H bonding
I + V II + V	Always + I + V frequently showing limited solubility	H bonds broken only
III + IV	Always −	H bonds formed only
I + IV II + IV	Always +, I + IV limited solubility	H bonds both formed and broken (dissociation common)
I + I I + II I + III II + II II + III	Usually + Some − giving maximum boiling azeotropes	H bonds both broken and formed
III + III III + V IV + IV IV + V V + V	Quasi-ideal always + or ideal Minimum azeotropes Minimum azeotropes (if any)	No H bonds involved

The same authors [9] classified maximum boiling azeotropes as follows:

1. Water + strong acids: HCl, HNO_3, HBr, etc.

2. Water + associated liquids: formic acid, hydrazine

3. Donor (III) liquids + nonassociated liquids having active H (IV) atoms: acetone + chloroform

4. Organic acids + amines

5. Phenols + amines

6. Organic acids + donor liquids containing O_2: formic acid + diethylketone

7. Phenols + donor liquids containing O_2: phenol + methyl hexylketone

8. Phenols + alcohols: phenol + *n*-octanol

Methods for prediction of selectivity The most dependable and accurate method of evaluating selectivity of added components or solvents is through use of complete or partially complete experimental data on the mixture. Since a complete experimental study is time consuming and expensive and since it is usually desirable to study a number of solvents, this approach, particularly for preliminary study, is not economically feasible. The next most dependable method involves the use of experimental binary data, with the key components and the component considered as the selective agent comprising three binary systems, and the prediction of ternary activity coefficients by some theoretical or empirical method, using the binary data. This is usually satisfactory if the required binary data are available. If they must be determined in the laboratory, the objections are the same as for the first method mentioned previously. Next in order of decreasing dependability is the prediction of relative volatility change by using van Laar, Margules, or other binary system constants determined experimentally from one or two points of data on each of the component binary systems. This is more economical and the method is commonly used, particularly for preliminary economic studies. The least dependable method but the most desirable from an engineering viewpoint is the evaluation of selectivity from individual pure-component data of which there is an abundance. Thus from the data on the pure components, if the mixture behavior could be predicted, the complete mathematical analysis of added component selectivity as a function of concentration, temperature, and pressure could be made. This would enable consideration and evaluation of a large number of selective agents with a considerable saving in time and money. Unfortunately, no accurate dependable method of general applicability has been developed. There are some, however, which apply to certain types of systems.

Mathematical Methods of Evaluating Selectivity

Quantitatively, selectivity is defined as the ratio of the relative volatility of the key components in the mixture which are to be separated in the presence of the separating agent to their relative volatility before the addition of the agent. A number of different expressions have been used to define selectivity (see Eqs. (10.3) through (10.5)).

$$S_{ij} = \frac{(\alpha_{ij})_T}{(\alpha_{ij})_B} = \frac{[(y_i/x_i)/(y_j/x_j)]_T}{[(y_i/x_i)/(y_j/x_j)]_B} = \frac{(\gamma_i P_i \nu_j/\gamma_j P_j \nu_i)_T}{(\gamma_i P_i \nu_j/\gamma_j P_j \nu_i)_B} \tag{10.3}$$

At normal pressures the fugacity coefficients are approximately unity and the selectivity can be expressed as

$$S_{ij} = \frac{(\alpha_{ij})_T}{(\alpha_{ij})_B} = \frac{(\gamma_i P_i/\gamma_j P_j)_T}{(\gamma_i P_i/\gamma_j P_j)_B} \tag{10.3a}$$

Where the relative volatilities of the key components i and j in the presence of the solvent and alone are being compared at constant temperature, the vapor pressure ratios are constant and, thus,

$$S_{ij}{}^t = \frac{(\gamma_i/\gamma_j)_T}{(\gamma_i/\gamma_j)_B} \tag{10.4}$$

Where it is assumed that the ratio of the terminal activity coefficients, $\gamma_i{}^\infty$ and $\gamma_j{}^\infty$ in the binary $i - S$ and $j - S$ systems represents a relative selectivity

$$S_{ij}^\infty = \frac{\gamma_i{}^\infty}{\gamma_j{}^\infty} \tag{10.5}$$

This follows directly because the value of γ_S is unity when the solute concentration is infinitely dilute.

The infinite-dilution selectivity, although having some value in the case of normal systems for screening solvents, is not always applicable because of the various types of behavior of the selectivity term as the solvent fraction is increased (see Figs. 10.6 through 10.10).

It is necessary to be able to evaluate the selectivity of a solvent at some practical value of solvent concentration to arrive at a meaningful comparison.

At low or moderate pressures where the fugacity coefficient is essentially unity, it is only necessary to evaluate the vapor pressures of the components and their activity coefficients with and without the added agent to evaluate S. To obtain the selectivity on a strictly comparable basis, it should be evaluated for the same relative liquid composition of the key components, and if the temperature is widely different, the activity coefficients and vapor pressures of the components should be corrected to the same basis.

In an azeotropic distillation where the entrainer is added to the feed, the boiling point of the entrainer is usually close to that of the components to be separated, and the correction for temperature is slight and may be neglected. However, where an extractive agent is used, as in extractive distillation, its boiling point is somewhat higher than those of the key components, boiling temperature of the mixture is higher, and the correction for temperature may be large and, therefore, must be evaluated. If the entrainer is added in the column at the bubble point of the entrainer-free mixture and if the temperature is far below the boiling point of the entrainer (i.e., its vapor pressure is low), the correction is small.

Robinson and Gilliland Method [24] This method was based upon the Bonham [4] equations (which have been shown to be the same as the Li and Coull equations [15]), relating the activity coefficients of the components in a mixture through the binary van Laar constants. Mathematical development of the selectivity relations is as follows:

$$T \ln \gamma_i = \frac{[x_j(B_{ij})^{0.5} + x_k A_{kj}(B_{ik})^{0.5}]^2}{(x_i A_{ij} + x_j + x_k A_{kj})^2} \tag{10.6a}$$

$$T \ln \gamma_j = \frac{[x_i A_{ij}(B_{ji})^{0.5} + x_k A_{kj}(B_{jk})^{0.5}]^2}{(x_i A_{ij} + x_j + x_k A_{kj})^2} \tag{10.6b}$$

Subtracting Eq. (10.6*b*) from (10.6*a*),

$$T \ln \frac{\gamma_i}{\gamma_j} = \frac{[x_j(B_{ij})^{0.5} + x_k A_{kj}(B_{ik})^{0.5}]^2 - [x_i A_{ij}(B_{ji})^{0.5} + x_k A_{kj}(B_{jk})^{0.5}]^2}{(x_i A_{ij} + x_j + x_k A_{kj})^2} \tag{10.7}$$

If i and j are similar in nature, B_{ij} is small and A_{ij} is approximately 1.0.

Consider only the numerator of the right-hand side of Eq. (10.7) and only the right-hand term and using the restriction in Eq. (10.8),

$$\left(\frac{B_{ij}}{A_{ij}}\right)^{0.5} + \left(\frac{B_{jk}}{A_{jj}}\right)^{0.5} + \left(\frac{B_{ki}}{A_{kj}}\right)^{0.5} = 0 \tag{10.8}$$

$$\begin{aligned} (B_{ji})^{0.5} &= -(B_{ij}A_{ji})^{0.5} \\ (B_{ki})^{0.5} &= -(B_{ik}A_{ki})^{0.5} \end{aligned} \tag{10.9}$$

$$\frac{A_{ki}}{A_{kj}} = A_{ji} \tag{10.10}$$

$$A_{ij} = \frac{1}{A_{ji}} \tag{10.11}$$

$$A_{jj} = 1.0 \tag{10.12}$$

$$-\left\{x_i A_{ij}(B_{ji})^{0.5} + x_k A_{kj}\left[-\left(\frac{B_{ij}}{A_{ij}}\right)^{0.5} - \left(\frac{B_{ki}}{A_{kj}}\right)^{0.5}\right]\right\}^2 \tag{10.13}$$

$$= -\left\{x_i A_{ij}(B_{ji})^{0.5} + x_k A_{kj}\left[-\left(\frac{B_{ij}}{A_{ij}}\right)^{0.5} - \left(\frac{B_{ik}A_{ki}}{A_{kj}}\right)^{0.5}\right]\right\}^2 \tag{10.14}$$

$$= -\left\{x_i A_{ij}(B_{ji})^{0.5} + x_k A_{kj}\left[-\left(\frac{B_{ij}}{A_{ij}}\right)^{0.5} - \left(\frac{B_{ik}}{A_{ij}}\right)^{0.5}\right]\right\}^2 \tag{10.15}$$

$$= -\left\{-x_i(A_{ij})^{0.5}(B_{ij})^{0.5} - x_k A_{kj}\left[\left(\frac{B_{ij}}{A_{ij}}\right)^{0.5} - \left(\frac{B_{ik}}{A_{ij}}\right)^{0.5}\right]\right\}^2 \tag{10.16}$$

Expanding the entire numerator of the right-hand side of Eq. (10.7),

$$x_j^2B_{ij} + 2x_kx_jA_{kj}(B_{ij})^{0.5}(B_{ik})^{0.5} + x_k^2A_{kj}^2B_{ik} - x_i^2A_{ij}B_{ij} + 2x_ix_kA_{kj}(A_{ij})^{0.5}(B_{ij})^{0.5}Z - x_k^2A_{kj}^2Z^2$$

$$Z^2 = \frac{B_{ij}}{A_{ij}} - 2\frac{(B_{ij})^{0.5}(B_{ik})^{0.5}}{A_{ij}} + \frac{B_{ik}}{A_{ij}} \tag{10.17}$$

$$\underline{x_j^2B_{ij}} + 2x_kx_jA_{kj}(B_{ij})^{0.5}(B_{ik})^{0.5} + \underline{x_k^2A_{kj}^2B_{ik}} - \underline{x_i^2A_{ij}B_{ij}} - 2x_ix_kA_{kj}(A_{ij})^{0.5}(B_{ij})^{0.5}\frac{(B_{ij})^{0.5}}{(A_{ij})^{0.5}} + 2x_ix_kA_{kj}(A_{ij})^{0.5}(B_{ij})^{0.5}\frac{(B_{ik})^{0.5}}{(A_{ij})^{0.5}} \underline{- x_k^2A_{kj}^2\frac{(B_{ij})^{0.5}}{A_{ij}}} + 2x_k^2A_{kj}^2\frac{(B_{ij})^{0.5}(B_{ik})^{0.5}}{A_{ij}} \underline{- x_k^2A_{kj}^2\frac{B_{ik}}{A_{ij}}}$$

All underlined terms in the numerator of the right-hand side containing B_{ij} contain second-order concentration terms (x_i^2, x_ix_j, etc.). Since B_{ij} is small (assumed) and the second-order concentration terms are less than unity, they are neglected.

In addition, since the A_{ij} terms are approximately unity (assumed), and B_{ij} and B_{ik} are similar because i and j are similar, the $x_k^2A_{kj}^2B_{ik}$ terms cancel.

The underlined terms in the above expression are eliminated, leaving

$$2x_kA_{kj}(B_{ij})^{0.5}(B_{ik})^{0.5}x_i + 2x_kA_{kj}(B_{ij})^{0.5}(B_{ik})^{0.5}x_j + 2x_kA_{kj}(B_{ij})^{0.5}(B_{ik})^{0.5}x_k\frac{A_{kj}}{A_{ij}}$$

Factoring,

$$2x_kA_{kj}(B_{ij})^{0.5}(B_{ik})^{0.5}\left(x_i + x_j + x_k\frac{A_{kj}}{A_{ij}}\right)$$

Equation (10.7) then becomes

$$T\ln\frac{\gamma_i}{\gamma_j} = \frac{2x_kA_{kj}(B_{ij})^{0.5}(B_{ik})^{0.5}\left(x_i + x_j + x_k\dfrac{A_{kj}}{A_{ij}}\right)}{(x_iA_{ij} + x_j + A_{kj}x_k)^2} \tag{10.18}$$

With

$$A_{ij} = 1.0 \qquad \text{and} \qquad A_{kj} = \frac{b_k}{b_j} = \frac{V_k}{V_j} \tag{10.19}$$

$$T\ln\frac{\gamma_i}{\gamma_j} = \frac{2x_kA_{kj}(B_{ij})^{0.5}(B_{ik})^{0.5}\left(x_i + x_j + x_k\dfrac{V_k}{V_j}\right)}{\left(x_i + x_j + x_k\dfrac{V_k}{V_j}\right)^2} \tag{10.20}$$

$$T\ln\frac{\gamma_i}{\gamma_j} = \frac{2x_kV_k(B_{ij})^{0.5}(B_{ik})^{0.5}}{x_i + x_j + x_k\dfrac{V_k}{V_j}} \tag{10.21}$$

$$T\ln\frac{\gamma_i}{\gamma_j} = \frac{2x_kV_k(B_{ij})^{0.5}(B_{ik})^{0.5}}{x_iV_j + x_jV_j + x_kV_k} \tag{10.22}$$

$$x_iV_j + x_jV_j + x_kV_k \cong V_{\text{mix}} \tag{10.23}$$

Thus,

$$T \ln \frac{\gamma_i}{\gamma_j} \simeq 2(B_{ij})^{0.5}(B_{ik})^{0.5} v_k \tag{10.24}$$

where v_k is the volume fraction of k in the mixture.

For the binary $i - j$,

$$\ln \gamma_i = \frac{\dfrac{b_i}{RT}\left[\dfrac{(a_i)^{0.5}}{b_i} - \dfrac{(a_j)^{0.5}}{b_j}\right]^2}{\left(1 + \dfrac{b_i}{b_j}\dfrac{x_i}{x_j}\right)^2} \tag{10.25}$$

Substituting

$$B_{ij} = \frac{b_i}{R}\left[\frac{(a_i)^{0.5}}{b_i} - \frac{(a_j)^{0.5}}{b_j}\right]^2$$

and (10.26)

$$A_{ij} = \frac{b_i}{b_j}$$

$$\ln \gamma_i = \frac{B_{ij}/T}{[1 + A_{ij}(x_i/x_j)]^2} \qquad \ln \gamma_j = \frac{A_{ij}B_{ij}/T}{[A_{ij} + (x_j/x_i)]^2} \tag{10.27}$$

Subtracting,

$$T \ln \frac{\gamma_i}{\gamma_j} = B_{ij}(v_j^2 - A_{ij}v_i^2) \tag{10.28}$$

Since i and j are similar in nature, $x_i \simeq v_j$.
Combining Eqs. (10.24) and (10.28),

$$T \ln \frac{(\gamma_i/\gamma_j)_T}{(\gamma_i/\gamma_j)_B} = B_{ij}\left[2\left(\frac{B_{ik}}{B_{ij}}\right)^{0.5} v_k - v_j^2 + A_{ij}v_i^2\right] \tag{10.29}$$

For any effective agent, $B_{ij}[2(B_{ik}/B_{ij})v_k]$ is larger than $B_{ij}(-v_j^2 + A_{ij}v_i^2)$. Therefore the latter term can be neglected for purposes of estimation and the approximate relation, Eq. (10.30), results.

$$T \ln \frac{(\gamma_i/\gamma_j)_T}{(\gamma_i/\gamma_j)_B} = 2(B_{ij})^{0.5}(B_{ik})^{0.5} v_k \tag{10.30}$$

The ratio $T \ln [(\gamma_i/\gamma_j)_T/(\gamma_i/\gamma_j)_B]$ varies as $(\alpha_{ij})_T/(\alpha_{ij})_B$ and, therefore, is an indication of selectivity. If the ratio varies appreciably from 1.0, either above or below, it indicates that the relative volatility can be modified by the added compound k, and separation is enhanced.

The effectiveness of the component k is increased by large values of $(B_{ik})^{0.5}$ and v_k [the derivation was based on the assumption that $(B_{ij})^{0.5}$ was small]. Under these conditions the type of system to which the equation is applicable is that shown in Fig. 10.6, where a large value of v_k gives a relatively large change in relative volatility, but where the effectiveness falls off at very high concentrations. For this type of system four to nine volumes of selective agent represent an economical limit. Nineteen volumes of k per volume of $i + j$ would be required for $v_k = 0.95$, while only nine would be needed for $v_k = 0.9$.

Effectiveness of the selective agent is indicated to be higher for higher values of $(B_{ik})^{0.5}$ and v_k. Because the terms $(B_{ij})^{0.5}$ and $(B_{ik})^{0.5}$ are square roots, the sign is important. The rule to follow is that when, e.g., component k is *more* polar than component i, the sign of the term $(B_{ik})^{0.5}$ is *negative*, and when k is *less* polar than i, the sign of $(B_{ik})^{0.5}$ is *positive*.

For a multicomponent system using a generalized form of Eq. (10.31), the following expansion results. For six components, $n = 6$, $LK = 3$, $j = HK = 4$,

$$T \ln \gamma_k = \frac{\left[\sum_{i=1}^{i=n}(x_i A_{ij} B_{ki}^{0.5})\right]^2}{\left[\sum_{i=1}^{i=n}(x_i A_{ij})\right]^2} \tag{10.31}$$

$$T \ln \gamma_3 = \frac{[x_1A_{14}(B_{31})^{0.5} + x_2A_{24}(B_{32})^{0.5} + x_3A_{34}(B_{33})^{0.5} + x_4A_{44}(B_{34})^{0.5} + x_5A_{54}(B_{35})^{0.5} + x_6A_{64}(B_{36})^{0.5}]^2}{(x_1A_{14} + x_2A_{24} + x_3A_{34} + x_4A_{44} + x_5A_{54} + x_6A_{64})^2} \tag{10.32}$$

$$T \ln \gamma_4 = \frac{[x_1A_{14}(B_{41})^{0.5} + x_2A_{24}(B_{42})^{0.5} + x_3A_{34}(B_{43})^{0.5} + x_4A_{44}(B_{44})^{0.5} + x_5A_{54}(B_{45})^{0.5} + x_6A_{64}(B_{46})^{0.5}]^2}{(x_1A_{14} + x_2A_{24} + x_3A_{34} + x_4A_{44} + x_5A_{54} + x_6A_{64})^2} \tag{10.33}$$

With the added selective component, $n = 7$, $LK = 3$, $j = HK = 4$. Add to numerator of Eq. (10.32) $x_7A_{74}(B_{37})^{0.5}$ and to denominator of Eq. (10.32) x_7A_{74}, add to numerator of Eq. (10.33) $x_7A_{74}(B_{47})^{0.5}$ and to the denominator of Eq. (10.33), x_7A_{74}, remembering to change the compositions of all components to the new basis.

To calculate the selectivity of the added component, calculate $T \ln \gamma_3$ and $T \ln \gamma_4$ from Eqs. (10.32) and (10.33) and determine the value of $T \ln (\gamma_3/\gamma_4)$. Then calculate $T \ln \gamma_3$ and $T \ln \gamma_4$ with the component 7 terms added and evaluate $T \ln (\gamma_3/\gamma_4)$. The selectivity is calculated by Eq. (10.5).

APPLICABILITY OF EQ. (10.30) The assumptions involved in the development of this equation were:

1. The van Laar assumptions.

2. The molecular character of components i and j are similar; primarily the molal volumes are the same or nearly so.

3. The second-order concentration terms were of such a small value to be neglected, mainly because x_i and x_j were small in the ternary mixture.

4. The ratio of the molal volumes of components k and j was assumed to be the same as the ratio of the *effective* molal volumes.

These assumptions limit the strict applicability of the equations to key components i, j, which are very similar with regard to their molecular volumes, whose concentrations in the ternary mixture are small compared to the concentration of the selective component k.

Separation Factors Method Black et al. [3] presented a method for predicting vapor-liquid equilibria based on certain empirical rules which were derived from observation of the behavior of compounds in nonideal mixtures using structural similarity as a parameter. The method requires some experimental data on some members of a series in the same solvent. Both screening and design are considered.

SCREENING The screening of solvent and solvent systems involves the semi-quantitative selection of solvents based upon determination of activity coefficients at infinite dilution. In these evaluations the nonideality of the vapor phase is neglected. Three experimental correlation parameters are used to construct volatility bands of the type shown in Fig. 10.12. Since in any system γ_i will vary from $\gamma_i = 1.0$ for the pure component to $\gamma_i = \gamma_i^{\infty}$ at infinite dilution of the component, the volatilities for a component in the system will fall between the limits, $P_i\gamma_i = P_i$, and $\gamma_i P_i = \gamma_i^{\infty} P_i$. Determining these limits for a range of temperatures establishes the volatility bands as shown in Fig. 10.12. Where the bands lap the separability is poor, and

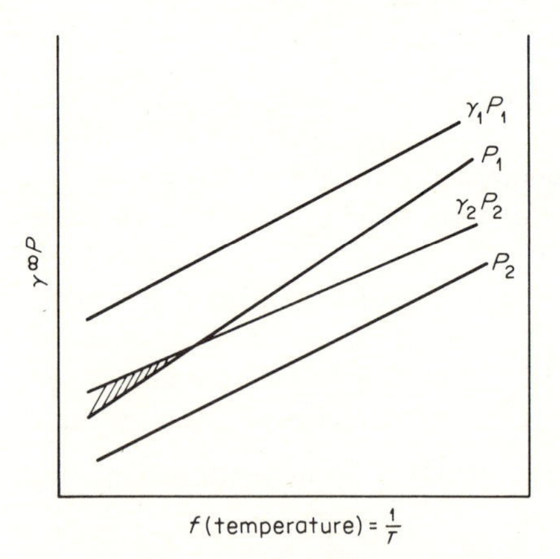

FIGURE 10.12 Volatility bands. [*From Black et al., Ind. Eng. Chem.*, *55*(*8*)*:40;* (*9*)*:38* (*1963*).]

the best separability is indicated by the greatest separation of the bands over the temperature range.

They recommend activity coefficients be determined from isothermal conditions by

$$\gamma_i = \frac{y_i P_t}{x_i P_i} \qquad \text{(vapor-liquid)}$$

$$\gamma_i = \frac{\gamma'_i x'_i}{x_i} \qquad \text{(liquid-liquid)}$$

$$\gamma_i = \frac{x_{L,\text{ideal}}}{x_i} \qquad \text{(solid-liquid)}$$

The infinite-dilution activity coefficients are calculated by the unmodified van Laar equation,

$$\log \gamma_{ij}^{\infty} = \log \gamma_{ij}\left[1 + \frac{\log \gamma_{ij}^{\infty}}{\log \gamma_{ij}^{\infty}}\left(\frac{x_i}{x_j}\right)\right]^2 \tag{10.34}$$

The ratio $\log \gamma_{ij}^{\infty}/\log \gamma_{ji}^{\infty}$ can be considered equal to unity unless sufficient data are known to provide more accurate evaluation. For nonisothermal data extrapolation over relatively short temperature intervals to the temperature desired may be done by assuming $\log \gamma_i^{\infty} = C(1/T)$.

STRUCTURAL CORRELATIONS Black et al. [3] established empirical correlations based on the observed systematic variation of infinite-dilution activity coefficients with molecular weight, molecular configuration, and functional group type. They show that, for a given series of solvents, e.g., paraffin hydrocarbons in a given series of solvents such as, e.g., aliphatic alcohols, when the data are plotted as in Fig. 10.13, the family of lines represent a

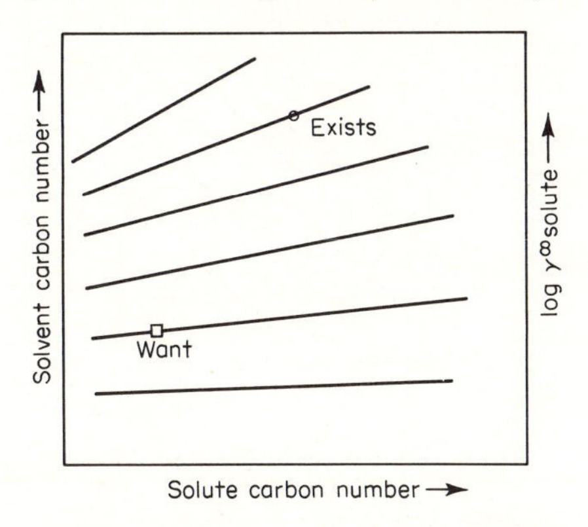

FIGURE 10.13 Correlation grid. [*From Black et al., Ind. Eng. Chem., 55(8):40; (9):38 (1963).*]

system with each line representing the activity-coefficient variation of paraffins in a given alcohol. The lines are said to be essentially straight and only two points of data are needed to establish the line. The slope of each line, B, is considered to be the property of the solvent. Thus, other solute series in aliphatic alcohols would have the same slopes as the paraffin solutes.

FIXED SOLVENT For fixed-solvent types,

$$\log \gamma_i^\infty = K + Bn_i + \frac{C}{n_i} \tag{10.35}$$

where K = a function of both solvent and solute series
B = a function of the solvent series alone
C = characteristic of the particular functional group alone

The authors regarded all hydrocarbon solutes as methylated series of differing characteristic groups, e.g., double bonds—olefins, saturated ring—cyclanes, unsaturated ring—aromatics. With Eq. (10.35) it is possible to evaluate the effectiveness of a solvent on a series of hydrocarbon solutes if activity coefficients for three solutes in a solvent are determined and if data for the series are available in a reference solvent.

$$\log \gamma_{i_r}{}^\infty - \log \gamma_{i_s}{}^\infty = K_{r_s} + B_{r_s} n_i + z_i \tag{10.36}$$

where r = reference solvent
s = solvent to be evaluated
K_{r_s} and B_{r_s} = solvent parameters
z_i = hydrogen deficiency of the hydrocarbon solute

Data calculated for a solvent and a series of solutes are plotted as shown in Fig. 10.14.

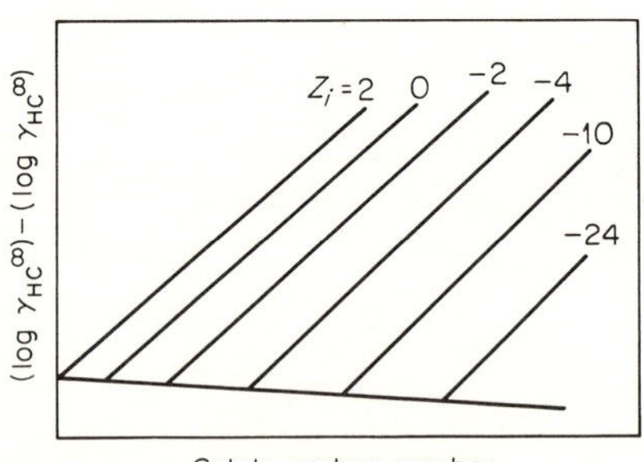

FIGURE 10.14 Estimation of infinite-dilution activity coefficients for a series of solutes in a solvent from data on another solvent and the same solutes. [*From Black et al., Ind. Eng. Chem.*, *55*(*8*)*:40;* (*9*)*:38* (*1963*).]

VARYING SOLVENT Where the solvent type is varied homologously, Black et al. [3] proposed

$$\log \gamma_i^{\infty} = K + \frac{B}{n_k} n_i + \frac{F}{n_k} \tag{10.37}$$

The change in γ_i^{∞} resulting in the change in carbon number of the solvent from n_k to n'_k can be shown by

$$\log \frac{\gamma_i^{\infty}}{\gamma_i^{\infty}} = n_i B\left(\frac{1}{n_k} - \frac{1}{n'_k}\right) + F\left(\frac{1}{n_k} - \frac{1}{n'_k}\right) \tag{10.38}$$

B is the same as in Eq. (10.35) and F is characteristic of the polar group of the solvent series. It may be estimated from data on members of the same solvent series (F can be ignored at moderate solvent molecular weights). F can be determined from water-solvent data (since water has no hydrocarbon grouping, $B = 0$). The originators of this method also presented a general equation for the case of varying solute type, varying solvent type which involves the evaluation of more constants, and they indicate that this equation might be less precise than the simpler forms [Eqs. (10.34) to (10.36)]. The values of the constants for many systems may be found in [20].

After the solvents have been screened and the selection has been made for the one or ones considered to be the more suitable, experimental study may be conducted to determine the vapor-liquid equilibria for the solvent and system, and only enough data need be determined to enable evaluation of the necessary constants in the correlating equations to establish the desired vapor-liquid equilibrium data. On such data the fractionator design can be based.

STEPWISE PROCEDURE FOR SCREENING

1. Select the key components to be separated and for which the solvents will be screened.

2. From available vapor-liquid equilibrium data on binaries composed of a solvent or solvents which are members of a homologous series and each of the keys or other compounds which are members of homologous series of each of the keys, determine the binary activity coefficient at some dilute concentration of each of the solutes and solvents using the defining equation.

3. Then determine the activity coefficient at infinite dilution for the solutes by Eq. (10.34). Extrapolate to the same temperature if necessary, assuming $\log \gamma = C(1/T)$.

4. Plot the $\log \gamma^{\infty}$ of the solute against the carbon number and associate the point with the parameter of carbon number of solvent. In order to establish the slope of the lines through the points, at least two points of data must be available for one solvent and solutes from a homologous series.

This enables the evaluation of the slope B which is a characteristic of the solvent. For paraffins, Black et al. state that $B \simeq \log \gamma_i^\infty/(n_i + 7)$.

4a. Alternatively, if enough binary data are available to enable evaluation of K (a function of the solvent and solute series), B (a function of solvent alone), and C (a function of solute series alone) in Eq. (10.35), the $\log \gamma^\infty$ values can be calculated for the whole series of solutes and solvents.

5. Calculate relative volatilities for the keys between the limits for two temperatures, as illustrated in Fig. 10.12. Where several sets of vapor-liquid equilibrium data are available involving a hydrocarbon solute and a number of solvents, Eq. (10.36) can be used to establish the values of the constants, and a plot of the type illustrated in Fig. 10.14 can be drawn.

Equation (10.37) can be used to evaluate constants from data wherein solvent and solute types are different, and Eq. (10.38) can be used to evaluate the constants when the solvent is varied homologously.

Solvent Effectiveness Predicted by the Wilson [31] *Equation* Flory [10] and Huggins [13] developed a theoretical equation based on the assumption that the excess enthaply of mixing, $\Delta H^E = 0$ (for athermal solutions)

$$\frac{G^E}{RT} = \sum^n x_i \ln \frac{\phi_i}{x_i} \tag{10.39}$$

where the volume fraction ϕ_i for component i is determined by

$$\phi_i = \frac{x_i V_{iL}}{\sum_1^n x_i V_{iL}} \tag{10.40}$$

where V_{iL} is the molar liquid volume of pure component i.

Wilson [31] considered molecules of different size and attractive forces in binary mixtures and extended the Flory-Huggins equation semiempirically to Eq. (10.41):

$$\frac{G^E}{RT} = -x_1 \ln (x_1 + \Lambda_{12} x_2) - x_2 \ln (\Lambda_{21} x_1 + x_2) \tag{10.41}$$

For any number of components Orye and Prausnitz [18] give the following generalized equation:

$$\frac{G^E}{RT} = -\sum_i^n x_i \ln \left(\sum_j^n x_j \Lambda_{ij} \right) \tag{10.42}$$

where

$$\Lambda_{ij} \simeq \frac{V_{jL}}{V_{iL}} \exp \left(- \frac{\lambda_{ij} - \lambda_{ii}}{RT} \right) \tag{10.43}$$

and

$$\Lambda_{ji} \cong \frac{V_{iL}}{V_{jL}} \exp\left(\frac{-\lambda_{ji} - \lambda_{jj}}{RT}\right) \tag{10.44}$$

The general equation for evaluation of activity coefficients is

$$\ln \gamma_k = -\ln\left(\sum_j^n x_j \Lambda_{kj}\right) + 1 - \sum_i^n \frac{x_i \Lambda_{ik}}{\sum_j^n x_j \Lambda_{ij}} \tag{10.45}$$

For a binary solution,

$$\ln \gamma_1 = -\ln (x_1 + \Lambda_{12} x_2) + x_2 \left(\frac{\Lambda_{12}}{x_1 + \Lambda_{12} x_2} - \frac{\Lambda_{21}}{\Lambda_{21} x_1 + x_2}\right) \tag{10.46}$$

$$\ln \gamma_2 = -\ln (\Lambda_{21} x_1 + x_2) + x_1 \left(\frac{\Lambda_{12}}{x_1 + \Lambda_{12} x_2} - \frac{\Lambda_{21}}{\Lambda_{21} x_1 + x_2}\right) \tag{10.47}$$

According to Orye and Prausnitz there are two principal advantages to the Wilson-type equation.

1. If it is assumed that $\lambda_{ij} - \lambda_{ii}$ and $\lambda_{ji} - \lambda_{jj}$ are independent of temperature, the heat of mixing can be calculated, e.g., for a binary,

$$\Delta h_m = x_1 \frac{x_2 \Lambda_{12}}{x_1 + \Lambda_{12} x_2} (\lambda_{12} - \lambda_{11}) + x_2 \frac{x_1 \Lambda_{21}}{x_2 + \Lambda_{21} x_1} (\lambda_{21} - \lambda_{22}) \tag{10.48}$$

2. Multicomponent solutions can be characterized by use of only binary data.

The equations may be used to calculate multicomponent vapor-liquid equilibrium data from binary data as follows:

1. From the x, y, t, P_t data on the binary systems comprising the multicomponent system and using Eqs. (10.46) and (10.47) evaluate the Λ_{12} and Λ_{21} terms or the Λ_{ij} and Λ_{ji} terms.
2. Using the values thus calculated, determine the activity coefficients for the components in the multicomponent mixture by using Eq. (10.45).

Orye and Prausnitz used the Wilson two-parameter equation to evaluate the parameters on 65 binary systems and found that the Wilson equation fits the data equally as well as the van Laar and Margules equations, and in many cases was better. They also compared the Wilson equation with others in fitting ternary data and found that it was better in this respect than conventional methods.

The disadvantages of this equation are that it does not describe adequately partially immiscible systems and it does not predict maxima or minima in activity coefficient.

Kyle and Leng [14] used the Scatchard-Hildebrand [12] regular solution equations to evaluate the vapor-liquid equilibria for 94 different ternary systems from the literature, which contained selective solvents to determine the selectivity. The Hildebrand solubility parameter for the components was calculated by Eq. (10.49).

$$\delta_i = \left(\frac{\Delta U_{\text{vap}}}{V_i}\right)^{0.5} \tag{10.49}$$

where ΔU_{vap} is the internal energy of vaporization, a measure of internal pressure, and V_i the liquid molar volume.

The solubility parameter results from dispersion forces. When a mixture of a nonpolar and a polar compound is formed, there is no hydrogen bonding, the induction forces are neglected (these are less than 6% of the total), and a polar contribution is added to the relation, resulting in

$$\frac{\Delta U_{\text{vap}}}{V_L} = \delta_i^2 + \omega_i^2 \tag{10.49a}$$

where δ_i^2 represents the dispersion energy contribution and ω_i^2 the polar energy contribution. The volume fraction was calculated by

$$\phi_i = \frac{x_i V_i}{\sum_1^n x_i V_i} \tag{10.50}$$

and these values were used in the Scatchard-Hildebrand equation,

$$RT \ln \frac{\gamma_1}{\gamma_2} = [V_1(\delta_1 - \delta_3)^2 - V_2(\delta_2 - \delta_3)^2]\phi_3^2 \tag{10.51}$$

where component 3 is the solvent. This equation is based on the assumption that the excess entropy of mixing is 0 and that the forces between molecules are dispersion forces. Where nonpolar solvents and systems were encountered, the regular solution equations predicted the selectivity well. With the polar-polar systems the agreement was not too good.

Weimer and Prausnitz Method Weimer and Prausnitz [30] developed a method for screening solvents for selectivity in the separation of hydrocarbons, which is based on pure-component properties. The method is strictly applicable to polar solvents which are nonassociating and does not apply to strongly associated solvents such as alcohols or organic acids. The selectivity criteria are based on the ratios of the activity coefficients at infinite dilution, and thus the correlation should apply to mixtures wherein the polar solvent is present in excess or in higher concentrations.

The Scatchard-Hildebrand concept of polar and nonpolar solubility parameters forms the basis for the correlation. Equation (10.52) is a modification of the Scatchard-Hildebrand equation, including dipole-dipole and dipole-induced dipole interactions.

$$G^E = \phi_i\phi_j(x_iV_i + x_jV_j)[(\lambda_i - \lambda_j)^2 + \tau_i^2 - 2\psi_{ij}] + \left(x_i \ln \frac{\phi_i}{x_i} + x_j \ln \frac{\phi_j}{x_j}\right)RT \quad (10.52)$$

where ϕ_i, ϕ_j = volume fractions
V_i, V_j = molar liquid volumes
λ = nonpolar solubility parameter
τ = polar solubility parameter
ψ_{ij} = energy of interaction between the polar and nonpolar species

Using the "homomorph" concept that compounds having the same size, molar volume, and structure at the same reduced temperature are homomorphs, plots of molar volume versus the ratio $\Delta U/V_L$ were prepared for *n*-paraffin, cycloparaffin, and aromatic hydrocarbons (Figs. 10.15 to 10.17) with parameters of reduced temperature.

Tables of molar volume and nonpolar and polar solubility parameters for various temperatures are presented for a number of polar solvents. These are reproduced here as Tables 10.3 through 10.8. While the tabular data are

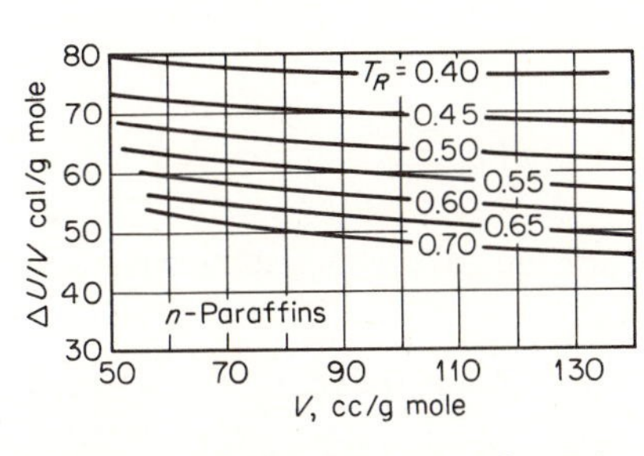

FIGURE 10.15 Homomorph relationship, *n*-paraffins. [*From Weimer and Prausnitz, Hydrocarbon Process. Petrol. Refiner (September, 1965), p. 237.*]

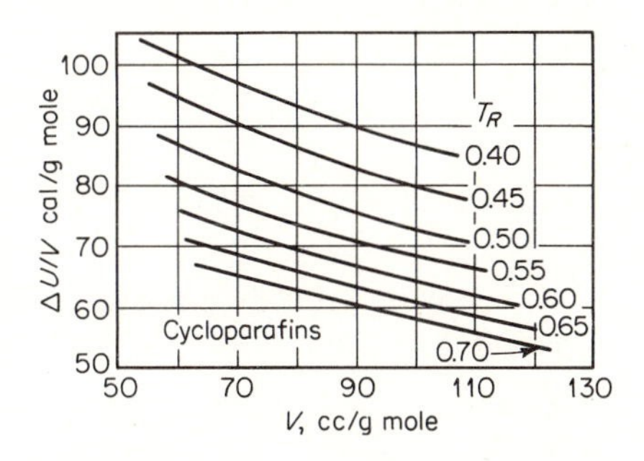

FIGURE 10.16 Homomorph relationship, cycloparaffins. [*From Weimer and Prausnitz, Hydrocarbon Process. Petrol. Refiner (September, 1965), p. 237.*]

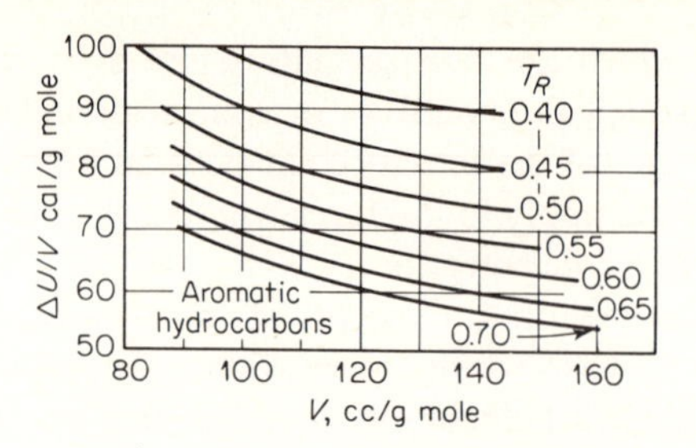

FIGURE 10.17 Homomorph relationship, aromatics. [*From Weimer and Prausnitz, Hydrocarbon Process. Petrol. Refiner (September, 1965), p. 237.*]

general in nature, the correlation presented is intended only for the case where solvent is present in excess.

According to the authors the values of ψ can be estimated for saturated paraffin hydrocarbons in nonassociating polar solvents to within $\pm 10\%$ by the empirical relation

$$\psi_{ij\mathrm{P}} = 0.396\tau_i^2 \tag{10.53}$$

For 1-pentane in various polar solvents the relation is

$$\psi_{ij\mathrm{O}} = 0.415\tau_i^2 \tag{10.54}$$

For benzene in polar solvents,

$$\psi_{ij\mathrm{B}} = 0.450\tau_i^2 \tag{10.55}$$

The values of the constants in Eqs. (10.53) through (10.55) were obtained from data where the hydrocarbon concentration was small. When the hydrocarbon concentration is larger and appreciable, the values of the constants would probably be different from those shown.

At infinite dilution, Eq. (10.52) for the activity coefficient for the nonpolar component becomes

$$RT \ln \gamma_{np}^{\infty} = V_{np}[(\lambda_p - \lambda_{np})^2 + \tau_p^2 - 2\psi_{p-np}] + RT\left(\ln \frac{V_{np}}{V_p} + 1 - \frac{V_{np}}{V_p}\right) \tag{10.56}$$

The stepwise procedure for screening solvents using the method proposed is as follows:

1. Obtain the molar volume V_L and the polar and nonpolar solubility parameters λ and τ from Tables 10.3 through 10.8 at the temperature desired (interpolate if necessary).

2. Substitute the values into Eq. (10.56) to calculate the activity coefficients at infinite dilution.

3. Calculate the selectivity by Eq. (10.5).

TABLE 10.3 **Volumes and Solubility Parameters of Hydrocarbons**

Hydrocarbon	v / cu cm/g mole	λ / (cal/cu cm)$^{1/2}$
$T = 0°C$:		
n-Pentane	111.83	7.45
n-Hexane	127.30	7.62
Cyclohexane	105.60	8.56
Methylcyclohexane	124.80	8.16
1-Pentene	106.06	7.52
$T = 25°C$:		
Propane	89.47	6.56
n-Butane	101.43	6.94
n-Pentane	116.16	7.17
n-Hexane	131.56	7.33
n-Heptane	147.44	7.47
n-Decane	195.92	7.85
n-Hexadecane	294.08	8.75
Cyclopentane	94.71	8.20
Methylcyclopentane	113.11	7.91
Cyclohexane	108.74	8.23
Methylcyclohexane	128.33	7.85
Ethylcyclohexane	143.13	8.01
1-Butene	95.28	7.08
1-Pentene	110.37	7.23
1,3-Butadiene	87.97	7.42
Benzene	89.40	9.20
Toluene	106.84	8.95
Ethylbenzene	123.06	9.01
p-Xylene	123.92	8.83
Mesitylene	139.58	8.88
$T = 45°C$:		
n-Pentane	120.25	6.94
n-Hexane	135.40	7.11
Cyclohexane	111.48	8.00
Methylcyclohexane	131.33	7.64
1-Pentene	113.11	7.04
Benzene	91.65	8.93
Toluene	109.18	8.71
$T = 60°C$:		
n-Hexane	138.67	6.95
n-Heptane	154.32	7.08
n-Decane	203.37	7.39
n-Hexadecane	303.57	8.02
Cyclohexane	113.64	7.80
Methylcyclohexane	133.69	7.48
Ethylcyclohexane	148.45	7.62
Benzene	93.44	8.74
Toluene	111.00	8.53

From Weimer and Prausnitz [30].

TABLE 10.3 (*cont.*)

Hydrocarbon	v, cu cm/g mole	λ, (cal/cu cm)$^{1/2}$
$T = 100°C$:		
n-Hexane	148.00	6.55
n-Heptane	163.99	6.67
n-Decane	212.96	6.95
n-Hexadecane	315.38	7.40
Cyclohexane	119.38	7.44
Methylcyclohexane	140.42	7.11
Benzene	98.80	8.27
Toluene	116.16	8.12

The homomorph plots may be used to evaluate the term $\Delta U/V_L$ at the reduced temperature of the system for the polar solvent. The term represents the sum of the dispersion energy and the polar energy densities, i.e.,

$$\left(\frac{\Delta U}{V_L}\right)_{\text{total}} = \left(\frac{\Delta U}{V_L}\right)_{\text{nonpolar}} + \left(\frac{\Delta U}{V_L}\right)_{\text{polar}} \equiv \lambda^2 + \tau^2 \tag{10.57}$$

The term $(\Delta U/V_L)_{\text{total}}$ can be evaluated from the molar volume V_L and the enthalpy of vaporization for the solvent at the temperature because

$$\left(\frac{\Delta U}{V_L}\right)_{\text{total}} = \frac{\Delta H_{\text{vap}} - RT}{V_L} = \text{total cohesive energy} \tag{10.58}$$

The value $(\Delta U/V_L)_{\text{nonpolar}}$ read from the homomorph plot for the homomorph of the solvent represents λ^2 ($\tau^2 = 0$). Then the value of τ^2 is calculated from

$$\left(\frac{\Delta U}{V_L}\right)_{\text{total}} - \lambda^2 = \tau^2 \tag{10.59}$$

$$\tau = \left(\frac{\Delta U}{V_L} - \lambda^2\right)^{0.5} \tag{10.60}$$

The cycloparaffinic homomorph plot is used for solvents having saturated ring structure and the aromatic homomorph plot for those having the aromatic ring structure.

TABLE 10.4 **Molar Volume and Nonpolar, Polar, and Solubility Parameters for Polar Solvents at 25°C**

Solvent	v cu cm/g mole	λ (cal/cu cm)$^{1/2}$	τ (cal/cu cm)$^{1/2}$
1. Acetophenone	117.4	9.44	3.69
2. Tetrahydrofuran	81.7	8.32	3.71
3. Pyridine	80.9	9.88	3.71
4. Cyclohexanone	104.2	8.84	4.04
5. Chloroethane	74.1	7.38	4.32
6. Diethyl ketone	106.4	7.75	4.44
7. Diethyl carbonate	121.9	7.89	4.49
8. Bromoethane	75.1	7.63	4.83
9. Nitrobenzene	102.7	9.70	4.89
10. di-(2-Chloroethyl)-ether	117.8	8.34	5.22
11. Trimethyl phosphate	116.2	8.46	5.22
12. Iodoethane	81.1	7.66	5.24
13. Methyl ethyl ketone	90.1	7.64	5.33
14. Cyclopentanone	89.5	8.70	5.37
15. 2,4-Pentanedione	103.0	8.06	5.69
16. 2,5-Hexanedione	117.7	8.45	5.88
17. Diethyl oxalate	136.2	8.37	5.94
18. 2-Nitropropane	90.7	7.95	6.02
19. Methoxyacetone	93.2	7.91	6.11
20. Acetone	74.0	7.66	6.14
21. Dimethyl carbonate	85.0	7.77	6.20
22. Butyronitrile	87.9	7.96	6.28
23. 2,3-Butanedione	87.8	7.73	6.35
24. Aniline	91.5	9.85	6.37
25. 1-Nitro propane	89.5	8.06	6.40
26. N-Methyl pyrrolidone	96.6	9.15	6.55
27. Acetic anhydride	95.0	7.85	7.11
28. Propionitrile	70.9	7.97	7.17
29. Citraconic anhydride	89.7	9.42	7.22
30. Methoxyacetonitrile	75.2	8.06	7.33
31. Furfural	83.2	9.04	7.62
32. Nitroethane	72.1	8.04	7.66
33. Dimethylacetamide	93.2	8.29	7.69
34. γ-Butyrolactone	77.1	9.50	8.01
35. Dimethylformamide	77.4	8.29	8.07
36. 3-Chloropropionitrile	77.7	8.44	8.73
37. Acetonitrile	52.6	8.03	8.98
38. Ethylenediamine	67.3	8.10	9.40
39. Nitromethane	54.3	8.08	9.44
40. Dimethyl sulfoxide	71.3	8.56	9.47

From Weimer and Prausnitz [30].

TABLE 10.5 **Molar Volume and Solubility Parameters for Polar Solvents at 0°C**

Solvent	Solvent	v cu cm/g mole	λ (cal/cu cm)$^{1/2}$	τ (cal/cu cm)$^{1/2}$
2.	Tetrahydrofuran	79.3	8.61	3.86
6.	Diethyl	103.3	8.04	4.54
10.	di-(2-Chloroethyl)-ether	109.7	8.68	5.72
13.	Methyl ethyl ketone	87.3	7.91	5.53
17.	Diethyl oxalate	132.7	8.72	6.02
20.	Acetone	72.3	7.91	6.32
22.	Butyronitrile	85.4	8.23	6.50
28.	Propionitrile	68.7	8.25	7.44
31.	Furfural	81.4	9.39	7.75
33.	Dimethylacetamide	90.7	8.60	7.90
35.	Dimethylformamide	75.5	8.56	8.32
37.	Acetonitrile	51.1	8.28	9.32

From Weimer and Prausnitz [30].

TABLE 10.6 **Molar Volume and Solubility Parameters for Polar Solvents at 45°C**

Solvent	v cu cm/g mole	λ (cal/cu cm)$^{1/2}$	τ (cal/cu cm)$^{1/2}$
2. Tetrahydrofuran	85.0	8.09	3.46
6. Diethyl ketone	108.9	7.55	4.34
8. Bromoethane	77.3	7.42	4.66
12. Iodoethane	83.0	7.44	5.13
13. Methyl ethyl ketone	92.6	7.43	5.19
17. Diethyl oxalate	139.3	8.10	5.91
19. Methoxyacetone	96.0	7.60	5.94
20. Acetone	76.2	7.45	5.95
21. Dimethyl carbonate	87.5	7.56	6.00
22. Butyronitrile	90.1	7.75	6.14
23. 2,3-Butanedione	90.4	7.52	6.17
28. Propionitrile	72.9	7.78	6.94
31. Furfural	84.8	8.81	7.40
32. Nitroethane	73.6	7.85	7.48
33. Dimethylacetamide	95.3	8.09	7.52
35. Dimethylformamide	79.0	8.07	7.91
37. Acetonitrile	54.4	7.85	8.65
39. Nitromethane	55.3	7.90	9.23

From Weimer and Prausnitz [30].

TABLE 10.7 **Molar Volume and Solubility Parameters for Polar Solvents at 60°C**

Solvent	v cu cm/g mole	λ (cal/cu cm)$^{1/2}$	τ (cal/cu cm)$^{1/2}$
3. Pyridine	83.9	9.33	3.81
6. Diethyl ketone	110.7	7.39	4.30
13. Methyl ethyl ketone	94.5	7.28	5.10
20. Acetone	78.1	7.31	5.80
24. Aniline	94.3	9.41	6.27
28. Propionitrile	74.4	7.62	6.81
31. Furfural	85.9	8.63	7.42
35. Dimethylformamide	80.2	7.94	7.77
37. Acetonitrile	55.5	7.72	8.46
38. Ethylenediamine	69.7	7.76	9.04
39. Nitromethane	56.3	7.77	9.05

From Weimer and Prausnitz [30].

TABLE 10.8 **Molar Volume and Solubility Parameters for Polar Solvents at 100°C**

Solvent	v cu cm/g mole	λ (cal/cu cm)$^{1/2}$	τ (cal/cu cm)$^{1/2}$
3. Pyridine	87.4	8.84	3.89
6. Diethyl ketone	116.1	7.02	4.17
13. Methyl ethyl ketone	99.9	6.90	4.87
20. Acetone	83.4	6.90	5.49
24. Aniline	97.9	8.94	6.20
28. Propionitrile	79.0	7.27	6.43
31. Furfural	89.4	8.28	7.14
35. Dimethylformamide	83.8	7.59	7.46
37. Acetonitrile	58.9	7.39	7.97
38. Ethylenediamine	72.9	7.43	8.67
39. Nitromethane	59.3	7.44	8.59

From Weimer and Prausnitz [30].

Nomenclature

A = constant in Wohl, van Laar, or Margules equations
B = constant in Wohl, van Laar, or Margules equations
C = ternary constant, Wohl equations
ΔE_{vap} = molal internal energy of vaporization
G = Gibbs' free energy
ΔG = change in free energy

ΔG^E = change in excess free energy
$\Delta \bar{G}^E$ = change in partial molal excess free energy
ΔH = change in enthalpy
N = number of components
P = vapor pressure
R = gas-law proportionality factor
T = temperature, absolute
ΔU = change in internal energy
V_L = molal volume, liquid
a, b = force constants
b_1, b_2 = van der Waals, volume constants
n = number of moles, number of components
q_1 = effective molal volume, component i
z_i = effective molal volume fraction, component i
v = volume fraction

Greek letters

α = relative volatility
Δ = difference
δ = partial sign, solubility parameter
ν = fugacity coefficient
γ = activity coefficient
Σ = summation
τ = solubility parameter, polar
ϕ = volume fraction
Λ = defined Eqs. (10.43) and (10.44)
λ = solubility parameter, nonpolar
ψ_{12} = energy of interaction
ω = polar solubility parameter

Subscripts

B = binary
HK = heavy key
LK = light key
Tern = ternary
i, j, k = component designation

REFERENCES

1. Anderson, R., R. Cambio, and J. Prausnitz, *A.I.Ch.E. J.*, **8**:66 (1962).
2. Benedict, M., C. A. Johnson, E. Solomon, and L. C. Rubin: *Trans. A.I.Ch.E.*, **41**:371 (1945).
3. Black, Cline, E. L. Derr, and M. N. Papadopoulos: *Ind. Eng. Chem.*, **55**(8):40–49 (1963); (9):38–47 (1963).
4. Bonham, M. S.: M.S. Thesis in Chemical Engineering, MIT, 1941.

5. Carlson, H. C., and A. P. Colburn: *Ind. Eng. Chem.*, **34**:581 (1942).
6. Chao, K. C., and O. A. Hougen: *Chem. Eng. Sci.*, **7**:246 (1958).
7. Chears, M. N., and E. C. Makin: *Am. Chem. Soc. Paper*, Regional Meeting, St. Louis (December 1–3, 1960).
8. Colburn, A. P., and E. M. Schoenborn: *Trans. A.I.Ch.E.*, **41**:42 (1945).
9. Ewell, R. H., J. M. Harrison, and Lloyd Berg: *Ind. Eng. Chem.*, **36**:871 (1944).
10. Flory, P. J.: *J. Chem. Phys.*, **10**:51 (1942).
11. Garner F. H., and S. R. M. Ellis: *Trans. Ind. Chem. Engrs.* (*London*), **29**:45 (1951).
12. Hildebrand, J. H., and R. L. Scott: "Solubility of Nonelectrolytes," 3d ed., *Am. Chem. Soc.*, Monograph Series, Reinhold, New York, 1955.
13. Huggins, M. L.: *Ann. N.Y. Acad. Sci.*, **43**:1 (1942).
14. Kyle, B. G., and D. E. Leng: *Ind. Eng. Chem.*, **57**:43 (1965).
15. Li, Y. M., and J. Coull: *J. Inst. Petrol.*, **34**:692 (1948).
16. London, F.: *Trans. Faraday Soc.*, **33**:8 (1937).
17. Margules, M.: *Sitzber. Akad. Wiss. Wien. Naturw. Kl. II*, **104**:1243 (1895).
18. Orye, R. V., and J. M. Prausnitz: *Ind. Eng. Chem.*, **57**:18 (1965).
19. Pauli, S.: *Z. Phys.*, **31**:765 (1925).
20. Pierrotti, G. J., C. A. Deal, and E. L. Derr: *Ind. Eng. Chem.*, **51**:95 (1959).
21. Prabhu, P. S., and M. Van Winkle: *J. Chem. Eng. Data*, **7**:210 (1962).
22. Prausnitz, J. M., and Ralph Anderson: *A.I.Ch.E. J.*, **7**:96 (1961).
23. Qozati, A., and M. Van Winkle: *J. Chem. Eng. Data*, **5**:269 (1960).
24. Robinson, C. S., and E. R. Gilliland: "Elements of Fractional Distillation," 4th ed., McGraw-Hill, New York, 1958.
25. Redlich, O., and A. T. Kister: *Ind. Eng. Chem.*, **30**:345 (1948).
26. Severns, W. H., Jr., Alexander Sesonske, R. H. Perry, and R. L. Pigford: *A.I.Ch.E. J.*, **1**:401 (1955).
27. Stephenson, R. L., and M. Van Winkle: *J. Chem. Eng. Data*, **7**:510 (1962).
28. Van Arkel, A. E.: *Trans. Faraday Soc.*, **428**:81 (1946).
29. Van Laar, J. J.: *Z. Phys. Chem.*, **72**:723 (1910); **185**:35 (1929).
30. Weimer, R. F., and J. M. Prausnitz: *Hydrocarbon Process. Petrol. Refiner* (September, 1966), p. 237.
31. Wilson, G. M.: *J. Am. Chem. Soc.*, **86**:127 (1964).
32. Wohl, Kurt: *Trans. A.I.Ch.E.*, **42**:215 (1946).
33. Wohl, Kurt: *Chem. Eng. Progr.*, **49**:219 (1953).

chapter

11

Azeotropic and extractive fractional distillation

Azeotropic distillation and extractive distillation are methods of separating hard to separate materials by fractional distillation with the addition of some component called an azeotrope former, entrainer, additive, or extractive solvent. The designations *entrainer* and *solvent* will be used for azeotropic and extractive solvents, respectively. The entrainer is used to modify the relative volatility of the key components to some value greater or less than unity which will enable them to be separated. The classes of compound mixtures for which separation by means of azeotropic distillation can be considered are close boiling compounds and azeotrope formers, where in each case the relative volatilities are at or near unity.

The principal difference between the processes of azeotropic distillation and extractive distillation is that the entrainer is almost entirely recovered in the distillate in azeotropic distillation, and in extractive distillation the solvent is recovered in the residue or bottoms. Also, the optimum point of addition of the entrainer and of the solvent to the column is different for the two types of processes.

11.1. MECHANISM OF RELATIVE VOLATILITY CHANGE

Although it is commonly accepted that the entrainer in azeotropic distillation forms a minimum boiling azeotrope with one of the key components and not with the other or it forms a ternary azeotrope having a widely different composition from that of an azeotrope between the keys, the assumption of a combination between the entrainer and one or more of the components resulting in a larger complex molecule does not satisfactorily explain the increased volatility of the composition under consideration. If hydrogen bonding is assumed to be the cause of complexing and if it is assumed that larger molecules have lower volatilities, then obviously the assumption of complexing to explain azeotropism (minimum boiling type) is erroneous. It would be better to assume that one or more of the original components in the mixture exists in a loosely bound, complex form, either with itself or with other types of molecules in the mixture. For many types of molecules the theory of hydrogen bonding can support this postulate, although, unfortunately, it does not apply to some systems which form azeotropes yet are deemed incapable of hydrogen bonding. On the basis of this assumption of molecular aggregation, when the entrainer is added, it is possible that it *breaks* bonds or destroys the original complexes of like or unlike molecules existing in the original solution. By doing this, the new molecular species is lower in molecular weight and, therefore, has a greater volatility, and the relative volatility of the key components changes and separation is enhanced. By the same reasoning, if an entrainer having a higher boiling point than the component with which it associates in the distillate is recovered in the distillate, it must have a higher volatility and, therefore, it must have existed in a more complex state as a pure liquid than it does in the mixture. Thus, when it was added to the mixture, the original larger molecular complexes must have been reduced in size so that the new molecular species' volatility is greater than that when the material existed alone.

Another possible explanation for the increase of volatility of a key component and the entrainer, when the entrainer is added to the mixture, is the tendency toward immiscibility of the entrainer and the associated key component with respect to the other key component. If immiscibility does occur or is approached, i.e., if two liquid phases are likely to be formed, one containing the entrainer and the light key and the other the heavy key, the entrainer would act as a stripping agent.

A maximum boiling azeotrope can be said to consist of a more or less loosely bound complex of a somewhat higher order of molecular aggregation than that of either (or any) of the pure components in the complex. Because of the higher order of aggregation, the boiling point is greater than that of either of the materials included in the combination. The existence of the maximum boiling point at a reproducible system composition merely indicates that the combination of physical and chemical attraction between the components is at a maximum at the particular composition at which the azeotrope occurs.

In an extractive distillation tower the solvent is introduced into the column near the top, associates with one or more of the components in the mixture (and certainly with one of the keys), and leaves the column in the residual material or bottoms. Although the solvent may have some dilution effect in breaking loosely bonded complexes, it must have bonding or other association characteristics so that it can decrease the volatility of one of the key components and make it more easily separable from the other key.

11.2. CHOICE OF ENTRAINERS OR SOLVENTS

An entrainer or solvent must have some or all of the following characteristics:

1. It must change the relative volatility of the key components in a mixture.

2. It must have volatility characteristics in the mixture so that it will distill into the distillate in the case of azeotropic entrainer. In the case of an extractive solvent its volatility will be very low compared to that of the mixture so that it will associate with the residue.

3. As an entrainer or as a solvent it must have a low molar latent heat since it is to be vaporized.

4. It has to be thermally stable.

5. It must be nonreactive with the other components in the mixture to which it is to be added.

6. It has to be available and inexpensive.

7. It must be noncorrosive.

8. It must be nontoxic.

9. It has to be easily separable from the components with which it associates.

10. It must be completely soluble with the components in the distilling system at the temperatures and concentrations in the column.

Compounds satisfying the specifications for entrainers for azeotropic distillation purposes are not as numerous as those which serve effectively as extractive distillation solvents.

The selection of the solvent can be based upon its ability to modify the relative volatility of the system by one of the methods shown in Chap. 10. However, its final selection must be determined through economic evaluation wherein all variables and criteria are considered in conjunction with selectivity to determine the minimum operating and investment costs for the process. Colburn and Schoenborn [6] and others have analyzed the selection of solvents on the cost basis.

It is necessary to emphasize that the whole process must be considered in the selection of an effective solvent, and the recovery process must be included in the evaluation. A solvent might have the best selectivity of all of those of an entire group considered, but it might be more difficult to separate from the components with which it associates than a solvent having a poorer selectivity, and economic considerations might prove the use of the latter to be more feasible.

11.3. SELECTION OF AN AZEOTROPIC OR EXTRACTIVE PROCESS

In many industrial mixtures the key components under consideration can be separated either by extractive or azeotropic distillation by the selection of the proper solvent or entrainer, and the economic comparison of the processes will usually indicate the one more suitable for the purpose. Another important consideration in the selection of a process is the thermal stability of the components, particularly that of the heavy key and heavier key. If one of these components is unstable or tends to decompose, polymerize, or otherwise react at a higher temperature level, it is better to use azeotropic distillation to keep the bubble-point temperature of the bottoms at its lowest point. A heavy solvent would increase the bubble point above this temperature and could have a deleterious effect on the components in residue.

11.4. DESIGN OF AN AZEOTROPIC DISTILLATION PROCESS

The factors to be considered in designing an azeotropic process are described in the following.

Selection of an Entrainer

This involves examination of vapor-liquid equilibrium data or selectivity data to determine which entrainers will enable the desired separation to be made. Complete evaluation of the process must be carried out before the final entrainer is selected.

Consideration of Entrainer Recovery from the Products

If an azeotrope is formed between the entrainer and one of the components to be separated, which is the usual case in azeotropic distillation, distillation will not serve as a satisfactory means of separating the entrainer from the distillate. Therefore, process methods for entrainer separation must be considered as well as the primary separation process.

Entrainer Concentration

Since the maximum effectiveness in the modification of the relative volatility of the components to be separated is generally obtained through the use of larger proportions of entrainer, it is considered better to maintain an appreciable concentration of entrainer in the liquid on all plates in the column. However, the advantage is somewhat lessened if the entrainer appears in the bottom product as well as in the distillate because an additional solvent recovery step is required.

Optimum Point of Entrainer Introduction

The point of addition of entrainer into the fractionator influences its concentration gradient throughout the tower. Generally, when the entrainer has essentially the same volatility as that of the feed, it should be added to or with the feed; when it is less volatile than the feed, it should be added above the feed near the top of the column; and when it is more volatile it should be added below the point of feed introduction. Benedict and Rubin [2] made a mathematical study of the optimum point of feed addition, using some simplifying assumptions, and concluded the addition of the entrainer to the upper section of the column is more desirable than with the feed.

A qualitative graphical examination using ternary diagrams is helpful in determining approximately the process limitations with regard to solvent-to-feed ratio related to the type of azeotropic system formed and the point of addition of the entrainer. In such a simplified study, the feed consisting of light and heavy key components could comprise a minimum boiling azeotrope or have essentially the same boiling point. The entrainer could form a binary azeotrope with one or both key components and possibly form a ternary azeotrope.

Figure 11.1 illustrates schematically the azeotropic distillation column where (*a*) the entrainer is added with the feed and (*b*) added above the feed.

The difference-point equations in terms of the over-all material balance are shown as follows. (Component material balance equations may also be written for each case as illustrated in Chaps. 6 and 7.)

Entrainer with feed	*Entrainer above feed*	
	$V_{S+1} - L_S = D = d$	(11.1*a*)
	$V_{S+1} - L_S = E + F - B$	(11.1*b*)
$V_{m+1} - L_m = D = d'$	$V_{m+1} - L_m = D - E = d'$	(11.2*a*)
$V_{m+1} - L_m = E + F - B$	$V_{m+1} - L_m = F - B$	(11.2*b*)
$\bar{V}_{p+1} - \bar{L}_p = D - (F + E) = d''$	$\bar{V}_{p+1} - \bar{L}_p = D - (F + E) = d''$	(11.3*a*)
$\bar{V}_{p+1} - \bar{L}_p = -B$	$\bar{V}_{p+1} - \bar{L}_p = -B$	(11.3*b*)

These equations indicate that the distillate location on the ternary diagram is one difference point and the bottoms location is another for both columns.

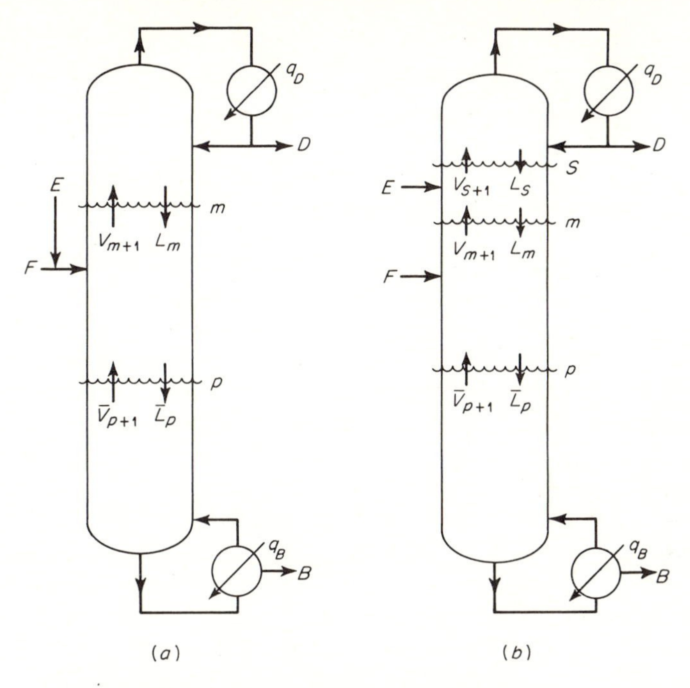

FIGURE 11.1 Location of entrainer introduction. (*a*) Entrainer with feed. (*b*) Entrainer above feed.

The difference point for the section between the point of entrainer addition and the feed is located with the use of Eqs. (11.4*a*) and (11.4*b*), which show that it lies on a line with the entrainer and top difference point

$$d' = d - E = D - E \tag{11.4a}$$

and also on a line with the bottom difference point and the feed

$$d' = F - d'' = F - (-B) \tag{11.4b}$$

Using these material balance relations and examining two types of azeotrope formation resulting from addition of an entrainer to a binary mixture, either with the feed or above its point of introduction to the column, certain qualitative pertinent conclusions can be drawn. Two types of systems will be considered for each type of entrainer addition: (1) components i and j are close boiling and the entrainer E forms an azeotrope with i; (2) i and j form a binary azeotrope and the entrainer forms binary azeotropes with both components and also a ternary azeotrope. Figure 11.2 illustrates the four cases.

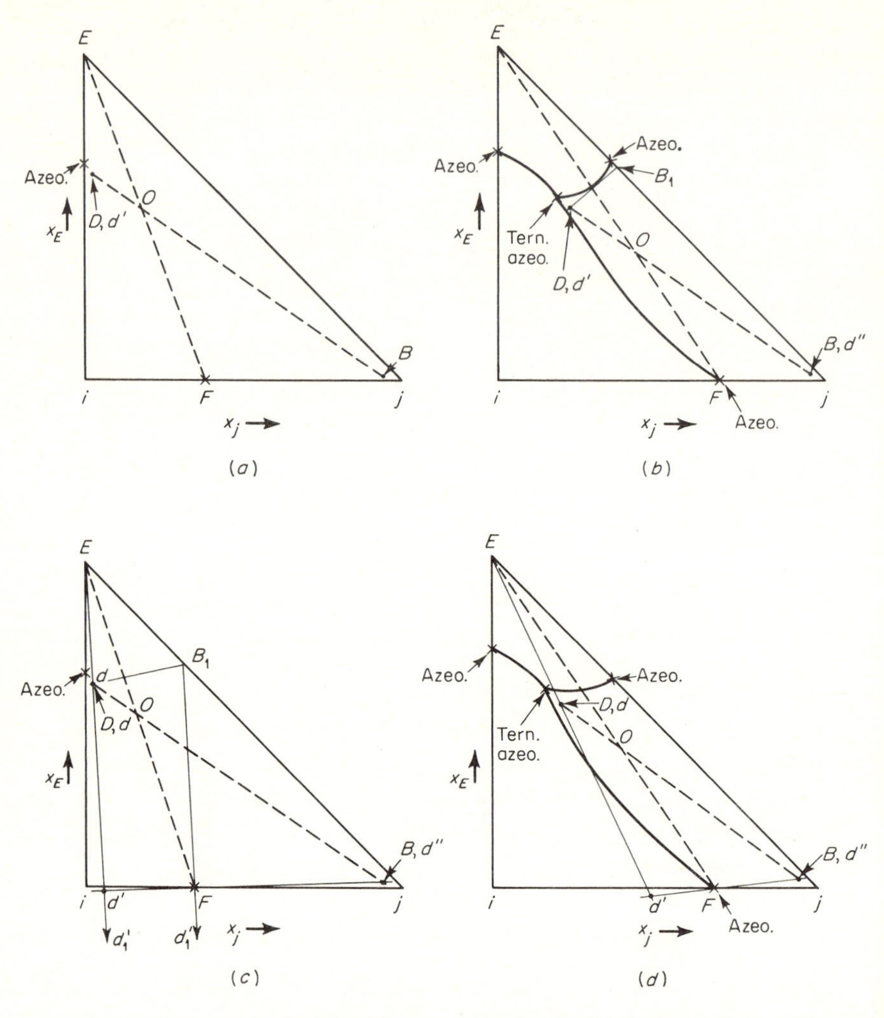

FIGURE 11.2 Azeotropic distillation cases. (*a*) and (*b*) Entrainer added to feed. (*c*) and (*d*) Entrainer added above feed. [*Note:* In (*b*) and (*d*) the solid, curved lines represent valleys connecting the azeotropic composition.]

Limitations: Entrainer-to-Feed Ratio

From the lever-arm principle the ratio of entrainer to feed is shown in all cases by the ratio of the distances along the line *EOF*.

$$\frac{E}{F} = \frac{\overline{OF}}{\overline{OE}} \tag{11.5}$$

For a fixed distillate and bottoms composition the entrainer-to-feed ratio is fixed. Because it is desirable to obtain the best separation between *i* and *j*, the bottoms composition should range from the *j* corner of the diagram to some value on the *jE* side of the diagram. It is desirable to eliminate the entrainer from the bottoms so that an additional solvent-recovery column will not be required. Thus the bottoms composition should be close to the *j* corner. Within these limitations, the entrainer-to-feed ratio will of necessity vary but slightly from that resulting from fixing the bottoms composition at the *j* corner.

In the case of the ternary azeotrope, the upper limitation on solvent-to-feed ratio would be that where the bottoms composition would lie at B_1 or some value where a straight line drawn between the distillate and bottoms compositions did not cross a valley with respect to temperature. However, high entrainer-to-*j* ratios would require an additional recovery step and, in the cases shown, the distillate from the entrainer recovery would approximate the composition of the *Ej* azeotrope. Unless some other method of separating the entrainer was available, this type of azeotropic distillation process would be unsatisfactory.

In the scheme of Fig. 11.2*c* the upper limit of entrainer-to-feed ratio is fixed economically by a bottoms composition at B_1. At this point the difference point d_1' lies at infinity, and approximately a minimum number of stages would be required in the section of the column between the feed and the point of entrainer addition. However, an additional entrainer recovery column to remove the entrainer from the bottoms would be required.

For almost all azeotropic distillations the ideal bottoms composition with respect to entrainer-to-feed ratio is at the entrainer-free composition. Unfortunately, as we shall see later, few commercial azeotropic distillations meet this requirement.

Azeotropic Fractional Distillation Column Design—Number of Equilibrium Stages

Because in the general case of azeotropic distillation the entrainer is required to be mutually soluble over the distillation temperature range with all components in the mixture including the lighter than light and heavier than heavy key components as well as the key components, and because the entrainer composition varies widely throughout the column, it is not generally feasible to estimate the number of theoretical stages by using a method based on the pseudobinary system of light key and heavy key component on an entrainer-free basis. If the system is composed essentially of the two keys and the entrainer, a graphical ternary method may be used. Usually, however, greater accuracy can be obtained from plate-to-plate methods, or empirical short-cut methods may be used to study a wide range of variables for preliminary design evaluation.

If a multicomponent system containing the two components to be separated can be prefractionated to produce a system composed essentially of the two key components which will be subjected to azeotropic distillation, the ternary method can be used. However, if a multicomponent system is to be subjected to azeotropic distillation with an added entrainer, it will be necessary to resort to the conventional multicomponent fractionation methods, either short-cut empirical methods or plate-to-plate methods.

Pseudoternary Method—Graphical

The general method for graphical evaluation of number of equilibrium stages used in azeotropic distillation is the same method as that described in Chap. 7 for the general ternary case.

Multicomponent Methods

Short-cut methods The short-cut methods described in Chaps. 5 and 6 are based on empirical correlations which relate minimum reflux, minimum stages at total reflux, and the number of equilibrium stages at selected operating reflux. Minimum reflux may be determined by the Colburn, Underwood, or Brown-Martin method; minimum stages may be determined by the Fenske or Underwood method, and either Gilliland's method or the Brown-Martin method can be used for the evaluation of the number of stages from some selected operating reflux ratio.

Plate-to-plate method If a greater accuracy is required and if the effort is justified, plate-to-plate calculations such as outlined in Chap. 6 are used. These require trial-and-error calculations to determine the temperatures and compositions from plate to plate throughout the column.

Processes

Essentially Ternary Azeotropic Distillation

Guinot and Clark [12] describe a process by which ethanol containing water is dehydrated by the addition of benzene as an entrainer. In this process the benzene forms a ternary azeotrope with ethanol and water which has a higher ratio of water to ethanol than the ethanol–water azeotrope. The schematic flow diagram for this process is shown in Fig. 11.3*a*.

In this system the azeotropic compositions and bubble-point temperatures are as follows:

System	Composition, wt %			Temperature, °C
	EtOH	H_2O	C_6H_6	
Ethanol–water	96.0	4.0	—	78.14
Ethanol–benzene	32.4	—	67.6	68.24
Benzene–water	—	8.8	91.2	99.25
Ethanol–benzene–water	18.5	7.4	74.1	64.86

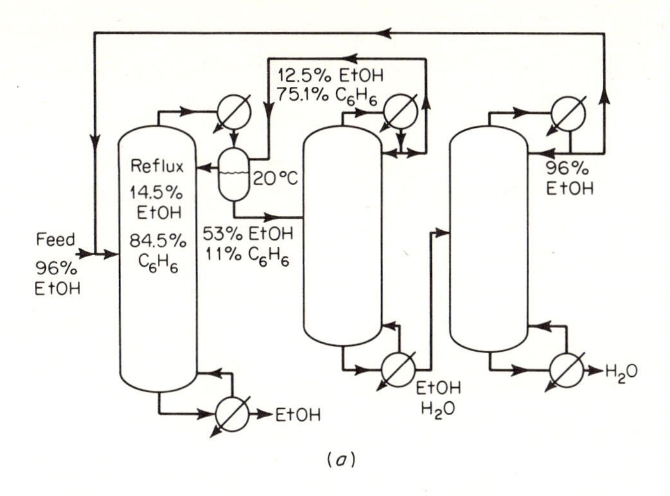

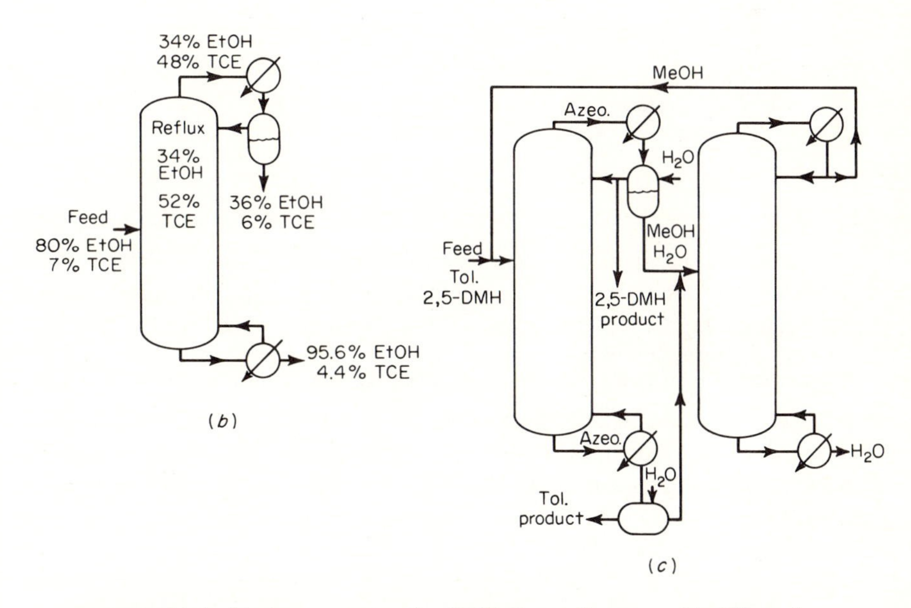

FIGURE 11.3 Various azeotropic distillation schemes. (*a*) Ethanol dehydration with benzene. (*b*) Ethanol dehydration with TCE. (*c*) Two binary azeotropes.

This azeotropic process utilizes the two-liquid-phase separation at 20°C in the decanter to concentrate the benzene entrainer in the reflux to the primary column. Here the bottoms product is essentially pure ethanol containing no entrainer.

Othmer [16] describes an azeotropic distillation separation of acetic acid and water using butyl acetate as an entrainer. Again, two-liquid-phase separation in a decanter is utilized to concentrate the entrainer in the reflux and a pure acetic acid is obtained as a bottoms product from the primary column.

Colburn and Phillips [5] describe a process for dehydration of ethanol utilizing trichloroethylene as an entrainer. Again, as in the above processes, a decanter is used to separate the two liquid phases formed at 20°C and to provide a concentrated trichloroethylene phase as reflux to the primary column. The ethanol product from this process contains some of the entrainer which must be removed by an additional fractionation. The primary column is shown in Fig. 11.3*b*.

The azeotrope compositions and bubble-point temperatures are as follows:

System	Composition, wt %			Temperature, °C
	EtOH	H_2O	C_2HCl_3	
EtOH–H_2O	96.0	4.0	—	78.7
EtOH–C_2HCl_3	27.5	—	72.5	70.9
C_2HCl_3–H_2O	—	5.4	94.6	73.6
Ternary	16.1	5.5	78.4	67.0
(Mole %)	(28.5)	(24.8)	(46.8)	

The water layer from the decanter can be distilled to produce an overhead distillate containing a lower percentage of water and a bottoms product containing a higher water content. These streams are fed into the separator or feed stream as recycle.

An azeotropic distillation described by Benedict and Rubin [2] involves the use of an entrainer which does not form a ternary azeotrope but which forms binary azeotropes with both components. The azeotrope compositions and temperatures are shown as follows:

System	Composition, wt %			Temperature, °C
	MeOH	2,5-Dimethyl hexane	Toluene	
2,5-DMH–toluene	—	65	35	107
Toluene–MeOH	69	—	31	63.8
2,5-DMH–MeOH	60	40	—	61.0

In this case methanol is added to the feed composed of toluene and 2,5-dimethyl hexane. The bottoms product is composed essentially of toluene and methanol, and the distillate is composed essentially of 2,5-dimethyl hexane and methanol. Figure 11.3*c* is the schematic flow diagram.

Multicomponent Azeotropic Distillation

Poffenberger et al. [18] describe a multicomponent azeotropic distillation which involves the addition of liquid ammonia to a C_4 hydrocarbon column to separate the butenes and butadienes. The operation is carried out at 230 psia. The schematic flow diagram is shown on Fig. 11.4.

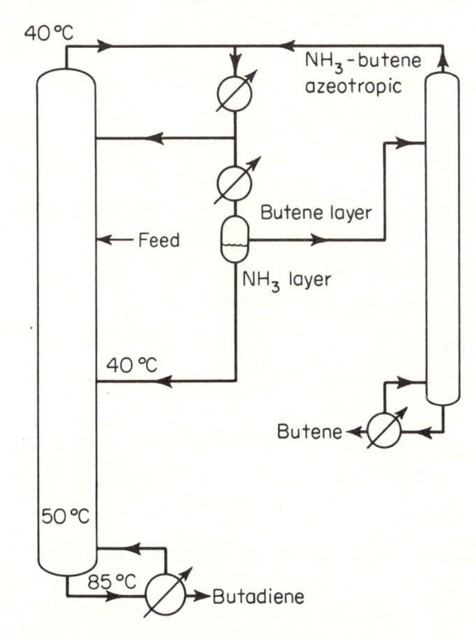

FIGURE 11.4 Azeotropic distillation of light hydrocarbons with ammonia. [*From Poffenberger et al., Trans. A.I.Ch.E.*, ***42****:815 (1946)*.]

Although this shows the two-liquid-phase separation by cooling, it is also possible to accomplish this by adding water to the butylene–ammonia azeotrope to extract the mixture with water and subsequently recover the NH_3 by distillation.

The following table from [18] lists the pertinent data concerning the compound ammonia azeotropes.

Apparently the separation of butylenes and butadiene is not one of temperature-difference effect on vapor pressure, since the column from the top plate to about four plates from the bottom was essentially isothermal at around 40°C. The effect probably was one of change in relative volatility through a large change in activity coefficients.

Hydrocarbon–Ammonia Azeotropes (P_t = 230 psia)

Compound	Boiling point, °C	NH_3 azeotrope, boiling point, °C	NH_3 in azeotrope, wt %
Propene	38	31	8
Propane	46	24	15
Methyl acetylene	61	40	85
Isobutane	90	39	40
Isobutene	93	39	60
1-Butene	94	39	65
1,3-Butadiene	98	39	75
n-Butane	101	39	60
Vinylacetylene	107	None	—
NH_3	41	—	—

11.5. DESIGN OF AN EXTRACTIVE DISTILLATION PROCESS

The factors which must be considered in the design of an extractive fractional distillation process are as follows:

Selection of a Suitable Solvent

Usually three or four of the most promising solvents examined are selected for complete evaluation with regard to the complete separation process.

Selection of the Method of Solvent Recovery for Recycle

Some solvents can be separated by simple fractionation. Others require extraction with some selective solvent such as, e.g., glycols extracted from hydrocarbons with water. Still others may be separated by cooling below the solubility limit so that two liquid phases are formed.

Selection of Conditions under Which the Process Components Are to Be Operated

The conditions for the extractive distillation column which must be selected are the pressure and temperature, solvent-to-feed ratio, and reflux ratio. The variables are not independent in that the higher the pressure the higher the bubble-point temperature for any composition. Also the greater the proportion of solvent the greater the temperature, and the greater the temperature the less selectivity, in general, of the solvent for one of the key components.

The pressure normally selected is that which will be most consistent with the requirements of the system to be distilled. Generally, atmospheric pressure (or slightly above) is used although, if the temperature level of the distillation, particularly in the reboiler, is such that polymerization or reaction might take place, the total pressure can be lowered so that the undesirable effects caused by the higher bubble-point temperature can be eliminated. If immiscibility and liquid-phase separation could take place because the temperature level was too low, the pressure can be increased to increase the temperature level to a point where the problem is eliminated.

With regard to solvent-to-feed ratio, in general, the greater the solvent proportion the greater will be the selectivity, unless the temperature increase (or bubble-point increase) is so great that it adversely affects the selectivity because of general over-all increase in solubility or a reduction of complex formation. If the system does not follow the "normal" pattern relating selectivity to solvent concentration (see Chap. 10), the specific behavior must be considered in selecting the solvent-to-feed ratio range.

Reflux ratio increase tends to decrease the number of stages necessary to make a given separation. However, on the other hand, it increases the nonsolvent liquid rate in the column, which for the "normal" systems tends to dilute the solvent, reducing its relative quantity and, therefore, reducing the relative volatility of the key components, and this is an adverse effect. Where the relative volatility-solvent composition curve is a maximum, for example, the increase in reflux ratio may be advantageous with respect to relative volatility or may be harmful, depending upon which side of the maximum on the selectivity-solvent concentration curve the reflux ratios fall (in terms of solvent concentration). Where the relative volatility-solvent composition curve is concave upward because of solubility decrease with respect to the nonselected component, the relative volatility is reduced with higher internal reflux rate (the reflux, particularly in the upper section of the column, is composed mainly of the nonselected component). On the other hand, because the solubility of the nonselected component is bettered, the possibility of encountering two-liquid-phase separation on the plates in the column is decreased. This will make the operation of the column easier and give a wider range of reflux ratios for satisfactory operation.

Where the curves for different parameters of selected-to-nonselected-component compositions cross or invert with respect to relative volatility as the solvent concentration is increased, increasing reflux ratio may be helpful, harmful, or have no effect. With a positive slope, increasing reflux ratio is harmful; with a negative slope it is helpful and with zero slope it makes no difference.

Extractive Distillation Column Design—Number of Theoretical Plates

In the design of an extractive distillation column the usual factors are to be evaluated: the size of the column (diameter, number of plates), the

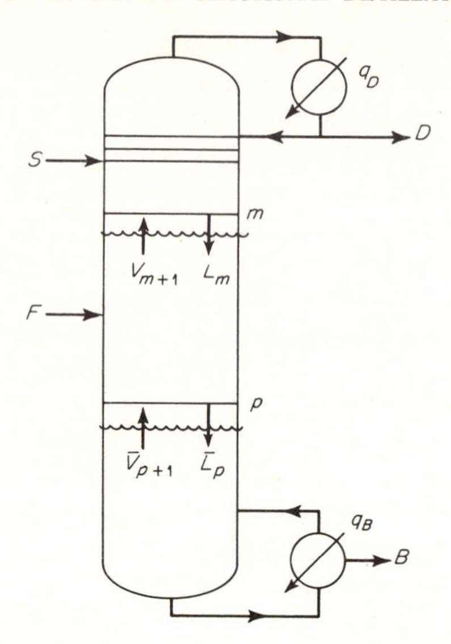

FIGURE 11.5 Extractive distillation column.

reboiler and condenser, controls, and accessories. The size of the column and number of plates depend upon the feed rate, reflux ratio, solvent rate, and efficiency. The number of theoretical plates or equilibrium stages may be determined by a number of methods, three of which are discussed here. These are the *pseudobinary* method, the *pseudoternary* method, and the *multicomponent* method. A modification of the pseudobinary method wherein a different L/V is used for each plate when occasion requires it is discussed by Chambers [4]. He used a McCabe-Thiele diagram to establish the number of theoretical stages.

Pseudobinary Method (Fig. 11.5)

The equations for the pseudobinary method are as follows: For the rectifying section, m, assume $Dx_{D_S} \cong 0$.

$$x_{S_m} = x_{S_{m+1}} = x_{S_{mS+2}} \text{ to } x_{S_F}$$

The material balances are:

$$V + S = L + D \tag{11.6}$$

$$Vy_S + S = Lx_S + Dx_{D_S} \tag{11.7}$$

$$Vy_N + Sx_N = Lx_N + Dx_{D_N} \tag{11.8}$$

Substituting into Eq. (11.7),

$$\alpha_{SN} = \frac{K_S}{K_N} \qquad K_S = \frac{y_S}{x_S}$$

$$VK_S x_S + S = Lx_S + Dx_{D_S} \tag{11.9}$$

$$VK_N(1 - x_S) = L(1 - x_S) + D(1 - x_{D_S}) \tag{11.10}$$

Solving for V in Eqs. (11.9) and (11.10) and equating,

$$V = \frac{Lx_S - S}{K_S x_S} = \frac{L(1 - x_S) + D(1 - x_{D_S})}{K_N(1 - x_S)} \tag{11.11}$$

$$-S = \frac{[L(1 - x_S) + D(1 - x_{D_S})]K_S x_S}{K_N(1 - x_S)} - Lx_S \tag{11.12}$$

$$-S = \frac{L - Lx_S + D - Dx_{D_S}}{1 - x_S}\alpha_{SN}x_S - Lx_S \tag{11.13}$$

$$Dx_{D_S} = 0$$

$$-S = \frac{\alpha_{SN}Lx_S - \alpha_{SN}Lx_S^2 + D\alpha_{SN}x_S - Lx_S(1 - x_S)}{1 - x_S} \tag{11.14}$$

$$-S(1 - x_S) = \alpha_{SN}Lx_S - \alpha_{SN}Lx_S^2 + \alpha_{SN}Dx_S - Lx_S + Lx_S^2 \tag{11.15}$$

$$-S(1 - x_S) = \alpha_{SN}Lx_S(1 - x_S) - Lx_S(1 - x_S) + \alpha_{SN}Dx_S \tag{11.16}$$

$$-S(1 - x_S) = \alpha_{SN}Lx_S(1 - x_S) + \left(-Lx_S + \frac{\alpha_{SN}Dx_S}{1 - x_S}\right)(1 - x_S) \tag{11.17}$$

$$-S = \alpha_{SN}Lx_S - Lx_S + \frac{\alpha_{SN}Dx_S}{1 - x_S} \tag{11.18}$$

$$x_S = \frac{S}{(1 - \alpha_{SN})L + \dfrac{\alpha_{SN}D}{1 - x_S}} \tag{11.19}$$

By similar balances around the stripping section p (see Fig. 11.5),

$$\bar{x}_S = \frac{S}{(1 - \alpha_{SN})\bar{L} + \dfrac{\alpha_{SN}B}{1 - \bar{x}_S}} \tag{11.20}$$

Equations (11.19) and (11.20) can be used to determine the solvent concentrations above and below the feed from the quantities of D, B, S, or S/F

ratio and the temperature-relative volatility relations. The quantity of S is selected on the basis of the type of selectivity curve as discussed in Chap. 10. The relative volatility of solvent referred to nonsolvent normally will be small because one of the criteria for selection is that the solvent be nonvolatile. If this be the case, the terms

$$\frac{\alpha_{SN} D}{1 - x_S} \quad \text{and} \quad \frac{\alpha_{SN} B}{1 - \bar{x}_S}$$

are small and the solvent compositions above and below the feed are calculated by

$$x_S = \frac{S}{(1 - \alpha_{SN})L} \quad \text{or} \quad x_S \cong \frac{S}{L} \tag{11.21}$$

and

$$\bar{x}_S = \frac{S}{(1 - \alpha_{SN})\bar{L}} \quad \text{or} \quad x_S \cong \frac{S}{L} \tag{11.22}$$

If the solvent is sufficiently volatile to produce an $\alpha_{SN} > 0.05$, the values of x_S and $\bar{x}_S$ must be calculated by trial and error after determining the relative volatility of the solvent referred to nonsolvent at some representative average temperature in both the rectifying and in the stripping sections.

The equations for the operating lines are as follows: For the rectifying section,

$$L_m x_m + D x_D = V_{m+1} y_{m+1} \tag{11.23}$$

$$V_{m+1} = D - S + L_m \tag{11.24}$$

$$L_m x_m + D x_D = D y_{m+1} - S y_{m+1} + L_m y_{m+1} \tag{11.25}$$

$$L_m (x_m - y_{m+1}) = y_{m+1}(D - S) - D x_D \tag{11.26}$$

$$L_m = S - D + V_{m+1} \tag{11.27}$$

$$S x_m - D x_m + V_{m+1} x_m + D x_D = V_{m+1} y_{m+1} \tag{11.28}$$

$$V_{m+1}(y_{m+1} - x_m) = x_m(S - D) + D x_D \tag{11.29}$$

$$V_{m+1}(x_m - y_{m+1}) = x_m(D - S) - D x_D \tag{11.30}$$

Dividing Eq. (11.26) by Eq. (11.30),

$$\frac{L_m}{V_{m+1}} = \frac{y_{m+1} - \dfrac{D}{D - S} x_D}{x_m - \dfrac{D}{D - S} x_D} \tag{11.31}$$

By the same method for the stripping section,

$$\frac{\bar{L}_p}{\bar{V}_{p+1}} = \frac{y_{p+1} - x_B}{x_p - x_B} \tag{11.32}$$

In terms of enthalpy balance for the rectifying section,

$$V_{m+1}H_{m+1} + Sh_S + q_D = Dh_D + L_m h_m \tag{11.33}$$

Defining $Q_D = q_D/D$ and rearranging and solving for $L_m/(V_{m+1})$,

$$\frac{L_m}{V_{m+1}} = \frac{H_{m+1} - \dfrac{D}{D-S}\left[(h_D + Q_D) - \dfrac{Sh_S}{D}\right]}{h_m - \dfrac{D}{D-S}\left[(h_D + Q_D) - \dfrac{Sh_S}{D}\right]} \tag{11.34}$$

and for the stripping section,

$$\bar{L}_p h_p + q_B = Bh_B + \bar{V}_{p+1}H_{p+1} \tag{11.35}$$

Defining $Q_B = q_B/B$ and solving for $\bar{L}_p/\bar{V}_{p+1}$,

$$\frac{\bar{L}_p}{\bar{V}_{p+1}} = \frac{H_{p+1} - (h_B - Q_B)}{h_p - (h_B - Q_B)} \tag{11.36}$$

Pseudobinary approximate method for number of theoretical plates The following are known:

Feed quantity and composition, F, x_F
Thermal condition of feed
Distillate composition (S-free) and quantity (by specification)
Bottoms composition (S-free) and quantity
Cooling-water temperature
Vapor-liquid equilibria in i, j, S system
Solubility of i, j in S at distillation temperatures

1. Select an operating pressure and calculate the dew-point temperature of the vapor from the top plate, assuming only the light key i, and lighter, and the heavy key present if S is nonvolatile. If S has sufficient volatility to appear in the distillate, the dew-point calculation must include it. Where a volatile system is being distilled and the distillation must be operated under pressure to attain a condensing temperature comparable with the cooling-water temperature, a temperature approximately 20°F above the highest expected cooling-water-inlet temperature is assumed and the dew-point temperature of the vapor is calculated.

2. Estimate the pressure drop across the column to obtain the pressure in the reboiler.

3. Calculate the bubble-point temperature in the reboiler, assuming the light key, heavy key and heavier, and solvent present.

4. At a temperature 5°F higher than the calculated dew-point temperature of V_1 determine

$$\alpha_{Si} = \frac{\gamma_S P_S}{\gamma_i P_i} \tag{11.37}$$

where the activity coefficients are determined from the binary iS data. If solubility is a problem, a temperature 5°F above the solubility temperature of i in S should be selected for the determination of α_{SN}.

5. At a temperature 5°F lower than the reboiler temperature calculate

$$\alpha_{Si} = \frac{\gamma_S P_S}{\gamma_j P_j} \tag{11.38}$$

where the activity coefficients are determined from binary jS data.

6. Determine the minimum reflux ratio $(L_0/D)_{\text{min}}$ by the Colburn or Underwood method (or other), using the average relative volatility of the light to heavy key at average temperature in the rectifying section, $(t_1 + 5 + t_F)/2$, and in the stripping section, $(t_n - 5 + t_F)/2$,

$$\alpha = \frac{\alpha_{\text{rect}} + \alpha_{\text{strip}}}{2}$$

7. For an operating reflux ratio = a factor times $(L_0/D)_{\text{min}}$, L_0 is determined. Since this stream is solvent-free or essentially so, this represents L_N leaving the top plate.

8. Since the solvent is added to the column at the bubble point of the liquid on the plate, no condensation of vapor caused by latent heat loss of the vapor to the sensible heat of the solvent takes place. However, as the solvent

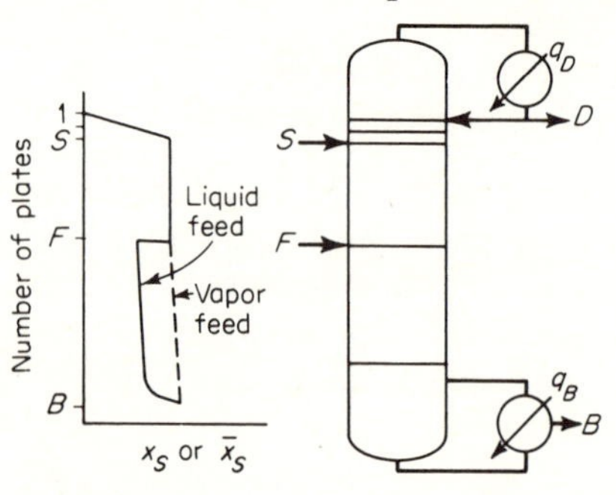

FIGURE 11.6 Solvent concentration in the column.

progresses down the column, its temperature rises until it reaches the reboiler (see Fig. 11.6). This heat absorption and associated vapor condensation increases the amount of liquid downflow. Also, the solubility effect reduces the relative volatility of i with respect to j normally (S/L ratio is decreased). The increase in the liquid downflow caused by condensation is calculated by trial and error. The temperature of the plate above the feed plate t_{F-1} is assumed and the sensible heat gain of the solvent from the point of its addition to plate $F - 1$ is calculated by

$$h_{S_{F-1}} - h_S = SC_{p_{S,\mathrm{av}}}(t_{S_{F-1}} - t_{S_F}) \tag{11.39}$$

Divide the molar heat of solution of the nonsolvent mixture (if known) or the average molar latent heat of the nonsolvent vapor into the heat quantity, $h_{S_{F-1}} - h_{S_F}$, to determine the moles of vapor condensed and absorbed, L_{abs}. The total liquid downflow from the plate above the feed plate is represented by

$$L_{F-1} = S + L_0 + L_{\mathrm{abs}} \tag{11.40}$$

$$x_{S_{F-1}} = \frac{S}{S + L_0 + L_{\mathrm{abs}}} \tag{11.41}$$

$$x_{i_{F-1}} = \frac{i}{S + L_0 + L_{\mathrm{abs}}}$$

$$x_{j_{F-1}} = 1.0 - x_{S_{F-1}} - x_{i_{F-1}}$$

9. Using the vapor pressure of the components at t_{F-1}, the compositions, and the activity coefficients, calculate the total pressure

$$P_t = \gamma_i x_i P_i + \gamma_j x_j P_j + \gamma_S x_S P_S \tag{11.42}$$

If the calculated and actual values differ, assume a new value of t_{F-1} and repeat until agreement is reached.

10. Calculate the temperature t_{n-5} in the same manner to evaluate the amount of $\bar{L}_{n_{t-5}}$.

11. α_{ij} is determined at plate $F - 1$ from equilibrium data or αt data and α_{ij} at plate t_{n-5} is determined similarly.

12. The average relative volatilities for the rectifying section and stripping section are evaluated as

$$\alpha_{\mathrm{rect}} = \frac{\alpha_{L1} + \alpha_{F-1}}{2} \qquad \alpha_{\mathrm{strip}} \frac{\alpha_{F-1} + \alpha_{n_{t+5}}}{2}$$

13. Calculate the pseudobinary equilibrium curves for the rectifying section and the stripping section by using the average relative volatilities

$$y = \frac{\alpha x}{1 + (\alpha - 1)x}$$

and plot on a conventional y versus x plot. (*Note:* Because different α's were used for the two sections, the curves do not coincide at the same composition.)

14. The operating lines are drawn using the slopes,

$$\left(\frac{L_0 + L_{\text{abs}} + S}{V_{F-1}}\right)_{\text{rect}} \quad \text{and} \quad \left(\frac{L_0 + L_{\text{abs}} + S + F_L}{\bar{V}_{n_{t+5}}}\right)_{\text{strip}}$$

More conservatively, the reflux ratios $(L_0 + S)/V_{F-1}$ and $(L_0 + S + F_L)/\bar{V}_{n_{t-5}}$ can be used since these are smaller. Equations (11.31) and (11.32) may also be used.

15. The number of plates is determined by graphical construction (see Fig. 11.7).

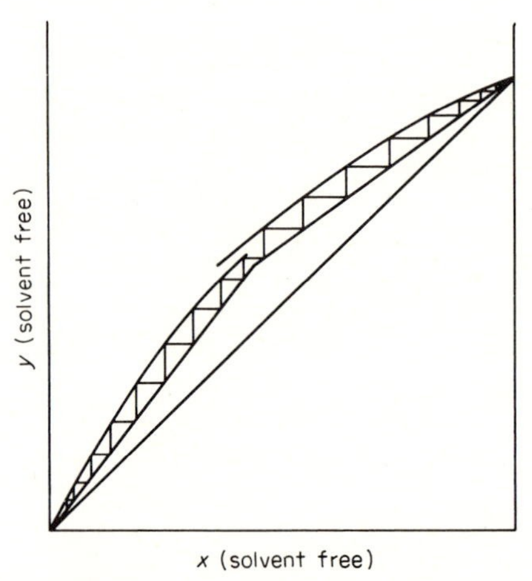

FIGURE 11.7 Theoretical plates determined by graphical construction—pseudobinary.

16. The reboiler heat load is calculated by an enthalpy balance around the stripping section of the column including the plate above the feed plate:

$$q_B = V_F H_F + B h_B - L_{F-1} h_{F-1} - [q h_F + (1 - q) H_F] F \quad (11.43)$$

17. The condenser heat load is calculated by heat balance around the rectifying section of the column including the plate above the feed plate:

$$q_D = L_{F-1} h_{F-1} + D h_D - S h_S - V_F H_F \quad (11.44)$$

Ternary Graphical Method

Where the system in the presence of a solvent can be considered a ternary system composed of the two components to be separated and the solvent, ternary graphical methods for determination of the number of equilibrium stages may be used. Commonly, Bonilla's method [3] is used and, if ternary experimental or calculated vapor-equilibrium data are available, a useful method of plotting them for ease in interpolation is that of Griswold and Dinwiddie [11]. The equations for the difference points are written as described in Chap. 7 and the method of construction for the number of theoretical stages is the same. A representative construction is shown on Fig. 11.8.

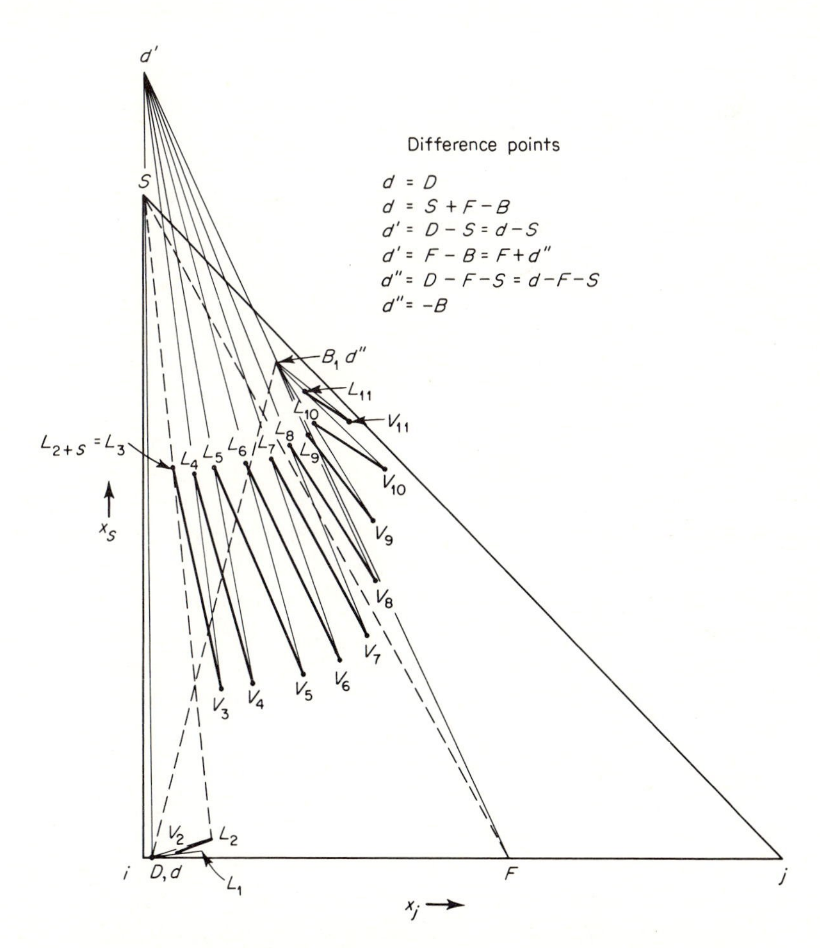

FIGURE 11.8 Ternary graphical construction—extractive distillation.

Multicomponent Method for Determining the Number of Stages in Extractive Distillation

Industrially the realistic extractive distillation process is one involving multicomponent and, in some instances, complex mixtures, and simple ternary systems are encountered infrequently. Calculations involving more than three components become more complex and some accuracy is sacrificed when simplifying assumptions are made to shorten the calculations. However, at least for preliminary design comparison, it is necessary to simplify the methods considerably. For determination of minimum reflux the method of Brown-Martin, Colburn, or Underwood may be used; for the minimum number of plates the methods of Fenske or Underwood are used; and for the actual evaluation of the number of theoretical stages the methods of Brown-Martin or Gilliland are satisfactory.

In these methods, the values of α_{iN} or α_{iS} are used, and the values of the relative volatilities selected must be averages which will represent some sort of mean values applicable throughout the column, or at least over a section of the column. By disregarding the knock-back plates above the solvent feed plate, average α's may be selected for the section between the solvent feed and the feed plate, and another set of average α's are selected for the section between the feed plate and the bottom. This is possible, particularly where large volumes of solvent are used since

$$\alpha_{ij} = \frac{y_i/x_j}{y_j/x_j} = \frac{K_i}{K_j} = \frac{\gamma_i P_i}{\gamma_j P_j}$$

Where large volumes of solvent are used, the bubble-point temperature of the liquids will vary only slightly between the solvent feed and the feed. Thus the vapor pressure ratios are reasonably constant. The γ ratio will remain reasonably constant since the compositions of the light key and lighter than light keys are small in the presence of the solvent and are essentially those calculated by the selectivity equations. Similarly, the value of γ for the heavy key is essentially constant. If the feed is a liquid, the composition changes below the feed are appreciable, and different average relative volatilities are used. The stepwise method for the evaluation of number of stages is as follows:

1. Determine the degree of separation of the mixture and thus the solvent-free distillate composition and solvent-free bottoms composition. This may not be a simple straightforward calculation because of the modification of relative volatilities of the various components by the presence of the solvent. It is possible for some of the light materials to be included in the bottoms product because they are capable of interacting with the solvent. Likewise it is possible for some of the normally heavier components to be removed in the distillate because of relative volatility rearrangement. Because of this possibility the selection of the key components as well as the estimation of the distillate and bottoms composition must be based upon selectivity characteristics of the solvent for each of the components present in the feed mixture. These can be estimated by the methods in Chap. 10.

2. Using this composition and fixing the solvent addition quantity, calculate the bubble-point temperature of the liquid leaving plate S.

3. By material balance determine the composition of the liquid leaving the feed plate.

4. Determine the bubble-point temperature of this liquid at the concentrations calculated in step 3.

5. Determine the bubble-point temperature for the bottoms product based on the estimated composition (relative to solvent selectivity).

6. Calculate the α's, K_i/K_j for all components in the presence of the solvent at the solvent feed-plate temperature, at the feed temperature, and at the bottoms temperature.

7. Average the first two and the second two to get relative volatilities for the top and bottom sections of the column.

8. Determine the minimum reflux ratio for the top and bottom section.

9. Determine the minimum plates for the top and bottom section.

10. Determine the number of plates for each section for some selected value of reflux ratio, $L_0/D = (L_0/D)_{min} \times$ factor.

Example 11.1

The following system is to be subjected to extractive distillation at 760 mm Hg pressure at the condenser to produce a bottoms product containing 99% (mole basis) of the ethylbenzene and 1% (mole basis) of the ethylcyclohexane. Use the following van Laar constants and assume 900 mm Hg pressure in the reboiler:

$A_{12} = 1.07$	$A_{13} = 1.38$	$B_{21} = 74.5$	$B_{31} = 71.5$	$B_{12} = 5.49^-$
$A_{32} = 0.775$	$A_{23} = 1.29$	$B_{23} = 52.6^-$	$B_{32} = 39.8$	$B_{13} = 65^-$
	$A_{43} = 0.685$	$B_{24} = 468^-$	$B_{34} = 1.61^-$	$B_{42} = 25.0^+$
		$B_{41} = 309^+$	$B_{14} = 613^-$	$B_{43} = 112^+$

The feed composition is as follows:

(1) *n*-Octane	$x_1 = 0.20$
(2) Ethylcyclohexane	$x_2 = 0.40$
(3) Ethylbenzene	$x_3 = 0.40$

Three moles of phenol (component 4) are to be used per mole of feed. A reflux ratio $L_0/D = 6.0$ is to be used. Determine the number of theoretical stages needed.

Solution

Short-cut methods will be used for determination of the number of stages. Therefore the relative volatility values for the components will be determined at the point of solvent addition, at the point of feed addition, and at the bottoms composition.

Stream Compositions: Basis—1 Mole of Feed

Components	Feed		Solvent		Distillate		Bottoms	
	Moles	x_F	Moles	x_S	Moles	x_D	Moles	x_B
(1) *n*-Octane	0.20	0.20	—	—	0.200	0.333	0.000	0.00
(2) ECH	0.40	0.40	—	—	0.396	0.660	0.004	0.0012
(3) EB	0.40	0.40	—	—	0.004	0.007	0.396	0.1165
(4) Phenol	—	—	3.0	1.0	0.000	0.000	3.000	0.8822
					0.600	1.000	3.400	0.9999

Vapor composition—solvent feed plate: Assume the vapor composition rising from the solvent feed plate to be that of the distillate. This assumes that there is little if any fractionation above the solvent feed plate (i.e., in the solvent knock-back section).

Temperature at solvent feed plate: Assume the solvent is added at the temperature of the dew point of the solvent-free vapor to minimize the thermal upset in the column. The $T \ln \gamma$ values are first determined for the components, t is assumed, and $\Sigma x = \Sigma y/K = \Sigma(yP_t/\gamma P)$ is calculated

$$T \ln \gamma_1 = \frac{[x_2A_{23}(B_{12})^{0.5} + x_3A_{33}(B_{13})^{0.5}]^2}{(x_1A_{13} + x_2A_{23} + x_3A_{33})^2}$$

$$T \ln \gamma_2 = \frac{[x_1A_{13}(B_{21})^{0.5} + x_3A_{33}(B_{23}{}^{0.5})]^2}{(x_1A_{13} + x_2A_{23} + x_3A_{33})^2}$$

$$T \ln \gamma_3 = \frac{[x_1A_{13}(B_{31})^{0.5} + x_2A_{23}(B_{32})^{0.5}]^2}{(x_1A_{13} + x_2A_{23} + x_3A_{33})^2}$$

$$T \ln \gamma_1 = \frac{[(0.660)(1.29)(2.45) + (0.007)(1.0)(8.1)]^2}{[(0.333)(1.38) + (0.660)(1.29) + (0.007)(1.0)]^2}$$

$$T \ln \gamma_1 = \frac{[(-2.08) + (0.0567)]^2}{(1.318)^2} = \frac{4.1}{2.64} = 1.55$$

$$T \ln \gamma_2 = \frac{[(0.333)(1.38)(8.3) + (0.007)(1.0)(-7.3)]^2}{2.64}$$

$$T \ln \gamma_2 = \frac{(3.81 - 0.051)^2}{2.64} = \frac{14.1}{2.64} = 5.35$$

$$T \ln \gamma_3 = \frac{[(0.333)(1.38)(8.2) + (0.660)(1.29)(6.35)]^2}{2.64}$$

$$T \ln \gamma_3 = \frac{(3.77 + 5.4)^2}{2.64} = \frac{84}{2.64} = 31.8$$

Assume $t = 133°\text{C} = 403°\text{K}$:

$$P_1 = 650 \text{ mm Hg} \qquad P_2 = 710 \text{ mm Hg} \qquad P_3 = 630 \text{ mm Hg}$$

$$\log \gamma_1 = \frac{1.55}{(403)(2.3)} = 0.00167 \qquad \gamma_1 = 1.00$$

$$\log \gamma_2 = \frac{5.35}{(403)(2.3)} = 0.00571 \qquad \gamma_2 = 1.01$$

$$\log \gamma_3 = \frac{31.8}{(403)(2.3)} = 0.034 \qquad \gamma_3 = 1.08$$

Check for dew point,

$$\Sigma x = 1.0 = \sum \frac{y_i P_t}{\gamma_i P_i}$$

Octane: $\dfrac{(0.333)(760)}{(1.00)(850)} = 0.296$

ECH: $\dfrac{(0.66)(760)}{(1.01)(710)} = 0.700$

EB: $\dfrac{(0.007)(760)}{(1.08)(630)} = 0.008$

$$\Sigma x = 1.004$$

Temperature at the bottom of the column:

$$T \ln \gamma_2 = \frac{[x_2 A_{23}(B_{22})^{0.5} + x_3 A_{33}(B_{23})^{0.5} + x_4 A_{43}(B_{24})^{0.5}]^2}{(x_2 A_{23} + x_3 A_{33} + x_4 A_{43})^2}$$

$$T \ln \gamma_3 = \frac{[x_2 A_{23}(B_{32})^{0.5} + x_3 A_{33}(B_{33})^{0.5} + x_4 A_{43}(B_{34})^{0.5}]^2}{(x_2 A_{23} + x_3 A_{33} + x_4 A_{43})^2}$$

$$T \ln \gamma_4 = \frac{[x_2 A_{23}(B_{42})^{0.5} + x_3 A_{33}(B_{43})^{0.5} + x_4 A_{43}(B_{44})^{0.5}]^2}{(x_2 A_{23} + x_3 A_{33} + x_4 A_{43})^2}$$

$$T \ln \gamma_2 = \frac{[(0.0012)(1.29)(0) + (0.1165)(1.0)(-7.3) + (0.8822)(0.685)(-21.7)]^2}{[(0.0012)(1.29) + (0.1165)(1.0) + (0.8822)(0.685)]^2}$$

$$T \ln \gamma_2 = \frac{(-0.85 - 13.1)^2}{(0.723)^2} = \frac{194.5}{0.521} = 373$$

$$T \ln \gamma_3 = \frac{[(0.0012)(1.29)(6.35) + (0.8822)(0.685)(-12.6)]^2}{0.521}$$

$$= \frac{(0.0098 - 7.6)^2}{0.521} = \frac{57.5}{0.521} = 111$$

$$T \ln \gamma_4 = \frac{[(0.0012)(1.29)(15.8) + (0.1165)(1.0)(10.3)]^2}{0.521}$$

$$= \frac{(0.0244 + 1.2)^2}{0.521} = \frac{1.5}{0.521} = 2.88$$

Assume that the total pressure is 900 mm Hg. Assume $t = 175°\text{C} = 448°\text{K}$:

$$P_2 = 2160 \qquad P_3 = 1970 \qquad P_4 = 610$$

$$\log \gamma_2 = \frac{373}{(2.3)(448)} = 0.366 \qquad \gamma_2 = 2.32$$

$$\log \gamma_3 = \frac{111}{(2.3)(448)} = 0.109 \qquad \gamma_3 = 1.28$$

$$\log \gamma_4 = \frac{2.88}{(2.3)(448)} = 0.0028 \qquad \gamma_4 = 1.01$$

$$\Sigma y_i = 1.0 = \Sigma x_i \frac{\gamma_i P_i}{P_t}$$

ECH: $$\frac{(0.0012)(2.32)(2160)}{900} = 0.067$$

EB: $$\frac{(0.1165)(1.28)(1970)}{900} = 0.329$$

PhOH: $$\frac{(0.8822)(1.01)(610)}{900} = \frac{0.610}{1.006}$$

Temperature on the plate above the feed plate: $P_t = 800$ mm Hg. Assume the heat necessary to cool the vapors to the saturation temperature to be negligible compared to the heat of vaporization plus the heat of mixing in heating the solvent.

Assume $C_{p_{S,\text{av}}} = 0.58$ cal/(g) (°C), 58.5 cal/(g mole)(°C):

$$\lambda_{\text{HC}} = 8650 \text{ cal/g mole}$$

Assume $t = 150°\text{C} = 423°\text{K}$:

$$\text{Moles HC condensed} = \frac{S(C_{p_S})(t_{F-1} - t_S)}{\lambda_{\text{HC}}} = \frac{3(58.5)(150 - 133)}{8650} = 0.345 \text{ mole}$$

$$x_{S_{F-1}} = \frac{S}{S + L + L_{\text{cond}}}$$

$$\frac{L_0}{D} = 6.0 \qquad D = 0.6 \qquad L_0 = 3.6 \qquad V_1 = 4.2$$

$$V_{F-1} = V_1 + \text{condensed hydrocarbon} = 4.71$$

$$x_{S_{F-1}} = \frac{3}{3 + 3.6 + 0.35} = \frac{3}{6.95} = 0.432$$

Assume the component distribution to be the same as in the feed:

Octane: $x_{F-1} = (0.2)(1 - 0.432) = 0.115$

ECH: $= (0.4)(1 - 0.432) = 0.230$

EB: $= (0.4)(1 - 0.432) = 0.230$

PhOH: $= 0.432 \qquad 0.423$

$\qquad 0.998$

Check temperature assumed:

$$T \ln \gamma_1 = \frac{[x_1A_{13}(B_{11})^{0.5} + x_2A_{23}(B_{12})^{0.5} + x_3A_{33}(B_{13})^{0.5} + x_4A_{43}(B_{14})^{0.5}]^2}{(x_1A_{13} + x_2A_{23} + x_3A_{33} + x_4A_{43})^2}$$

$$= \frac{[(0.115)(1.38)(0) + (0.23)(1.29)(-2.34) + (0.23)(1.0)(-8.05) + (0.422)(0.685)(-24.8)]^2}{[(0.115)(1.38) + (0.23)(1.29) + (0.23)(1.0) + (0.422)(685)]^2}$$

$$T \ln \gamma_1 = \frac{(-0.68 - 1.85 - 7.2)^2}{(0.159 + 0.297 + 0.23 + 0.29)^2} = \frac{(9.75)^2}{(0.976)^2} = \frac{9.47}{0.95} = 99.5$$

$$T \ln \gamma_2 = \frac{[x_1A_{13}(B_{21})^{0.5} + x_2A_{23}(B_{22})^{0.5} + x_3A_{33}(B_{23})^{0.5} + x_4A_{43}(B_{24})^{0.5}]^2}{0.95}$$

$$= \frac{[(0.115)(1.38)(8.61) + (0.23)(1.0)(-7.3) + (0.422)(0.685)(-21.7)]^2}{0.95}$$

$$= \frac{(1.37 - 1.68 - 6.3)^2}{0.95} = \frac{44}{0.95} = 46.5$$

$$T \ln \gamma_3 = \frac{[x_1A_{13}(B_{31})^{0.5} + x_2A_{23}(B_{32})^{0.5} + x_3A_{33}(B_{33})^{0.5} + x_4A_{43}(B_{34})^{0.5}]^2}{0.95}$$

$$= \frac{[(0.115)(1.38)(8.2) + (0.23)(1.29)(6.35) + (0.422)(0.685)(-12.6)]^2}{0.95}$$

$$= \frac{(1.31 + 1.89 - 3.65)^2}{0.95} = \frac{(0.45)^2}{0.95} = \frac{0.203}{0.95} = 0.214$$

$$T \ln \gamma_4 = \frac{[x_1A_{13}(B_{41})^{0.5} + x_2A_{23}(B_{42})^{0.5} + x_3A_{33}(B_{43})^{0.5}]^2}{0.95}$$

$$= \frac{[(0.115)(1.38)(17.6) + (0.23)(1.29)(15.7) + (0.23)(1.0)(10.6)]^2}{0.95}$$

$$= \frac{(2.8 + 4.67 + 2.44)^2}{0.95} = \frac{(9.91)^2}{0.95} = \frac{98.5}{0.95} = 104$$

$$\ln \gamma_1 = \frac{99.5}{(2.3)(4.28)} = 0.101 \qquad \gamma_1 = 1.26$$

$$\ln \gamma_2 = \frac{46.5}{(2.3)(4.28)} = 0.0477 \qquad \gamma_2 = 1.12$$

$$\ln \gamma_3 = \frac{0.214}{(2.3)(4.28)} = 0.00022 \qquad \gamma_3 = 1.00$$

$$\ln \gamma_4 = \frac{104}{(2.3)(4.28)} = 0.107 \qquad \gamma_4 = 1.28$$

Check on temperature assumed: bubble point t_{F-1}; $\Sigma y_i = 1.0 = \Sigma(x_i \gamma_i P_i / P_t)$.

		$t = 150°C$	$t = 145°C$
Octane:	$\dfrac{(0.115)(1.26)P_1}{800} = 0.00182P_1$	0.264	0.233
ECH:	$\dfrac{(0.23)(1.12)P_2}{800} = 0.00324P_2$	0.390	0.349
EB:	$\dfrac{(0.23)(1.00)P_3}{800} = 0.00286P_3$	0.310	0.272
PhOH:	$\dfrac{(0.422)(1.20)P_4}{800} = 0.00676P_4$	0.189	0.158
		1.153	1.012

The temperature is closer to 145°C than to 150°C. Correcting for the new temperature use

$$\text{Moles HC condensed} = \frac{3(58.5)(145 - 133)}{8650} = 0.244$$

$$x_S = \frac{3}{3 + 3.6 + 0.244} = 0.44 \text{ instead of } 0.432$$

This correction for γ need not be made because of the uncertainty introduced by the many assumptions made.

Relative volatility based on ethylbenzene:

Relative volatility	Solvent feed, 133°C	Feed, 145°C	Reboiler, 175°C
α_{13}	1.23	1.70	—
α_{23}	1.06	1.25	2.00
α_{33}	1.00	1.00	1.00
α_{43}	—	0.31	0.23

Minimum reflux—Underwood: Use average $t = 152°C$; use γ at feed condition.

$$\alpha_{13} = \frac{(1.26)(1530)}{(1.0)(1120)} = 1.72$$

$$\alpha_{23} = \frac{(1.12)(1270)}{(1.0)(1120)} = 1.27$$

$$\alpha_{33} = \frac{(1.28)(875)}{(1.0)(1120)} = 1.00$$

$$\alpha_{43} = \frac{(1.28)(290)}{(1.0)(1120)} = 0.331$$

Saturated liquid feed $1 - q = 0$.

$$\Sigma = 0 = \frac{0.115}{(1.72 - \theta)/1.72} + \frac{0.23}{(1.27 - \theta)/1.27} + \frac{0.23}{(1.0 - \theta)/1.0} + \frac{0.423}{(0.331 - \theta)/0.331}$$

$$= \frac{0.198}{1.72 - \theta} + \frac{0.293}{1.27 - \theta} + \frac{0.23}{1.0 - \theta} + \frac{0.14}{0.331 - \theta}$$

$$\theta = 1.115$$

$$\left(\frac{L}{D}\right)_m + 1 = \frac{(0.333)(1.72)}{1.72 - 1.115} + \frac{(0.66)(1.27)}{1.27 - 1.115} + \frac{(0.007)(1.0)}{1.0 - 1.115}$$

$$= 0.95 + 5.4 - 0.06 = 6.3$$

$$\left(\frac{L}{D}\right)_m = 5.3$$

Minimum plates—Fenske:

$$N_m = \frac{\log (0.66/0.007)(0.1165/0.0012)}{\log (1.27)} = \frac{\log 9200}{0.105} = \frac{3.965}{0.105}$$

$$N_m = 38 \text{ stages}$$

Number of theoretical stages—Gilliland:

$$\frac{\dfrac{L}{D} - \left(\dfrac{L}{D}\right)_m}{\dfrac{L}{D} + 1} = \frac{6.0 - 5.3}{6.0 + 1.0} = \frac{0.7}{7} = 0.1$$

From Fig. 5.17,

$$\frac{N - N_m}{N + 1} = F = 0.52$$

$$N - 38 = (N + 1)F$$

$$N - NF = 38 + F$$

$$N = \frac{38 + 0.52}{1 - 0.52} = 80$$

Extractive Distillation Processes

Although the process of extractive distillation was known prior to World War II, industrial adoption of the process as a standard distillation separations method could be said to be the result of the great demand, both in quantity and purity, of two materials—toluene and butadiene. Although these materials had been produced in relatively small quantities with high purity, the wartime demand for nitration grade toluene for explosive production and for butadiene for synthetic rubber production increased tremendously. Previous sources for the raw materials were unable to supply the demand, and the new processes developed produced the desired material in mixtures which could not be separated by straight fractionation. Therefore extractive distillation (as well as azeotropic) quickly became a large-scale industrial process.

Solvents for Hydrocarbon Separations

Many solvents have been studied to determine their selectivity for hydrocarbon species. Dunn et al. [8], among others, listed a number of solvents selective for toluene and general aromatic recovery from paraffinic mixtures. A few of these solvents are listed in Table 11.1.

TABLE 11.1 **Solvents Selective for Aromatics**

Furfural	Hexyleneglycol
Acetonyl acetone	2-Ethyl hexylamine
Nitrobenzene	*o*-Chloroamine
Nitrotoluene	*o*-Phenetidine
Phenol	*o*-Chlorophenol
Aniline	*o*-Nitrophenol
Dichloroethylether	*o*-Phenyl phenol
Phenyl Cellosolve	Methyl salicylate
Phenol–cresol: 60/40	Nonanoic acid
Cresol	Sulfolane
Dimethyl sulfolane–H_2O	Sodium-*o*-xylene sulfonate + H_2O
Dimethyl aniline–aniline	Diphenyloxide
Phthalic anhydride	
2-Ethyl hexanol	
Propylene glycol	

This table shows by no means a complete list and is only given to indicate the extensive work done in this field. A specific process for recovery of toluene from a paraffin–toluene mixture was discussed by Drickamer and Hummel [7]. The schematic flow diagram is shown in Fig. 11.9. The solvent-to-feed rate was approximately 3 to 1. Benedict and Rubin [2] discussed this process and developed some equations which showed that increasing the solvent rate increases the separation while increasing the product rate (toluene) decreases the separation.

The process of extraction distillation has been applied to separation of

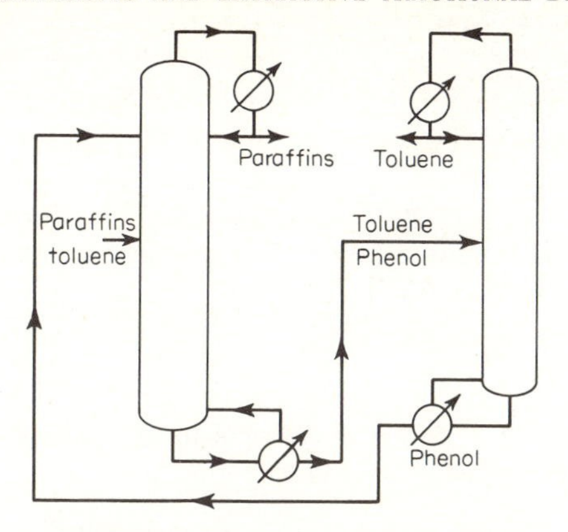

FIGURE 11.9 Separation of toluene from paraffin hydrocarbons using phenol in extractive distillation. [*From Dunn et al., Trans. A.I.Ch.E.*, ***41**:631* (*1945*).]

butenes from butanes, the butenes being intermediates in butadiene production. Atkins and Boyer [1] described a process in which the solvent consisted of a mixture of 85% acetone and 15% water. The flow diagram is shown in Fig. 11.10. The solvent to hydrocarbon feed is around 85% solvent.

Happel et al. [13], Mertes and Colburn [15], and Gerster et al. [10] describe an extractive distillation process (see also Perry [17] wherein the butanes and the butenes are separated by means of extractive distillation using furfural as a solvent).

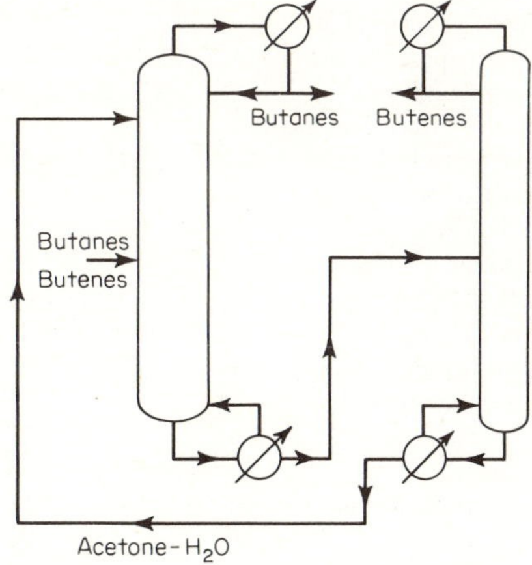

FIGURE 11.10 Separation of butanes from butenes using acetone–water as the extractive solvent. [*From Atkins and Boyer, Chem. Eng. Progr.*, ***45**:553* (*1949*).]

There have been many additional hydrocarbon separations by extractive distillation described in the literature, and investigations are continuing to seek solvents to separate hydrocarbon mixtures.

Swanson and Gerster [20] described the separation of isoprene and 2,methyl-2-butene with dimethyl formamide. The selectivity at infinite dilution, 2,methyl-2-butene relative to isoprene, ranged from 1.9 at 100°F (37.8°C) to around 1.4 at 200°F (93.3°C). Also, a number of patents have been issued on the separation of isoprene from other C_5 olefins and diolefins

TABLE 11.2 **Selective Solvents for Separation of Pentene–Pentane Mixtures***

Solvent	Selectivity, 25°C, $\gamma_{C_5}^{\infty}/\gamma_{C_5}^{\infty}$
Tetrahydrofuran	1.41
Diethylketone	1.43
Diethylcarbonate	1.52
Methylethyl ketone	1.62
Pentanedione	1.72
Cyclopentanone	1.65
Acetone	1.67
Butyronitrile	1.62
Acetylpiperidine	1.69
Acetophenone (45°C)	1.65
Pyridine	1.65
Diethyl oxalate	1.75
Propionitrile	1.85
Dimethyl acetamide	1.85
n-Methylpyrrolidone	1.96
Acetonyl acetone	1.87
Tetrahydrofurfural alcohol	1.62
Dimethylsulfolane	1.95
Dimethylcyanamide	1.96
Methylcarbitol	1.72
Dimethyl formamide	1.96
Methyl Cellosolve	1.69
Furfural	1.87
Acetonitrile	2.77
Ethylenechlorohydrin	1.79
n-Butyrolactone	2.17
Methanol	1.53
bis-Chloropropionitrile	2.18
Pyrrolidone	1.99
Propylene carbonate	2.10
Nitromethane	2.49
Ethylenediamine	2.11
Ethylene carbonate (36.8°C)	1.92

* From Gerster et al. [9].

TABLE 11.3 **Selective Solvents for Separation of Pentene–Pentane Mixtures—Mixed Solvents***

Solvent 1	Solvent 2	Selectivity, 25°C, $\gamma^{\infty}_{C_5}/\gamma^{\infty}_{C_5}$
Methyl Cellosolve	H_2O (5%)	1.69
	Tetramethylenesulfone (5%)	1.70
	Tetramethylenesulfone (20%)	1.81
	Nitromethane (5%)	1.70
Pyridine	H_2O (10%)	1.70
	-oxydipropionitrile (10%)	1.66
	-butyrolactone (32.1%)	1.79
Benzene	Tetramethylenesulfone (50%)	1.73
	-oxydipropionitrile (50%)	1.75
Methylethyl ketone	-butyrolactone (50%)	1.79

* From Gerster et al. [9].

involving the use as solvent, ketones, amines, nitriles, pyridine, and acetic anhydride.

Gerster et al. [9] studied the selectivity of various solvents for the pentene–pentane system at various temperatures. A comparison of the solvent selectivities at 25°C is shown in Tables 11.2 and 11.3.

TABLE 11.4 **Selectivity of Solvents for the Butane–Butene System***

Solvent	Vol/vol HC	Temperature, °F	$\alpha = \gamma_{C_4}/\gamma_{C_4-}$
Hydroxyethylacetate	4	133	1.54
Methylsalicylate	2	156	1.46
Dimethylphthalate	2	142	1.41
Ethyl oxalate	2	154	1.38
Carbitolacetate	2	160	1.35
Diethyl carbonate	2	172	1.28
Amylacetate	2	180	1.21
Acetonitrile	2	137	1.49
Butyronitrile	2	161	1.42
Acrylonitrile	2	156	1.23
Acetonyl acetone	2	141	1.43
Cyclohexanone	2	171	1.32
Acetophorone	2	145	1.31
Methylhexyl ketone	2	166	1.27
Methylamyl ketone	2	173	1.23
Methylisobutyl ketone	2	171	1.23
Methyldiisobutyl ketone	2	177	1.18
Nitromethane	1.8	134	1.60

TABLE 11.4 (*Continued*)

Solvent	Vol/vol HC	Temperature, °F	$\alpha = \gamma_{C_4}/\gamma_{C_4^-}$
Nitroethane	2	146	1.46
1-Chloro-1-nitropropane	2	155	1.46
Nitrobenzene	2	150	1.41
o-Nitrotoluene	2	155	1.38
o-Nitroanisole	2	130	1.30
n-Formylmorpholine	4.6	133	1.60
Morpholine	2	160	1.41
Pyridine	2	176	1.35
Quinoline	2	148	1.33
Picoline	2	188	1.29
Benzyl alcohol	3	144	1.48
Phenol	2	138	1.47
Diacetone alcohol	2	146	1.32
Butyl alcohol	2	152	1.21
2-Ethyl butyl alcohol	2	161	1.20
o-Hexanol	2	159	1.18
tert-Butyl alcohol	2	154	1.16
Benzaldehyde	2	145	1.42
Furfural	3	158	1.40
3,4-Diethoxybenzaldehyde	2	125	1.11
Butyraldehyde	2	165	1.09
Aniline	3	130	1.65
o-Chloroaniline	2	152	1.44
Methylaniline	2	146	1.42
o-Toluidine	2	148	1.38
Dimethyl aniline	2	169	1.37
n-Tributyl amine	2	176	1.09
Cellosolve	2	152	1.40
Dichloroethylether	2	152	1.39
Anisole	2	175	1.28
Butyl Cellosolve	2	163	1.24
Diethyl Cellosolve	2	179	1.23
Diethyl carbitol	2	173	1.23
n-Butylether	2	197	1.10
Solvents with water			
Furfural, 96 wt %	3.7	128	1.78
Aniline, 96.5 wt %	4.4	132	1.77
Methylacetoacetate (90 vol %)	3.6	134	1.67
Phenol (90 vol %)	2.5	133	1.66
Acetonylacetone (95 vol %)	3	128	1.58
Acetonitrile (90 vol %)	4	133	1.58
Benzyl alcohol (95 vol %)	2.5	133	1.51
o-Chlorophenol (90 vol %)	5.0	180	1.50

*From Hess et al. [14].

Hess et al. [14] studied the selectivity of various solvents for separation of 2-butene and butane while the selectivity in this case was the relative volatility at some concentration other than infinite dilution and the temperatures varied, a fairly good approximation of the selectivity of the various solvents is obtained. These are shown in Table 11.4.

Some of the more common extractive distillation hydrocarbon separations utilized in industry involve the separation of acetylene from substituted acetylenes and C_3 hydrocarbons, butadiene from other C_4 hydrocarbons, and aromatics from paraffin and cycloparaffin materials.

Nonhydrocarbon Extractive Distillation

Many nonhydrocarbon mixtures including alcohols, ketones, esters, chlorinated hydrocarbons, and other organic mixtures have been successfully

TABLE 11.5 **Some Common Commercial Extractive Distillation Processes**

System	Extractive agent
HCl–H_2O	H_2SO_4
HNO_3–H_2O	H_2SO_4
Ethanol–H_2O	Glycerine
Butadiene–butane	Acetone, furfural
Butane–butane	Acetone, furfural
Isoprene–pentane	Acetone
Toluene–paraffins	Phenol, furfural
Acetone–methanol	Aniline, water

processed by extractive distillation. Hydrochloric acid and water and nitric acid and water can be extractively distilled with sulfuric acid to make the pure acids. Robinson and Gilliland [19] describe a process for separation of 2-propanol–ethanol by extractive distillation using water as a solvent. The schematic flow diagram is shown in Fig. 11.11. The extractive distillation of nitric acid and water using sulfuric acid as a solvent is schematically shown in Fig. 11.12. The vapor feed is an azeotrope between nitric acid and water. The extractive distillation breaks the azeotrope.

11.6. SOLVENT RECOVERY

A factor of considerable importance in comparing the economics of a given azeotropic or extractive distillation process with that of another system involving another solvent is the cost of recovery of the solvent for recycling. A particular solvent may be very satisfactory from the standpoint of its

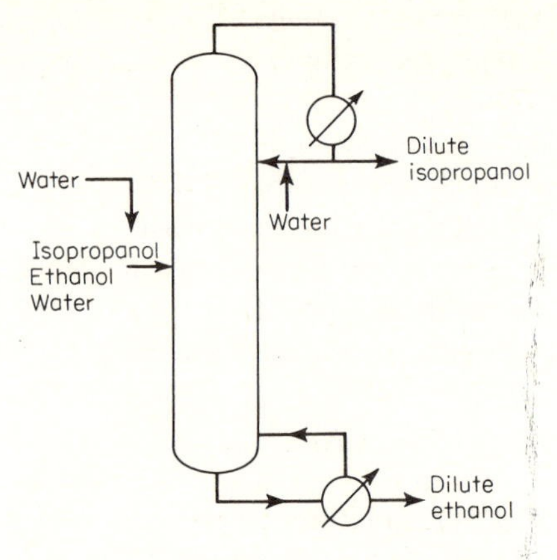

FIGURE 11.11 Extractive distillation of ethanol and isopropanol. (*From Robinson and Gilliland, "Elements of Fractional Distillation," McGraw-Hill, New York, 1950.*)

ability to modify the relative volatility of the key components and from the standpoint of many other criteria involving properties, availability, and cost. However, if it is difficult to separate from the material with which it associates, its separation cost may make it less desirable than some solvent which has less effect on the relative volatility of the keys and yet is easily separable.

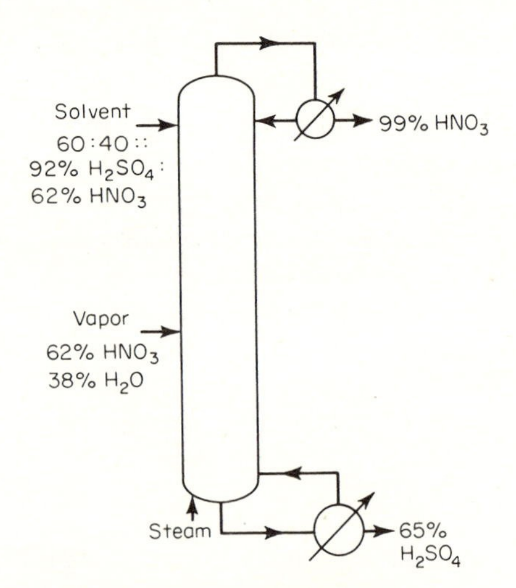

FIGURE 11.12 Extractive distillation of nitric acid water with sulfuric acid as a solvent.

Therefore, to make the best solvent selection, consideration of the solvent recovery system must be made to determine the cost of solvent recovery.

Methods for Solvent Recovery

Fractionation

Generally, a solvent which is nonvolatile would be used with an extractive distillation column designed to eliminate the solvent from the distillate, both by mechanical entrainment separation and by equilibrium fractionation stages. Therefore, the only stream from the extractive distillation column which contains solvent is the residue or bottoms stream. In many cases the solvent may be recovered from this stream by straight fractionation at atmospheric pressure. For azeotropic distillation an entrainer usually has a volatility similar to that of the keys. If it forms a homogeneous azeotrope, straight fractionation will not suffice to make the separation, and some other method must be used.

If a heterogeneous azeotrope is formed, i.e., the condensate will form two liquid phases, the separation of the entrainer from the nonentrainer-rich phase is easily accomplished by gravity separation. Because of the problems

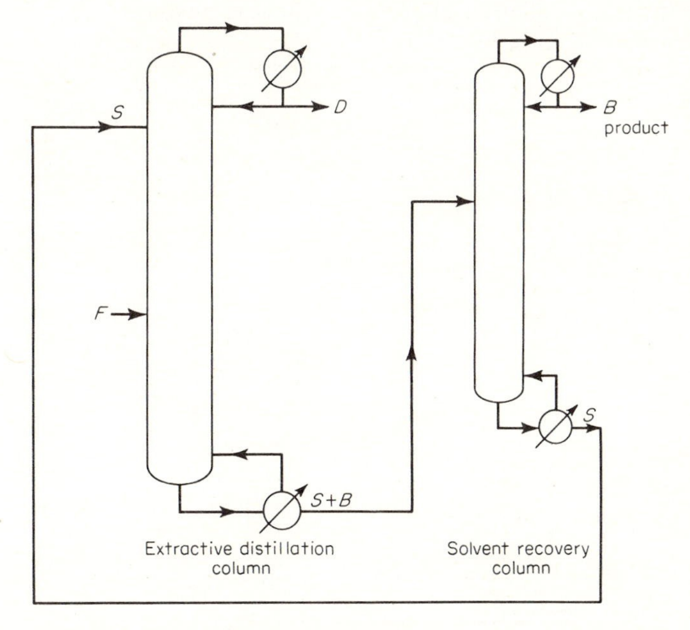

FIGURE 11.13 Extractive distillation with solvent recovery by fractionation.

of separating the entrainer from a homogeneous azeotrope, it is desirable to select an entrainer which will give a heterogeneous system. The following methods are commonly applied to solvent or entrainer recovery.

If the solvent tends to become unstable at its bubble point (essentially the pure-solvent boiling point), it is necessary to resort to some means of reducing the bubble-point temperature. This is done in the conventional manner by distilling at a lower system pressure or by use of a stripping agent such as stream or inert gas to reduce the effective partial pressure of the system. Figure 11.13 is a schematic diagram showing solvent recovery by fractional distillation.

Separation of Solvent by Extraction

A commonly used method for separation of the solvent or entrainer from the product with which it is associated is liquid-liquid solvent extraction wherein either the solvent or entrainer is extracted from the material with which it is associated or, in a few cases, where the associated material is extracted from the solvent. In either case it is necessary for the extraction solvent to be easily separable from the azeotropic entrainer or extractive distillation solvent and from the associated product. An example is the separation of a water-soluble glycol from an aromatic hydrocarbon with which it associates in extractive distillation of paraffin–aromatic mixtures. A schematic flow diagram is shown in Fig. 11.14.

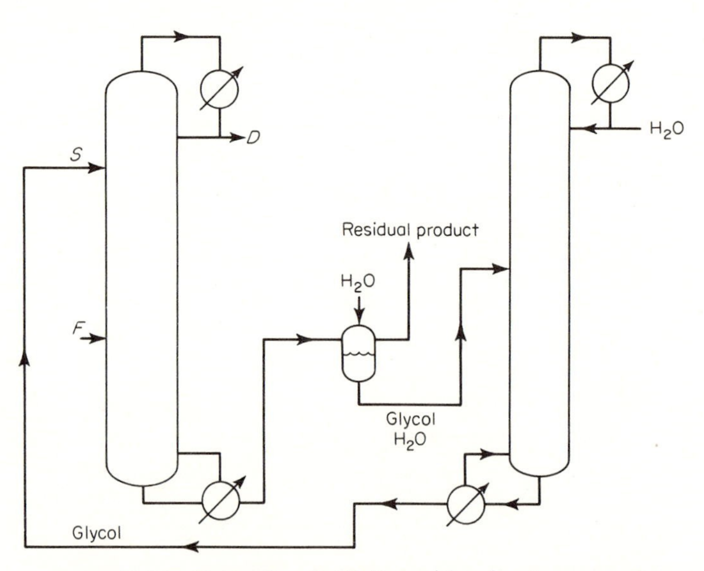

FIGURE 11.14 Extractive distillation with solvent recovery by extraction and distillation.

PROBLEMS

1. The following data are ternary vapor-liquid equilibrium data for the system methanol–acetone–water.

t, °C	Liquid, mole %		Vapor, mole %	
	Acetone	Methanol	Acetone	Methanol
70.0	10	10	61.0	13.0
70.0	10	20	52.0	24.0
69.4	10	30	43.0	35.5
68.8	10	40	37.0	45.0
68.0	10	50	32.0	54.0
66.5	10	60	26.0	63.0
65.0	10	70	22.0	71.0
63.5	10	80	18.5	78.0
65.0	20	10	72.0	77.0
65.5	20	20	63.5	18.0
65.5	20	30	56.0	28.0
64.5	20	40	50.0	37.0
63.5	20	50	44.0	47.0
62.5	20	60	38.5	55.0
61.0	20	70	34.0	62.5
62.8	30	10	76.0	6.5
62.7	30	20	70.0	15.0
62.3	30	30	62.5	25.5
61.5	30	40	57.0	34.5
60.5	30	50	52.0	41.5
59.5	30	60	46.5	50.0
61.5	40	10	79.0	6.0
60.8	40	20	72.5	15.5
60.0	40	30	67.0	24.2
59.4	40	40	61.5	32.2
59.4	40	50	56.5	40.2
60.0	50	10	81.0	66.5
59.4	50	20	75.0	16.0
58.6	50	30	70.0	23.2
57.0	50	40	65.5	31.7
59.1	60	10	82.0	8.0
58.3	60	20	77.0	16.3
57.3	60	30	72.0	24.5
58.2	70	10	82.7	9.5
57.2	70	20	77.8	18.4
57.2	80	10	86.0	10.0

From Griswold and Buford, *Ind. Eng. Chem.*, **41**:2347 (1949).

It is desired to dehydrate methanol containing water by azeotropic distillation with acetone. The distillate is to contain 9.5% of water or less, and the bottoms less than 5% methanol. The original feed consists of a 50:50 mole % mixture of methanol and water. Acetone is added to the feed to produce a mixture containing 40 mole % acetone. A reflux ratio of $L_0/D = 4.0$ is selected.

(*a*) Determine the maximum and minimum entrainer-to-feed ratio which would be feasible.

(*b*) For the above system determine the minimum reflux by the Underwood method, the minimum plates by the Fenske method, and the number of theoretical stages by the Gilliland correlation.

(*c*) Determine the number of stages needed by the ternary graphical method.

(*d*) According to Fordyce and Simonsend, *Ind. Eng. Chem.*, **41**:104 (1949), there is no azeotrope between methanol and acetone at 100 mm Hg. How would you separate the components of the distillate?

(*e*) If the entrainer were added by the top plate for the same conditions, what would be the number of theoretical stages required by the ternary graphical method?

2. A mixture composed of 35 mole % isooctane and 65 mole % toluene is to be separated by extractive distillation at 760 mm Hg pressure using phenol as a solvent in the ratio of 5 moles of phenol to 1 mole of hydrocarbon mixture. A reflux ratio of $L_0/D = 5$ is selected. Determine the number of theoretical stages, by the pseudobinary method, required to produce a distillate containing 99% of the isooctane with a composition of 99.0 mole % isooctane. The data are:

$$T \log \gamma_I = \left(\frac{-8.0 - 22.4x_P/x_T}{x_T/x_T + 1.00 + 0.84x_P/x_T}\right)^2$$

$$T \log \gamma_T = \left(\frac{8.0 - 14.4x_P/x_I}{x_T/x_I + 1.00 + 0.84x_P/x_I}\right)^2$$

$$T \log \gamma_P = \left(\frac{28.9 + 19.6x_T/x_I}{x_P/x_I + 1.18 + 1.18x_T/x_I}\right)^2$$

3. For the system and conditions in Prob. 2, using the ternary graphical method, determine the number of theoretical stages.

4. Using the following short-cut methods, determine the number of equilibrium stages for solvent-to-hydrocarbon ratios of 2, 4, 6, and 8 and L_0/D values of 3, 5, 7, 10 for the system and conditions of Prob. 2.

(*a*) For minimum reflux, use Underwood's method.

(*b*) For minimum stages, use Fenske's method.

(*c*) For theoretical stages at operating reflux, use Gilliland's method.

Plot the results of the calculations to show the effect of solvent-to-feed ratio and reflux ratio on the number of stages necessary.

Nomenclature

B = moles residue per unit time
C_p = heat capacity, Btu/(lb)(°F)
D = moles distillate per unit time
E = moles entrainer per unit time
F = moles feed per unit time
H = enthalpy of vapor, Btu/lb mole
K = equilibrium vaporization ratio

L = moles liquid per unit time, rectifying section
$\bar{L}$ = moles vapor per unit time, stripping section
N = moles of nonsolvent per unit time
N = number of stages
O = addition point
P_t = total pressure
Q_B = heat added in the reboiler per mole residue
Q_D = heat added in the condenser per mole distillate
S = moles solvent per unit time
V = moles vapor per unit time, rectifying section
$\bar{V}$ = moles vapor per unit time, stripping section
d, d, d'' = difference points
h = enthalpy of liquid, Btu/mole
i, j, k = components
q = mole fraction of feed as liquid
q_B = heat added in the reboiler, Btu/unit time
q_D = heat added in the condenser, Btu/unit time
t = temperature
x, y = mole fraction, liquid and vapor

Greek letters

α = relative volatility
γ = activity coefficient
λ = latent heat of vaporization, Btu/lb mole

Subscripts

B = residue
D = distillate
E = entrainer
F = feed
i, j, S = component i, j, and solvent
L = liquid
m, n, p = section of column
N = nonsolvent
S = solvent
1, 2, 3 = conditions

REFERENCES

1. Atkins, G. T., and C. M. Boyer: *Chem. Eng. Progr.*, **45**:553 (1949).
2. Benedict, M., and C. Rubin: *Trans. A.I.Ch.E.*, **41**:353 (1945).
3. Bonilla, A.: *Trans. A.I.Ch.E.*, **37**:669 (1941).
4. Chambers, J. M.: *Chem. Eng. Progr.*, **47**:555 (1951).
5. Colburn, A. P., and J. C. Phillips: *Trans. A.I.Ch.E.*, **40**:333 (1944).

6. Colburn, A. P., and E. M. Schoenborn: *Trans. A.I.Ch.E.*, **41**:421 (1945); Errata: **41**:645 (1945).
7. Drickamer, H. G., and H. H. Hummel: *Trans. A.I.Ch.E.*, **41**:607 (1945).
8. Dunn, C. L., R. W. Miller, C. J. Pierotti, R. N. Shiras, and Mott Souders, Jr.: *Trans. A.I.Ch.E.*, **41**:631 (1945).
9. Gerster, J. A., J. A. Gorton, and R. B. Eklund: *J. Chem. Eng. Data*, **5**:423 (1960).
10. Gerster, J. A., T. S. Mertes, and A. P. Colburn: *Ind. Eng. Chem.*, **39**:797 (1947).
11. Griswold, John, and J. Dinwiddie: *Ind. Eng. Chem.*, **34**:1188 (1942).
12. Guinot, H., and F. W. Clark: *Trans. Inst. Chem. Engrs.* (*London*), **16**:187 (1938).
13. Happel, J., P. W. Cornell, D. B. Eastman, M. J. Fowle, C. A. Porter, and A. H. Schutte: *Trans. A.I.Ch.E.*, **42**:189 (1946); Disc., 1001 (1946).
14. Hess, H. V., E. A. Narragon, and C. A. Coghlin: *Chem. Eng. Progr. Symp. Ser.*, no. 2:72–79 (1952).
15. Mertes, T. S., and A. P. Colburn: *Ind. Eng. Chem.*, **39**:787 (1947).
16. Othmer, D. F.: *Chem. Met. Engr.*, **42**:356 (1935).
17. Perry, J. H.: "Chemical Engineers' Handbook," 3d ed., McGraw-Hill, New York, 1950.
18. Poffenberger, N., L. H. Horsley, H. S. Nutting, and E. C. Britton: *Trans. A.I.Ch.E.*, **42**:815 (1946).
19. Robinson, C. S., and E. R. Gilliland: "Elements of Fractional Distillation," 4th ed., McGraw-Hill, New York, 1950.
20. Swanson, R. W., and J. A. Gerster: *J. Chem. Eng. Data*, **7**:132 (1962).

part 4

Fractionating plate and column design

chapter 12

Fractionation devices—equilibrium stages

Over the period of time that fractional distillation has been practiced, many vapor-liquid contacting methods and devices have been used in an attempt to attain equilibrium between the phases in contact. Since the rate of mass transfer is increased in proportion to the area of contact between the phases, the devices were designed to obtain the maximum interfacial area. For constant vapor and liquid rates the contact time between the phases is constant, and increased contact area increases the mass-transfer rate, thus providing a greater possibility for attainment of true equilibrium which represents 100% efficiency. However, the combination of design of contact device, operating conditions, and system characteristics rarely produces 100% efficiency. Some devices contribute toward attainment of a better efficiency more than others and, under different conditions, the order of effectiveness may be interchanged.

The fractionation process conventionally is conducted in columns. Those which use equilibrium-stage devices to obtain vapor and liquid contact are designated as *equilibrium-stage columns* and those which utilize differential contact between vapor and liquid on the surface of packing are designated as *differential columns.*

Equilibrium-stage columns produce a change in the concentration of the vapor and liquid on plates or trays, each one of which ideally represents an "equilibrium stage." Although there undoubtedly is some mass transfer taking place between the liquid spray and vapor between the plates, and between the vapor and liquid in the form of foam above the plate and in the downcomers, the principal mass transfer is generally attributed to the contact of the two phases on the plate or in the vicinity of the plate. Also, the equilibrium-stage column has certain other features which differentiate it from a differential column: a means of maintaining a definite liquid depth on the plate is provided in the form of a weir or dam; and a conduit or pipe is provided to conduct the liquid from a plate above to the next plate below.

Differential columns enable the mass transfer between phases to take place over each increment of height throughout the entire height of the column. Although this normally occurs on packing placed in the column shell to provide a large surface area, it can take place in "pseudostage" columns, both on the plate and in the space between the plates. The principal difference between the differential column and the plate column is that the vapor and liquid must pass countercurrent to one another through the same passages in the differential column while the two phases are provided with entirely separate paths in the plate or stage column, except, of course, when they are in contact on the plate.

Thus a column in which both vapor and liquid has to pass through the same passage ways must be considered a differential column when downcomers are not provided, even if plates such as perforated plates, grid plates, and barred plates are used for vapor-liquid contact. Such columns are considered types of packed columns and are subject to the same considerations as packed columns.

12.1. THE BUBBLE-CAP PLATE

The oldest widely used "equilibrium-stage plate" is the bubble-cap plate (see Fig. 12.1). Bubble-cap plates comprised the majority of equilibrium-stage fractionating devices during the initial development of the distillation industries. However, recent developments and improvements in the knowledge of principles of distillation and in the hardware design have resulted in the specification of a relatively small percentage of bubble-cap columns compared with other types for new installations. Nevertheless, a large number of bubble-cap columns are in use and will be in use for some time in the future, and engineers in the distillation field must be familiar with their design and characteristic behavior.

The bubble-cap plate consists of a flat plate which is bolted, riveted, or welded to the walls of the column and which is provided with holes, *chimneys* or *risers* over the holes, and caps in the form of inverted cups or troughs over the risers. The caps are provided with slots or holes through which at least the major portion of the vapor passes to be mixed with the liquid flowing across the plate. Each plate is provided with one or more conduits called *downcomers* or *downspouts* through which the liquid flowing across the plate is conducted to the plate below. A liquid head or seal is maintained on the

FIGURE 12.1 Bubble-cap plate. (*Courtesy Fritz W. Glitsch and Sons, Dallas, Texas.*)

plate by a dam placed on the outlet side of the plate near the downcomer(s). This is commonly called the *outlet weir*.

There are other fixtures added to the plate, in some cases for special purposes, but basically the bubble plate includes only the parts described. The flow pattern of vapor and liquid in the column is shown schematically in Fig. 12.2.

The bubble-cap plate can operate at low vapor and liquid rates because liquid and foam is trapped on the plate to a depth at least equal to that of the weir height. All plates are provided with weep holes so that the plates will drain when the column is not in operation. At extremely low rates much of the liquid will drain through the holes and the liquid head will be much less than the weir height.

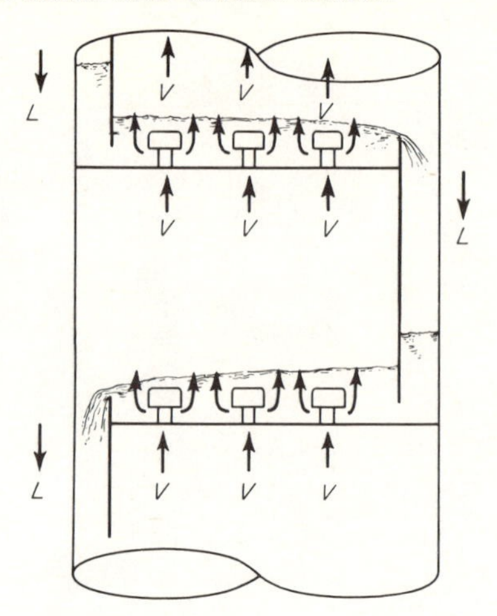

FIGURE 12.2 Schematic vapor and liquid flow in a bubble-cap column.

Bubble Caps—Types

Originally it was thought that the principal resistance to mass transfer in a distillation process was in the vapor phase. Thus a bubbling device seemed to be the most logical type of contact device because the greatest area per unit volume of vapor phase serving as contact area results from a discontinuous vapor phase in a continuous liquid phase. For this reason, for many years the bubble cap—and its modifications—was the standard vapor-liquid contacting device in fractionating distillation.

In general, a bubble cap is an inverted cup or trough which is provided with a mechanical support that fastens it to the plate on or at some distance from the plate. The fastening device may consist either of extensions from the edge of the cap which are bolted to the plate surface or of bolting arrangements which enable the cap to be bolted to the pipe or riser which forms a necessary part of the cap assembly. The riser conducts the vapor from the vapor space below the plate into the cap where it passes through the slots into the liquid. Typical cap assemblies are shown in Fig. 12.3.

There is a wide variation in cap shape and size, as well as in slot size, shape, and arrangements. Figure 12.4 shows a number of types manufactured by cap suppliers.

Various slot shapes have been used: rectangular, trapezoidal, square, diamond-shaped, triangular, oval, and round. They are located in the side of the cap or extending to the lower edge. In some instances, where extremely high vapor loads are encountered, no slots are provided and all of the vapor leaves the cap from under the edge or the skirt.

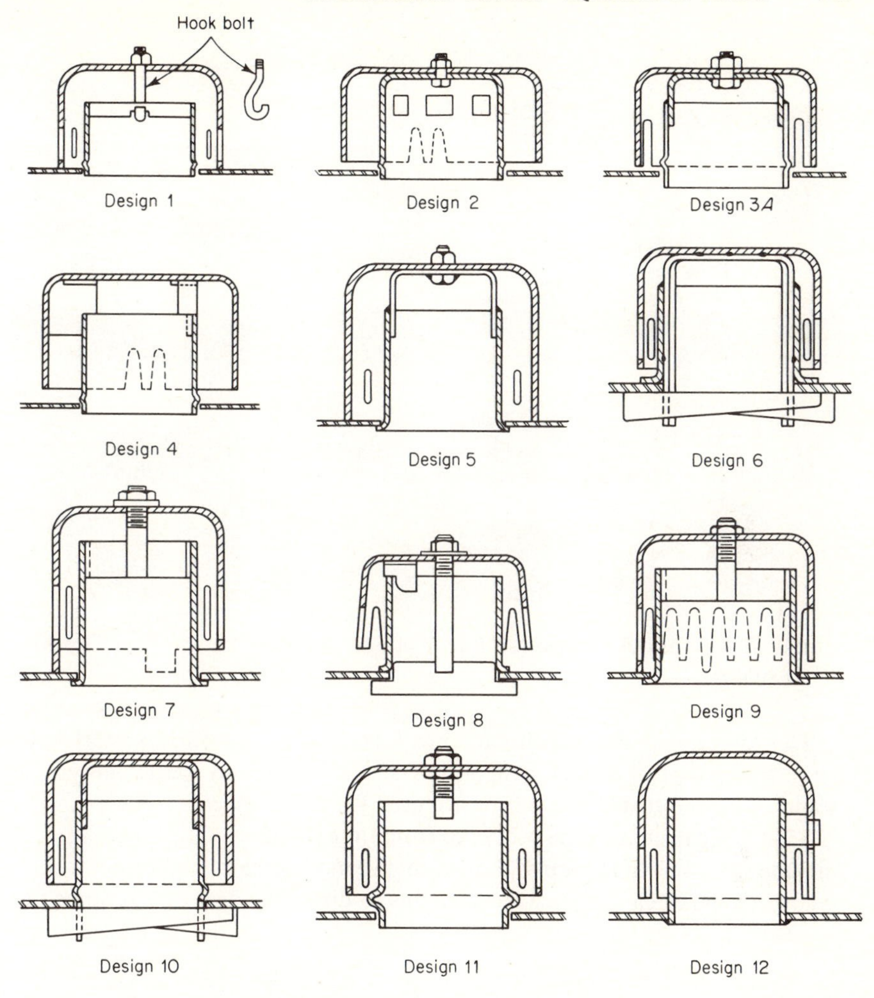

FIGURE 12.3 Typical bubble-cap and riser assembly designs. (*Courtesy The Pressed Steel Company, Wilkes-Barre, Pa.*)

Rectangular slots are the most commonly used at the present time, but trapezoidal slots and sawtooth slots are widely used also. Slot sizes found in commercial caps range from ⅛ to ⅜ in. wide and from ½ to 1½ in. long. The minimum slot spacing is by "rule of thumb" not less than 1.5 times the metal thickness. Commercial caps range in size from 1 in. in diameter used in small (6- to 12-in. diameter) columns up to 6 in. in diameter. Some caps are in the form of tunnel caps and range in width from 2 to 6 in. and are from 12 to 24 in. long. The number of slots is controlled by the dimensions of the cap and the slot size and spacing, and varies from 12 to 70.

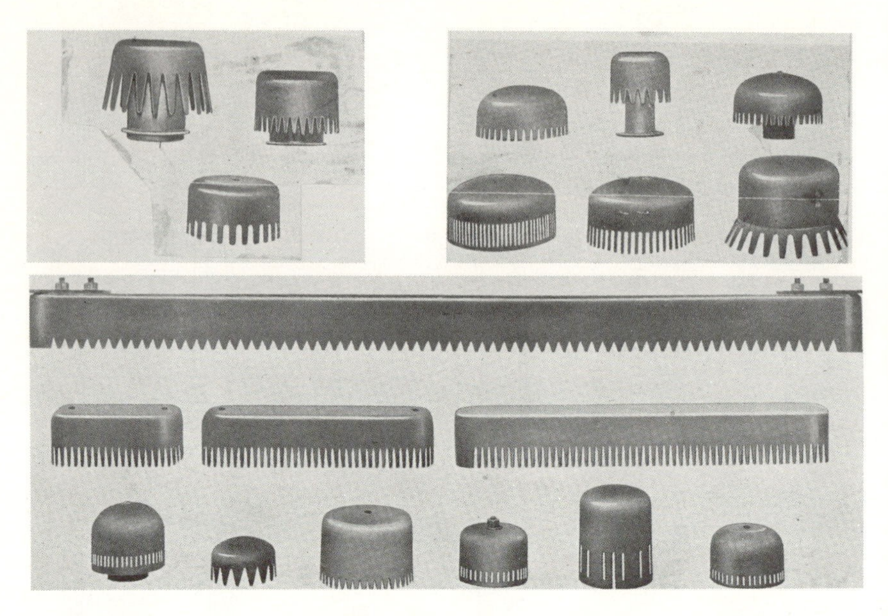

FIGURE 12.4 Various bubble-cap types. (*Courtesy The Pressed Steel Company, Wilkes-Barre, Pa.*)

The materials from which caps are formed are: low-carbon steel, alloy steel, stainless steel, nickel, copper, brass, aluminum, and special alloys. In some cases, molded ceramic caps are used for acid service.

Modified caps have been proposed from time to time. One type, Fig. 12.5, includes an insert in the top of the cap to give more streamlined vapor flow [1]. Another type of cap is provided with fins to direct the flow of the vapor into

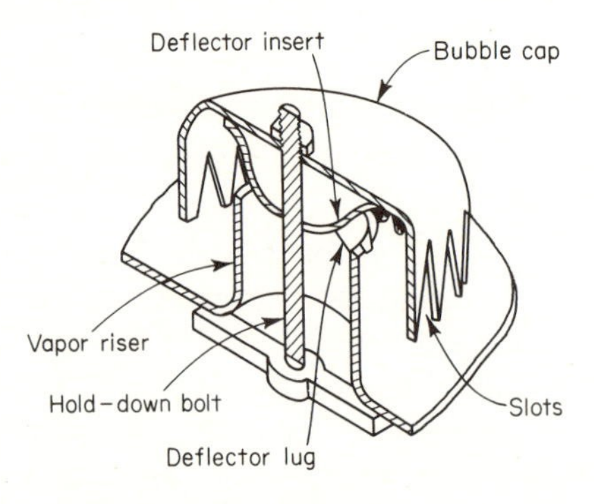

FIGURE 12.5 Bubble-cap design for streamlined vapor flow. [*From Chem. Eng. (January, 1955), p. 238.*]

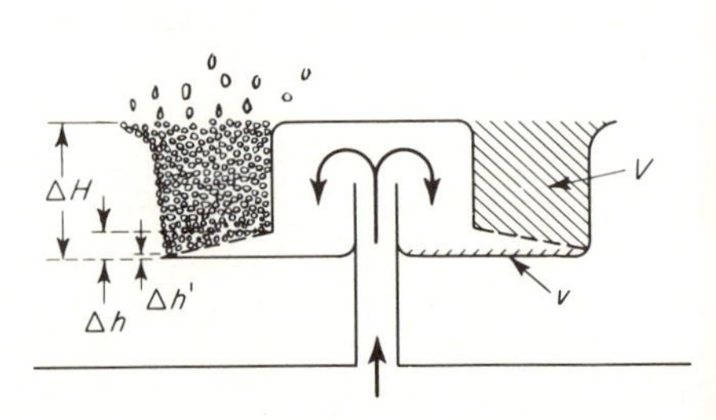

FIGURE 12.6 Lateral screen skirt cap. [*From Zenz, Petrol. Refiner (June, 1950), p. 103.*]

the liquid. The cap shown in Fig. 12.6 has a screened skirt provided to increase the area of contact between the vapor and liquid [10].

Risers

Risers are pipes, usually of circular cross section but may be rectangular for tunnel caps, which conduct the vapor from the vapor space below the plate to the annular space in the bubble cap above the slots. In most cases the riser acts as a support and anchor for the cap so that vapor impingement and turbulence in the liquid will not move it from its position above the riser. Risers may be bolted, welded, riveted, or wedged to the plate. The materials from which the risers are made are the same as those for the caps.

The riser area or cross-sectional area of the riser is usually selected so that approximately the ratios of slot area/riser area = riser area/annular area = 1.0 to 1.1. The selection of cap diameter therefore fixes the riser and annular area.

The caps are mounted on the plate in a number of ways relative to the riser. They may be bolted, welded, wedged, or pinned to the riser. Some of the arrangements are shown in Fig. 12.3.

Skirt Clearance

Caps are mounted on the plate so that the clearance between the bottom of the cap and plate surface ranges from zero to as much as 2½ in. in extreme cases. It is common practice to limit skirt clearances between 0.5 to 1.5 in. In some cases where vapor loads are low, the caps may be butted on the plate surface and the slots are extended to the lower edge of the cap. Most engineers consider this poor practice.

Bubble-cap Arrangement

Bubble caps are usually arranged on the plate at the corners of equilateral triangles with the rows oriented normal to the direction of flow across the plate, as shown in Fig. 12.7. The caps may be closely spaced or widely spaced from one-quarter of the cap diameter to one cap diameter between the walls.

Flow Arrangement

The fractionating plate must be designed to meet certain requirements:

1. It has to provide a high degree of contact between the vapor and liquid phases so that equilibrium between the vapor and liquid may be approximated.

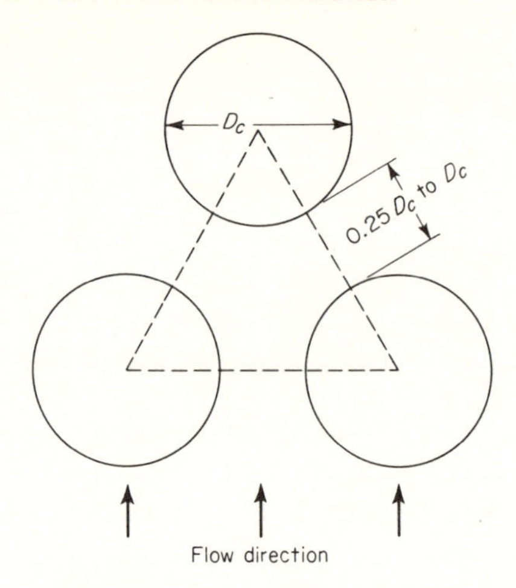

FIGURE 12.7 Equilateral triangular cap spacing.

2. It must cause a minimum of pressure drop throughout the column while providing the necessary contact between the phases.

3. It has to provide adequate liquid flow paths on the plate and in the downcomers such that the liquid-handling capacity of the column is well within required limits.

In order to satisfy these requirements, a number of plate layouts have been used conventionally with no particular attempt at standardization. The principal ones are shown in Fig. 12.8. These layouts serve for all types of equilibrium-stage plates.

The liquid flow pattern on a fractionator plate is chosen to give the best contact with the vapor, with the minimum hydraulic gradient or buildup of liquid across the plate. The "best" contact is that which provides the optimum contact time and area to give the closest approach to equilibrium. Each of the design layouts shown in Fig. 12.8 was devised to increase the contact time between vapor and liquid with a relatively small hydraulic gradient, and the selection of any one type is based primarily on column diameter or on the distance which the liquid must travel from entrance to exit on the plate.

Hydraulic gradient is a term applied to the difference in height of liquid on the plate at the point of entrance and at the outlet weir, as shown in Fig. 12.9. This liquid head is necessary to aid in causing the liquid to flow across the plate. Besides cascading the plate to avoid excessive hydraulic gradient, stepping bubble caps to maintain essentially the same submergence has been proposed [3]. Another arrangement is that of tilting the tray so the gradient

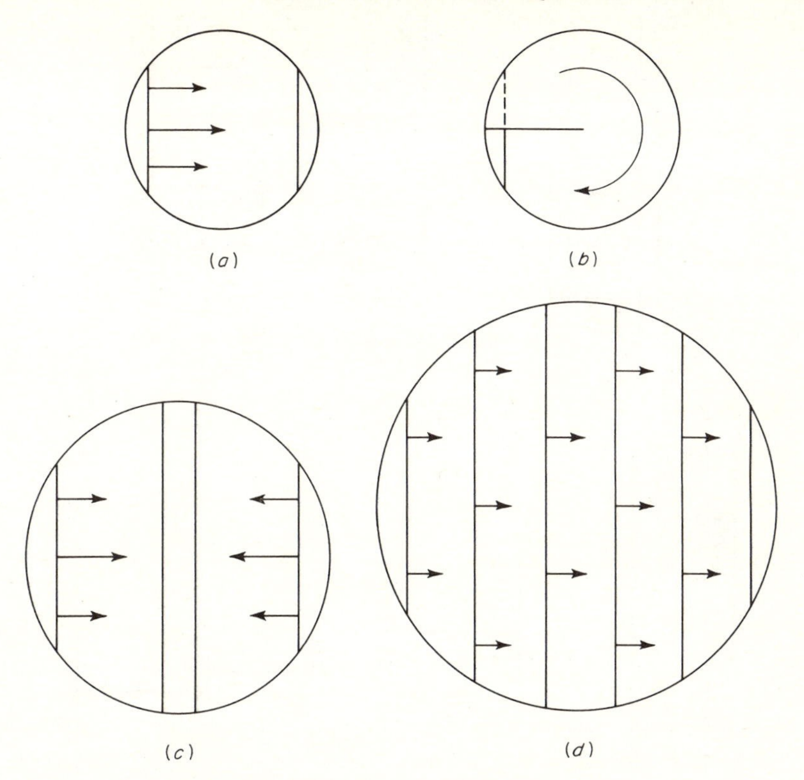

FIGURE 12.8 Various flow patterns on fractionator plates. (*a*) Cross flow, 1 to 6 ft. (*b*) Reverse flow, 6 to 9 ft. (*c*) Double-cross flow, 9 to 12 ft. (*d*) Multiple-dam cascade, 6 to 35 ft.

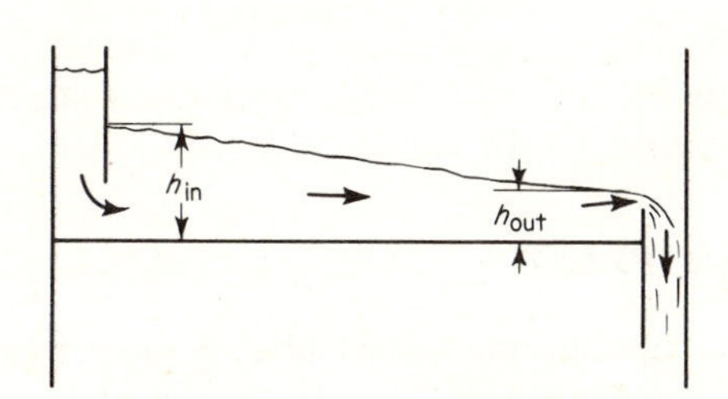

FIGURE 12.9 Hydraulic gradient.

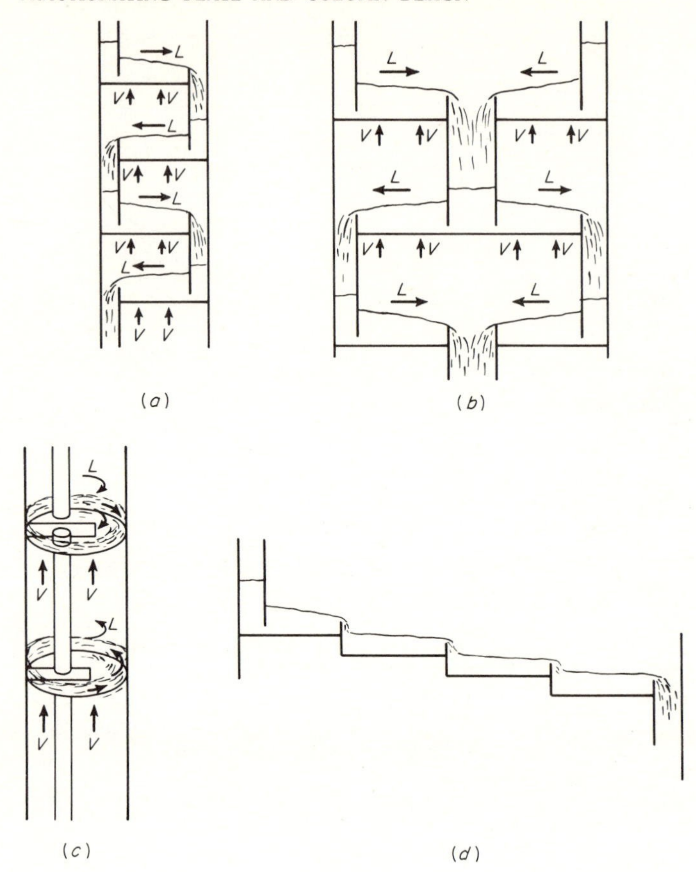

FIGURE 12.10 Vapor and liquid flow patterns in columns. (*a*) Cross flow. (*b*) Double-pass flow. (*c*) Single reverse. (*d*) Cascade.

is decreased and stepping the caps to maintain essentially the same submergence [2]. Schematic flow patterns for vapor and liquid in the various types of column arrangements are shown in Fig. 12.10.

12.2. PERFORATED PLATE

The least expensive equilibrium contact plate is the perforated plate (Fig. 12.11). The vapor flows through the holes in the plate and is dispersed through the liquid flowing across the plate. Downcomers and outlet weirs must be provided, and inlet weirs and splash baffles may be included as part of the plate assembly. Because the holes provide the means of vapor introduction into the liquid, and the holes may be punched in the plate material, the construction is inexpensive compared with bubble-cap trays. Also their cost is less than that of other types of perforated trays in which valves are provided for the holes to give a variable-orifice flow effect, or the plate is ridged, corrugated, or subjected to some type of machine or forge work.

Although the perforated plate has a higher capacity than a bubble-cap tray at essentially the same maintained efficiency—and under certain conditions,

FIGURE 12.11 Large-diameter perforated plate. (*Courtesy Fritz W. Glitsch and Sons, Dallas, Texas.*)

higher capacity at higher efficiency—it has one disadvantage which becomes serious at lower vapor rates than those for which the plate was designed to operate. Because the vapor flow prevents the liquid from flowing down through the holes in the plate to the plate below, each plate design has a minimum operable vapor velocity below which the plate "dumps" or "showers"—a situation wherein the liquid is flowing freely through the holes. This minimum vapor velocity of "weep point" seriously restricts the use of perforated plates in services where flexibility of operation with respect to the regions of low vapor flow rate—and to a lesser extent with respect to low liquid flow rate—is necessary. On the other hand, however, the capacity of the plate with respect to high vapor throughput more than compensates for the lower limit of operation in most industrial service. This results from the practice of operating columns under essentially steady-state conditions, and the tendency is always to increase rather than decrease loading.

Hole Size and Arrangement

Hole size and arrangements are varied depending upon the service and the designer's preference. Hole sizes range from $\frac{1}{16}$ in. in diameter in small columns using 16-gage metal to more than 1 in. in diameter in plate thicknesses up to $\frac{1}{4}$ in. The larger hole sizes are more self-cleaning but better vapor-liquid contact is obtained with smaller holes. Except in special cases the most widely specified hole sizes range from $\frac{1}{8}$ to $\frac{3}{8}$ in. in diameter and the commonly specified plate thicknesses are 12 to 16 gage. The general rule for punching plain carbon steel or copper plate is that the thickness of the plate may not be greater than the diameter of the holes. For stainless steel the thickness (for punching) is limited to $\frac{1}{2}$ to $\frac{2}{3}$ the hole diameter. The hole arrangement may be either on a triangular pitch or square pitch, as shown in Fig. 12.12.

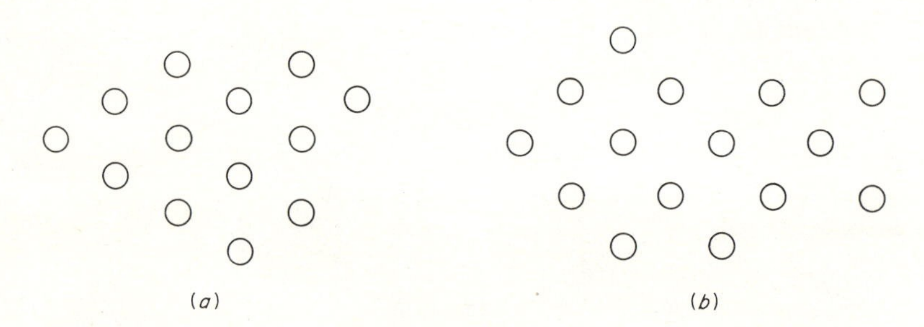

FIGURE 12.12 Hole arrangement, pitch/diameter = 3. (*a*) Triangular. (*b*) Square.

Hole spacing or pitch in terms of pitch-to-hole-diameter ratio ranges from 1.0 to 5.0. The selection of the percent-free area or total hole area divided by the column area, and the hole diameter fixes the pitch-to-diameter ratio. The percent-free area usually ranges from 5% to 15%, with 10% most commonly specified. The pitch-to-hole-diameter ratio usually falls between 2.5 and 4.0.

The plate thickness related to the hole diameter is a factor in pressure drop. The greater the thickness-to-diameter ratio the lower the pressure drop. Unfortunately, the punchability of the plate severely limits the plate thickness. In addition, the added cost of thicker plates is not justified. Where relatively large-diameter plates are needed, thin metal can still be used, and necessary support bars and structure can be added to support the plate and the liquid contained on it.

(*a*)

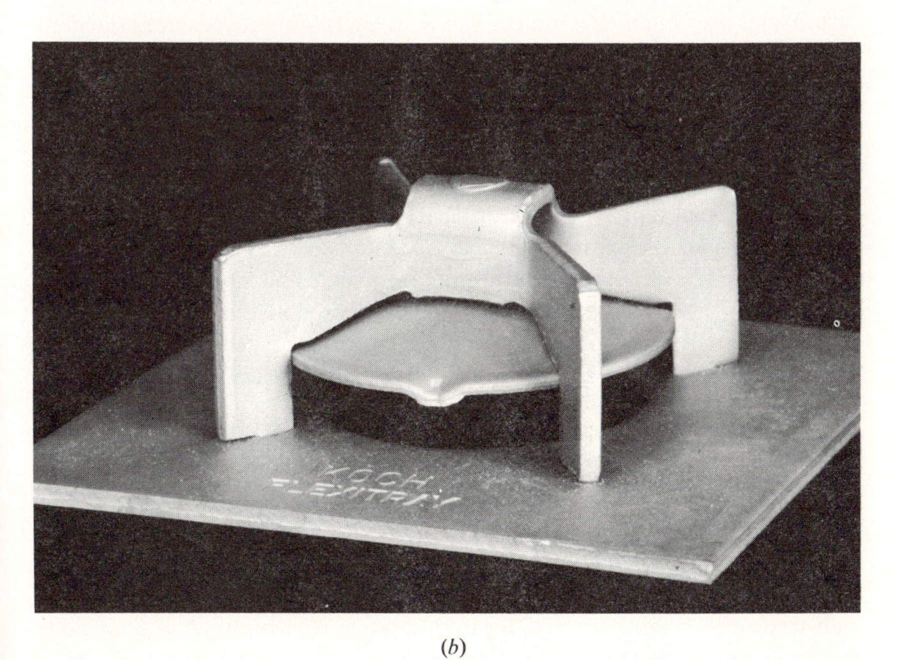

(*b*)

FIGURE 12.13 (*a*) Koch tray, type *T*. (*b*) Koch unit, type *T*. (*Courtesy Koch Engineering Company, Wichita, Kansas.*)

12.3. VARIABLE-ORIFICE PERFORATED PLATES

Because of the limitation on the operability of conventional perforated plates at lower vapor loadings, devices have been developed to enable the vapor loading to fix, within limits, the area of the holes available for vapor flow. These devices may be circular valve disks held in place over the holes in the tray by guides, either fastened to the plate surface or to the movable valve disk itself. These valve disks will rise as the vapor rate is increased and they remain open or settle intermittently over the holes as the vapor load becomes less. Such devices may also be of the tip-lift or trap-door type where the movable metal plate is heavier at one end than the other and opens proportionately with increased vapor flow rate. The upper limit of opening is controlled by restrictive bars or cages.

Trays of the former type, wherein the valve disks are circular and free floating, are the Koch Engineering Flexitray [7] and the Glitsch ballast tray [9]. The Flexitray has two distinct valve designs, the type *T* and type *A*. These are illustrated in Figs. 12.13 and 12.14. Generally two weights of valve disks are used, alternating in rows across the plate in the direction of liquid flow. This is for the purpose of giving better vapor distribution throughout the vapor flow range. In some instances, smaller diameter perforations are combined with the valves in alternating rows across the plate to give flexibility in the higher flow range combined with a more economical design. Figure 12.15 illustrates one type of Flexitray design.

The Glitsch ballast unit has a number of possible variations. There are principally two basic units—the *V* unit and the *A* unit. The type *V* unit consists of a circular valve disk to which guides are fastened that allow the valve disk to open to a fixed limit or partly to close. The type *A* unit has an orifice cover and a separate ballast disk contained in a fixed cage over the hole. Figure 12.16 shows a type *A* unit and a plate assembly. Figure 12.17 shows a type *V* unit and a plate assembly composed of these units.

The Nutter valve tray [5], shown in Fig. 12.18, is the type which opens from one side while partly open and lifts off the plate surface to its maximum opening limited by the restraining bracket.

12.4. OTHER EQUILIBRIUM-STAGE TRAYS

Uniflux

The Uniflux tray [4] was developed by Socony-Mobil Oil Company, Inc., as an improved bubble tray. The tray consists of a number of S sections extending across the tray. The interlocking S forms longitudinal vapor passage ways or risers, and the liquid flows across the tops of the sections where it is mixed with the vapors. The advantages claimed for this tray are

(*a*)

(*b*)

FIGURE 12.14 (*a*) Koch tray, type *A*. (*b*) Koch unit, type *A*. (*Courtesy Koch Engineering Company, Wichita, Kansas.*)

FIGURE 12.15 Koch tray, type *AF*. (*Courtesy Koch Engineering Company, Wichita, Kansas.*)

flexibility of operation without loss in efficiency. (These have been operated as low as 12% of maximum capacity without liquid backflow.)

Benturi [6, 8]

The Benturi tray, which has had limited use, was a tray designed for low pressure drop and high vapor capacity. Its design was unique in that it had built-in entrainment-separating devices which consisted of curved perforated baffles against which the vapor impinged as it left the liquid-vapor contact area on the plate.

12.5. PLATE COMPONENTS

As stated ealier, equilibrium-stage plates or trays have certain things in common which distinguish them from differential fractionation devices such as a packed column. These items consist primarily of liquid downcomers and weirs and vapor and liquid disengaging space.

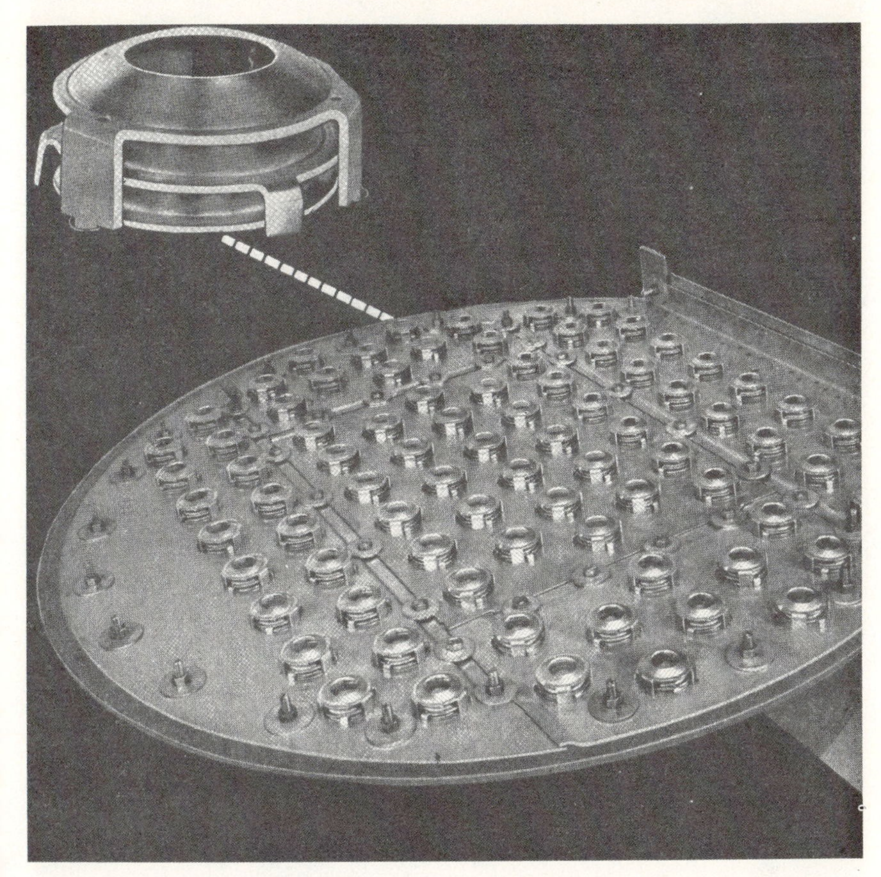

FIGURE 12.16 Glitsch ballast tray, type *A* unit. (*Courtesy Fritz W. Glitsch and Sons, Dallas, Texas.*)

Downcomers

Downcomers are provided on all equilibrium-stage trays to conduct the flow of liquid from the trays above to the trays below. They are designed to provide adequate liquid-handling capacity for the distillation column and, at the same time, occupy a minimum of tray cross-sectional area, so that the active area of the tray will be at a maximum. Downcomers may be circular or segmental in shape, single or multiple. Circular downcomers are used only in small columns or for some special reason.

The segmental downcomer whose upper edge forms the segmental weir is the most commonly used type of downcomer. In multiple-flow plate arrangements, i.e., double-pass flow, four pass-flow, etc., the (essentially)

FIGURE 12.17 Glitsch ballast tray, type *V*. (*Courtesy Fritz W. Glitsch and Sons, Dallas, Texas.*)

FIGURE 12.18 Nutter float-valve tray. (*Courtesy Nutter Engineering Company, Tulsa, Okla.*)

rectangular downcomer is used in conjunction with the segmental form. Figure 12.19 illustrates the type. Segmental downcomers may have vertical aprons or slanted or inclined aprons which provide a greater volume for the froth on top of the liquid in the downcomer. This allows more time for disengagement of the vapor and liquid.

The capacity of the downcomer should be designed for the maximum liquid load. This rating can be expressed in terms of the net average linear

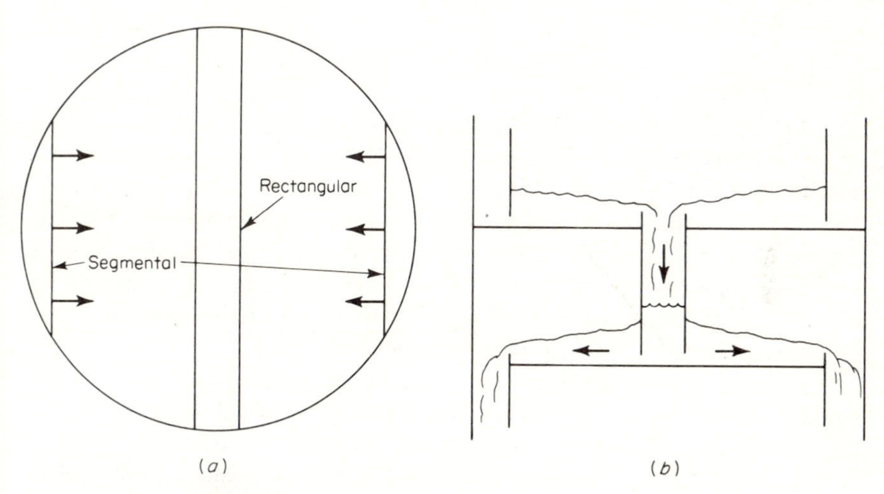

FIGURE 12.19 Segmental and rectangular downcomers.

velocity of liquid flow through the downcomer assembly by dividing the volume of liquid flowing per unit time by the cross-sectional area of the downcomer. Where foaming systems are encountered, clear liquid linear velocities as low as 0.05 to 0.1 fps are common. In non-foaming systems, clear liquid velocities as high as 0.3 fps are suitable.

Downcomer capacity can also be expressed in terms of mean residence time of the liquid in the downcomer. This is calculated by determining the volume of liquid held in residence in the downcomer and dividing the total volume of liquid flow per unit time into this. The usual practice is to design the downcomer for a residence time of 5 to 6 sec where foaming is encountered and for a residence time of around 3 sec where the system is nonfoaming.

The downcomer should extend to a distance above the surface of the plate below which will give at least the area of flow for the liquid between the edge of the downcomer and the plate surface equal to the area of flow in the downcomer. Also the downcomer should extend into the liquid to seal against vapor flow up the downcomer. For this purpose the height of weir minus the clearance under the downcomer is maintained between 1.0 and 1.5 in.

Plate Spacing

The plates are separated by some distance to enable the liquid droplets and foam to separate from the vapor before it reaches the plate above. It will be shown in the following chapters that plate spacing is a function of many variables which must be calculated. However, it can be said here that plate spacing will range from 12 in. in small columns to 48 in. in large vacuum columns, with the commonest spacing around 24 in. for most commercial fractionators.

Overflow or Outlet Weirs

Outlet weirs are merely dams which maintain a predetermined liquid head on the plate so that the vapor rising from the plate below must bubble or flow through the liquid on the plate before it escapes to the plate above. The weirs are usually segmental since they are merely an extension of the inner edge of the segmental downcomer. In some cases adjustable weirs are provided which enable adjustment of the weir height over a range of 1 to 2 in. The adjustable sections are slotted bars which are bolted to the edge of the downcomer. Notched weirs are sometimes used.

The height of the weir is generally determined by the requirement of the liquid depth on the plate which is controlled by the depth of seal over the vapor passages. This may be either static seal, h_{SS}, or dynamic seal, h_{dS}. Figure 12.20 illustrates the dimensions of the terms h_{SS} and h_{dS} for bubble-cap plates, perforated plates, and variable-orifice plates. The static seal varies with the allowable pressure drop. The dynamic seal is defined by

$$(h_{dS})_{\text{av}} = h_{SS} + h_{OW} + \tfrac{1}{2}\Delta$$

It is the equivalent clear-liquid seal over the point at which the vapor emerges onto the plate proper—the top of the slot in the case of bubble caps and the surface of the plate in perforated and valve plate designs. Because of the liquid gradient on the plate, the dynamic seal is greater on the inlet side of the plate than on the exit side.

The length of segmental weirs ranges from 0.5 times the tower diameter to around 0.8 times the tower diameter. For single-pass and split-flow trays, weir lengths are usually 0.6 to 0.75 times the tower diameter; for double-pass trays, they are 0.5 to 0.6 times the tower diameter.

Inlet Weirs

In some few cases inlet or distributing weirs are provided (Fig. 12.21). The inlet weir distributes the liquid flow evenly over the plate because it maintains somewhat of a calming section for the liquid falling from the plate above before it is introduced to the bubbling area in the plate. In addition,

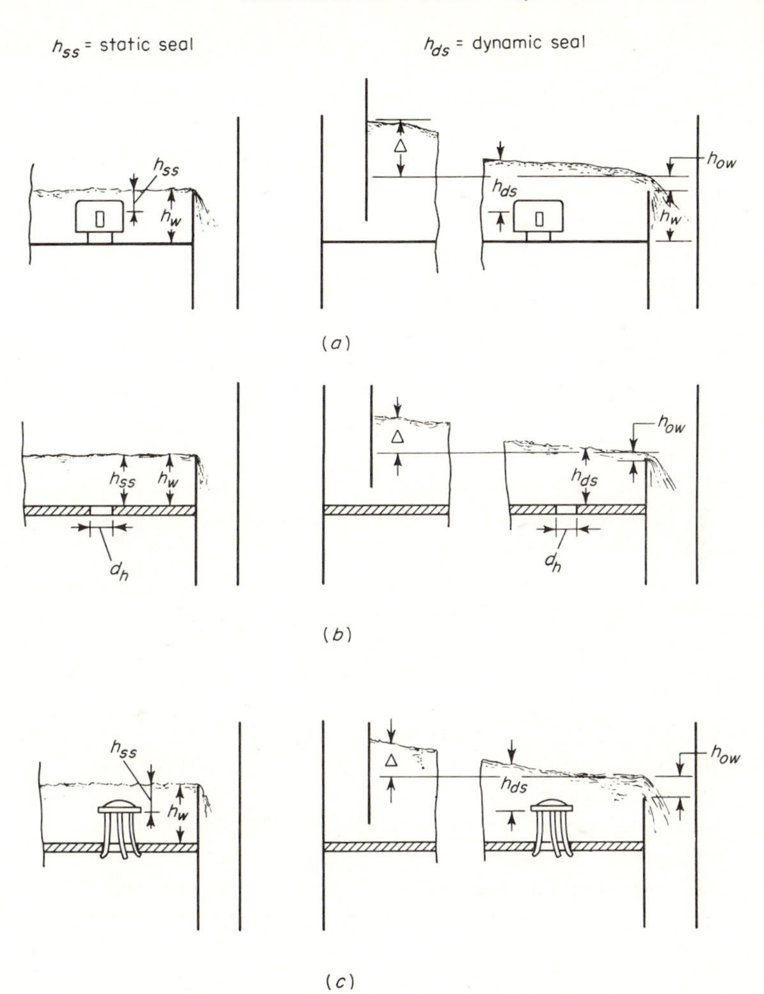

FIGURE 12.20 Static seal and dynamic liquid seal. (*a*) Bubble cap. (*b*) Perforated. (*c*) Valve or variable orifice.

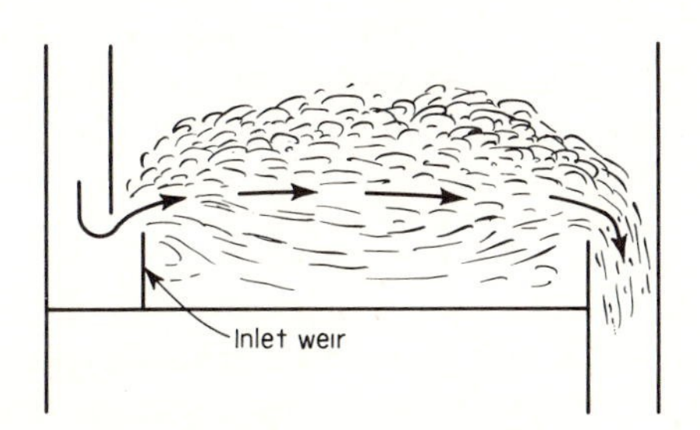

FIGURE 12.21 Inlet weir.

inlet weirs provide a barrier to the liquid, possessing a large amount of inertia or flow energy, impinging directly on the vapor passages or cap slots. This is sufficient in many cases to prevent the proper operation of the caps or perforations in the first row and, in some cases, in the first three or four rows because the pressure thus exerted shuts off the vapor flow.

To introduce the liquid coming on to the plate without using inlet weirs but still preventing direct impingement of the liquid on the caps or other vapor-liquid contacting devices, seal pots are sometimes provided. Essentially, a seal pot is a recess between the column wall and the plate into which the downcomer extends and from which the liquid flows up and over the plate surface. Examples of this type of arrangement are shown in Fig. 12.22.

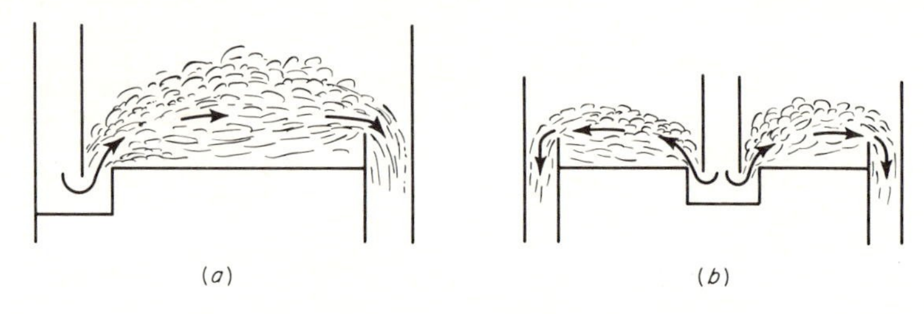

FIGURE 12.22 Seal pots.

Neither of the above arrangements is satisfactory where dirty liquids are being distilled or where solids from any source can accumulate behind the weirs or in the seal pots. Flooding of the column will result from the blocking of the liquid passage on to the plate from the downcomer.

Generally, seal pots are more expensive and difficult to fabricate than flat plates with weirs and, consequently, they are not widely used. If used, the seal depths are similar to those for the regular downcomer arrangement, i.e., 1 to 1.5 in.

Redistribution Baffles

Where the end caps in the row are placed so that the cap-to-wall clearance is 1 in. (or more) larger than cap-to-cap wall spacing, redistribution baffles can be placed so that the baffle to cap-to-wall distance equals cap spacing. The baffles should be twice the calculated liquid depth in height. These are only used in special cases (see Fig. 12.23).

Splash Baffles

Where the distilling system is of such a nature or the operating conditions are such that excessive splashing or foaming might take place, it is common practice to provide splash baffles near the outlet weir. The splash baffle

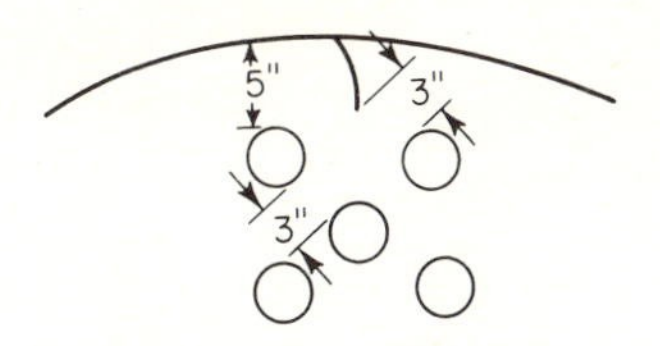

FIGURE 12.23 Redistribution baffle spacing.

usually consists of a flat plate extending either from the plate above or fastened to the plate or weir such that the lower edge extends a short distance into the liquid flowing over the weir. Any spray or foam which is thrown into the vapor space because of the vapor-liquid interaction on the plate hits the baffle and only clear liquid is allowed to flow over the weir. Location of splash baffles should be such that no obstruction to liquid flow across the plate will be introduced.

REFERENCES

1. Anon.: *Chem. Eng.* (January, 1955), p. 238.
2. Kelley, C. A., and S. D. Lawson: To Phillips Petrol. Co., U.S. 2,539,142 (January 25, 1957).
3. May, J. A., and J. C. Frank: *Chem. Eng. Progr.*, **51**:189 (1955).
4. Muller, H. M., and D. F. Othmer: *Ind. Eng. Chem.*, **51**:625 (1959).
5. Nutter, I. E.: *Oil Gas J.* (April 26, 1954), p. 165.
6. Thornton, D. P.: *Petrol. Process.* (May, 1952), p. 2.
7. Thrift, G. C.: *Chem. Eng.*, **61**:177 (1954).
8. Thrift, G. C.: *Petrol. Eng.* (May, 1954), p. C-26.
9. Winn, F. W.: *Petrol. Refiner* (October, 1960), p. 145.
10. Zenz, F. A.: *Petrol. Refiner* (June, 1950), p. 103.

chapter 13

Plate and column hydraulics and efficiency

In a fractionating column the flowing system of vapor and liquid following the paths provided develop frictional expansion and contraction as well as orifice and inertial losses best expressed as head losses or pressure drop. Excessive pressure drop will result in "flooding" wherein the liquid is forced back up the downcomers from where it flows on to the plates above. This represents an inoperable situation. Where excessive pressure drop is encountered, even without flooding, the change in pressure over the column may be great enough to affect seriously the vapor-liquid equilibrium relations so that predicted separations cannot be obtained and properties of the phases such as, e.g., liquid and vapor entrainment may be encountered, thereby reducing efficiency. In vacuum fractional distillation very little pressure drop can be tolerated because it would tend to raise the column pressure to a point where the effectiveness of vacuum operation would be lost. For these and other reasons the design engineer evaluates the pressure drop or head loss for the fractionating plate and column to be used for a given service.

13.1. FLOW-ENERGY LOSSES IN PLATE AND COLUMN

The flow-energy losses in both vapor and liquid flow regions are summarized as follows and are indicated schematically on Figs. 13.1 and 13.2.

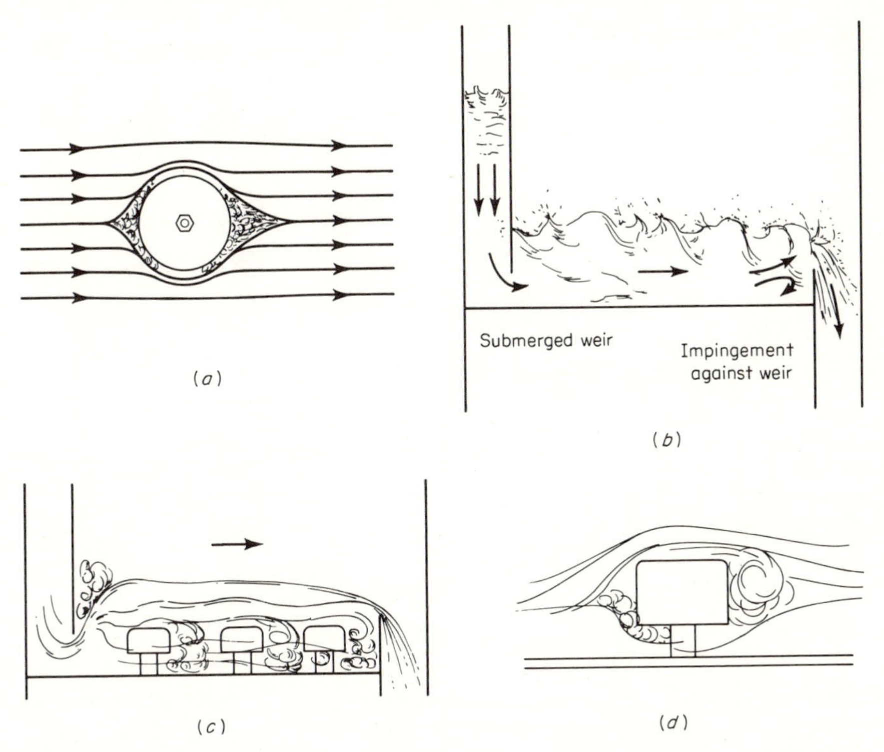

FIGURE 13.1 (*a*) Liquid flow pattern around cap. (*b*) Liquid flow resistance caused by weir and apron. (*c*) Liquid flow across bubble-cap plate. (*d*) Liquid flow pattern across cap.

Liquid

(*a*) Frictional losses caused by flow of the liquid through the downcomers

(*b*) Frictional losses caused by flow of the liquid under the edge of the downcomer, between the downcomer and inlet weir (if any), or between the downcomer and the side of the seal pot (if any)

(*c*) Inertial losses caused by the change in direction of the liquid under the edge of the downcomer and on to the plate

(*d*) Contraction and expansion losses caused by differences in cross-sectional area through which the liquid flows in the downcomer system

(*e*) Eddy losses caused by flow of liquid around bubble caps or around valve structures

(*f*) Eddy and frictional losses caused by the flow of the liquid around the inside periphery of the column

(*g*) Eddy losses caused by liquid flowing up over the weir

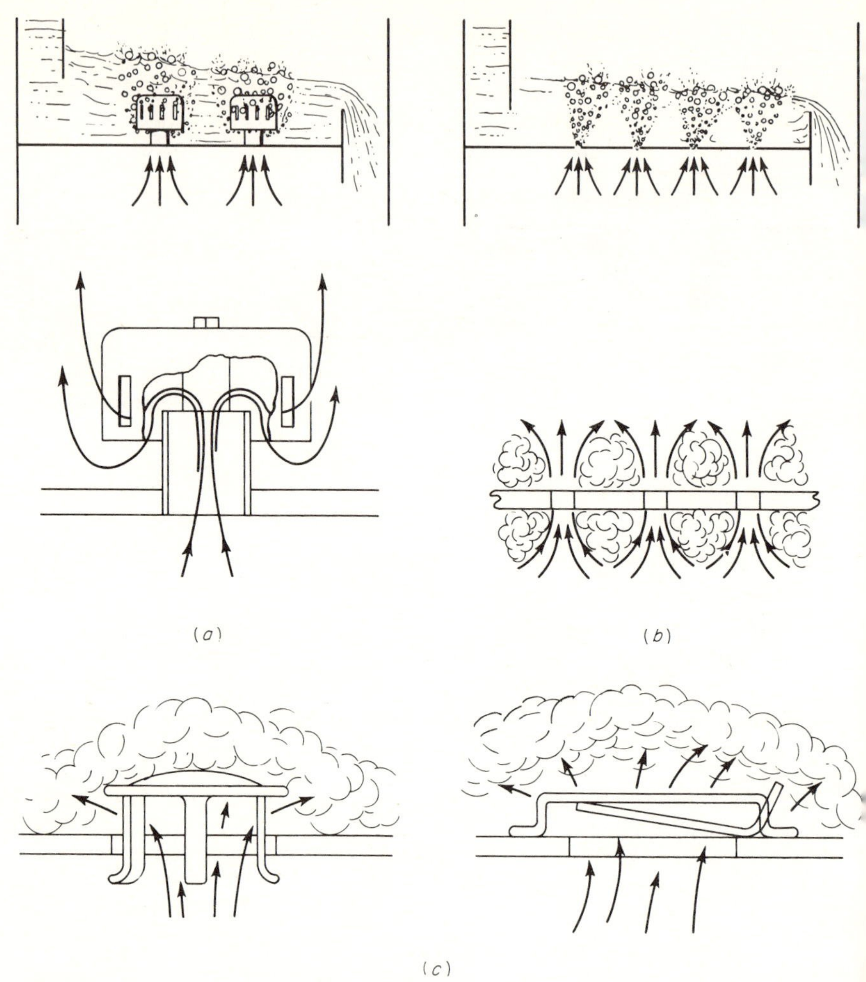

FIGURE 13.2 Flow through vapor passages. (*a*) Vapor flow through bubble cap. (*b*) Vapor flow through perforations. (*c*) Vapor flow through valves.

Vapor

1. Contraction losses of vapor flowing from space below the plate into the riser or into a perforation or hole

2. Expansion losses of vapor flowing from the riser into the annular space of bubble caps or under the edge of valves

3. Reversal and eddy losses caused by the vapor changing direction in the annular space in the cap assembly or valve assembly

4. Frictional losses through friction drag of the vapor in the riser and on the inside surface of the cap or inside surface of valve assembly

5. Expansion losses caused by the vapor flowing through the cap slots into the liquid, through the perforations into the liquid, or through the valve assembly into the liquid

6. Energy loss in overcoming the head of liquid above the openings leading into the liquid

7. Expansion energy losses in bubble expansion as it forms and rises through the liquid

8. Energy losses in the vapor overcoming the head of foam

9. Energy loss in raising the weight of the valve plate to the open position

In terms of column operation these losses are evidenced as pressure drop through the plate and through the column, which may result in abnormal temperature gradients from plate to plate throughout the column and as liquid backup in the downcomers which may become great enough to cause flooding. This would occur when the liquid from the plate below was forced back up the downcomer on to the plate above. Because of the importance of either or both of these effects on the performance of the column, it is necessary, if possible, to evaluate their extent quantitatively from consideration of the design, operating, and system variables.

Although it is recognized that the various frictional and other head losses exist in an operating column, not all of them can be calculated and some of the head-loss contributions are so small that neglecting them will introduce no serious error. Table 13.1 lists those head-loss terms commonly evaluated in pressure-drop and liquid backup calculations. Those not included are neglected.

Because there are differences in terminology encountered in the literature, it is considered necessary to define all terms used in this and the following discussions. The symbols to be used in the various equations utilized for

Table 13.1 Head-loss Terms Commonly Evaluated*

	Bubble cap	Valve	Sieve
h_d	*a, b, c, d*	*a, b, c, d*	*a, b, c, d*
Δ	*e, f, g*	*e, f, g*	—
h_0	—	—	5
h_{rc}	1, 2, 3, 4, 5	1, 4, 5	—
h_r	1	—	—
h_{ra}	2, 3, 4	—	—
h_{ds}	5	—	—
h_σ	—	—	7
$h_w + h_{ow} + f(\Delta)$	6	6	6
h_m	—	9	—

* Letters and numbers refer to the losses described under "Liquid" and "Vapor."

bubble-cap and perforated-plate pressure-drop calculations are defined as shown in Figs. 13.3 through 13.6. Calculations of pressure drop for valve plates use empirical factors supplied in the manufacturers' design manuals.

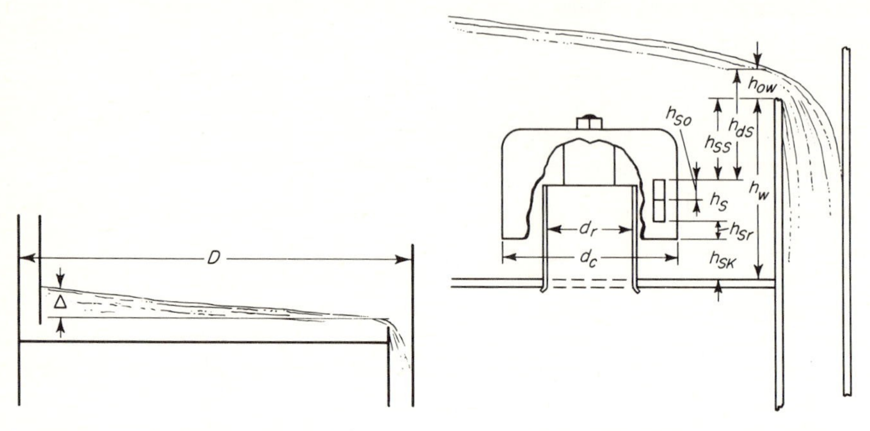

FIGURE 13.3 Definition of Δ.

FIGURE 13.4 Definition of symbols—bubble-cap plate.

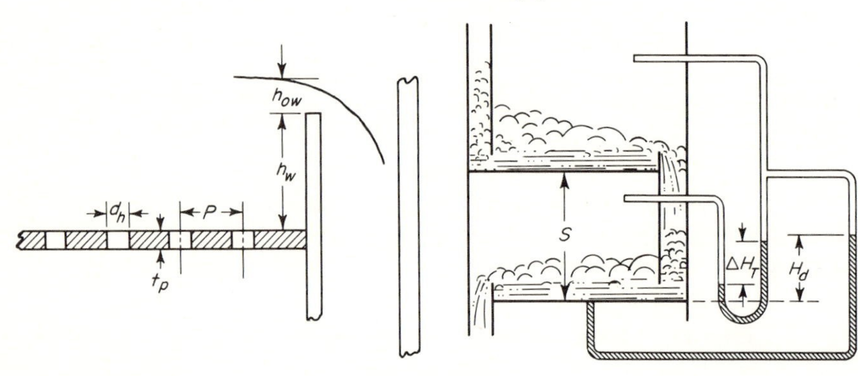

FIGURE 13.5 Definitions of symbols—perforated plate.

FIGURE 13.6 Total pressure drop and liquid backup.

13.2. EVALUATION OF PRESSURE DROP AND LIQUID BACKUP IN DOWNCOMERS

Numerous methods have been proposed for evaluation of pressure drop ΔH_T and the liquid backup H_D in fractionating columns. (For bubble-cap plates, see [11, 12, 17, 19, 24, and 55]; for perforated plates, see [45, 46, and 61]; for valve plates, see [38, 69, 80, 81, 82].) There are many similarities as well as differences in method for calculation of specific items of pressure drop and the calculated results from the various methods agree surprisingly well in most instances. Because of the similarity in results, only one method and the equations for calculation of pressure drop and liquid backup are given for each type of plate in Table 13.2.

Table 13.2 Pressure Drop and Liquid Backup Equations

	Pressure drop ΔH_T, inches of fluid	Liquid backup H_D, inches of aerated fluid
Bubble-cap plates	$\Delta H_{T,\,min} = h_{rc} + h_{so} + \beta h_{ds}$ (13.1*a*)	$H_{D,\,max} = (\Delta H_{T,\,max} + h_w + h_{ow} + \Delta + h_d)\dfrac{1}{\phi_d}$ (13.1*b*)
Perforated plates	$\Delta H_{T,\,min} = h_o + \beta\left(h_w + h_{ow} + \dfrac{\Delta}{2}\right) + h_\sigma$ (13.2*a*)	$H_{D,\,max} = \left[\Delta H_{T,\,max} + \left(h_w + h_{ow} + \Delta\right) + h_d\right]\dfrac{1}{\phi_d}$ (13.2*b*)
Valve plates	$\Delta H_{T,\,min} = h_v + \beta\left(h_w + h_{ow} + \dfrac{\Delta}{2}\right)$ (13.3*a*)	$H_{D,\,max} = \left[\Delta H_{T,\,max} + h_d + \beta\left(h_w + h_{ow} + \Delta\right)\right]\dfrac{1}{\phi_d}$ (13.3*b*)

Note: To convert inches of liquid to inches of water, multiply inches of liquid by $\rho_L/62.4$. To convert inches of liquid to pounds per square inch, multiply inches of liquid by $(\rho_L/62.4)(0.0361)$ or $\rho_L/1728$.

The terms in Eqs. (13.1) through (13.3) are defined and evaluated quantitatively by the following definitions and recommended equations.

General Pressure-drop Equations

h_w is the weir height in inches measured from the surface on the tray to the top of the weir. This is a design variable only.

h_{ow} designates the height of the liquid crest over the weir in inches of the fluid flowing. This is a function of the operating variable, liquid rate, and design variables, weir length, and column diameter. Equation (13.4) defines the term

$$h_{ow} = 0.48F_w\left(\frac{Q_L}{l_w}\right)^{0.67} = 30F_w\left(\frac{Q'_L}{l_w}\right)^{0.67} \tag{13.4}$$

F_w is read from Fig. 13.7 for segmental weirs; for circular weirs $F_w = 1.0$. Q_L is the liquid flow rate in gallons per minute, Q'_L is the liquid flow rate in cubic feet per second, l_w is the weir length in inches.

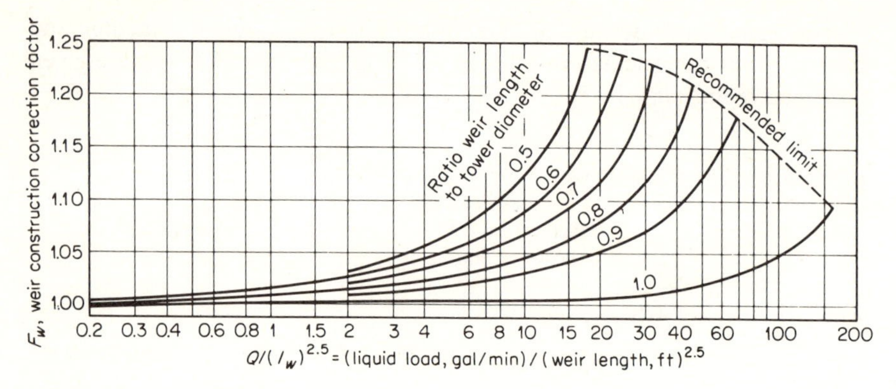

FIGURE 13.7 Correlation of F_w, weir formula correction factor, for constricting tower wall. [*From Bolles, Petrol. Refiner*, **25**:*613* (*1946*).]

h_{ss} is the head of liquid in inches of fluid system equal to the distance from the top of the slot to the top of the weir. This is a design variable.

h_d represents the liquid head in inches of fluid system equivalent to the pressure drop caused by the liquid flow in the downcomer. This is a function of liquid flow rate, characteristics of the liquid at the temperature of the downcomer and the downcomer design. It is evaluated by means of

$$h_d = 0.03\left[\frac{(Q_L)}{100A_{dm}}\right]^2 \tag{13.5}$$

where Q_L = liquid flow rate, gpm
A_{dm} = minimum area through which liquid flows in the downcomer assembly, sq ft

Bubble-cap Plate Equations

h_{rc} corresponds to the head loss in inches of fluid system caused by the vapor flow through the wet cap assembly. This is a function of both operating and design variables and is evaluated by Bolles [78]:

$$h_{rc} = K_c \frac{\rho_V}{\rho_L}\left(\frac{Q_V}{A_r}\right)^2 = K_c \frac{\rho_V}{\rho_L} U_{VC}^2 \tag{13.6}$$

where U_{VC} = maximum vapor velocity through cap assembly, i.e., at minimum cross section, fps
Q_V = tray vapor load, cu ft/sec
A_r = total riser area per tray, sq ft

K_c is obtained from Fig. 13.8:

$$h_{so} = 1.17 \left(\frac{\rho_V}{\rho_L - \rho_V}\right)^{0.33} \left(h_s \frac{Q_V}{A_s}\right)^{0.67} \tag{13.7}$$

for rectangular slots.

$$A_s = \text{total slot area per plate, sq ft}$$

h_{so} can be read directly as percent of slot height for various shaped slots as a function of vapor load from Fig. 13.9.

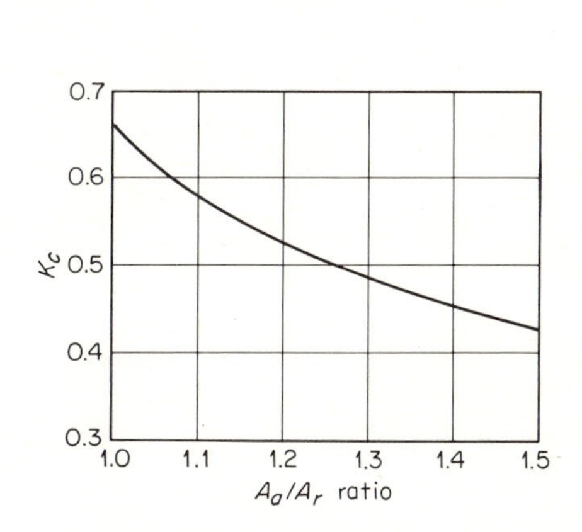

FIGURE 13.8 Bubble-cap pressure-drop constant. Reversal area = 1.35 $(A_a - A_r)/2$. [*From Bolles, Petrol. Process. (February, 1956), p. 64.*]

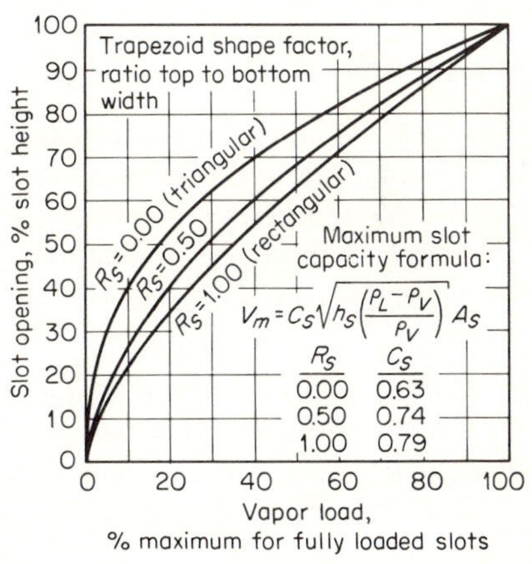

FIGURE 13.9 Trapezoidal slot, generalized correlation. [*From Bolles, Petrol. Process. (February, 1956), p. 64.*]

Δ is designated (erroneously) as a "gradient" which implies a slope. Actually it is the difference in the height of the liquid crest flowing over the weir and the height of the liquid on the inlet side of the tray—both measured from the same reference elevation, usually the tray surface. The term Δ is a somewhat complex function of the operating variables (liquid and vapor flow rates), the design variables of cap diameter, cap spacing, and arrangement, skirt clearance, weir height, weir length, and tower diameter, and system variables (density of the vapor and liquid).

Davies [20, 21] developed a general method of calculating liquid gradient uncorrected for vapor load based on mathematical analysis of a fluid hydraulics model. The equations which he developed require a trial-and-error solution. Bolles [11] rearranged the equations for caps spaced on equilateral triangular pitch and assumed the riser diameter to be 0.7 of the cap diameter. Equation (13.8) was developed through this modification and used to construct the charts shown in Figs. 13.10 through 13.13

$$\frac{Q_L/W_a}{C_d} = 25.8 \frac{\gamma}{1+\gamma} (\Delta_r')^{1/2} \left[1.6\Delta_r' + 3\left(h_c + \frac{0.3h_{sk}}{\gamma}\right)\right] \tag{13.8}$$

where Q_L = liquid load on the tray, gpm
W_a = total flow width across tray normal to flow, ft

$$W_a = \frac{D + l_w}{2} = \frac{\text{tower diameter + weir length}}{2} \tag{13.9}$$

C_d = liquid gradient factor obtained from Fig. 13.14
γ = ratio of distance between caps to cap diameter
Δ_r' = liquid gradient per row of caps normal to flow, uncorrected, in.
h_c = depth of clear liquid on the tray, in.
h_{sk} = cap skirt clearance, in.

The correction of liquid gradient for vapor load is made through use of Fig. 13.15 whereon the vapor load correction factor C_V is correlated with the Q_L/W_a term at parameters of superficial F factors $U_V(\rho_V)^{1/2}$.

Thus the corrected gradient is calculated by

$$\Delta = C_V\Delta' \tag{13.10}$$

h_{ds} represents the liquid head loss in inches of fluid caused by resistance to vapor flow through the liquid and froth above the slots. A maximum or conservative value is calculated with Eq. (13.11) which assumes no foaming.

$$h_{ds} = h_{ss} + h_{ow} + \frac{\Delta}{2} \tag{13.11}$$

The term Δ is the liquid gradient evaluated with the use of Figs. 13.10 through 13.15.

A smaller, less conservative head loss h_a is computed by inclusion of an aeration factor β in the equation to take into account the occlusion of trapped vapor in the expanded liquid bed on the plate.

$$h_a = \beta h_{ds} \tag{13.12}$$

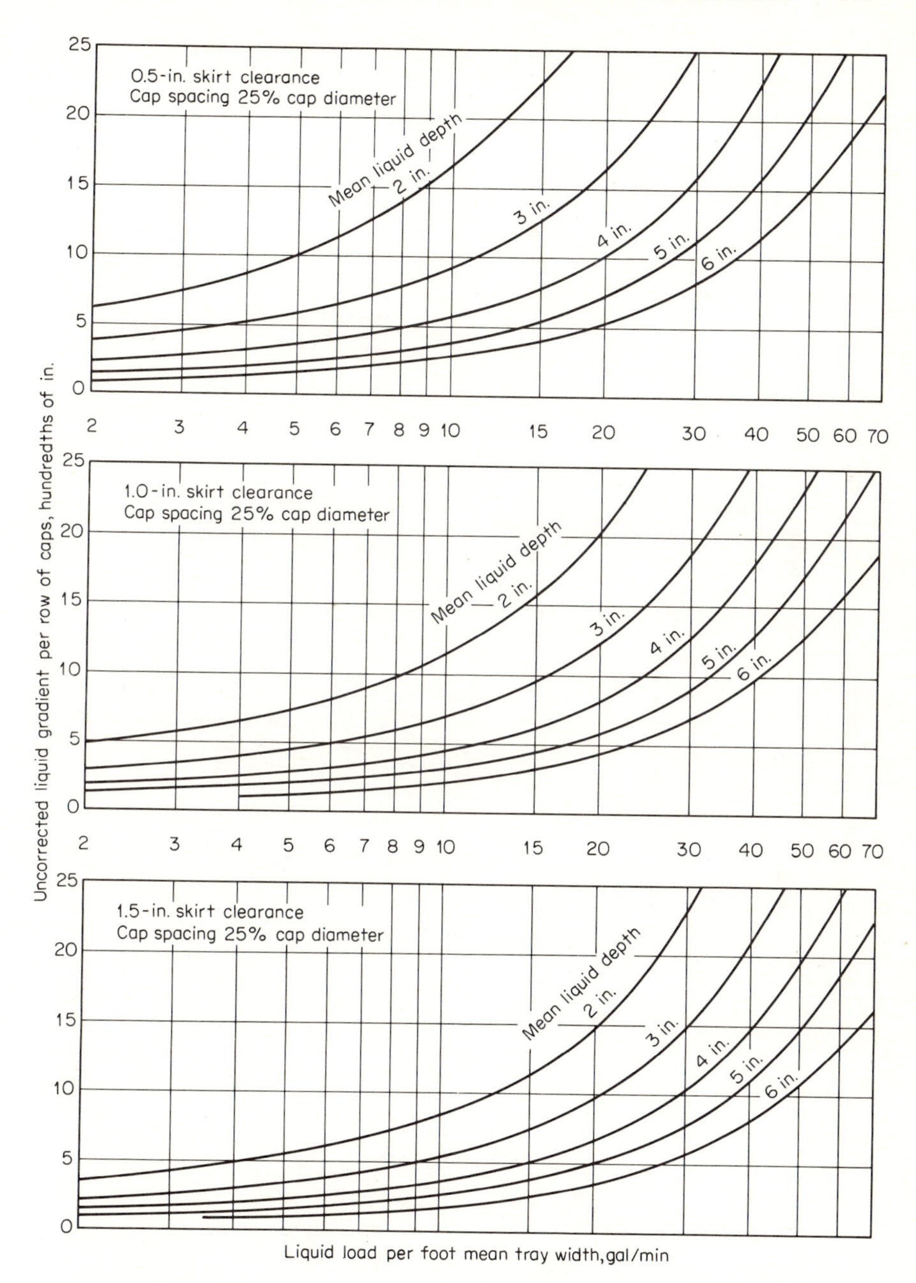

FIGURE 13.10 Liquid gradient chart—cap spacing 25% cap diameter—for equilateral triangular cap pitch. Mean liquid depth $= h_w + h_{ow} + \frac{1}{2}\Delta$. [*From Bolles, Petrol. Process. (February, 1956), p. 64.*]

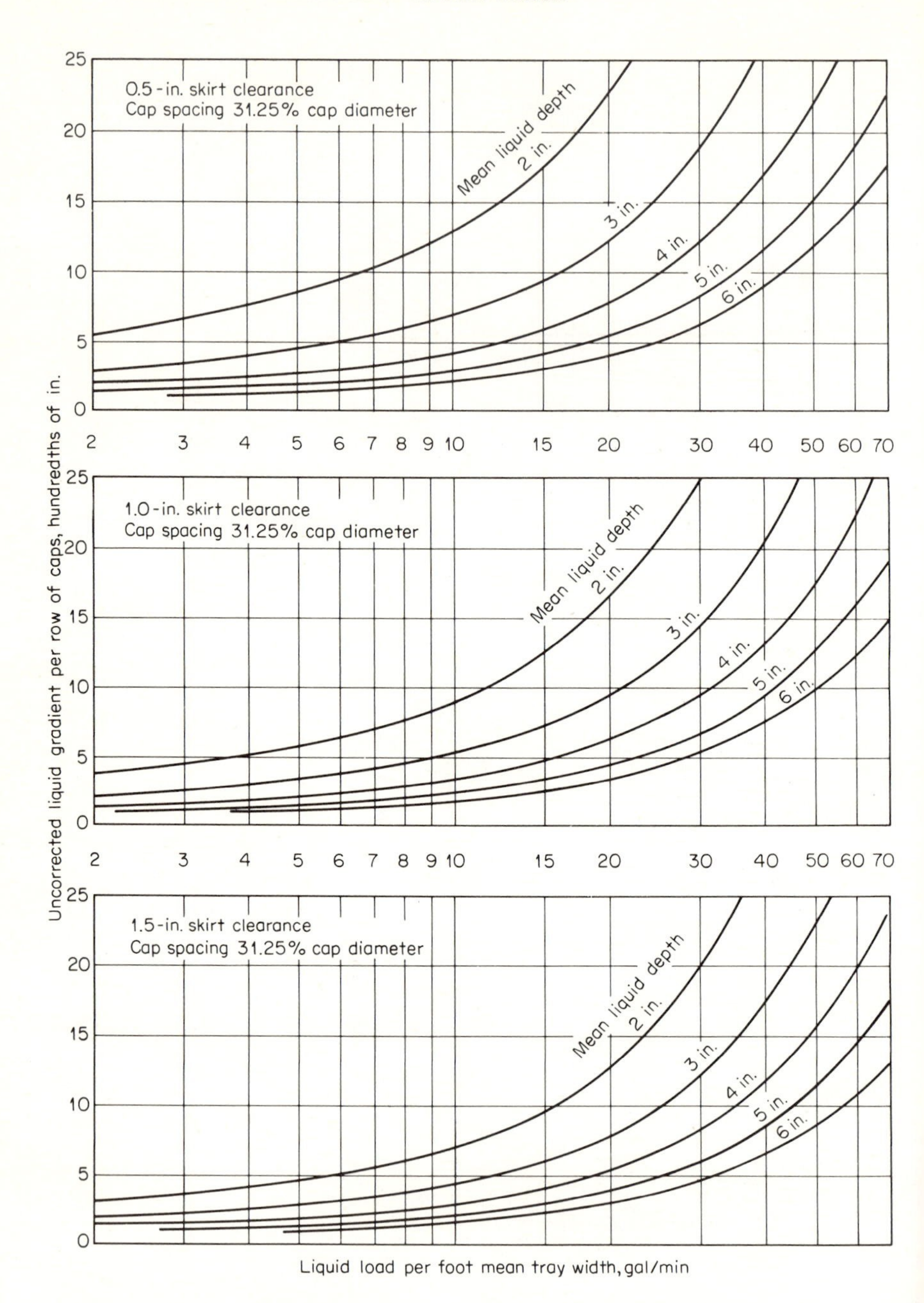

FIGURE 13.11 Liquid gradient chart—cap spacing 31.25% cap diameter—for equilateral triangular cap pitch. Mean liquid depth $= h_w + h_{ow} + \frac{1}{2}\Delta$. [*From Bolles, Petrol. Process.* (*February, 1956*), *p. 64.*]

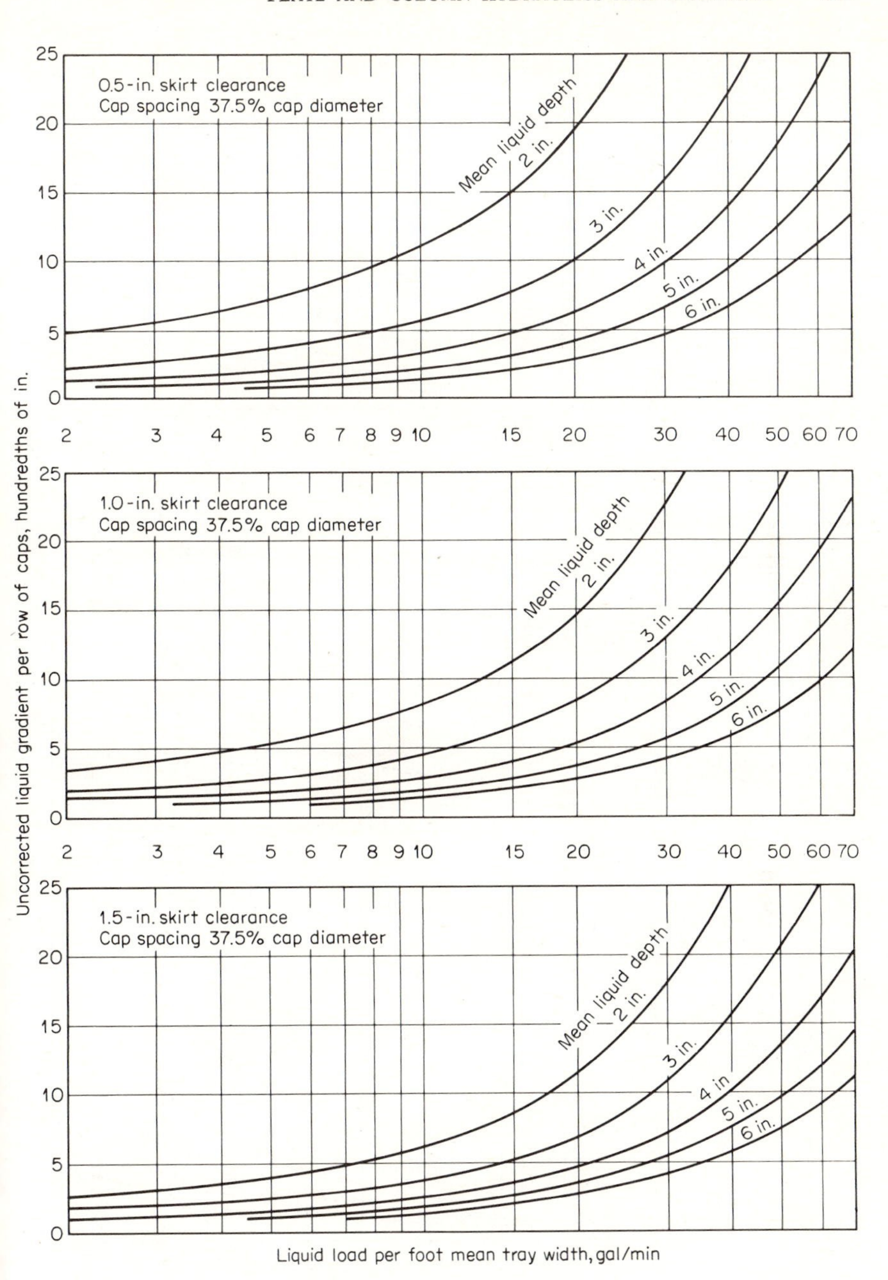

FIGURE 13.12 Liquid gradient chart—cap spacing 37.5% cap diameter—for equilateral triangular cap pitch. Mean liquid depth $= h_w + h_{ow} + \frac{1}{2}\Delta$. [*From Bolles, Petrol. Process.* (*February, 1956*), *p. 64.*]

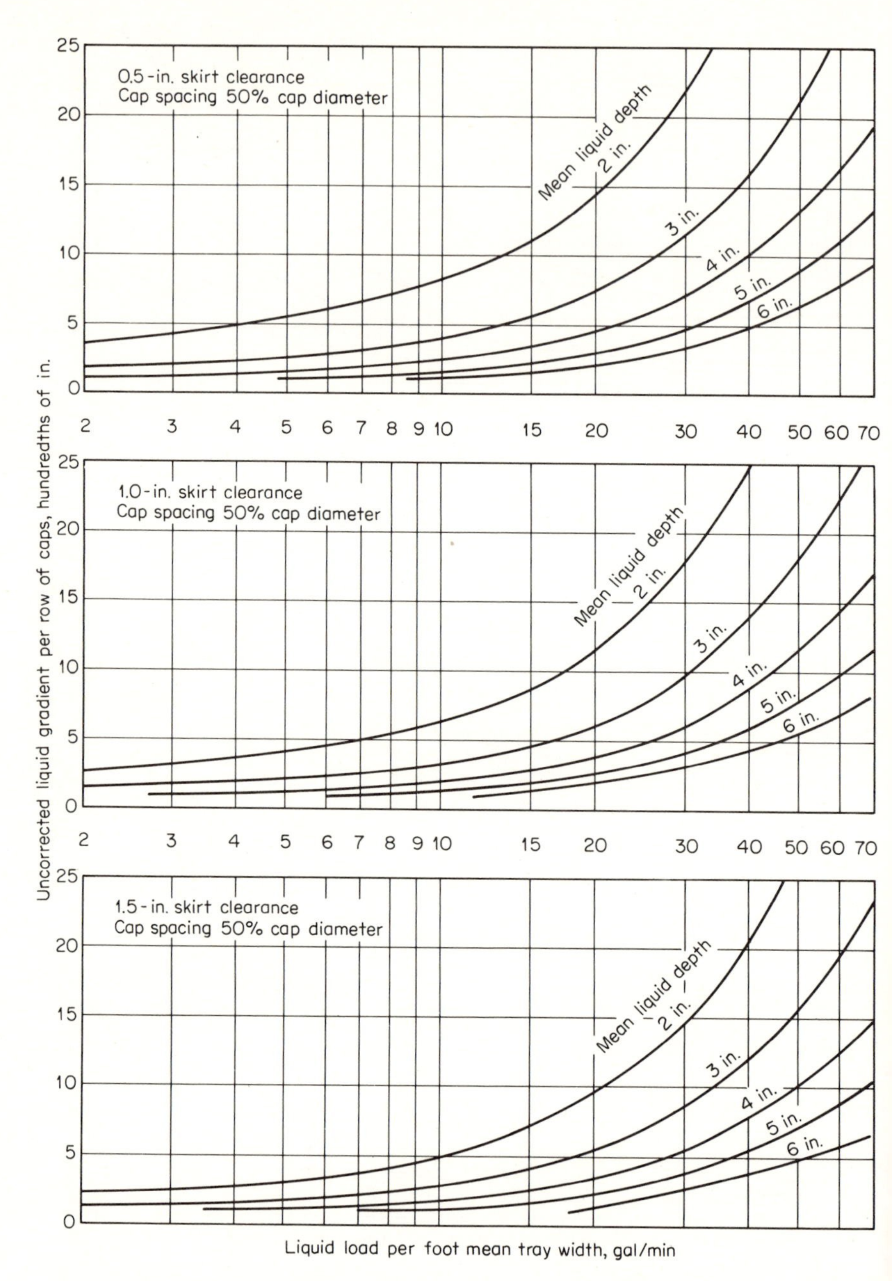

FIGURE 13.13 Liquid gradient chart—cap spacing 50% cap diameter—for equilateral triangular cap pitch. Mean liquid depth $= h_w + h_{ow} + \frac{1}{2}\Delta$. [*From Bolles, Petrol. Process. (February, 1956), p. 64.*]

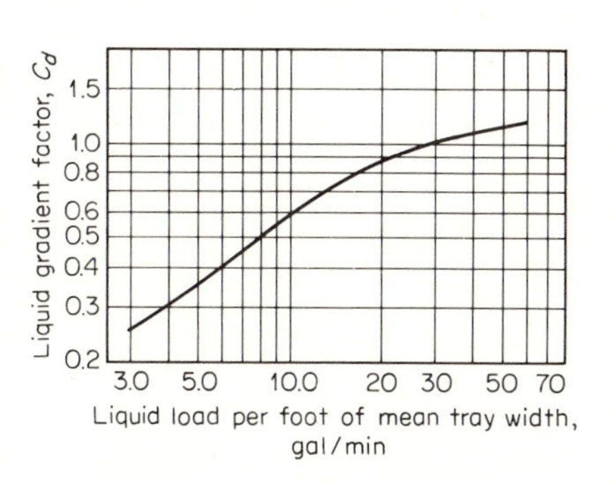

FIGURE 13.14 Liquid gradient correction factor. [*From Bolles, Petrol. Process.* (*February, 1956*), *p. 64.*]

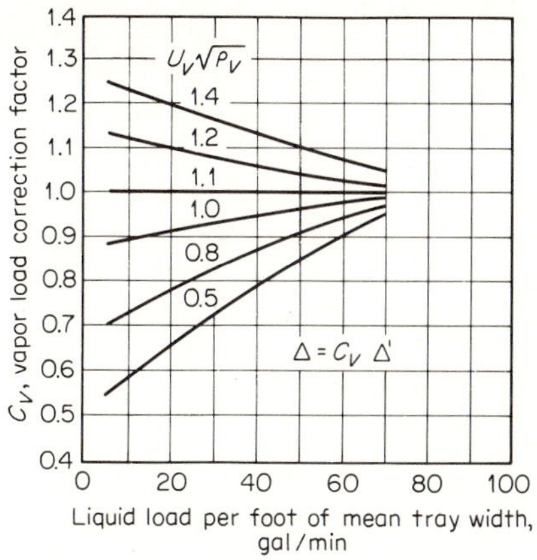

FIGURE 13.15 Correction of liquid gradient for vapor load. [*From Bolles, Petrol. Process.* (*February, 1956*), *p. 64.*]

Aeration Factor and Foam Density

The characteristics of the liquid-vapor mixture on an operating plate in a fractionating column range from those of a slightly aerated liquid having almost the same density as the liquid alone to those of a foam mass containing relatively large quantities of vapor occluded in the liquid having a density much less than the liquid alone. The liquid in the downcomer occludes foam also and has a density less than that of the liquid. There have been several experimental studies of aeration and foam density. Kemp and Pyle [54] and Gilbert [37] studied aeration and bubble-cap plate hydraulics and Hutchinson et al. [49], Mayfield et al. [63], and Lee [59] obtained experimental data on foaming and aeration with perforated plates. Bolles [78] plotted the data of Kemp and Pyle as a function of the active area F factor, $F_{VA} = U_{VA}(\rho_V)^{0.5}$, and this is shown on Fig. 13.16 as the B curve. Fair plotted the data of Foss and Gerster [30] against the same parameter, and this appears on Fig. 13.16 as curve P.

Foam density ϕ is the density of the aerated mass, or bulk density, and is defined by

$$\phi = \frac{h_c}{h_f} \tag{13.13}$$

the ratio of the height of clear liquid contained in the foam to the height of the foam. Hutchinson et al. [49] related the aeration factor β and foam density ϕ theoretically by means of Eq. (13.11),

$$\beta = \frac{\phi + 1}{2} \tag{13.14}$$

The data for curve P for aeration factor and for the curve B for froth density, Fig. 13.16, were calculated by means of Eq. (13.14).

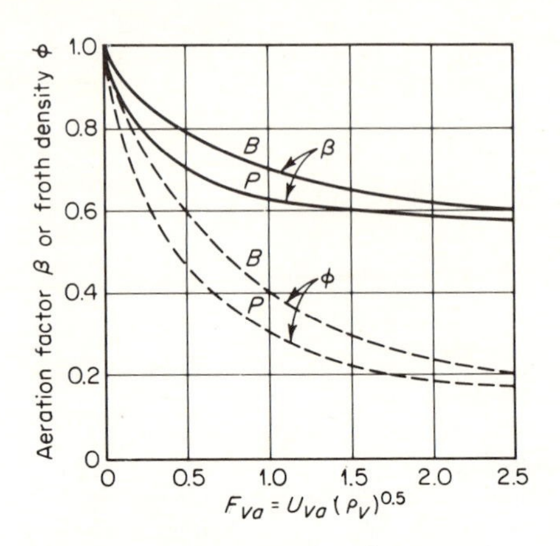

FIGURE 13.16 Aeration factor and froth density. B-bubble-cap plates, P-perforated plates.

Example 13.1

Calculate the maximum and minimum pressure drop per plate for the following column, system, and operating conditions.

Column data

Diameter = 10 ft
Bubble-cap, single cross flow
Weir length = 7.1 ft
Distance between weirs = 7.1 ft
Area downcomer = 7.4 sq ft
Area under apron = 1.8 sq ft
Weir height = 3 in.
Cap diameter = 4 in.
Distance between caps = 2 in.
No. of caps = 300
No. of rows = 18
Slot area/cap = 8.12 sq in.
Riser area/cap = 4.80 sq in.
Annular area/cap = 5.99 sq in.
Skirt height = 1.00 in.
Slot height = 1.25 in.
Riser ID = 2.5 in.
Plate spacing = 24 in.
Shroud ring height = 0.125 in.
Rectangular slots

System data

$\rho_V = 0.168$ lb/cu ft
$\rho_L = 52.0$ lb/cu ft
$Q_V = 153$ cu ft/sec
$Q_L = 108$ gpm
$P_t = 5$ psia

Solution

$$\Delta H_{T,\max} = h_{rc} + h_{so} + h_{ds}$$

$$h_{rc} = K_c\left(\frac{\rho_V}{\rho_L}\right) U_{VC}^2 \qquad U_{VC} = \frac{153}{4.8}\,\frac{144}{300} = 15.3 \text{ fps}$$

$$\frac{\rho_V}{\rho_L} = \frac{0.168}{52} = 0.00323 \qquad \frac{A_a}{A_r} = 1.25 \qquad K_c = 0.505$$

$$h_{rc} = (0.505)(0.00323)(15.3)^2 = 0.382 \text{ in.}$$

$$h_{so} = 1.17\left(\frac{\rho_V}{\rho_L - \rho_V}\right)^{0.33}\left(h_s\frac{Q_V}{A_s}\right)^{0.67}$$

$$\left(\frac{\rho_V}{\rho_L - \rho_V}\right)^{0.33} = 0.147 \qquad \left(h_s\frac{Q_V}{A_s}\right)^{0.67} = \left[1.25\,\frac{(153)(144)}{(8.12)(300)}\right]^{0.67} = 5.02$$

$$h_{so} = (1.17)(0.147)(5.02) = 0.865 \text{ in.}$$

$$h_{ds} = h_{ss} + h_{ow} + \frac{\Delta}{2} \qquad h_{ss} = 0.625$$

$$h_{ow} = 0.48F_w\left(\frac{Q_L}{l_w}\right)^{0.667} \qquad \frac{Q_L}{l_w^{2.5}} = 0.795 \qquad F_w = 1.01$$

$$h_{ow} = (0.48)(1.01)(1.175) = 0.568 \text{ in.}$$

$$2\left(\frac{108}{7.1 + 10}\right) = 12.6 \text{ gpm/ft mean width}$$

Estimate $h_w + h_{ow} + \Delta/2 = 4$ in. From Fig. 13.13, $\Delta' = 0.025$ in. per row; Δ' for 18 rows = (0.025)(18) = 0.45 in.

$$U_V(\rho_V)^{0.5} = \frac{153}{78.6}(0.168)^{0.5} = 0.80$$

From Fig. 13.15, $C_V = 0.75$.

$$\Delta = C_V\,\Delta' = (0.75)(0.45) = 0.338 \text{ in.}$$

$$h_{ds} = 0.625 + 0.568 + 0.169 = 1.362 \text{ in.}$$

$$\Delta H_{T,\max} = 0.382 + 0.865 + 1.362 = 2.609 \text{ in.}$$

$$\Delta H_{T,\min} = h_{rc} + h_{so} + \beta h_{ds}$$

$$U_{VA}(\rho_V)^{0.5} = \frac{153}{78.6 - (2)(7.4)}\,0.41 = 0.98$$

From Fig. 13.16, $\beta = 0.70$.

$$\Delta H_{T,\min} = 0.382 + 0.865 + (1.362)(0.70) = 2.205 \text{ in.}$$

Example 13.2

Determine the minimum and maximum liquid backup in the downcomers for the column and conditions in Example 13.1.

Solution

$$H_{D,\min} = \Delta H_{T,\min} + h_w + h_{ow} + \Delta + h_d$$

$$h_d = 0.03\left(\frac{Q_L}{100A_{dm}}\right)^2 = 0.03\left[\frac{108}{(100)(1.8)}\right]^2 = (0.03)(0.36)$$

$$= 0.0108 \text{ in.}$$

$$H_{D,\min} = 2.205 + 3.0 + 0.568 + 0.338 + 0.011 = 6.122 \text{ in.}$$

$$H_{D,\max} = (\Delta H_{T,\max} + h_w + h_{ow} + \Delta + h_d)\frac{1}{\phi_d}$$

$$= (2.609 + 3.0 + 0.568 + 0.338 + 0.011)\frac{1}{\phi_d}$$

$$= (6.526)\frac{1}{\phi_d} \qquad \text{(from Fig. 13.16, } \phi_d = 0.40)$$

$$H_{D,\max} = \frac{6.526}{0.40} = 16.3 \text{ in.}$$

Perforated Plate Equations

The dimensional symbols for perforated plates are defined in Fig. 13.5 and the equivalent head-loss terms appearing in Eq. (13.2*a*) and (13.2*b*) are defined quantitatively by the following equations and with the use of the indicated figures.

The term h_0 represents the equivalent dry-plate pressure drop in inches of liquid system caused by vapor passage through the holes or perforations. This head loss can be calculated as a dry-orifice loss by any one of a number of methods [45, 46, 57, 61, and 77] which differ only in the evaluation of the orifice or discharge coefficient. This coefficient is a function of a number of interrelated variables, many of which cannot be correlated satisfactorily. Smith, Kolodzie, and Van Winkle [57, 77] correlated the factors of vapor (hole) Reynolds number, plate thickness, hole diameter, hole pitch, and fractional area of the plate allotted to holes. The correlation is shown in Fig. 13.17 and the value of C_0 derived from the figure is used in

$$h_0 = 0.186\left(\frac{U_h}{C_0}\right)^2\frac{\rho_V}{\rho_L}\left[1 - \left(\frac{A_h}{A_a}\right)^2\right] \tag{13.15}$$

where U_h = vapor velocity through holes, fps
A_h = total plate hole area
A_a = total active area of plate $= A - 2A_d$

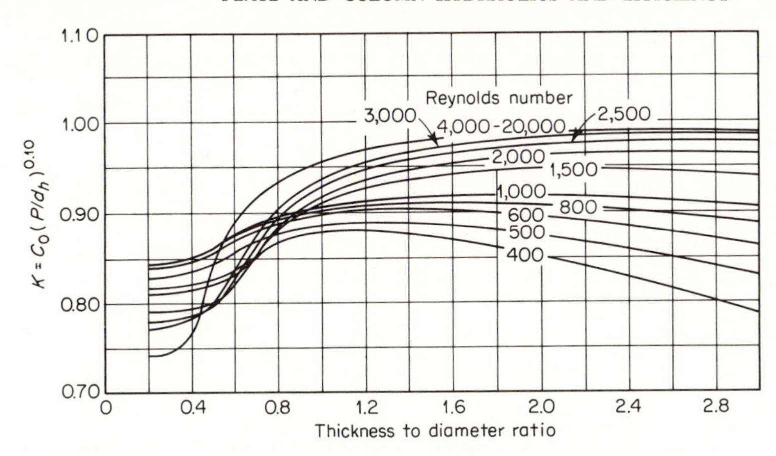

FIGURE 13.17 Orifice-coefficient perforated plate. [*From Van Winkle et al., A.I.Ch.E. J.,* ***3****:205 (1957);* ***4****:226 (1958).*]

Liebson et al. [61] developed a correlation shown on Fig. 13.18 which relates the orifice coefficient C_0 for 3⁄16-in.-diameter-hole plates, with the hole area-to-active area ratio and the plate thickness-to-hole diameter ratio. The coefficient from this relation is usually sufficiently accurate for design purposes. It is used in

$$h_0 = 0.186\left(\frac{U_h}{C_0}\right)^2 \frac{\rho_V}{\rho_L} \tag{13.16}$$

Fair [79] uses a correction factor to include the effect of liquid in vapor

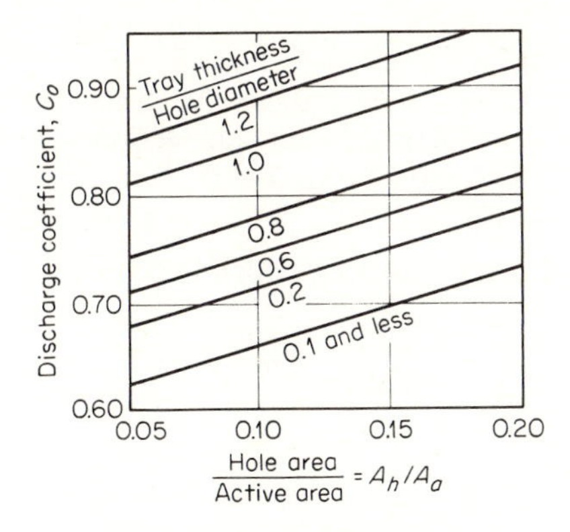

FIGURE 13.18 Discharge coefficients for vapor flow, sieve trays. [*From Liebson et al., Petrol. Refiner,* ***36****(2):127 (1957);* ***36****(3): 288 (1957).*]

entrainment when the entrainment factor is 10% or above

$$h_0' = h_0 \left\{ \left[\frac{\psi}{1-\psi} \frac{L}{V} \left(\frac{\rho_V}{\rho_L} \right)^{0.5} \frac{M_L}{M_V} \right] 15 + 1 \right\} \tag{13.17}$$

where ψ = moles entrained per mole liquid downflow (Fig. 13.26).

For perforated plates the hydraulic gradient Δ may be calculated using the method of Hughmark and O'Connell [46]:

$$\Delta = \frac{f_f U_f^2 Z_L}{12 g_c R_H} \tag{13.18}$$

f_f, the foam friction factor, is correlated in Fig. 13.19 (modified by Fair [79])

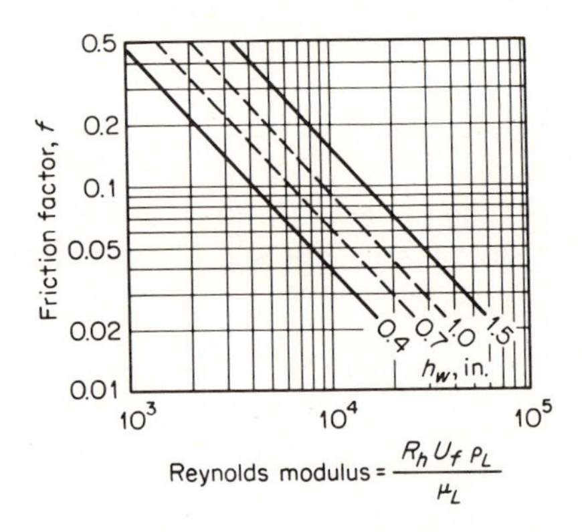

FIGURE 13.19 Friction factor for froth cross flow, sieve trays. *(From Smith, "Design of Equilibrium Stages," Chap. 15, McGraw-Hill, New York, 1963.)*

with the foam Reynolds number Re_f, which is defined by

$$Re_f = \frac{R_H U_f \rho_L}{\mu_L} \tag{13.19}$$

The hydraulic radius of the aerated liquid is defined by

$$R_H = \frac{W_a h_f}{W_a + 2h_f} \tag{13.20}$$

U_f is the velocity of the foam which is the same as the velocity of the clear liquid calculated by

$$U_f = \frac{144\ Q_L/W_a}{(60)(7.48)h_c} \tag{13.21}$$

In most cases the hydraulic gradient on perforated plates is small enough compared with the other pressure-drop terms so that it can be neglected.

The head loss required to overcome surface-tension effect in forming froth and foam and forcing the vapor through the aerated mass is designated h_σ in inches of liquid on the plate.

$$h_\sigma = \frac{0.04\sigma}{\rho_L d_h} \tag{13.22}$$

where σ = the surface tension of the liquid at the temperature of the plate, dynes/cm

ρ_L = liquid density, lb/cu ft

d_h = hole diameter, in.

This can also be represented as a pressure-drop residual equivalent head defined as

$$h_\sigma = \Delta H_T - \beta\left(h_w + h_{ow} + \frac{\Delta}{2}\right) - h_0 \tag{13.23}$$

13.3. FLOW CONDITIONS CAUSING INOPERABILITY

There are limits of operability imposed by flow and design characteristics for each type of plate and column. These are not necessarily just conditions where operation can continue under reduced efficiency, since probably the column will fail to function at all at or near these conditions. In all types of columns the conditions of flooding, priming, and intermittent or cyclic instability are critical, and in perforated-plate columns an additional limiting condition at the weep point or falloff point can be encountered.

Flooding

Flooding is the condition where the pressure drop across the plate is sufficient to cause the dynamic liquid head to be equivalent to the plate spacing S plus the weir height h_w. At this point the liquid backup in the downcomer is just at the point of overflowing the weir on the plate above. When this happens the column fills with foamy liquid and it becomes inoperable (Fig. 13.20).

Quantitatively the point of flooding occurs when

$$\frac{H_D}{\phi_d} = S + h_w \tag{13.24}$$

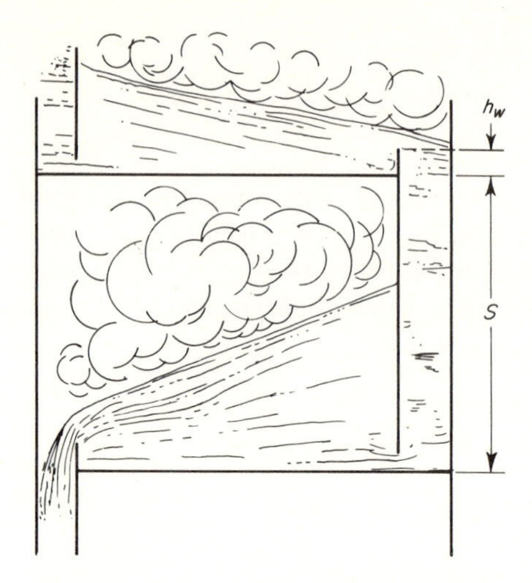

FIGURE 13.20 Condition of flooding.

where β_d is the average aeration factor of the aerated liquid in the downcomer.

If the plate spacing is computed on the basis of the relations given on the previous pages, which enable calculation of the liquid backup in the downcomer, and the sum of the plate spacing plus the weir height is greater than $2H_D$, the condition of flooding will rarely be encountered. The reason for this is that clear liquid heads with no foaming are inherently assumed. Since foaming or frothing takes place on all plates to some extent, the effect is to reduce the density of the liquid and, therefore, the pressure drop across the plate.

Example 13.3

Calculate the pressure drop for the following column and conditions. Determine both maximum and minimum.

Column data

Perforated plate
Single cross flow
Diameter of column = 10 ft
Weir length = 7.1 ft
Distance between weirs = 7.1 ft
Area of downcomer = 7.4 sq ft
Area under apron = 2.0 sq ft
Weir height = 2.0 in.
Hole diameter = 0.25 in., a_h = 0.0495 sq in.
Hole spacing = 0.5 in.
Free area = 10.2% A_N
Plate spacing = 24 in.
Plate thickness = 0.1875 in.

System data

$\sigma_L = 18$ dynes/cm
$\rho_V = 0.27$ lb/cu ft
$\rho_L = 52$ lb/cu ft
$\mu_L = (1.0)(6.72 \times 10^{-4})$, lb/fps
$Q_V = 200$ cu ft/sec
$Q_L = 108$ gpm
$P_T = 5$ psia

Solution

$$\Delta H_{T,\max} = h_0 + h_w + h_{ow} + \Delta + h_\sigma$$

$$U_h = \frac{200}{(78.6 - 7.4)(0.102)} = 27.5 \text{ fps}$$

$$h_0 = 0.186\left(\frac{U_h}{C_0}\right)^2\left(\frac{\rho_V}{\rho_L}\right)\left[1 - \left(\frac{A_h}{A_A}\right)^2\right]$$

$$\frac{A_h}{A_A} = \frac{(78.6 - 7.4)(0.102)}{78.6 - 2(7.4)} = 0.113 \qquad \frac{t_p}{d_h} = 0.75$$

C_0 from Fig. 13.17 = 0.77.

$$h_0 = 0.186\left(\frac{27.5}{0.77}\right)^2 \frac{0.27}{52} \quad [1 - (0.113)^2]$$

$$= 0.186(1272)(0.0052)(0.987) = 1.22 \text{ in.}$$

$$\Delta = \frac{f_f U_f^2 Z_L}{12 g_c R_H} \qquad \text{Re}_f = \frac{R_H U_f \rho_L}{\mu_L}$$

$$R_H = \frac{\dfrac{7.1 + 10}{2} h_f}{\dfrac{7.1 + 10}{2} + 2\,h_f} \qquad \text{(assume } h_f = 6 \text{ in.} = 0.5 \text{ ft)}$$

$$R_H = \frac{(8.55)(0.5)}{8.55 + (2)(0.5)} = 0.448 \text{ ft}$$

$$U_f = \frac{108}{(60)(7.48)(8.55)(0.5)} = 0.0565 \text{ fps } (\max, \phi = 1.0)$$

$$\text{Re}_f = \frac{(0.448)(0.0565)(52)}{(1.0)(6.72 \times 10^{-4})} = 1960$$

$$f_f = 0.12 \qquad \text{(Fig. 13.19)}$$

$$\Delta = \frac{(0.12)(0.0565)^2(7.1)}{(12)(32.2)(0.448)} = 0.0000154 \qquad \text{(neglect)}$$

$$h_\sigma = \frac{0.04\sigma}{\rho_L d_h} = \frac{0.04(18)}{(0.25)(52)} = 0.0555 \text{ in.}$$

$$h_{ow} = 0.568 \text{ in.} \qquad \text{(Example 13.1)}$$

$$\Delta H_{T,\max} = 1.22 + 2.0 + 0.568 + 0.055$$
$$= 3.84 \text{ in.}$$

$$\Delta H_{T,\min} = h_0 + \beta\left(h_w + h_{ow} + \Delta\right) + h_\sigma$$

β from Fig. 13.16 at $U_{VA}(\rho_V)^{0.5} = 1.83$.

$$\beta = 0.58$$

$$\Delta H_{T,\min} = 1.055 + (0.58)(2.568) = 2.545 \text{ in.}$$

Example 13.4

For the same column and conditions given in Example 13.3 calculate the maximum height of foam backup in the downcomer.

Solution

$$H_{D,\max} = \left(\Delta H_{T,\max} + h_w + h_{ow} + \frac{\Delta}{2} + h_d\right)\frac{1}{\phi_d}$$

$$= (3.623 + 2.0 + 0.568 + h_d)\frac{1}{\phi}$$

$$h_d = 0.03\left(\frac{Q_L}{100A_{dm}}\right)^2 = 0.03\left[\frac{108}{(100)(2)}\right]^2 = 0.0087 \text{ in.}$$

$$U_{VA}(\rho_V)^{0.5} = \frac{200}{78.6 - 2(7.4)}(0.27)^{0.5}$$

$$= (3.14)(0.52) = 1.83$$

$$\phi = 0.195 \qquad \text{(Fig. 13.16)}$$

$$H_{D,\max} = \frac{6.125}{0.195} = 31.8 \text{ in.}$$

This column would prime if the foam density in the downcomer were the same as that on the plate since $S + h_w = 26$ in.

It is customary to design a column so that at maximum throughput the approach to flooding does not exceed 85% based on the flooding correlations, such as the Fair and Matthews [28] for bubble-cap plates and Fair [27] for perforated plates (Fig. 13.21). Where foaming is expected, 50 to 60% of flooding based on such correlations is usually considered to be safe. If reliable data are available for foam density under a given set of operating and design conditions, design for as much as 95% of flooding can be used as a maximum.

Figure 13.21 represents a correlation developed by Fair and Matthews [28] using literature data on a number of bubble-plate and perforated-plate columns and systems. The flooding capacity of the tray is related through plate spacing and flow parameter $(L/V)(\rho_V/\rho_L)^{0.5}$, to a capacity parameter $U_{VN}[\rho_V/(\rho_L - \rho_V)]^{0.5}$

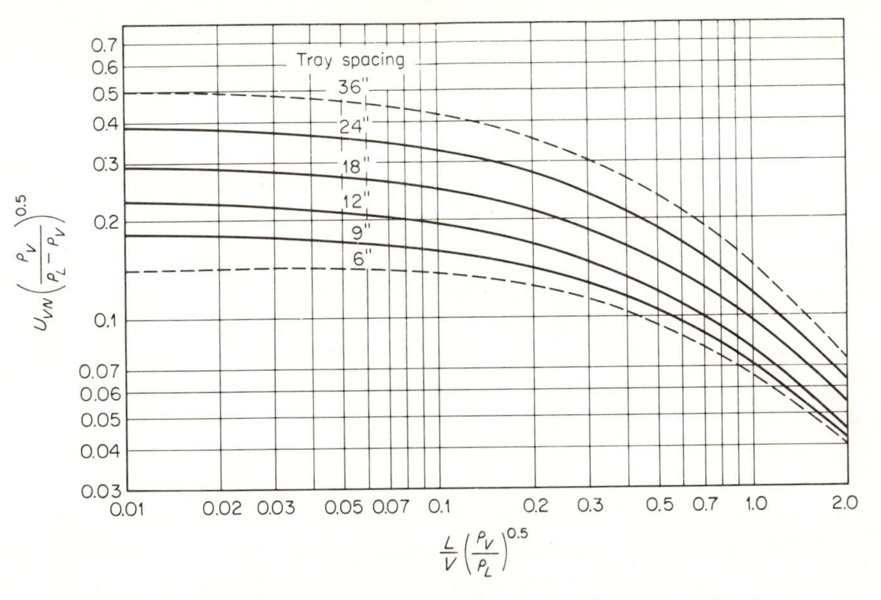

FIGURE 13.21 Flooding limits for bubble-cap and perforated plates. [*From Fair, Petro./Chem. Eng.* (*September, 1961*), *p. 45; Fair and Matthews, Petrol. Refiner* (*April, 1958*), *p. 153.*]

This chart may be used for both bubble-cap plates and perforated plates to determine the flooding vapor velocity for a given value of flow parameter and plate spacing within about 10%, with the following restrictions:

1. The distilling system is a low-foaming or nonfoaming type.

2. The weir height is less than 15% of the tray spacing.

3. The ratio of slot area, or hole area, to active plate area is equal to or greater than 0.1. For hole-to-active area ratios of less than 0.1, it is recommended by Fair that the flooding capacity be modified by the following factors

$\frac{A_h}{A_A}$	Multiply U_{VN} (from Fig. 13.21) by
0.10	1.00
0.08	0.90
0.06	0.80

4. The perforation diameters are equal to or less than 0.25 in.

5. The plate spacings of 6 and 9 in. do not apply to bubble-cap plates.

6. The system surface tension is 20 dynes/cm. For surface tensions different from 20 dynes, the capacity from the chart is modified by

$$\frac{U_{VN}}{U_{VN}^{\dagger}} = \left(\frac{\sigma}{20}\right)^{0.2} \tag{13.25}$$

Priming

The term *priming* refers to the condition where the foam height on the plate becomes great enough to pass through the risers in a bubble-cap column or perforations in a perforated-plate column to mix with the liquid on the plate above. This may be a serious matter if all plates are priming and the foam density is relatively low. Under this condition the change in concentration of the liquid on the plate above by mixing with the foam from the plate below could be appreciable and, in fact, reduce the separation effectiveness greatly. Design to eliminate this problem is almost impossible because the presence of surface-active contaminants might cause unforeseen foaming tendencies, and prediction of foam height as a function of design and operating variables has proved to be highly unreliable because no suitable correlations based on experimental observations of operating columns have been developed. In the absence of data and where foaming is expected, safe design will limit the vapor velocity to 50 to 60% of the flooding velocity, as indicated above.

Based upon observations of especially designed small bubble-plate columns in the A.I.Ch.E. Distillation Research Program and on some additional data from the Fractionation Research Incorporated Program, an approximate correlation [1] relating the visual froth (or foam) height to the active area F factor (based on the active area of the column) was suggested.

$$h_f = 2.53F_{VA}^2 + 1.89h_w - 1.6 \tag{13.26}$$

where h_f = foam height, in.
$F_V = U_V(\rho_V)^{0.5}$

Hughmark [47] correlated foam height with the variables, vapor velocity, weir height, height of the liquid crest over the weir, and density of vapor and liquid. His equation is

$$h_f = -0.61 + 115U_{VA}^2\frac{\rho_V}{\rho_L - \rho_V} + 1.64h_w + 1.49h_{ow} \tag{13.26a}$$

From Eq. (13.26) the priming F_V factor is at 2.8 although it is known that commercial columns have been noted to prime at F_V factors lower than 2.8. It was also noted that the liquid rate had some effect.

The Fair and Matthews [28] correlation gave a prime point of 2.28 under the same conditions as for the 2.8 value above (Fig. 13.21).

† U_{VN} from Fig. 13.21, σ = 20 dynes/cm.

Priming in perforated-plate columns has been studied by a few investigators, and foam or froth height has been observed without attempt at correlation by others. Fair [27] presented a correlation (Fig. 13.21) for perforated plates relating plate spacing, flow parameter, and flooding or priming capacity similar to that of Fair and Matthews [28].

Hunt et al. [48] investigated pressure drop on perforated plates using a static-liquid system and a number of different vapors and liquids. They assumed a foam density of $0.4\rho_L$ in the bubbling zone and an average foam density over the whole plate of $0.7\rho_L$ based upon visual observations. On this basis the relation between the pressure drop and priming was shown to be (at priming conditions)

$$0.7\,(S - h_f) = \Delta H_T \tag{13.27}$$

where S represents plate spacing, and h_f is height of foam.

Dumping or Excessive Weeping

Dumping or weeping takes place in perforated-plate columns when the liquid head on the plate becomes equal to the pressure holding it on the plate. At this point the liquid starts flowing through the holes or perforations to the plate below. Contrary to some observations, this is not a definite point. It has been observed in the author's laboratory that weeping takes place to some extent under practically all conditions and for all designs of plates except for very-small-diameter perforations ($\frac{1}{16}$ in. or less) and high-surface-tension liquids. The reason for this is that because of sloshing and oscillation of the liquid on the tray, the depth varies instantaneously at different locations on the tray and, therefore, the liquid head varies. With the pressure essentially constant in the vapor space below the tray, liquid will weep through the holes at those locations where temporarily the head is high. This takes place even though the average head across the plate is less than that required to equal or overcome the resisting pressure below the tray.

Because the weeping takes place to some extent over a range of conditions, the amount of weeping must be excessive and continuous in order for the operation of the column to be seriously affected. This excessive, continuous weeping is called *dumping* and it can be determined experimentally by either a sharp change in efficiency or pressure drop.

Umholtz and Van Winkle [84, 85], Jones and Van Winkle [52], and Hellums et al. [43] have indicated that the dumping or fall-off point is a function of the following variables: G or hole F factor F_h, hole diameter d_h, plate thickness t_p free area, $\%\,FA$, and weir height h_w, to a lesser extent. Liquid rate apparently had little effect in the ranges studied. Jones and Van Winkle reported also that the pressure drop at the fall-off point was a function of weir height, hole diameter, surface tension, and to a smaller extent, density of vapor and liquid. In general the falloff F factor F_h' ranged from around 1.5 for 1/16-in. holes to around 6 for 3/16-in. holes for a 1/8-in. plate and up to 8 for a 1/4-in. plate.

Attempts at correlation of dumping or falloff point in terms of all the variables known to affect it have been highly unsuccessful. Lee [59] and Mayfield et al. [63] related dry-plate pressure drop, and clear-liquid depth on the tray. Hunt et al. [48] studied plate stability and found that, in general, plates with hole spacing or pitch in the range of $3d_h$ to $4d_h$ gave much more stability than those with $p = 2d_h$. With the air-water system they found that for 1/8-in. holes, $p = 4d_h$ and $h_0 = 1.0$, the critical velocity of air through the holes was 5 fps. The critical velocity increased with increase in hole size and with increase in liquid head up to 70 fps. McAllister et al. [65, 66], Hughmark and O'Connell [46], and Zenz [90] correlated wet-plate pressure drop at weeping conditions with hole F factor. These different correlations agreed only over part of the range of F factors. Ellis and Moyade [25] and Arnold et al. [5] reported data on weeping related to pressure drop and vapor velocity. Liebson et al. [61] related weeping and dry-plate pressure drop and liquid seal for a 3/16-in.-hole plate, and Huang and Hodson [45] suggested an equation to calculate the minimum dry-plate pressure drop necessary for no weeping operation. Unfortunately, the calculated results from these relations

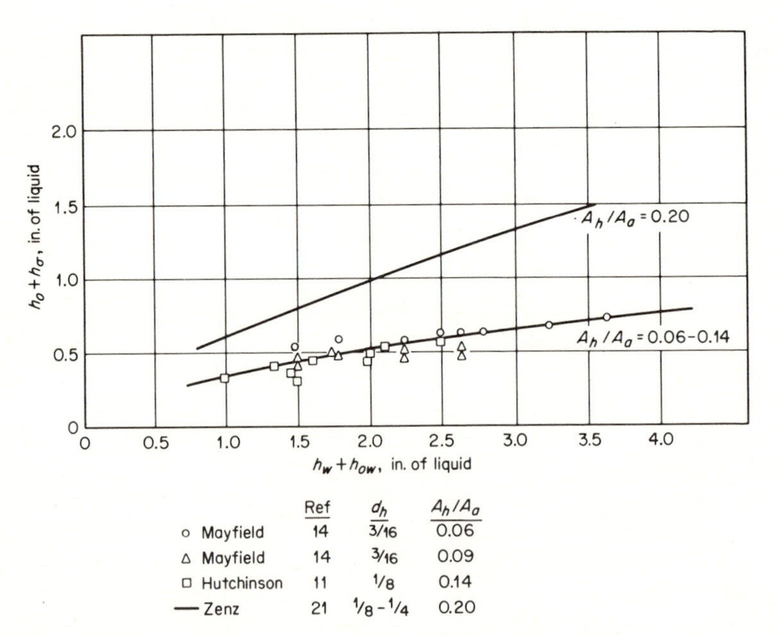

FIGURE 13.22 Weeping, sieve trays. *(From Smith, "Design of Equilibrium Stage Processes," Chap. 15, McGraw-Hill, New York, 1963.)*

scatter rather widely and insufficient reliable data are available to verify the applicability over a wide range of any correlation at this point.

The weeping point has been designated for a plate where no sloshing occurs at that point where

$$h_0 + h_\sigma > f\left(h_w + h_{ow} + \frac{\Delta}{2}\right) \tag{13.28}$$

Figure 13.22 shows a correlation by Fair [79] relating the weeping characteristic of a perforated tray with the terms in Eq. (13.28). This figure indicates that incipient weeping will take place when the calculated sum of the head loss through the perforations plus the surface-tension head just equals the ordinate value read from Fig. 13.22 corresponding to the calculated liquid head, $h_w + h_{ow}$, at the exit weir. If the calculated orifice plus surface-tension loss is greater than that read from Fig. 13.22, the plate is operating above the weep point.

Example 13.5

Determine the vapor flow rate for the system and column described in Example 13.3, which would result in weeping or dumping at the liquid rate specified.

Solution

Minimum vapor rate for a liquid rate of 108 gpm,

$$\frac{A_h}{A_A} = \frac{(78.6 - 7.4)(0.102)}{78.6 - 2(7.4)} = \frac{7.18}{63.8} = 0.113$$

$$h_w = 2.0 \text{ in.} \qquad h_\sigma = 0.056 \text{ in.} \qquad h_{ow} = 0.568 \text{ in.}$$

$$h_w + h_{ow} = 2.568 \text{ in.}$$

From Fig. 13.22, $h_0 + h_\sigma = 0.6$,

$$h_0 = 0.6 - 0.056 = 0.544 \text{ in.}$$

$$\left(\frac{U_h}{C_0}\right)^2 = \left(\frac{0.544}{0.186}\right)\left(\frac{52}{0.27}\right) = 562$$

$$C_0 = 0.76 \quad \text{(assumed)}$$

$$U_h^2 = (562)(0.577) = 325$$

$$U_h = 18 \text{ fps}$$

Cyclic Instability

Cyclic instability in a fractionating column may be one of several different types, and some of these may seriously affect the operation of the column.

With respect to bubble-cap plates and columns specifically, instability caused by excessive buildup or hydraulic gradient can be a serious problem. The behavior of the plate under such a condition of instability can be described as follows: When the plate is operating normally (Fig. 13.23*a*), the liquid is flowing evenly across the plate, the vapor is approximately evenly distributed among the caps, and only slight differences in vapor flow rates are encountered between the inlet row and outlet row of caps. The liquid gradient or buildup is relatively small. As time proceeds, the resistance to the flow of the liquid across the plate causes it to start "piling up" at the entrance weir to the plate (Fig. 13.23*b*). At this point, since the liquid head

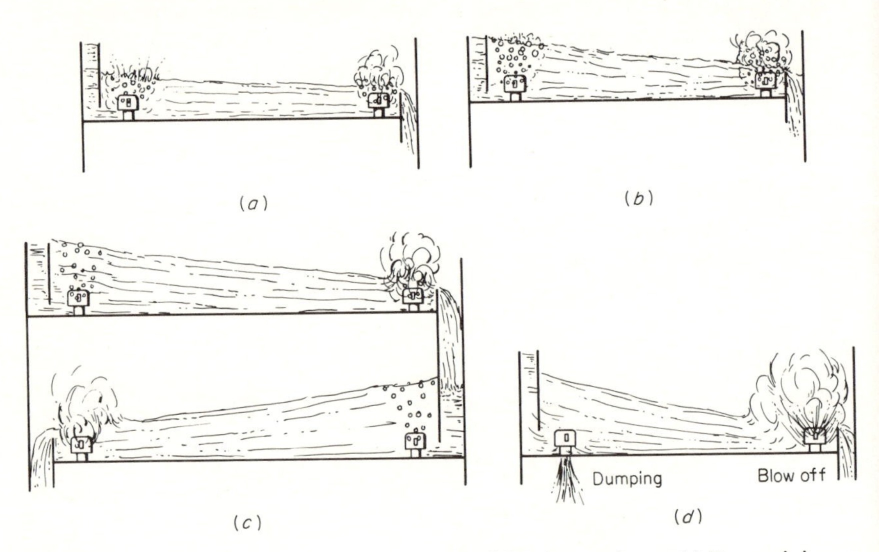

FIGURE 13.23 (*a*) Normal operation. (*b*) Buildup increasing. (*c*) Bare minimum stability. (*d*) Dumping.

is much greater over the entrance rows of caps than over the exit rows, the vapor rate increases through the exit rows of caps and decreases through the entrance rows. The condition becomes worse with the buildup increasing to the point where the entrance rows of caps are just barely bubbling. This is called the *condition of bare minimum stability* (Fig. 13.23*c*). The entire vapor flow is being carried by the other caps on the plate, and the ones near the exit weir handle so much vapor that they tend to "blow off," blowing the liquid from around them. When this condition arises, the pressures in the vapor spaces above and below the plate are essentially equalized and the high liquid head over the entrance rows of caps forces the liquid to flow through the slots and down through the riser to the plate below. This is called *dumping* (Fig. 13.23*d*). After the dumping takes place, the gradient becomes normal and all of the caps again start normal operation. This cycle is repeated until the condition(s) causing the instability is rectified.

Good et al. [39] studied this phenomenon using a square bubble-cap plate and the air-water system. They concluded that buildup or increase in liquid gradient leading to instability was caused by increase in liquid flow path, liquid rate, vapor rate, number of rows of caps normal to the liquid flow path, number of caps per square foot (cap concentration), and decrease in liquid level on the plate and cap skirt clearance.

Kemp and Pyle [54] came to similar conclusions based on their studies.

Maximum Liquid Load for Minimum Stability

Good et al. [39] presented a correlation relating the maximum liquid load with the design and operating variables shown on Fig. 13.24. This was

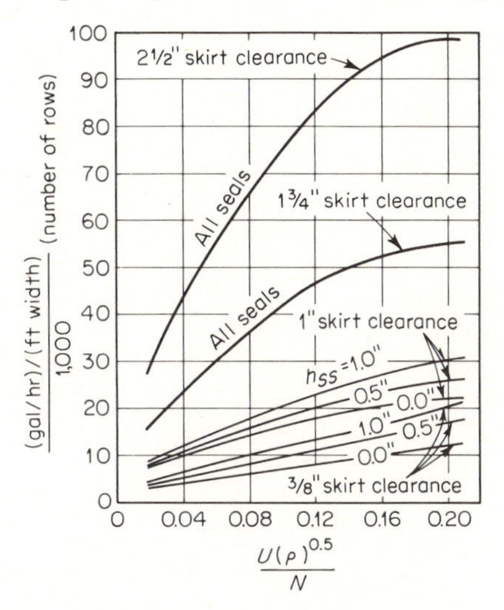

FIGURE 13.24 Liquid capacity at bare minimum stability. [*From Good et al., Ind. Eng. Chem.*, ***34**:1445* (*1942*).]

devised for 3-in.-diameter caps but also serves as a basis for estimation of maximum stable liquid loads for plates having other sizes of caps.

Davies [20] concluded from hydraulic gradient studies that when $\Delta/h_{rc} > 1.0$ the plate is unstable and dumping occurs when

$$\Delta H_T = h_{ow} + h_{ss} + \Delta - h'_{sr} \tag{13.29}$$

at the inlet row of caps, where $\Delta = h_{rc} + h'_{sr}$ and h'_{sr} is the distance from the top of the slots to the top of the riser opening.

Example 13.6

Determine for the column and system in Example 13.1 the maximum liquid load at bare minimum stability.

Solution

$$A_A = (100)(0.786) - 2(7.4) = 63.8 \text{ sq ft}$$

$$A_N = 78.6 - 7.4 = 71.2 \text{ sq ft}$$

$$N = \frac{300}{63.8} = 4.7 \text{ caps/sq ft}$$

$$U_{VN} = \frac{153}{71.2} = 2.15 \text{ fps}$$

$$F_{VN} = U_{VN}(\rho_V)^{0.5} = 2.15(0.168)^{0.5} = 0.885$$

$$U_{VN}(\rho_V)^{0.5}\frac{7.14}{N} = \frac{(0.885)(7.14)}{4.7} = 1.77$$

$$N_R = 18 \qquad h_{sk} = 1.0 \text{ in.} \qquad s_m = h_{ss}$$

$$h_{ss} = 3.0 - 1.0 - 0.125 - 1.25 = 0.625 \text{ in.}$$

From Fig. 13.24,

$$\frac{\text{gph/ft}}{1000} N_R = 30$$

$$W_a = \frac{10.0 + 7.1}{2} = 8.55$$

$$\text{gph} = \frac{(1000)(8.55)(30)}{18} = 14{,}200 \text{ gph}$$

$$L_{\text{actual}} = (108)(60) = 6480$$

Therefore the plate is stable.

Pulsations

Pulsations in the liquid and vapor on both bubble-cap and perforated-plate trays have been noted by a number of investigators. These pulsations have been studied by McAllister and Plank [66] in two columns, one bubble-cap column and one perforated-tray column. They found the oscillation to be correlated by

$$f = k\frac{C}{2n}\frac{F_L}{F_V}\left(\frac{V_S}{l_{vs}}\right)^{0.5} \tag{13.30}$$

where f = frequency of pulsations, cps
$k = 2.5 \pm 0.2$
C = velocity of sound in the fluid, fps
$F_L = U(\rho_L)^{0.5}$
$F_V = U(\rho_V)^{0.5}$
l_{vs} = length of system (containing vapor, including reboiler, condenser, and piping), ft
V_S = volume of system, cu ft

The equation describes only the phenomena for the columns studied and these involved cross flow, alternating with successive plates.

Such pulsations are claimed to increase the efficiency through backmixing. However, it is possible that conditions could be encountered in design and operation such that harmonic oscillation of the equipment could cause destructive effects on the piping or the column. To the author's knowledge, this has not been reported. However, the possibility has not been definitely eliminated by information concerning these phenomena at the present time.

13.4. EFFICIENCY EVALUATION AND PREDICTION

It was pointed out in Chap. 5 that efficiency of a fractionating device, over-all or local, with regard to a plate or stage is affected by a large number of variables. Most of these recognized variables have been studied and their effects are fairly well defined. Unfortunately, insufficient quantitative study has been made of the interaction effects of the variables, and the interaction effects may be as important in some cases as the variables themselves. The individual variables affecting efficiency of bubble-cap valve and perforated-plate columns are shown in Table 13.3.

Experimental Evaluation of Efficiency

Murphree Efficiency [68]

The Murphree dry-vapor-plate efficiency is calculated by

$$E_0 = \frac{y_n - y_{n+1}}{y_n^* - y_{n+1}} \tag{13.31}$$

where E_0 = plate efficiency factor
y_n^* = vapor composition in equilibrium with L_n
y_{n+1} = actual vapor composition

Table 13.3 Variables Affecting Efficiency

Bubble-cap columns	Perforated-plate and valve-plate columns
Operating variables:	
1. Temperature	Temperature
2. Pressure	Pressure
3. Liquid rate, L/V	Liquid rate, L/V
4. Vapor rate	Vapor rate
System variables:	
5. Density of liquid	Density of liquid
6. Viscosity of liquid	Viscosity of liquid
7. Surface tension of liquid (at boiling temp.)	Surface tension of liquid (at boiling temp.)
8. Density of vapor	Density of vapor
9. Viscosity of vapor	Viscosity of vapor
10. Relative volatility of components	Relative volatility of components
Design variables—plate:	
11. Free area, A_s/A	Free area, A_h/A
12. Slot design	Hole diameter
12a	Weight of valve element
13. Skirt clearance	Hole pitch
14. No. of caps/sq ft	Plate thickness
15. Cap arrangement	Hole arrangement
16. Overflow weir height	Overflow weir height
17. Inlet weir	Inlet weir
18. Placement of weirs with respect to caps	Placement of weirs with respect to holes
19. Splash baffles	Splash baffles
20. Flow pattern	Flow pattern
Design variables—column:	
21. Downcomer area	Downcomer area
22. Plate spacing	Plate spacing
23. Downcomer arrangement	Downcomer arrangement

Thus the actual vapor enrichment is divided by the theoretical maximum enrichment to obtain the efficiency factor. Experimentally, vapor samples above and below the plate are taken simultaneously with the liquid sample from the downcomer. The Murphree *point* efficiency is calculated by Eq. (13.32) which applies to only one point on the plate. A number of values of E_p are averaged to get E_0.

$$E_p = \left(\frac{y_n - y_{n+1}}{y_n^* - y_{n+1}}\right)_P \tag{13.32}$$

In this case also the vapor and liquid samples are taken simultaneously from adjacent locations on the plate.

Plate and Column Efficiency

In equilibrium-stage fractionation there are two efficiencies considered—*plate efficiency* and *column efficiency*. It has been common practice to determine by prediction or scale-up methods a plate efficiency under certain operating conditions for a particular type of plate, and then to evaluate the number of actual plates needed by

No. of theoretical stages/plate efficiency factor = no. of actual stages

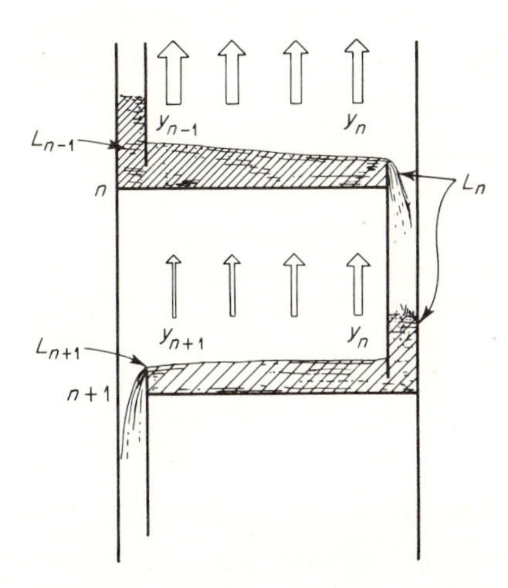

FIGURE 13.25 Schematic illustration of concentration change. (The size of the arrows in Fig. 13.25 indicates the relative concentration of the lighter material in the vapor; the density of cross hatching shows the relative concentration of the same material in the liquid from an idealized standpoint. Actually there is mixing in both phases.)

This is not valid in most cases because the plate efficiency is evaluated for one particular plate and this rarely represents an average for all of the plates in the column. The mass-transfer characteristics for that plate might be very different from those of a plate at some other location in the column. Even if such an "average" plate could be selected, use of the above relation would not be satisfactory in the strict sense. For example, if a plate efficiency were evaluated for a plate at the average temperature, pressure, flow rates, and compositions in the column, the efficiency found to be 50%, and the number of theoretical stages were 30, then 60 actual plates would be required. For a binary system plotted on a McCabe-Thiele or Ponchon-Savarit diagram, it is indicated that more stages are required near the top and bottom of the column where the change in concentration with plates is usually much smaller. If the efficiency evaluated at the end condition, e.g., at the top, were 40%, and at the bottom were 35%, 87 and 75 plates, respectively, would be required. Based on this consideration it is evident that predicting efficiency at three widely different locations in the column and applying the efficiency to the sections selected would be better practice since it would enable a more realistic approach to determination of the actual number of plates.

Where point efficiency is evaluated at a point on a plate where mass transfer might be good or might be poor and this value is used to evaluate plate efficiency and then column efficiency, serious error could result.

Factors Affecting Efficiency

There are three principal factors affecting efficiency related to design, operating, and system variables in the operation of a fractionating column: liquid-in-vapor entrainment, vapor-in-liquid entrainment, and effective contact of vapor and liquid on the plate.

Liquid-in-Vapor Entrainment

Where liquid-in-vapor entrainment is encountered, liquid is carried by the vapor to the plate above through the risers or perforations, and the liquid of lower concentration from the plate below reduces the concentration of the liquid on the plate above (with respect to the more volatile materials). The vapor rising from the plate, therefore, will be of lower concentration, the net amount of mass transfer is less, and the efficiency is reduced. This situation requires more contact stages than would be necessary for equilibrium separation.

A number of studies of entrainment from fractionating plates as a function of operating, design, and system variables have been made. Some of these are listed in Table 13.4.

Bubble-cap trays were studied by Sherwood and Jenny [74], Holbrook and Baker [44], Atterig et al. [7], Simkin et al. [76], and Fair and Matthews [28]. Fair [27], Bain and Van Winkle [8], Arnold et al. [5], and Hunt et al. [48] investigated entrainment with perforated-plate columns. These investigations indicate that entrainment is increased for all vapor-liquid contacting devices by (*a*) decrease in plate spacing; (*b*) increase in superficial vapor velocity; (*c*) increase in weir height; (*d*) increase in liquid flow rate; (*e*) increase in vapor density; (*f*) decrease in liquid surface tension; (*g*) increase in cap spacing; (*h*) decrease in liquid flow path; and (*i*) increase in hole diameter.

Table 13.4 Entrainment Studies

References	Tray type	Diameter, ft	System	Variables
[74]	BC	1.5	Air–H_2O	G, L, S
[44]	BC	0.67	Steam–H_2O (brine)	G, L, h_w, S
[7]	BC	1.5 × 7.0	Air–H_2O	G, L, h_w, S, Z_L
[76]	BC	1.3	Air–H_2O	$G, L, h_w, S, \sigma, \rho_V, \rho_L$
[28]	BC	0.67–3.0	Air–H_2O	G, L, S, % flood, h_w, σ
			Air–D.E.G.	
			Air–G.O.	
			HOAc–H_2O	
			EtOH–H_2O	
[27]	P	1.5–13.0	Air–D.E.G.	G, L, S, % flood, h_w, σ
			Air–Oil	
			EtOH–H_2O	
			Benzene–toluene	
			Air–H_2O	
			HOAc–H_2O	
[8]	P	0.5 × 2.0	Air–H_2O	G, L, S, d_h, pitch, h_w, σ
[48]	P	0.5	Air–H_2O	$G, S, d_h, \rho_V, \sigma$, pitch
			F_{12}–H_2O	
			CO_2–H_2O	
			Ar–H_2O	
			CH_4–H_2O	
[5]	P	1.0–5.5	Air–H_2O	G, L, S, d_h

The entrainment correlation presented in Fig. 13.26 was developed by Fair and Matthews [28] for bubble-cap plates and by Fair for perforated plates using literature data for a number of columns and systems. This correlation was developed upon the basis of a flow parameter

$$\frac{L}{V}\left(\frac{\rho_V}{\rho_L}\right)^{0.5}$$

and a capacity parameter (Brown-Sounders [13]),

$$U_{VN}\left(\frac{\rho_V}{\rho_L - \rho_V}\right)^{0.5}$$

The correlation relates percent of flood, fractional entrainment (moles of liquid entrained in the vapor per mole of gross liquid downflow), liquid and vapor flow rate, and the densities of the vapor and liquid, and it is claimed to be accurate within ±15%. The restrictions applying to the use of Fig. 13.26 are the same as those for Fig. 13.21 (see p. 525).

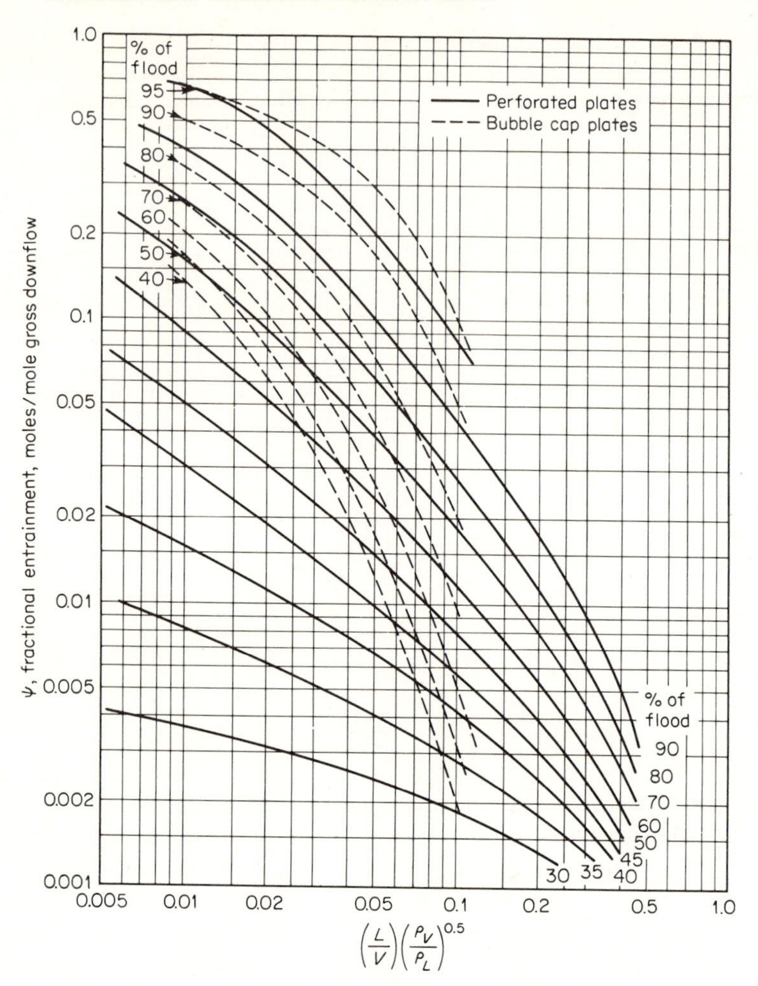

FIGURE 13.26 Entrainment correlation. [*From Fair, Petro./Chem. Eng. (September, 1961), p. 45; Fair and Matthews, Petrol. Refiner (April, 1958), p. 153.*]

Example 13.7

Determine the entrainment predicted for an ethanol–water fractionation by the Fair and Matthews correlation. The data are:

Bubble-cap column
Single cross flow
Diameter = 6 ft
Area downcomer = 3.2 sq ft
Weir height = 2.0 in.
Weir length = 5 ft

Plate spacing = 18 in.
$\rho_V = 0.075$ lb/cu ft
$\rho_L = 45$ lb/cu ft
$\sigma = 39$ dynes/cm
$L = 2000$ lb/(hr)(sq ft of A)
$Q_V = 120$ cu ft/sec

Solution

$$V = \frac{(120)(3600)(0.075)}{(0.786)(36)} = 1145 \text{ lb/(hr)(sq ft of } A)$$

$$\frac{L}{V}\left(\frac{\rho_V}{\rho_L}\right)^{0.5} = \left(\frac{2000}{1145}\right)\left(\frac{0.075}{45}\right)^{0.5} = 0.0715$$

From Fig. 13.21 at 18-in. plate spacing,

$$U_{VN}\left(\frac{\rho_V}{\rho_L - \rho_V}\right)^{0.5} = 0.26$$

Correcting for surface tension,

$$0.26\left(\frac{39}{20}\right)^{0.2} = 0.297$$

$$U_{VN} = 0.297\left(\frac{45 - 0.075}{0.075}\right)^{0.5} = 7.25 \text{ fps}$$

$$A_N = (0.786)(36) - 3.2 = 25.1 \text{ sq ft}$$

$$U_{VN,\text{actual}} = \frac{120}{25.1} = 4.79 \text{ fps}$$

$$\text{Percent of flood} = \frac{4.79}{7.25} = 66\%$$

From Fig. 13.26, $\psi = 0.0185$ mole/mole,

$$\text{Fractional entrainment} = \frac{0.0185}{1 - 0.0185} = 0.0189$$

Bain and Van Winkle [8] studied entrainment from a section of a 24-in.-diameter perforated-plate column using the air–water system and considered the effect of plate spacing, hole diameter, hole spacing, weir height, liquid rate, and gas rate. An empirical correlation was developed which represented 90% of the experimental data within 25% maximum deviation. This correlation is shown in the following equation and in Figs. 13.27 through 13.29.

$$\ln E' = K \ln\left[\frac{d_h}{S}\left(\frac{l}{L}\right)^g G^f\right] + B \tag{13.33}$$

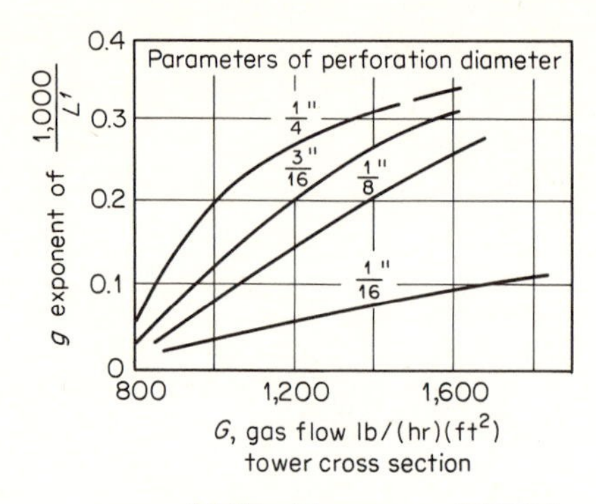

FIGURE 13.27 Exponent of 1000/L' as a function of d_h, G. [*From Bain and Van Winkle, A.I.Ch.E. J., 7:363 (1961).*]

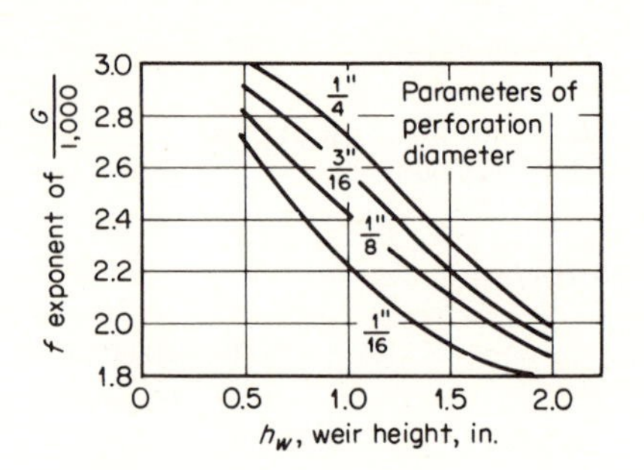

FIGURE 13.28 Exponent of G/1000 as a function of d_h and h_w. [*From Bain and Van Winkle, A.I.Ch.E. J., 7:363 (1961).*]

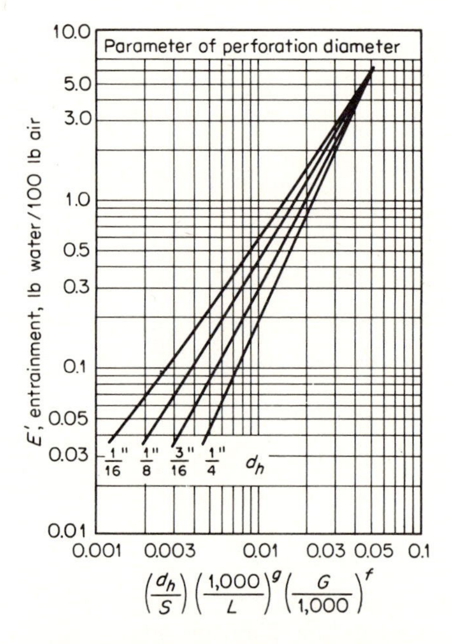

FIGURE 13.29 Entrainment correlation. [*From Bain and Van Winkle, A.I.Ch.E. J., 7:363 (1961).*]

Because the air–water system was used in the experiments, a suggested correction for surface tension and density of the vapor and liquid is presented by

$$E' = E'\left(\frac{73}{\sigma}\right)^{0.2}\left(\frac{\rho_V/0.08}{\rho_L/62.4}\right)^{0.5} \tag{13.34}$$

Example 13.8

Calculate the entrainment for the system and the column in Example 13.7 but assume perforated plates with hole diameters of 0.25 in. rather than bubble-cap plates and $A_h/A_A = 0.06$. Determine the entrainment by the (*a*) Fair correlation; (*b*) Bain and Van Winkle correlation.

Solution

(*a*) The flow parameter is

$$\frac{L}{V}\left(\frac{\rho_V}{\rho_L}\right)^{0.5} = \frac{2000}{1145}\left(\frac{0.075}{45}\right)^{0.5} = 0.0715$$

From Fig. 13.21 at 18-in. plate spacing,

$$C = U_{VN}\left(\frac{\rho_V}{\rho_L - \rho_V}\right)^{0.5} = 0.26$$

Correcting for surface tension and for $A_h/A_A = 0.06$,

$$C_{\text{cor}} = 0.26\left(\frac{39}{20}\right)^{0.2} 0.8 = 0.237$$

$$U_{VN} = \frac{0.237}{0.041} = 5.8 \text{ fps}$$

$$\text{Percent flood} = \frac{120/25.1}{5.8} = \frac{4.79}{5.8} = 82.5\%$$

From Fig. 13.26, $\psi = 0.047$,

$$\text{Fractional entrainment} = \frac{0.047}{1 - 0.047} = 0.0492$$

(*b*)

$$G = \frac{(120)(3600)(0.075)}{28.3} = 1145 \text{ lb/(hr)(sq ft of area)}$$

From Fig. 13.27 at $d_h = 0.25$, $G = 1145$, $g = 0.25$. From Fig. 13.28 at $d_h = 0.25$, $h_w = 2.0, f = 2.0$.

$$\frac{d_h}{S} = \frac{0.25}{18} = 0.0139$$

$$\left(\frac{1000}{L'}\right)^g = \left[\frac{1000}{\dfrac{(2000)(28.3)}{1_w}}\right]^{0.25} = (0.088)^{0.25} = 0.545$$

$$\left(\frac{G}{1000}\right)^f = \left(\frac{1145}{1000}\right)^{2.0} = 1.31$$

$$(0.0139)(0.545)(1.31) = 0.0099$$

From Fig. 13.29, $E' = 0.185$ lb/100 lb

$$E'_{\text{cor}} = E'\left(\frac{73}{\sigma}\right)^{0.2}\left(\frac{\rho_V/0.08}{\rho_L/62.4}\right)^{0.5}$$

$$= (0.185)(1.135)\left(\frac{0.935}{0.721}\right)^{0.5}$$

$$= 0.24 \text{ lb/100 lb vapor}$$

In terms of pounds of liquid entrainment per pound of liquid, ψ', this represents

$$\psi' = 0.0024(1145/2000) = 0.00137$$

Assuming that the molecular weights are the same, $\psi' = 0.00137$ mole of entrained liquid per mole of liquid, and the fractional entrainment is

$$\frac{0.00137}{1 - 0.00137} = 0.00133$$

Vapor-in-Liquid Entrainment

Another factor that affects efficiency of a fractionating plate is the occlusion of vapor in the liquid as foam which can be carried to the plate below, or trapping of vapor in the downcomer by the liquid which carries it to the plate below. (If the foam is sufficiently stable it can be carried to the plate above by the vapor. This is known as *priming*.) The vapor, containing a higher proportion of volatile components than the liquid on the plate, "dilutes" the liquid with respect to the heavier components and reduces the effectiveness of separation, and more plates are required to make the separation because of a reduction in efficiency. Since the quantity of vapor entrained in the liquid is related to the factors which cause froth height to increase or cause liquid throw over the outlet weir, vapor-in-liquid entrainment can be said to be a function of the same factors. Entrainment is increased by increase in vapor load, increase in liquid load, increase in viscosity of the liquid, decrease in density difference between that of the liquid and that of the vapor, increase in weir height, increase in liquid flow path, decrease in surface tension of liquid, increase in density of the vapor, and decrease in distance between the last row of holes or caps and the overflow weir. It is possible to reduce the quantity of vapor in liquid entrainment by locating the last row of caps, valves, or holes at some distance from the overflow weir so that there is less tendency for the vapor to be trapped as the liquid overflows. Also sufficient volume in the downcomers is provided to allow time for the vapor to disengage and escape up the downcomer.

Foam or splash baffles are provided to prevent splashing of the foam or froth into the downcomer and to serve as a "skimmer" to enable only essentially clear liquid to flow into the downcomer. The baffles should be placed above and in front of the overflow weir and located such that the area of flow path of the liquid and froth is never less than the smallest area for flow in the downcomer.

Effectiveness of Vapor-Liquid Contact

When the plate or stage is so designed and operated that sufficient interfacial area between the vapor and liquid phases and a sufficient time of contact (for the area) are provided for equilibrium to be reached between the liquid and vapor compositions at all points, an efficiency of 100% or an efficiency factor of 1.0 is reached. Thus, the actual plate becomes an ideal plate or equilibrium stage. It is possible in some instances to attain an indicated efficiency greater than 100%. The reasons for this are that efficiency calculations are based on compositions of nonrepresentative samples of vapor and/or liquid or the column is operated at such a superficial vapor velocity that a spray of liquid droplets and/or a foam or froth is maintained above the plate which provides additional mass-transfer area and contact time to that provided on the plate. As the result of such factors, plate efficiencies as high as 150% have been reported.

The design and operating factors involved in the mechanism of formation of interfacial area and provision of contact time between phases are liquid depth, length of liquid path, liquid rate, liquid distribution across the plate, vapor rate, size of vapor bubbles or area of phase, and vapor distribution relative to the liquid phase. The system properties which affect the contact between phases are generally assumed to be vapor and liquid density, vapor and liquid viscosity, liquid surface tension, diffusivity, and relative volatility.

Because possibly all of the factors involved have numerous interrelated effects on efficiency, it is difficult to separate them and evaluate their effects individually. For example, increased weir height increases the liquid depth, thus increasing the length of path for the rising vapor bubble and increasing contact time. It also increases liquid holdup, thereby increasing time of contact. It may be increased to the point where the effective plate spacing is reduced and some entrainment takes place. This would reduce efficiency. Because of the complexity of the interaction effects of the variables, experimental studies have succeeded in producing valuable qualitative information on the problem as a whole and relatively little quantitative information, except for particular designs, conditions, and systems. Table 13.5 includes a general summary of observed effects resulting from reported experimental work on bubble-cap and perforated plates.

Table 13.5 Observations of Effects of Variables on Efficiency

Variable increase	Bubble cap		Perforated	
	Effect on efficiency	Reference	Effect on efficiency	Reference
Vapor rate, G				
$G = 0–200$	Increases	[6, 13, 23]	No effect or slight decrease	[25, 30, 42, 50, 63, 73, 85]
$G = 200–500$	Little effect	[32, 41]		
$G = 500–1000$	Decreases	[72, 86]		
Liquid rate, L	Increases	[32, 36, 37, 41]	Increases	[30, 37, 43, 50]
At very high rates	Decreases	[6, 35]		[63, 73, 84]
Reflux ratio, L/V	No effect	[16, 58, 72, 75, 89]	No effect at $L/V > 0.7$; decreases at $L/V < 0.7$	[43]
Weir height, h_w	Increases	[11, 14, 34, 35, 36, 71, 72]	Increases	[25, 50, 73, 85]
Liquid path, Z_L	Increases	[31, 33, 37]		
Plate spacing, S	Increases or no effect	[11, 14, 76]	Increases or no effect	[43, 51]
Cap concentration or hole free area	Decreases	[1, 76]	Increases or no effect	[20, 25, 43]
Hole diameter			Not defined	[25, 42, 43, 84]
Hole pitch			Increases slightly	[84, 85]
Pressure	Increases	[35, 62, 71]	Decreases	[85]
System properties				
Viscosity	Decreases	[22, 58, 70, 76]	Decreases	[50]
Alpha	Increases	[11, 68]		
Surface tension	Increases	[11, 13, 32]	Increases	[15]

Flow Regime and Efficiency

With respect to perforated-plate operation, examination of efficiency curves plotted as a function of numerous different operating variables from experimental data, as well as examination of various correlations which have been developed to relate these variables, indicates some very interesting and apparently contradictory behavior.

For example, at low-column and hole F factors the efficiency increases rapidly until the vapor rate is above that of the dumping point. At this point the efficiency levels off and remains essentially constant with perhaps a slight increase or decrease indicated by the data. These deviations from constancy are well within the experimental limit of error. At higher rates, around an F factor (superficial) of 1.8 to 2.0, efficiency is seen to rise in one case, fall in another, and remain constant in another. Finally, at flooding or priming the efficiency curve turns downward. The two ends of the curve, below the point of dumping and above the point of incipient flooding and priming, are easily explained. The behavior of the curve between these points requires special analysis.

Effect of weir height Above the falloff or dumping point to some intermediate F factor, efficiency increases with increase in weir height in some cases and stays constant in others. This may be explained by the type of controlling resistance to diffusion. If the gas-phase resistance controls, and the liquid phase is the continuous phase (which is usually the case at lower vapor F factors), increase in depth of liquid increases the time of contact and this increases the amount of transfer, resulting in greater efficiency. If the liquid-phase resistance is controlling, increasing the depth of liquid will have little effect since the surface area-to-volume ratio is much smaller. At higher vapor F factors the increase in area per unit volume of liquid is increased through foaming, and the liquid phase becomes the discontinuous phase (liquid droplets in vapor entrainment) and the vapor phase becomes the continuous phase. If the vapor-phase resistance is controlling, little or no increase in transfer rate will be noted and, indeed, it may be lowered by entrainment. Increase in weir height will have little effect. If the liquid-phase resistance controls, increase in dispersion of the liquid phase will increase the transfer and increase in weir height will have no effect.

Bubble-forming mechanism or phase contact The form of interfacial contact between phases affects the rate of mass transfer and the efficiency attained on the plate. If the gas-phase resistance controls, it is important to obtain the maximum area of interface per unit volume of gas phase—bubbling or foaming. Where the liquid-phase resistance controls, the interfacial area per unit volume of the liquid phase should be the greater area. This would represent a spray of liquid droplets; the gas phase would be the continuous phase and the liquid phase would be the discontinuous phase. Experimental studies made under the same operating conditions have resulted in the following conclusions.

Slot Size or Hole Size

Where the column is operating normally there is little effect of slot size or hole size on efficiency in perforated plates. Gouveia et al. [40], Karim and Nandi [53], Hellums et al. [43], Umholtz and Van Winkle [84] all showed this to be true for perforated plates.

Free Area or Bubbling Area

Volland [87] and Foss and Gerster [30] showed that there was a small effect of free area on efficiency, while Umholtz and Van Winkle [85], Karim and Nandi [53], and Hellums et al. [43] all showed little if any effect of free area on efficiency on perforated plates operating between the upper limit of entrainment and the lower limit of flow (where insufficient vapor liquid contact was encountered in the bubble-cap plate or where weeping, dumping, or falloff was encountered with the perforated plate).

Relative vapor and liquid rates The relative vapor and liquid rates should have an effect on efficiency since without either vapor or liquid there would be no mass transfer and, therefore, no efficiency. Investigators have shown this effect experimentally in efficiency versus either liquid or vapor rate, with the other held constant, or in terms of reflux ratio. Gerster et al. [35] show that the efficiency tends to increase with increase in gas rate at some liquid rates and to decrease with others. Similarly, the efficiency decreases with increase in liquid rate and then increases at the higher liquid rates. Ellis and Moyade [25] have shown the same effect. That is, there seems to be a slight optimum efficiency with regard to vapor and liquid flow rate. In perforated-plate studies, Hellums et al. [43], Umholtz and Van Winkle [85], Karim and Nandi [53], etc., show that efficiency increases with decrease in reflux ratio after $L/D = 2.0$ is reached. The change in efficiency is appreciable between $L/D = 2.0$ and $L/D = 1.0$. However, between total reflux and a reflux ratio of around 2.0 the effect is very slight.

Length of liquid path The length of liquid flow path depends upon the diameter of the column as well as on the flow pattern. Essentially it is the distance the liquid travels from the point of entrance on the plate to the point of overflow over the outlet weir. Since there is a concentration gradient across a plate in the liquid, the gradient must be a function of the length of path. Also the efficiency would be at a maximum if the liquid were completely unmixed, that is, there is no forward or back mixing. Kirschbaum [56] assumed that a plate could be divided (for the purposes of mathematical analysis) into a series of completely mixed pools of liquid. Lewis [60] assumed three different cases and derived equations for each. The cases were as follows:

1. The liquid flows across the plate unmixed and the vapor enters the plate completely mixed.
2. The liquid flows across the plate unmixed, vapors enter the plate unmixed, and the liquid flow is the same direction on all plates.
3. The liquid is unmixed and the vapors are unmixed, and the liquid flows in opposite directions on alternate plates.

For the first case, if L/V is constant and equilibrium line slope $m = y/x$ for each plate is constant, and the local efficiency across each plate is constant,

$$E_0 = a(e^{E_v/a} - 1) \tag{13.35}$$

Actually there is mixing, and Gautreaux and O'Connell [33] developed an equation to estimate the number of stages or completely mixed pools:

$$E_0 = a\left(1 + \frac{E_p}{N'a}\right)^{N'} - 1 \tag{13.36}$$

where E_0 = over-all plate efficiency

$$E_0 = \frac{y_{av} - y}{y_n^* - y}$$

$$a = \frac{L}{Vm}$$

$$E_p = \frac{y_i - y}{y_i^* - y} = \text{point efficiency}$$

N' = number of stages in series

Equation (13.36) indicates that the plate efficiency E increases as the number of stages in series is increased or, since the number of stages increases with the length of travel of the liquid Z_L, plate efficiency increases with increase in liquid path (see Fig. 13.30).

Gilbert [37] studied the mixing of liquid on plates and found that the number of mixing stages was roughly equal to the number of rows of bubble caps traversed by the liquid across the plate. For purposes of mathematical analysis he assumed a diffusion model and used frequency response to determine a characteristic mixing parameter for a continuous-flow system.

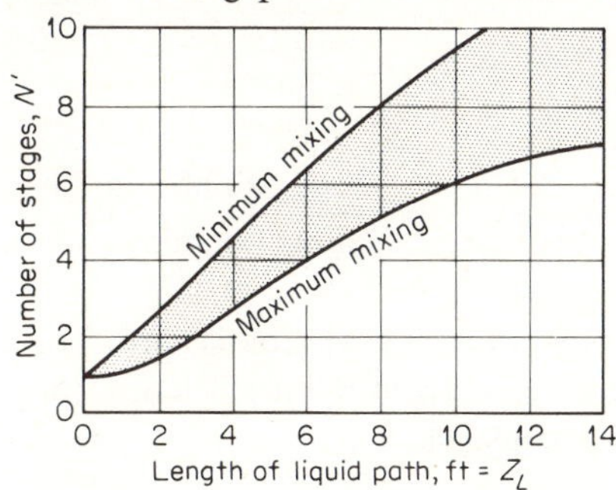

FIGURE 13.30 Correlation for number of liquid stages with length of liquid path and degree of mixing. [*From Gautreaux and O'Connell, Chem. Eng. Progr.*, ***51****:232 (1955)*.]

The longitudinal eddy diffusivity was correlated with liquid velocity, liquid holdup, and froth density for both bubble-cap data and perforated-plate data by

$$\frac{2D_e}{U_L} = C_1(\rho_f U_L)^{-C_2} \tag{13.37}$$

Bubble-cap plate:

$$C_1 = 0.9 \qquad C_2 = 2.4$$

Perforated plate:

$$C_1 = 0.5 \qquad C_2 = 3.0$$

Thus, for a bubble-cap plate,

$$\frac{D_e}{U_L h_c} = 0.9\left(\frac{h_c}{h_f}\right)^{-2.4}\left(\frac{1}{U_L}\right)^{0.4} \tag{13.38}$$

and for a perforated plate,

$$\frac{D_e}{U_L h_c} = 0.25\left(\frac{h_c}{h_f}\right)^{-3.0}\frac{1}{U_L} \tag{13.39}$$

where U_L is the linear velocity of the liquid across the plate between the weirs. Thus the eddy diffusivity increases with distance between weirs or length of liquid travel.

E_{OG}, the over-all plate efficiency based on the gas phase, is related to the point efficiency using the parameter of modified Peclet number,

$$\alpha' Z_L = \frac{L}{h_c \rho_L D_e} Z_L \tag{13.40}$$

and as a function of β',

$$\beta' = \frac{y/x}{L/GZ_L} = \frac{\text{slope equilibrium line}}{\text{slope operating line}} \tag{13.41}$$

A plot based on the Peclet number relation is shown in Fig. 13.31.

To calculate plate efficiency from point efficiency, foam height h_f must be evaluated. For bubble caps [1],

$$h_f = 2.53 F_{VA}{}^2 + 1.89 h_w - 1.6 \tag{13.42}$$

No reliable relationship has been developed for determination of foam height for perforated plates. However, it is commonly assumed that $h_f = 2h_c$, and

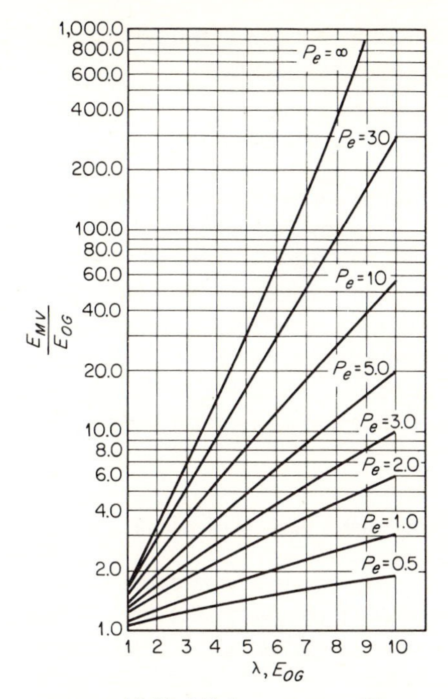

FIGURE 13.31 Mixing curves. (*From A.I.Ch.E. Bubble Tray Design Manual, 1958.*)

since this is just an estimate, it can be greatly in error. Foam heights can be estimated by methods described in the foregoing discussion in this chapter.

Gerster et al. [35] examined the efficiency obtained on the methanol–water system for a small-diameter (13 in.) column and a large-diameter (15 ft) column with split flow (45-in. flow paths), and showed that for equivalent slot velocities and foam densities the efficiencies are comparable. The corrected basis, however, tends to eliminate the mixing effect. Foss et al. [31] studied specifically the effect of liquid mixing on efficiency of bubble trays. They utilized the residence time concept to characterize the degree of mixing on bubble trays and gave equations which require detailed knowledge of the distribution function. They also give a rapid method for estimation of efficiencies by using approximate parameters of the distribution function.

Oliver and Watson [71] studied a number of factors affecting efficiency on a bubble-plate column, 18 in. in diameter, using the ethylene dichloride–toluene system. They developed a mixing parameter equation

$$F = \frac{x_n - x_e}{x_n - x_{n+1}} = f\left(\frac{U_V{}^2 w_v U_L}{2Q_L'^2 \mu_L}\right) \qquad (13.43)$$

where F is the fractional mixing concentration change across the inlet weir divided by total change across the plate.

13.5. PREDICTION OF EFFICIENCY

In order to arrive at the number of actual stages to specify for a given separation, it is necessary to have some knowledge of the tray efficiency which can be expected. Experience data or observed efficiencies on similar systems in similar column and plate designs help greatly and are used whenever available. Where data are unavailable, it is necessary to predict efficiencies from theoretical or empirical relationships. It has become common practice to predict "dry vapor" plate efficiencies or efficiencies, excluding liquid in vapor entrainment and then correcting the dry plate efficiency for the adverse effect on entrainment.

"Dry-vapor" Efficiency and Phase Resistances

The actual rate of mass transfer compared to the theoretical rate of transfer is an expression of efficiency of contact between phases. If the actual rate of transfer were equivalent to the theoretical rate, the efficiency would be 100%. Gerster et al. [36] defined local efficiencies or point efficiencies in terms of a vapor-phase basis and in terms of a liquid-phase basis

$$E_{OG} = \frac{y_n - y_{n-1}}{y_n^* - y_{n-1}} \tag{13.44}$$

$$N_{OG} = -2.3 \log (1 - E_{OG}) = \frac{Z_V K_{OG} a'}{G} \tag{13.45}$$

Since the over-all mass-transfer coefficient and the individual film coefficients are related by

$$\frac{1}{K_{OG} a'} = \frac{1}{k_G a'} + \frac{m}{k_L a'} \tag{13.46}$$

When the slope of the equilibrium curve m is constant,

$$\frac{1}{-2.3 \log (1 - E_{OG})} = \frac{1}{N_G} + \frac{mG/L}{N_L} \tag{13.47}$$

$$N_{OG} = \frac{Z_V k_G a'}{G} \tag{13.48}$$

$$N_G = \frac{Z_V k_G a'}{G} \tag{13.49}$$

$$N_L = \frac{Z_V k_L a'}{L} \tag{13.50}$$

They further express efficiency as follows:

$$\frac{1}{-2.3 \log (1 - E_{OG})} = \frac{1}{N_{OG}} = \frac{1}{N_G} + \frac{mG}{Z_V k_L a'} \tag{13.51}$$

Analogously E_G is defined by

$$-2.3 \log (1 - E_G) = \frac{Z_V k_G a'}{G} \tag{13.52}$$

$$-2.3 \log (1 - E_G) = N_G \tag{13.53}$$

$$N_G = \frac{Z_V}{H_G} \tag{13.54}$$

The pure-liquid-film local or point efficiency is defined by

$$E_L = \frac{x_{n+1} - x_n}{x_{n+1} - x_n^*} \tag{13.55}$$

$$-2.3 \log (1 - E_L) = \frac{Z_V k_L a}{L_m} \tag{13.56}$$

$$-2.3 \log (1 - E_L) = N_L \tag{13.57}$$

$$N_L = \frac{Z_L}{H_L} \tag{13.58}$$

N_L may also be defined as

$$N_L = \frac{x_{n+1} - x_n}{(x - x^*)_m} \tag{13.59}$$

where

$$(x - x^*)_m = \frac{(x_{n+1} - x_{n+1}^*) - (x_n - x_n^*)}{\log \dfrac{x_{n+1} - x_{n+1}^*}{x_n - x_n^*}} \tag{13.60}$$

Since the efficiency at a point on a plate or the average efficiency for the whole plate is the efficiency of mass transfer between phases, there can be no difference in actual efficiency, regardless of the basis on which it is expressed, any more than the number of moles transferred across the gas film per unit time can be different from the number transferred across the liquid film or across both films. Thus the evaluation of efficiency in terms of E_G, E_L, E_{OL}, or E_{OG} is merely arbitrary, and when different values are obtained the so-called *efficiencies* are not efficiencies but defined transfer factors. These are useful in evaluating the major resistances to mass transfer with regard to phases and aid in studying various flow and design characteristics.

Empirical Dry-vapor Efficiency Prediction

Geddes [34] applied mass-transfer theory to the Murphree point-efficiency equations and developed a semitheoretical relationship for prediction of dry-vapor efficiency. He gave empirical equations for evaluation of film and over-all mass-transfer coefficients in terms of bubble size, rate of bubble rise, flow rates, and design variables.

Wet-vapor Efficiency (Fig. 13.32)

Wet-vapor efficiency is the efficiency of the plate and column when the effect of entrainment of liquid in vapor is included. Colburn [18] developed

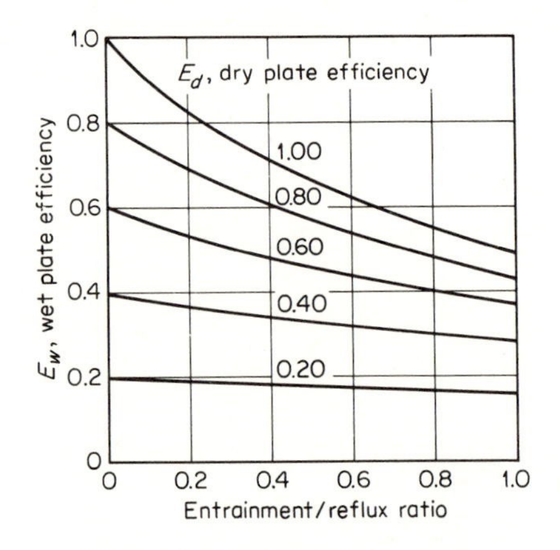

FIGURE 13.32 Wet vapor efficiency and entrainment. Effect of entrainment on plate efficiency. [*From Bolles, Petrol. Process.* (*February, 1956*), *p. 64*.]

a correlation in which wet-vapor plate efficiency is related to reflux ratio and amount of entrainment.

$$E_w = \frac{E_d}{1 + \dfrac{E''E_d}{L/V}} \tag{13.61}$$

where E_w = wet-vapor plate efficiency
E_d = dry-vapor plate efficiency
E'' = entrainment moles liquid per mole dry vapor

A.I.Ch.E. Method—Bubble Tray [1]

The Distillation Subcommittee of the American Institute of Chemical Engineers recommends the following method (included in "Bubble Tray Design Manual") for efficiency. This method involves fixing a plate design and calculating or fixing the operating variables and the system variables. These variables are total pressure, tray temperature, total vapor load, total liquid load, vapor and liquid densities, vapor and liquid viscosities, vapor and liquid molecular weights, vapor diffusivity (molar), liquid diffusivity (molar), and liquid surface tension. The method of calculation is as follows:

1. Calculate the vapor load Q_V and liquid load Q_L in cubic feet per second and gallons per minute, respectively.

2. Determine the vapor velocity based on bubbling area in feet per second:

$$U_{VA} = \frac{Q_V}{A_A} \tag{13.62}$$

3. Calculate $F_{VA} = U_{VA}(\rho_V)^{0.5}$.

4. Calculate liquid rate per foot of average flow width, Q_L/W_a.

5. Calculate the absorption factor $\lambda = mG_m/L_m$ by

$$\lambda = \frac{mQ_V\rho_V M_L C}{Q_L\rho_L M_V} \tag{13.63}$$

6. Calculate the foam height (inches) by

$$h_f = 2.53F_{VA}^2 + 1.89h_w - 1.6 \tag{13.64}$$

7. Calculate the clear liquid height (inches) by

$$h_c = \frac{103 + 11.8h_w - 40.5F_{VA} + 1.25Q_L/W_a}{\rho_L} \tag{13.65}$$

8. Calculate the liquid contact time in seconds by

$$\theta_L = \frac{37.4h_c A_A}{Q_L} \tag{13.66}$$

9. The number of liquid-phase transfer units are determined by

$$N_L = 103D_L^{0.5}(0.26F_{VA} + 0.15)\theta_L \tag{13.67}$$

10. The number of gas-phase transfer units are calculated by

$$N_G = \left(\frac{\rho_V D_V}{\mu_V}\right)^{0.5}\left(0.776 + 0.116h_w - 0.290F_{VA} + 0.0217\frac{Q_L}{W_a} + 0.200\Delta\right) \tag{13.68}$$

11. The over-all gas-phase efficiency E_{OG} is evaluated by

$$-\log(1 - E_{OG}) = 0.434 \frac{N_L N_G}{N_L + \lambda N_G} \tag{13.69}$$

12. The percent liquid-phase resistance is determined by

$$\% R_L = \frac{\lambda N_G 100}{N_L + \lambda N_G} \tag{13.70}$$

13. The eddy diffusion coefficient D_e is calculated by

$$D_e = [1.0 + 0.044(d_c - 3)]^2 \times \left(0.0124 + 0.0150 h_w + 0.017 U_{VA} + 0.00250 \frac{Q_L}{W_a}\right)^2 \tag{13.71}$$

14. The Peclet number is next evaluated

$$\text{Pe} = \frac{Z_L^2}{D_e \theta_L} \tag{13.72}$$

15. The term λE_{OG} is evaluated so that Fig. 13.31 can be used to determine E_{MV}/E_{OG}.

16. Evaluate E_{MV}.

17. Calculate the apparent plate spacing

$$S' = S - h_f \tag{13.73}$$

18. Using the value of U_{VA}/S' determine $e_w\sigma$ from Fig. 13.33.

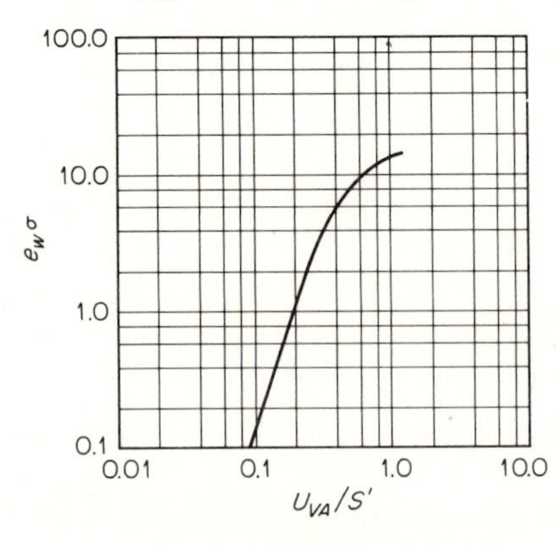

FIGURE 13.33 Entrainment relation. (*From A.I.Ch.E. Bubble Tray Design Manual, 1958.*)

19. Calculate the net fraction of liquid entrained r_e by

$$r_e = \frac{449 e_w Q_V \rho_V}{Q_L \rho_L} \tag{13.74}$$

20. Determine wet-plate efficiency E_w and the over-all efficiency E_0 by

$$E_w = \frac{E_{MV}}{1 + r_e E_{MV}} \tag{13.75}$$

and

$$E_0 = \log \frac{1 + E_w(\lambda - 1)}{\log \lambda} \tag{13.76}$$

Example 13.9

Calculate the plate efficiency E_0 by the A.I.Ch.E. Bubble Tray Design Manual method for the following column and condition:

Column diameter, $D = 6$ ft

$h_w = 1.5$ in.

Average flow width, $W_a = 5.4$ ft

$l_w = 4.8$

Distance between weirs, $Z_L = 4.5$ ft

Bubbling area, $A_A = 24.4$ sq ft

Slot area, $A_S = 1.955$ sq ft

Tray spacing, $S = 24$ in.

Liquid gradient, $\Delta = 4.0$ in.

Cap diameter $= 4.0$ in.

Cap clearance $= 3.0$ in.

Total vapor load, $Q = 170$ cu ft/sec

Total liquid load, $Q_L = 180$ gpm

$\dfrac{L}{V} = 0.458$ mole/mole

$\mu_V = 0.03$ lb/(ft)(hr)

$\mu'_L = 0.30$ centipoise

$P_T = 10$ psia

Tray temperature $= 190°$F

$M_V = 68$

$M_L = 72$

$\rho_V = 0.095$ lb/ft

$\rho_L = 45.2$ lb/ft

$m = 0.85 = \dfrac{dx}{dy}$

$D_V = 8.6 \times 10^{-5}$ sq ft/hr

$D_V = 0.0535$ sq ft/hr

$\alpha = 1.8$

$\sigma = 14.5$

Solution

$$U_{VA} = \frac{170 \text{ cu ft/sec}}{24.4 \text{ sq ft}} = 7.0 \text{ fps}$$

$$F_{VA} = U_{VA}(\rho_V)^{0.5} = (7.0)(0.095)^{0.5} = 2.14$$

$$\frac{Q_L}{W_a} = \frac{180}{5.4 \text{ ft}} \text{ gpm} = 33.4 \text{ gpm/ft average plate width}$$

$$\lambda = \frac{(170)(0.095)(72)}{(180)(45.2)(68)}(449)(0.85) = m\frac{G_m}{L_m} = 0.802$$

$$h_f = 2.53F_{VA}^2 + 1.89h_w - 1.6 = (2.53)(2.14)^2 + (1.89)(1.5) - 1.6 = 12.83 \text{ in.}$$

$$h_c = \frac{103 + (11.8)(1.5) - (40.5)(2.14) + (1.25)(33.4)}{45.2} = 1.7 \text{ in.}$$

$$D_e = [1.0 + 0.044(d_c - 3)]^2\left(0.0124 + 0.0150h_w + 0.017U_{VA} + 0.0025\frac{Q_L}{W_a}\right)^2$$

$$= [1.0 + 0.044(1)]^2[0.0124 + 0.0150(1.5) + 0.017(7.0) + 0.0025(33.4)]^2$$

$$= (1.09)[(0.0124) + (0.0225) + (0.119) + 0.0835]^2$$

$$= 1.09(0.2374)^2 = 0.0615$$

$$\theta_L = \frac{37.4h_cA_A}{Q_L} = 37.4\frac{(24.4)(1.7)}{180} = 8.6 \text{ sec}$$

$$\text{Pe} = \frac{Z_L^2}{D_e\theta_L} = \frac{(4.5)^2}{(0.0615)(8.6)} = 38.4$$

$$N_L = 103(D_L)^{0.5}(0.26F_{VA} + 0.15)\Phi_L$$

$$= 103(86 \times 10^{-6})^{0.5}[(0.26)(2.14) + 0.15](8.6)$$

$$= (103 \times 10^{-3})(9.2)[(0.555) + 0.15]8.6$$

$$= 8.160(0.7) = 5.74$$

$$N_G = \left(\frac{\rho_V D_V}{\mu_V}\right)^{0.5}\left(0.776 + 0.116h_w - 0.290F_{VA} + 0.217\frac{Q_L}{W_a} + 0.200\Delta\right)$$

$$= \left[\frac{(0.095)(0.0535)}{0.03}\right]^{0.5}[0.776 + (0.116)(1.5) - (0.290)(2.14)$$

$$+ (0.0217)(33.4) + 0.2(4)]$$

$$= (0.169)^{0.5}(0.776 + 0.174 - 0.62 + 0.725 + 0.8)$$

$$= 0.76$$

$$-\log(1 - E_{OG}) = 0.434\frac{N_LN_G}{N_L + \lambda N_G} = 0.434\frac{(5.74)(0.76)}{(5.74) + (0.802)(0.76)}$$

$$= 0.434\frac{4.35}{6.32} = 0.305$$

$$1 - E_{OG} = 0.495 \qquad E_{OG} = 0.505$$

$$\lambda E_{OG} = (0.802)(0.505) = 0.405$$

From Fig. 13.31, $E_{MV}/E_{OG} = 1.18$,

$$E_{MV} = (0.505)\ 1.18 = 0.595$$

$$S' = S - h_f = 24.0 - 12.83 = 11.17$$

$$\frac{U_{VA}}{S'} = \frac{7}{11.17} = 0.6 \qquad e_w\sigma = 9.0$$

$$e_w = \frac{9}{14.5} = 0.62 \qquad r_e = 4490.62\left(\frac{170}{180}\right)\left(\frac{0.095}{45.2}\right) = 0.551$$

$$E_w = \frac{E_{MV}}{1 + r_e E_{MV}} = \frac{0.595}{1 + 0.551(0.595)} = 0.446$$

$$E_0 = \frac{\log[1 + E_w(\lambda - 1)]}{\log \lambda} = \frac{\log[1 + 0.446(0.802 - 1)]}{\log 0.802}$$

$$E_0 = \frac{\log 0.911)}{\log (0.802)} = \frac{-0.041}{-0.096} = 0.427 \text{ or } 42.7\%$$

Empirical Methods

Hughmark [47] proposed empirical correlations for interfacial area and liquid residence time for estimation of gas and liquid transfer units for binary system distillations. Drickamer and Bradford [22] presented a correlation, which was entirely empirical, relating plate efficiency with a pseudomolal viscosity of the feed to the fractionating column. The correlation resulted from a study of a large number of industrial fractionators including those in hydrocarbon service as well as nonhydrocarbon applications. The authors implied restriction of the correlation to columns of standard or customary design and to operation near the maximum capacity.

The correlation equation is

$$E_0 = 0.17 - 0.616 \log (m\mu'_L) \tag{13.77}$$

O'Connell [70] presented an empirical correlation of plate efficiency as a function of feed viscosity and the relative volatility of the key components which was based on the study of the performance of 32 commercial columns and 5 laboratory columns. He presents correlations for both fractionators and absorbers. The fractionator plate efficiency was correlated as a function of relative volatility of the light key to the heavy key times the molal average viscosity of the feed at the arithmetic average temperature of the top- and bottom-column conditions.

Chaiyavech and Van Winkle [15] studied a number of systems selected for their wide variation in physical properties of vapor and liquid density, vapor and liquid viscosity, vapor and liquid diffusivity, and liquid surface tension.

Plate efficiencies were determined in a laboratory column at essentially the same operating conditions.

Dimensional analysis gave the following grouping:

$$E = A\left(\frac{\sigma}{\mu'_L U'_V}\right)^B \left(\frac{\mu'_L}{\rho'_L D'_L}\right)^C \left(\frac{\mu'_V}{\rho'_V D'_V}\right)^D \left(\frac{\mu'_L}{\mu'_V}\right)^E \left(\frac{\rho'_L}{\rho'_G}\right)^F \alpha^G \tag{13.78}$$

where A, B, C, D, E, F, G are constants. Mathematical analysis of the data indicated that the exponents D, E, and F were so small the terms could be neglected.

Thus the equation became

$$E = A\left(\frac{\sigma}{\mu'_L U'_V}\right)^{0.643} \left(\frac{\mu'_L}{\rho'_L D'_L}\right)^{0.19} \alpha^{0.056} \tag{13.79}$$

For the column used, $A = 0.0691$. Analysis of literature data gave values of A, which incorporates operating and design variables, ranging from 0.0101 to 0.236. Using the correct A value correlated the data with a maximum deviation of 12 efficiency units and an average deviation of around 2.4 efficiency units.

English and Van Winkle [26], using the basic correlation of properties developed by Chaiyavech and Van Winkle, studied the efficiency data reported in the literature in order to determine the relationship between efficiency and operating and design variables.

From mathematical analysis of these data [2, 3, 4, 9, 10, 13, 67, 83, 88] an empirical correlation was developed which gave an average absolute deviation between calculated and experimental data of 4.6 efficiency units. Equation (13.80) relates the operating, design, and system variables for both perforated and bubble-cap columns.

$$E = 10.84(FFA)^{-0.280}\left(\frac{L}{V}\right)^{0.024} h_w^{0.241} G^{-0.013} \left(\frac{\sigma}{\mu'_L U'_V}\right)^{0.044} \left(\frac{\mu'_L}{\rho'_L D'_L}\right)^{0.137} \alpha^{-0.028} \tag{13.80}$$

The data calculated from the equation had a maximum deviation of 20% from the experimental results. The range of applicability of the equation in terms of operating and design variables is as follows:

Column type—bubble cap, perforated plate
Column diameter, 1 to 24 in.
Percent free area or slot area, 2.7 to 18.5
Plate spacing, 2 to 36 in.

Hole diameter, 1/16 to 7/8 in.
Weir height, 1/2 to 6 in.
$G = 100 - 1000$ lb/(hr)(sq ft)
$L = 100 - 1000$ lb/(hr)(sq ft)
$L/V = 0.6$ to 1.0

In addition, the data on which this analysis was carried out included different sizes and types of caps and cap arrangement and nine different systems.

Example 13.10

For the column and system shown in Example 13.9 calculate the efficiency by (*a*) the Drickamer and Bradford method; (*b*) the English and Van Winkle method.

Solution

(*a*)

$$\begin{aligned} E &= 0.17 - 0.616 \log \mu'_{L,\mathrm{av}} \\ &= 0.17 - 0.616 \log 0.30 \\ &= 0.17 + 0.323 = 0.493 \text{ or } 49.3\% \end{aligned}$$

(*b*)

$$E = 10.84(FFA)^{-0.28}\left(\frac{L}{V}\right)^{0.024} h_w^{0.241} G^{-0.013}$$

$$\left(\frac{\sigma}{\mu'_L U'_V}\right)^{0.044}\left(\frac{\mu'_L}{\rho'_L D'_L}\right)^{0.137} \alpha^{-0.028}$$

$$= 10.84(0.08)^{-0.28}(0.458)^{-0.224}(1.5)^{0.241}(2060)^{-0.013}$$

$$(26.2)^{0.44}(1.88 \times 10^3)^{0.137}(1.8)^{-0.028}$$

$$= (10.84)(2.06)(0.998)(1.125)(0.873)(1.15)(2.81)(0.985)$$

$$= 69.5\%$$

Scale-up of Efficiency from Laboratory Data

A useful method for extrapolating efficiency data determined in the laboratory on bench scale or pilot scale equipment to plant scale equipment was developed by Finch and Van Winkle [29]. The procedure in general is as follows:

1. Use a test column, preferably 12 in. or more in diameter (although a 4-in.-diameter plate may be satisfactory if the measurements are precise), with the plate design being a prototype of the proposed large-scale plate having the same percent-free area, hole diameter, pitch, percent downcomer area, weir height, and preferably, 12-in. plate spacing. Furthermore, use total reflux and at least three values of vapor (liquid) loading in the range of proposed operating loading.

2. Take two vapor samples above and two below the test plate and three liquid samples, one from the inlet downcomer, one from the center of the plate, and one from the outlet downcomer.

3. Take the total pressure drop from taps provided above the test plate and below it.

4. Measure the average total foam (vapor-liquid mixture) height above the plate surface.

5. If possible make runs on liquid concentrations corresponding to feed, upper column, and lower column concentrations.

6. Measure the plate temperatures for each condition.

7. Calculate efficiency by

$$E_{MV} = \frac{y_n - y_{n+1}}{y_n^* - y_{n+1}} \tag{13.81}$$

where y_n^* = vapor composition in equilibrium with the average liquid composition on the test plate

y_n = average vapor composition above the test plate

y_{n+1} = average vapor composition below the test plate

8. Determine the dry-plate pressure drop h_0 by the Kolodzie, Smith, Van Winkle method [Eq. (13.15)] and Fig. 13.17.

9. Determine the equivalent clear-liquid head-pressure drop h_l

$$h_l = \Delta H_T - h_0 \tag{13.82}$$

where ΔH_T is the total pressure drop across the plate.

10. Calculate the clear-liquid holdup B_L on the tray in feet of liquid at the plate temperature.

11. Calculate the tray vapor holdup B_G

$$B_G = B_T - B_L \tag{13.83}$$

where B_T is the average foam height on the tray in feet.

12. Determine the gas contact time θ_G. Calculate the density of the vapor at the temperature of the plate and the total pressure and the rate of vaporization in terms of G', pounds per hour per square foot of bubbling area (between inlet downcomer and outlet weir):

$$\theta_G = \frac{B_G \rho_G}{G'} \tag{13.84}$$

13. Determine the liquid contact time θ_L. Calculate ρ_L at the plate temperature and the liquid flow rate L' in terms of pounds of liquid per hour per foot of average width of flow path, and the distance between weirs, Z_L,

$$\theta_L = \frac{B_L Z_L \rho_L}{L} \tag{13.85}$$

14. Plot on rectangular coordinate paper values of E_{MV}/θ_L versus θ_G/θ_L for each of the three rates studied and draw a straight line through the three points. The intercept is A_L and the slope is A_G in the followinge quation:

$$\frac{E_{MV}}{\theta_L} = A_G \frac{\theta_G}{\theta_L} + A_L \tag{13.86}$$

15. From the values of A_G and A_L thus established the efficiency can be determined for a larger-diameter column through θ_L since A_G and A_L have been shown to be primarily functions of the diffusive and surface-tension characteristics of the system.

PROBLEMS

1. For the same system and column described in Example 13.1 determine the pressure drop if 6-in.-diameter caps having the characteristics listed in Table 14.6 were used on equilateral triangular spacing with 2 in. between caps. Use the Bolles method.
2. Determine the liquid backup H_D.
3. If the holes were spaced on ¾-in. centers for the column and conditions shown in Example 13.3, determine the pressure drop.
4. Calculate the liquid backup H_D.
5. Using the Fair and Matthews correlation determine whether the bubble-cap column in Prob. 1 would flood or prime.
6. Determine the dumping flow (liquid) rate for the conditions in Prob. 3 using the Liebson correlation, p. 528.
7. For the column in Prob. 1 determine the maximum liquid flow rate at bare minimum stability.
8. Calculate the quantity of entrainment to be expected for a bubble-cap column of the following characteristics:

   Column diameter = 4 ft
   Bubble-cap diameter = 3 in.
   Slot height = 1.0 in.
   Slot opening = 0.7 in.
   Height of slop top above plate = 1.5 in.
   Weir length, l_w = 3.0 ft
   Weir height = 3.0 in.
   h_{ow} = 0.6 in.
   ρ_v = 0.11 lb/cu ft
   ρ_L = 48.0 lb/cu ft
   Tray spacing, in. = 15 in.
   Vapor load = 120 cu ft/sec
   Liquid load = 220 gpm
   σ = 50 dynes/cm

   (*a*) Use the Fair relationship.
   (*b*) Use the Bain and Van Winkle method.

9. Calculate the efficiency to be expected for the column in Prob. 8 with the A.I.Ch.E. method. Additional data are: $\mu_L = 0.8$ centipoise, slot width = 0.125 in., compressibility factor = 1.0, distance between weirs = 3.0 ft, 48 caps, 2½ in. from wall to wall, 40 slots per cap, plate temperature = 250°F, $M_V = 65$, $M_L = 68$, $y/x = m = 1.3$, $P_t = 1$ atm, $\mu_V = 0.029$ centipoise, $D_V = 0.03$ sq ft/hr, $D_L = 1.1 \times 10^{-4}$ sq ft/hr.
10. Determine the plate efficiency by the Drickamer and Bradford method.
11. Determine the plate efficiency by the English and Van Winkle method.

Nomenclature

A = constant
A = cross-sectional area of column, sq ft
A_A = bubbling area, sq ft = $A - 2A_d$
A_a = total annular area per tray, sq ft
A_d = area of downcomer, sq ft
A_{dm} = minimum area for liquid flow in downcomer assembly, sq ft
A_G = slope, Eq. (13.86)
A_h = area of holes per tray, sq ft
A_L = intercept, Eq. (13.86)
A_N = net tray area for vapor flow = $A - A_d$
A_r = total riser area per tray, sq ft
A_s = total slot area per tray, sq ft
a = $\dfrac{L}{Vm}$
a' = phase contact area, sq ft/cu ft
a_a = annulus area per cap, sq in.
a_c = cap area per cap, sq in.
a_r = area of the riser per cap, sq in.
a_s = slot area per cap, sq in.
B = constant
B_G = tray vapor holdup, Eq. (13.83)
B_L = liquid holdup, ft
B_T = average foam height, ft
C = velocity of sound in the fluid, fps
C = compressibility factor
C, C_1, C_2 = constants
C_d = liquid gradient factor, Fig. 13.44
C_0 = orifice coefficient
C_V = vapor load correction factor, Fig. 13.15
D = tower diameter, ft or in.
D_e = longitudinal eddy diffusivity, sq cm/sec or sq ft/sec
D_L = volumetric liquid diffusion coefficient, sq ft/hr
D'_L = volumetric liquid diffusion coefficient, sq cm/sec
D_V = volumetric vapor diffusion coefficient, sq ft/hr

D'_V = volumetric vapor diffusion coefficient, sq cm/sec
d_c = cap diameter, in.
d_h = hole diameter, in.
d_r = diameter of riser, in.
E = entrainment, lb/min
E' = entrainment, lb liquid/100 lb vapor
E'' = entrainment, moles liquid/mole dry vapor
E_d = dry-vapor plate efficiency
E_G = vapor efficiency
E_{MV} = Murphree vapor efficiency
E_{OG} = over-all vapor phase efficiency
E_L = liquid efficiency
E_0 = over-all plate efficiency
E_p = Murphree point efficiency
E_w = actual efficiency corrected for entrainment
e_d = dry-vapor efficiency
e_w = entrainment ratio, lb liquid/lb vapor
F = mixing parameter.
FA = % free area
FFA = fractional free area
F_h = hole F factor
F'_h = hole falloff factor
F_L = liquid F factor
F_V = F factor based on free cross-sectional area
F_{VA} = F factor based on active area
F_{VN} = F factor based on net area
F_w = function for h_{ow} correction, Fig. 13.7
f = function
f = frequency, cps
f = friction factor
f_f = foam friction factor
G = vapor mass velocity, lb/(hr) (sq ft of area)
G' = vapor mass velocity, lb/(hr) (sq ft A_A)
G_e = vapor entrainment velocity, lb/(min) (sq ft of area)
G_m = molar vapor velocity, moles/(hr) (sq ft of area)
g = constant
g_c = gravitational constant
H_D = liquid backup in downcomer, in. (from plate surface)
H_G = height of gas transfer unit
H_L = height of liquid transfer unit
ΔH_T = total pressure drop across plate equivalent head, in. of fluid
h_c = clear-liquid height, in.
h_d = equivalent head loss through downcomer assembly, in. of fluid

h_{ds} = dynamic seal, in., Fig. 13.4
h_e = equivalent head loss caused by vapor passage through liquid and formation of bubbles, in. of fluid
h_f = height of foam, in.
h_l = clear-liquid head, defined Eq. (13.82)
h_L = height of liquid, ft or in.
h_m = equivalent head loss to lift metal valve, in. of fluid
h_0 = equivalent dry-plate pressure drop through holes, in. of fluid
h_0' = equivalent dry-plate pressure drop through holes, in. of fluid, corrected for entrainment
h_{ow} = height of crest over weir, in.
h_r = equivalent riser pressure drop, in. of fluid
h_{ra} = equivalent riser annulus, reversal pressure drop, in. of fluid
h_{rc} = equivalent head loss through wet cap, in. of fluid
h_s = slot length, in.
h_{sk} = skirt clearance, surface of plate to bottom of cap
h_{so} = slot opening measured from top of slot, in.
h_{sr} = height of shroud ring, in.
h_{ss} = static seal distance from top of slot to top of weir, in.
h_σ = equivalent surface-tension head loss, in. of fluid
h_w = weir height, in.
K_c = correction factor, Fig. 13.8
K_{OG} = over-all mass-transfer coefficient based on vapor
K_{OL} = over-all mass-transfer coefficient based on liquid
k = constant
K = constant
k_G = vapor film mass-transfer coefficient
k_L = liquid film mass-transfer coefficient
L = liquid flow rate, lb/(hr)/(sq ft of area)
L' = liquid flow rate, lb/(hr) (ft W_a)
L_m = moles liquid/(hr)(sq ft of area)
l_{vs} = length of system, ft
l_w = length of weir, in. or ft
M = molecular weight
m = slope of equilibrium curve y/x
N = number of caps per sq ft
N' = number of stages in series
N_G = number of vapor-phase transfer units
N_L = number of liquid-phase transfer units
N_{OG} = number of over-all transfer units (vapor)
N_{OL} = number of over-all transfer units (liquid)
N_R = number of rows of caps normal to liquid flow
P = pitch, distance between hole centers, in.

Pe = Peclet number = $Z_L^2/D_e\theta_L$
P_t = pressure, total
Q_L = liquid flow rate, gpm
Q'_L = liquid flow rate, cu ft/hr or cu ft/sec
Q_V = vapor load, cu ft/sec
R = gas-law constant
Re_f = Reynolds number of foam, Eq. (13.19)
R_H = hydraulic radius, Eq. (13.20)
r_e = net fraction of liquid entrained, Eq. (13.74)
S = tray spacing, in.
S' = corrected tray spacing, in., $S - Z_f$
S_m = minimum submergence, Fig. 13.24
T = absolute temperature, °R or °K
t = temperature, °F or °C
t_p = plate thickness, in.
U_h = vapor velocity through holes, fps
U_f = velocity of foam, consistent units
U_L = liquid velocity fps across plate based on distance between weirs
U_V = vapor velocity, fps based on A
U'_V = vapor velocity, cm/sec based on A
U_{VA} = vapor velocity based on active area, fps
U_{VC} = maximum vapor velocity through cap assembly, fps
U_{VN} = velocity of vapor based on net area
V = vapor flow rate, lb/(hr)(sq ft of area) or moles/(hr)(sq ft of area)
V_S = volume of system, cu ft
W_a = $(D + l_w)/2$, average width of liquid flow across plate, in.
w_s = width of slot, ft or in.
w_v = mass rate of flow, lb/sec
X = mole ratio in liquid, $x/(1 - x)$
x = mole fraction in liquid
Y = mole ratio in vapor
y = mole fraction in vapor
y^* = mole fraction in vapor in equilibrium with x
Z_L = distance between weirs, ft, in., or cm
Z_V = distance of vapor travel through liquid, ft or in.

Greek Letters

α = relative volatility
α' = relative parameter, Eq. (13.40)
β = aeration factor, Eq. (13.14)
β_d = aeration factor for liquid in downcomer
β' = slope ratio, Eq. (13.41)
γ = ratio of distance between caps to cap diameter

Δ = difference
Δ' = liquid gradient across plate, in. of fluid, uncorrected
Δ = liquid gradient across plate, in. of fluid, corrected
Δ'_r = liquid gradient per row of caps normal to flow, uncorrected, in.
λ = mG_m/L_m, absorption factor
μ_L = viscosity of liquid, lb/(ft)(hr)
μ'_L = viscosity of liquid, centipoises
μ_V = viscosity of vapor, lb/(ft)(hr)
μ'_V = viscosity of vapor, centipoises
ϕ = relative foam density, Eq. (13.13)
ϕ_d = relative foam density, downcomer
ψ = moles entrainment per mole of gross downflow
$\Delta\rho$ = $\rho_L - \rho_V$
ρ_f = h_c/h_f
ρ_L = density of liquid, lb/cu ft
ρ'_L = density of liquid, g/cu cm
ρ_V = density of vapor, lb/cu ft
ρ'_V = density of vapor, g/cu cm
σ = surface tension, dynes/cm or lb/ft
θ = time, sec
θ_G = vapor contact time, sec
θ_L = liquid contact time, sec
Σ = summation

Subscripts

b = normal boiling point
c = critical
$n, n + 1$ = plate numbers
i, j = components
1, 2 = components

REFERENCES

1. American Institute of Chemical Engineers Distillation Committee: *A.I.Ch.E. Bubble Tray Design Manual*, 1958.
2. *A.I.Ch.E. Res. Comm. Addenda to Third Annual Progr. Rept.* (1956).
3. *A.I.Ch.E. Res. Comm. Final Rept. North Carolina State College* (1956).
4. *A.I.Ch.E. Res. Comm. Final Rept. Univ. Delaware* (1958).
5. Arnold, D. S., C. A. Plank, and E. M. Schoenborn: *Chem. Eng. Prog.*, **48**:633 (1952).
6. Atkins, G. T.: *Chem. Eng. Progr.*, **5e**:116 (1954).
7. Atterig, P. T., E. J. Lemieux, W. C. Schreine, and R. A. Sundback: *A.I.Ch.E. J.*, **2**:3 (1956).
8. Bain, J. L., and M. Van Winkle: *A.I.Ch.E. J.*, **7**:363 (1961).
9. Barker, P. E., and M. H. Choudbury: *Brit. Chem. Eng. J.* (June, 1959), p. 348.

10. Berg, L., and D. O. Popovac: *Chem. Eng. Progr.*, **45**:683 (1949).
11. Bolles, W. L.: *Petrol. Process.* (February, 1956), p. 64; (March, 1956), p. 82; (April, 1956), p. 72; (May, 1956), p. 109.
12. Brown, G. G., and Associates: "Unit Operations," Wiley, New York, 1950.
13. Brown, G. G., and Mott Sounders: *Ind. Eng. Chem.*, **26**:98 (1934).
14. Carey, J. S., J. Griswold, W. K. Lewis, and W. H. McAdams: *Trans. A.I.Ch.E.*, **30**:504 (1933).
15. Chaiyavech, P., and M. Van Winkle: *Ind. Eng. Chem.*, **53**:187 (1961).
16. Chu, Ju Chin: *Petrol. Process.* (January, 1951), p. 39.
17. Cicalese, J. J., J. A. Davies, P. J. Harrington, G. S. Houghland, A. J. L. Hutchinson, and T. J. Walsh: *Petrol. Refiner* (April, 1947), p. 141; (May, 1947), p. 127; *Trans. Am. Petrol. Inst.* (Div. of Refining), **26**(III):180 (1946).
18. Colburn, A. P.: *Ind. Eng. Chem.*, **28**:526 (1936).
19. Dauphine, T. C.: "Pressure Drop in Bubble Trays," Sc.D. Thesis, MIT, 1959.
20. Davies, James A.: *Petrol. Refiner* (August, 1950), p. 93; (September, 1950), p. 21.
21. Davies, James A.: *Ind. Eng. Chem.*, **39**:774 (1947).
22. Drickamer, H. G., and J. R. Bradford: *Trans. A.I.Ch.E.*, **39**:319 (1943).
23. Edulgee, H. E.: *Trans. Inst. Chem. Engrs. (London)*, **24**:128 (1946)..
24. Eld, A. C.: *Petrol. Refiner* (October, 1948), p. 119.
25. Ellis, S. R. M., and H. D. Moyade: *Brit. Chem. Eng.* (June, 1959), p. 342.
26. English, G. E., and M. Van Winkle: *Chem. Eng.* (November 11, 1963), pp. 241–245.
27. Fair, J. R.; *Petro./Chem. Eng.* (September, 1961), p. 45.
28. Fair J. R., and R. L. Matthews: *Petrol. Refiner* (April, 1958), p. 153.
29. Finch, R. N., and Matthew Van Winkle: *Ind. Eng. Chem., Process Design Develop.*, **3**:106–116 (1964).
30. Foss, A. S., and J. A. Gerster: *Chem. Eng. Progr.*, **52**:28-j (1956).
31. Foss, A. S., J. A. Gerster, and R. L. Pigford: *A.I.Ch.E. J.*, **4**:231 (1958).
32. Garner, F. H., and D. C. Freshwater: *Trans. Inst. Chem. Engrs. (London)*, **33**:280, (1955).
33. Gautreaux, M. F., and H. E. O'Connell: *Chem. Eng. Prog.*, **51**:232 (1955).
34. Geddes, R. L.: *Trans. A.I.Ch.E.*, **42**:79 (1946).
35. Gerster, J. A., W. E. Bonnet, and I. Hess: *Chem. Eng. Progr.*, **47**:573, 621 (1951).
36. Gerster, J. A., A. P. Colburn, W. E. Bonnet, and T. W. Carmody: *Chem. Eng. Progr.*, **45**:716 (1949).
37. Gilbert, J. T.: *Chem. Eng. Sci.*, **10**:243 (1959).
38. Glitsch, F. W.: Ballast Tray Design Manual, *Bull.* 4900, Fritz W. Glitsch & Sons, Dallas, 1961.
39. Good, A. J., M. H. Hutchinson, and W. C. Rousseau: *Ind. Eng. Chem.*, **34**:1445 (1942).
40. Gouveia, W. R., Ju C. Chu, and O. P. Kharbanda: Abstracts of Papers, given at the Am. Chem. Soc. Meeting, New York, September, 1954.
41. Grohse, E. W., R. F. McCartney, H. J. Hauer, J. A. Gerster, and A. P. Colburn: *Chem. Eng. Progr.*, **45**:725 (1949).
42. Gunness, R. C., and J. C. Baker: *Ind. Eng. Chem.*, **30**:1394 (1938).

43. Hellums, J. D., D. J. Braulick, C. D. Lyda, and M. Van Winkle: *A.I.Ch.E. J.*, **4**:465 (1958).
44. Holbrook, G. E., and E. M. Baker: *Trans. A.I.Ch.E.*, **30**:520 (1934).
45. Huang, C. J., and J. R. Hodson: *Petrol. Refiner* (February, 1958), p. 104.
46. Hughmark, G. A., and H. E. O'Connell: *Chem. Eng. Progr.*, **53**:127M (1957).
47. Hughmark, G. A.: *Chem. Eng. Progr.*, **61**(7):97 (1965).
48. Hunt, C. d'A., D. N. Hanson, and C. R. Wilke: *A.I.Ch.E. J.*, **1**:441 (1955).
49. Hutchinson, M. H., A. G. Buron, and B. B. Miller: Aerated Flow Principle Applied to Sieve Plates, *A.I.Ch.E. Paper*, Los Angeles Meeting (March, 1949).
50. Johnson, A. I., and J. Marangozis: *Can. J. Chem. Eng.* (August, 1958), p. 61.
51. Jones, J. B., and C. Pyle: *Chem. Eng. Progr.*, **51**:424 (1955).
52. Jones, P. D., and M. Van Winkle: *Ind. Eng. Chem.*, **49**:232 (1952).
53. Karim, B., and S. K. Nandi: *J. Indian Chem. Soc., Ind. News Ed.*, **11**(1): 3 (1948).
54. Kemp, H. S., and Cyrus Pyle: *Chem. Eng. Progr.*, **45**:435 (1949).
55. Kirkbride, C. G.: *Petrol. Refiner* (September, 1944), p. 321.
56. Kirschbaum, E.: "Distillation and Rectification," Chemical Publishing, New York, 1948.
57. Kolodzie, P. A., and M. Van Winkle: *A.I.Ch.E. J.*, **3**:305 (1957).
58. Langdon, W. M., and D. B. Keyes: *Ind. Eng. Chem.*, **35**:464 (1943).
59. Lee, D. C.: *Chem. Eng.* (May, 1954), p. 179.
60. Lewis W. K., Jr.: *Ind. Eng. Chem.*, **28**:399 (1936).
61. Liebson, I., R. E. Kelley, and L. A. Bullington: *Petrol. Refiner* (February, 1957), p. 127.
62. Marek, J., and Z. Novasad: *Collection Czech. Chem. Commun.*, **21** (1956).
63. Mayfield, F. D., W. L. Church, A. C. Green, D. C. Lee, and R. W. Rasmussen: *Ind. Eng. Chem.*, **44**:2238 (1952).
64. McAllister, R. A., P. H. McGuinnis, Jr., and C. A. Plank: *Chem. Eng. Sci.*, **9**:25 (1958).
65. McAllister, R. A., P. H. McGuinnis, Jr., and C. A. Plank: *Chem. Eng. Sci.*, **13**:269 (1961).
66. McAllister, R. A., and C. A. Plank: *A.I.Ch.E. J.*, **4**:282 (1958).
67. Miller, R. H.: Plate Efficiencies and Mass Transfer for Valve Trays and Trays with Large Perforations, Dissertation, Univ. of Michigan, 1958.
68. Murphree, E. V.: *Ind. Eng. Chem.*, **17**:747, 960 (1925).
69. Nutter, I. E.: *Oil Gas J.* (April 26, 1954), p. 165.
70. O'Connell, H. E.: *Trans. A.I.Ch.E.*, **42**:741 (1946).
71. Oliver, E. D., and C. C. Watson: *A.I.Ch.E. J.*, **2**:18 (1956).
72. Peavy, C. C., and E. M. Baker: *Ind. Eng. Chem.*, **29**:1956 (1937).
73. Rush, F. E., Jr., and C. Stirba: *A.I.Ch.E. J.*, **3**:336 (1951).
74. Sherwood, T. K., and F. J. Jenny: *Ind. Eng. Chem.*, **27**:265 (1935).
75. Shilling, G. D., G. H. Beyer, and C. C. Watson: *Chem. Eng. Progr.*, **49**:128 (1953).
76. Simkin, D. J., and C. P. Strand, and R. B. Olney: *Chem. Eng. Progr.*, **50**:565 (1954).
77. Smith, P. L., M. Van Winkle: *A.I.Ch.E. J.*, **4**:266 (1958).
78. Smith, B.: "Design of Equilibrium Stage Processes," chap. 17, McGraw-Hill, New York, 1963.

79. Smith, B.: "Design of Equilibrium Stage Processes," chap. 15, McGraw-Hill, New York, 1963.
80. Thrift, G. C.: *Chem. Eng.*, **61**:177 (1954).
81. Thrift, G. C.: *Petrol. Eng.* (May, 1954), p. c-26.
82. Thrift, G. C.: *Petrol. Refiner*, **39**(8):93 (1960).
83. Toor, H. L., and J. K. Burchard: *A.I.Ch.E. J.*, **6**:202 (1960).
84. Umholtz, C. L., and M. Van Winkle: *Petrol. Refiner* (July, 1955), p. 114.
85. Umholtz, C. L., and M. Van Winkle: *Ind. Eng. Chem.*, **49**:226 (1957).
86. Van Wijk, W. R., and H. A. C. Thiessen: *Chem. Eng. Sci.*, **3**:153 (1954).
87. Volland, G.: *Chem. Fabrik*, **8**:5 (1935).
88. Warzel, L. A.: Plate Efficiencies for Absorption and Desorption in a Bubble Cap Column, Dissertation, Univ. of Michigan, 1955.
89. Williams, G. C., E. K. Stigger, and J. H. Nichols: *Chem. Eng. Progr.*, **46**:7 (1950).
90. Zenz, F. A.: *Petrol. Refiner* (February, 1954), p. 99.

chapter 14

Plate-fractionating-column design methods

Design procedures for equilibrium-stage or plate-fractionating columns have not been generally standardized, and within a particular company and associated companies it is common practice to follow a design method based on standards which appear to be the most reliable and satisfactory on the basis of the company's experience. The philosophy of design related to the selection of items to be evaluated, the order in which the various items are to be evaluated, and the methods of evaluation of the different items vary widely among engineers engaged in distillation column design. Only three of the many available design methods have been selected for discussion in detail in this chapter, and each applies only to the design of one type of column, namely, bubble-cap, perforated-tray, and valve-tray columns. In selecting the particular method described there is no intention whatever, either implied or actual, to advocate the use of either the method or the particular hardware to which the design is applied. Rather the selection is the personal preference of the author based on experience and the availability of the necessary design correlations and information.

Because bubble-cap trays are more expensive for the same service than perforated trays, and in most instances more expensive than the valve trays, the question arises as to the advisability of devoting space to bubble-cap tray design. This is justifiable from at least two viewpoints. First, there are literally hundreds of bubble-tray columns already in operation (most of them paid for) which will remain in operation in the same service or in different service for some time and there is no better way to understand the operation of the column than to know how to design it. Also, in order to adapt the column to a new service requires a knowledge of the design method. Secondly, many engineers prefer the bubble-cap columns to others because of their experience with that type of column and because of its effectiveness at low to medium vapor rates.

14.1. GENERAL COLUMN DESIGN

After the number of trays and reflux ratio have been determined, the column design involves the evaluation of the column diameter, plate type and design, plate spacing, and certain miscellaneous design factors involving tolerances, materials of construction, etc. To serve as a guide in the stepwise design procedure, Table 14.1 was constructed to include the items to be evaluated in

TABLE 14.1 **Check list of Design Items for Bubble-cap, Perforated, and Valve Tray**

Column
1. Operating temperature and pressure
2. Reflux ratio
3. Number of trays
4. Feed and draw off trays location
5. Column diameter
6. Tray spacing

Tray
7. Liquid-flow arrangement or tray type
8. Active area
9. Downcomer type, area, and clearance
10. Tray outlet weir type, height, and length
11. Tray inlet weir, type, height, and length (if any)
12. Tray outlet splash baffle, antijump baffles
13. Tray and weir level tolerances
14. Materials of construction

For specific types of trays

Bubble cap	Perforated	Valve
15. Bubble-cap diameter, number	15. Free hole area	15. Size holes and valve type
16. Cap layout, pitch, and spacing	16. Hole size, pitch pattern	16. Number of valves and spacing
17. Skirt clearance	17. Tray thickness	17. Tray thickness
18. Static seal	18. Hole blanking	
19. Riser dimensions		
20. Tray baffles		
21. Tray drain holes		
22. Leakage		

TABLE 14.2 **Recommended Limits: Tray and Column Design**

	Bubble-cap trays [3, 11]	Perforated trays	Valve trays [4]
Column			
Diameter	1–24 ft	1–24 ft	1–24 ft
Basis: % flood	80–85% NF* 70–75% F†	80–85% NF 70–75% F	82% NF 60% F 77% vacuum service 65–75%, col. diam., 3.0 ft
Tray spacing	Fig. 13.21 12–48 in. Col. diam., 2.5–4.0 ft; 18 in. Col. diam., 5.0–24 ft; 24–36 in. Check: liquid backup, entrainment	Fig. 13.21 12–36 in. Col. diam., 2.0–4.0 ft; 12–18 in. Col. diam., 5.0–24.0 ft; 24–36 in. Check: liquid backup, entrainment	12–36 in. Col. diam., 2.0–4.0 ft; 12–18 in. Col. diam., 5.0–24 ft; 24–36 in. Check: liquid backup, entrainment
Tray			
Flow arrangement	(Table 14.3)	(Table 14.3)	
General	Cross flow	Cross flow	Cross flow
Med. diam., 6–12 ft	DP‡, cascade	DP	5–6 ft DP
Large diam., 12–24 ft	Multiple pass, cascade	Multiple pass	8–15 ft multiple pass
Tray layout	Bubble-cap diam. / Tray diam.: 3 in. / 2.5–4.0 ft 4 in. / 5–16 ft 6 in. / 16 ft and over Slot area = 10–20% col. area Spacing 1–3 in. between caps Pattern-equilateral triangular Skirt clearance 0.5–1.5 in. Cap clearance: Cap-tower wall, 1.5 in., min Cap-weir, 3.0 in., min Cap-apron, 3.0 in., min Ave. dynamic seal, h_{ds}: Vacuum, 0.5–1.5 in. Atmospheric, 1.0–2.5 in. 50–100 psig, 1.5–3.0 in. 200–500 psig, 2.0–4.0 in. Slot opening, $w_s = 0.25–0.5$ in., $h_s = 1.0–1.5$ in.	Hole diam.: 1/8–1/2 in. Hole area: 6–15% col. area Spacing: pitch/hole diam., 2–4 Tray thickness: 16 gage to 1/4 in. Hole clearance: Hole-tower wall, 1.5 in. Hole-weir, 2.0 in., min Hole-apron, 2.0 in., min Ave. dynamic seal h_{ds}: Vacuum, 0.5–0.60 in. Atmospheric, 0.5–1.5 in. Pressure, 1.5–3.0 in.	Standard manufacture Specification, 1 7/8 in. diam. holes Special up to 6 in. Tray thickness: 1/16–1/4 in. Ave. dynamic seal h_{ds}: Vacuum, 0.5–0.75 in. Atmospheric, 1.0–2.5 in. Pressure, 2.0–6.0 in.
Downcomers			
Type	Segmental	Segmental	Segmental
Apron	Vertical	Vertical	Vertical

Liquid			
Residence			
Time	5 sec, min, F 3 sec, min, NF	5 sec, min, F 3 sec, min, NF	5 sec, min, F 3 sec, min, NF
Liquid velocity	0.3–1.0 fps	0.3–1.0 fps	0.5–3.0 fps
Apron clearance	Weir to baffle distance, to 6 ft-0.5 in. Weir to baffle distance, 6–12 ft-1.0 in. Weir to baffle distance, over 12 ft-1.5 in.	½ weir height	Seal area: ½–⅓ downcomer area
Weirs—outlet			
Type	Segmental	Segmental	Segmental
Height	2.0–6.0 in.	1.0–3.0 in.	0.75–3 in.
Adjustment	1.0–2.0 in.	1.0–2.0 in.	1.0–2.0 in.
Length, % col. diam.	Cross flow, 60–75% DP, 50–60% Center, 8–12 in. wide	Cross flow, 60–75% DP, 50–60% Center, 8–12 in. wide	Cross flow, 60–75% DP, 50–60% Center, 8–12 in. wide
Inlet	Not recommended	Recommended only for low liquid rates	Not recommended Seal sumps recommended
Intermediate	Optional, $h_w \gtrless h$ liquid downstream	Not recommended	Not recommended
Splash baffles	Optional, bottom 2–3 in. above outlet weir	Optional, bottom 2–3 in. above outlet weir	Optional, recommended for high vapor rates
Antijump baffles	Recommended for DP trays Extend to elevation of top of weir	Recommended for DP trays Extend to elevation of top of weir	Recommended for DP trays Extend to elevation of top of weir
Redistributive baffles	All rows of caps where end space is 1 in. cap spacing. Clearance same as caps. Height 2 × clear liquid head, min	Not recommended	Not recommended
Miscellaneous			
Drainholes	⅜–⅝ in. diam., 4 sq in./100 sq ft tray area	Not recommended	Not recommended
Leakage	Max. fall 1 in. below top of weir in 20 min with drain holes plugged	—	—
Construction tolerances			
Tray level	¼ in. max ⅛ in. max	¼ in. max ⅛ in. max	¼ in. max ⅛ in. max

* NF—nonfoaming
† F—foaming
‡ DP—double pass

TABLE 14.3 **Tray-type Selection Based on Liquid-handling Capacity [3]***

Tower diameter, ft	Reverse	Cross	Double pass	Cascade double pass
3	0–30	30–200	—	—
4	0–40	40–300	—	—
6	0–50	50–400	400–700	—
8	0–50	50–500	500–800	—
10	0–50	50–500	500–900	900–1400
12	0–50	50–500	500–1000	1000–1600
15	0–50	50–500	500–1100	1100–1800
20	0–50	50–500	500–1100	1100–2000

* Range of liquid capacity, gallons per minute.

TABLE 14.4 **Approximate Distribution of Areas as Percent of Tower Area [3]***

Tower diameter, ft	Downflow area†		Liquid dist. area†			End wastage†
	Cross	Double pass	Cross	Double pass	Cascade double pass	
3	10–20	—	10–25	—	—	10–30
4	10–20	—	8–20	—	—	7–22
6	10–20	20–30	5–12	15–20	—	5–18
8	10–20	18–27	4–10	12–16	—	4–15
10	10–20	16–24	3–8	9–13	20–30	3–12
12	10–20	14–21	3–6	8–11	15–25	3–10
15	10–20	12–18	2–5	6–9	12–20	2–8
20	—	10–15	—	5–7	9–15	2–6

* Allocated cap area (active area) = tower area − (downflow area + liquid distribution area + end wastage).

† From Fig. 14.1.

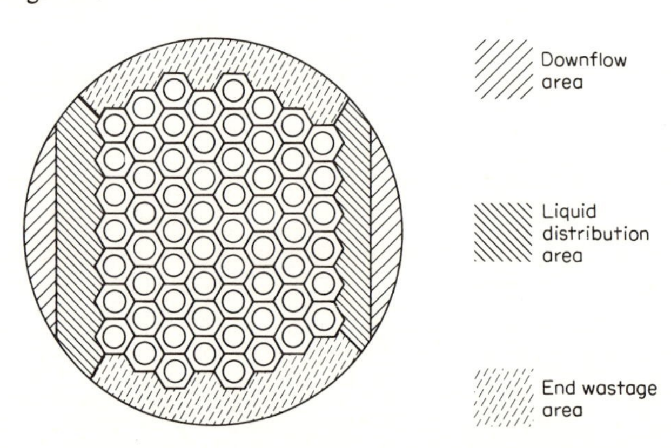

FIGURE 14.1 Area designations—bubble-cap plate.

the design of each of three types of columns: bubble-cap, perforated, and valve tray.

Experience in both research and operation combined with manufacturers' recommendations has enabled some standardization of general design limits on the various types of trays. In an attempt to summarize these limits for the three types of trays discussed here, Table 14.2 was composed to serve as a guide in designing new towers and trays. Although the limits indicated in Table 14.2 are generally applicable, undoubtedly there are many exceptions necessary to cover all possible situations encountered.

14.2. BUBBLE-CAP TRAY DESIGN

Bolles [3] presented a design method for bubble-cap trays, subsequently slightly modified [11], which provides a reasonably "tight" design, i.e., one which produces a column and internals that will operate satisfactorily at the design loading for the required separation, but which does not provide quite as wide a margin of safety with regard to increased load as the former, less precise, methods. In addition, it offers a method of optimization which the author claims "ensures required performance, provides required flexibility to handle varying loads, and does this at minimum cost."

The recommended stepwise method of design is as follows:

1. Estimate the tower diameter based upon flooding considerations using Fig. 13.21 and an assumed tray spacing. The capacity factor P_c read from the chart is for 100% of flooding. For foaming systems use 70 to 75% of this value; for nonfoaming systems use 80 to 85%.

2. Check the effect of surface tension by Eq. (13.25).

3. Determine the net vapor velocity U_{VN} from the corrected capacity factor.

4. Calculate the net flow area for vapor by dividing the volumetric vapor load by the vapor velocity $Q_V/U_{VN} = A - A_d = A_N$.

5. Select the type of tray based upon the liquid load Q_L in gallons per minute from Table 14.3.

6. From Table 14.4 select a value for the net flow area, $A_N = A - A_d$, from the average downflow area given in the column corresponding to the selected tower diameter. For example, for a 10-ft-diameter column, the downflow area for a cross-flow arrangement is 10 to 20% of tower area. If 15% is used for the first trial, the net area $A_N = 85\%$.

7. Calculate the column area A by dividing the net area by 0.85.

8. Calculate the tower diameter D based upon A.

9. Check the assumed tray spacing from the recommendations in Table 14.2.

10. Select a bubble-cap size and type from Table 14.6 based on diameter recommendations in Table 14.2.

TABLE 14.5 **Bubble-cap Size and Slot Area [3]***

Cap size, in.	S_c/d_c			
	0.25	0.3125	0.375	0.50
3	0.39	0.35	0.32	0.27
4	0.56	0.33	0.30	0.25
6	0.29	0.26	0.24	0.20

* Ratio: slot/allocated cap area.

$$\frac{\text{Distance between cap walls}}{\text{Cap diameter}} = \frac{s_c}{d_c}$$

TABLE 14.6 **Proposed Standard Bubble-cap Designs [3]**

	Carbon steel, in.			Alloy steel, in.		
	3	4	6	3	4	6
Cap						
U.S. Standard gage	12	12	12	16	16	16
OD, in.	3.093	4.093	6.093	2.999	3.999	5.999
ID, in.	2.875	3.875	5.875	2.875	3.875	5.875
Height over-all, in.	2.500	3.000	3.750	2.500	3.000	3.750
Number of slots	20	26	39	20	26	39
Type of slots	Trpzl.	Trpzl.	Trpzl.	Trpzl.	Trpzl.	Trpzl.
Slot width, in.						
Bottom	0.333	0.333	0.333	0.333	0.333	0.333
Top	0.167	0.167	0.167	0.167	0.167	0.167
Slot height, in.	1.000	1.250	1.500	1.000	1.250	1.500
Height shroud ring, in.	0.250	0.250	0.250	0.250	0.250	0.250
Riser						
U.S. Standard gage	12	12	12	16	16	16
OD, in.	2.093	2.718	4.093	1.999	2.624	3.999
ID, in.	1.875	2.500	3.875	1.875	2.500	3.875
Standard heights, in.						
0.5-in. skirt height	2.250	2.500	2.750	2.250	2.500	2.750
1.0-in. skirt height	2.750	3.000	3.250	2.750	3.000	3.250
1.5-in. skirt height	3.250	3.500	3.750	3.250	3.500	3.750
Riser-slot seal, in.	0.500	0.500	0.500	0.500	0.500	0.500
Cap areas, sq. in.						
Riser*	2.65	4.80	11.68	2.65	4.80	11.68
Reversal	3.99	7.30	17.40	4.18	7.55	17.80
Annular	3.05	5.99	13.95	3.35	6.38	14.55
Slot	5.00	8.12	14.64	5.00	8.12	14.64
Cap	7.50	13.15	29.0	7.07	12.60	28.3
Area ratios						
Reversal/riser	1.50	1.52	1.49	1.58	1.57	1.52
Annular/riser	1.15	1.25	1.20	1.26	1.33	1.25
Slot/riser	1.89	1.69	1.25	1.89	1.69	1.25
Slot/cap	0.67	0.62	0.50	0.71	0.65	0.52

* Allowing for ⅜-in. hold-down bolt.

11. Assume a tray having characteristics based on the recommendations in Table 14.7, values for the first design evaluation.

A_r = riser area = % tower area = % A
A_d = downflow area = % tower area = % A
A_{Am} = maximum active area = % tower area = % A
l_w = weir length = % tower diameter = % D
Z_{Lm} = maximum liquid flow path = % tower diameter = % D

12. Assume a tray layout based upon recommended practice by first scaling the column, selecting the weir(s), and drawing them on the column plan. Then select a cap spacing, e.g., $d_c/s_c = 0.25$, and draw in the caps to determine the number which will fit into the plate area.

TABLE 14.7 **Recommended Standard Trays [11]***

Tower diameter and flow arrangement, %	Up to 6 ft—cross	Over 6 ft—double pass
A_r/A	10	10
A_s/A	17	17
A_d/A	12	12
A_A/A	76	76
l_w/D, side	77	62
l_w/D, center	—	97
Z_{Lm}/D	64	30

* Caps have 4-in. diameter.

13. Select a skirt clearance and weir height and calculate the hydraulic factors utilizing Table 14.5 and the information in Chap. 13.

14. After several layouts have been examined so that the optimum arrangement and sizing can be approximated, a final layout is decided upon which should include

l_w = weir length, % D
P_c = cap pitch
s_c = cap spacing
N_c = number of caps per tray
h_{sk} = cap skirt clearance
h_{ss} = static slot seal

15. Locate the downcomers and make a plan drawing, including the cap layout.

16. Determine the weir height, $h_w = h_{sk} + h_{sr} + h_s + h_{ss}$.

17. Assume the apron seal for the downcomer and calculate the area for liquid flow under the edge of the apron, A_{AP}.

18. Determine the following both as areas and as percents of tower area.

Riser, A_r
Slot, A_s
Downcomer, A_d
Active area, A_A
Area under apron, A_{AP}
Tower, A
Net area, A_N

If a multiple-pass tray, e.g., two-pass, is used, the arithmetic average areas for successive trays are used, i.e., one with two side downcomers and one with a center downcomer.

19. The tray dynamics are checked for the trays having the most widely different loadings and operating conditions.

(*a*) *Foaming or foam height:* Use Eq. (13.26); $h_f = 2.53F_{VA}^2 + 1.89h_w - 1.6$.

(*b*) *Flooding:* Calculate the percent flooding based upon the final tray arrangement using the net area vapor velocity and compare with the velocity calculated in steps 1 through 3. This velocity divided into the velocity for the final tray design gives the percent flood.

(*c*) *Entrainment:* Calculate the fractional entrainment using the percent flood from (*b*), P_F, and Fig. 13.26. This should be less than 0.1.

(*d*) *Slot opening:* Calculate shape factor and use Fig. 13.9 to get maximum slot capacity. Calculate percent of maximum slot capacity and use Fig. 13.9 to get the percent of slot opening and h_{so}. This should be greater than 0.5 in.

(*e*) Crest over weir, h_{ow}, is calculated by Eq. (13.4). F_w is read from Fig. 13.7.

(*f*) Liquid gradient is estimated by the use of Figs. 13.10 through 13.13 and corrected for vapor rate by Fig. 13.15.

(*g*) Tray pressure drop is calculated by means of Eq. (13.1*a*) and the defining equations and figures in Chap. 13.

(*h*) Vapor distribution ratio is calculated by

$$R_V = \frac{\Delta}{h_{rc} + h_{so}}$$

20. Downcomer hydraulics:

(*a*) Head loss in the downcomer is based on the most restrictive area for liquid flow. This usually is under the apron. Equation (13.5) is used to calculate h_d.

(*b*) Liquid backup is calculated by Eq. (13.1*b*).

(*c*) The height of the aerated liquid is calculated by dividing the H_D by the relative froth density, h_c/h_f. Unless known, it is estimated to be 0.5 minimum.

(*d*) The residence time is obtained by dividing the volumetric flow rate of clear liquid by the clear liquid volume in the downcomer, i.e., $A_d \times H_D$.

21. The final evaluation is made by comparing the actual calculated values with the so-called *standard* or *predetermined* limits of the values. Bolles compares the following:

Data	Standard
Percent flooding, max	80
Entrainment ratio, max	0.15
Slot opening, % h_s, max	100
Dynamic seal, h_{ds}, in.	1.0–2.5
Vapor distribution ratio, max	0.5
Height, froth in downcomer, %$(S + h_w)_{max}$	100
Downcomer residence time, sec, min	3

Example 14.1

A benzene distillate product containing 99.8 wt % benzene is to be fractionated from a feed material composed of benzene, 23.5 wt %, cyclohexane, 10.0 wt %, ethylcyclohexane, 1.5 wt %, ethylbenzene, 60.0 wt %, and 1,4-diethylbenzene, 5.0 wt %, in a 43-plate tower. It is to operate at essentially atmospheric pressure at the top of the tower.

Data	Top tray	Bottom tray
Temperature, °F,	181	291
Pressure, psia,	15.7	Approx. 20
Reflux ratio, L_0/D	1.3	
Reflux ratio, L/V	0.64	
Liquid load L, lb/hr	3080	
Q_L, gpm	7.58	
Q'_L, cu ft/sec	0.0169	
Vapor load V, lb/hr	4835	
Q_V, cu ft/sec	7.52	
Liquid density ρ_L, lb/cu ft	50.7	
Vapor density ρ_V, lb/cu ft	0.178	
Surface tension σ, dynes/cm	21	
Dry-plate efficiency, %	45	
No. of plates, n	43	
Plate spacing S, in.	18	
Assumed operation, % flood	60	

Design a bubble-tray column whose characteristics fall within the recommended limits in Table 14.2.

Solution

The design will be illustrated for the top tray only. Complete design calculations require evaluation of tray requirements at the top, bottom, and feed locations to include the widest range of loads and conditions.

Tower diameter. Liquid-vapor flow parameter,

$$P_F = \frac{L}{V}\left(\frac{\rho_V}{\rho_L}\right)^{0.5} = \frac{3080}{4835}\left(\frac{0.178}{50.7}\right)^{0.5} = 0.0378$$

From Fig. 13.21, $P_c = 0.272$,

$$U_{VN} = \frac{P_c}{\left(\dfrac{\rho_V}{\rho_L - \rho_V}\right)^{0.5}} = 4.54 \text{ fps}$$

For 60% flood,

$$U_{VN} = (0.6)(4.54) = 2.73 \text{ fps}$$

Select $A_d/A = 0.08$. Assume no splash baffle and

$$A_N = A - A_d = A - 0.08A$$

$$= 0.92A$$

$$A = \frac{A_N}{0.92} = \frac{Q_V}{0.92U_{VN}} = \frac{7.52}{(0.92)(2.73)}$$

$$= 3.0 \text{ sq ft}$$

Select column diameter $D = 2.0$ ft,

$$A = 3.14 \text{ sq ft}$$

Tray type. Select cross flow.

Caps (refer to Table 14.5)
- Cap diameter, $d_c = 3$ in.
- Slot shape, trapezoidal, $R_s = 0.5$
- Slot height, $h_s = 1.0$ in.
- Slot area, $a_s = 5.0$ sq in. cap
- Cap area, $a_c = 7.5$ sq in. cap
- Shroud ring, $h_{sr} = 0.25$ in.
- Riser diameter/cap diameter = 0.675

Minimum slot area—total

$$A_{sm} = \frac{Q_V}{C_s\left(h_s \dfrac{\rho_L - \rho_V}{\rho_V}\right)^{0.5}}$$

From Fig. 13.9,

$$C_s = 0.74$$

$$A_{sm} = \frac{7.52}{0.74\left(1.0\,\dfrac{50.7 - 0.178}{0.178}\right)^{0.5}}$$

$$= 0.604 \text{ sq ft}$$

Minimum number of caps operating at 100% slot capacity,

$$\frac{(0.604)(144)}{5.0} = 18$$

Tray layout

Tower diameter $D = 2.0$ ft
Tray spacing $S = 18$ in.
Bubble-cap arrangement, equilateral triangle
Bubble-cap spacing $= 0.25d_c = 0.75$ in.
Skirt clearance $h_{sk} = 0.50$ in.
Static seal $h_{ss} = 1.0$ in.
Total caps—selecting best fit from trial layouts = 22
Weir type, segmental
Weir length $l_w = (0.680)D = 16.4$ in. $= 1.365$ ft

From Table 14.10 corresponding to $A_d/A = 0.08$

Downcomer area $A_d = (0.08)(3.14) = 0.251$ sq ft
Weir height $h_w = h_{sk} + h_{sr} + h_s + h_{ss} = 2.75$ in.
Downcomer seal $= 0.5$ in.

$$\text{Area under apron} = \frac{(2.75 - 0.5)(16.4)}{144} = 0.255 \text{ sq ft}$$

Flow areas	sq in./cap	sq ft/tray	Percent area
Riser	2.65	0.41	13
Annular	3.05	—	—
Slot	5.00	0.76	24
Downcomer	—	0.251	8
Apron seal	—	0.255	8
Tower	—	3.14	100

Tray dynamics

(*a*) Maximum slot capacity:

$$Q_{V,\max} = (0.74)0.76\left(1.0\,\frac{50.7 - 0.178}{0.178}\right)^{0.5}$$

$$= 9.0 \text{ cu ft/sec}$$

$$\frac{\text{Operating vapor load}}{\text{Maximum vapor load}} = \frac{7.52}{9.0} = 0.835$$

Mean slot opening h_{so} (Fig. 13.9) $= (0.90)(1.0) = 0.9$ in.

(*b*) Liquid height over weir:

$$\frac{Q_L}{(l_w)^{2.5}} = \frac{7.58}{(1.365)^{2.5}} = 3.5$$

From Fig. 13.7, $F_w = 1.035$,

$$h_{ow} = 0.092F_w\left(\frac{Q_L}{l_w}\right)^{0.67} = 0.3 \text{ in.}$$

(*c*) Cap pressure drop:

$$h_c = h_{rc} + h_{so} \qquad \frac{a_a}{a_r} = 1.15$$

From Fig. 13.8, $K_c = 0.55$,

$$h_{rc} = K_c \frac{\rho_V}{\rho_L}\left(\frac{Q_V}{A_r}\right)^2 = 0.55\,\frac{0.178}{50.7}\left(\frac{7.52}{0.41}\right)^2 = 0.65$$

$$h_c = 1.55 \text{ in.}$$

(*d*) Liquid gradient:

$$\text{Mean flow width} = \frac{D + l_w}{2} = \frac{2.0 + 1.36}{2} = 1.7 \text{ ft}$$

$$\text{Liquid flow loading} = \frac{Q_L}{W_a} = \frac{7.6}{1.7} = 4.46 \text{ gpm/ft}$$

Skirt clearance $h_{sk} = 0.5$ in.
Assume mean liquid depth = 3.0 in.
Uncorrected liquid gradient per row of caps Δ' (Fig. 13.10) = 0.055 in.
Δ' total for four rows = 0.220 in.
Superficial vapor velocity,

$$U_V = \frac{7.52}{3.14} = 2.4 \text{ fps}$$

$$F_V = U_V(\rho_V)^{0.5} = (2.4)(0.178)^{0.5} = 1.01$$

$$C_V \text{ (Fig. 13.15)} = 0.88$$

Corrected liquid gradient $\Delta = C_V\Delta'$,

$$\Delta = (0.88)(0.220) = 0.194 \text{ in.}$$

Check mean liquid depth assumption of 3 in.

$$h_w + h_{ow} + \frac{\Delta}{2} = 2.75 + 0.3 + \frac{0.194}{2} = 3.15 \text{ in.}$$

(*e*) Vapor distribution ratio:

$$R_V = \frac{\Delta}{h_{rc} + h_{so}} = \frac{0.194}{1.55} = 0.125$$

(*f*) Mean dynamic slot submergence:

$$h_{ds} = h_{ss} + h_{ow} + \frac{\Delta}{2}$$

$$= 1.0 + 0.3 + 0.097 = 1.397 \text{ in.}$$

(*g*) Tray pressure drop:

$$\Delta H_T = h_{rc} + h_{so} + h_{ds}$$

$$= h_{rc} + h_{so} + \beta\left(h_{ss} + h_{ow} + \frac{\Delta}{2}\right)$$

$$= 0.65 + 0.9 + \beta(1.0 + 0.3 + 0.097)$$

$$= 1.55 + \beta(1.40)$$

$$F_{VA} = U_{VA}(\rho_V)^{0.5} = 2.64(0.178)^{0.5} = 1.1$$

$$\beta = 0.68 \quad \text{(Fig. 13.16)} \qquad \Delta H_T = 1.55 + 0.68(1.40) = 2.50 \text{ in.}$$

(*h*) Clear liquid backup in downcomer: Pressure drop through downcomer,

$$h_d = 0.03\left(\frac{Q_L}{100A_d}\right)^2 = 0.03\left(\frac{7.6}{25.1}\right)^2 = 0.027 \text{ in.}$$

$$H_D = h_w + h_{ow} + h_d + \Delta + \Delta H_T = 2.75 + 0.3 + 0.194 + 2.5 = 5.74 \text{ in.}$$

For height of aerated liquid assume relative froth density $\phi = 0.5$,

$$H_{D_{al}} = \frac{5.73}{\phi} = 11.46 \text{ in.}$$

$$\text{Available downcomer height} = S + h_w = 18 + 2.75 = 20.75$$

$$\text{Percent backup, clear liquid} = \left(\frac{5.73}{20.75}\right) 100 = 25.6$$

$$\text{Percent backup, aerated liquid} = \left(\frac{11.46}{20.75}\right) 100 = 51.2$$

(*i*) Downcomer residence time:

$$A_d = 0.251 \text{ sq ft} \qquad H_d = 5.74 \text{ in.} = 0.48 \text{ ft}$$

$$\text{Downcomer volume} = (0.251)(0.48) = 0.12 \text{ cu ft}$$

$$\text{Residence time} = \frac{V_d}{Q'_L} = \frac{0.12}{0.0169} = 7.1 \text{ sec}$$

(*j*) Entrainment:

$$A_N = A - A_d = 2.89 \text{ sq ft} \qquad U_{VN} = \frac{7.52}{0.289} = 2.6 \text{ fps}$$

$$P_F = \frac{3080}{4835}\left(\frac{0.178}{50.7}\right)^{0.5} = 0.038$$

From Fig. 13.26, $\psi = 0.044$ mole/mole net downflow or (0.044)(3080/4835) = 0.027 mole/mole vapor.

Summary	Top tray	Allowable
Mean slot opening, % h_s	90	100
Mean dynamic slot seal, h_{ds}	1.40	1.0–2.0
Vapor distribution ratio	0.125	0.5 max
$(H_d/S + h_w)100$, height liquid in downcomer, %	25.6	50 max
Downflow residence time, sec	7.1	5 min
Entrainment, mole/mole dry vapor	0.027	0.10 max
Liquid height over weir, in.	0.30	—
Liquid gradient, in.	0.194	—
Tray pressure drop, in. liquid	2.50	—

Other Design Methods for Bubble Plates

Many other design methods for bubble-cap trays have been used extensively. Although most methods are basically oriented toward entrainment limitations, others are concerned with vapor-liquid contact area, flexibility with regard to loading, and maximum mixing.

The Souders-Brown method [13] is based on the effect of entrainment on efficiency and the vapor capacity chart is somewhat conservative. The A.I.Ch.E. bubble-tray method [1] can be used to design trays for a given efficiency by trial-and-error calculations. The Chemical Engineers' Handbook [10] gives methods for evaluation of the various design features for bubble-tray columns as well as others. Atkins [2] proposed a method based on experience factors which in essence divides the cross section of the tower into zones as follows: a vapor contacting zone, a liquid-in-vapor entrainment elimination zone, an outlet zone for liquid collection, and an inlet zone for liquid from the tray above. He considers each of the zones to have an average loading for a well-designed tray. The proper allocation of areas in the zones including plate spacing and downcomer areas is based heavily on the experience of the design engineer. Because of this the method cannot be used by the novice.

14.3. PERFORATED-TRAY AND COLUMN DESIGN

Many of the items to be sized and determined are the same for perforated-tray design as for bubble-tray design, as indicated in Table 14.1. A stepwise outline of the design procedure recommended is as follows:

1. The tower diameter is determined on the basis of a specified allowable percent of flooding, e.g., 80%, and an assumed plate spacing (see Table 14.2). The capacity factor for 100% of flooding is read from Fig. 13.21 and corrected for the surface-tension effect by Eq. (13.25). U_{VN} is then calculated. The vapor velocity for 80% of flooding is calculated as U_{VN} (0.8), and the net area of the tower A_N is calculated by dividing the total volumetric vapor flow rate by the velocity. Calculate the diameter D from the area. ($A = A_N + A_D$, no splash baffle; $A = A_n + 2A_D$ with splash baffle; A_D is usually selected to be $0.1A$.)

2. Select a flow arrangement for the tray based upon liquid-handling capacity, as indicated in Table 14.3. The upper limit shown in Table 14.3 may be increased about 20% for perforated trays.

3. Select a tower diameter to the nearest 0.5 ft to that calculated in step 1, assume a tentative tray layout, and calculate the various design lengths and areas in accordance with the values recommended in Table 14.8.

4. Calculate the new U_{VN} if the diameter is different from that calculated in step 1.

TABLE 14.8 **Recommended Limits for Sieve Trays**

Data	Cross flow	Double pass
Tray diameter, ft	1–8	8–12
d_h, hole diameter, in.	⅛–⅜	⅛–⅜
p, hole pitch	2.5–3.0	2.5–3
h_w, weir height, in.	1.0–2.0	1.0–2.0
l_w/D, weir length/tower diam., side	0.68–0.76	0.55–0.63
l_w/D, center	—	0.97
A_d/A	0.08–0.12	0.08–0.12
A_h/A	0.06–0.12	0.06–0.12
A_N/A	0.92–0.88	0.92–0.88
A_N/A (splash baffle)	0.84–0.76	0.84–0.76
A_A/A	0.84–0.76	0.84–0.76
t_p gage	12–14	12–14

5. Calculate the approach to flooding by dividing the new U_{VN} by that read from Fig. 13.21.

6. Using Fig. 13.26 and the calculated flow parameter and percent of flood calculated in step 5, determine the entrainment. This should be not more than 10% for conservative design.

7. Using Eq. (13.2*a*), determine the total pressure drop as follows:

h_{ow} by Eq. (13.4) and Fig. 13.7
h_w selected as, e.g., 2 in.
h_σ by Eq. (13.22)
h_0 by Eq. (13.16) and C_0 from Fig. 13.18, or Eq. (13.15) and C_0 from Fig. 13.17

Calculate $h_{ow} + h_w$: From Fig. 13.16 and the F factor for the active area, $F_{VA} = U_{VA}(\rho_V)^{0.5}$, determine the froth density.

8. Calculate the weep point to be that where the clear-liquid head just balances the dry-plate pressure-drop equivalent head plus the h_σ, using Eq. (13.28).

9. Calculate the hydraulic gradient by Eqs. (13.18) through (13.21).

10. Calculate the liquid backup H_D in the downcomer by Eq. (13.2*b*). The height of foamy liquid is H_D/ϕ. Assume $\phi = 0.5$ if data are not available. Calculate the residence time by using the clear-liquid backup in the downcomer times the downcomer area A_d and divide this into the volumetric liquid flow rate. The linear velocity of the liquid in the downcomer is the volumetric liquid flow rate divided by the downcomer area.

11. Check the calculated tray operating characteristics against the recommendations in Table 14.2. Adjust for optimization within the recommended limits by successive trial designs.

TABLE 14.9 **Relation between Area, Hole Diameter, Pitch, and Column Diameter—Perforated Plates**

Hole diam., d_h, in.	1/16	1/8	3/16	1/4	5/16	3/8	1/2	3/4	1.0
a_h, sq in. per hole	0.0214	0.0867	0.193	0.343	0.534	0.770	1.37	3.08	5.49

$$\text{Maximum, } FFA = A_h/A = \frac{\text{total hole area}}{\text{cross-sectional area of column}}$$

Arrangement		p/d_h			
		2.0	3.0	4.0	5.0
	60° △	0.227	0.101	0.0555	0.0363
	□	0.197	0.0875	0.0490	0.0315
		Hole area for 1-sq-ft column area with 80% effective bubbling area, sq ft*			
	60° △	0.182	0.0808	0.0444	0.0262
	□	0.158	0.0700	0.0392	0.0252
		Hole area for various diameter columns with 80% effective bubbling area, sq ft*			
D	△	0.144	0.0635	0.0350	0.0228
1	□	0.111	0.055	0.0307	0.0178
2	△	0.565	0.254	0.140	0.0920
	□	0.445	0.221	0.124	0.0711
3	△	1.28	0.571	0.315	0.207
	□	1.00	0.496	0.278	0.160
4	△	2.30	1.02	0.560	0.366
	□	1.78	0.882	0.495	0.285
5	△	3.58	1.59	0.971	0.561
	□	2.78	1.38	0.770	0.445
6	△	5.16	2.29	1.260	0.823
	□	4.00	1.98	1.141	0.675

* The areas listed apply to perforated metal area allotting 20% of the cross-sectional area of the column to downcomer and onflow area but not allowing for plate supports, blanking strip areas, or other design features which reduce hole area. These factors must be taken into consideration for each design.

For other diameters multiply the number across from 1-ft diameter in the correct p/d_h column times D^2.

Example 14.2

Design a perforated plate column for the following conditions:

Data	Top	Bottom
Pressure, psia	20.0	—
Temperature, °F	181	191
Surface tension σ, dynes/cm	13	13
ρ_L, lb/cu ft	50.7	44.6
ρ_V, lb/cu ft	0.178	0.190
Internal reflux, L/V or $\bar{L}/\bar{V}$	0.638	1.88
Maximum vapor, lb/hr	9670	6670
Maximum liquid, lb/hr	6160	18,160
Maximum vapor rate, cu ft/sec	15.0	9.76
Maximum liquid rate, cu ft/sec	0.034	0.113
Maximum liquid rate, gpm	15.2	50.8
Tray spacing, ft	1.5	1.5

Solution

The calculations are made for the top and bottom trays, and the system is assumed to be nonfoaming under the conditions specified. Where two calculations appear on the same line, the left-hand calculation is for top trays and the right-hand one is for bottom trays.

1. *Tower diameter.* Based upon 80% of flood and the use of a splash baffle. The active area $A_A = A_N = A - 2A_d$,

$$P_F = \frac{L}{V}\left(\frac{\rho_V}{\rho_L}\right)^{0.5} = \left(\frac{6160}{9670}\right)\left(\frac{0.178}{50.7}\right)^{0.5} \qquad P_F = \left(\frac{18160}{6670}\right)\left(\frac{0.190}{44.6}\right)^{0.5}$$

$$= 0.0376 \qquad\qquad = 0.177$$

$$P_C = 0.274 \qquad\qquad P_C = 0.22 \quad \text{(Fig. 13.21)}$$

$$(P_C)_{\text{corr. for }\sigma} = 0.274\left(\frac{13}{20}\right)^{0.2} \qquad (P_C)_{\text{corr. for }\sigma} = 0.22\left(\frac{13}{20}\right)^{0.2}$$

$$= 0.272 \qquad\qquad = 0.218$$

For 100% of flood,

$$U_{VN} = P_C\left(\frac{\rho_L - \rho_V}{\rho_V}\right)^{0.5}$$

$$U_{VN} = 0.272\left(\frac{50.7 - 0.178}{0.178}\right)^{0.5} \qquad U_{VN} = 0.218\left(\frac{44.6 - 0.19}{0.19}\right)^{0.5}$$

$$= 4.60 \text{ fps} \qquad\qquad = 3.36 \text{ fps}$$

For 80% of flood,

$$A_N = \frac{Q_V}{U_{VN}(0.8)}$$

$$A_N = \frac{15}{(4.60)(0.8)} = 4.17 \text{ sq ft} \qquad A_N = \frac{9.76}{(3.36)(0.8)} = 3.64 \text{ sq ft}$$

Assume:

$$A_d = 0.1A$$

$$A = A_N + 2A_d = 4.17 \text{ sq ft} + 2(0.1)A$$

$$0.8A = 4.17$$

$$A = 4.56 \text{ sq ft for the bottom section}$$

$$A = 5.22 \text{ sq ft for the top section}$$

$$D = \left(\frac{4.56}{0.786}\right)^{0.5} = 2.41 \text{ ft (bottom)}$$

$$D = 2.58 \text{ ft (top)}$$

Select a diameter of 2.5 ft.
Check percent flood.

$$\text{Percent flood} = \frac{15/3.94}{4.6} = 83\% \qquad \text{Percent flood} = \frac{9.76/3.94}{3.64} = 73.8\%$$

$$A = 4.92 \text{ sq ft}$$

$$A_d = 0.1A = 0.49 \text{ sq ft}$$

$$A_A = A - 2A_d = 3.94 \text{ sq ft}$$

$$t_p = 12 \text{ gage} = 0.0825\text{-in. plate thickness}$$

$$d_h = 0.25 \text{ in.}$$

$$P = (3)(0.25) = 0.75\text{-in. pitch}$$

Use equilateral triangular pattern.
Select $h_w = 1.5$ in. From Table 14.10,

$$l_w = (0.727)(2.5) = 1.82 \text{ ft or } 21.8 \text{ in.}$$

2. *Entrainment*

$$P_F = 0.0376 \qquad P_F = 0.177$$

$$\psi = 0.096 \qquad \psi = 0.0105$$

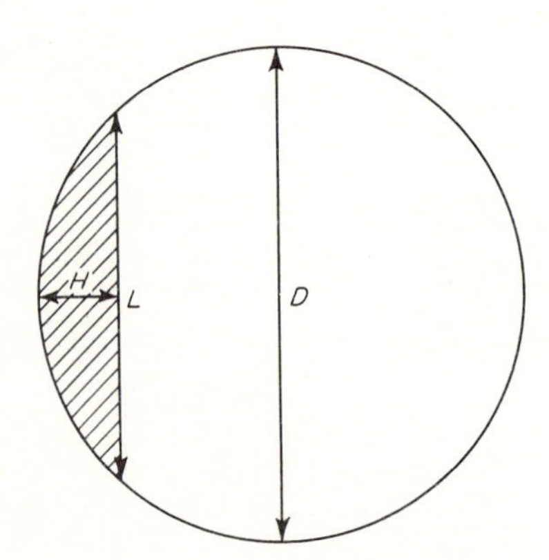

TABLE 14.10 **Segmental Functions [4]**

H/D from 0.0 to 0.1														
H/D	L/D	A_D/A_T	H/D	L/D	A_D/A_T	H/D	L/D	A_D/A_T	H/D	L/D	A_D/A_T	H/D	L/D	A_D/A_T
0.0000	0.0000	0.0000	0.0200	0.2800	0.0048	0.0400	0.3919	0.0134	0.0600	0.4750	0.0245	0.0800	0.5426	0.0375
0.0005	0.0447	0.0000	0.0205	0.2834	0.0050	0.0405	0.3943	0.0137	0.0605	0.4768	0.0248	0.0805	0.5441	0.0378
0.0010	0.0632	0.0001	0.0210	0.2868	0.0051	0.0410	0.3966	0.0139	0.0610	0.4787	0.0251	0.0810	0.5457	0.0382
0.0015	0.0774	0.0001	0.0215	0.2901	0.0053	0.0415	0.3989	0.0142	0.0615	0.4805	0.0254	0.0815	0.5472	0.0385
0.0020	0.0894	0.0002	0.0220	0.2934	0.0055	0.0420	0.4012	0.0144	0.0620	0.4823	0.0257	0.0820	0.5487	0.0389
0.0025	0.0999	0.0002	0.0225	0.2966	0.0057	0.0425	0.4035	0.0147	0.0625	0.4841	0.0260	0.0825	0.5502	0.0392
0.0030	0.1094	0.0003	0.0230	0.2998	0.0059	0.0430	0.4057	0.0149	0.0630	0.4859	0.0263	0.0830	0.5518	0.0396
0.0035	0.1181	0.0004	0.0235	0.3030	0.0061	0.0435	0.4080	0.0152	0.0635	0.4877	0.0266	0.0835	0.5533	0.0399
0.0040	0.1262	0.0004	0.0240	0.3061	0.0063	0.0440	0.4102	0.0155	0.0640	0.4895	0.0270	0.0840	0.5548	0.0403
0.0045	0.1339	0.0005	0.0245	0.3092	0.0065	0.0445	0.4124	0.0157	0.0645	0.4913	0.0273	0.0845	0.5563	0.0406
0.0050	0.1411	0.0006	0.0250	0.3122	0.0067	0.0450	0.4146	0.0160	0.0650	0.4931	0.0276	0.0850	0.5578	0.0410
0.0055	0.1479	0.0007	0.0255	0.3153	0.0069	0.0455	0.4168	0.0162	0.0655	0.4948	0.0279	0.0855	0.5592	0.0413
0.0060	0.1545	0.0008	0.0260	0.3183	0.0071	0.0460	0.4190	0.0165	0.0660	0.4966	0.0282	0.0860	0.5607	0.0417
0.0065	0.1607	0.0009	0.0265	0.3212	0.0073	0.0465	0.4211	0.0168	0.0665	0.4983	0.0285	0.0865	0.5622	0.0421
0.0070	0.1667	0.0010	0.0270	0.3242	0.0075	0.0470	0.4233	0.0171	0.0670	0.5000	0.0288	0.0870	0.5637	0.0424
0.0075	0.1726	0.0011	0.0275	0.3271	0.0077	0.0475	0.4254	0.0173	0.0675	0.5018	0.0292	0.0875	0.5651	0.0428
0.0080	0.1782	0.0012	0.0280	0.3299	0.0079	0.0480	0.4275	0.0176	0.0680	0.5035	0.0295	0.0880	0.5666	0.0431
0.0085	0.1836	0.0013	0.0285	0.3328	0.0081	0.0485	0.4296	0.0179	0.0685	0.5052	0.0298	0.0885	0.5680	0.0435
0.0090	0.1889	0.0014	0.0290	0.3356	0.0083	0.0490	0.4317	0.0181	0.0690	0.5069	0.0301	0.0890	0.5695	0.0439
0.0095	0.1940	0.0016	0.0295	0.3384	0.0085	0.0495	0.4338	0.0184	0.0695	0.5086	0.0304	0.0895	0.5709	0.0442
0.0100	0.1990	0.0017	0.0300	0.3412	0.0087	0.0500	0.4359	0.0187	0.0700	0.5103	0.0308	0.0900	0.5724	0.0446
0.0105	0.2039	0.0018	0.0305	0.3439	0.0090	0.0505	0.4379	0.0190	0.0705	0.5120	0.0311	0.0905	0.5738	0.0449
0.0110	0.2086	0.0020	0.0310	0.3466	0.0092	0.0510	0.4400	0.0193	0.0710	0.5136	0.0314	0.0910	0.5752	0.0453
0.0115	0.2132	0.0021	0.0315	0.3493	0.0094	0.0515	0.4420	0.0195	0.0715	0.5153	0.0318	0.0915	0.5766	0.0457
0.0120	0.2178	0.0022	0.0320	0.3520	0.0096	0.0520	0.4441	0.0198	0.0720	0.5170	0.0321	0.0920	0.5781	0.0460
0.0125	0.2222	0.0024	0.0325	0.3546	0.0098	0.0525	0.4461	0.0201	0.0725	0.5186	0.0324	0.0925	0.5795	0.0464
0.0130	0.2265	0.0025	0.0330	0.3573	0.0101	0.0530	0.4481	0.0204	0.0730	0.5203	0.0327	0.0930	0.5809	0.0468
0.0135	0.2308	0.0027	0.0335	0.3599	0.0103	0.0535	0.4501	0.0207	0.0735	0.5219	0.0331	0.0935	0.5823	0.0472
0.0140	0.2350	0.0028	0.0340	0.3625	0.0105	0.0540	0.4520	0.0210	0.0740	0.5235	0.0334	0.0940	0.5837	0.0475
0.0145	0.2391	0.0030	0.0345	0.3650	0.0108	0.0545	0.4540	0.0212	0.0745	0.5252	0.0337	0.0945	0.5850	0.0479
0.0150	0.2431	0.0031	0.0350	0.3676	0.0110	0.0550	0.4560	0.0215	0.0750	0.5268	0.0341	0.0950	0.5864	0.0483
0.0155	0.2471	0.0033	0.0355	0.3701	0.0112	0.0555	0.4579	0.0218	0.0755	0.5284	0.0344	0.0955	0.5878	0.0486
0.0160	0.2510	0.0034	0.0360	0.3726	0.0115	0.0560	0.4598	0.0221	0.0760	0.5300	0.0347	0.0960	0.5892	0.0490
0.0165	0.2548	0.0036	0.0365	0.3751	0.0117	0.0565	0.4618	0.0224	0.0765	0.5316	0.0351	0.0965	0.5906	0.0494
0.0170	0.2585	0.0037	0.0370	0.3775	0.0119	0.0570	0.4637	0.0227	0.0770	0.5332	0.0354	0.0970	0.5919	0.0498
0.0175	0.2622	0.0039	0.0375	0.3800	0.0122	0.0575	0.4656	0.0230	0.0775	0.5348	0.0358	0.0975	0.5933	0.0501
0.0180	0.2659	0.0041	0.0380	0.3824	0.0124	0.0580	0.4675	0.0233	0.0780	0.5363	0.0361	0.0980	0.5946	0.0505
0.0185	0.2695	0.0042	0.0385	0.3848	0.0127	0.0585	0.4694	0.0236	0.0785	0.5379	0.0364	0.0985	0.5960	0.0509
0.0190	0.2730	0.0044	0.0390	0.3872	0.0129	0.0590	0.4712	0.0239	0.0790	0.5395	0.0368	0.0990	0.5973	0.0513
0.0195	0.2765	0.0046	0.0395	0.3896	0.0132	0.0595	0.4731	0.0242	0.0795	0.5410	0.0371	0.0995	0.5987	0.0517

TABLE 14.10 (*Continued*)

H/D from 0.1 to 0.2														
H/D	L/D	A_D/A_T	H/D	L/D	A_D/A_T	H/D	L/D	A_D/A_T	H/D	L/D	A_D/A_T	H/D	L/D	A_D/A_T
0.1000	0.6000	0.0520	0.1200	0.6499	0.0680	0.1400	0.6940	0.0851	0.1600	0.7332	0.1033	0.1800	0.7684	0.1224
0.1005	0.6013	0.0524	0.1205	0.6511	0.0684	0.1405	0.6950	0.0855	0.1605	0.7341	0.1037	0.1805	0.7692	0.1229
0.1010	0.6027	0.0528	0.1210	0.6523	0.0688	0.1410	0.6960	0.0860	0.1610	0.7351	0.1042	0.1810	0.7700	0.1234
0.1015	0.6040	0.0532	0.1215	0.6534	0.0692	0.1415	0.6971	0.0864	0.1615	0.7360	0.1047	0.1815	0.7709	0.1239
0.1020	0.6053	0.0536	0.1220	0.6546	0.0696	0.1420	0.6981	0.0869	0.1620	0.7369	0.1051	0.1820	0.7717	0.1244
0.1025	0.6066	0.0540	0.1225	0.6557	0.0701	0.1425	0.6991	0.0873	0.1625	0.7378	0.1056	0.1825	0.7725	0.1249
0.1030	0.6079	0.0544	0.1230	0.6569	0.0705	0.1430	0.7001	0.0878	0.1630	0.7387	0.1061	0.1830	0.7733	0.1253
0.1035	0.6092	0.0547	0.1235	0.6580	0.0709	0.1435	0.7012	0.0882	0.1635	0.7396	0.1066	0.1835	0.7742	0.1258
0.1040	0.6105	0.0551	0.1240	0.6592	0.0713	0.1440	0.7022	0.0886	0.1640	0.7406	0.1070	0.1840	0.7750	0.1263
0.1045	0.6118	0.0555	0.1245	0.6603	0.0717	0.1445	0.7032	0.0891	0.1645	0.7415	0.1075	0.1845	0.7758	0.1268
0.1050	0.6131	0.0559	0.1250	0.6614	0.0721	0.1450	0.7042	0.0895	0.1650	0.7424	0.1080	0.1850	0.7766	0.1273
0.1055	0.6144	0.0563	0.1255	0.6626	0.0726	0.1455	0.7052	0.0900	0.1655	0.7433	0.1084	0.1855	0.7774	0.1278
0.1060	0.6157	0.0567	0.1260	0.6637	0.0730	0.1460	0.7062	0.0904	0.1660	0.7442	0.1089	0.1860	0.7782	0.1283
0.1065	0.6170	0.0571	0.1265	0.6648	0.0734	0.1465	0.7072	0.0909	0.1665	0.7451	0.1094	0.1865	0.7790	0.1288
0.1070	0.6182	0.0575	0.1270	0.6659	0.0738	0.1470	0.7082	0.0913	0.1670	0.7460	0.1099	0.1870	0.7798	0.1293
0.1075	0.6195	0.0579	0.1275	0.6671	0.0743	0.1475	0.7092	0.0918	0.1675	0.7468	0.1103	0.1875	0.7806	0.1298
0.1080	0.6208	0.0583	0.1280	0.6682	0.0747	0.1480	0.7102	0.0922	0.1680	0.7477	0.1108	0.1880	0.7814	0.1303
0.1085	0.6220	0.0587	0.1285	0.6693	0.0751	0.1485	0.7112	0.0927	0.1685	0.7486	0.1113	0.1885	0.7822	0.1308
0.1090	0.6233	0.0591	0.1290	0.6704	0.0755	0.1490	0.7122	0.0932	0.1690	0.7495	0.1118	0.1890	0.7830	0.1313
0.1095	0.6245	0.0595	0.1295	0.6715	0.0760	0.1495	0.7132	0.0936	0.1695	0.7504	0.1122	0.1895	0.7838	0.1318
0.1100	0.6258	0.0598	0.1300	0.6726	0.0764	0.1500	0.7141	0.0941	0.1700	0.7513	0.1127	0.1900	0.7846	0.1323
0.1105	0.6270	0.0602	0.1305	0.6737	0.0768	0.1505	0.7151	0.0945	0.1705	0.7521	0.1132	0.1905	0.7854	0.1328
0.1110	0.6283	0.0606	0.1310	0.6748	0.0773	0.1510	0.7161	0.0950	0.1710	0.7530	0.1137	0.1910	0.7862	0.1333
0.1115	0.6295	0.0610	0.1315	0.6759	0.0777	0.1515	0.7171	0.0954	0.1715	0.7539	0.1142	0.1915	0.7870	0.1338
0.1120	0.6307	0.0614	0.1320	0.6770	0.0781	0.1520	0.7180	0.0959	0.1720	0.7548	0.1146	0.1920	0.7877	0.1343
0.1125	0.6320	0.0619	0.1325	0.6781	0.0785	0.1525	0.7190	0.0963	0.1725	0.7556	0.1151	0.1925	0.7885	0.1348
0.1130	0.6332	0.0623	0.1330	0.6791	0.0790	0.1530	0.7200	0.0968	0.1730	0.7565	0.1156	0.1930	0.7893	0.1353
0.1135	0.6344	0.0627	0.1335	0.6802	0.0794	0.1535	0.7209	0.0973	0.1735	0.7574	0.1161	0.1935	0.7901	0.1358
0.1140	0.6356	0.0631	0.1340	0.6813	0.0798	0.1540	0.7219	0.0977	0.1740	0.7582	0.1166	0.1940	0.7909	0.1363
0.1145	0.6368	0.0635	0.1345	0.6824	0.0803	0.1545	0.7229	0.0982	0.1745	0.7591	0.1171	0.1945	0.7916	0.1368
0.1150	0.6380	0.0639	0.1350	0.6834	0.0807	0.1550	0.7238	0.0986	0.1750	0.7599	0.1175	0.1950	0.7924	0.1373
0.1155	0.6392	0.0643	0.1355	0.6845	0.0811	0.1555	0.7248	0.0991	0.1755	0.7608	0.1180	0.1955	0.7932	0.1378
0.1160	0.6404	0.0647	0.1360	0.6856	0.0816	0.1560	0.7257	0.0996	0.1760	0.7616	0.1185	0.1960	0.7939	0.1383
0.1165	0.6416	0.0651	0.1365	0.6866	0.0820	0.1565	0.7267	0.1000	0.1765	0.7625	0.1190	0.1965	0.7947	0.1388
0.1170	0.6428	0.0655	0.1370	0.6877	0.0825	0.1570	0.7276	0.1005	0.1770	0.7633	0.1195	0.1970	0.7955	0.1393
0.1175	0.6440	0.0659	0.1375	0.6887	0.0829	0.1575	0.7285	0.1009	0.1775	0.7642	0.1200	0.1975	0.7962	0.1398
0.1180	0.6452	0.0663	0.1380	0.6898	0.0833	0.1580	0.7295	0.1014	0.1780	0.7650	0.1204	0.1980	0.7970	0.1403
0.1185	0.6464	0.0667	0.1385	0.6908	0.0838	0.1585	0.7304	0.1019	0.1785	0.7659	0.1209	0.1985	0.7977	0.1409
0.1190	0.6476	0.0671	0.1390	0.6919	0.0842	0.1590	0.7314	0.1023	0.1790	0.7667	0.1214	0.1990	0.7985	0.1414

H/D	L/D	A_D/A_T	H/D	L/D	A_D/A_T	H/D	L/D	A_D/A_T	H/D	L/D	A_D/A_T	H/D	L/D	A_D/A_T
0.2000	0.8000	0.1424	0.2200	0.8285	0.1631	0.2400	0.8542	0.1845	0.2600	0.8773	0.2066	0.2800	0.8980	0.2292
0.2005	0.8007	0.1429	0.2205	0.8292	0.1636	0.2405	0.8548	0.1851	0.2605	0.8778	0.2072	0.2805	0.8985	0.2298
0.2010	0.8015	0.1434	0.2210	0.8298	0.1642	0.2410	0.8554	0.1856	0.2610	0.8784	0.2077	0.2810	0.8990	0.2304
0.2015	0.8022	0.1439	0.2215	0.8305	0.1647	0.2415	0.8560	0.1862	0.2615	0.8789	0.2083	0.2815	0.8995	0.2309
0.2020	0.8030	0.1444	0.2220	0.8312	0.1652	0.2420	0.8566	0.1867	0.2620	0.8794	0.2088	0.2820	0.8999	0.2315
0.2025	0.8037	0.1449	0.2225	0.8319	0.1658	0.2425	0.8572	0.1873	0.2625	0.8800	0.2094	0.2825	0.9004	0.2321
0.2030	0.8045	0.1454	0.2230	0.8325	0.1663	0.2430	0.8578	0.1878	0.2630	0.8805	0.2100	0.2830	0.9009	0.2326
0.2035	0.8052	0.1460	0.2235	0.8332	0.1668	0.2435	0.8584	0.1884	0.2635	0.8811	0.2105	0.2835	0.9014	0.2332
0.2040	0.8059	0.1465	0.2240	0.8338	0.1674	0.2440	0.8590	0.1889	0.2640	0.8816	0.2111	0.2840	0.9019	0.2338
0.2045	0.8067	0.1470	0.2245	0.8345	0.1679	0.2445	0.8596	0.1895	0.2645	0.8821	0.2116	0.2845	0.9024	0.2344
0.2050	0.8074	0.1475	0.2250	0.8352	0.1684	0.2450	0.8602	0.1900	0.2650	0.8827	0.2122	0.2850	0.9028	0.2349
0.2055	0.8081	0.1480	0.2255	0.8358	0.1689	0.2455	0.8608	0.1906	0.2655	0.8832	0.2128	0.2855	0.9033	0.2355
0.2060	0.8089	0.1485	0.2260	0.8365	0.1695	0.2460	0.8614	0.1911	0.2660	0.8837	0.2133	0.2860	0.9038	0.2361
0.2065	0.8096	0.1490	0.2265	0.8371	0.1700	0.2465	0.8619	0.1917	0.2665	0.8843	0.2139	0.2865	0.9043	0.2367
0.2070	0.8103	0.1496	0.2270	0.8378	0.1705	0.2470	0.8625	0.1922	0.2670	0.8848	0.2145	0.2870	0.9047	0.2372
0.2075	0.8110	0.1501	0.2275	0.8384	0.1711	0.2475	0.8631	0.1927	0.2675	0.8853	0.2150	0.2875	0.9052	0.2378
0.2080	0.8118	0.1506	0.2280	0.8391	0.1716	0.2480	0.8637	0.1933	0.2680	0.8858	0.2156	0.2880	0.9057	0.2384
0.2085	0.8125	0.1511	0.2285	0.8397	0.1721	0.2485	0.8643	0.1938	0.2685	0.8864	0.2161	0.2885	0.9061	0.2390
0.2090	0.8132	0.1516	0.2290	0.8404	0.1727	0.2490	0.8649	0.1944	0.2690	0.8869	0.2167	0.2890	0.9066	0.2395
0.2095	0.8139	0.1521	0.2295	0.8410	0.1732	0.2495	0.8654	0.1949	0.2695	0.8874	0.2173	0.2895	0.9071	0.2401
0.2100	0.8146	0.1527	0.2300	0.8417	0.1738	0.2500	0.8660	0.1955	0.2700	0.8879	0.2178	0.2900	0.9075	0.2407
0.2105	0.8153	0.1532	0.2305	0.8423	0.1743	0.2505	0.8666	0.1961	0.2705	0.8884	0.2184	0.2905	0.9080	0.2413
0.2110	0.8160	0.1537	0.2310	0.8429	0.1748	0.2510	0.8672	0.1966	0.2710	0.8890	0.2190	0.2910	0.9084	0.2419
0.2115	0.8167	0.1542	0.2315	0.8436	0.1754	0.2515	0.8678	0.1972	0.2715	0.8895	0.2195	0.2915	0.9089	0.2424
0.2120	0.8174	0.1547	0.2320	0.8442	0.1759	0.2520	0.8683	0.1977	0.2720	0.8900	0.2201	0.2920	0.9094	0.2430
0.2125	0.8182	0.1553	0.2325	0.8449	0.1764	0.2525	0.8689	0.1983	0.2725	0.8905	0.2207	0.2925	0.9098	0.2436
0.2130	0.8189	0.1558	0.2330	0.8455	0.1770	0.2530	0.8695	0.1988	0.2730	0.8910	0.2212	0.2930	0.9103	0.2442
0.2135	0.8196	0.1563	0.2335	0.8461	0.1775	0.2535	0.8700	0.1994	0.2735	0.8915	0.2218	0.2935	0.9107	0.2448
0.2140	0.8203	0.1568	0.2340	0.8467	0.1781	0.2540	0.8706	0.1999	0.2740	0.8920	0.2224	0.2940	0.9112	0.2453
0.2145	0.8210	0.1573	0.2345	0.8474	0.1786	0.2545	0.8712	0.2005	0.2745	0.8925	0.2229	0.2945	0.9116	0.2459
0.2150	0.8216	0.1579	0.2350	0.8480	0.1791	0.2550	0.8717	0.2010	0.2750	0.8930	0.2235	0.2950	0.9121	0.2465
0.2155	0.8223	0.1584	0.2355	0.8486	0.1797	0.2555	0.8723	0.2016	0.2755	0.8935	0.2241	0.2955	0.9125	0.2471
0.2160	0.8230	0.1589	0.2360	0.8492	0.1802	0.2560	0.8728	0.2021	0.2760	0.8940	0.2246	0.2960	0.9130	0.2477
0.2165	0.8237	0.1594	0.2365	0.8499	0.1808	0.2565	0.8734	0.2027	0.2765	0.8945	0.2252	0.2965	0.9134	0.2482
0.2170	0.8244	0.1600	0.2370	0.8505	0.1813	0.2570	0.8740	0.2033	0.2770	0.8950	0.2258	0.2970	0.9139	0.2488
0.2175	0.8251	0.1605	0.2375	0.8511	0.1818	0.2575	0.8745	0.2038	0.2775	0.8955	0.2264	0.2975	0.9143	0.2494
0.2180	0.8258	0.1610	0.2380	0.8517	0.1824	0.2580	0.8751	0.2044	0.2780	0.8960	0.2269	0.2980	0.9148	0.2500
0.2185	0.8265	0.1615	0.2385	0.8523	0.1829	0.2585	0.8756	0.2049	0.2785	0.8965	0.2275	0.2985	0.9152	0.2506
0.2190	0.8271	0.1621	0.2390	0.8529	0.1835	0.2590	0.8762	0.2055	0.2790	0.8970	0.2281	0.2990	0.9156	0.2511
0.2195	0.8278	0.1626	0.2395	0.8536	0.1840	0.2595	0.8767	0.2060	0.2795	0.8975	0.2286	0.2995	0.9161	0.2517

TABLE 14.10 (*Continued*)

H/D from 0.3 to 0.4														
H/D	L/D	A_D/A_T	H/D	L/D	A_D/A_T	H/D	L/D	A_D/A_T	H/D	L/D	A_D/A_T	H/D	L/D	A_D/A_T
0.3000	0.9165	0.2523	0.3200	0.9330	0.2759	0.3400	0.9474	0.2998	0.3600	0.9600	0.3241	0.3800	0.9708	0.3487
0.3005	0.9170	0.2529	0.3205	0.9333	0.2765	0.3405	0.9478	0.3004	0.3605	0.9603	0.3247	0.3805	0.9710	0.3493
0.3010	0.9174	0.2535	0.3210	0.9337	0.2771	0.3410	0.9481	0.3010	0.3610	0.9606	0.3253	0.3810	0.9713	0.3499
0.3015	0.9178	0.2541	0.3215	0.9341	0.2777	0.3415	0.9484	0.3016	0.3615	0.9609	0.3259	0.3815	0.9715	0.3505
0.3020	0.9183	0.2547	0.3220	0.9345	0.2782	0.3420	0.9488	0.3022	0.3620	0.9612	0.3265	0.3820	0.9718	0.3512
0.3025	0.9187	0.2552	0.3225	0.9349	0.2788	0.3425	0.9491	0.3028	0.3625	0.9614	0.3272	0.3825	0.9720	0.3518
0.3030	0.9191	0.2558	0.3230	0.9352	0.2794	0.3430	0.9494	0.3034	0.3630	0.9617	0.3278	0.3830	0.9722	0.3524
0.3035	0.9195	0.2564	0.3235	0.9356	0.2800	0.3435	0.9498	0.3040	0.3635	0.9620	0.3284	0.3835	0.9725	0.3530
0.3040	0.9200	0.2570	0.3240	0.9360	0.2806	0.3440	0.9501	0.3046	0.3640	0.9623	0.3290	0.3840	0.9727	0.3536
0.3045	0.9204	0.2576	0.3245	0.9364	0.2812	0.3445	0.9504	0.3053	0.3645	0.9626	0.3296	0.3845	0.9730	0.3543
0.3050	0.9208	0.2582	0.3250	0.9367	0.2818	0.3450	0.9507	0.3059	0.3650	0.9629	0.3302	0.3850	0.9732	0.3549
0.3055	0.9212	0.2588	0.3255	0.9371	0.2824	0.3455	0.9511	0.3065	0.3655	0.9631	0.3308	0.3855	0.9734	0.3555
0.3060	0.9217	0.2593	0.3260	0.9375	0.2830	0.3460	0.9514	0.3071	0.3660	0.9634	0.3315	0.3860	0.9737	0.3561
0.3065	0.9221	0.2599	0.3265	0.9379	0.2836	0.3465	0.9517	0.3077	0.3665	0.9637	0.3321	0.3865	0.9739	0.3567
0.3070	0.9225	0.2605	0.3270	0.9382	0.2842	0.3470	0.9520	0.3083	0.3670	0.9640	0.3327	0.3870	0.9741	0.3574
0.3075	0.9229	0.2611	0.3275	0.9386	0.2848	0.3475	0.9524	0.3089	0.3675	0.9642	0.3333	0.3875	0.9744	0.3580
0.3080	0.9233	0.2617	0.3280	0.9390	0.2854	0.3480	0.9527	0.3095	0.3680	0.9645	0.3339	0.3880	0.9746	0.3586
0.3085	0.9237	0.2623	0.3285	0.9393	0.2860	0.3485	0.9530	0.3101	0.3685	0.9648	0.3345	0.3885	0.9748	0.3592
0.3090	0.9242	0.2629	0.3290	0.9397	0.2866	0.3490	0.9533	0.3107	0.3690	0.9651	0.3351	0.3890	0.9750	0.3598
0.3095	0.9246	0.2635	0.3295	0.9401	0.2872	0.3495	0.9536	0.3113	0.3695	0.9653	0.3357	0.3895	0.9753	0.3605
0.3100	0.9250	0.2640	0.3300	0.9404	0.2878	0.3500	0.9539	0.3119	0.3700	0.9656	0.3364	0.3900	0.9755	0.3611
0.3105	0.9254	0.2646	0.3305	0.9408	0.2884	0.3505	0.9543	0.3125	0.3705	0.9659	0.3370	0.3905	0.9757	0.3617
0.3110	0.9258	0.2652	0.3310	0.9411	0.2890	0.3510	0.9546	0.3131	0.3710	0.9661	0.3376	0.3910	0.9759	0.3623
0.3115	0.9262	0.2658	0.3315	0.9415	0.2896	0.3515	0.9549	0.3137	0.3715	0.9664	0.3382	0.3915	0.9762	0.3629
0.3120	0.9266	0.2664	0.3320	0.9419	0.2902	0.3520	0.9552	0.3143	0.3720	0.9667	0.3388	0.3920	0.9764	0.3636
0.3125	0.9270	0.2670	0.3325	0.9422	0.2908	0.3525	0.9555	0.3150	0.3725	0.9669	0.3394	0.3925	0.9766	0.3642
0.3130	0.9274	0.2676	0.3330	0.9426	0.2914	0.3530	0.9558	0.3156	0.3730	0.9672	0.3401	0.3930	0.9768	0.3648
0.3135	0.9278	0.2682	0.3335	0.9429	0.2920	0.3535	0.9561	0.3162	0.3735	0.9675	0.3407	0.3935	0.9771	0.3654
0.3140	0.9282	0.2688	0.3340	0.9433	0.2926	0.3540	0.9564	0.3168	0.3740	0.9677	0.3413	0.3940	0.9773	0.3661
0.3145	0.9286	0.2693	0.3345	0.9436	0.2932	0.3545	0.9567	0.3174	0.3745	0.9680	0.3419	0.3945	0.9775	0.3667
0.3150	0.9290	0.2699	0.3350	0.9440	0.2938	0.3550	0.9570	0.3180	0.3750	0.9682	0.3425	0.3950	0.9777	0.3673
0.3155	0.9294	0.2705	0.3355	0.9443	0.2944	0.3555	0.9573	0.3186	0.3755	0.9685	0.3431	0.3955	0.9779	0.3679
0.3160	0.9298	0.2711	0.3360	0.9447	0.2950	0.3560	0.9576	0.3192	0.3760	0.9688	0.3438	0.3960	0.9781	0.3685
0.3165	0.9302	0.2717	0.3365	0.9450	0.2956	0.3565	0.9579	0.3198	0.3765	0.9690	0.3444	0.3965	0.9783	0.3692
0.3170	0.9306	0.2723	0.3370	0.9454	0.2962	0.3570	0.9582	0.3204	0.3770	0.9693	0.3450	0.3970	0.9786	0.3698
0.3175	0.9310	0.2729	0.3375	0.9457	0.2968	0.3575	0.9585	0.3211	0.3775	0.9695	0.3456	0.3975	0.9788	0.3704
0.3180	0.9314	0.2735	0.3380	0.9461	0.2974	0.3580	0.9588	0.3217	0.3780	0.9698	0.3462	0.3980	0.9790	0.3710
0.3185	0.9318	0.2741	0.3385	0.9464	0.2980	0.3585	0.9591	0.3223	0.3785	0.9700	0.3468	0.3985	0.9792	0.3717
0.3190	0.9322	0.2747	0.3390	0.9467	0.2986	0.3590	0.9594	0.3229	0.3790	0.9703	0.3475	0.3990	0.9794	0.3723
0.3195	0.9326	0.2753	0.3395	0.9471	0.2992	0.3595	0.9597	0.3235	0.3795	0.9705	0.3481	0.3995	0.9796	0.3729

H/D from 0.4 to 0.5

H/D	L/D	A_D/A_T	H/D	L/D	A_D/A_T	H/D	L/D	A_D/A_T	H/D	L/D	A_D/A_T	H/D	L/D	A_D/A_T
0.4000	0.9798	0.3735	0.4200	0.9871	0.3986	0.4400	0.9928	0.4238	0.4600	0.9968	0.4491	0.4800	0.9992	0.4745
0.4005	0.9800	0.3742	0.4205	0.9873	0.3992	0.4405	0.9929	0.4244	0.4605	0.9969	0.4498	0.4805	0.9992	0.4752
0.4010	0.9802	0.3748	0.4210	0.9874	0.3998	0.4410	0.9930	0.4251	0.4610	0.9970	0.4504	0.4810	0.9993	0.4758
0.4015	0.9804	0.3754	0.4215	0.9876	0.4005	0.4415	0.9931	0.4257	0.4615	0.9970	0.4510	0.4815	0.9993	0.4765
0.4020	0.9806	0.3760	0.4220	0.9878	0.4011	0.4420	0.9932	0.4263	0.4620	0.9971	0.4517	0.4820	0.9994	0.4771
0.4025	0.9808	0.3767	0.4225	0.9879	0.4017	0.4425	0.9934	0.4270	0.4625	0.9972	0.4523	0.4825	0.9994	0.4777
0.4030	0.9810	0.3773	0.4230	0.9881	0.4023	0.4430	0.9935	0.4276	0.4630	0.9973	0.4529	0.4830	0.9994	0.4784
0.4035	0.9812	0.3779	0.4235	0.9882	0.4030	0.4435	0.9936	0.4282	0.4635	0.9973	0.4536	0.4835	0.9995	0.4790
0.4040	0.9814	0.3785	0.4240	0.9884	0.4036	0.4440	0.9937	0.4288	0.4640	0.9974	0.4542	0.4840	0.9995	0.4796
0.4045	0.9816	0.3791	0.4245	0.9885	0.4042	0.4445	0.9938	0.4295	0.4645	0.9975	0.4548	0.4845	0.9995	0.4803
0.4050	0.9818	0.3798	0.4250	0.9887	0.4049	0.4450	0.9939	0.4301	0.4650	0.9975	0.4555	0.4850	0.9995	0.4809
0.4055	0.9820	0.3804	0.4255	0.9888	0.4055	0.4455	0.9940	0.4307	0.4655	0.9976	0.4561	0.4855	0.9996	0.4815
0.4060	0.9822	0.3810	0.4260	0.9890	0.4061	0.4460	0.9942	0.4314	0.4660	0.9977	0.4567	0.4860	0.9996	0.4822
0.4065	0.9824	0.3816	0.4265	0.9891	0.4068	0.4465	0.9943	0.4320	0.4665	0.9978	0.4574	0.4865	0.9996	0.4828
0.4070	0.9825	0.3823	0.4270	0.9893	0.4074	0.4470	0.9944	0.4326	0.4670	0.9978	0.4580	0.4870	0.9997	0.4834
0.4075	0.9827	0.3829	0.4275	0.9894	0.4080	0.4475	0.9945	0.4333	0.4675	0.9979	0.4586	0.4875	0.9997	0.4841
0.4080	0.9829	0.3835	0.4280	0.9896	0.4086	0.4480	0.9946	0.4339	0.4680	0.9979	0.4593	0.4880	0.9997	0.4847
0.4085	0.9831	0.3842	0.4285	0.9897	0.4093	0.4485	0.9947	0.4345	0.4685	0.9980	0.4599	0.4885	0.9997	0.4854
0.4090	0.9833	0.3848	0.4290	0.9899	0.4099	0.4490	0.9948	0.4352	0.4690	0.9981	0.4606	0.4890	0.9998	0.4860
0.4095	0.9835	0.3854	0.4295	0.9900	0.4105	0.4495	0.9949	0.4358	0.4695	0.9981	0.4612	0.4895	0.9998	0.4866
0.4100	0.9837	0.3860	0.4300	0.9902	0.4112	0.4500	0.9950	0.4364	0.4700	0.9982	0.4618	0.4900	0.9998	0.4873
0.4105	0.9838	0.3867	0.4305	0.9903	0.4118	0.4505	0.9951	0.4371	0.4705	0.9983	0.4625	0.4905	0.9998	0.4879
0.4110	0.9840	0.3873	0.4310	0.9904	0.4124	0.4510	0.9952	0.4377	0.4710	0.9983	0.4631	0.4910	0.9998	0.4885
0.4115	0.9842	0.3879	0.4315	0.9906	0.4131	0.4515	0.9953	0.4383	0.4715	0.9984	0.4637	0.4915	0.9999	0.4892
0.4120	0.9844	0.3885	0.4320	0.9907	0.4137	0.4520	0.9954	0.4390	0.4720	0.9984	0.4644	0.4920	0.9999	0.4898
0.4125	0.9846	0.3892	0.4325	0.9908	0.4143	0.4525	0.9955	0.4396	0.4725	0.9985	0.4650	0.4925	0.9999	0.4905
0.4130	0.9847	0.3898	0.4330	0.9910	0.4149	0.4530	0.9956	0.4402	0.4730	0.9985	0.4656	0.4930	0.9999	0.4911
0.4135	0.9849	0.3904	0.4335	0.9911	0.4156	0.4535	0.9957	0.4409	0.4735	0.9986	0.4663	0.4935	0.9999	0.4917
0.4140	0.9851	0.3910	0.4340	0.9912	0.4162	0.4540	0.9958	0.4415	0.4740	0.9986	0.4669	0.4940	0.9999	0.4924
0.4145	0.9853	0.3917	0.4345	0.9914	0.4168	0.4545	0.9959	0.4421	0.4745	0.9987	0.4675	0.4945	0.9999	0.4930
0.4150	0.9854	0.3923	0.4350	0.9915	0.4175	0.4550	0.9959	0.4428	0.4750	0.9987	0.4682	0.4950	1.0000	0.4936
0.4155	0.9856	0.3929	0.4355	0.9916	0.4181	0.4555	0.9960	0.4434	0.4755	0.9988	0.4688	0.4955	1.0000	0.4943
0.4160	0.9858	0.3936	0.4360	0.9918	0.4187	0.4560	0.9961	0.4440	0.4760	0.9988	0.4695	0.4960	1.0000	0.4949
0.4165	0.9860	0.3942	0.4365	0.9919	0.4194	0.4565	0.9962	0.4447	0.4765	0.9989	0.4701	0.4965	1.0000	0.4955
0.4170	0.9861	0.3948	0.4370	0.9920	0.4200	0.4570	0.9963	0.4453	0.4770	0.9989	0.4707	0.4970	1.0000	0.4962
0.4175	0.9863	0.3954	0.4375	0.9922	0.4206	0.4575	0.9964	0.4460	0.4775	0.9990	0.4714	0.4975	1.0000	0.4968
0.4180	0.9865	0.3961	0.4380	0.9923	0.4213	0.4580	0.9965	0.4466	0.4780	0.9990	0.4720	0.4980	1.0000	0.4975
0.4185	0.9866	0.3967	0.4385	0.9924	0.4219	0.4585	0.9965	0.4472	0.4785	0.9991	0.4726	0.4985	1.0000	0.4981
0.4190	0.9868	0.3973	0.4390	0.9925	0.4225	0.4590	0.9966	0.4479	0.4790	0.9991	0.4733	0.4990	1.0000	0.4987
0.4195	0.9870	0.3979	0.4395	0.9927	0.4232	0.4595	0.9967	0.4485	0.4795	0.9992	0.4739	0.4995	1.0000	0.4994
												0.5000	1.0000	0.5000

3. *Pressure drop*

$$h_{ow} = 0.48F_w\left(\frac{Q_L}{l_w}\right)^{0.67}$$

$$\frac{Q_L}{(l_w)^{2.5}} = \frac{15.2}{(1.82)^{2.5}} = \frac{15.2}{4.4} \qquad \frac{Q_L}{(l_w)^{2.5}} = \frac{50.8}{4.4}$$

$$= 3.46 \qquad\qquad = 11.5$$

From Fig. 13.7 at $l_w/D = 0.73$,

$$F_w = 1.035 \qquad F_w = 1.08$$

$$h_{ow} = (0.48)(1.035)\left(\frac{15.2}{21.8}\right)^{0.67} \qquad h_{ow} = (0.48)(1.08)\left(\frac{50.8}{21.8}\right)^{0.67}$$

$$= 0.39 \text{ in.} \qquad\qquad = 0.91 \text{ in.}$$

$$h_w = 1.5 \text{ in.} \qquad h_w = 1.5 \text{ in.}$$

$$h_\sigma = \frac{(0.04)(13)}{(50.7)(0.25)} \qquad h_\sigma = \frac{(0.04)(13)}{(44.6)(0.25)}$$

$$= 0.041 \text{ in.} \qquad\qquad = 0.0465 \text{ in.}$$

$$h_0 = 0.186\,\frac{\rho_V}{\rho_L}\left(\frac{U_h}{C_0}\right)^2$$

$$t_p = 0.0825 \qquad d_h = 0.25 \qquad t_p/d_h = 0.33$$

Select $A_h/A = 8/80 = 0.10$, $C_0 = 0.73$. From Fig. 13.18,

$$A_h = 0.10(3.94) = 0.394 \text{ sq ft}$$

$$U_h = 38.0 \text{ fps} \qquad U_h = 24.8 \text{ fps}$$

$$h_0 = 0.186\left(\frac{0.178}{50.7}\right)\left(\frac{38.0}{0.73}\right)^2 \qquad h_0 = 0.186\left(\frac{0.190}{44.6}\right)\left(\frac{24.8}{0.73}\right)^2$$

$$= 1.77 \text{ in.} \qquad\qquad = 0.915 \text{ in.}$$

$$F_{VA} = U_{VA}(\rho_V)^{0.5}$$

$$F_{VA} = (3.8)(0.178)^{0.5} \qquad F_{AV} = (2.48)(0.190)^{0.5}$$

$$= 1.6 \qquad\qquad = 1.08$$

From Fig. 13.16,

$$\beta = 0.59 \qquad \beta = 0.62$$

The total pressure drop is

$$\Delta H_T = \beta(h_w + h_{ow}) + h_0 + h_\sigma$$

$$\Delta H_T = 0.59(1.5 + 0.390) + 1.77 + 0.04 \qquad \Delta H_T = 0.62(1.5 + 0.910) + 0.915 + 0.047$$

$$= 2.92 \text{ in.} \qquad\qquad = 2.46 \text{ in.}$$

4. *Check for weep point.* Weep point (Fig. 13.22), $A_h/A_A = 0.100$.

$$h_w + h_{ow} = 1.890 \qquad h_w + h_{ow} = 2.410$$

From Fig. 13.22,

$$h_0 + h_\sigma = 0.5 \qquad h_0 + h_\sigma = 0.65$$

Calculated:

$$h_0 + h_\sigma = 1.81 \qquad h_0 + h_\sigma = 0.962$$

The sum of the hole plus surface-tension head terms is greater than the value read from Fig. 13.22 at the calculated $h_w + h_{ow}$ quantities and, therefore, both sections will operate well above the weep point.

5. *Liquid backup in downcomer*

$$H_D = \left[\Delta H_t + h_w + h_{ow} + \frac{\Delta}{2} + h_d\right]\frac{1}{\phi_d}$$

Assume 1.5-in. clearance under apron,

$$A_{AP} = \frac{1.5 \times 21.8}{144} = 0.227 \text{ sq ft}$$

$$h_d = 0.03\left(\frac{Q_L}{100A_d}\right)^2 \qquad h_d = 0.03\left[\frac{50.8}{(100)(0.227)}\right]^2$$

$$= 0.03\left[\frac{15.2}{(100)(0.227)}\right]^2 \qquad = 0.03\left[\frac{50.8}{(100)(0.227)}\right]^2$$

$$= 0.0135 \text{ in.} \qquad\qquad = 0.15 \text{ in.}$$

For top,

$$H_D = [2(0.59)(1.5 + 0.390) + 1.77 + 0.04 + 0.0135]\frac{1}{\phi_d}$$

$$= 4.05 \text{ in.}$$

For bottom,

$$H_D = [2(0.62)(0.910 + 1.5) + 0.915 + 0.047 + 0.15]\frac{1}{\phi_d}$$

$$= 4.102 \text{ in.}$$

Assume $\phi_d = 0.5$,

$$H_{D_{al}} = \frac{4.05}{0.5} = 8.1 \text{ in.} \qquad H_{D_{al}} = \frac{4.10}{0.5} = 8.2 \text{ in.}$$

6. *Liquid gradient.* The liquid gradient is usually negligible for small-diameter perforated trays, but because of the liquid loading it is checked here.

$$\Delta = \frac{fU_f^2 Z_L}{12gR_h}$$

$$R_h = \frac{h_f W_a}{2h_f + W_a}$$

$$W_a = \frac{D + l_w}{2} = \frac{2.5 + 1.82}{2} = 2.16 \text{ ft}$$

$$h_f = \frac{(0.59)(1.89)}{1.18 - 1} \qquad h_f = \frac{(0.62)(2.41)}{1.24 - 1}$$

$$= 6.2 \text{ in. or } 0.516 \text{ ft} \qquad = 6.22 \text{ in. or } 0.52 \text{ ft}$$

$$R_h = \frac{(0.516)(2.16)}{(1.032) + (2.16)} \qquad R_h = \frac{(0.52)(2.16)}{(1.04) + (2.16)}$$

$$= \frac{1.113}{3.192} = 0.35 \text{ ft} \qquad = \frac{1.122}{3.20} = 0.352 \text{ ft}$$

$$U_f = \frac{12Q'_L}{h_c W_a}$$

$$U_f = \frac{(12)(0.034)}{(0.59)(1.89)(2.16)} \qquad U_f = \frac{(12)(0.113)}{(0.62)(2.41)(2.16)}$$

$$= 0.17 \text{ fps} \qquad = 0.42 \text{ fps}$$

$$\text{Re}_f = \frac{R_h U_f \rho_L}{\mu_L}$$

$$\text{Re}_f = \frac{(0.35)(0.17)(50.7)}{(0.4)(6.72 \times 10^{-4})} \qquad \text{Re}_f = \frac{(0.352)(0.42)(44.6)}{(0.4)(6.72 \times 10^{-4})}$$

$$= 11210 \qquad = 24500$$

$$f = 0.15 \qquad f = 0.11 \quad \text{(Fig. 13.19)}$$

$$\Delta = \frac{(0.15)(0.17)^2(1.72)}{(12)(32.2)(0.35)} \qquad \Delta = \frac{(0.11)(0.42)^2(1.72)}{(12)(32.2)(0.352)}$$

$$= 0.000055 \text{ in.} \qquad = 0.00025 \text{ in.}$$

7. *Liquid residence time in downcomer*

$$A_d = 0.49 \text{ sq ft}$$

$$H_D = 4.05 \text{ in.} \qquad H_D = 4.10 \text{ in.}$$

The flow rate is

$$Q'_{Ld} = 0.034 \text{ cu ft/sec} \qquad Q'_{Ld} = 0.113 \text{ cu ft/sec}$$

$$\text{Residence time} = \frac{\text{volume downcomer}}{\text{flow rate}}$$

$$\text{Residence time} = \frac{(0.49)(4.05/12)}{0.034} \qquad \text{Residence time} = \frac{(0.49)(4.10/12)}{0.113}$$

$$= 4.85 \text{ sec} \qquad = 1.48 \text{ sec}$$

The residence time in the downcomer in the stripping section is below the recommended time for vapor disengagement. Some adjustment to increase the backup could be made such as decreasing apron clearance or reducing downcomer area. If the apron clearance were reduced to 0.5 in., h_d would increase to 1.35 in., H_D to 5.39 in., and the residence time would increase to 1.95 sec. Reduction of the downcomer volume by sloping the apron would also increase residence time.

8. *Summary*

Tower diameter = 2.5 ft

Tray spacing = 1.5 ft

Active area = 3.94 sq ft

Area of holes = 0.394 sq ft

Area downcomer = 0.49 sq ft

$A_h/A = 0.08$

$A_d/A = 0.10$

$A_h/A_A = 0.10$

$d_h = 0.25$ in.

$l_w = 21.8$ in.

$h_w = 1.5$ in.

Downcomer clearance = 1.5 in. for rectifying section, 0.5 in. for stripping section
Tray thickness = 12 gage

Several tray designs should be made to optimize the performance with regard to efficiency, capacity, and cost. No one particular design will prove particularly advantageous over others from all standpoints.

Other Methods for Sieve-tray Design

Besides the method of design for perforated trays given in the foregoing, there have been several methods, all somewhat similar, presented in the literature. In general, all methods of design involve devising a perforated plate which will neither dump, prime, nor flood under the conditions for which it is expected to operate. Therefore, the designs attempt to minimize the pressure drop to get high capacity, yet maintain sufficient pressure drop across the tray to prevent dumping or excessive weeping.

Hughmark and O'Connell [6] consider the following:

1. Tower diameter and plate spacing are selected on the basis of the Souders-Brown [13] capacity-factor chart and on the evaluation of the factors which cause backup of liquid in the downcomer to the plate above or foam carryover. The dry-plate pressure drop, hydrostatic head above the holes including the effect of hydraulic gradient, and the head equivalent for bubble formation are calculated.

2. Plate stability is related to the minimum flow condition where dumping occurs.

Liebson et al. [8] consider the limits of design to be flooding and priming at one of the capacity ranges and dumping and pulsation at the other extreme, and their design method hinges upon the evaluation of maximum and minimum vapor loadings.

Huang and Hodson [5] discuss general perforated-plate and column design as well as present quantitative relations for design evaluation. Their method evaluates the minimum and maximum operating capacity with respect to priming, flooding, and dumping.

The methods for evaluation of the various design variables are given in Chap. 13.

14.4. VALVE-TRAY DESIGN METHODS

The manufacturers of specific types of valve trays have developed design methods for their particular proprietary type of tray. All manufacturers prefer to do the engineering design of fractionators using their specific valve types to meet specifications for efficiency and capacity established by the potential purchaser. One detailed design procedure [4] is made available to engineers who wish to make their own preliminary detailed studies. Discussions of valve-tray designs of a partial nature have appeared in the literature [7, 9, 14], and more or less detailed procedures are made available by all of the manufacturers of proprietary trays upon request. The broad, general procedure followed in valve-plate design and an indication of the special data necessary to obtain reasonably accurate results are discussed in the following section.

General Considerations—Valve-tray Design

Although specific performance test data are necessary for reasonably accurate valve-tray design, and these data are usually unpublished except by the manufacturer in design bulletins, certain generalizations can be considered to be valid in relation to the information developed in this chapter and Chap. 13.

1. The diameter of the tower can be estimated by the Souders-Brown [13] capacity factor from the surface tension, densities of vapor and liquid, and plate spacing, or by use of a limiting percent of flood (see Table 14.2). The same variables as in the foregoing can be used to determine vapor velocity by Fig. 13.21 from which the diameter is calculated.

2. The tray type and downcomer areas can be selected by following the recommendations in Tables 14.2 through 14.4.

3. The total pressure drop through the valve tray can be shown in the same form as in the pressure-drop equations for the bubble and perforated trays:

$$\Delta H_T = h_V + \beta_V\left(h_{ow} + h_w + \frac{\Delta}{2}\right) \tag{14.1}$$

where h_V is the dry-plate pressure drop through the valves expressed as a function of the velocity of the vapor through the valve, the density of vapor and liquid, density of metal, and volume of metal valve plate to be lifted by the vapor pressure in passing through the valve. Thus, in one form [4],

$$h_V = f_1\left(t_m \frac{\rho_m}{\rho_L}\right) + f_2\left(U_{VV}{}^2 \frac{\rho_V}{\rho_L}\right) \tag{14.2}$$

where β_V = aeration factor for the valve-plate design in question and is obtained from experimental data
h_w = the weir height
h_{ow} = the crest over the weir and is a function of $(Q_L/l_w)^{0.67}$

and the values of the functions f_1 and f_2 are obtained from experimental data. Δ is evaluated from experimental correlations for the particular type of valve tray.

4. The backup in the downcomers can be evaluated from relations similar to those for the bubble and valve trays.

$$H_D = [\Delta H_T + h_d + \beta_V(h_w + h_{ow})]\frac{1}{\phi_d}$$

where h_d is calculated by Eq. (13.5), h_{ow} by Eq. (13.4) using the recommended coefficient, ΔH_T from the functions and experimental data as previously described. ϕ_d, the foam density in the downcomer, can be estimated as 0.5 if no data are available. Residence time is computed by the same method used for any type of tray.

5. The selection of valve area, number, and type of valves must be based on the experience of the manufacturers and their recommended practices. The maximum and minimum liquid and vapor capacity and plate stability characteristics are determined also with the use of experimental performance data.

6. Experimental correlations of entrainment as a function of the operating conditions and particular valve plate are necessary for quantitative entrainment calculations. The same observation can be made in the case of efficiency evaluation.

PROBLEMS

1. A mixture composed of the following materials is charged at the rate of 6000 gph to a fractionator to make the products indicated. The column is operated at 100 psia.

Component	Feed	Distillate	Bottoms
	Mole fraction		
C_2H_6	0.04	0.21	—
C_3H_8	0.15	0.71	0.018
C_4H_{10}	0.15	0.08	0.166
C_4H_{10}	0.59	—	0.730
C_5H_{12}	0.07	—	0.086

Determine the bubble-cap plate design for this column operated at a reflux ratio $L_0/D = 4.0$ by using the Bolles method.

2. For the conditions in Prob. 1, design a perforated plate for the column.

3. If the quantity fed to the column were increased to 10,000 gph, determine if the bubble-cap column design in Prob. 1 would still operate satisfactorily?

4. Would the perforated-plate column designed in Prob. 2 operate satisfactorily with the increase in feed to 10,000 gph?

5. If the quantity of feed were reduced to 5000 gph, would the bubble-cap plate and perforated plate designed in Probs. 1 and 2 be satisfactory?

Nomenclature

A = area of column, sq ft
A_A = active area for bubbling, sq ft
A_{Am} = maximum active area, sq ft
A_{AP} = area for liquid flow under apron, sq ft
A_d = area of downcomer, sq ft
A_h = hole area, total, sq ft
A_N = net vapor flow area between plates, $A - A_d$
A_r = total riser area, sq ft
A_s = total slot area, sq ft
A_{sm} = minimum slot area
a_a = area of annulus per cap
a_c = cap area, sq in. per cap
a_h = hole area, sq in. per hole
a_r = riser area per cap
a_s = slot area per cap, sq in.
C_0 = orifice coefficient
C_S = correction factor, Fig. 13.9

C_V = vapor loading factor
D = tower diameter, feet or distillate rate, moles/hr or lb/hr
d_c = cap diameter, in.
d_h = hole diameter, in.
F_{VA} = $U_{VA}(\rho_V)^{0.5}$
F_{VN} = $U_{VN}(\rho_V)^{0.5}$
F_w = factor, Fig. 13.7
f = friction factor
f_1, f_2 = functions from experimental data
H_D = height of liquid backup in downcomer, in.
$H_{D_{al}}$ = height of foam backup in downcomer, in.
ΔH_T = total pressure drop, in. of fluid
h_d = equivalent head loss in downcomer, in.
h_{ds} = dynamic seal, in.
h_c = clear liquid head, in.
h_f = foam height, in.
h_0 = equivalent head loss through holes, in. of fluid
h_{ow} = liquid crest over weir, in.
h_{rc} = equivalent head loss, risers and caps, in.
h_s = slot height, in.
h_{sk} = skirt clearance, in.
h_{so} = slot opening, in.
h_{ss} = static seal, in.
h_σ = equivalent surface-tension head loss, in.
h_{sr} = height of shroud ring, in.
h_V = dry pressure drop through valve, in. of liquid
h_w = weir height, in.
K_c = correction, Fig. 13.8
L = liquid rate, moles/hr or lb/hr
l_w = weir length, in.
L_0 = reflux rate, moles/hr or lb/hr
N_C = number of caps per tray
n = number of plates
p = pitch, in.
P_C = capacity factor
P_F = flow parameter
P_c = cap pitch, in.
Q_L = liquid load, gpm
Q'_L = liquid load, cu ft/sec
Q_{L_d} = liquid rate in downcomer, gpm/(min)(sq ft A_d)
Q_V = vapor load, cu ft/sec
Q'_V = corrected vapor load, cu ft/sec
R_h = hydraulic radius

R_V = vapor distribution ratio, $\Delta/(h_{re} + h_{so})$
S = tray spacing, in.
s_c = cap spacing, distance between walls, in.
t_p = plate thickness, in.
t_m = valve metal thickness, in.
U_d = velocity of liquid on downcomer, fps
U_f = foam velocity, fps
U_h = velocity of vapor through holes, fps
U_{VN} = vapor velocity, fps, based on A_N
U_{VV} = velocity of vapor through valves, fps
V = vapor load, lb/hr or moles/hr
V_d = volume of downcomer, cu ft
W_a = average width of flow path, in.
W_s = slot width, in.
Z_L = liquid flow path, in. or ft
Z_{Lm} = maximum liquid flow path, in.

Greek letters

β = relative foam density
Δ = liquid gradient inches
Δ' = uncorrected liquid gradient, in.
μ_L = liquid viscosity, lb/fps
ϕ = froth density
ϕ_d = froth density, downcomer
ψ = entrainment, moles per mole of downflow
ρ_L = density of liquid, lb/cu ft
ρ_m = density of metal, lb/cu ft
ρ_V = density of vapor, lb/cu ft
σ = surface tension

REFERENCES

1. American Institute of Chemical Engineers: A.I.Ch.E. Bubble Design Manual, 1958.
2. Atkins, G. T.: *Chem. Eng. Progr.*, **50**:119 (1954).
3. Bolles, W. L.: *Petrol. Process.* (February, 1956, p. 65); (March, 1956), p. 82; (April, 1956), p. 72.
4. Glitsch, F. W., and Sons: Ballast Tray Design Manual, *Bull.* 4900, Dallas.
5. Huang, C. J., and J. R. Hodson: *Petrol. Refiner* (February, 1958), p. 104.
6. Hughmark, D. A., and O'Connell: *Chem. Eng. Progr.*, **53**:127m (1957).
7. Koch Engineering Co.: "Koch Flexitray—Preliminary Sizing," Wichita, Kansas, 1960.
8. Liebson, I., R. E. Kelley, and L. A. Bullington: *Petrol. Refiner* (February 19, 1957), p. 127.
9. Nutter, I. E.: *Oil Gas J.*, **52**(51):165 (1954).
10. Perry, J. H.: "Chemical Engineers' Handbook," McGraw-Hill, New York, 1963.

11. Smith, B. O.: "Design of Equilibrium Stage Processes," Chap. 14, McGraw-Hill, New York, 1963.
12. Smith, B. O.: "Design of Equilibrium Stage Processes," Chap. 15, McGraw-Hill, New York, 1963.
13. Souders, M., and G. G. Brown: *Ind. Eng. Chem.*, **25**:98 (1934).
14. Thrift, G. C.: *Petrol. Refiner*, **39**(8):93 (1960).
15. Thrift, G. C.: *Oil Gas J.*, **52**(51):163 (1954).

chapter 15

Packed column design

Discussion of fractionating column design has to this point been restricted to the equilibrium-stage or equilibrium-plate design concept because the great majority of fractionating columns in use today consists of plate columns. However, there is a class of columns used for many of the same applications as plate columns functioning on the principle of obtaining contact between the phases, not at specific levels throughout the column but differentially at all levels. These are "packed" columns, and the contact between the phases is obtained on the surface provided by the packing.

15.1. PACKINGS

There are generally three principal types of packed columns:

1. The conventional dumped-packed or random-packed columns, wherein discrete pieces of packing of a specific design are "dumped" or random packed into a shell or section of a shell.

2. The systematically geometrically packed packing, wherein manufactured sections of packing are placed by hand in particular reference to other sections to insure a completely uniform area of contact between the phases for any point or cross section in the column.

3. The "pseudoplate" columns, wherein essentially the contact area for the phases is that provided on plates at definite locations in the column. Since the vapor and liquid passageways are the same in this type of contact device because no downcomers are provided, these columns have the same operating limitations as packed columns but with less contact area per unit volume of shell.

The ring packings used in the first category of packed columns are Raschig rings, Lessing single-partition rings, cross-partition rings, splined rings, spiral rings, and Pall rings. The saddle packings utilized are for the most part Berl saddles, Intalox saddles, and the McMahon metal packing. Certain other miscellaneous types such as the protruded packing and helical packing are used primarily for laboratory and pilot-plant columns.

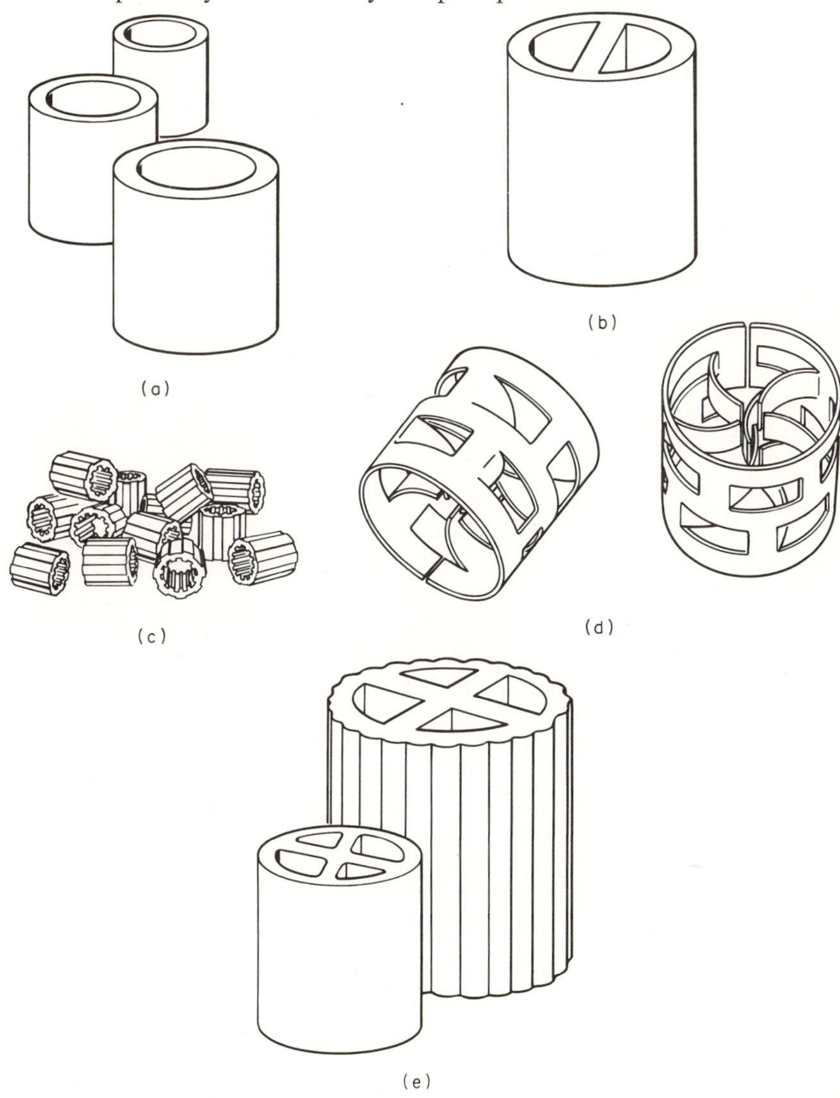

FIGURE 15.1 Ring packings. (*a*) Raschig ring. (*b*) Lessing ring. (*c*) Splined ring. (*d*) Pall ring. (*e*) Cross-partition rings. (*Courtesy U.S. Stoneware Company, Akron, Ohio.*)

Spraypak (Panapak), Stedman, Drippoint, wire mesh, wire cloth, and Multifil packings are employed in the second category of packed columns.

The third category of pseudopacked columns includes Turbogrid trays, grid trays, Ripple trays, Kittel trays, and shower deck trays.

Ring Packings

There are a number of packings available which are classed as ring packings (see Figs. 15.1 and 15.2). Raschig rings are hollow cylinders where the

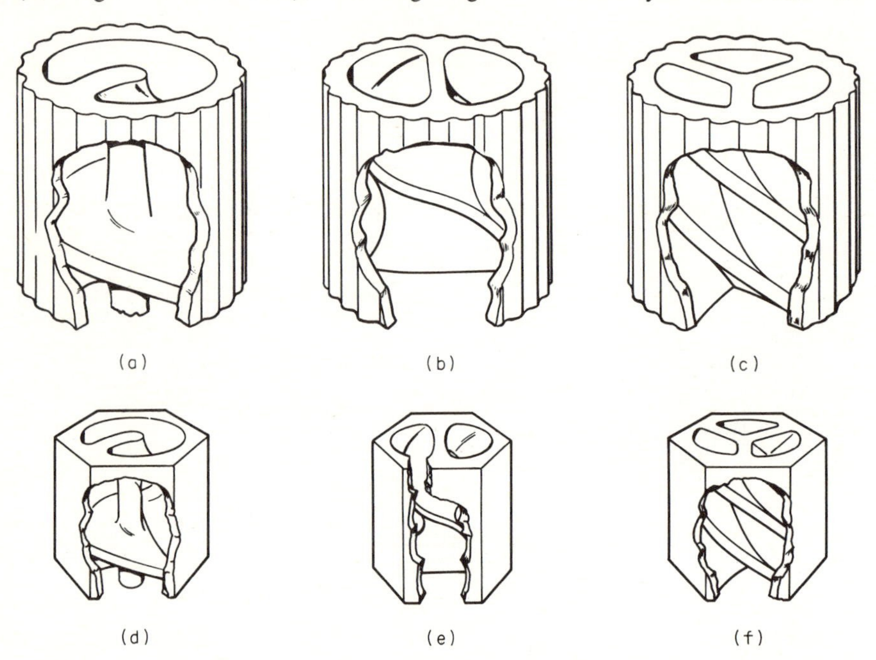

FIGURE 15.2 Spiral ring packing. (*a*) Cyclohelix single-spiral ring (with center post). (*b*) Cyclohelix double-spiral ring. (*c*) Cyclohelix triple-spiral ring. (*d*) Hexahelix single-spiral packing. (*e*) Hexahelix double-spiral packing. (*f*) Hexahelix triple-spiral packing. (*Courtesy U.S. Stoneware Company, Akron, Ohio.*)

diameter and the length of the cylinder are the same and the wall thickness varies with the type of ring from $\frac{1}{32}$ in. to $\frac{3}{8}$ in. Lessing rings are rings with a partition in the center to give added strength. Because of the partition there is a slight increase in surface area over the Raschig ring of the same nominal size. Cross-partition rings are made with a double partition and are sometimes ribbed to give greater surface area. Splined rings are made with the ribs on the inside to give increased surface. Pall rings [12] are metal, plastic, or ceramic rings of special manufacture to give very high surface area and low pressure drop.

Cyclohelix and Hexahelix rings are a specialized ring type containing a helical passage inside. The Cyclohelix has a cylindrical form and the Hexahelix is hexagonal in shape.

Generally, the rings are available in a variety of materials including ceramic materials, porcelain and stoneware, carbon, plastic, and metals including stainless steel, copper, and aluminum. The helix types are available only in ceramic materials.

Data on the size, surface area, weight, free space, and other characteristics are available from the manufacturers and in many handbooks. Tables 15.1 through 15.4 include information supplied by the manufacturers for the most commonly used ring types.

TABLE 15.1 **Ceramic Raschig Rings (Chemical Stoneware and White Porcelain)**

Size, in.	Wall thickness, in.	OD and length, in.	Approximate number per cu ft	Approximate weight per cu ft, lb*	Approximate surface area, sq ft/cu ft	Percent free gas space
1/4†	1/32	1/4	88,000	46	240	73
5/16†	1/16	5/16	40,000	56	145	64
3/8	1/16	3/8	24,000	52	155	68
1/2	3/32	1/2	10,500	54	111	63
1/2†	1/16	1/2	10,600	48	114	74
5/8	3/32	5/8	5,600	48	100	68
3/4	3/32	3/4	3,140	44	80	73
1	1/8	1	1,350	40	58	73
1 1/4	3/16	1 1/4	680	43	45	74
1 1/2	1/4	1 1/2	375	46	35	68
1 1/2†	3/16	1 1/2	385	42	38	71
2	1/4	2	162	38	28	74
2†	3/16	2	164	35	29	78
3‡	3/8	3	48	40	19	74
3(M)§	3/8	3	74	67	29	60
3(R)§	3/8	3	64	58	25	66

* Porcelain rings are about 5% heavier than stoneware. These weights are the average for both.

† Offered in porcelain only.

‡ The 4-in. OD and 6-in. OD sizes are also standard. The 3-in., 4-in., and 6-in. OD sizes are also made with ribbed or corrugated outside surfaces. The 3-in., 4-in., and 6-in. sizes can be made in lengths up to 12 in., on special order.

§ Data for stacked arrangement. M indicates diamond pattern. R indicates square pattern.

(*Courtesy U.S. Stoneware Company.*)

TABLE 15.2 **Metal Raschig Rings**

Size, in.	Wall thickness, in.	OD and length, in.	Approximate number per cu ft	Approximate weight per cu ft, lb*	Approximate surface area, sq ft/cu ft	Percent free gas space
1/4	1/32	1/4	88,000	150	236	69
5/16	1/32	5/16	45,000	120	190	75
5/16	1/16	5/16	43,000	198	176	60
1/2	1/32	1/2	11,800	77	128	84
1/2	1/16	1/2	11,000	132	118	73
19/32	1/32	19/32	7,300	66	112	86
19/32	1/16	19/32	7,000	120	106.5	75
3/4	1/32	3/4	3,410	55	83.5	88
3/4	1/16	3/4	3,190	100	71.8	78
1	1/32	1	1,440	40	62.7	92
1	1/16	1	1,345	73	56.7	85
1 1/4	1/16	1 1/4	725	62	49.3	87
1 1/2	1/16	1 1/2	420	50	41.2	90
2	1/16	2	180	38	31.4	92
3	1/16	3	53	25	20.6	95
3 (M)†	1/16	3	74	35	29	93
3 (R)†	1/16	3	64	30	25	94

* Weights shown are for carbon steel. Metal Raschig rings are also made in stainless steel, copper, and aluminum.

† Data for stacked arrangement. M indicates diamond pattern. R indicates square pattern.

(*Courtesy U.S. Stoneware Company.*)

TABLE 15.3 **Carbon Raschig Rings**

Size, in.	Wall thickness, in.	OD and length, in.	Approximate number per cu ft	Approximate weight per cu ft, lbs	Approximate surface area, sq ft/cu ft	Percent free gas space
1/4	1/16	1/4	85,000	46	212	55
1/2	1/16	1/2	10,600	27	114	74
3/4	1/8	3/4	3,140	34	75	67
1	1/8	1	1,325	27	57	74
1 1/4	3/16	1 1/4	678	31	45	69
1 1/2	1/4	1 1/2	392	34	37.5	67
2	1/4	2	166	27	28.5	74
3	5/16	3	49	33	19	78
3(M)*	5/16	3	74	49.5	29	66
3 (R)*	5/16	3	64	43	25	71

* Data for stacked arrangement. M indicates diamond pattern. R indicates square pattern.

(*Courtesy U.S. Stoneware Company.*)

TABLE 15.4 **Pall-ring Data**

OD and length, in.	Wall thickness, gage	Approximate number per cu ft	Approximate weight per cu ft, lb (carbon steel)	Approximate surface area, sq ft/cu ft	Percent free gas space
5/8	26	7700	50	131.2	90.2
1	24	1520	33	66.3	93.4
1½	22	490	30	48.1	94.0
2	20	210	27.5	36.6	94.0

Also available in stainless steels, aluminum, and copper.

Saddle Packings

Saddle packings were devised to give a large surface area, low pressure drop through the packed beds, and better vapor and liquid distribution to prevent channeling and bypassing. Berl saddles and Intalox saddles are illustrated in Fig. 15.3. The Intalox saddle gives a higher surface area than the Berl

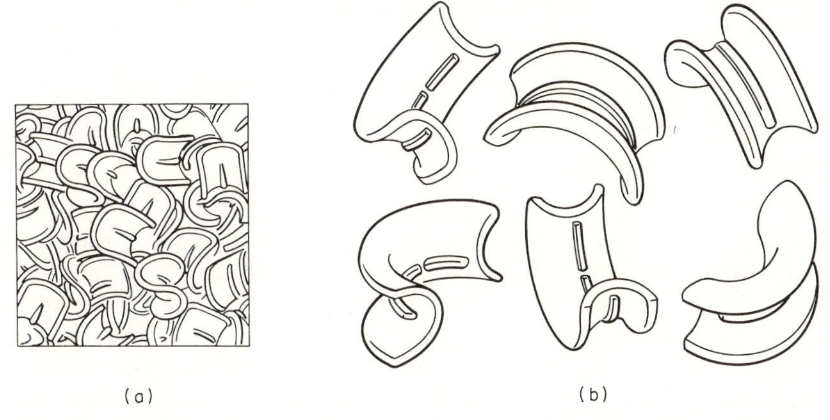

FIGURE 15.3 Saddle packings. (*a*) Berl saddles. (*b*) Intalox saddles. (*Courtesy U.S. Stoneware Company, Akron, Ohio.*)

saddle for the same nominal size. These packings are made from porcelain or stoneware, and the nominal size designation applies to the over-all symmetrical dimension. Pertinent design information is given in Tables 15.5 and 15.6.

Other Packings

A number of different packings have been developed primarily to increase the surface area of the packing per unit packed volume and to maintain or

TABLE 15.5 **Berl-Saddle Packings**

Nominal size, in.	Approximate number per cu ft	Approximate weight per cu ft, lb	Approximate surface area, sq ft/cu ft	Percent free gas space
¼	113,000	56	274	60
½	16,200	54	142	63
¾	5,000	48	82	66
1	2,200	45	76	69
1½	580	38	44	75
2	250	40	32	72

TABLE 15.6 **Average Data and Specifications for Intalox Saddles**

Nominal size, in.	Approximate number per cu ft	Approximate weight per cu ft, lb	Approximate surface area, sq ft/cu ft	Percent free gas space
¼	117,500	54	300	75
½	20,700	47	190	78
¾	6,500	44	102	77
1	2,385	42	78	77.5
1½	709	37	59.5	81
2	265	38	36	79

(*Courtesy U.S. Stoneware Company.*)

decrease the pressure drop through the packed bed. Reduction in packed bed weight, reduction of channeling tendencies of the liquid and vapor passing through the bed, increased strength, and other factors of special application have promoted development of a wide variety of packing forms.

Some of the packings are shown on Fig. 15.4. The McMahon stainless-steel packing [17] is made of wire mesh formed into a Berl-saddle shape. The Rosette or Tellerette packing [44] has a high interstitial holdup and gives high efficiency. It is made of polyethylene and cannot be used where solubility or reactivity is a problem. Cannon [5] developed a protruded packing for laboratory use which is made in small sizes. It has a high efficiency and is used widely in laboratory columns. Glass helices are widely used also in laboratory columns because of high surface area and chemical unreactivity.

Symmetrically Arranged Packing

A number of packing materials are available consisting of uniform sections which can be placed in a column in such a manner that the interstitial space provided for vapor and liquid passage is uniform, or essentially uniform,

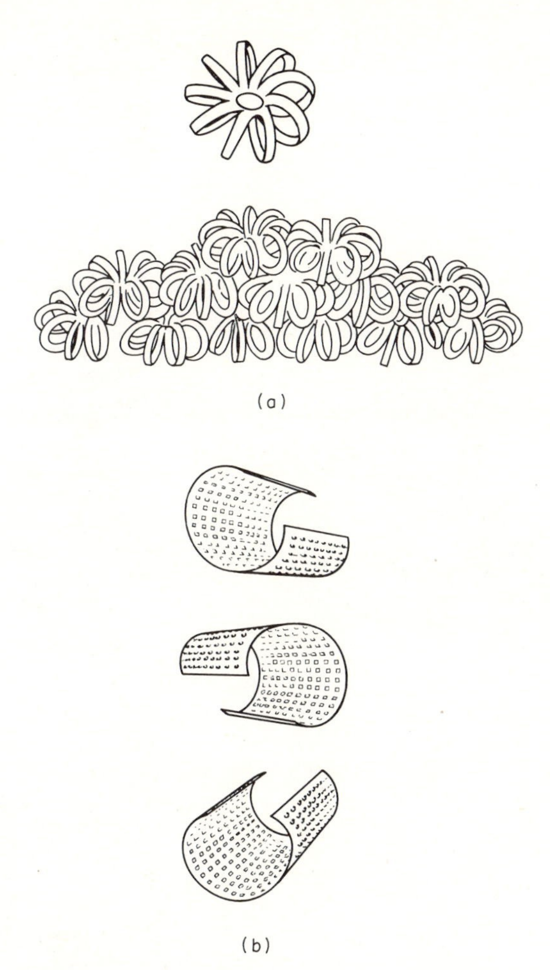

FIGURE 15.4 Packings. (*a*) Tellerette or Rosette packing. (*Courtesy Dr. A. J. Teller.*) (*b*) Protruded packing. (*Courtesy Scientific Development Company, State College, Pa.*)

throughout the whole bed. This method of packing prevents to a large extent bypassing and channeling and provides good vapor-liquid contact. Among these packings are Spraypak, Stedman, drip point [28–30, 33], and various wire-mesh packing materials such as those manufactured by York, Metal Goods, Metex, and others.

Spraypak [18], Fig. 15.5, is made from layers of expanded metal screen fastened together and pressed into a corrugated form, with the corrugation angle being 90° or less. The corrugated material is then bolted together

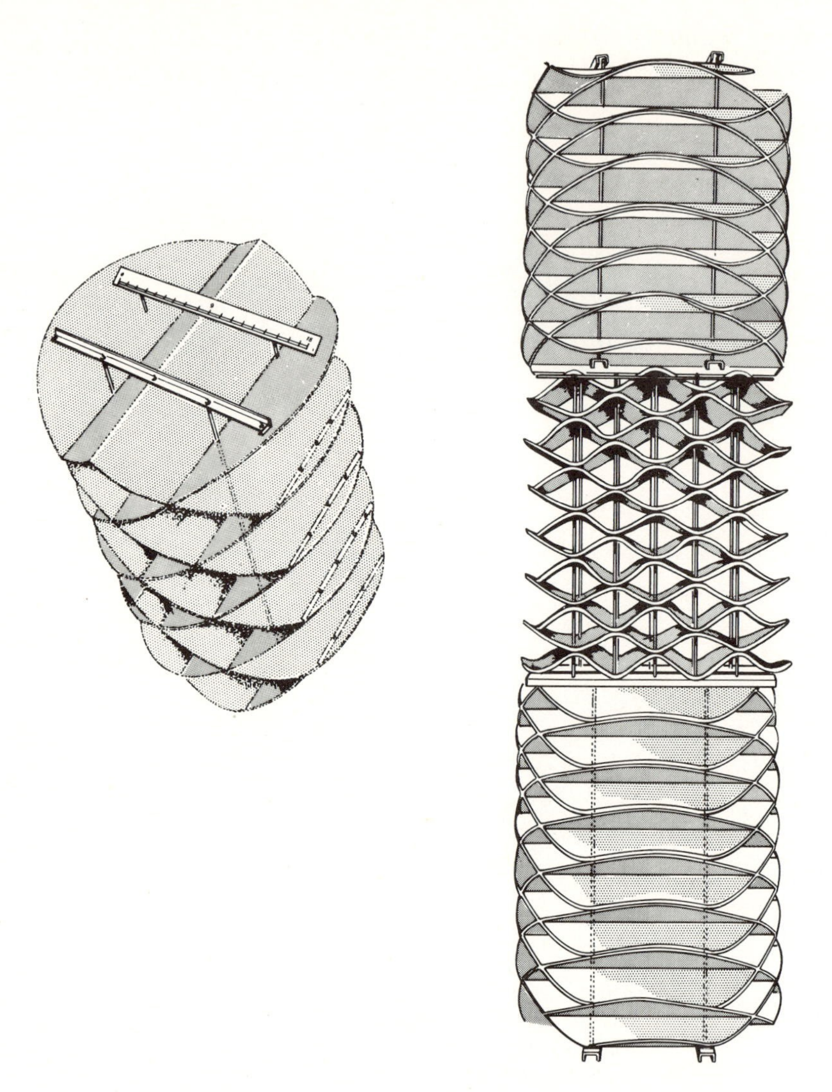

FIGURE 15.5 Spraypak. (*Courtesy Denholme, Inc., Process Equipment Division, Englewood, N.J.*)

through the apexes of the corrugations into sections of 10 to 20 layers. These are then trimmed to fit the circular-column cross section and placed into the column, with each section alternately at right angles (with respect to the corrugations) to the other. This packing has a good contact area, low pressure drop, and provides a uniform flow pattern.

Stedman packing is made from layers of wire mesh formed into pie-shaped or rectangular segments to fit various sizes of column. It is packed by hand

and the oriented packing furnishes a uniform cross section for vapor and liquid flow. It has a large contact area and gives low pressure drop.

15.2. "PSEUDOEQUILIBRIUM STAGE" DEVICES

Where the vapor and liquid pass countercurrently through the same passages in a plate and their flow paths are not separated as is the case where downcomers are used, a form of packed column results, even though discrete plates are provided in the column. The behavior of this type of column is governed by the same factors governing the behavior of a packed column. Several of this type are in use.

Turbogrid Trays

Turbogrid trays [32, 34, 35] consist of a flat grating which extends uniformly over the cross section of the column. This grating may be continuous or a series of slots, as shown in Fig. 15.6. It is claimed that these trays gave 20

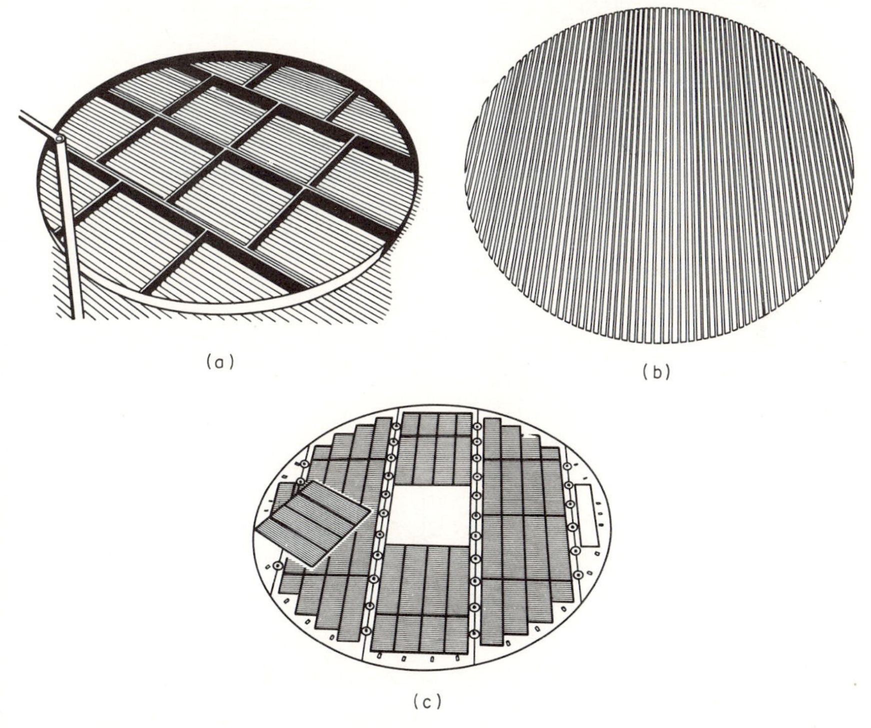

FIGURE 15.6 Turbogrid trays. (*a*) Bottom side of partly completed tray (welding and trimming of bars and sectioning of tray incomplete). (*b*) Top side of completed tray. (*c*) Turbogrid tray, with one section removed. (*From Petroleum Engineer, May, 1954, p. C-17.*)

to 100% greater capacity and 40 to 80% less pressure drop than the conventional bubble tray design. The trays were primarily designed for high-capacity service in the petroleum industry and they have the additional advantage of low cost of construction.

The Kittel Tray

The Kittel tray [1] was designed to give a horizontal flow of vapor across the contact plate surface. The liquid is carried along with the vapor and mixed with it. Two plates are mounted together, as shown in Fig. 15.7, and the pairs of plates are oriented so that one plate throws the liquid outward and the other throws it in toward the center. The liquid flows through the

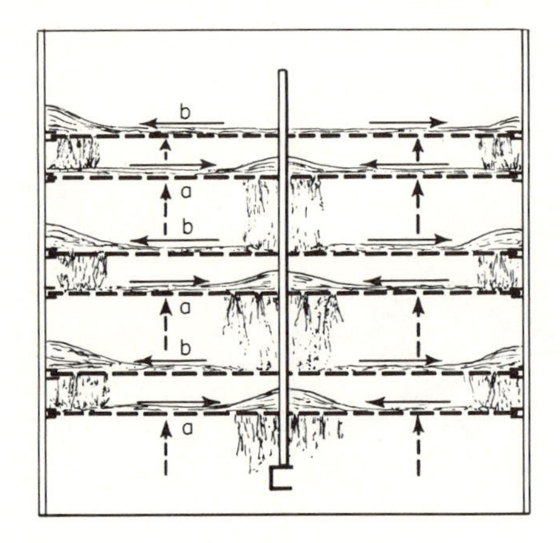

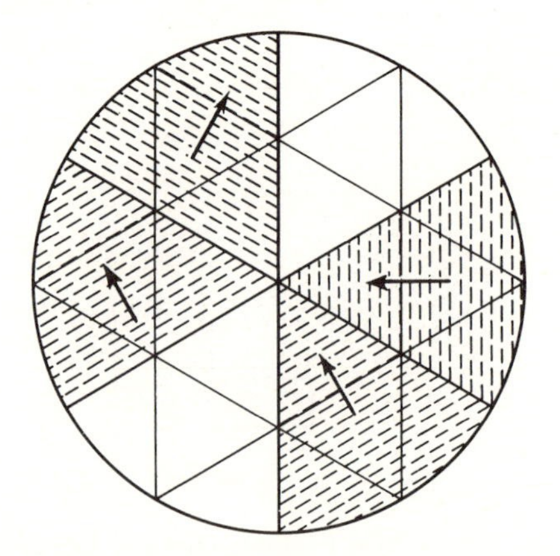

FIGURE 15.7 Kittel tray. Plate opening direct flow. [*From Chem. Eng.* (*April, 1953*), *p. 242.*]

slots on the sides of the plates where it flows to the plates below. These plates have been widely used in Europe in distillation and absorption service.

The Ripple Tray

The Ripple tray [22] illustrated in Fig. 15.8 is a perforated or sieve tray which has been bent into sinusoidal waves. The purpose of the waves is to give predetermined discharge points for the liquid flowing through the tray and to give structural strength. The hole diameters range from $\frac{1}{8}$ to $\frac{1}{2}$ in.

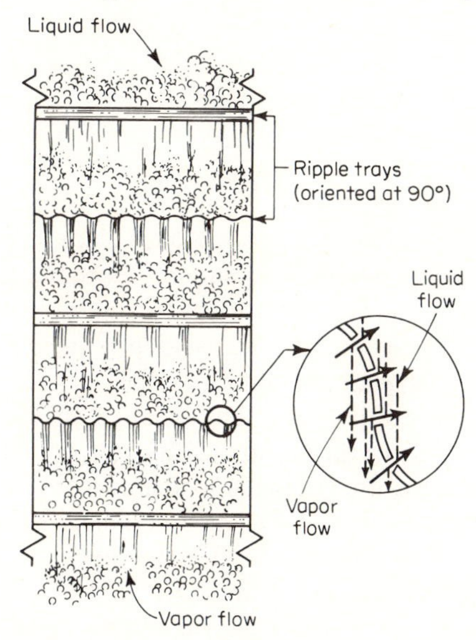

FIGURE 15.8 Ripple tray. [*From Hutchinson and Baddour, Refining Eng. (February, 1957), p. C-6.*]

and the free area from 15 to 30% of the tower cross section. The depth of the waves in the plate can be varied: shallow for liquid loads of 500 gph/sq ft, deep for 2000 gph/sq ft. The trays are placed in the column alternately oriented at 90° (with respect to the waves) with one another. These trays are supposed to have lower pressure drops, equivalent efficiencies, and greater capacity than bubble trays.

15.3. USES OF PACKED COLUMNS

Packed columns lend themselves well to (*a*) a high vapor-low liquid loading or low vapor-high liquid loading service; (*b*) where highly corrosive materials are to be handled; (*c*) where harmful contamination from metals might be encountered, such as in food and drug materials; and (*d*) where a small size would make the design of conventional fractionating plates difficult and the construction would be relatively more expensive, e.g., for laboratory and small-scale pilot-plant columns.

Since packing material made of ceramic substances, porcelain, glass, clay tile, and carbon is essentially chemically inert, one of the most important applications is in the area of handling corrosive materials.

Packed columns can range in cost from relatively inexpensive towers consisting of a cheap packing material in a steel shell to very expensive arrangements utilizing expensive packings, distribution devices, and supports. Economic considerations govern the selection of a particular packed tower much on the same basis as for other equipment.

15.4. DIFFERENTIAL MASS TRANSFER

The transfer of mass in a packed column takes place throughout the column at the interface between the liquid and vapor. Both liquid and vapor phases are moving generally countercurrently, with the liquid making its way through the maze of passages in and around the packing to the bottom of the column and the vapor rising through the same passages to the top of the column. The compositions of the vapor and liquid change differentially as they pass through the column. Since the liquid is at its bubble point and the vapor at its dew point, the temperature gradually increases from the top of the column to the bottom. With respect to the flow characteristics, highly turbulent conditions may exist in one passageway for the vapor and liquid contact and viscous flow conditions may exist in another.

Mass-transfer mechanisms of at least two types are considered important where two phases are in contact. These are *molecular diffusion* and *eddy diffusion*. Under conditions of viscous flow, mass transfer by molecular diffusion is the controlling process. Under turbulent conditions, the controlling process is mass transfer by eddy diffusion or convective diffusion, including penetration and surface renewal effects. In the flow ranges between viscous and turbulent flow, both mechanisms contribute to the mass transfer.

It has been shown that the Schmidt number $\mu/\rho D$, which includes molecular diffusivity, represents the controlling variables in the case of viscous flow, while the terms $E_m\rho/E$, representing the ratio of eddy diffusivity to eddy viscosity, control the mass-transfer rate under turbulent conditions.

The rate of mass transfer in conventional film (molecular) diffusion can be expressed as

$$\begin{aligned} Q &= k_y(y - y_i)A = k_x(x_i - x)A \\ &= K_y(y - y_L^*)A = K_y(x_G^* - x)A \end{aligned} \tag{15.1}$$

where Q = moles of material transferred per unit time, unit area of interface
k_y = mass-transfer coefficient moles per unit time, Δy
k_x = mass-transfer coefficient moles per unit time, Δx
K_y = over-all mass-transfer coefficient based on gas concentrations
K_x = over-all mass-transfer coefficient based on liquid concentrations
y,x = gas- and liquid-phase concentrations, respectively
y_i,x_i = gas- and liquid-phase concentrations at the interface
y^*,x^* = gas- and liquid-phase concentrations corresponding to the liquid- and gas-phase concentration (equilibrium values)

In applying the relations for mass transfer at point conditions to a packed column, the differential equations for the point condition must be integrated over the entire packed section (Fig. 15.9).

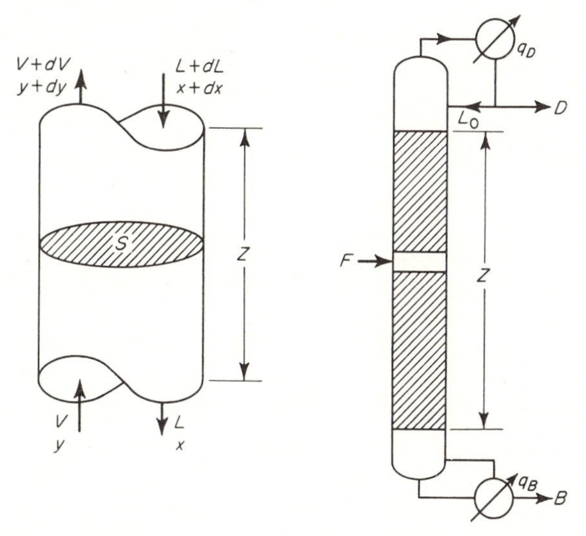

FIGURE 15.9 Schematic of packed column and differential section.

A component material balance around the differential section gives

$$Vy + Lx + dL\,dx + L\,dx + x\,dL = Vy + V\,dy + y\,dV + dV\,dy + Lx \tag{15.2}$$

Neglecting the product of the differentials,

$$L\,dx + x\,dL = V\,dy + y\,dV \tag{15.2a}$$

The over-all balance gives

$$V + L + dL = V + dV + L \tag{15.3}$$

$$dL = dV \tag{15.4}$$

Combining,

$$L\,dx + dV = V\,dy + y\,dV \tag{15.5}$$

$$L\,dx - V\,dy = dV(y - x) \tag{15.6}$$

Assuming that the moles of vapor remain constant in the rectifying section and are also constant in the stripping section,

$$\frac{L}{V} = \frac{dy}{dx} \qquad \text{operating line (rectifying section)} \tag{15.7}$$

and

$$\frac{\bar{L}}{\bar{V}} = \frac{dy}{dx} \qquad \text{operating line (stripping section)} \tag{15.8}$$

If the diffusion equations are written in differential form,

$$dQ = k_y(y - y_i)\,dA \tag{15.9a}$$

or

$$dQ = k_x(x_i - x)\,dA \tag{15.9b}$$

or

$$dQ = K_y(y - y_L^*)\,dA \tag{15.9c}$$

or

$$dQ = K_x(x_V^* - x)\,dA \tag{15.9d}$$

$$dA = aS\,dZ \tag{15.10}$$

where a = area of interface per cubic foot of packing
S = cross-sectional area of packed column
Z = height of packed section, ft

Combining the material balance and rate equations results in the following:
For the vapor phase,

$$dQ = k_y(y - y_i)aS\,dZ = V\,dy \tag{15.11}$$

For the liquid phase,

$$dQ = k_x(x_i - x)aS\,dZ = L\,dx \tag{15.12}$$

Over-all gas,

$$dQ = K_y(y - y_L^*)aS\,dZ = V\,dy \tag{15.13}$$

Over-all liquid,

$$dQ = K_x(x_V^* - x)aS\,dZ = L\,dx \tag{15.14}$$

If the mass-transfer constants can be considered constant over the conditions in the packed section to which they are applied,

$$\int_0^Z \frac{K_y aS\,dZ}{V} = \int_{y_1}^{y_2} \frac{dy}{y - y_L^*} \tag{15.15}$$

$$\frac{K_y aSZ}{V} = \int_{y_1}^{y_2} \frac{dy}{y - y_L^*} \tag{15.16}$$

$$\frac{K_x aSZ}{L} = \int_{x_1}^{x_2} \frac{dx}{x_V^* - x} \tag{15.17}$$

The numerical value of the right-hand side of the equation is designated as the *number of transfer units* based upon vapor phase, liquid phase, over-all vapor, or over-all liquid, i.e., N_V, N_L, N_{OV}, N_{OL} [8, 9].

Similarly, the value of the integral divided into the height of the packed section is called the *height of a transfer unit* $H_V = Z/N_V$,

$$H_L = \frac{Z}{N_L} \qquad H_{OV} = \frac{Z}{N_{OV}} \qquad H_{OL} = \frac{Z}{N_{OL}} \tag{15.18}$$

Colburn [8] related the number of transfer units and the height of transfer unit to composition in terms of an *absorption factor* defined as

$$\frac{mG_m}{L_m} \quad \text{or} \quad \frac{mV}{L}$$

where m is the slope of the equilibrium line and G_m, L_m, V, and L are total moles of the two phases. Figures 15.10 and 15.11 relate Colburn's N_{OG} and N_p to the absorption factor for dilute solutions only.

Since the K values vary with temperature and the flow rates, the H values, which are less sensitive to temperature and flow rates, form a better design basis. The relations between the over-all and film H's are as follows:

$$H_{OV} = H_V + m\frac{V}{L}H_L \tag{15.19}$$

$$H_{OL} = H_L + \frac{L}{mV}H_V \tag{15.20}$$

$$H_{OV} = H_{OL}\frac{mV}{L} \tag{15.21}$$

Equations (15.19) to (15.21) apply only to dilute concentrations.

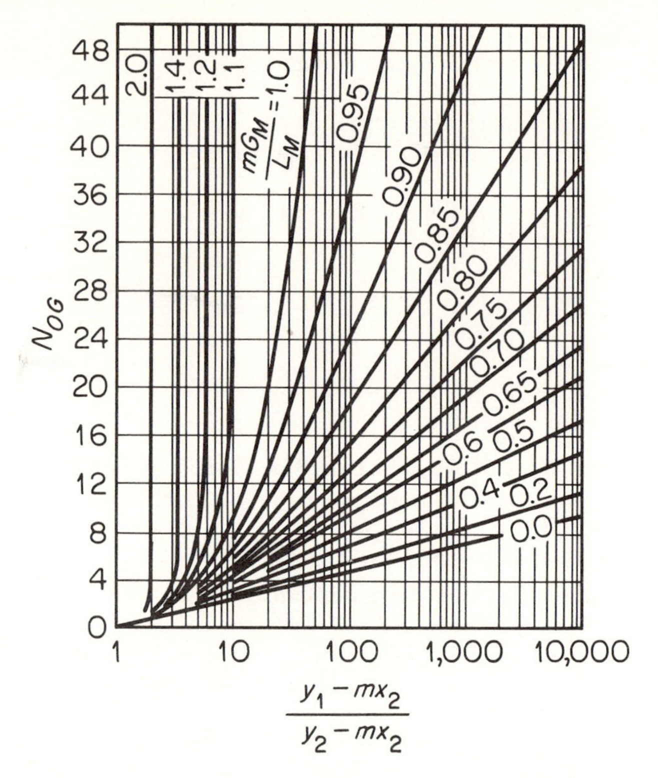

FIGURE 15.10 Number of transfer units in an absorption column. [*From Colburn, Ind. Eng. Chem.*, **33**:459 (1941).]

For concentrated solutions,

$$H_{OV} = H_V + \frac{mV}{L} \frac{x_{B_m}}{y_{B_m}} H_L \tag{15.22}$$

$$H_{OL} = H_L + \frac{L}{mV} \frac{y_{B_m}}{x_{B_m}} H_V \tag{15.23}$$

In any case the use of any relation involving m, the slope of the equilibrium line, is restricted to the range of concentrations wherein the slope is constant or where some kind of an average slope will serve satisfactorily.

15.5. EMPIRICAL DESIGN RELATIONSHIPS

Yu and Coull [45] developed a relationship, empirical in nature, which relates minimum reflux, finite reflux, minimum N_{OV}, and operating N_{OV},

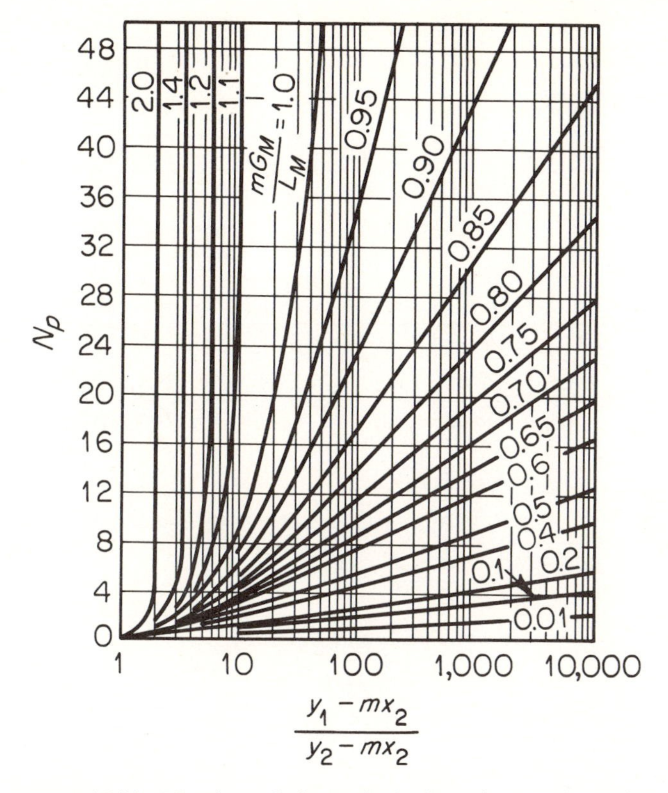

FIGURE 15.11 Number of theoretical plates in an absorption column. [*From Colburn, Ind. Eng. Chem., 33:459 (1941).*]

similar to the Brown and Martin correlation for plates or stages. They claim a maximum deviation of not more than 6%. The relation is strictly for systems having a constant relative volatility over the range of temperatures encountered in the column.

$$N_{OV} = \int_{y_1}^{y_2} \frac{dy}{y^* - y} \tag{15.24}$$

Equation (15.24) was integrated to give Eq. (15.25), assuming total reflux and constant relative volatility.

$$N_{OV,\min} = \frac{1}{\alpha - 1}\left(\ln \frac{x_D}{x_B} + \alpha \ln \frac{1 - x_B}{1 - x_D}\right) \tag{15.25}$$

Equation (15.24) was also integrated for the rectifying and for the stripping sections, respectively, assuming a saturated liquid feed.

$$N_{OV} = \frac{2R + B}{2(B^2 + 4Ax_D)^{1/2}}$$
$$\times \ln \frac{[2Ax_D + B - (B^2 + 4Ax_D)][2Ax_F + B + (B^2 + 4Ax_D)^{1/2}]}{[2Ax_F + B - (B^2 + 4Ax_D)][2Ax_D + B + (B^2 + 4Ax_D)^{1/2}]}$$
$$+ \frac{2R' + B'}{2(B'^2 - 4A'x_B)^{1/2}}$$
$$+ \ln \frac{[2A'x_F + B' - (B'^2 - 4Ax_B)^{1/2}][2A'x_F + B' + (B'^2 - 4A'x_B)^{1/2}]}{[2A'x_B + B' + (B'^2 - 4Ax_B)^{1/2}][2A'x_B + B' - (B'^2 - 4Ax_B)^{1/2}]}$$
$$+ \tfrac{1}{2} \ln \frac{Ax_F{}^2 + Bx_F - x_D}{Ax_D{}^2 + Bx_D - x_D} + \tfrac{1}{2} \ln \frac{A'x_B{}^2 + B'x_B + x_B}{A'x_F{}^2 + B'x_F + x_B} \tag{15.26}$$

where

$$A = R(1 - \alpha) \tag{15.27}$$

$$B = R(\alpha - 1) - x_D(\alpha - 1) + \alpha \tag{15.28}$$

$$R' = \frac{F - R}{F - 1} \tag{15.29}$$

$$A' = R'(1 - \alpha) \tag{15.30}$$

$$B' = R'(\alpha - 1) + x_B(\alpha - 1) - \alpha \tag{15.31}$$

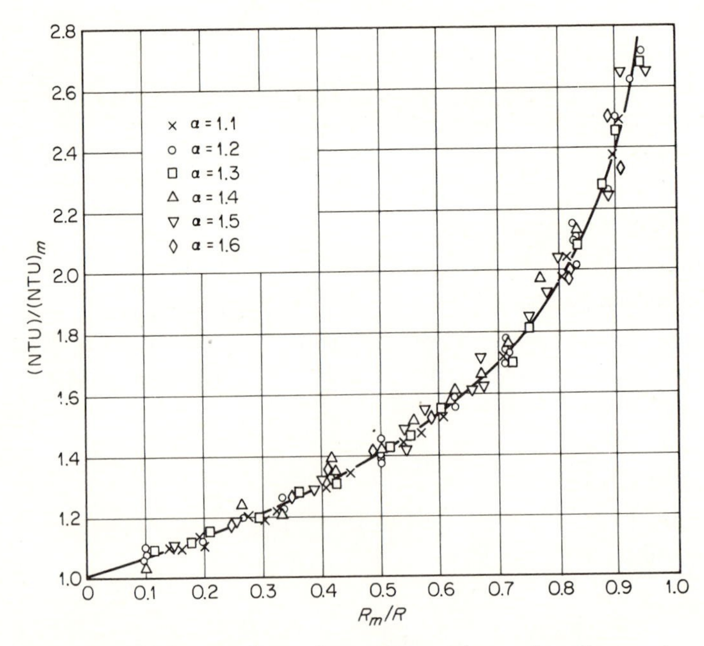

FIGURE 15.12 Number of transfer units and reflux ratio. [*From Yu and Coull, Chem. Eng. Progr.*, **46**:89 (1950).]

The correlation is based on the following ranges:

$$\alpha = 1.1\text{–}1.6 \qquad x_F = 0.3\text{–}0.6 \qquad x_D = 0.7\text{–}0.99$$

$$x_B = 0.01\text{–}0.3 \qquad R = 3\text{–}200 \qquad R_m = 2.47 \text{ to } 28.4$$

$$N_{OV,\min} = 8.43\text{–}53.7 \qquad N_{OV} = 9.09\text{–}155$$

For nonideal systems where relative volatility is not constant, an empirical relationship was developed by Yu and Coull [45] which represented the data with less than a 2% maximum deviation between a concentration range of 0.04 to 0.96 mole fraction.

$$\frac{y}{1-y} = \alpha\left(\frac{x}{1-x}\right)^b \tag{15.32}$$

If $b = 1.0$, α is constant.

Based on the assumption that the number of equilibrium stages and the number of transfer units are the same (only strictly true when $mV/L = 1.0$), in order to calculate $N_{OV,\min}$ when the system is nonideal, it was assumed that α was constant over one theoretical plate. This was done for a series of n plates and the results were summed for $N_{\min} = N_{OV,\min}$.

Thus,

$$N_{OV} = -\ln\frac{x_D}{x_B} + 2\sum_1^n \frac{\alpha_B{}^{b^n}}{\alpha_B^{b^n+1}} \tag{15.33}$$

This is not an exact solution. For a small number of transfer units, a stepwise summation is possible. For a large number of transfer units they recommend the approximation

$$\sum_1^n = n\,\frac{\alpha_B(1+b+b^2+b^3+\cdots+b^{n-1})^{1/n}}{\alpha_B[(1+b+b^2+b^3+\cdots+b^{n-1})^{1/n}+1]} \tag{15.34}$$

As an alternative to the use of the graphical correlation, the N_{OV} may be calculated through a method described as follows:

By defining an average volatility α_{av} as

$$\frac{x_D}{1-x_D} = \alpha_{\text{av}}^n\,\frac{x_B}{1-x_B} \tag{15.35}$$

which relates the terminal compositions with an average α and number of plates, Eq. (15.35) can be written for N_{OV} as follows:

$$N_{OV} = \ln\frac{x_D}{x_B} + \frac{2n\alpha_{\text{av}}}{\alpha_{\text{av}}+1} \tag{15.36}$$

$$\alpha_{\text{av}} = \left(\frac{x_D}{1-x_D}\,\frac{1-x_B}{x_B}\right)^{1/n} \tag{15.37}$$

where n is number of equilibrium stages.

$$n = \frac{1}{\ln b} \ln \frac{\dfrac{1}{b-1} \ln a' + \ln \dfrac{x_D}{1 - x_D}}{\dfrac{1}{b-1} \ln a' - \ln \dfrac{1 - x_B}{x_B}} \tag{15.38}$$

where a', b' are constants for the particular system.

The agreement between calculated results and the results obtained graphically is very good.

Height Equivalent to a Theoretical Plate

The concept of a height of packed section equivalent to one theoretical plate with regard to a particular separation was introduced to enable comparison of the efficiency of packed columns and plate columns. Also it serves as a means of comparison through basic mass-transfer theory. Height equivalent to a theoretical plate (HETP) is defined by

$$\text{HETP} = \frac{Z}{n} \tag{15.39}$$

For a binary separation when the equilibrium and operating lines are straight and parallel, i.e., $mG_m/L_m = 1.0$, HETP $= H_{OV}$, or the height equivalent to a theoretical plate equals the height of a transfer unit. When they are straight but not parallel,

$$\frac{\text{HETP}}{H_{OV}} = \frac{(mG_m/L_m) - 1}{\ln (mG_m/L_m)} \tag{15.40}$$

Gerster et al. [19] studied the factors relating Murphree plate and over-all column efficiency and the concept of height of a transfer unit H. They introduced the concept of plates equivalent to a theoretical plate (PETP) and plates per transfer unit (PTU).

Empirical formulas were developed by a number of investigators for the prediction of HETP from experimental data and properties of the packing and system. Murch [31], Ellis [16], Hands and Whitt [20], Cornell et al. [10], and Clay et al. [7] considered various factors affecting the HETP in packed distillation columns. Eckert and Walter [15] discussed some of the methods proposed for prediction of HETP and compared the calculated and experimental results [12, 13] developed in their laboratories on the distillation of the isooctane–toluene system with two sizes and types of packing. They concluded that no method was completely adequate to predict accurately the values of HETP for all packing systems and conditions.

The method of Cornell et al. [10] proposes Eqs. (15.41) and (15.42) to compute H_V and H_L:

$$H_V = \frac{\psi \mathrm{Sc}_V^{0.5}}{(L, f_1, f_2, f_3)^n} \left(\frac{D}{12}\right)^x \frac{Z^{0.33}}{10} \tag{15.41}$$

$$H_L = \varphi \mathrm{Sc}_L^{0.5} C \left(\frac{Z}{10}\right)^{0.15} \tag{15.42}$$

where C = correction factor for H_L at high rates
D = diameter of column, in.
Z = packed height between distributors, ft

$\mathrm{Sc}_V = \dfrac{\mu_V}{\rho_V D_V}$ = vapor Schmidt number

$\mathrm{Sc}_L = \dfrac{\mu_L}{\rho_L D_L}$ = liquid Schmidt number

L = liquid mass velocity, lb/(sq ft)(hr)

$f_1 = \left(\dfrac{\mu_L}{\mu_W}\right)^{0.16}$ = ratio of liquid viscosity at column conditions to viscosity at water at 20°C

$f_2 = \left(\dfrac{\rho_W}{\rho_L}\right)^{1.25}$ = ratio of water density at 20°C to density of liquid at column conditions

$f_3 = \left(\dfrac{\sigma_W}{\sigma_L}\right)^{0.8}$ = ratio of water surface tension at 20°C to liquid surface tension at column conditions

x = 1.24 for Raschig rings

x = 1.11 for Berl saddles

$\psi = \dfrac{H_V (L, f_1, f_2, f_3)^n}{\mathrm{Sc}_V^{0.5}}$ determined from experimental data or from Figs. 15.13 and 15.14

$\phi = \left(\dfrac{H_L}{\mathrm{Sc}_L}\right)^{0.5}$ determined from experimental data or from Figs. 15.15 and 15.16

n = 0.6 for Raschig rings

n = 0.5 for Berl saddles

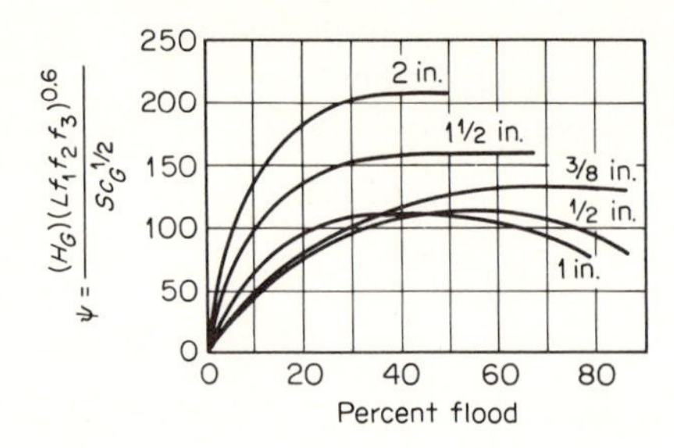

FIGURE 15.13 H_G correlation for various size Raschig rings; 10-ft packed height, 1-ft column diameter. [*From Cornell et al., Chem. Eng. Progr.*, ***56***(*7*)*:68; 49* (*1960*).]

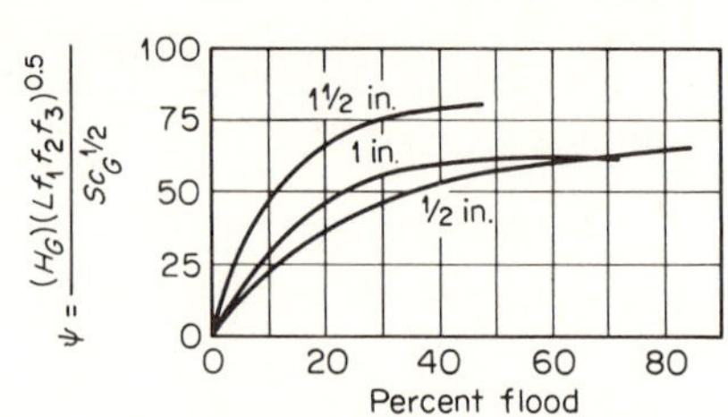

FIGURE 15.14 H_G correlation for various sizes of Berl saddles; 10-ft packed height, 1-ft column diameter. [*From Cornell et al., Chem. Eng. Progr.*, ***56***(*7*)*:68; 49* (*1960*).]

Figures 15.13 and 15.14 are plots relating experimental H_G data, mostly air and water, 1.0 ft of diameter column with percent flood, and various sizes of ring and saddle packings in terms of ψ. Figures 15.15 and 15.16 are similar relationships for ϕ, based on liquid transfer unit data.

The authors recommend that values of ϕ or φ be read from the suitable figure, to substitute the value into Eq. (15.11) or (15.42) and to calculate H_V or H_L.

The authors show the average absolute deviations of predicted and observed values of H_V for the different packing sizes and types range from 5.2 to 34.5%.

Chilton and Colburn [6] used the following relation to calculate the number of transfer units for a packed column. In use of this equation α was assumed to be constant and represented an average value between the top and bottom

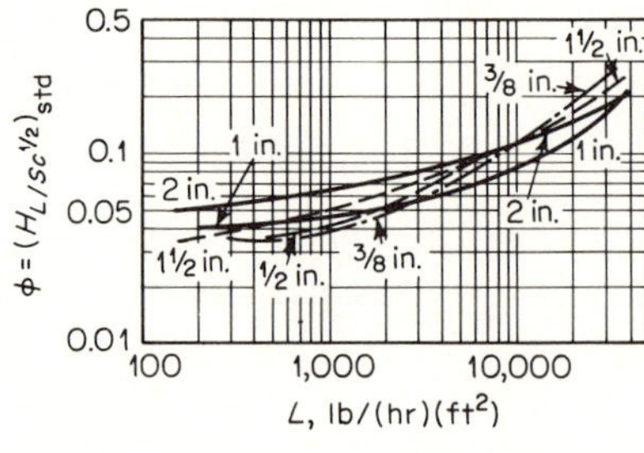

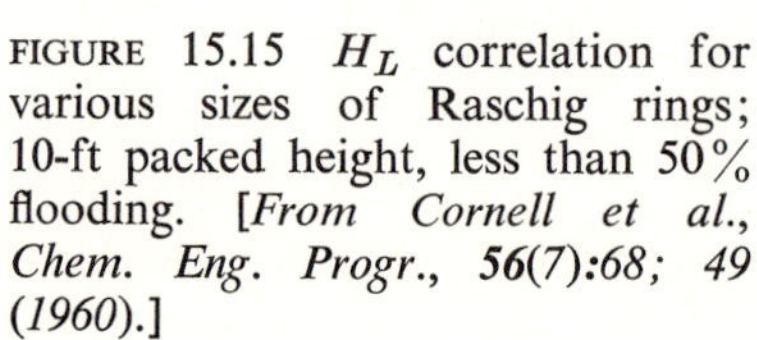

FIGURE 15.15 H_L correlation for various sizes of Raschig rings; 10-ft packed height, less than 50% flooding. [*From Cornell et al., Chem. Eng. Progr.*, ***56***(*7*)*:68; 49* (*1960*).]

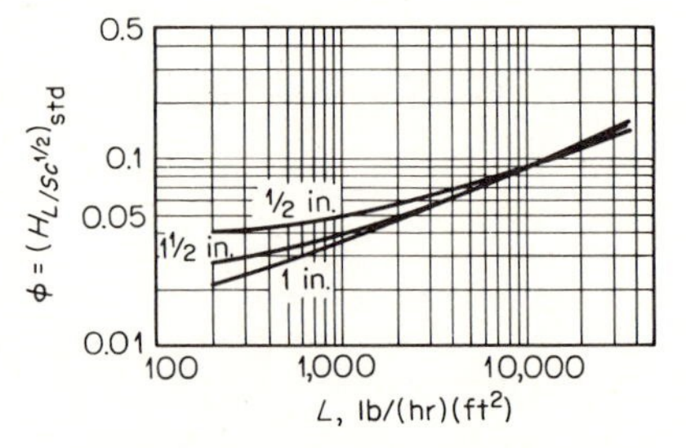

FIGURE 15.16 H_L correlation for various sizes of Berl saddles; 10-ft packed height, less than 50% flooding. [*From Cornell et al., Chem. Eng. Progr.*, ***56***(*7*)*:68; 49* (*1960*).]

conditions of the column. Also it applies only to total reflux

$$N_{OV} = \frac{1}{\alpha - 1} \ln \frac{y_D(1 - y_B)}{y_B(1 - y_D)} + \ln \frac{1 - y_B}{1 - y_D} \tag{15.43}$$

Simon and Rau [43] derived an equation for evaluation of the height of packing needed for a given service.

$$Z = \frac{G}{k_G a} N_{G_1} + \frac{L}{k_L a} N_{G_2} = z_1 + z_2 \tag{15.44}$$

$$N_{G_1} = \int_{y_1}^{y_D} \frac{dy}{y^* - y} \qquad N_{G_2} = \int_{y_1^*}^{y_D^*} \frac{dy}{y^* - y} \tag{15.45}$$

where y_1 = composition of the vapors at the bottom of the rectifying section of the column (above feed plate)

y_1^* = composition of vapors in equilibrium with the liquid leaving the rectifying section, i.e., with x_1

y_D = composition of distillate = x_D

y_D^* = composition of vapors in equilibrium with x_D

This method involves a knowledge of $k_G a$ and $k_L a$. Since these are not always available for a given system and packing, the method can be used only for the systems and packings for which data are available. Shulman et al. [37–41] have reported $k_G a$ and $k_L a$ data on a number of systems and packings which aid in the use of the above method and others involving film transfer coefficients.

Example 15.1

For the binary system hexane–benzene at atmospheric pressure, α varies from 1.14 at the top of the column to 1.54 at the bottom. Assuming the feed to consist of 60% hexane and 40% benzene and making a distillate product containing 95% of the hexane and 5% of the benzene in the feed, calculate the following:

(*a*) The number of transfer units at total reflux [Eq. (15.43)].

(*b*) Determine the height of packing for total reflux.

(*c*) For a feed of 160 moles/hr and a reflux ratio $L/D = 2.5$, $(L/D)_{\min} = 1.05$, calculate the height of the packed section Z, for the separation using Eq. (15.43).

(*d*) Determine H_{OV} for both total and actual reflux.

(*e*) Determine H_{OL} for both total and actual reflux.

The following data apply:

Packing = 1-in. Raschig rings

$a = 62.7$ sq ft/cu ft

Column diameter = 55 ft

$K_{0v} = 0.106$ mole/(hr)(sq ft), Δy (assumed)

Solution

Material Balances

Compound	Feed		Distillate		Bottoms	
	Moles	x_F	Moles	x_D	Moles	x_B
Hexane	96.0	0.60	91.2	0.966	5.8	0.087
Benzene	64.0	0.40	3.2	0.034	60.8	0.913
			94.4	1.000	66.6	1.000

$$\frac{L}{D} = 2.5 \qquad D = 94.4 \qquad L = 236 \text{ moles}$$

$$V = 236 + 94.4 = 330.4 \qquad \frac{L}{V} = \frac{236}{330.4} = 0.715$$

$$\bar{L} = 160 + 236 = 396 \qquad \frac{\bar{L}}{\bar{V}} = \frac{396}{330.4} = 1.2$$

For the rectifying section:

$$m = \left(\frac{y}{x}\right)_{\text{av}}$$

$$y = \frac{\alpha x}{1 + (\alpha - 1)x} = \frac{(1.14)(0.8)}{1 + (0.14)(0.8)} = \frac{0.912}{1.12} = 0.815$$

$$\frac{y}{x} = \frac{0.815}{0.8} = 1.02 \qquad y = \frac{(1.34)(0.6)}{1 + (0.34)(0.6)} = \frac{0.805}{1.202} = 0.67$$

$$\frac{y}{x} = \frac{0.67}{0.60} = 1.11$$

$$\alpha_{\text{av}} = 1.24 \qquad m_{\text{av}} = 1.06$$

For the stripping section:

$$y = \frac{(1.54)(0.10)}{1 + (0.54)(0.10)} = \frac{0.154}{1.054} = 0.146$$

$$\frac{y}{x} = \frac{0.146}{0.10} = 1.46$$

$$y = \frac{(1.54)(0.40)}{1 + (0.54)(0.4)} = \frac{0.615}{1.216} = 0.505$$

$$\frac{y}{x} = \frac{0.505}{0.4} = 1.26$$

$$\alpha_{\text{av}} = 1.44 \qquad m_{\text{av}} = 1.36$$

(*a*) Assume $\alpha = 1.34$:

$$N_{OV} = \frac{1}{\alpha - 1} 2.303 \log \frac{y_D(1 - y_B)}{y_B(1 - y_D)} + 2.303 \log \frac{1 - y_B}{1 - y_D}$$

$$= \frac{1}{1.34 - 1} 2.303 \log \frac{0.966}{0.087} \frac{0.913}{0.034} + 2.303 \log \frac{0.913}{0.034}$$

$$= 6.79 \log 208 + 2.303 \log 26.8$$

$$= (6.79)(2.474) + (2.303)(1.428) = 16.8 + 3.3 = 20.1$$

(*b*)
$$Z = \frac{VN_{OV}}{K_{Oy}aS} = \frac{(20.1)(330.4)}{(0.106)(62.7)(23.8)} = 42 \text{ ft}$$

(*c*) $N_{V,\text{actual}}$ will be determined in two parts, assuming constant m and α for the rectifying section and the stripping section.

Rectifying section: $m = 1.06$, $\alpha = 1.24$.

$$N_{OV} = \frac{1}{1.24 - 1} 2.303 \log \frac{0.965(0.40)}{0.60(0.034)} + 2.303 \log \frac{0.4}{0.034}$$

$$= 9.6 \log 18.9 + 2.303 \log 11.75$$

$$= (9.6)(1.276) + (2.303)(1.07) = 12.3 + 2.47 = 14.77$$

Stripping section: $m = 1.36$, $\alpha = 1.44$.

$$N_{OV} = \frac{1}{1.44 - 1} 2.303 \log \frac{0.60(0.913)}{0.087(0.40)} + 2.303 \log \frac{0.913}{0.40}$$

$$= 5.23 \log 15.8 + 2.303 \log 2.28$$

$$= (5.23)(1.198) + (2.303)(0.358) = 6.25 + 0.825 = 7.75$$

$$N_{OV,\text{total}} = 14.77 + 7.75 = 22.52$$

$$Z = 42 \frac{N_\infty}{N_a} = 42 \frac{20.1}{22.52} = 37.5 \text{ ft}$$

(*d*)
$$H_{OV,\text{total}} = \frac{Z}{N_{OV}} = \frac{42}{20.1} = 2.09 \qquad N_{OV,\text{actual}} = \frac{37.5}{22.52} = 1.67$$

(*e*)
$$H_{OL,\text{actual}} = H_{OV} \frac{L_m}{mV_m}$$

$$= 2.09 \frac{1}{m} = 2.09 \frac{1}{(1.06 + 1.36)/2} = 1.74$$

$$H_{OL,\text{actual}} = H_{OV,\text{actual}} \frac{1}{1.36} 0.82 = \frac{1.67}{1.36} 0.82 = 1.010$$

Example 15.2

Determine the value of N_{OV} for the system in Example 15.1, where α varies linearly between 0% hexane ($\alpha = 1.14$) to 100% hexane ($\alpha = 1.54$), using (*a*) Graphical correlation; (*b*) Eqs. (15.35) and (15.36), $(L/D)_m = R_m = 1.05$.

Solution

(*a*)
$$\frac{R_m}{R} = \frac{1.05}{2.5} = 0.42 \qquad \alpha_{av} = 1.34$$

From Fig. 15.12, $N_{OV}/N_{OV,\min} = 1.33$. From Eq. (15.25),

$$N_{OV,\min} = \frac{1}{\alpha - 1}\left(\ln\frac{x_D}{x_B} + \alpha \ln\frac{1 - x_B}{1 - x_D}\right)$$

$$= \frac{1}{0.34}\left(\ln\frac{0.965}{0.087} + 1.34 \ln\frac{1 - 0.087}{1 - 0.965}\right)$$

$$= \frac{1}{0.34}[2.41 + (1.34)(2.33)] = \frac{5.54}{0.34} = 16.3$$

$$N_{OV,\min} = 16.3 \qquad N_{OV} = (16.3)(1.33) = 21.7$$

(*b*)
$$\frac{X_D}{1 - x_D} = \alpha_{av}{}^n \frac{x_B}{1 - x_B}$$

$$\alpha_{av}{}^n = \frac{x_D/(1 - x_D)}{y_B/(1 - x_B)} = \frac{0.965/0.034}{0.087/0.913}$$

$$\alpha_{av}{}^n = \frac{28.4}{0.0955} = 298$$

$$\alpha_{av}{}^n = \frac{1.14 + 1.54}{2} = \frac{2.68}{2} = 1.34$$

$$n = 19.5$$

$$N_{OV} = \ln\frac{x_D}{x_B} + \frac{2_n\alpha_{av}}{\alpha_{av} + 1} = \ln\frac{0.965}{0.087} + \frac{2(19.5)(1.34)}{1.34 + 1}$$

$$= 2.42 + 22.3 = 24.72$$

15.6. CAPACITY DESIGN OF PACKED TOWERS

Packed towers, like all other phase-contacting equipment, have limitations in their capacity to handle vapor and liquid loads. In distillation, the ratio of vapor and liquid is fixed by reflux ratio. The operating limits, therefore, lie between the minimum reflux ratio and total reflux. For a given product rate (e.g., distillate rate), the actual quantities of vapor and liquid are fixed in relation to the minimum or total reflux. This limitation is based upon producing the specified product.

For a given column and packing, for a given system and separation, the limits may be shown as illustrated in Fig. 15.17.

In addition to the operating limits imposed by separation of the desired components, in terms of interrelated liquid and vapor quantities, there are

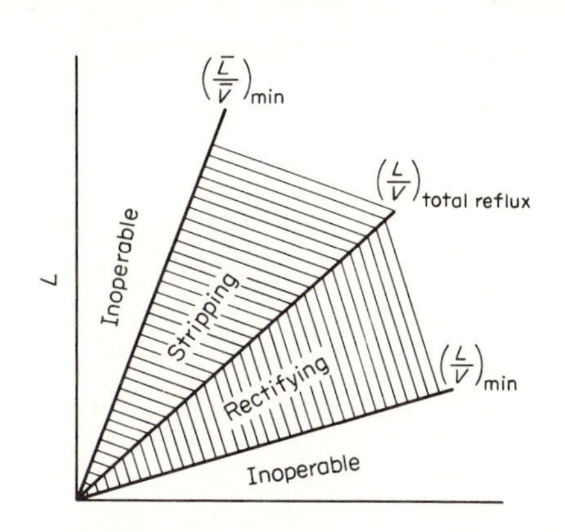

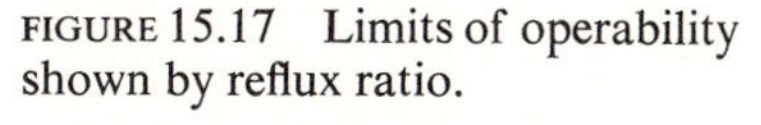
FIGURE 15.17 Limits of operability shown by reflux ratio.

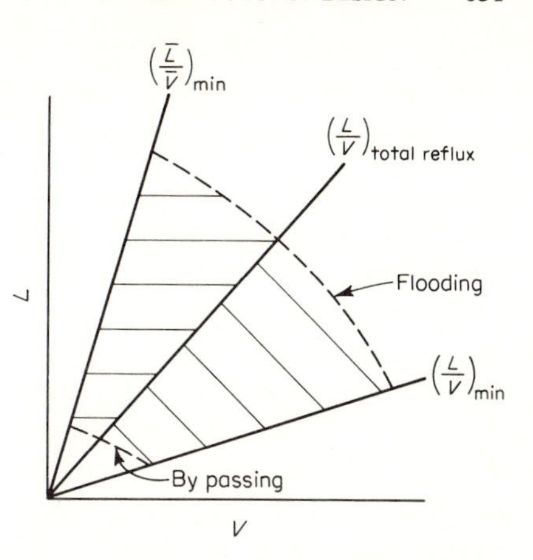

FIGURE 15.18 Limits of operability based on flooding and bypassing.

other limits to capacity. On the high end of liquid and vapor flow rates the condition of flooding is encountered. In this situation the liquid backs up the column and fills all of the flow space between the packing. On the low end of packed-column operation in terms of L or V, or both, there is insufficient flow to cover the packing with liquid, and bypassing of liquid and vapor takes place with very low efficiency of mass transfer. Including these limits, Fig. 15.18 shows schematically the maximum and minimum operating range of any packed column or (similar device) for distillation.

There are still further capacity limits imposed by poor efficiency of contact at the low liquid and vapor flow rates and by excessive pressure drop at the high rates of flow. When these are plotted on a performance chart, the area A indicated on Fig. 15.19 is the performance area.

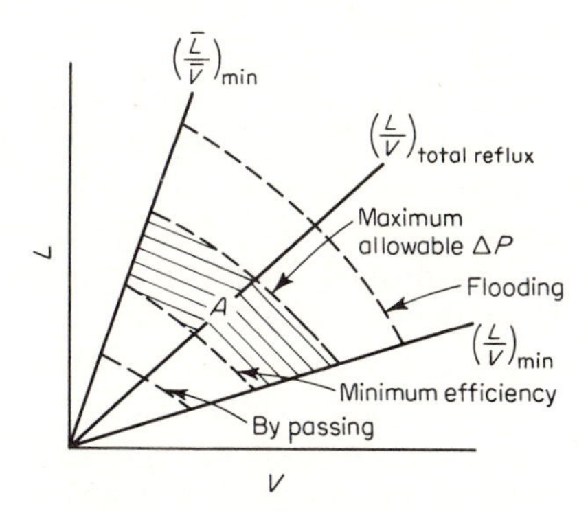

FIGURE 15.19 Performance chart for packed columns.

Since packed-column design is dependent on the flooding and pressure drop characteristics of the column, packing, system, and operating conditions, it is necessary to evaluate these factors quantitatively.

Flooding

Flooding will occur in any device wherein the liquid and vapor flow countercurrently through the same space. In a packed distillation column the energy potential causing the vapor to flow up the column is derived from the expansion of the distilling system in changing from the liquid to the vapor state. That which causes the liquid to flow down the column is the hydraulic head or height potential. When the energy potentials causing the flow upward and downward are balanced, the point of flooding is reached. When the energy potential of the vapor is greater than that of the liquid, the liquid backs up and the column floods (Fig. 15.20).

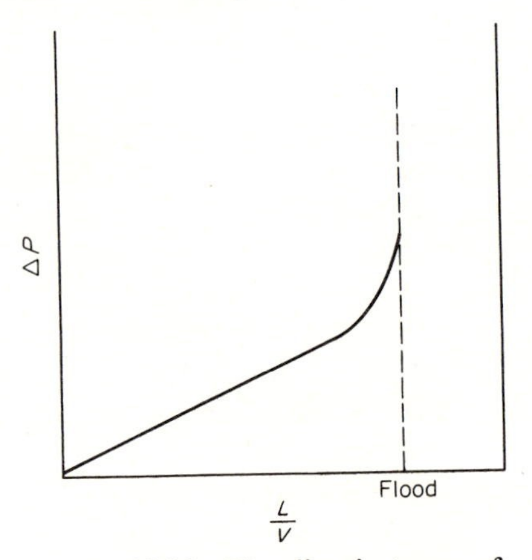

FIGURE 15.20 Flooding in terms of Δp and reflux ratio.

The point of flooding in terms of pressure drop is indicated by plotting the pressure drop across a packed-column section, for a given system, as a function of, e.g., reflux ratio, at a constant vapor or liquid rate. Flooding is a function of a number of variables and no satisfactory correlation of these variables with flooding based upon fundamental theory has been offered. Empirical correlations of the variables have been presented by Bain and Hougen [2], Bertetti [4], Lerner and Grove [25], Lobo et al. [26], Sherwood et al. [36], Leva [24], Zenz [46], and Eckert [11].

The form of the Sherwood et al. correlation has generally been accepted as the most useful. It was modified by Eckert et al. [13] and is shown in Fig. 15.21.

Hoffing and Lockhart [21] developed a single generalized flooding correlation for both liquid-liquid and liquid-vapor countercurrent flow in packed

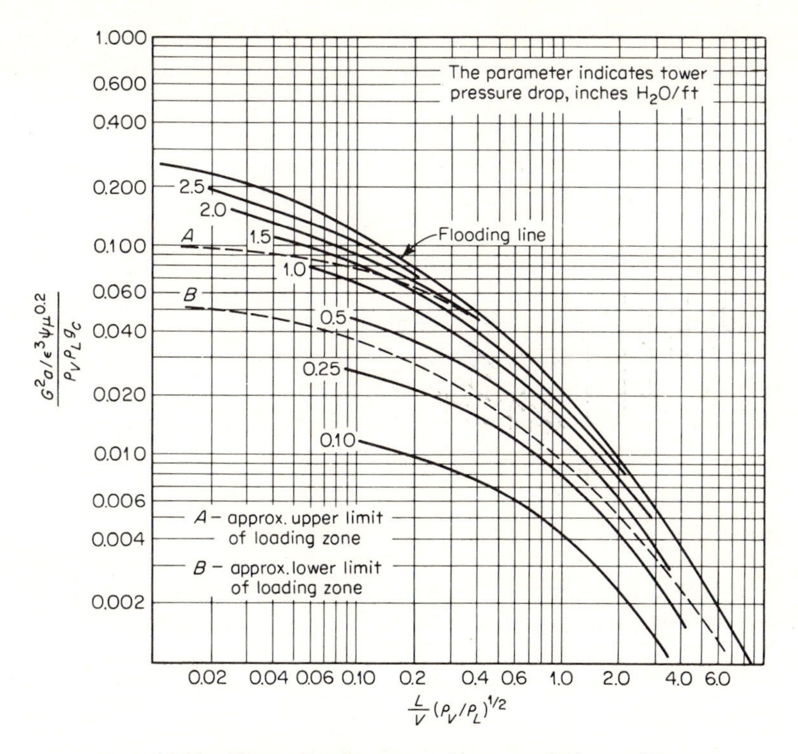

FIGURE 15.21 Generalized pressure-drop correlation. (*Courtesy U.S. Stoneware Company, Akron, Ohio.*)

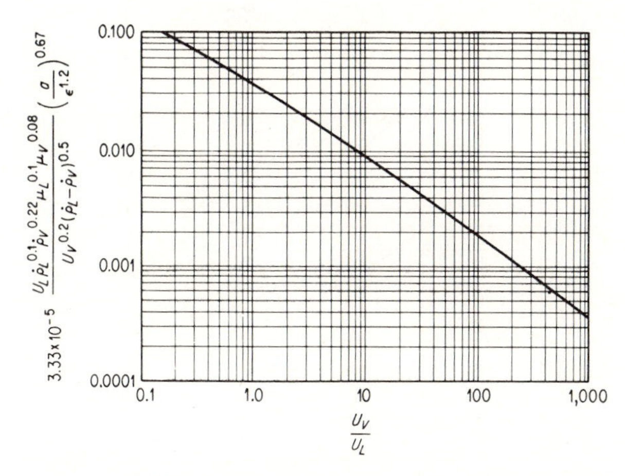

FIGURE 15.22 Correlation for two-phase flow-packed towers; curve is limiting locus for flow rates. [*Courtesy F. J. Lockhart Fluor Corporation, Chem. Eng. Progr.*, **50:**94 (*1954*).]

beds. The curve is shown on Fig. 15.22. When the correlation is used for liquid-liquid extraction, a term which is the ratio of the interfacial tension between the two liquids is used as a multiplier in the ordinate. This is not applicable in vapor-liquid applications since the ratio is unity.

Zenz and Eckert [47] developed a simplified design chart for packed-tower flooding rates based upon the experimental flooding rate = pressure-drop relation. The chart relates the density of vapor and liquid, the volumetric flow rates of vapor and liquid, the column diameter, viscosity of liquid, and packing factor.

The flooding line on Fig. 15.21 is used as a limit curve for the various packings. The other lines have parameters of either pressure drop or loading as designated. The value of this type of correlation is that it enables selection of conditions which will reasonably ensure operability with respect to flow rates. The line on Fig. 15.22 is the flooding line for all packings.

Pressure Drop

Pressure drop in packed distillation columns is generally related to the column operation in the same manner as it is in plate columns.

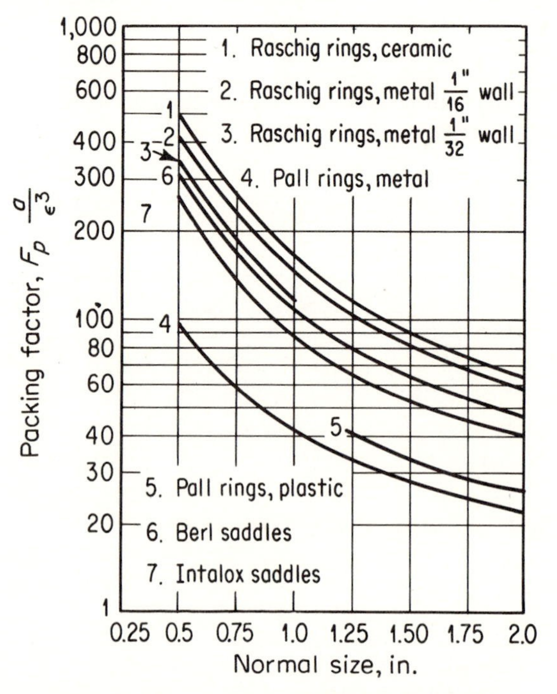

FIGURE 15.23 Average packing factors, a/ϵ^3.

TABLE 15.7 **Pressure-drop Correlations for Packed Towers***

Packing		Constants		
Type	Nominal size, in.	α	β	$\frac{a}{\epsilon^3}$ †
Raschig rings Ceramic	⅜	4.70	0.41	750
	½	3.10	0.41	640
	⅝	2.35	0.26	380
	¾	1.34	0.26	255
	1	0.97	0.25	160
	1¼	0.57	0.23	125
	1½	0.39	0.23	95
	2	0.24	0.17	65
Raschig rings Metal, 1/32 in. Wall, 1/16 in.	⅝	1.20	0.28	258 (290)
	1	0.42	0.21	115 (145)
	1½	0.29	0.20	— (83)
	2	0.23	0.135	— (57)
Pall rings, metal	⅝	0.43	0.17	71
	1	0.15	0.16	45
	1½	0.08	0.15	24
	2	0.06	0.12	17
Berl saddles	½	1.2	0.21	380
	¾	0.62	0.17	170
	1	0.39	0.17	110
	1½	0.21	0.13	65
Intalox saddles	½	0.82	0.20	265
	¾	0.28	0.16	130
	1	0.31	0.16	98
	1½	0.14	0.14	52

* Valid for liquid-gas systems below the loading point.
† a/ϵ^3 = packing factor F_p.
(*Courtesy U.S. Stoneware Company.*)

High pressure drop causes flooding (and loading), it increases the temperature level at which the column operates, and increases the temperature gradient through the column. Generally, it would be expected that, because of its effect on equilibrium relationships, efficiency of separation would be reduced. However, Berg and Popovac [3] concluded that pressure had little, if any, effect on packed-column efficiency.

Pressure drop may be estimated from Fig. 15.21 for the particular packing, system properties, and operating conditions, or it may be calculated by Eq. (15.46), presented by Leva [24], and by Table 15.7.

$$\Delta P = \alpha 10^{\beta L'} \frac{G'^2}{\rho_V} \qquad (15.46)$$

This equation was developed for air–water systems. For other liquids multiply L by ρ_{H_2O}/ρ_L.

Figure 15.23 is a plot of packing factors F_p based on experimental data. The individual curves represent different types of packing and are smoothed averages of the data. Clay et al. [7], Eckert [13, 14], and Silvey and Keller [42] found that the experimental packing factors deviated from those indicated by the averaged data in Fig. 15.23 with different systems, different packing materials, and different wall thicknesses of the packing. This indicates that pressure-drop variation is a function of more than just packing size, and experimental data should be used if possible. The packing factors from Fig. 15.23 can be used for approximation purposes.

Example 15.3

(*a*) Determine the flooding vapor rate for the system in Example 15.1 for 1-in. Intalox saddles in terms of internal reflux ratio.

(*b*) Determine the pressure drop for the system described in Example 15.1 by Fig. 15.21 and use 1-in. Intalox saddles.

(*c*) Determine the pressure drop by Eq. (15.46).

Solution

(*a*) For 1.0-in. Intalox saddles, $a = 78$ sq ft/cu ft. $\epsilon = 0.775$ cu ft/cu ft.

$$\frac{G^2(a/\epsilon^3)\psi\mu^{0.2}}{\rho_G\rho_L g_c} = \frac{G^2(78)/(0.775)^3(1/0.711)^2(0.73)^{0.2}}{(0.184)(44.2)(32.2)}$$

$$= G^2\frac{(78)(1.41)(0.994)}{(0.465)(0.184)(44.2)(32.2)} = 0.90G^2$$

$$\frac{L}{G}\left(\frac{\rho_G}{\rho_L}\right)^{0.5} = \left(\frac{0.184}{44.2}\right)^{0.5}\frac{L}{G} = 0.065\frac{L}{G}$$

From Fig. 15.21 the following results:

$\frac{L}{G}$	$0.065\frac{L}{G}$	Ordinate	G^2	G, lb/(sec)(sq ft)	G, lb/(hr)(sq ft)	G_{total}, lb/hr
0.7	0.0455	0.185	0.206	0.455	1640	39,000
0.8	0.052	0.175	0.195	0.442	1590	37,800
0.9	0.0585	0.170	0.189	0.432	1558	37,100

(*b*) From Fig. 15.21,

$$\frac{L}{G} = 0.715 \qquad 0.065\,\frac{L}{G} = 0.0465$$

$$G = \frac{(330.4)(83)}{23.8} = 1150 \text{ lb/(hr)(sq ft)} \qquad \frac{1150}{3600} = 0.32 \text{ lb/(sec)(sq ft)}$$

$$\text{Ordinate} = 0.9G^2 = (0.10)(0.9) = 0.09$$

$$\text{Abscissa} = 0.0465$$

$$\Delta P = 1.2 \text{ in water}$$

(*c*)

$$\Delta P = \alpha 10^{\beta L}\,\frac{G^2}{\rho_G}$$

$$\Delta P = (0.82)(10)^{(0.2)L}\,\frac{0.104}{0.184}$$

$$L = (0.715)(0.32) = 0.229$$

$$\Delta P = (0.82)(10)^{(0.2)(0.229)}\,\frac{0.104}{0.184}$$

$$= (0.82)(1.012)(0.565)$$

$$= 0.47 \text{ in water per foot of packing}$$

$$Z = \text{adjusted} = \frac{(20.1)(330.4)}{(0.106)(23.8)(78)} = 34.8 \text{ ft}$$

$$\Delta P_{\text{total}} = (34.8)(0.47) = 15.9 \text{ in water}$$

Liquid Holdup

Because of increased weight of the packing and liquid, and because of reduced temperature differential, liquid holdup should be kept at a minimum. For dumped packings a generalized relation has been presented by Leva [24].

$$h_w = 0.0004\left(\frac{L}{D_b}\right)^{0.6} \tag{15.47}$$

where h_w = cu ft liquid/cu ft packed tower volume
L = liquid rate, lb/(hr)(sq ft)
D_b = equivalent spherical packing diameter, in. (see tables for the respective packings)

For liquids other than water,

$$h_{\text{liquid}} = h_w \mu_L^{0.1}\left(\frac{62.3}{\rho_L}\right)^{0.78}\left(\frac{73}{\sigma}\right)^n \tag{15.48}$$

where h_{liquid} = cu ft liquid/cu ft tower packed space
h_w = cu ft water/cu ft tower packed space [Eq. (15.47)]
μ_L = viscosity of liquid, centipoise
ρ_L = density of liquid, lb/cu ft
σ = surface tension, dynes/cm

15.7. DESIGN OF PACKED TOWERS FOR DISTILLATION

Design of a packed distillation tower involves (*a*) the selection of the packing and the tower diameter to get reasonable flow rates, good contact of the liquid and vapor phases and permissible pressure drop; (*b*) the selection of reflux ratio; and (*c*) the determination of the height of the packed section. Optimization involves getting the greatest throughput of a given product with the least height of packing.

According to Eckert [11], a suggested method of procedure is as follows:

1. Knowing the feed rate or product rate desired and having established a reflux ratio, select a packing and calculate $(L/V)(\rho_L/\rho_V)^{0.5}$.

2. Determine the flow rate per unit area, either V or L, from Fig. 15.21 using the approximate pressure-drop line and the experimental packing factor, a/ϵ^3, for the packing if available; if unavailable, use the factor from Fig. 15.23. This tends to be low.

3. Get the cross-sectional area of the column by dividing the rate of flow or V or L by the rate per unit area. Solve for the diameter.

4. Determine the N_{OV} with the use of the equilibrium data and Eq. (15.16), or (15.26), or (15.36).

5. Determine HTU by the method described on page 625.

6. Get the height of packed section $Z = H_V N_V$.

7. *Selection of packing.* The packing is selected on the basis of economics through optimization of the design. Generally however, as a guide, the largest size packing consistent with the tower diameter is selected.

Recommended Packing Sizes

Type	Column diameter, D, ft
Raschig rings	$< \dfrac{D}{30}$
Saddles	$< \dfrac{D}{15}$
Pall rings	$< \dfrac{D}{10}$

The material from which the packing is made should be selected upon the basis of cost, strength, and corrosion resistance.

8. *Pressure drop.* The column should be designed to operate at the maximum economical pressure drop, although the pressure drop should never be more than 90% of the $(\Delta P)_{flood}$. Many designers insist that it should not be more than $0.6(\Delta P)_{flood}$.

The pressure of the operation will in many cases dictate the allowable pressure drop per foot. Under atmospheric conditions 0.2–0.6 in. of H_2O per foot of packing is usual. Under vacuum, less than 0.1 in. of H_2O per foot of packing may be the maximum allowable.

9. *Vapor and liquid distribution—redistribution.* Eckert [11] recommends the following rules pertaining to distribution or redistribution of vapor and liquid in packed towers.

(*a*) Raschig rings: Redistribution should be provided at each 2½ to 3 column diameters or at 20 ft maximum.

(*b*) Intalox and Berl saddles: Redistribution should be provided at each 5 to 8 column diameters with a maximum of 20 ft.

(*c*) Pall rings: Redistribution should be provided at each 5 to 10 column diameters or a maximum of 20 ft. (Generally all columns should be packed wet. This means the column is full of liquid when packing is poured into the shell.)

"Good" liquid distribution according to Eckert is indicated by the following tabulation:

Tower diameter	Liquid drip or flow points per sq ft of cross-sectional area
1.0–1.25	32
2.5	16
4.0–4.0+	4

Liquid and vapor distribution or redistribution is attained by the support plates for the packing. These are discussed in the following section. Eckert recommends gas injection plates for all services except with stacked packing.

Support Plates

Generally, the support plates are for the purpose of supporting the packing in the column shell and also serving as liquid and vapor redistribution plates.

It has been found [23] that the design of support plates will seriously affect the flooding rate of the columns. Because the design is important, there have been a number of special types of plates developed for this purpose:

Grid bars
Perforated plates
Interdistributor screens
Gas injection support plates
Multibeam support plates
Orifice plates
Weir flow distributors
Trough-type distributors

Perforated plates are sometimes used as redistributor plates. Their use has been successful where the free area is as large or larger than that of the packing, the spacing of the holes is uniform, and the plate is maintained level. Grid bars are merely grids made of bars welded to a circular insert which can be supported within the column shell. In this case (and in all cases) the vapor-flow free area should be equal to or larger than that of the packing. Interdistributor screens are advocated by Manning and Cannon [27]. Those described were made of 10-mesh screen, and consist of screen cones mounted between horizontal screens. Higher efficiencies were claimed for packed columns using this type of distributor.

Example 15.4

Design a packed tower to handle the system in Example 15.3.

Solution

1. Operating line slope (rectifying section) = 0.715.

$$x_B = 0.966 \qquad x_F = 0.60$$

Operating line slope (stripping section) = 1.2.

$$x_B = 0.087 \qquad x_F = 0.60$$

For rectifying section,

$$\frac{L}{G}\left(\frac{\rho_V}{\rho_L}\right)^{0.5} = (0.715)(0.065) = 0.0465$$

For stripping sections,

$$\frac{L}{G}\left(\frac{\rho_V}{\rho_L}\right)^{0.5} = (1.2)(0.065) = 0.78$$

2. Select 1-in. Pall rings, $a/\epsilon^3 = 45$; ordinate $= 0.56G^2$.
3. $\Delta P = 0.5$ in./ft packing.

For rectifying section, ordinate = 0.06; G = 0.326 lb/(sec)(sq ft) or 1175 lb/(hr)(sq ft).
For stripping section, ordinate = 0.17; G = 0.55 lb/(sec)(sq ft) or 1980 lb/(hr)(sq ft).
Vapor loss is the same for both sections = (330.4)(83) = 27,400 lb/hr.

4. Area of column based on limiting section, i.e., rectifying, 27,400/1175 = 23.3 sq ft. Diameter = $(23.3/0.786)^{0.5}$ = 5.43 ft; use 5.5 ft.

5. $K_{0y}a = (0.106)(66) = 7.0 = N/ZS\,\Delta y$; $\Delta y_{\text{av}} = 0.045$.

$$Z = \frac{330.4}{(7.0)(23.8)(0.045)} = 44 \text{ ft}$$

The column will be 5.5 ft in diameter, packed height = 44 ft, 1-in. Pall rings, 3- to 15-ft sections of packing.

Gas injection support plates were designed to give a good vapor and liquid redistribution and a relatively high free area. Figure 15.24 illustrates some of the types of gas injection plates. Orifice distributor support plates include gas risers and are supported by means of lugs which leave an annular space around the plate. These are made to handle a variety of different flow rates and are shown in Fig. 15.25. The weir flow distributor support plate gives essentially an equal or uniform liquid distribution over the cross section of the column. It is shown in Fig. 15.25. Trough-type distributor plates

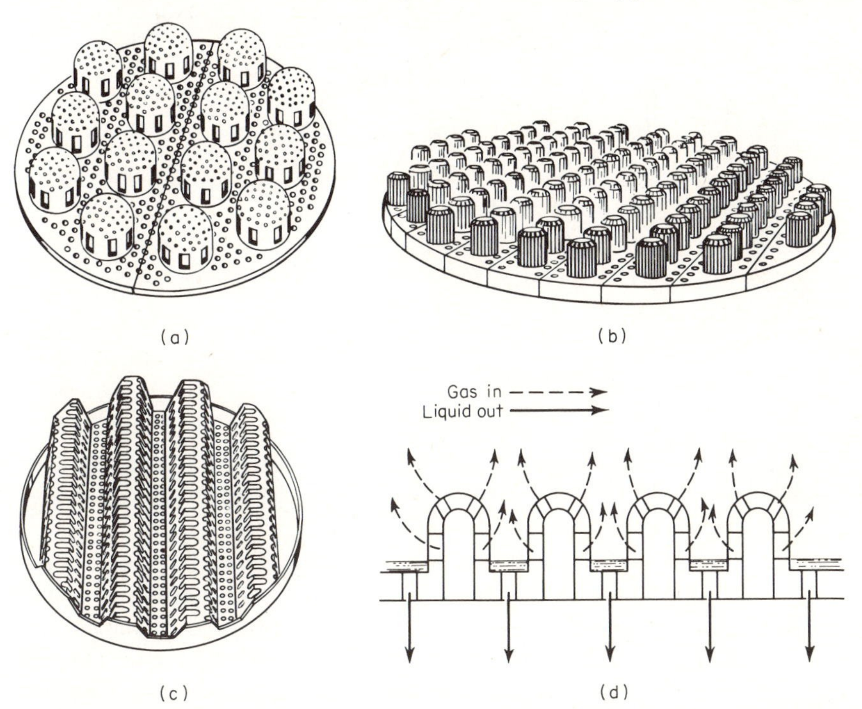

FIGURE 15.24 Gas injection support plates. (*a*) Chemical ceramic. (*b*) Carbon. (*c*) Metal. (*d*) Flow pattern. (*Courtesy U.S. Stoneware Company, Akron, Ohio.*)

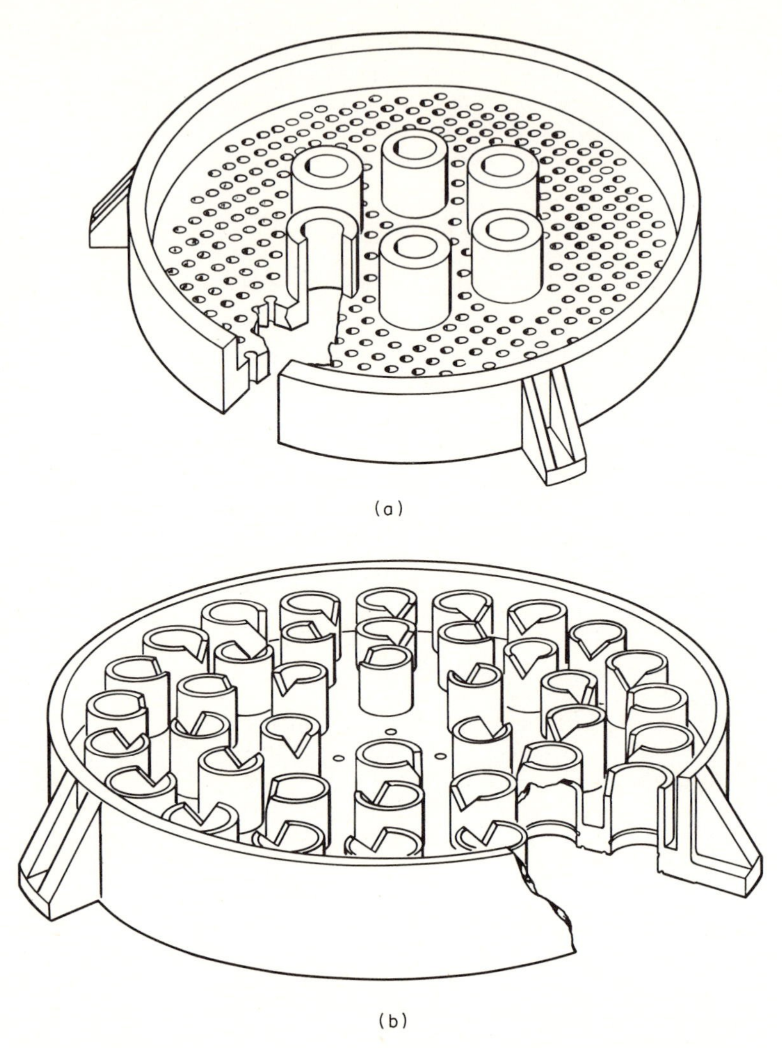

FIGURE 15.25 Support plates. (*a*) Orifice distributor. (*b*) Weir flow distributor. (*Courtesy U.S. Stoneware Company, Akron, Ohio.*)

represent another type of design which attempts to provide a uniform liquid distribution with low pressure drop. Figure 15.26 is an illustration of this type. Hold-down plates are commonly provided for packed columns where flow conditions and the character of the packing might be such that the packing would move about with possible injurious effects on the discrete pieces of packing. Particularly with ceramic packing this leads to cracking and chipping with resultant decrease in surface and increase in pressure drop.

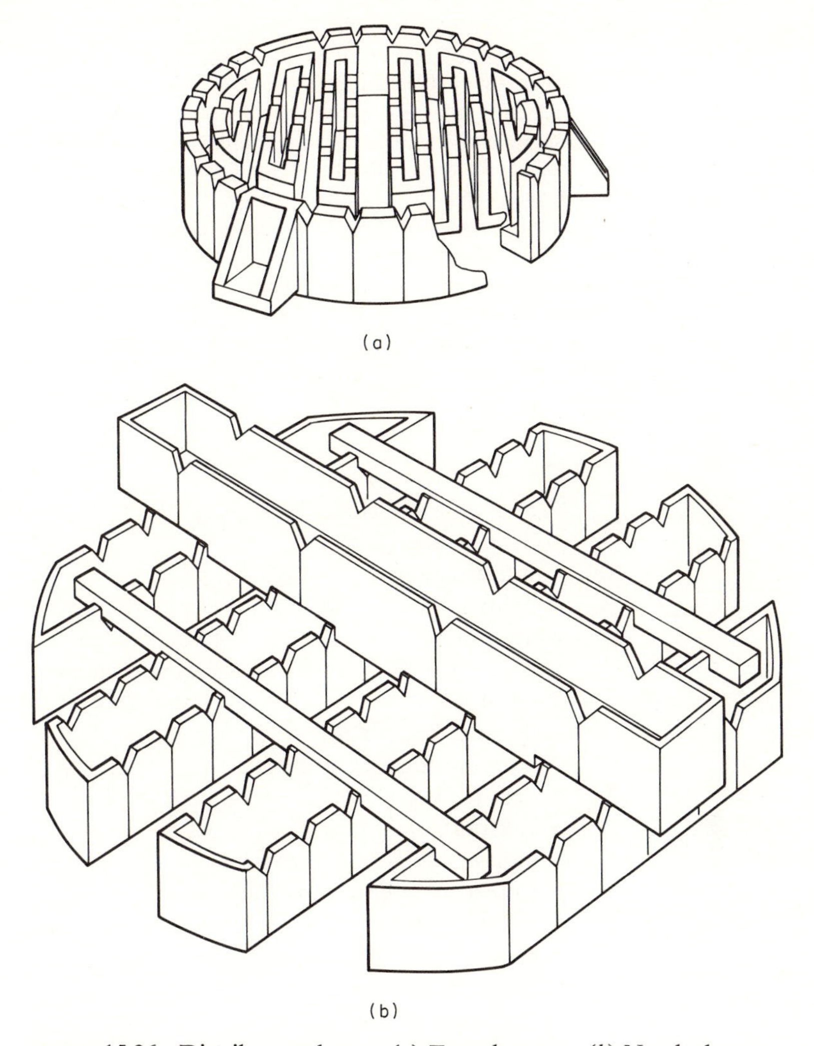

FIGURE 15.26 Distributor plates. (*a*) Trough type. (*b*) Notched type. (*Courtesy U.S. Stoneware Company, Akron, Ohio.*)

Hold-down plates are available either in ceramic or metal construction and consist essentially of grid construction, with high free area, and with sufficient weight to hold the packing in place.

PROBLEMS

1. Determine the number of transfer units, N_{OV}, at total reflux for the separation of a 50/50 mixture of benzene–toluene into a distillate consisting of 99 mole % benzene and a bottoms consisting of 99 mole % toluene. The reflux ratio $L_0/D = 2.0$ and the feed quantity is 2000 gph. Use Eq. (15.36).

2. Determine the number of transfer units, N_{OV}, by the graphical method (Fig. 15.12). Determine R_{min} by the conventional method.
3. A feed of 60% methanol, 40% water is to be distilled in a packed column using 1-in. Berl saddles and a reflux ratio of 3.5. The separation produces a distillate of 99.9% methanol and 99% water as bottoms. Determine the number of transfer units by using Eqs. (15.35) through (15.38).
4. Determine the height of the packed section by means of Eq. (15.39) and the N_{OV} calculated in Prob. 2.
5. Calculate the flooding reflux ratio for 1.0-in. Raschig rings, and for 1.0-in. Berl saddles for the system in Prob. 3 by means of Fig. 15.21.
6. Determine the pressure drop for the system in Prob. 1 (for 1-in. saddles) using Fig. 15.23 and Eq. (15.46).

Nomenclature

A = area of interface between phases
A = defined, Eq. (15.27)
A_0 = interfacial area
A' = defined Eq. (15.30)
a = area of interface per cu ft of packing or area of packing surface per cu ft of packed column
a, a' = constant
B = defined, Eq. (15.28)
B = moles residue per unit time
B' = defined, Eq. (15.31)
b, b' = constant
D = moles distillate per unit time
D_b = equivalent spherical packing factor
D_c = tower diameter, in.
D_m = molecular diffusivity, sq ft/hr
d = differential
E = eddy viscosity, lb/(ft)(hr)
E_m = eddy diffusivity, sq ft/hr
F = moles feed per unit time
G = lb of vapor/(hr)(sq ft) tower cross section
G' = lb of vapor/(sec)(sq ft) tower cross section
G_m = moles of vapor/(hr)(sq ft) tower cross section
g_c = gravitational constant, (lb mass)(ft)/(lb force)(sec^2)
H_V, H_L, H_{OV}, H_{OL} = height of transfer unit based on vapor phase, liquid phase, over-all vapor, and over-all liquid phases, respectively
h_w = liquid holdup, cu ft/cu ft packed tower

HETP = height equivalent to a theoretical plate, ft
K_y, K_2, K_3 = constants
K_1, K_x = over-all mass-transfer coefficients, moles/(hr)(sq ft) for y (or x)
k_y, k_x = film mass-transfer coefficients, moles/(hr)(sq ft) for y (or x)
L = moles liquid per unit time
L' = lb liquid/(sec)(sq ft) tower cross section
L_m = moles liquid/(hr)(sq ft) tower cross section
L_0 = moles of reflux per unit time
m = slope of equilibrium curve, y/x
N_V, N_L, N_{OV}, N_{OL} = number of transfer units based on vapor phase, liquid phase, over-all vapor, over-all liquid, respectively
n = number of equilibrium stages
ΔP = pressure drop, in H_2O/ft packed height
Q = moles transferred per unit time per unit area of interface
q_B = reboiler duty, Btu/hr
q_D = condenser duty, Btu/hr
R = reflux ratio, L/D
R' = defined, Eq. (15.29)
R_m = minimum reflux ratio $(L_0/D)_{min}$
S = cross-sectional area of column, sq ft
Sc = Schmidt number, $\mu/\rho D$
U_V, U_L = velocity of vapor or liquid, cu ft/(sq ft)(hr)
V = moles vapor per unit time
x, y = mole fraction in liquid, vapor
x_i, y_i = mole fraction in liquid, vapor at interface
x^*, y^* = equilibrium mole fractions
x_{BM}, y_{BM} = log mean average concentration in mole fraction of the nontransferred component
Z = height of packed section, ft

Greek letters

α = relative volatility
α = constant
β = constant
ψ = ρ_{H_2O}/ρ_{fluid}
ϵ = fractional voids, cu ft/cu ft packing
ρ = density, lb/cu ft
μ = viscosity, lb/(ft)(hr)
μ_L = viscosity liquid, centipoise

μ_V = viscosity vapor, centipoise
Σ = summation
σ = surface tension, dynes/cm

REFERENCES

1. Anon.: *Chem. Eng.* (April, 1953), p. 242.
2. Bain, W. A., Jr., and O. A. Hougen: *Trans. A.I.Ch.E.*, **40**:29(1944).
3. Berg, L., and D. O. Popovac: *Chem. Eng. Progr.*, **46**:683 (1949).
4. Bertetti, J. W.: *Trans. A.I.Ch.E.*, **38**:1023 (1942).
5. Cannon, M. R.: *Ind. Chem.*, **41**:1953 (1949).
6. Chilton, T. B., and A. P. Colburn: *Ind. Eng. Chem.*, **27**:255 (1935).
7. Clay, H. A., J. W. Clark, and B. L. Munro: *Chem. Eng. Progr.*, **12**(1):51 (1966).
8. Colburn, A. P.: *Ind. Eng. Chem.*, **33**:459 (1941).
9. Colburn, A. P.: *Trans. A.I.Ch.E.*, **35**:211 (1939).
10. Cornell, D., W. G. Knapp, and J. R. Fair: *Chem. Eng. Progr.*, **56**(7):68, 49 (1960).
11. Eckert, J. S.: *Chem. Eng. Progr.*, **57**:64 (1961).
12. Eckert, J. S.: *Chem. Eng. Progr.*, **59**(5):75 (1963).
13. Eckert, J. S., E. H. Foote, and R. L. Huntington: *Chem. Eng. Progr.*, **54**(1):70 (1958).
14. Eckert, J. S., E. H. Foote, and L. F. Walter: *Chem. Eng. Progr.*, **62**(1):59 (1966).
15. Eckert, J. S., and L. F. Walter: *Hydrocarbon Process., Petrol. Refiner*, **43**(2):107 (1964).
16. Ellis, S. R. M.: *Chem. Eng. News*, **3**(44):4613 (1953).
17. Fisher, A. W., Jr., and R. J. Bowen: *Chem. Eng. Progr.*, **45**:359 (1949).
18. Fractionating Towers, Inc., Englewood, N.J.: Spraypak, *Bull.* 8.
19. Gerster, J. A., J. H. Koffolt, and J. R. Withrow: *Trans. A.I.Ch.E.*, **41**:693 (1945).
20. Hands, C. H. G., and F. R. Whitt: *J. Appl. Chem.* (March, 1951), p. 135.
21. Hoffing, E. H., and F. J. Lockhart: *Chem. Eng. Progr.*, **50**:94 (1954).
22. Hutchinson, M. H., and R. F. Baddour: *Refining Eng.* (February, 1957), p. C-6.
23. Leva, M., J. M. Lucas, and H. H. Frahme: *Ind. Eng. Chem.*, **46**:1225 (1954).
24. Leva, M.: "Tower Packing and Packed Tower Design," U.S. Stoneware Company.
25. Lerner, B. J., and C. S. Grove, Jr.: *Ind. Eng. Chem.*, **43**:216 (1951).
26. Lobo, W. E., F. Hashmall, Leo Friend, and F. A. Zenz: *Trans. A.I.Ch.E.*, **41**:693 (1945).
27. Manning, R. E., and M. R. Cannon: *Ind. Eng. Chem.*, **49**:347 (1957).
28. Molstad, M. C., R. G. Abbey, A. R. Thompson, and J. F. McKinney: *Trans. A.I.Ch.E.*, **38**:387, 410 (1942).
29. Molstad, M. C., and L. F. Parsly: *Chem. Eng. Progr.*, **46**:20 (1950).
30. Molstad, M. C., J. F. McKinney, and R. G. Abbey: *Trans. A.I.Ch.E.*, **39**:605 (1943).
31. Murch, D. P.: *Ind. Eng. Chem.*, **45**:2616 (1953).
32. Olney, R. B.: *A.I.Ch.E. Paper*, Annual Meeting, St. Louis (December 13–16, 1953).
33. Parsly, L. F., M. C. Molstad, H. Cress, and L. G. Bauer: *Chem. Eng. Progr.*, **46**:17 (1950).

34. Shell Development Co.: *Petrol. Refiner* (November, 1952), p. 105.
35. Shell Development Co.: *Petrol. Eng.* (May, 1954), p. C-17.
36. Sherwood, T. K., G. H. Shipley, and F. A. L. Holloway: *Ind. Eng. Chem.*, **30**:765 (1938).
37. Shulman, H. L., C. F. Ulrich, A. Z. Prouls, and J. D. Zimmerman: *A.I.Ch.E. J.*, **1**:247, 253, 259 (1955).
38. Shulman, H. L., and J. E. Margolis: *A.I.Ch.E. J.*, **3**:157 (1957).
39. Shulman, H. L., and L. J. Delancy: *A.I.Ch.E. J.*, **5**:290 (1959).
40. Shulman, H. L., and W. G. Whitehouse: *A.I.Ch.E. J.*, **6**:174, 469 (1960).
41. Shulman, H. L., C. G. Saviani, and R. V. Edwin: *A.I.Ch.E. J.*, **9**:479 (1963).
42. Silvey, F. C., and C. J. Keller: *Chem. Eng. Progr.*, **62**(1):68 (1966).
43. Simon, M. J., and M. A. G. Rau: *Ind. Eng. Chem.*, **40**:93 (1948).
44. Teller, A. J.: *Chem. Eng. Progr.*, **50**:65 (1954).
45. Yu, K. T., and J. Coull: *Chem. Eng. Progr.*, **46**:89 (1950).
46. Zenz, F. A.: *Chem. Eng. Progr.*, **43**:415 (1947).
47. Zenz, F. A., and R. A. Eckert: *Petrol. Refiner* (February, 1961), p. 130.

Appendix

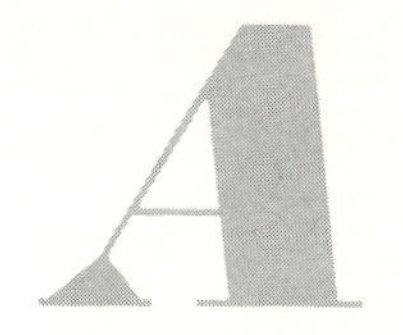

Tables

TABLE A.1 **API Gravity—Density***

API gravity, 60/60°F	Specific gravity, 60/60°F	Density, lb/cu ft	Density, lb/gal
Ethylene	0.338	21.10	2.8149
Ethane	0.374	23.35	3.1147
Propylene	0.522	32.59	4.3472
Propane	0.509	31.78	4.2390
i-Butylene	0.600	37.46	4.9970
n-1-Butene	0.601	37.52	5.0051
i-Butane	0.563	35.15	4.6887
n-Butane	0.584	36.46	4.8636
i-Pentane	0.625	39.02	5.2050
n-Pentane	0.631	39.39	5.2550
0	1.0760	67.18	8.962
1.0	1.0679	66.68	8.895
2.0	1.0599	66.18	8.828
3.0	1.0526	65.68	8.762
4.0	1.0443	65.20	8.698
5.0	1.0366	64.72	8.634
6.0	1.0291	64.25	8.571
7.0	1.0217	63.79	8.509
8.0	1.0143	63.33	8.448
9.0	1.0071	62.88	8.388
10.0	1.0000	62.43	8.328
11.0	0.9930	62.00	8.270
12.0	0.9861	61.56	8.212
13.0	0.9792	61.14	8.155
14.0	0.9725	60.72	8.099
15.0	0.9659	60.30	8.044

TABLE A.1 (*Continued*)

API gravity, 60/60°F	Specific gravity, 60/60°F	Density, lb/cu ft	Density, lb/gal
16.0	0.9593	59.90	7.989
17.0	0.9529	59.50	7.935
18.0	0.9465	59.10	7.882
19.0	0.9402	58.71	7.830
20.0	0.9340	58.31	7.778
21.0	0.9279	57.93	7.727
22.0	0.9218	57.55	7.676
23.0	0.9159	57.18	7.627
24.0	0.9100	56.82	7.578
25.0	0.9042	56.45	7.529
26.0	0.8984	56.09	7.481
27.0	0.8927	55.73	7.434
28.0	0.8871	55.38	7.387
29.0	0.8816	55.04	7.341
30.0	0.8762	54.70	7.296
31.0	0.8708	54.36	7.251
32.0	0.8645	54.03	7.206
33.0	0.8602	53.70	7.163
34.0	0.8550	53.37	7.119
35.0	0.8498	53.05	7.076
36.0	0.8448	52.74	7.034
37.0	0.8398	52.43	6.993
38.0	0.8348	52.12	6.951
39.0	0.8299	51.81	6.910
40.0	0.8251	51.51	6.870
41.0	0.8203	51.21	6.830
42.0	0.8155	50.91	6.790
43.0	0.8109	50.62	6.752
44.0	0.8063	50.33	6.713
45.0	0.8017	50.05	6.675
46.0	0.7972	49.77	6.637
47.0	0.7927	49.49	6.600
48.0	0.7883	49.21	6.563
49.0	0.7839	48.94	6.526
50.0	0.7796	48.67	6.490
51.0	0.7753	48.40	6.455
52.0	0.7711	48.14	6.420
53.0	0.7669	47.88	6.385
54.0	0.7628	47.62	6.350
55.0	0.7587	47.36	6.316
56.0	0.7547	47.12	6.283
57.0	0.7507	46.87	6.249
58.0	0.7467	46.62	6.216
59.0	0.7428	46.37	6.184
60.0	0.7389	46.13	6.151

TABLE A.1 (*Continued*)

API gravity, 60/60°F	Specific gravity, 60/60°F	Density, lb/cu ft	Density, lb/gal
61.0	0.7351	45.89	6.119
62.0	0.7313	45.66	6.087
63.0	0.7275	45.42	6.056
64.0	0.7238	45.19	6.025
65.0	0.7201	44.96	5.994
66.0	0.7165	44.73	5.964
67.0	0.7128	44.51	5.934
68.0	0.7093	44.28	5.904
69.0	0.7057	44.06	5.874
70.0	0.7022	43.84	5.845
71.0	0.6988	43.63	5.817
72.0	0.6953	43.41	5.788
73.0	0.6919	43.20	5.759
74.0	0.6886	42.99	5.731
75.0	0.6857	42.78	5.703
76.0	0.6819	42.57	5.676
77.0	0.6787	42.37	5.649
78.0	0.6754	42.17	5.622
79.0	0.6722	41.97	5.595
80.0	0.6690	41.77	5.568
81.0	0.6659	41.57	5.542
82.0	0.6628	41.38	5.516
83.0	0.6597	41.19	5.491
84.0	0.6566	41.00	5.465
85.0	0.6536	40.81	5.440
86.0	0.6506	40.62	5.415
87.0	0.6476	40.43	5.390
88.0	0.6446	40.25	5.365
89.0	0.6417	40.07	5.341
90.0	0.6388	39.88	5.316
91.0	0.6360	39.71	5.293
92.0	0.6331	39.53	5.269
93.0	0.6303	39.35	5.246
94.0	0.6275	39.18	5.222
95.0	0.6247	39.00	5.199
96.0	0.6220	38.83	5.176
97.0	0.6193	38.66	5.154
98.0	0.6166	38.49	5.131
99.0	0.6139	38.33	5.109
100.0	0.6112	38.16	5.086

* From National Standard Petroleum Oil Tables, *Cir. 410*, National Bureau of Standards.

TABLE A.2 **Volume Correction***

Range, °API	Coefficient of expansion, C_e, at 60°F, vol/(vol)(°F)
0–14.9	0.00035
15–34.9	0.00040
35–50.9	0.00050
51–63.9	0.00060
64–78.9	0.00070
79–88.9	0.00080
89–93.9	0.00085
94–99.9	0.00090

* From Abridged Volume Correction Table Supplement to *Cir. 410*, National Bureau of Standards.

Note: For approximate calculation of liquid volume from 30 to 120°F, the volume at 60°F can be corrected by $V_{60} + (t - 60)C_e$.

Appendix

Vapor pressures at various temperatures

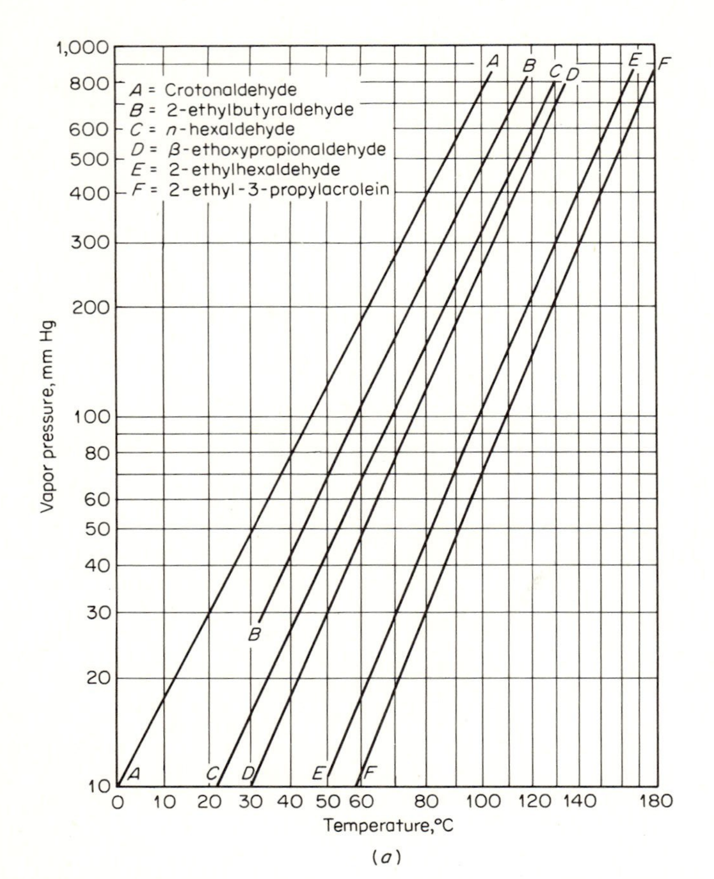

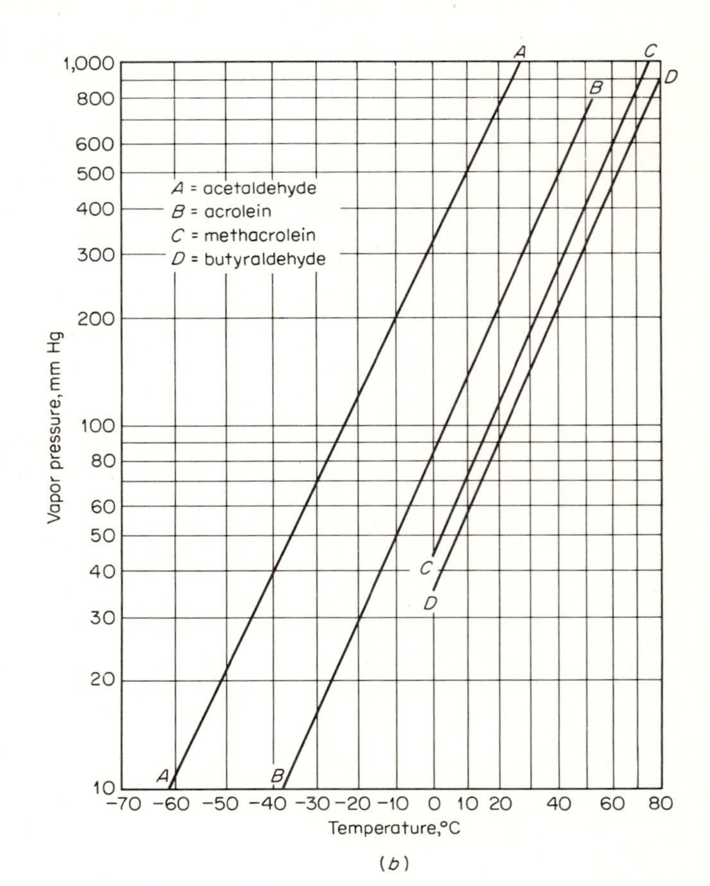

FIGURE B.1 Aldehydes. (Courtesy Union Carbide Corporation.)

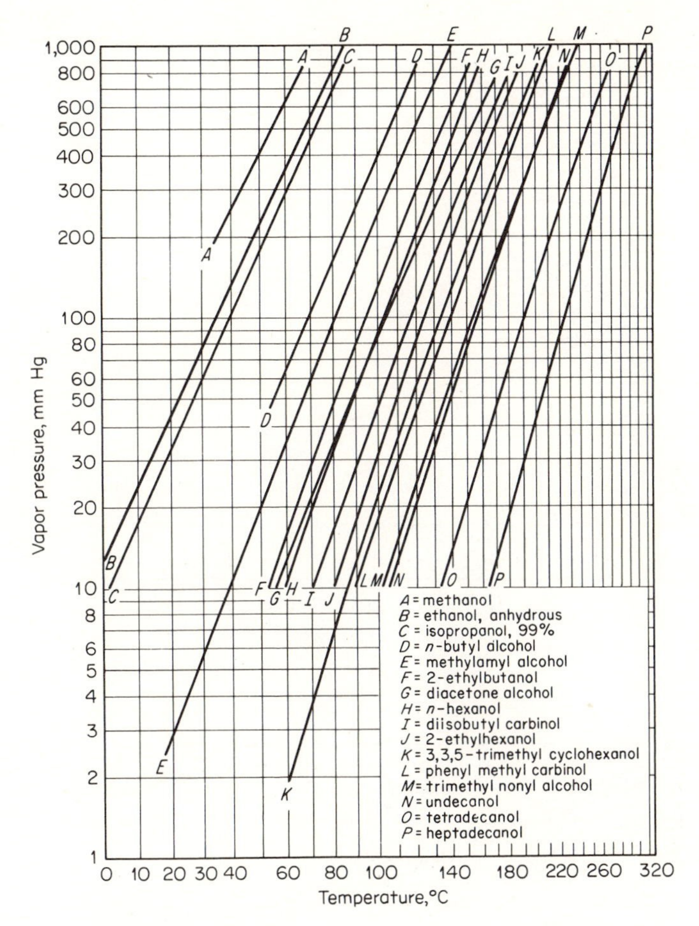

FIGURE B.2 Alcohols. (*Courtesy Union Carbide Corporation.*)

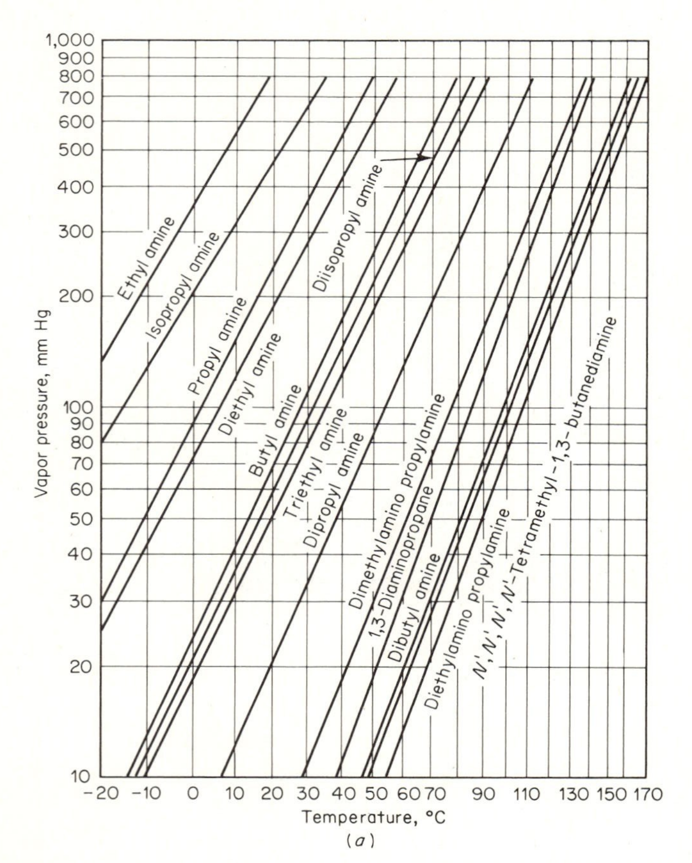

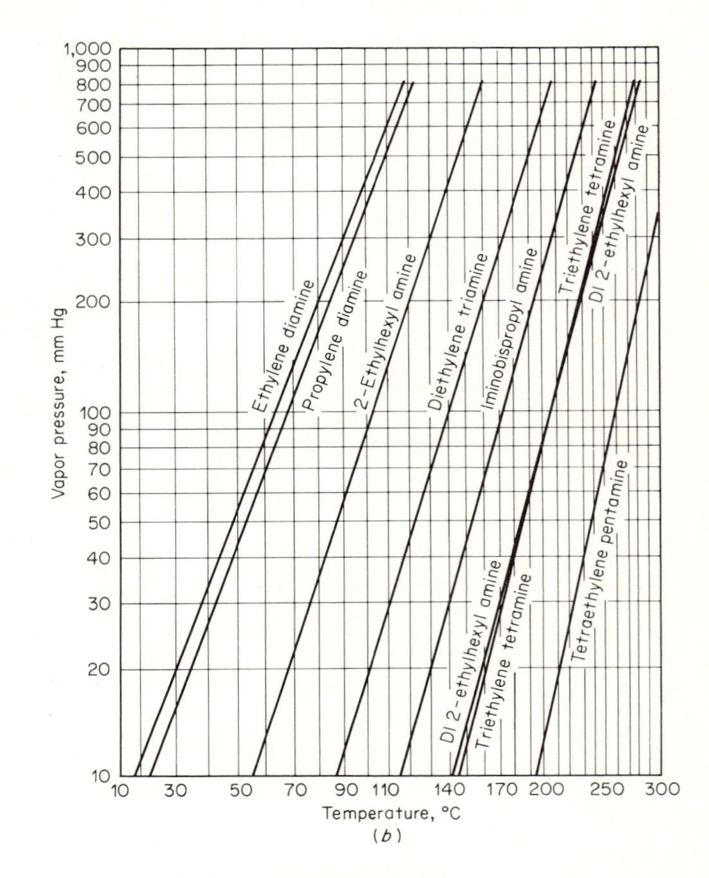

FIGURE B.3 Amines. (*Courtesy Union Carbide Corporation.*)

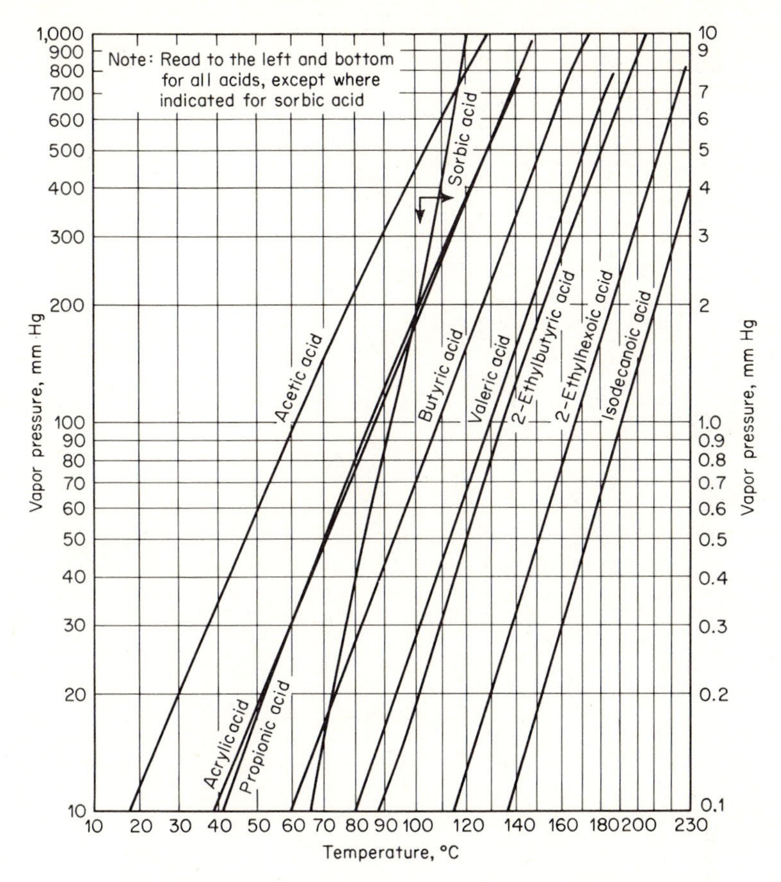

FIGURE B.4 Acids. (*Courtesy Union Carbide Corporation.*)

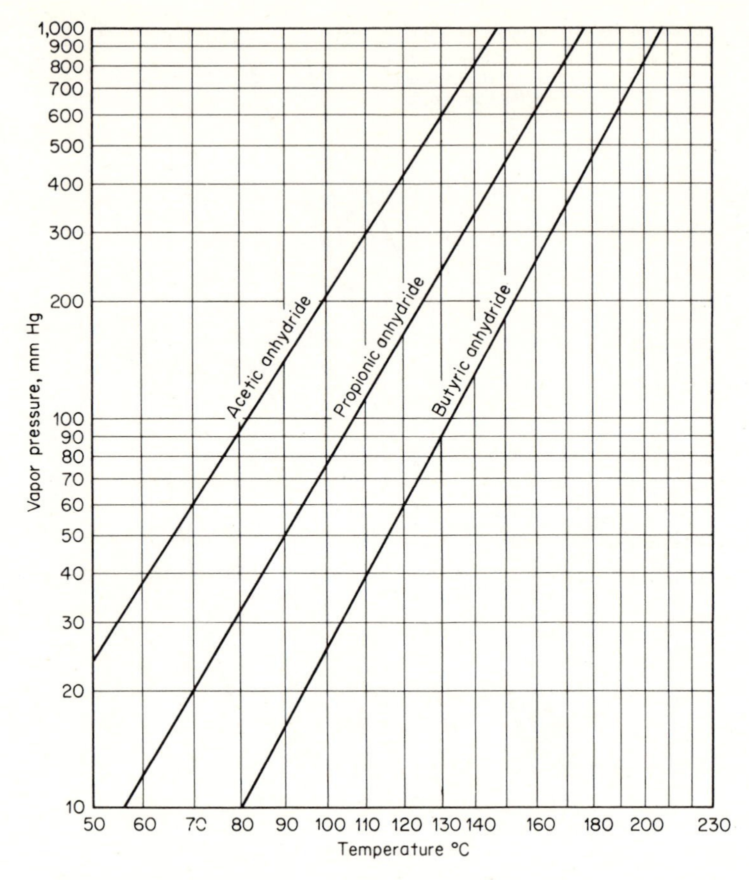

FIGURE B.5 Anhydrides. (*Courtesy Union Carbide Corporation.*)

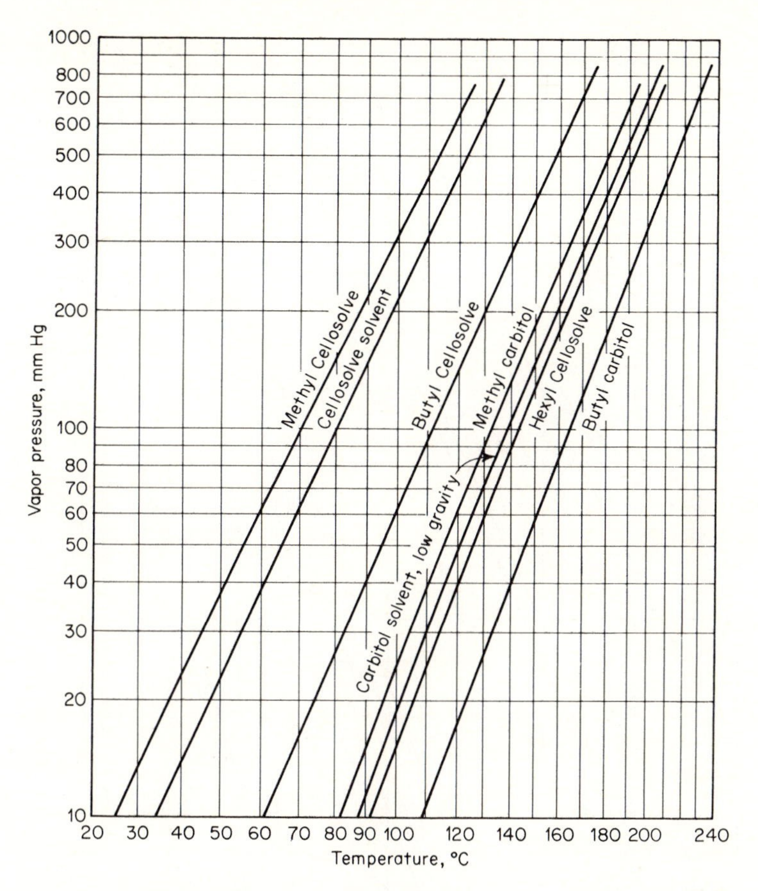

FIGURE B.6 Cellosolve-carbitols. (*Courtesy Union Carbide Corporation.*)

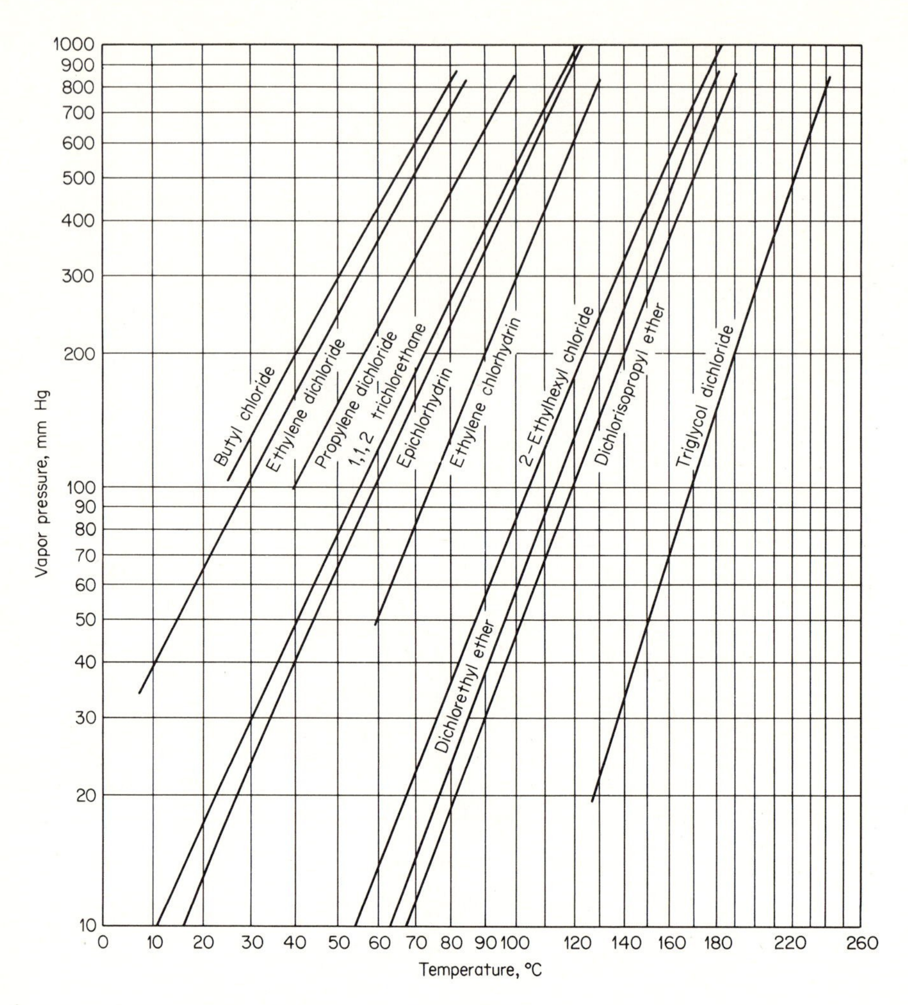

FIGURE B.7 Chlorides. (*Courtesy Union Carbide Corporation.*)

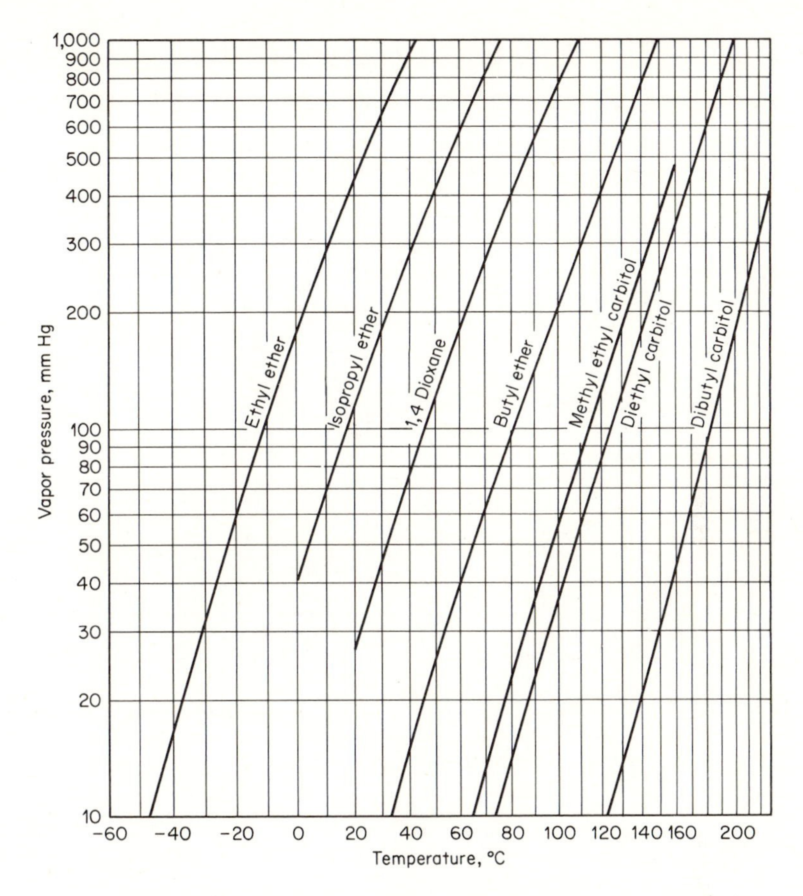

FIGURE B.8 Ethers. (*Courtesy Union Carbide Corporation.*)

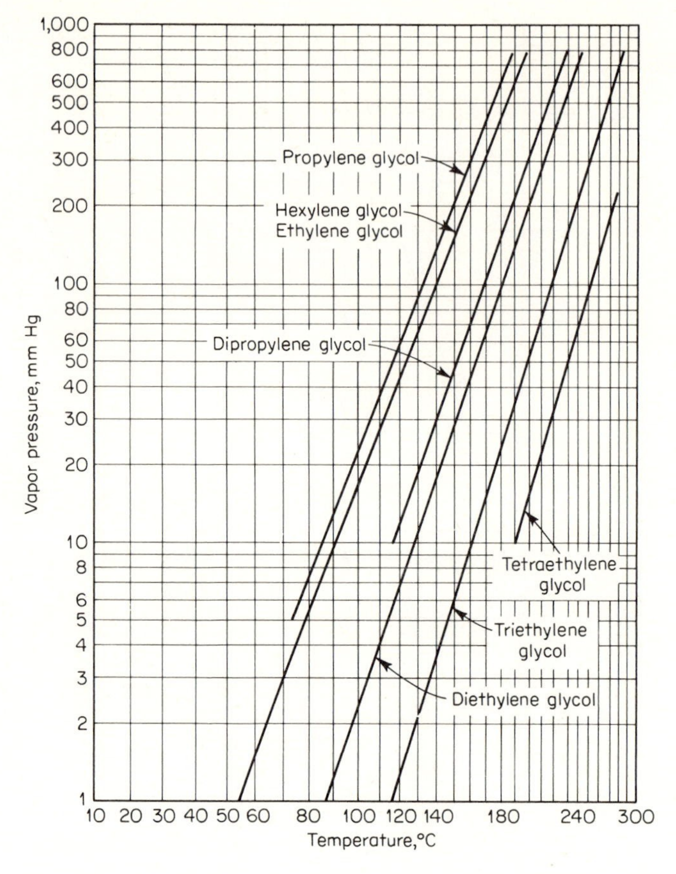

FIGURE B.9 Glycols. (*Courtesy Union Carbide Corporation.*)

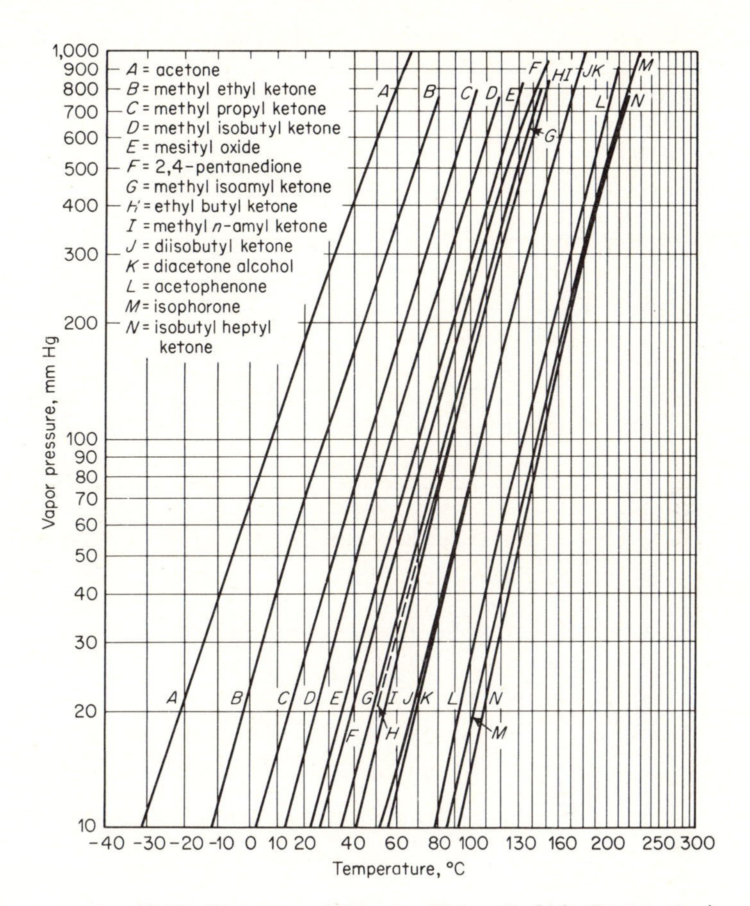

FIGURE B.10 Ketones. (*Courtesy Union Carbide Corporation.*)

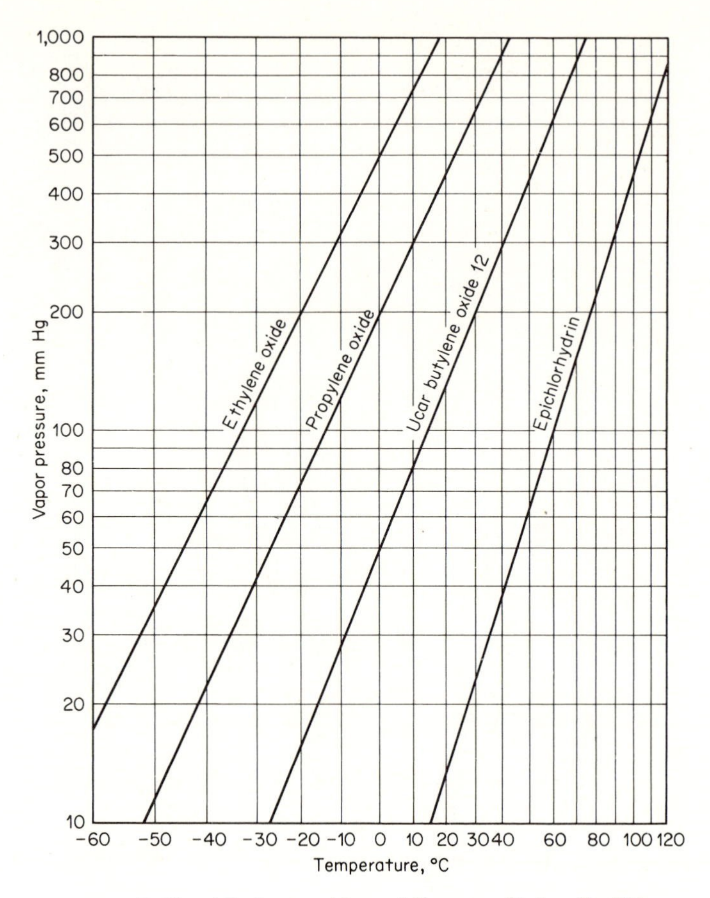

FIGURE B.11 Alkylene oxides. (*Courtesy Union Carbide Corporation.*)

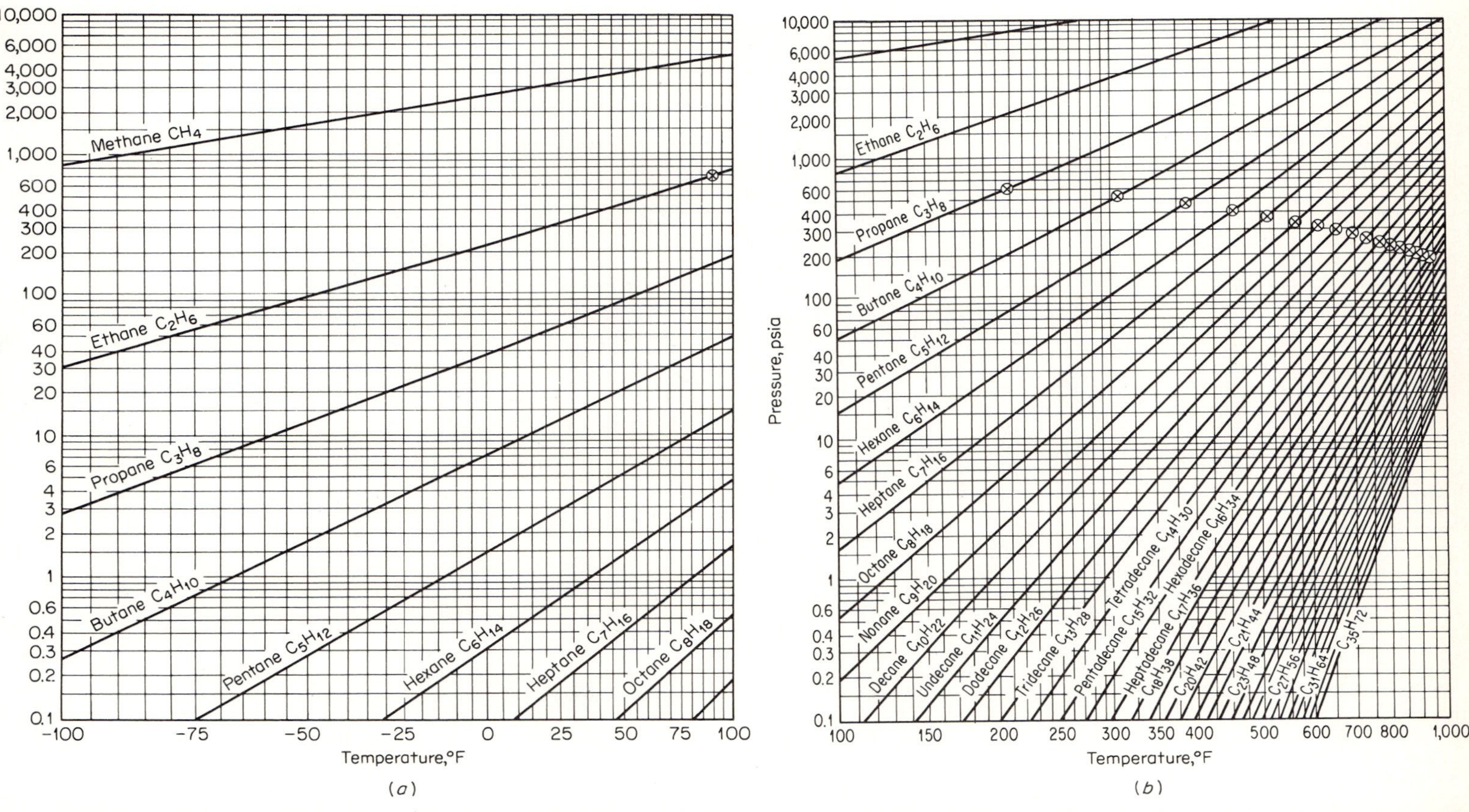

FIGURE B.12 Normal paraffins. (*From Katz et al., "Handbook of Natural Gas Engineering," McGraw-Hill, New York, 1959. Data collected by M. W. Kellogg Company.*)

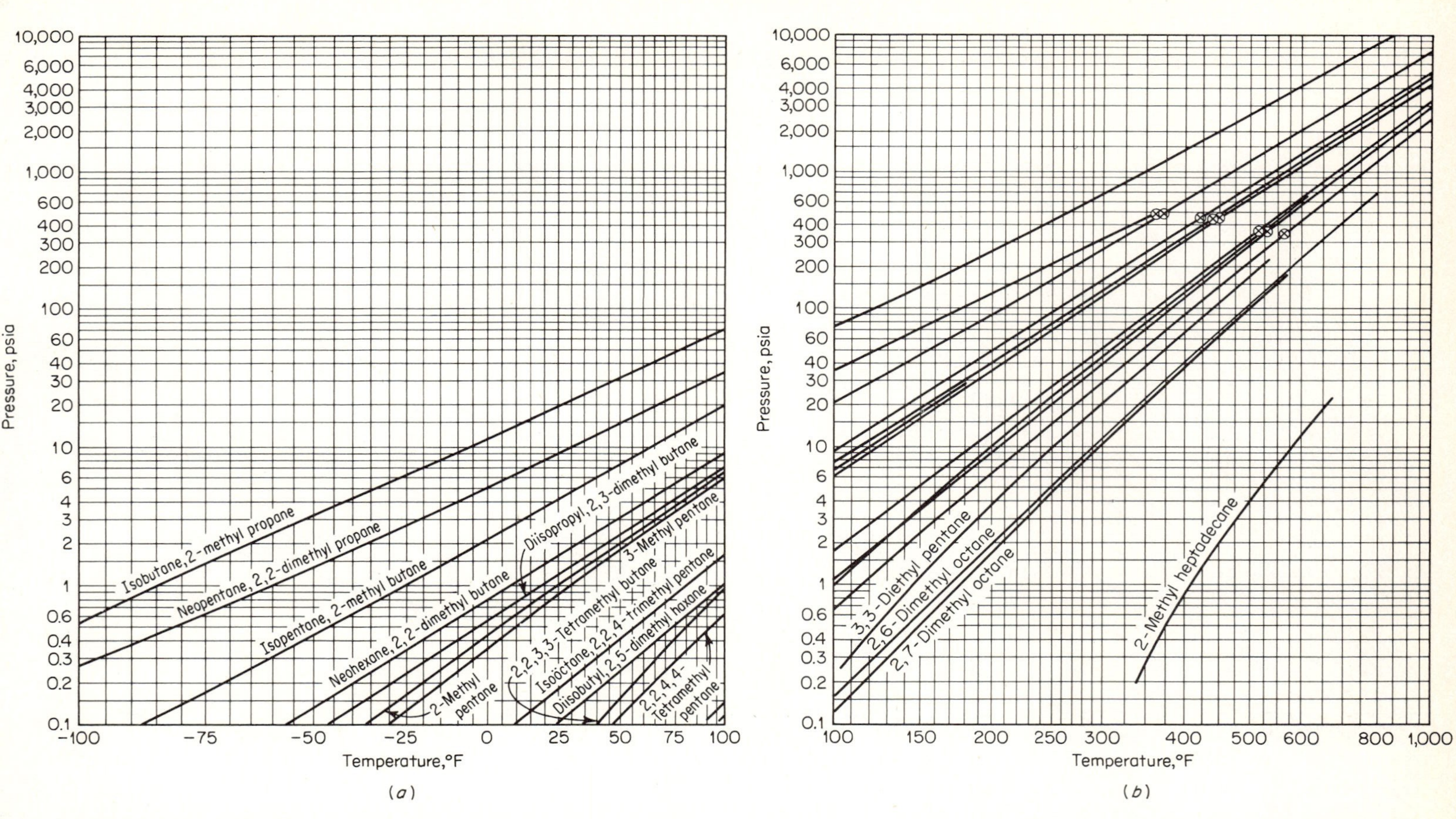

FIGURE B.13 Isomeric paraffins. (*From Katz et al., "Handbook of Natural Gas Engineering," McGraw-Hill, New York, 1959. Data collected by M. W. Kellogg Company.*)

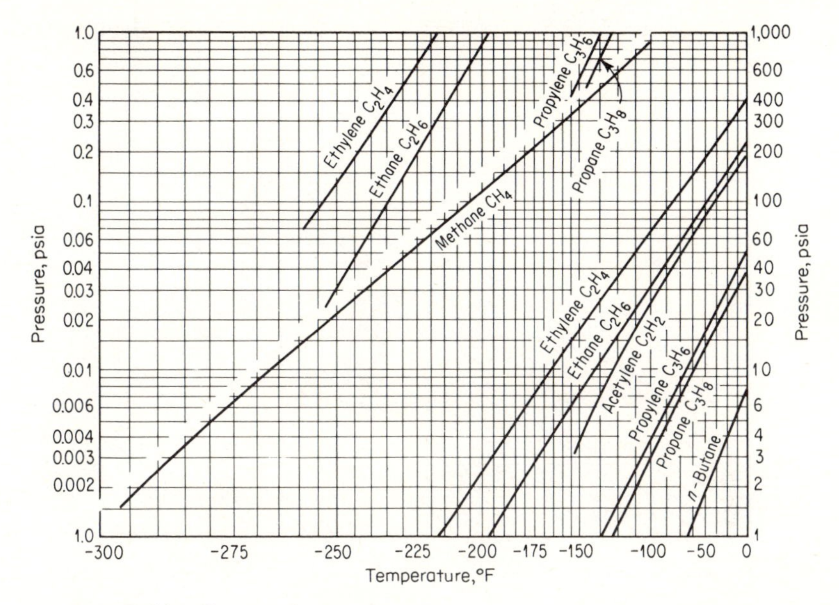

FIGURE B.14 Gases at low temperature. (*From Katz et al., "Handbook of Natural Gas Engineering," McGraw-Hill, New York, 1959. Data collected by M. W. Kellogg Company.*)

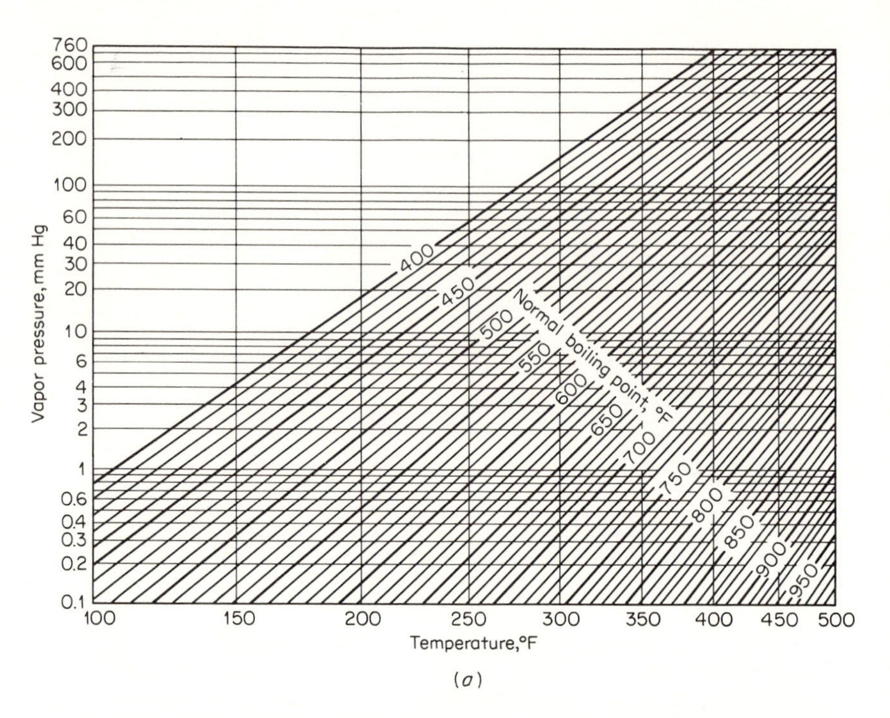
Vapor pressure, mm Hg
760
600
400
300
200
100
60
40
30
20
10
6
4
3
2
1
0.6
0.4
0.3
0.2
0.1
100
150
200
250
300
350
400
450
500
Temperature, °F
400
450
500
550
600
650
700
750
800
850
900
950
Normal boiling point, °F

(*a*)

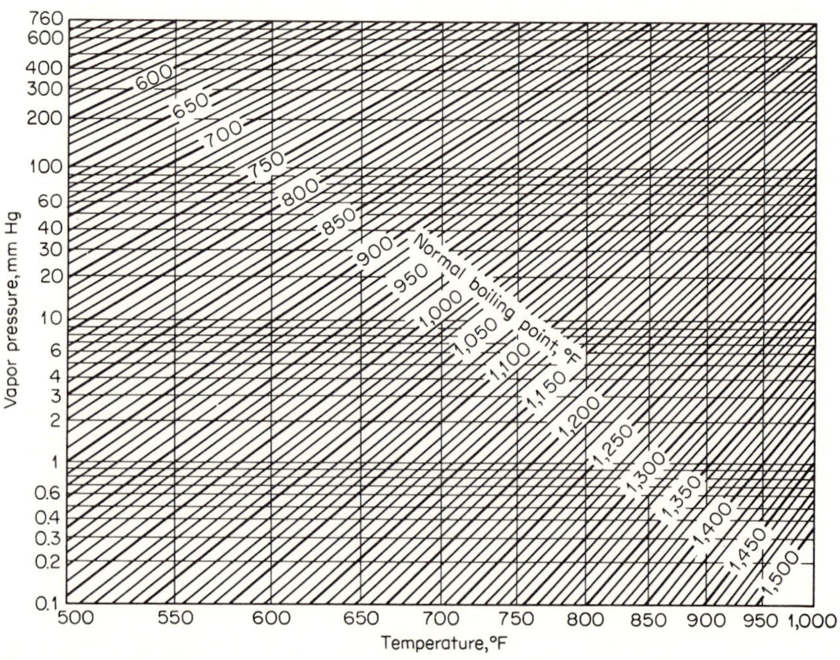
Vapor pressure, mm Hg
760
600
400
300
200
100
60
40
30
20
10
6
4
3
2
1
0.6
0.4
0.3
0.2
0.1
500
550
600
650
700
750
800
850
900
950
1,000
Temperature, °F
600
650
700
750
800
850
900
950
1,000
1,050
1,100
1,150
1,200
1,250
1,300
1,350
1,400
1,450
1,500
Normal boiling point, °F

(*b*)

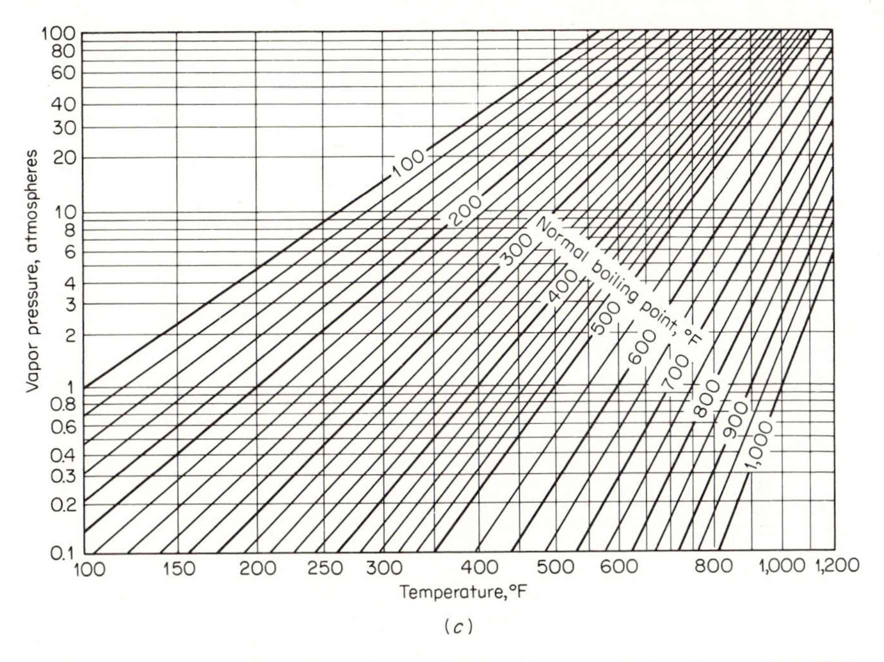

FIGURE B.15 Petroleum hydrocarbons. Basis—characterization factor $K = 12.0$. Use Fig. B.16 to correct for K other than 12.0. In Fig. B.15c correction for variation in K is so small that it can be neglected. (*From J. B. Maxwell and L. L. Bonnell, "Vapor Pressure Charts for Petroleum Hydrocarbons," Esso Research and Engineering Company, 1955.*)

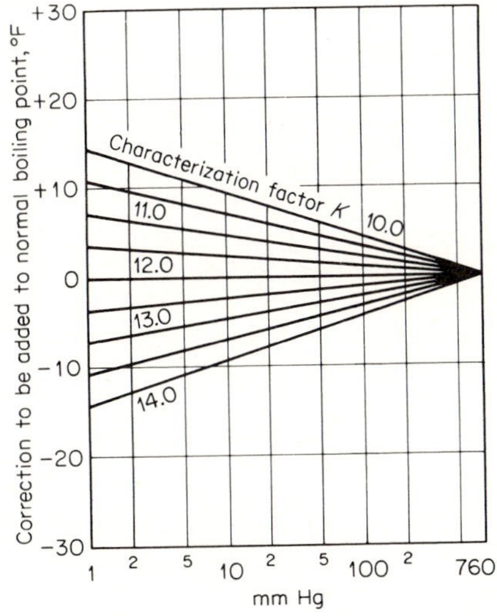

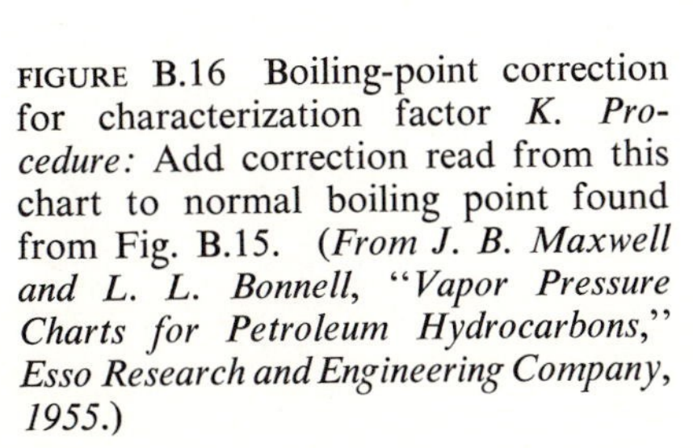
FIGURE B.16 Boiling-point correction for characterization factor K. *Procedure:* Add correction read from this chart to normal boiling point found from Fig. B.15. (*From J. B. Maxwell and L. L. Bonnell, "Vapor Pressure Charts for Petroleum Hydrocarbons," Esso Research and Engineering Company, 1955.*)

Name Index

Abbey, R. C., 611, 646
Acrivos, A., 306
Adler, S. B., 32
Amundson, N. R., 306
Anderson, R., 396–398, 408, 409, 433
Andrew, L., 1
Antoine, C., 9
Arnold, D. S., 528, 536, 566
Atkins, G. T., 465, 475, 584, 602
Atterig, P. T., 536, 566

Bachelor, J. B., 169, 189, 217, 247
Baddour, R. F., 615, 646
Baer, R. M., 306
Bain, J. L., 536, 539–541, 566
Bain, W. A., Jr., 632, 646
Baker, E. M., 536, 544, 568
Baker, J. C., 544, 567
Bard, Y., 306
Barker, P. E., 558, 566
Bauer, C. R., 351, 378
Bauer, L. G., 611, 646
Beattie, J. A., 74, 124
Beiswenger, G. A., 131, 158
Benedict, M. G., 63, 72, 393, 433, 438, 444, 464, 475
Berg, L., 410, 433, 635, 646
Bertetti, J. W., 632, 646
Bertholet, D., 1
Beyer, G. H., 544, 568
Billingsly, D. S., 306
Black, C., 28, 32, 70–72, 393, 418–420, 433
Bolles, W. L., 506, 508, 509, 511–515, 552, 567, 572, 574–576, 579, 602
Bonham, M. S., 28, 61, 62, 72, 393, 414, 433
Bonilla, A., 455, 475
Bonilla, C. F., 322, 332
Bonnell, L. L., 667–669
Bonner, J. S., 301, 305, 313
Bonnet, W. E., 546, 549, 550, 567
Bowen, R. J., 610, 646
Bowman, J. R., 157, 158
Boyd, C. W., 353, 354, 378
Boyer, C. M., 465, 475
Boyle, R., 2
Bradford, J. R., 557, 567
Braulick, D. J., 527, 545, 568
Brearly, C. S., 32
Bridgeman, O. C., 74, 124
Britton, E. C., 445, 476
Broughton, D. B., 32
Brown, G. G., 131, 154, 155, 158, 217, 218, 229, 231, 241, 242, 247, 304, 340, 349, 375, 376, 378, 506, 537, 567, 584, 597, 598, 603
Brunshwig, H., 1
Buford, R., 473
Bullington, L. A., 506, 518, 519, 528, 568, 598, 602
Burchard, J. K., 544, 569
Buron, A. G., 515, 568

Cajander, B. C., 76, 105, 109–116, 124
Cambio, R., 396–398, 409, 433
Cannon, M. R., 610, 640, 646
Carey, J. S., 544, 567
Carlson, H. C., 24, 25, 59, 306, 393, 433
Carmody, T. W., 550, 567
Chaiyavech, P., 557, 567
Chambers, J. M., 448, 475
Chao, K. C., 68, 72, 306, 393, 433
Chears, M. N., 405, 433
Chilton, T. B., 626, 646
Choudbury, M. H., 558, 566
Chu, J. C., 16, 131, 158, 545, 567
Church, W. L., 515, 528, 568
Cicalese, J. J., 506, 567

Clark, F. W., 442, 476
Clark, J. W., 624, 636, 646
Clay, H. A., 624, 636, 646
Coffee, A., 2
Coghlin, C. A., 468, 469, 476
Colburn, A. P., 24, 25, 59, 130, 158, 216, 218, 224, 226, 227, 229, 247, 303, 340, 375, 378, 390, 393, 433, 436, 444, 465, 475, 476, 550, 552, 567, 620, 621, 626, 646
Cooper, A., 2
Cornell, D., 624–626, 646
Cornell, P. W., 465, 476
Coull, J., 28, 59, 61, 62, 72, 393, 414, 433, 620, 622, 623, 647
Cox, E. R., 11
Cress, H., 611, 646
Crozier, R. D., 306

Dalton, J., 9, 18, 46, 49
Dauphine, T. C., 506, 567
Davies, J. A., 506, 509, 531, 567
Davisson, R. R., 169, 189
Deal, C. A., 396, 433
Delancy, L. J., 647
DePriester, C. L., 74, 125
Derr, E. L., 28, 393, 396, 418-420, 433
Dinwiddie, J., 455, 476
Dodge, B. F., 19, 68, 283–285
Donata d'Eremita, 2
Dreisbach, R. R., 10, 60
Drickamer, H. G., 464, 476, 557, 567
Duffin, J. H., 306
Duhem, P., 19, 31
Dunbar, A. K., 284
Dunn, C. L., 464, 476

Eastman, D. B., 465, 476
Eckert, J. S., 606, 624, 632, 634, 636, 638, 639, 646
Eckert, R. A., 634, 647
Edmister, W. C., 131–145, 156–158
Edulgee, H. E., 544, 567
Edwards, B. S., 63, 72
Edwin, R. V., 647
Egloff, G., 1
Eklund, R. B., 467, 476
Eld, A. C., 506, 567
Elenbaas, J. R., 11, 44
Ellis, S. R. M., 400, 433, 528, 546, 567, 624, 646
English, G. E., 558, 567
Ewell, R. H., 410, 433

Fair, J. R., 515, 520, 524–527, 536–538, 567, 624, 646
Felt, R. E., 75, 125
Fenske, M. R., 236, 238, 247, 304, 340, 375, 378
Filak, G. A., 131, 132, 158
Finch, R. N., 31, 559, 567
Fisher, A. W., Jr., 610, 646
Flory, P. J., 71, 72, 422, 433
Foote, E. H., 624, 636, 646
Foss, A. S., 515, 545, 549, 567
Fowle, M. J., 465, 476
Fractionating Towers, Inc., 611, 646
Frahme, H. H., 639, 646
Frank, J. C., 486, 501
Franks, R. G. E., 306
Frazier, J. P., 306
Freshwater, D. C., 544, 567
Fried, V., 28
Friedland, D., 63, 72
Friend, L., 32, 632, 646

Gamson, B. W., 107, 125
Garner, F. H., 400, 433, 567
Gautreaux, M. F., 547, 567
Geddes, R. L., 291, 301, 305, 318, 319, 338, 378, 552, 567
Gerster, J. A., 306, 465–467, 476, 515, 545, 546, 549, 550, 567, 624, 646
Gibbs, J. W., 8, 19, 31, 47
Gibson, C. H., 217, 247
Gilbert, J. T., 515, 547, 567
Gilliland, E. R., 200, 217, 241–243, 247, 292, 304, 340, 376, 378, 414, 433, 469, 470, 476
Gilmont, R., 28, 63
Glitsch, F. W., 506, 567, 572, 589–593, 598, 602
Good, A. J., 531, 567
Gorton, J. A., 467, 476
Gouveia, W. R., 545, 567
Graydon, W. F., 28
Grayson, H. G., 157, 158, 349, 378
Green, A. C., 515, 528, 568
Greenkorn, R. A., 30, 44, 60
Greenstadt, J., 306
Griswold, J., 455, 473, 476

Grohse, E. W., 544, 567
Grove, C. S., Jr., 632, 646
Guinot, H., 442, 476
Gunness, R. C., 544, 567

Hackmuth, K., 76, 125
Hadden, S. T., 157, 158, 349, 378
Hala, E., 28, 56
Hamer, W. J., 28
Hands, C. H. G., 624, 646
Hanson, D. N., 217, 247, 306, 527, 528, 536, 568
Happel, J., 465, 476
Harrington, P. G., 506, 567
Harrison, J. M., 410, 433
Harvey, R. J., 131, 158
Hashmall, F., 28, 63, 632, 646
Hauer, H. J., 544, 567
Hausbrand, E., 2
Haylett, R. E., 131, 159
Hellums, J. D., 527, 545, 546, 568
Hengstebeck, R. J., 291
Herington, E. F. G., 38
Hess, H. V., 468, 469, 476
Hess, I., 546, 549, 567
Hildebrand, J. H., 31, 72, 394, 424, 433
Hipkin, H. G., 76, 105, 108–116, 125
Hodson, J. R., 506, 518, 528, 568, 598, 602
Hoefer, F., 1
Hoffing, E. H., 632, 646
Hoffman, P. S., 75, 125
Holbrook, G. E., 536, 568
Holland, C. D., 169, 184, 189, 211, 305, 306, 378
Holloway, F. A. L., 632, 647
Horsley, L. H., 445, 476
Hougen, O. A., 12, 13, 68, 72, 393, 433, 632, 646
Houghland, G. S., 506, 567
Houser, H. F., 28
Huang, C. J., 506, 518, 528, 568, 598, 602
Huggins, M. L., 71, 72, 422, 433
Hughmark, G. A., 506, 518, 520, 526, 528, 557, 568, 597, 602
Hummel, H. H., 301, 464, 476
Hunt, C. d'A, 527, 528, 536, 568
Hunter, T. G., 322, 332
Huntington, R. L., 624, 636, 646
Hutchinson, A. J. L., 506, 515, 568
Hutchinson, M. H., 531, 567, 615, 646

Ibl, N. V., 19, 68

Jenny, F. J., 536, 568
Joffe, J., 28
Johnson, A. I., 544, 568
Johnson, C. A., 63, 393, 433
Jones, C. A., 130, 158
Jones, J. B., 515, 568
Jones, P. D., 527, 568

Karim, B., 545, 546, 568
Katz, D. L., 11, 131, 154, 155, 158, 349, 378, 665–667
Keller, C. J., 636, 646
Kelley, C. A., 488, 501, 506, 518, 519, 528, 568
Kelley, R. E., 598, 602
Kellogg, M. W., Co., 76, 125
Kemp, H. S., 515, 568
Keyes, D. B., 544, 568
Kharbanda, O. P., 545, 567
Kirkbride, C. G., 293, 506, 568
Kirschbaum, E., 322, 333, 546, 568
Kister, A. T., 28, 32, 69, 72, 393, 433
Knapp, W. G., 624, 646
Kobayashi, R., 11, 44
Koch Engineering Co., 598, 602
Koffolt, J. H., 624, 646
Kolodzie, P. A., 518, 568
Kopp, H., 1
Kramer, F., 28, 63
Kwong, J. N., 306
Kyle, B. G., 424, 433

Langdon, W. M., 544, 568
Lawson, S. D., 488, 501
Lee, D. C., 515, 528, 568
Lemieux, E. J., 536, 566
Leng, D. E., 393, 424, 433
Lenoir, J. M., 77–94, 105, 108–116, 124, 125
Lerner, B. J., 632, 646
Leva, M., 632, 635, 637, 639, 646
Lewis, G. N., 13
Lewis, W. K., 156, 158, 285, 300, 305, 314, 316
Lewis, W. K., Jr., 546, 567
Li, Y. M., 28, 59, 61, 72, 393, 414, 433
Libavius, A., 1, 2
Liebson, I., 506, 518, 519, 528, 568, 598, 602
Lobo, W. E., 632, 646

Lockhart, F. J., 632, 633, 646
London, F., 31, 394, 433
Longtin, B., 266, 285
Lonicer, A., 1
Lowry, C. D., 1
Lu, B. C-Y, 28
Lucas, J. M., 639, 646
Lyda, C. D., 527, 545, 568
Lyderson, A. L., 12, 30, 44, 60
Lyster, W. N., 306

McAdams, W. H., 544, 567
McAllister, R. A., 528, 532, 568
McCabe, W. L., 255, 256, 260, 285, 535
McCartney, E. R., 131, 159
McCartney, R. F., 544, 567
McGuinnis, P. H., Jr., 528, 568
McIntyre, R. L., 305
McKinney, J. F., 611, 646
McReynolds, E. E., 169, 189
Makin, E. C., 405, 443
Manning, R. E., 640, 646
Marangozis, J., 544, 568
Marek, J., 544, 568
Margolis, J. E., 647
Margules, M., 28, 62, 63, 72, 393, 433
Martin, H. Z., 217, 218, 229, 231, 241, 242, 247, 304, 340, 375, 376, 378
Matheson, G. L., 300, 305, 314, 316
Matthews, R. L., 524–527, 536–538, 567
Maxwell, J. B., 667–669
May, J. A., 217, 247, 486, 501
Mayfield, F. D., 217, 247, 515, 528, 568
Mayland, B. J., 157, 159, 349, 378
Mertes, T. S., 465, 476
Middleton, J. F., 351, 378
Miller, B. B., 515, 568
Miller, E., 28, 63
Miller, R. H., 544, 568
Miller, R. W., 464, 476
Mills, A. K., 306
Molstad, M. C., 611, 646
Montrose, C. F., 217, 247
Morse, B., 306
Moyade, H. D., 528, 546, 567
Muller, H. M., 492, 501
Munro, B. L., 624, 636, 646
Murch, D. P., 624, 646
Murphree, E. V., 533, 568
Murphy, G. B., 337, 339, 371, 372, 378
Myers, H. S., 75, 77, 125

Nandi, S. K., 545, 546, 568
Narragon, E. A., 468, 469, 476
Nash, A. W., 322, 332
Nelson, E. F., 131, 159, 337, 339, 371, 372, 378
Nelson, W. L., 131, 132, 158, 353, 354, 359, 370, 378
Newman, J. S., 305
Nichols, J. H., 544, 569
Norrish, R. S., 32
Novasad, Z., 544, 568
Nutter, I. E., 492, 501, 506, 568, 598, 602
Nutting, H. S., 445, 476

Obrien, N. G., 306
Obryadchakoff, S. N., 131, 154, 158, 349, 378
O'Connell, H. E., 506, 518, 520, 528, 547, 557, 567, 568, 597, 602
Oehler, H. A., 131, 153, 158
Okamoto, K. K., 130–132, 136, 143, 144, 158
Oliver, E. D., 549, 568
Olney, R. B., 536, 568, 613, 646
Organick, E. I., 76, 125
Orye, R. V., 71, 393, 422, 423, 433
Othmer, D. F., 28, 63, 130, 157–159, 443, 476, 492, 501

Packie, J. W., 131, 158, 353, 354, 378
Papadopoulos, M. N., 28, 393, 418–420, 433
Parsly, L. F., 611, 646
Pauli, S., 409, 433
Peavy, C. C., 544, 568
Perrier, A., 2
Perry, J. H., 10, 465, 476, 584, 602
Perry, R. H., 62, 393, 433
Phillips, J. C., 444, 475
Pick, J., 28
Pierrotti, G. J., 393, 396, 433, 464, 476
Pigford, R. L., 32, 62, 393, 433, 549, 567
Pike, M. A., Jr., 28
Piromoov, R. S., 131, 158
Plank, C. A., 528, 532, 536, 566, 568
Poettmann, F. H., 157, 159, 169, 189, 349, 378
Poffenberger, N., 445, 476
Pollock, D. H., 131–133, 143, 158
Ponchon, M., 255, 266, 285, 535

Pontinen, A. J., 306
Popovac, D. D., 635, 646
Porter, C. A., 465, 476
Prabhu, P. S., 406, 407, 433
Prater, N. H., 353, 354, 378
Prausnitz, J. M., 71, 72, 393, 396-398, 408, 409, 422–427, 429–431, 433
Prengle, H. W., 28
Prouls, A. Z., 647
Pyle, C., 515, 568

Qozati, A., 405, 433

Ragatz, E. G., 131, 159
Ragatz, R. A., 12, 13
Randall, M., 13, 266, 285
Raoult, F. M., 9, 18, 46, 49
Rasmussen, R. W., 515, 528, 568
Rau, M. A. G., 627, 647
Rayleigh, L., 171, 175, 182, 189
Redlich, O. A., 28, 32, 69, 72, 306, 393, 433
Reed, C. E., 200, 247
Reid, R. C., 60
Reilly, P. M., 169, 189
Rice, R. B., 169, 189
Robinson, C. S., 414, 433, 469, 470, 476
Rose, A., 306
Rosseli, R. M., 32
Rousseau, W. C., 531, 567
Rubin, L. C., 63, 393, 433, 438, 444, 464, 475
Ruheman, M., 283, 285
Rumford, B. T., 2
Rush, F. E., Jr., 544, 568

Sandlin, H. L., 131, 132, 158
Savarit, R., 255, 266, 285, 535
Saviani, C. G., 647
Scatchard, G., 28, 424, 433
Schechter, R. S., 25
Scheibel, E. G., 63, 72, 217, 247
Schoenborn, E. M., 390, 433, 436, 476, 528, 536, 566
Schreine, W. C., 536, 566
Schrodt, V., 306
Schutte, A. H., 465, 476
Scott, R. L., 394, 433
Seader, J. D., 306
Sesonske, A., 62, 393, 433
Severns, W. H., 62, 393, 433
Shell Development Co., 613, 647
Shelton, R. O., 305
Sherwood, F. K., 60, 536, 568, 632, 647
Shilling, G. D., 544, 568
Shipley, G. H., 632, 647
Shiras, R. N., 217, 247, 464, 476
Shulman, H. L., 647
Silvey, F. C., 636, 647
Simkin, D. J., 536, 568
Simon, M. J., 627, 647
Smith, B. D., 528, 568, 572, 575, 577, 603
Smith, K. A., 349, 378
Smith, P. L., 518, 520, 568
Smith, R. B., 349, 378
Smith, R. L., 338, 378
Smoker, E. H., 261, 285
Solomon, E., 63, 393, 433
Somerville, G. F., 306, 333
Sorel, E., 2, 249, 285, 293, 300, 313
Souders, Mott., Jr., 131, 158, 464, 476, 537 567, 584, 597, 598, 603
Staffel, E. J., 131, 158
Stephenson, R. L., 405, 433
Stigger, E. K., 544, 569
Stillman, R. E., 306
Stirba, C., 544, 568
Stockholm, C. J., 131, 132, 158
Strand, C. P., 536, 568
Sullivan, S. L., 306
Sundback, R. A., 536, 566
Swanson, R. W., 466, 476
Sweeny, R. F., 306

Teller, A. J., 610, 611, 647
Teneyck, E. H., 157, 159
Thiele, E. W., 255, 256, 260, 285, 301, 305, 318, 319, 535
Thiessen, H. A. C., 544, 569
Thompson, A. R., 611, 646
Thornton, D. P., 494, 501
Thrift, G. C., 492, 494, 501, 506, 568, 598, 603
Tierney, J. W., 306
Toor, H. L., 544, 569
Twigg, G. H., 32

Ulrich, C. F., 647
Umholtz, C. L., 527, 545, 546, 569
Underwood, A. J. V., 216, 217, 222, 234, 236–238, 247, 304, 340, 375, 378

Union Carbide, 654–664
U.S. Stoneware Company, 605–610, 633, 635, 641–643

van Arkel, A. E., 31, 394, 433
van der Waals, 21, 24
Van Laar, J. J., 21, 24, 25, 30, 58, 59, 61, 72, 393, 433
Van Wijk, W. R., 544, 569
Van Winkle, M., 25, 28, 31, 130–132, 146–150, 153, 158, 159, 358, 378, 405–407, 433, 518, 519, 527, 536, 539–541, 545, 546, 557–559, 568, 569
Vary, J. A., 11, 44
Vaswani, N. R., 156, 159
Vilim, O., 28
Volland, G., 545

Walker, J. R., 75, 125
Walsh, T. J., 506, 567
Walter, L. F., 624, 636, 646
Wang, W. J., 16
Warzel, L. A., 544, 569
Waterman, W. W., 306
Watson, C. C., 544, 549, 568
Watson, K. M., 12, 13, 131, 159, 337–339, 371, 372, 378
Webb, G. B., 72, 124
Weber, J. H., 75, 125
Weimer, R. F., 72, 393, 424–427, 429–431, 433
Weinaug, C. F., 11, 44
Weinman, E. A., 28, 63
Welch, N. E., 184, 189
White, G. A., 77–87, 125
White, R. R., 28, 61, 72
Whitehouse, W. G., 647
Whitt, F. R., 624, 646
Wilde, D. L., 156, 158
Wilke, C. R., 527, 528, 536, 568
Williams, G. C., 544, 569
Williams, T. J., 306
Williams, V. C., 283, 285
Wilson, C. L., 169, 189
Wilson, G. M., 71, 72, 422, 433
Winn, F. W., 492, 501
Withrow, J. R., 624, 646
Wohl, K., 23, 58, 62, 72, 393, 433
Wood, S. E., 28

Yu, K. T., 28, 620, 622, 623, 647

Zenz, F. A., 484, 501, 528, 569, 632, 634, 646, 647
Zimmerman, J. D., 647
Zudkevitch, D., 28

Subject Index

Absorption factor, 619
Activity coefficient, 13
  binary correlations, 19–24
    Carlson-Colburn, 24
    Gibbs-Duhem, 19
    Ibl-Dodge, 19
    Margules, 28
    Van Laar, 22
    Wohl, 23, 24
  infinite dilution, 413
    correlations, 418–431
      Black *et al.*, 418–422
      Kyle-Leng, 424
      Orye-Prausnitz, 422–424
      Weimer-Prausnita, 424–431
      Wilson, 422
    definition, 413
  multicomponent correlations, 58–63, 68, 69, 71–73, 393
    Benedict *et al.*, 63
    Black, 70, 71
    Bonham, 61, 62
    Chao-Hougen, 68
    Li-Coull, 59
    Redlich-Kister, 69
    summary, 72, 393
    Wilson, 71–73
    Wohl, 58, 62, 63
  variables affecting, 19, 20
    composition, 19
    molecular size, 396
    pressure, 20
    temperature, 20
    unsaturation, 396
Addition agents (*see* Solvent)
Aeration factor, 515, 516, 521, 522
American Institute of Chemical Engineers (A.I.Ch.E.), 306, 378, 526, 553–555, 566
  Distillation Committee, 526, 566, 584, 602
  Research Committee, 526, 556, 558
American Society for Testing Materials (ASTM), 129, 132, 134, 137, 142, 145–147, 151, 152, 155, 158
  distillation curves, 129, 132
    EFV and TBP, atmospheric, 131, 132, 136–140, 145–153
      pressure, 132–136
      vacuum, 137, 139–143, 145–150
    equilibrium, 155
    interrelationship with other, 129, 131–153
Antoine equation, 9
API gravity-density, 649–652
  hydrocarbons, 649–651
  temperature correction, 652
API Project Report No. 44, 10
Azeotrope, binary, 14, 15
    techniques of separation, 381–389
Azeotropic distillation, 441–446
  column design stages, 385–387, 441, 442
    multicomponent method, 442
    pseudoternary method, 442
  entrainers in, 410–412
    feed ratio to, 440, 441
    optimum point of introduction, 438–440
  processes, 442–445
    selection of, 437
  two-column, 385–387

Baffles, 500, 501
  redistribution, 500, 501
  splash, 500
Ballast tray, 492, 494, 496
Benturi tray, 494
Boiling point, average, 134, 337–339
    correlations, 337–339
      cubic, 338
      mean, 339

Boiling point, average, correlations, molal, 339
volumetric, 337, 338
weight, 339
Brown-Martin correlation, short-cut method, 241, 242
Bubble cap plate, 480, 485–488
aeration factor, 515, 516
area designation, 574
backup in downcomer, 507
cap arrangement, 483
design, 575–584
standards, 576, 577
dumping, 529–533
efficiency, 533–549
prediction, 550–559
scale up, 559–561
entrainment, 536–541
flooding, 537, 538
flow pattern on, 485–488
froth density, 515, 516
hydraulics, 502–533
liquid gradient, 509–515
pressure drop, 508–515
priming, 526
pulsations, 532, 533
vapor distribution, 578
Bubble caps, 482–485
arrangement, 485
design, 482–483
materials, 484
pressure drop through, 508, 509
risers, 482, 485
selection, 576
skirt clearance, 485
slots, 482, 483
types, 484
Bubble point, 47, 161–169, 175, 253, 287, 291, 296, 299
ideal system, 47
nonideal system, 161–169, 175, 253, 287, 291, 296, 299

Complex system, 127–157, 334–376
density mid-percent curve, 337
fractionation, stages, 334–376
pseudo components, 336, 337
vapor-liquid equilibrium, 127–157
Components, 153–155, 197, 198, 287, 290, 291, 293, 336–338, 341
key, 197, 198, 287, 290, 291, 293, 338
pseudo, 153–155, 336, 337, 341
Compressibility factor, 11, 13
Concentration profile in fractionaters, 241, 314, 315, 317, 318, 320, 321, 452
Condensers, 204–207, 289, 290, 294, 295, 369
equations, 205–207
partial, 204
selection, 289, 290, 369
total, 204
Consistency tests, thermodynamic, 31–38
Adler *et al.*, 32, 38
Black, 32, 33, 37
Broughton-Brearly, 32
Herington, 38, 39
Norrish-Twigg, 32, 33, 37
Redlich-Kister, 32, 35
Contact stage or plate, 196, 197
Convergence pressure, 75–101, 104
charts to determine, 78, 81–87, 89, 95–97
correction for K values, 92–94
hydrogen, 92
methane, 93, 94
definition, 76
effective properties for, 79, 80
grid relationships, 104
pressure function, 80
Critical pressure, 133, 134
petroleum, 134
Critical temperature, 133
Cyclic instability, 529–533
dumping, 529–531
pulsations, 532, 533

Capacity parameter, 537
Carlson-Colburn equation, 24
constants, 25
Clausius-Clapeyron equation, 36
Cohesive energy density, 428
Column design methods, 570–603, 638–643
equilibrium stage, 570–603
packed, 638–643

Dalton's law, 9, 18, 46, 49
Degrees of freedom, fractionation, 199, 200
Density, hydrocarbons, 649–651
temperature correction, 652
Dewpoint, 47, 162, 164, 166, 167, 288, 291, 294
definition, 47, 166

Dewpoint, ideal system, 47, 48
nonideal system, 162, 164, 166, 167, 288, 291, 294
Difference point equations, 274–278, 324, 385–388
Differential column, 479
Diffusivity, 548, 553
eddy, 548
molar, 553
Discharge coefficients, 518–520
correction factor, 520
Distillation, 129–472
analytical curves, 129–155
ASTM or Engler-type, 129
EFV, 130
interrelationship between, 131–155
slope of, 131, 132
correction for curvature, 132
TBP, 127, 128, 130, 131
transposition with pressure, 141, 146, 147
complex system, 334–376
component distribution, 338, 339, 345
key components, 338, 344
overflash, 349, 358, 371
reflux, 369, 370
side stream operation, 347–376
effective distillation pressure, 348, 350
feed vapor composition, 349
plate temperatures, 354
plates between side streams, 370
products, 353
stripping steam required, 359
TBP gap-overlap, 353
yields, theoretical, 352–357
stages, pseudo component method, 334–376
conventional fractionation, 336–347
side stream, 347–376
differential batch, 170–180
binary system, 170–173
multicomponent system, 173–180
constant pressure, 173–177
constant temperature, 177–180
equilibrium, EFV, 153, 160–169
binary system, 162, 165
complex system, 349, 350
composition of phases, 153, 162, 165
fraction vaporized in, 161, 166, 169
Distillation, equilibrium, EFV, limiting conditions, 166, 169
multicomponent system, 165, 166
fractional, 249–279
azeotropic, 385–388
entrainers, 436–441
processes, 442–445
solvent recovery in, 469–472
stages, 434
binary system, stages, 249–279
equations for, 249–255
Lewis, 255
Sorel, 249–254
graphical methods, 256–279
McCabe-Thiele, 256–265
Ponchon-Savarit, 266–279
extractive, 436–438, 446–472
processes, 463–469
solvent recovery in, 469–472
solvents, 436–438
stages, 446–472
multicomponent, 241–243, 286–321
component distribution, 290, 291
computer calculations, 304–321
feed plate location, 292, 293
general considerations, 286–293
pressure, selection of, 288–290
stages, calculation method, 241–243, 293–304
Brown-Martin, 241, 242
Gilliland, 242, 243
Lewis-Matheson, 300, 301
short-cut, 303, 304
Sorel, 293–300
Thiele-Geddes, 301–303
ternary, 300–304, 321–329, 441, 442, 455
stages, calculation method, 303, 304, 321–329, 441, 442, 455
analytical, 300–303, 441
graphical, 321–329, 442, 455
short-cut, 303, 304
steam, 178–185
batch, 178–182
binary, 182
constant and varying steam rate, 182
multicomponent, 184, 185
continuous, 183
cocurrent and counter current, 183
rates, 180–182
Downcomers, 480, 495, 497
residence time in, 579, 585

Dumping, 527–531
  conditions causing, 527, 529, 530
  definition, 527
  limits, evaluation of, 527, 528, 531

Effective boiling point, 78–80
Effective distillation pressure, 348, 350
Efficiency stage or plate, 502–561
  calculation from data, 533, 534
  experimental, 533, 534
  factors affecting, 534, 536–550
  Murphree plate and point, 533, 534
  prediction of, 550–558
  relation to actual stages, 535
  scale up, 559–561
  vapor, 550–552
  vaporization, 178–181
Energy of interaction, 425
Engler-type distillation, 129
Enthalpy mixtures, 351
  balances, 205, 209, 272, 274–276, 295, 351, 372–375
  calculation of, 209, 266–268, 296
  diagrams-concentration, 267, 268, 270, 271, 274–278
Entrainers, 432–440
  choice of, 432
  concentration, 437
  feed ratio, limitations, 440
  introduction point, optimum, 438
Entrainment, 515, 516, 537–542, 545, 554
  correlations, 537–540, 554
  liquid in vapor, 537–541
  vapor in liquid, 515, 516, 542, 545
Equilibrium characteristics, prediction of, 153–167
  liquid and vapor, 153–167
Equilibrium data, 8–39, 46–121, 127–157
  consistency of, 31–39
  correlation of, 8–31, 46–121, 127–157
    binary, 8–31
    complex system, 127–157
    multicomponent, 46–121
Equilibrium flash distillation, 130, 131, 160–169
Equilibrium flash vaporization (EFV) curves, 127–157
  interrelationship with other distillation curves, 131–157
  prediction from analytical data, 131–157
Equilibrium stages, 202, 241–243, 249–279, 286–329, 334–376, 442, 455, 479
  calculation methods, 241–243, 249–279, 286–329, 334–376, 442, 455
    analytical, 249–279, 286–321, 334–376
      binary, 249–279
      complex system, 334–376
      multicomponent, 286–321
    graphical, 256–279, 321–329, 442, 455
      binary, 256–279
      ternary, 321–329, 442, 455
    short-cut, 241–243
  definition, 479
  minimum, 202
Equilibrium vaporization ratios (*K*'s), 74–116, 157
  hydrocarbons, 74–116
    calculation methods, 74–76, 94–116
      empirical, 75, 76
        Hadden *et al.*, 94–105
        Lenoir *et al.*, 110–116
      thermodynamic, 74, 75
    corrections for, 76–97
      convergence pressure, 76–91, 94–97
      hydrogen content, 88–95
      methane content, 92, 93
    extarpolation of, 108, 117
    K_{10} charts, 109–116
    selection of, 97
  hydrogen, 108
  petroleum oils, 105–107, 157
    volatility exponent, 105
Excess volume change, 397
Extractive distillation, 446–472
  column design stages, 447–464
    multicomponent method, 456–464
    pseudo-binary method, 448–454
    pseudo-ternary method, 455
  conditions, selection of, 446
  processes, 464–470
  selection of type, 437
  solvents, 446, 464, 467, 468, 471, 472
    aromatics for, 464
    butenes-butanes for, 467, 468
    pentenes-pentanes for, 467
    recovery, 471, 472
    selection, 446

Feed plate location, 251, 252, 258, 259, 276, 277, 292, 293
  analytical, 251, 252, 258, 292, 293
  graphical, 258, 259, 276, 277

Flexitray, 491–493
Flooding, 521–526
  correlations, 524–526
  evaluation, 521, 522
Flow parameter, 524, 525
Foam, 515, 516, 520, 548
  density, 515, 516
  friction factor, 520
  height, 548
  Reynold's number, 520
Focal pressure, 135
Focal temperature, 135
Fractionation, general considerations in, 193–205
    column, 195, 196
    contact stage, 196
    efficiency and number of stages, 199–201
    limits of operability, 203, 204
    pressure of operation, 198, 199, 298–300
    reflux, 194, 195
  general equations, 205–210
  (*See also* Distillation, fractional)
Free energy, excess, 23, 58, 422, 423, 425
Froth, 515, 516, 548
  density, 515, 516
  height, 548
Fugacity, 13
  coefficient, 11
  Lewis-Randall rule, 13

Gas mixtures, 8, 11, 12
  ideal, 8
  nonideal, 11, 12
Gibbs-Duhem equation, 19, 31
Gilliland correlation, short-cut method, 242, 243
Graphical methods, 256–279, 321–329
  Hunter-Nash, 321–329
  McCabe-Thiele, 256–265
  Ponchon-Savarit, 266–279
Grid pressure, 104

H bonding, 409–411
  azeotropes in, 410
Heat of mixing, 396, 397, 423
  equilibrium data from, 423
  molecular size, effect of, 396
  polarity, effect of, 396
  unsaturation, effect of, 397
Height of a transfer unit (HTU), 619, 620
Height equivalent to a theoretical plate (HETP), 624–627
Homomorph, 425, 426
  plots for hydrocarbons, 425, 426
Hydraulic gradient, 486
  bubble cap tray, 509–515
  perforated tray, 520
Hydraulic radius, 520, 521

Instability in fractionator operation, 529–533

K charts, 98, 99, 102, 103, 108, 109–116
  hydrocarbons, 98, 99, 102, 103, 109–116
    correction charts, 92, 93, 94
  hydrogen, 108
  petroleum oils, 105–107, 157
Key components, 197, 287
  distributed, 287
  heavy, 287
  intermediate, 287
  light, 287
  selection of, 287
Kittel tray, 614

Lever arm principle, 271
Liquid backup in downcomers, 506, 507
Liquid gradient (*see* Hydraulic gradient)
Liquids, classification, based on nonideality, 411

McCabe-Thiele graphical method, 256–265
McMahon packing, 610
Margules equations, 28, 62, 63
Mass transfer, differential, 615–619
    coefficients, 618, 619
Minimum reflux, 202, 203, 211–213, 216, 217, 260, 277
  equations, 211–213
  graphical evaluation, 260, 277
    McCabe-Thiele, 260
    Ponchon-Savarit, 277
  rigorous method, 215, 216
  short-cut methods, 217–223
    Brown-Martin, 217–222
    Colburn, 217, 224–228
    Underwood, 217, 222–223

Molar liquid volume, 426–431
  hydrocarbons, 427, 428
  solvents, 429, 431
Murphree efficiencies, 533, 534

National Bureau of Standards, 649–652
NGPA Engineering Data Book, 157
Nonideal mixtures, 11, 13, 14, 18–29, 57–73
Number of transfer units (NTU), 619, 621–624, 627
Nutter tray, 492, 496

Obryadchakoff equilibrium curves, 154

Packed column, 604–644
  distributor plates, 640–643
  flooding, 632
  liquid holdup, 637, 638
  packings, 604–613
  pressure drop, 634–636
    and reflux ratio, 632–634
  support plates, 639, 643
Perforated plate or tray, 488–490, 504–507, 515–529, 536–543, 557–561, 571–573, 584–587
  aeration factor, 515–516
  area terms, 586
  backup in downcomer, 506–507
  capacity based on flooding, 525, 526
  design, 571–573, 584–587
    standards, 571–573, 586
  dumping-weeping, 489, 527–529
  efficiency, 557–561
    prediction, 557–559
    scale up, 559–561
  entrainment, 536–543
  fall-off point, 489, 527–529
  flooding, 524, 525
  flow pattern, 504
  foam density, 515, 516
  hole area, 490
  hole arrangement, 490
  hole size, 490
  hydraulics, 515–527
  liquid gradient, 520, 521
  pressure drop, 505–507, 518–521
  priming, 526, 527
  surface tension pressure drop, 521
  weeping, 527–529
Petroleum oils, 133, 134, 331, 339, 351, 358, 372
  characterization factor, 339
  critical pressure, 134
  critical temperature, 133, 339
  enthalpy, 351
  flash point, 358
  latent heat, 372
  molecular weight, 331
Phase diagrams, 14, 15, 133, 135
  binary, 14, 15
  naphtha-kerosene, 133
  petroleum fractions, 135
Phase equilibria, 718
Phase rule, Gibbs, 8
Plate hydraulics, 502–533
Plate spacing, 498
  related to flooding, 536–540
Ponchon-Savarit method, 266–279
Pressure, critical, petroleum, 134
effect on distillation, 198, 199, 298, 299
effect on phase equilibria, 14 grid, 104
vapor (*see* Vapor pressures)
Pressure convergence (*see* Convergence pressure)
Pressure drop evaluation, 507–521
  bubble cap plate, 508–518
  downcomer, 508
  dynamic submergence, 610
  liquid crest, 507
  liquid gradient, 509, 510–515, 520, 521
  orifice or hole, 518, 519
  riser and cap, 508
  slot, 509
  static submergence, 508
  surface tension, 521
  weir, 507
Pressure drop factors, 502–505
  liquid, 502, 503
  vapor, 504, 505
Pressure selection, 288, 289
Priming, 526, 527
  conditions causing, 526
  evaluation of, 526, 527
Protruded packing, 611
Pulsations, 532, 533

q-line, 259

Raoult's law, 9, 18, 46, 49
Reflux, 195, 196, 211–213, 215–228, 236–243, 260, 261, 277, 278
internal, 195
minimum, evaluation methods, 215–223, 260, 277
analytical, 215–223
rigorous, 215
short-cut, 216–223
Brown-Martin, 217–222
Colburn, 224–228
Underwood, 222–223
graphical, 260, 277
McCabe-Thiele, 260
Ponchon-Savarit, 277
related to operating reflux and stages, 241–243
Brown-Martin, 241, 242
Gilliland, 242, 243
variables affecting, 196
operating, 195, 241–243
minimum and total reflux in terms of, 241–243
ratio, 211–213
internal and operating equations for, 211–213
total, evaluation, 236, 237, 239–241, 261, 277, 278
graphical, 261, 277, 278
McCabe-Thiele, 261
Ponchon-Savarit, 277, 278
rigorous, 239–241
short-cut, 236, 237
Fenske-Underwood, 236, 237
Relative volatility, 11, 14, 48, 51, 390–394, 434
alteration of, methods, 391–394
change, mechanism of, 391, 392, 394, 434
economic factor, as, 390
nonideal, 11, 14
stages and, 391
ternary, 48, 51
Residence time, downcomers, 579, 585
Ripple tray, 615
Risers, 485

Scatchard-Hamer equations, 28
Seal, 498, 499
dynamic, 498, 499
static, 498, 499
Seal pots, 500
Selectivity, 412, 413–431, 464–468
defining equations, 412, 413–431
Black *et al.*, 417–421
Orye-Prausnitz, 422–424
Weimer-Prausnitz, 424–431
Wilson, 422–424
effective agents, 399, 416, 417
factors affecting, 407–409
concentration of solvent, 408, 409
H bonding, 409
molecular size, 409
pressure, 408
temperature, 407
infinite dilution, 413
molecular interaction and, 394
solvents for, 464–468
aromatics, 464
butenes-butanes, 467, 468
pentenes-pentanes, 466, 467
system type and, 400–407
Side stream columns, 347–376
Sieve trays (*see* Perforated plate or tray)
Skirt clearance, 485
Smoker equation, 261
Solvent, 421–422, 434–436, 464–468, 471, 472
hydrocarbon separation for, 464–468
recovery of, 471, 472
screening procedure, 421, 422
selection of, 434–436
Solubility parameter, 424–428
Scatchard-Hildebrand, 424, 425
tabulation, 427–431
hydrocarbons, 427, 428
polar solvents, 429–431
Solution, 395–400, 409–411
nonideal, 395–400, 409–411
classification of, 410, 411
H bonding, 409, 410
heat of mixing, 396, 397
molecular interaction, 410
types, polarity, 395, 396
volume change, excess, 397–400
Spraypack packing, 611, 612
Stages, 210
reflux ratio and, 210
Stedman packing, 612
Stripping, 358, 359
steam required, 358, 359
Submergence, 506, 510
dynamic and static, 506, 510

Tellerette packing, 610, 611
Thermodynamic consistency tests, 31–39
Thiele-Geddes method, 301, 303
Total reflux (*see* Reflux, total)
Transfer units, 619–627
  height of, 619, 620, 625, 626
  HETP, 624
  number of, 619, 621–623, 627
    and reflux ratio, 622, 623
    and theoretical plates, 623, 624
Trays (*see under* Plate)
True boiling point (TBP) curves, 127–132, 136–140, 154–157
  binary system, 128
  complex mixture, 128
  equilibrium, 154
  flash distillation resulting from, 130, 131
  interrelationship with EFV and ASTM, 136, 137, 139, 140
  pseudo component method of determination, 156, 157
Turbogrid plates, 613, 614

Underwood equations, 222, 223, 236, 237
Uniflux tray, 492
Unstable operation, 529–533

Valve trays, 492, 598–600
  ballast, 492
  design, 598–600
  Flexitray, 492
  Nutter, 492
Van Laar equations, 21, 22, 24, 30
Vapor-liquid equilibrium data, 13, 16, 18, 22–25, 28, 30, 49–51, 70–73, 107, 117–121, 127–157, 392, 393
  binary system, 13, 25, 28, 30
    prediction, 13, 25, 28, 30
  complex system, 127–157
  correlations, 16, 18, 22, 23, 24, 28
  index of equations, 393
  multicomponent, 49–51, 70, 71–73, 107, 117–121, 392
    prediction, 71–73
  ternary, 52–54, 58–63, 68–69
    plotting, 69
    prediction, 58–63, 68, 69
      graphical, 52–54
Vapor pressures, 9
  chemical compounds of, 653–669
Vaporization efficiency, 178
Volatility exponent, 104, 105
Volume molar (*see* Molar liquid volume)

Weeping, 527–529
  conditions causing, 527
  evaluation, 528, 529
Weir, 481, 498, 508
  formula, 508
  inlet, 498
  liquid crest, 508
  outlet, 481, 498
Wilson equation, 71–73
Winn homograph, 98
Wohl equations, 23, 24, 58, 59, 62, 63